Benchmark Papers in Inorganic Chemistry

Series Editor: Harry H. Sisler
University of Florida at Gainesville

Published Volumes and Volumes in Preparation

METAL–AMMONIA SOLUTIONS
William L. Jolly
COMPOUNDS CONTAINING PHOSPHORUS–PHOSPHORUS BONDS
Alan H. Cowley
HARD AND SOFT ACIDS AND BASES
Ralph G. Pearson
SYMMETRY IN CHEMICAL THEORY
John P. Fackler, Jr.
CHLORAMINATION REACTIONS
Stephen E. Frazier
LIGAND FIELD THEORY
R. Carl Stoufer
COMPOUNDS CONTAINING AS–N BONDS
Larry K. Krannich
QUANTUM THEORY OF INORGANIC MOLECULES
Joyce Kaufman
TRANSPORTATION OF ALKALI AND ALKALINE EARTH METAL IONS IN BIOLOGICAL SYSTEMS
Dan W. Urry and Marianna Long

Benchmark Papers in Inorganic Chemistry

A *BENCHMARK* ™ Books Series

SYMMETRY IN CHEMICAL THEORY: Application of Group Theoretical Techniques to the Solution of Chemical Problems

Edited by
JOHN P. FACKLER, JR.
Case Western Reserve University

Dowden, Hutchinson & Ross, Inc.
Stroudsburg, Pennsylvania

Benchmark Papers in Inorganic Chemistry, Volume 4
ISBN: 0-87933-018-X

Library of Congress Cataloging in Publication Data

Fackler, John P comp.
Symmetry in chemical theory.

(Benchmark papers in inorganic chemistry, v. 4)
Bibliography: p.
1. Symmetry (Physics)--Addresses, essays, lectures. 2. Groups, Theory of--Addresses, essays, lectures. 3. Chemistry, Physical and theoretical--Addresses, essays, lectures. I. Title.
QD461.F23 541'.2 73-12620
ISBN 0-87933-018-X

Manufactured in the United States of America.

Exclusive distributor outside the United States and Canada: John Wiley & Sons, Inc.

Acknowledgments and Permissions

ACKNOWLEDGMENTS

The American Physical Society—*Physical Review*

"Theory of the Variations in Paramagnetic Anisotropy Among Different Salts of the Iron Group"

"Influence of Crystalline Fields on the Susceptibilities of Salts of Paramagnetic Ions. II. The Iron Group, Especially Ni, Cr, and Co"

"Note on the Electric Field in Paramagnetic Crystals"

"Information Concerning Crystal Structures from Data on Magnetic Susceptibilities"

"Electronic Structures of Polyatomic Molecules and Valence"

"Electronic Structures of Polyatomic Molecules and Valence. IV. Electronic States, Quantum Theory of the Double Bond"

"The Normal Modes and Frequencies of Vibrations of the Regular Plane Hexagon Model of the Benzene Molecule"

PERMISSIONS

The following papers have been reprinted with the permission of the authors and the copyright owners.

Academic Press, Inc.

Annals of Physics

"Complete Theory of Ni(II) and V(III) in Cubic Crystalline Fields"

Journal of Molecular Spectroscopy

"Intensities of Inorganic Complexes. Part II. Tetrahedral Complexes"

American Chemical Society

Inorganic Chemistry

"Molecular Orbital Calculations for Complexes of Heavier Transition Elements. III. The Metal–Metal Bonding and Electronic Structure of $Re_2Cl_8^{2-}$"

Journal of the American Chemical Society

"The Nature of the Chemical Bond. Application of Results Obtained from the Quantum Mechanics and from a Theory of Paramagnetic Susceptibility to the Structure of Molecules"

"The Nature of the Chemical Bond. III. The Transition from One Extreme Bond Type to Another"

"New Tetrahedral Complexes of Nickel(II)"

"Electronic Structure and Molecular Association of Some Bis-(β-diketone)nickel(II) Complexes"

"Vibrational Spectra and Structure of Xenon Tetrafluoride"

"Stereochemistry of Electrocyclic Reactions"

"Electronic Mechanism of Electrocyclic Reactions"

"Selection Rules for Concerted Cycloaddition Reactions"

"Selection Rules for Sigmatropic Reactions"

American Institute of Physics

American Journal of Physics

"Interpretation of the Spectra of Polyatomic Molecules by Use of Group Theory"

"Application of Group Theory to the Calculation of Vibrational Frequencies of Polyatomic Molecules"

Journal of Chemical Physics
"The Group Relation Between the Mulliken and Slater–Pauling Theories of Valence"
"Valence Strength and the Magnetism of Complex Salts"
"Directed Valence"
"Spectra of Transition Metal Complexes"
"Optical Spectra of Hydrated Ions of the Transition Metals"
"Effect of Pressure on the Spectra of Certain Transition Metal Complexes"
"The Degeneracy, Selection Rules, and Other Properties of the Normal Vibrations of Certain Polyatomic Molecules"
"Symmetry Considerations Concerning the Splitting of Vibration–Rotation Levels in Polyatomic Molecules"

Annual Reviews Inc.—*Annual Review of Physical Chemistry*
"Quantum Theory"

Johann Ambrosius Barth, Leipzig—*Annalen der Physik*
"Termaufspaltung in Kristallen"

The Faraday Society—*Transactions of the Faraday Society*
"A Note on the Bonding Powers of Groups of *d* Electrons"

Macmillan (Journals) Ltd.—*Nature*
"Stereochemistry of Complex Halides of the Transition Metals"

Microforms International Marketing Corporation—*Journal of Inorganic and Nuclear Chemistry*
"On the Stabilities of Transition Metal Complexes. I."
"The Nephelauxetic Series of Ligands Corresponding to Increasing Tendency of Partly Covalent Bonding"

The Physical Society of Japan—*Journal of the Physical Society of Japan*
"On the Absorption Spectra of Complex Ions. I."
"On the Absorption Spectra of Complex Ions. II."

Royal Academy of Sciences, Holland—*Proceedings of the Section of Sciences*
"Théorie générale de la rotation paramagnétique dans les cristaux"

The Royal Society, London—*Proceedings of the Royal Society, London*
"Stability of Polyatomic Molecules in Degenerate Electronic States. I. Orbital Degeneracy"
"The Colours and Magnetic Properties of Hydrated Iron Group Salts, and Evidence for Covalent Bonding"

Taylor & Francis Ltd—*Molecular Physics*
"The Symmetry Groups of Non-rigid Molecules"

Series Editor's Preface

The Benchmark Series in Inorganic Chemistry has as its objective the selection by a knowledgable editor of the outstanding original research publications in an important area of inorganic chemistry and the assembling of these publications with appropriate comments, discussion, and explanation into a single volume. Such volumes will assist the teacher, student, researcher, or professional scientist in bringing himself up to date in various specific fields in a manner which would never be possible with the use of only secondary literature.

Certainly the applications of symmetry theory constitute a most important field in modern inorganic chemistry and few concepts have had a more significant effect on the development of chemical theory. John P. Fackler, Jr., is the ideal editor for this volume. His contributions in this area have been major and his reputation as a teacher has been most impressive. As Series Editor I am proud to recommend this volume of Benchmark Papers in Inorganic Chemistry.

Harry H. Sisler

To
John H. Van Vleck

Preface

Group theory has enjoyed a particularly important place in the development of chemical theory. With this premise in mind, articles for this volume on symmetry were selected. Geometric concepts and theories of symmetry which do not utilize group theory have not been included, although some of them have made important contributions to chemical thought.

The first direct application of group theory to what today might be considered chemical theory was concerned with the description of crystals. However, it is with the development of quantum mechanics that the power of group theory as applied to chemistry became manifest. Consequently, the articles chosen for inclusion in this volume all assume an acceptance of the fundamental concepts of quantum mechanics, and none predate its development. In fact, symmetry has been used to define basic quantum relationships in several of the more important papers.

I have emphasized three distinct areas of chemical theory in choosing articles for this volume: electronic structure, vibrational behavior, and molecular rearrangements and reactivity. The first two began to develop in the late 1920s and early 1930s, while the last remains in its infancy in the 1970s. I have generally relied on my rather imperfect background in the application of symmetry to chemical theory to help select articles from each area. Consequently, I take full blame for any and all omissions. I also take full credit for having included particularly appropriate articles, although anyone even moderately familiar with the field would have made some of the same choices. Experts such as F. A. Cotton, R. H. Holm, R. Hoffman, and others also have helped me, directly and indirectly. I am particularly delighted with the way John H. Van Vleck of Harvard University responded to my questions. It is to him that I have dedicated this volume. Special credit also goes to my secretary, Helen Bircher, without whose help this undertaking would not have been completed.

Finally, I apologize to those numerous inorganic chemists who have contributed excellent work relevant to the development of symmetry in chemical theory but who, for various reasons, do not find their material cited. Certainly none of the articles included here would be of any significant value had its concepts failed under extensive testing. Without the work of the many scientists doing this testing, symmetry in chemical theory would be an elegant nothingness.

John P. Fackler, Jr.

Contents

III. MOLECULAR ORBITAL THEORY

IV. THE SYNTHESIS–LIGAND FIELD THEORY

V. DEVELOPMENT AND APPLICATION OF THE BONDING THEORIES

Contents by Author

Introduction

The commonly accepted beginning of our modern theory of valence occurred in Berkeley with G. N. Lewis [1]. This "nearly completely empirical" [2] theory preceded the development of wave mechanics by a decade. Yet the concepts of electron pair bonding and stability of rare gas electronic shells remain useful to the practicing chemist today. We now think of these electron pair bonds (and nonbonded lone pairs) as having specific directional characteristics. We also think of them in terms of the symmetry of molecules. In fact, rather useful structural theories [3] have been developed from those concepts which are able to correlate molecular geometries. The valence shell electron pair repulsion model has given a sufficiently consistent account of the details of molecular structure to allow prediction (prior to structure determination) that the coordination of fluorine atoms to xenon in XeF_4 is square, for example.

Very shortly after wave mechanics gained acceptance among physicists, it was recognized by some, particularly Eugene P. Wigner, that "the actual solution of quantum mechanical equations is, in general, so difficult that one obtains by direct calculations only crude approximations to the real solutions. It is gratifying, therefore, that a large part of the relevant results can be deducted by considering the fundamental symmetry operations [4]. Wigner suggests that what he originally thought to be the most important result of symmetry, the concept of parity, has been superseded by the "recognition that almost all rules of spectroscopy follow from the symmetry of the problem."

The idea of symmetry has had a profound impact on the development of the chemical sciences. This fact certainly needs no documentation to those familiar with the stereochemical literature in organic chemistry [5, 6]. L. Pasteur, J. H. van't Hoff [7], and J. A. LeBel [8] stand out as giants for having recognized early in the development of organic chemistry the relationship between optical activity and molecular asymmetry. Prior to 1950, chemists, with few exceptions, did not apply group theoretical techniques to symmetry-related problems. Indeed, only recently have these techniques been applied to problems dealing with molecular asymmetry [9]. Consequently, chemical theory that relied on the concepts of mathematical groups was relegated to the realm of the theoretician.

Two events appear to have changed all of this. The impact of the book *Chemical Applications of Group Theory* by Frank Albert Cotton [10] has been enormous. Cotton's translation of the mathematics of abstract group theory to a language readable by the average chemist has set the stage for the use of symmetry concepts in all branches of structural and reaction chemistry. Prior to this book, few texts were available to pique the chemist's interest in the topic. E. P. Wigner's 1931 classic *Gruppentheorie* [11] unfortunately was not translated into English until 1960.

The electronic revolution and its child, the high-speed digital computer, have brought a detailed understanding of molecular structure within the grasp of all branches of the chemical sciences. Symmetry by its very nature plays an important role in any complete description of ordered arrays—crystals suitable for diffraction of x-rays or neutrons. Space groups that describe the symmetry present in crystalline materials contain most of the chemically important point groups (32) as a subset. While the mathematical concepts of groups [12] were used very early by practicing crystallographers [13], the chemist was shielded from the details until modern computers brought the techniques into nearly every university or industrial research laboratory. The names C. Hermann, K. Lonsdale, C. Mauguin, M. von Laue, M. J. Buerger, W. H. and W. L. Bragg, and others are meaningful to chemists today as a result of the widespread use of x-ray crystallography.

Laue writes in his historical introduction to the 1965 edition of the *International Tables for X-Ray Crystallography* [14] that the concept of three-dimensional symmetry in crystals was introduced as early as 1824 by L. A. Seeber. The mathematical problems raised by ordered arrays led to significant studies by M. L. Frankenheim and A. Bravais in 1850, L. Sohncke in 1879, and finally culminated in the work of E. S. Federov and A. Schoenflies in 1890. Thus, before the end of the nineteenth century all of the mathematical concepts relating to the description of the 230 three-dimensional space groups, 32 crystallographic point groups, and 14 Bravais lattices had been discovered. The descriptive languages of Hermann, Mauguin, and Schoenflies are employed today by chemists [15] to describe the symmetry properties of molecules and crystals.

This collection of Benchmark papers will not deal with the concepts of symmetry as applied to the theory of ordered arrays. It is a matter of some regret that the size of this volume prevents their inclusion.

Symmetry has been applied with considerable success to the development of three fairly distinct areas of chemical theory. In the broadest sense these areas are (1) electronic structure, (2) vibrational behavior, and (3) chemical reactivity. The first two areas developed simultaneously and are fairly complete, while research in the latter area is still in its early stages of evolution, particularly with regard to inorganic materials. The articles presented here have been grouped to conform to this pattern.

Bethe's classic on term splitting in crystals introduces the part on electronic structure. Papers 1–7 deal with the origins of crystal field theory and its development, largely by Van Vleck. Papers 8–11 develop the covalent bond and molecular orbital theories of Pauling and Mulliken. Papers 12–29 synthesize these ideas into ligand field theory (Papers 12–15) and apply the concepts to "real" molecular systems (Papers 16–29).

With Wigner's *Göttingen Nachrichten* article, Paper 31, on vibrational symmetry, published in 1930, the part on symmetry applied to vibrational problems begins. Wilson's classic papers, Papers 32–34, are followed by some well-written review articles. The Meister–Cleveland review articles, Papers 35 and 36, bring vibrational theory to the practicing chemist with the same degree of forcefulness that the Tanabe–Sugano articles, Papers 17 and 18, and the Orgel article, Paper 19, inform the chemist of the power of symmetry in describing the electronic spectra and colors of coordination compounds.

Beginning with the 1963 Longuet-Higgins article, Paper 38, symmetry is used to deal with the nonvibronic motional behavior of molecules. The Woodward–Hoffman articles, Nos. 39, 41, and 42, and the Longuet-Higgins–Abrahamson article, Paper 40, initiate the use of symmetry rules to understand mechanisms of certain chemical reactions, notably concerted organic reactions. It can be expected that similar relationships will be developed which are readily applicable to inorganic reactions. Indeed, some such papers have already appeared [16, 17]. However, owing to the breadth of chemistry encompassed by inorganic reaction mechanisms, the resultant questionable generality [18] of any all-encompassing theory, and the fact that these theories have not yet been adequately tested, no article has been reproduced here which deals specifically with inorganic compounds.

I am aware of the fact that optical activity and the concepts dealing with chirality in molecules represent an important area of symmetry in chemical theory not covered by the articles included here. This area has been purposely shunned. Certainly I may be accused of neglect by adopting this approach, since much significant work has appeared, particularly from the Illinois School of John C. Bailar, which deals with optical activity and the stereochemistry of inorganic complexes. However, group theory has played only a minor part (if any) in this work to date, although there are indications that its impact will be substantially felt in the near future. Indeed, for purposes of selection of papers for this volume, I have excluded those not directly recognized to be an outgrowth of the intimate relationships which exist between molecular symmetry and the theory of groups.

In order for the reader to be cognizant of the underlying relationships which exist in mathematical groups, a thorough study of a text [10, 15, 19] is essential. As a reminder that many of these principles have been available to the English-speaking scientific community since 1936 in an easily readable paper, two pages from a review by Rosenthal and Murphy, *Vibrations of Polyatomic Molecules* [20], are reproduced.

General Concepts

1. Definition of a group

An ensemble of elements, E, A, B, G, $\cdots$ is said to form a *group* if the following four postulates are obeyed:

I. There is a rule of combination such that the combination of any two elements A and B of the group will give a third element of the group, C, called the product of A and B and written $C = AB$.

II. The associative law holds: $(AB)C = A(BC)$.

III. Every group contains a unit element, E, for which $AE = EA = A$.

IV. Every element of the group has an inverse, $X = A^{-1}$ such that $XA = A^{-1}A = E$. An element and its inverse may be identical and obviously, $E^{-1} = E$.

The members of a group may be considered as abstract elements to which a meaning is to be assigned later. They may be identified with real or complex numbers, matrices or the motions of a geometrical figure in space.

In what follows, the rule of combination is multiplication and if necessary matrix multiplication. In the case of ordinary multiplication, the four numbers ± 1, $\pm i$ form a group and it is easily verified that the four group postulates as given above are fulfilled.

If the group contains a finite number of elements, it is called a *finite group* and the number of its elements, h is called the *order* of the group. Infinite groups are of considerable interest in quantum mechanics but here we are mainly concerned with finite groups. If the commutative law holds, $AB = BA$, and the group is said to be *Abelian* but in general $AB \neq BA$.

A simple group has been given above. An example of a slightly more complex group of order 6 will now be given and some additional group properties presented. A group is completely defined if all of its products are known. Let the elements of the group, E, A, B, C, D, F be arranged in rows and columns in such a way that the products stand at the intersections. For example, the product AB will stand at the intersection of the row headed by A and the column by B. Let the multiplication table for this group be:

	E	A	B	C	D	F
E	E	A	B	C	D	F
A	A	B	E	D	F	C
B	B	E	A	F	C	D
C	C	F	D	E	B	A
D	D	C	F	A	E	B
F	F	D	C	B	A	E

(7)

It is easy to see that the six elements form a group and that the group postulates are obeyed. Every product is contained in the group and every element has an inverse. This group is not Abelian for $AC = CB \neq CA$.

2. Sub-groups

Consider some one element of the group, say X and form successive powers of this element. Since the element chosen and all of its powers are members of the group and since the group is finite, this series will eventually repeat itself. Let $X^n = E$, then

$$X,\ X^2,\ X^3,\ \cdots X^{n-1},\ X^n = E$$

is called the *period of* X and is indicated by the symbol $\{X\}$. If n is the smallest number for which $X^n = E$, n is called the order of X. The period of A in the group above is:

$$A,\ A^2 = B,\ A^3 = E, \text{ its order is 3.}$$

The period of B is B, $B^2 = A$, $B^3 = E$ and its order is also 3. The order of C is 2, however, since $C^2 = E$. The period of X forms a group itself since all the group postulates are fulfilled and it is said to be a sub-group of the original group, $\mathfrak{G}$. In the above group, the sub-groups are:

$$\{A\} = \{B\} = E,\ A,\ B,$$
$$\{C\} = E,\ C,$$
$$\{D\} = E,\ D,$$
$$\{F\} = E,\ F.$$

The order of a sub-group may be shown to be a divisor of the order of the whole group. Since the only divisors of 6 are 2 and 3 (besides 1 and 6) these are the only possible sub-groups.

3. Conjugate elements

If A, B, and X are elements of a group and

$$B = XAX^{-1}, \tag{8}$$

A and B are said to be conjugate to one another. The following laws about conjugate elements are almost self-evident and are readily verified using the group multiplication table given above:

I. Every element is conjugate with itself.
II. If A is conjugate with B, then B is conjugate with A.
III. If A is conjugate with B and also with C, then B and C are conjugate.

The elements conjugate with each other are called a *class*. The whole group then splits into a number of classes none of which contains elements in common. For the group (7), the class of A is A and B. For,

$$EAE^{-1}=A,\ AAA^{-1}=A,\ BAB^{-1}=A,\ CAC^{-1}=B,\ DAD^{-1}=B,\ FAF^{-1}=B.$$

Similarly, it can be shown that the class of C (also D and F) is C, D, F.

The unit element always forms a class by itself. There are then 3 classes in $\mathfrak{G}$: E; A, B; C, D, F. Elements of the same class have the same order, in the above example, 1, 3 and 2, respectively. The number of elements in any class is always a factor of the order of the whole group.

A *complex* is a set of elements from a group and is designated by German script letters. The complex of elements forming a class is always indicated by $\mathfrak{C}$. If the complex $\mathfrak{A}$ contains A, B, C, the product $C\mathfrak{A}$ will contain CA, CB, C^2. By the product of two complexes $\mathfrak{A}\mathfrak{B}$ is meant the product of every element of $\mathfrak{A}$ with every element of $\mathfrak{B}$ but products occurring more than once are only taken once. If the complex $\mathfrak{H}$ is a sub-group

$$\mathfrak{H}\mathfrak{H}=\mathfrak{H}^2=\mathfrak{H}$$

$\mathfrak{H}$ and $X\mathfrak{H}X^{-1}$ are called *conjugate sub-groups*.

4. Isomorphism

Two groups are said to be isomorphous if elements A, B and C of $\mathfrak{G}$ are associated with elements A', B' and C' of $\mathfrak{G}'$ in such a way that if $AB=C$, also $A'B'=C'$. If the multiplication tables of the two groups are known, there will be a one-to-one correspondence with the elements of the isomorphous groups throughout the tables, although the meaning of the elements will be different in each case. Instead of a simple isomorphism as illustrated here, it is possible to extend this concept to the case of general or multiple isomorphism where instead of having a one-to-one correspondence of elements in two groups, there may be several elements of one group isomorphous to a single element of another group.

Crystal Field Theory

I

Editor's Comments on Papers 1 Through 7

1 **Bethe:** *Termaufspaltung in Kristallen*

2 **Kramers:** *Théorie générale de la rotation paramagnétique dans les cristaux*

3 **Van Vleck:** *Theory of the Variations in Paramagnetic Anisotropy Among Different Salts of the Iron Group*

4 **Schlapp and Penney:** *Influence of Crystalline Fields on the Susceptibilities of Salts of Paramagnetic Ions. II. The Iron Group, Especially Ni, Cr, and Co*

5 **Gorter:** *Note on the Electric Field in Paramagnetic Crystals*

6 **Jordahl:** *Information Concerning Crystal Structures from Data on Magnetic Susceptibilities*

7 **Jahn and Teller:** *Stability of Polyatomic Molecules in Degenerate Electronic States. I. Orbital Degeneracy*

The idea that the electronic structure of an atom is perturbed by the symmetry of the arrangement of ligands surrounding it establishes the base for crystal field theory—the earliest theory to successfully explain colors, magnetism, and related electronic properties of transition metal salts. The theory simply recognizes that the potential-energy function, which is spherically symmetric for a free ion, must take on some characteristics in a molecule that are demanded by the reduced symmetry. Crystal field theory itself assumes that the atoms bound to the metal ion can be represented by point charges. However, even with this assumption, now known to be naive, many of the spectral and magnetic properties of transition metal compounds could be understood.

Credit for the concept of term splitting by reduction from spherical symmetry clearly goes to Hans A. Bethe. However, it was John H. Van Vleck and his students who nurtured the growth of the theory and established its applicability. The first seven articles presented here develop this theory to its practical limit.

Bethe (Paper 1) was the first person to point out the power of symmetry as applied to the electronic structure of atoms in crystals. His paper became the basis for almost all later developments in crystal field theory (CFT) and its supplement, ligand field theory (LFT). With this paper crystal field theory was born. Bethe's application of group theory to the electronic spectra of ions paves the way for the application of the theory to all other symmetry-related quantum mechanical problems. Other quantitive theories dealing with the effect of symmetry on the electronic properties of molecules, such as ligand field theory and molecular orbital theory, take cognizance of Bethe's contribution.

Bethe received the Nobel Prize in physics in 1967 for his contributions to atomic and nuclear physics. The paper presented here has received much less recognition in the physics community than his contributions to nuclear theory; however, its impact on chemistry has been paramount.

Van Vleck and his students in the early 1930s were concerned with magnetic problems relating to transition metal salts. Expanding on the principles enumerated by Bethe, they developed CFT to a point where it dealt satisfactorily with the splitting of spectroscopic terms of metal ions in crystals. Magnetic susceptibilities for ions in field of cubic symmetry could now be understood (Papers 2–6). The work by H. A. Kramers [21] was particularly important to this development, as was the book *The Theory of Electric and Magnetic Susceptibilities* by Van Vleck [22]. References to other significant contributions can be found in the articles reproduced.

Paper 3, one of three by Van Vleck included here, directly applies Bethe's theory to the magnetic properties of some first-row transition-metal ions in crystals. Van Vleck and his students at Harvard University placed the crystal field theory on a firm basis with their contributions to both theory and experiment.

Van Vleck's contributions were so significant that it is often suggested that crystal field theory be called the "Bethe–Van Vleck theory."

The part on crystal field theory is completed with the presentation of the 1937 paper by H. A. Jahn and E. Teller (Paper 7). The results [23] of the remarkable theorem developed in this paper proved that unless all atoms of a molecule having orbital degeneracy lie in a straight line, the nuclear configuration is unstable. Since the underlying assumptions are based on group theory, accidental degeneracy is excluded. Spin degeneracy in even-electron systems (non-Kramers multiplets) should, in principle, also produce an instability, although the effect "will be small." It was Van Vleck who presented the first concrete illustration of the "Jahn–Teller effect" a couple of years later [24]. A new rigorous derivation of the Jahn–Teller theorem appeared in 1959 [25]. The Jahn–Teller theory has been widely applied to the interpretation of electronic spectra and structure in organic and inorganic systems. Studies of the six-coordinate complexes of copper(II), high-spin chromium(II) and manganese(III), and low-spin cobalt(II) are particularly significant in this context.

Reprinted from *Ann. Phys.*, **5**, 133–208 (1929)

1

(A translation of § 1 and 2 of this paper will be found on pages 86–90)

Termaufspaltung in Kristallen

Von H. Bethe

(Mit 8 Figuren)

Der Einfluß eines elektrischen Feldes von vorgegebener Symmetrie (Kristallfeld) auf ein Atom wird wellenmechanisch behandelt. Die Terme des Atoms spalten auf in einer Weise, die von der Symmetrie des Feldes und vom Drehimpuls l (bzw. j) des Atoms abhängt. s-Terme spalten allgemein, p-Terme in Feldern von kubischer Symmetrie *nicht* auf. Für den Fall, daß die einzelnen Elektronen des Atoms separat behandelt werden dürfen (aufgehobene Wechselwirkung im Atom), werden zu jedem Term im Kristall die Eigenfunktionen nullter Näherung angegeben; aus diesen ergibt sich eine für den Term charakteristische Gruppierung der Elektronendichte nach den Symmetrieachsen des Kristalls. — Die *Größe* der Termaufspaltung bewegt sich in der Ordnung einiger hundert cm^{-1}. — Bei tetragonaler Symmetrie kann ein quantitatives Maß für die Abweichung von der kubischen Symmetrie definiert werden, welches eindeutig die stabilste Elektronenanordnung im Kristall bestimmt.

§ 1. Einleitung

Für die Wellenmechanik der Kristalle bieten sich von vornherein zwei Ausgangspunkte: Entweder wir behandeln den Kristall als fertiges Ganzes, beschreiben ihn also etwa durch ein räumlich-periodisches Potentialfeld und Eigenfunktionen von derselben räumlichen Periodizität, auf deren Modulation durch die feinere Struktur der im Kristall enthaltenen Atome wir erst in zweiter Linie eingehen. Dieser Weg ist vor allem von Bloch[1]) zur Behandlung der elektrischen Leitfähigkeit verwendet worden und erscheint als besonders angemessen im Falle weitgehend „freier" Elektronen. Andererseits kann man vom freien Atom ausgehen und dessen Störung im Kristall analog zum Verfahren von London und Heitler[2]) behandeln, man wird dies Verfahren vor allem zur Berechnung von Gitterenergie, Gitterabständen usw. benutzen können.

1) F. Bloch, Ztschr. f. Phys. **52.** S. 555. 1929.

2) F. London u. W. Heitler, Ztschr. f. Phys. **44.** S. 455. 1927.

Eine Störung des freien Atoms beim Einfügen in einen Kristall ist aus zweierlei Ursachen zu erwarten: Einerseits wird das Atom in Elektronenaustausch mit den anderen Atomen des Kristalls treten, d. h. seine Permutationsgruppe wird geändert. Dieser Austauscheffekt wird ganz analog zu behandeln sein wie in einem Molekül, höchstens besteht ein rein quantitativer Unterschied bezüglich der Anzahl der Nachbarn, mit denen ein Austausch stattfinden kann. *Andererseits* wirkt auf das Atom im Kristall ein von den übrigen Atomen herrührendes elektrisches Feld von bestimmter Symmetrie, welches die Richtungsentartung des freien Atoms aufhebt. In dieser verschiedenen räumlichen Symmetrie ist der *qualitative* Unterschied zwischen dem Molekülproblem und dem Kristallproblem zu suchen. Wir wollen in dieser Arbeit die durch die Symmetrie des Kristalls bedingten spezifischen Störungen des freien Atoms behandeln, ohne vorerst auf die Elektronenaustauschphänomene zwischen den Atomen des Kristalls einzugehen; gleichzeitig wollen wir damit einen Ausgangspunkt für die Behandlung der letzteren schaffen.

Ein elektrisches Feld von bestimmter Symmetrie wird eine Aufspaltung der Terme des ungestörten Atoms verursachen, die analog ist zur Starkeffektaufspaltung und charakteristisch sein wird eben für die Symmetrie des Feldes, d. h. für die Symmetrie der Lage des Atoms im Kristall. Die Anzahl der Komponenten, in die ein Term des freien Atoms aufspaltet, wächst mit abnehmender Symmetrie. Der Betrag der Aufspaltung kann von sehr verschiedener Größenordnung sein, und man wird je nachdem drei Fälle zu unterscheiden haben.

1. Starkeffektaufspaltung im Kristallfeld *groß* gegen den Abstand *verschiedener* Multipletts: Die Beeinflussung des Atoms durch das Kristallfeld überwiegt die Wechselwirkung der Elektronen im Atom und hebt ihre Kopplung in erster Näherung auf. Wir gehen dann aus von einem Modell des freien Atoms, bei dem nur die Besetzungszahlen der durch Hauptquantenzahl n_i und Azimutalquantenzahl l_i (des Einzelelektrons) charakterisierten Quantenzellen vorgegeben, dagegen die Termaufspaltung infolge der Austauschentartung außer acht gelassen ist. (Die elektrostatischen Wirkungen der Elektronen aufeinander können wir etwa nach der Hartreeschen Methode des self-consistent field einbeziehen.)

In *erster* Näherung berücksichtigen wir dann die Störung der einzelnen, außerhalb von abgeschlossenen Schalen befindlichen Elektronen durch das Feld der anderen Atome im Kristall, d. h. die möglichen Einstellungen des Bahndrehimpulses l_i des einzelnen Elektrons gegenüber den Kristallachsen, und die hierdurch bedingte „Starkeffektaufspaltung" der Terme des Atoms. In *zweiter* Näherung wäre dem Elektronenaustausch innerhalb des Atoms Rechnung zu tragen; hierdurch findet im allgemeinen eine weitere Termaufspaltung statt, deren Größenordnung natürlich dem Abstand verschiedener Multipletts des freien Atoms entspricht. Schließlich bedingt die Wechselwirkung zwischen Bahndrehimpuls und Spin nochmals in üblicher Weise die Multiplettaufspaltung.

2. Kristallaufspaltung von *mittlerer* Größe, d. h. *klein* gegen den Abstand *verschiedener* Multipletts, aber *groß* gegen die Termdifferenzen eines und desselben Multipletts. In diesem (wohl meist zutreffenden) Fall ist auszugehen von dem freien Atom *mit* Berücksichtigung der Termaufspaltung durch Austauschentartung, aber *ohne* Berücksichtigung der Wechselwirkung von Spin und Bahn. Sodann wäre dieses Atom in den Kristall einzusetzen und die Einstellungen seines Gesamt-Bahndrehimpulses l zu den Achsen des Kristalls sowie die zugehörigen Termwerte zu untersuchen, und schließlich die Wechselwirkung zwischen Spin und Bahn bei festgehaltener Orientierung der letzteren im Kristall.

3. Kristallaufspaltung *klein* gegen den Termabstand *innerhalb* eines Multipletts. Das vollkommen „fertige" Atom mit Berücksichtigung der Wechselwirkung zwischen Elektronenbahnen und Spin wird der Störung durch das elektrische Feld des Kristalls unterworfen. Dabei stellt sich der Gesamtdrehimpuls j des Atoms relativ zu den Kristallachsen ein, und nicht mehr, wie in Fall 1 und 2, der *Bahn*drehimpuls, weil dieser jetzt mit dem Spin fest verkoppelt bleibt, während die Kopplung in Fall 1 und 2 durch das Kristallfeld aufgehoben wird.

Wir werden folgende Quantenzahlen zur Beschreibung der Starkeffekt-Aufspaltung benutzen:

λ azimutale Kristallquantenzahl zur Kennzeichnung der Einstellung des Bahndrehimpulses des Atoms im Kristall.

λ_i azimutale Kristallquantenzahl des einzelnen ($i$ten), Elektrons.

l_i gewöhnliche azimutale Quantenzahl des einzelnen Elektrons im freien Atom.

μ innere Kristallquantenzahl (Einstellung des Gesamtdrehimpulses j des Atoms im Kristall).

Da es sich bei den Kristallquantenzahlen im allgemeinen um keine „echten" Quantenzahlen handelt, die etwa als Drehimpuls um eine Achse oder ähnlich gedeutet werden könnten, sondern nur um die Unterscheidung verschiedener gruppentheoretischer Darstellungseigenschaften der einzelnen Terme, so schreiben wir sie nicht als Zahlen, sondern als griechische Buchstaben.

Ein Atom ist dann in den drei oben erwähnten Fällen charakterisiert durch folgende Quantenzahlen:

1. Große Kristallaufspaltung: n_i, l_i, λ_i, λ, μ.
2. Mittlere Kristallaufspaltung: n_i, l_i, l, λ, μ.
3. Kleine Kristallaufspaltung: n_i, l_i, l, j, μ.

Zunächst werden wir die drei möglichen Fälle — Einstellung des Bahnimpulses des Einzelelektrons, des gesamten Bahnimpulses oder des Gesamtdrehimpulses des Atoms im Kristall — gemeinsam behandeln und die jeweilige Anzahl der Aufspaltungskomponenten eines Terms bei gegebenem Drehimpuls (Darstellung der Drehgruppe) und gegebener Symmetrie der Lage des Atoms im Kristall gruppentheoretisch berechnen. Dabei wird der Fall der Einstellung eines halbzahligen Gesamtdrehimpulses (zweideutige Darstellung der Drehgruppe) gesondert zu behandeln sein. Hierauf werden wir die zu den einzelnen Aufspaltungskomponenten gehörigen winkelabhängigen Faktoren der Eigenfunktionen nullter Näherung angeben, wie sie für Austauschrechnungen benötigt werden. Dann werden wir die drei durch die Größenordnung der Kristallaufspaltung unterschiedenen Fälle einzeln diskutieren, und dabei besondere Rücksicht auf die Symmetrie der Elektronendichteverteilung nehmen. Schließlich werden wir für Ionenkristalle die Größe der Aufspaltung berechnen.

I. Gruppentheoretische Lösung

§ 2. Gang der Lösung

Die Schrödingersche Differentialgleichung des freien Atoms ist bekanntlich invariant gegenüber beliebigen Drehungen des Koordinatensystems sowie gegenüber einer Inversion am

Kern (und gegenüber Vertauschung der Elektronen, die uns aber vorerst nicht interessiert). Ihre Substitutionsgruppe[1]) ist u. a. die Drehungsgruppe der Kugel, welche je eine irreduzible Darstellung von der Dimension $2l + 1$ besitzt ($l = 0, 1, 2 \ldots$).

Setzen wir nun das Atom in einen Kristall hinein, so reduziert sich die Symmetrie des Potentialansatzes von Kugelsymmetrie auf die Symmetrie der Lage, welche das Atom im Kristall einnimmt, z. B. bei Einfügung in einen Kristall vom NaCl-Typ auf kubisch-holoedrische, beim ZnS-Typ auf kubisch-hemiedrische Symmetrie. Die Substitutionsgruppe der Schrödingerschen Differentialgleichung für das betrachtete Atom umfaßt nur mehr solche Symmetrieoperationen, welche die Lage des Kerns des betrachteten Atoms ungeändert lassen und gleichzeitig den ganzen Kristall in sich überführen (Symmetriegruppe des Kristallatoms). Der Weg, eine solche Verminderung der Symmetrie zu behandeln, ist allgemein von Wigner angegeben worden. Man hat einfach die zu einem bestimmten Term des ungestörten Atoms gehörige Darstellung der ursprünglichen Substitutionsgruppe als Darstellung der neuen Substitutionsgruppe, welche ja eine Untergruppe der früheren ist, auszureduzieren, und erhält damit Anzahl und Vielfachheit der Terme, in welche der betrachtete ungestörte Term bei der vorgenommenen Verminderung der Symmetrie zerfällt. Dabei ist es bekanntlich ganz belanglos, ob die Störung klein oder groß und überhaupt von welcher Form sie im einzelnen ist, ob man die Störungsenergie nur in erster oder in beliebig hoher Ordnung berechnet: Die Feststellung einer bestimmten *Symmetrie* des Störungspotentials genügt vollkommen.

Um die $2l + 1$-dimensionale Darstellung der Drehgruppe praktisch als Darstellung einer bestimmten Kristallsymmetriegruppe (z. B. der Oktaedergruppe) auszureduzieren, verwenden wir den fundamentalen Satz der Gruppentheorie: Jede reduzible Darstellung einer Gruppe läßt sich auf eine und nur eine Weise in ihre irreduziblen Bestandteile zerlegen, dabei ist der Charakter jedes Gruppenelements in der reduziblen Darstellung gleich der Summe der Charaktere, die dem Element in den irreduziblen Bestandteilen zukommen. Wir müssen also den Charakter jeder Symmetrieoperation des Kristalls, welche den Atomkern in seiner Lage beläßt, kennen: einerseits in den

1) E. Wigner, Ztschr. f. Phys. **43.** S. 624. 1927.

irreduziblen Darstellungen der Substitutionsgruppe des Kristallatoms; andererseits in den irreduziblen Darstellungen der Substitutionsgruppe des freien Atoms.

Jede Symmetrieoperation des Kristallatoms läßt sich nun zusammensetzen aus einer reinen Drehung und evtl. einer Inversion am Kern, ebenso wie jede Symmetrieoperation des freien Atoms. Wir brauchen aber nur den Charakter einer vorgegebenen *Drehung* in der $2l+1$-dimensionalen Darstellung der Drehgruppe zu berechnen, fügt man der Drehung eine Inversion hinzu, so multipliziert sich der Charakter einfach mit $+1$ oder -1, je nachdem ob es sich um einen positiven oder einen negativen Term handelt.[1]) Die allgemeinste Drehung ist diejenige um eine beliebige Achse um den Winkel Φ. Ihre einfachste Darstellung wird vermittelt durch die Transformation der auf die Drehachse bezogenen Kugelfunktionen mit unterem Index l:

$$f_\mu(x) = P_l^\mu(\cos\vartheta)\, e^{i\mu\varphi}.$$

Unsere Drehung R soll ϑ in sich, φ in $\varphi + \Phi$ überführen, also $f_\mu(x)$ in:

$$f_\mu(R\,x) = P_l^\mu(\cos\vartheta)\, e^{i\mu(\varphi+\Phi)},$$

d. h. unsere Drehung wird dargestellt durch die Matrix[1]):

$$\begin{pmatrix} e^{-il\Phi} & 0 & \ldots\ldots\ldots & 0 \\ 0 & e^{-i(l-1)\Phi} & \ldots & 0 \\ \ldots & \ldots & \ldots & \ldots \\ 0 & 0 & \ldots\ldots & e^{il\Phi} \end{pmatrix}$$

mit dem Charakter:

$$\chi(\Phi) = \frac{\sin(l+\frac{1}{2})\,\Phi}{\sin\frac{1}{2}\,\Phi}. \tag{1}$$

§ 3. Holoedrische und hemiedrische Symmetrie

Wenn ein Atom im Kristall holoedrische Symmetrie besitzt, so muß es Symmetriezentrum des ganzen Kristalles sein. Dann lassen sich seine Symmetrieoperationen in zwei Kategorien von je gleichviel Elementen zusammenfassen, die reinen Drehungen einerseits und die Spiegelungen und Drehspiegelungen andererseits

---

1) Vgl. E. Wigner u. J. v. Neumann, Ztschr. f. Phys. **49.** S. 73, 91. 1928.

seits. Dabei bilden die reinen Drehungen einen Normalteiler der Gruppe. Aus jeder Klasse reiner Drehungen geht durch Multiplikation mit der Inversion am Atomkern eine Klasse von Spiegelungen oder Drehspiegelungen hervor.

Denn, nehmen wir an, C sei eine Klasse des aus den reinen Drehungen bestehenden Normalteilers N, d. h., wenn X irgendein Element des Normalteilers:

$$X^{-1} C X = C .$$

J sei die Inversion, welche natürlich mit jeder Drehung vertauschbar ist ($J^2 = E$). Dann folgt zunächst, daß C auch eine Klasse in der *ganzen* Gruppe bildet:

$$(J X)^{-1} C J X = C .$$

Ferner bildet aber auch $J C$ eine Klasse der Gruppe:

$$X^{-1} J C X = J C ,$$

w. z. b. w. (Vgl. das Beispiel im folgenden Paragraphen.)

Die Gruppe besteht also aus genau doppelt so vielen Elementen in genau doppelt so vielen Klassen wie der erwähnte Normalteiler, und besitzt demnach doppelt so viele Darstellungen wie dieser, nämlich erstens die positiven Darstellungen, in welchen die Inversion durch die Einheitsmatrix dargestellt wird und die Spiegelungsklasse $J C$ denselben Charakter hat wie die Drehklasse C, zweitens die negativen Darstellungen, bei denen die Inversion durch die negative Einheitsmatrix dargestellt wird und bei Multiplikation einer Klasse C mit der Inversion der Charakter sich mit -1 multipliziert. Man sieht sofort durch Vergleich mit der Definition Wigners und v. Neumanns[1]): Ein positiver Term des freien Atoms zerfällt in einer Lage von holoedrischer Symmetrie in lauter positive, ein negativer in lauter negative Kristallterme. Im übrigen braucht man sich um die Inversion nicht weiter zu kümmern, sondern nur die Darstellungen der Kugeldrehgruppe als Darstellungen der aus den reinen Drehungen des Kristalls um den Atomkern bestehenden Gruppe auszureduzieren, welche wir als Kristalldrehgruppe (tetragonale, hexagonale Drehgruppe) bezeichnen wollen.

Durch Fortfall der Hälfte aller Symmetrieelemente und der Hälfte aller Symmetrieklassen entsteht aus der holoedrischen

---

1) E. Wigner u. J. v. Neumann a. a. O. S. 91.

eine hemiedrische Symmetriegruppe, vorausgesetzt, daß die übrig bleibenden Elemente eine Gruppe bilden. Gehört die Inversion auch zu den Elementen der *hemiedrischen* Symmetriegruppe, so kann man wie bei holoedrischer Symmetrie schließen und die Betrachtung auf die aus reinen Drehungen bestehenden Elemente der Gruppe beschränken. Die Aufspaltung eines Terms des freien Atoms liefert dann im allgemeinen *mehr* Komponenten als bei holoedrischer Symmetrie. Ist die Inversion *nicht* Element der hemiedrischen Symmetriegruppe, so kann man die hemiedrische Gruppe durch Multiplikation mit der Inversion zur holoedrischen Symmetriegruppe ergänzen, wobei wie oben jeder Klasse der hemiedrischen Gruppe *zwei* Klassen der holoedrischen entsprechen. Also entspricht jede hemiedrische Symmetriegruppe, welche die Inversion nicht enthält, Klasse für Klasse dem aus den reinen Drehungen bestehenden Normalteiler der holoedrischen Gruppe, hat also auch genau dieselben Darstellungen wie dieser. (Der Normalteiler selbst bildet natürlich auch eine hemiedrische Symmetriegruppe, die wir als Dreh-Hemiedrie bezeichnen können, daneben sind aber auch andere Hemiedrien ohne Inversionszentrum denkbar.)

Nun wollen wir eine Darstellung einer hemiedrischen Symmetriegruppe positiv-gleich einer Darstellung der Drehhemiedriegruppe nennen, wenn entsprechende Klassen der beiden Gruppen durchweg denselben Charakter besitzen. Negativ gleich wollen wir zwei Darstellungen der Gruppen nennen, wenn nur die Drehklassen der hemiedrischen Symmetriegruppe genau den gleichen Charakter haben wie die entsprechenden Klassen der Drehhemiedriegruppe, die Charaktere der Spiegelungsklassen der hemiedrischen Symmetriegruppe dagegen entgegengesetzt gleich sind den Charakteren der entsprechenden Drehklassen in der vergleichsweise herangezogenen Darstellung der Drehhemiedriegruppe.

Ein Term des freien Atoms spaltet bei einer hemiedrischen Symmetrie der Lage des Atoms im Kristall, bei welcher dieses kein Inversionszentrum bildet, in genau ebensoviele Komponenten auf wie bei der zugehörigen holoedrischen Symmetrie. Die Darstellungen der Terme, die aus einem positiven Terms des freien Atoms in einer Lage von hemiedrischer Symmetrie entstehen, sind dabei positiv-gleich den irreduziblen Darstellungen der Kristalldrehgruppe (Drehhemiedrie), welche man durch Ausreduktion derjenigen Darstellung der Kugeldrehgruppe erhält, die zu dem Term des freien Atoms gehört. Aus einem negativen Term des freien Atoms dagegen entstehen in einer Lage von hemiedrischer Symmetrie Terme, deren Darstellungen negativ-gleich den entsprechenden Darstellungen der Kristall-Drehgruppe sind.

Nunmehr gehen wir an die Betrachtung einzelner Symmetriegruppen.

§ 4. Kubisch-holoedrische Symmetrie (z. B. NaCl, Ca in CaF_2)

Wir nehmen an, daß sämtliche Drehachsen und Spiegelebenen der kubisch-holoedrischen Symmetrie durch den Kern des betrachteten Atoms hindurchgelegt werden können. Die 24 reinen Drehungen der kubisch-holoedrischen Symmetriegruppe, welche mit den Drehungen der Oktaedergruppe identisch sind, ordnen sich in 5 Klassen:

E Identität.

C_2 Drehungen um die drei 4zähligen Achsen (Würfelkanten) um π (3 Elemente).

C_3 Drehung um die 4zähligen Achsen um $\pm\frac{\pi}{2}$ (6 Elemente).

C_4 Drehungen um die sechs 2zähligen Achsen (Flächendiagonalen des Würfels) um π (6 Elemente).

C_5 Drehungen um die vier 3zähligen Achsen (Würfeldiagonalen) um $\pm\frac{2\pi}{3}$ (8 Elemente).

Durch Multiplikation mit der Inversion entstehen hieraus:

J Inversion.

JC_2 Spiegelung an den zu den 4zähligen Achsen senkrechten Ebenen (3 Elemente).

JC_3 Drehspiegelung an den 4zähligen Achsen (Drehung um $\pm\frac{\pi}{2}+$ Spiegelung an der zur Achse senkrechten Ebene) (6 Elemente).

JC_4 Spiegelung an den durch je eine 4zählige und eine dazu senkrechte 2zählige Achse gelegten Ebenen (6 Elemente).

JC_5 Drehung um die 3zählige Achse (6zählige Drehspiegelachse) um $\pm\frac{\pi}{3}$ mit gleichzeitiger Spiegelung an der zur Achse senkrechten Ebene (8 Elemente).

Die aus 24 Elementen bestehende Oktaedergruppe besitzt also fünf irreduzible Darstellungen. Die Quadratsumme der Dimensionen aller Darstellungen muß gleich der Anzahl der Gruppenelemente sein; wir müssen also 24 als Summe von 5 Quadraten darstellen. Die einzige Möglichkeit ist die Zerlegung:

$$24 = 3^2 + 3^2 + 2^2 + 1^2 + 1^2 .$$

D. h. die Oktaedergruppe besitzt zwei dreidimensionale, eine zweidimensionale und zwei eindimensionale Darstellungen. Jeder Term des ungestörten Atoms von höherer als dreifacher Richtungsentartung *muß* also im Kristall von kubisch-holoedrischer Symmetrie aufspalten. Um nun die Charaktere der Darstellungen der Oktaedergruppe zu berechnen, gehen wir aus von der Formel[1]):

$$h_i h_k \chi_i \chi_k = \chi_1 \Sigma c_{ikl} h_l \chi_l .$$

Die Produkte je zweier Klassen sind:

$$\begin{aligned}
C_2^2 &= 3E + 2C_2 & C_2 C_3 &= C_3 + 2C_4 \\
C_3^2 &= 6E + 2C_2 + 3C_5 & C_2 C_4 &= 2C_3 + C_4 \\
C_4^2 &= 6E + 2C_2 + 3C_5 & C_2 C_5 &= 3C_5 \\
C_5^2 &= 8E + 8C_2 + 4C_5 & C_3 C_4 &= 4C_2 + 3C_5 \\
& & C_3 C_5 = C_4 C_5 &= 4C_3 + 4C_4
\end{aligned}$$

Daraus ergeben sich die folgenden Charakterensysteme.

Tabelle 1

Darstellung	Klasse				
	E	C_2	C_3	C_4	C_5
Γ_1	1	1	1	1	1
Γ_2	1	1	−1	−1	1
Γ_3	2	2	0	0	−1
Γ_4	3	−1	1	−1	0
Γ_5	3	−1	−1	1	0

1) A. Speiser, Theorie der Gruppen. 2. Aufl. S. 171.

In der $2l+1$-dimensionalen Darstellung der Drehgruppe entsprechen den Klassen der Oktaedergruppe gemäß dem Ende von § 2 die folgenden Charaktere:

Klasse E Drehung um $\Phi = 0$ $\qquad \chi_E = 2l+1$

$$\text{„}\ C_2 \text{ u. } C_4 \quad \text{„} \quad \text{„} \quad \Phi = \pi \quad \chi_2 = \chi_4 = \frac{\sin\left(l+\frac{1}{2}\right)\pi}{\sin\frac{\pi}{2}} = (-)^l$$

$$\text{„}\ C_3 \quad \text{„} \quad \text{„} \quad \Phi = \frac{\pi}{2} \quad \chi_3 = \frac{\sin\left(\frac{l}{2}+\frac{1}{4}\right)\pi}{\sin\frac{\pi}{4}} = (-)^{\left[\frac{l}{2}\right]}$$

$$\text{„}\ C_5 \quad \text{„} \quad \text{„} \quad \Phi = \frac{2\pi}{3} \quad \chi_5 = \frac{\sin\left(\frac{2l}{3}+\frac{1}{3}\right)\pi}{\sin\frac{\pi}{3}}$$

$$= \begin{cases} 1, & \text{wenn } l = 3m \\ 0, & \text{„} \quad l = 3m+1 \\ -1, & \text{„} \quad l = 3m+2. \end{cases}$$

Im einzelnen entsprechen den Darstellungen der Kugeldrehgruppe, die in der folgenden Tabelle aufgeführten Charaktere der einzelnen Klassen der Oktaedergruppe, aus denen sich die rechts verzeichneten Zerlegungen in irreduzible Darstellungen der Oktaedergruppe ergeben:

Tabelle 2

Charaktere der Klassen der Oktaedergruppe in der $2l+1$-dimensionalen Darstellung der Drehgruppe						Zerlegung der $2l+1$-dimensionalen Darstellung der Drehgruppe in irreduzible Darstellungen der Oktaedergruppe	Anzahl der Terme
l	E	C_2	C_3	C_4	C_5		
0	1	1	1	1	1	Γ_1	1
1	3	−1	1	−1	0	Γ_4	1
2	5	1	−1	1	−1	$\Gamma_3+\Gamma_5$	2
3	7	−1	−1	−1	1	$\Gamma_2+\Gamma_4+\Gamma_5$	3
4	9	1	1	1	0	$\Gamma_1+\Gamma_3+\Gamma_4+\Gamma_5$	4
5	11	−1	1	−1	−1	$\Gamma_3+2\Gamma_4+\Gamma_5$	4
6	13	1	−1	1	1	$\Gamma_1+\Gamma_2+\Gamma_3+\Gamma_4+2\Gamma_5$	6
12	25	1	1	1	1	$2\Gamma_1+\Gamma_2+2\Gamma_3+3\Gamma_4+3\Gamma_5$	11

S- und P-Term des ungestörten Atoms spalten demnach *nicht auf*, wenn man das Atom in eine Lage von kubisch-holoedrischer Symmetrie in einem Kristall bringt. Von D-Term ab spaltet *jeder* Term auf, und zwar allgemein in höchstens so viele Komponenten wie die azimutale Quantenzahl l angibt, also in weniger Komponenten, als beim Starkeffekt im homogenen Feld auftreten.[1]) Für $l = 12$ (allgemein für $l = 12\,m$) sind die Charaktere aller Klassen der Oktaedergruppen gleich $+1$, nur der Charakter des Einheitselements $= 25$ $(24\,m + 1)$. Also enthält die 25-dimensionale Darstellung der Drehgruppe, als Darstellung der Oktaedergruppe ausreduziert, genau einmal die reguläre Darstellung und noch einmal die identische Darstellung dieser Gruppe extra. (In der regulären Darstellung sind ja alle Charaktere Null, nur der des Einheitselements gleich der Anzahl der Gruppenelemente, hier 24.) Generell folgt, daß die $2\,(12\,m + k) + 1$-dimensionale Darstellung der Drehgruppe m mal die reguläre Darstellung der Oktaedergruppe enthält und dazu die gleichen Darstellungen wie die $2\,k + 1$-dimensionale Darstellung der Drehgruppe. Reduziert man die $24\,m - (2\,k + 1)$-dimensionale Darstellung aus, so fehlen gerade die Bestandteile der $2\,k + 1$-dimensionalen Darstellung an der m-fachen regulären Darstellung.

Anmerkung: Fast das gleiche Problem wie das in diesem Paragraphen und in §§ 12, 13 behandelte hat bereits Ehlert[2]) von einer ganz anderen Seite her gelöst. Es handelt sich bei ihm um die Aufsuchung von Kernschwingungsfunktionen mit vorgegebenem Hundschen Symmetriecharakter bezüglich Vertauschung der Protonen eines Moleküls vom Typ CH_4. Dies führt u. a. auf die Untersuchung der Transformation von Kugelflächenfunktionen bei Vertauschung der Eckpunkte eines Tetraeders. Diese Vertauschungen bilden eine kubisch-hemiedrische Gruppe, welche die Klassen E, C_2, C_5, JC_3 und JC_4 der holoedrischen Gruppe umfaßt. Sieht man mit Ehlert von den Vertauschungen ab, welche einer Spiegelung entsprechen, so ergeben sich drei mögliche Symmetriecharaktere $\{\overline{1234}\}$ bzw. $\{\overline{123}\}4$ bzw. $\{12\}\{34\}$ der Eigenfunktion, welche den Darstellungen Γ_1 und Γ_2 bzw. Γ_4 und Γ_5 bzw. Γ_3 der Oktaedergruppe korrespondieren. Die Anzahl der Kugelfunktionen $l$ter Ordnung vom Symmetriecharakter $\{\overline{1234}\}$ ergibt sich z. B., wenn wir die Anzahlen der Terme mit den Darstellungseigenschaften Γ_1 und Γ_2 (bezüglich Drehungen des Oktaeders) addieren, welche aus einem Term des ungestörten

1) Wenn bei der Ausreduktion zwei Terme mit gleicher Darstellung auftreten, so sind diese natürlich trotzdem verschieden, ebenso wie ihre Eigenfunktionen verschiedene Winkelabhängigkeit im einzelnen besitzen und nur bei den speziellen Symmetrieoperationen des Kristallatoms sich in gleicher Weise transformieren. Darüber vgl. §§ 11, 18, 22.

2) W. Ehlert, Ztschr. f. Phys. 51. S. 8. 1928.

Atoms mit der azimutalen Quantenzahl l bei kubischer Symmetrie entstehen. Ebenso ist die Anzahl der Funktionen mit dem Charakter $\{12\}\{34\}$ gleich der doppelten Anzahl der Terme Γ_3, weil zu jedem solchen Term zwei Eigenfunktionen gehören und die Anzahl der Funktionen mit dem Symmetriecharakter $\{\overline{123}\}4$ gleich dreimal der Gesamtzahl der Terme mit den Darstellungen Γ_4 und Γ_5. Auf diese Weise erhält man *sofort* die bei Ehlert angegebenen Anzahlen der Eigenfunktionen mit bestimmtem Symmetriecharakter. Für die Aufstellung der Eigenfunktionen selbst ist allerdings die Ehlertsche Methode wohl vorzuziehen (vgl. §§ 12, 13).

§ 5. Hexagonale, tetragonale und rhombische (holoedrische) Symmetrie

a) Die *hexagonale* Symmetriegruppe umfaßt 12 reine Drehungen, die sich in 6 Klassen ordnen:

E Identität.

C_2 Drehung um die hexagonale Achse um π (1 Element).

C_3 Drehung um die hexagonale Achse um $\pm\frac{2\pi}{3}$ (2 Elemente).

C_4 Drehung um die hexagonale Achse um $\pm\frac{\pi}{3}$ (2 Elemente).

C_5 Drehung um eine der drei 2zähligen Kristallachsen um π (3 Elemente).

C_6 Drehung um eine der drei zu den vorigen senkrechten 2zähligen Achsen um π (3 Elemente).

Dazu kämen die durch Multiplikation mit der Inversion entstehenden 6 Spiegelungs- bzw. Drehspiegelungsklassen, um die wir uns aber nicht zu kümmern brauchen (§ 3). Die Dimensionen der 6 Darstellungen ergeben sich aus der Zerlegung von 12 in die Summe von 6 Quadraten, die nur in der Form

$$12 = 2^2 + 2^2 + 1^2 + 1^2 + 1^2 + 1^2$$

möglich ist: Es gibt vier eindimensionale und zwei zweidimensionale Darstellungen der hexagonalen Drehgruppe; also muß bereits der P-Term eines freien Atoms aufspalten, wenn das Atom in eine Lage von hexagonal-holoedrischer Symmetrie in einem Kristall gebracht wird.

Die Produkte von je zwei Klassen ergeben:

$$C_2^2 = E \qquad C_2 C_3 = C_4$$
$$C_3^2 = C_4^2 = 2E + C_3 \qquad C_2 C_4 = C_3$$
$$C_5^2 = C_6^2 = 3E + 3C_3 \qquad C_2 C_5 = C_6$$
$$C_3 C_4 = 2C_2 + C_4 \qquad C_2 C_6 = C_5$$
$$C_5 C_6 = 3C_2 + 3C_4 \qquad C_3 C_5 = C_4 C_6 = 2C_5$$
$$C_3 C_6 = C_4 C_5 = 2C_6$$

Daraus erhält man als Charaktere der irreduziblen Darstellungen der hexagonal-holoedrischen Gruppe:

Tabelle 3

	E	C_2	C_3	C_4	C_5	C_6
Γ_1	1	1	1	1	1	1
Γ_2	1	1	1	1	-1	-1
Γ_3	1	-1	1	-1	1	-1
Γ_4	1	-1	1	-1	-1	1
Γ_5	2	2	-1	-1	0	0
Γ_6	2	-2	-1	1	0	0

In der $2l+1$-dimensionalen Darstellung der Kugeldrehgruppe kommen den Klassen der hexagonalen Drehgruppe gemäß § 2, Ende, die folgenden Charaktere zu:

$$\chi_E = 2l + 1$$

$$\chi_2 = \chi_5 = \chi_6 = (-)^l$$

$$\chi_3 = \frac{\sin \frac{2l+1}{3}\pi}{\sin \frac{\pi}{3}} = \begin{cases} 1, & \text{wenn } l = 3m \\ 0, & 3m+1 \\ -1, & 3m+2 \end{cases}$$

$$\chi_4 = \frac{\sin \frac{2l+1}{6}\pi}{\sin \frac{\pi}{6}} = \begin{cases} 1, & \text{wenn } l \equiv 0 \text{ oder } 2 \pmod 6 \\ 2, & \text{,, } l \equiv 1 \quad (\text{,, } 6) \\ -1, & \text{,, } l \equiv 3 \text{ oder } 5 \quad (\text{,, } 6) \\ -2, & \text{,, } l \equiv 4 \quad (\text{,, } 6) \end{cases}$$

Im einzelnen ergibt sich

Tabelle 4

l	Charaktere der Klassen der hexagonalen Drehgruppe in der $2l+1$-dimensionalen Darstellung der Kugeldrehgruppe						Zerlegung der $2l+1$-dimensionalen Darstellung der Drehgruppe in irreduzible Darstellungen der hexagonalen Drehgruppe	Anzahl der Terme
	E	C_2	C_3	C_4	C_5	C_6		
0	1	1	1	1	1	1	Γ_1	1
1	3	-1	0	2	-1	-1	$\Gamma_2+\Gamma_6$	2
2	5	1	-1	1	1	1	$\Gamma_1+\Gamma_5+\Gamma_6$	3
3	7	-1	1	-1	-1	-1	$\Gamma_2+\Gamma_3+\Gamma_4+\Gamma_5+\Gamma_6$	5
4	9	1	0	-2	1	1	$\Gamma_1+\Gamma_3+\Gamma_4+2\Gamma_5+\Gamma_6$	6
5	11	-1	-1	-1	-1	-1	$\Gamma_2+\Gamma_3+\Gamma_4+2\Gamma_5+2\Gamma_6$	7
6	13	1	1	1	1	1	$2\Gamma_1+\Gamma_2+\Gamma_3+\Gamma_4+2\Gamma_5+2\Gamma_6$	9

Die $2(6m+k)+1$-dimensionale Darstellung der Drehgruppe enthält m mal die reguläre und außerdem dieselben Darstellungen der hexagonalen Gruppe wie die $2k+1$-dimensionale Darstellung der Kugeldrehgruppe. Der $2l+1$fach entartete Term des freien Atoms spaltet, wie man leicht nachrechnet, im hexagonalen Kristall in $\left[\frac{4}{3}l\right]+1$ Terme auf, wovon $2\left[\frac{l}{3}\right]+1$ Terme einfach, die übrigen $\left[\frac{2l-1}{3}\right]+1$ zweifach entartet sind.

b) Die *tetragonale* Symmetriegruppe enthält 8 Drehungen in 5 Klassen:

E Identität.

C_2 Drehung um die tetragonale Achse um π.

C_3 Drehung um die tetragonale Achse um $\pm\frac{\pi}{2}$ (2 Elemente)

C_4 Drehung um eine dazu senkrechte 2zählige Achse um π (2 Elemente).

C_5 Drehung um die Winkelhalbierenden der vorigen Achsen um π (2 Elemente).

Die Gruppe besitzt vier eindimensionale und eine zweidimensionale Darstellung, die Charaktere sind:

Tabelle 5

Charaktere der tetragonalen Drehgruppe

	E	C_2	C_3	C_4	C_5
Γ_1	1	1	1	1	1
Γ_2	1	1	1	−1	−1
Γ_3	1	1	−1	1	−1
Γ_4	1	1	−1	−1	1
Γ_5	2	−2	0	0	0

Die Charaktere der tetragonalen Drehungen in der $2\,l+1$-dimensionalen Darstellung der Kugeldrehgruppe sind (§ 2):

$$\chi_1 = 2\,l+1$$

$$\chi_2 = \chi_4 = \chi_5 = (-)^l$$

$$\chi_3 = (-)^{\left[\frac{l}{2}\right]}.$$

Daraus ergibt sich die Zerlegung der $2\,l+1$-dimensionalen Darstellung der Kugeldrehgruppe in irreduzible Darstellungen der tetragonalen Drehgruppe allgemein wie folgt:

Tabelle 6

Wenn $l =$	4λ	$4\lambda+1$	$4\lambda+2$	$4\lambda+3$,
so ist enthalten die Darstellung $(\Gamma_i)\ldots(n_i)$ mal:				
Γ_1	$\lambda+1$	λ	$\lambda+1$	λ
Γ_2	λ	$\lambda+1$	λ	$\lambda+1$
Γ_3	λ	λ	$\lambda+1$	$\lambda+1$
Γ_4	λ	λ	$\lambda+1$	$\lambda+1$
Γ_5	2λ	$2\lambda+1$	$2\lambda+1$	$2\lambda+2$

Man überzeugt sich leicht, daß hierdurch die Charakterbeziehungen erfüllt sind. Allgemein spaltet der $2\,l+1$-fach entartete Term des freien Atoms in $\left[\frac{3}{2}\,l\right]+1$ Terme auf, von denen $\left[\frac{l+1}{2}\right]$-Terme zweifach entartet und $2\left[\frac{l}{2}\right]+1$-Terme einfach sind.

c) *Rhombische Symmetrie.* Die 4 Drehungen (Identität und Drehung um jede der drei Achsen um π) bilden je eine Klasse für sich. Die Gruppe besitzt vier eindimensionale Darstellungen, jeder Term des freien Atoms spaltet *vollkommen* auf in $2l+1$ *einfache* Terme. Dasselbe geschieht bei noch niedrigerer Symmetrie.

§ 6. Einstellung eines halbzahligen Drehimpulses im Kristall. Zweideutige Gruppendarstellungen

Wenn die Starkeffektaufspaltung im Kristall klein ist gegen den Abstand der Terme eines und desselben Multipletts des freien Atoms, so stellt sich der *Gesamt*drehimpuls j des Atoms im Kristall ein (jede Komponente eines Multipletts spaltet einzeln auf) (§ 1, Fall 3). Besitzt nun das in den Kristall einzufügende Atom *gerade* Termmultiplizität (ungerade Ordnungszahl), so ist der Drehimpuls j halbzahlig, ebenso wie auch für ein einzelnes Elektron, und es gehört zu ihm eine *zweideutige* Darstellung der räumlichen Drehgruppe. Der Drehung um eine beliebige Achse um den Winkel Φ kommt nämlich der Charakter:

$$\chi(\Phi) = \frac{\sin(j+\frac{1}{2})\,\Phi}{\sin\frac{1}{2}\,\Phi} \tag{2}$$

zu, ganz analog zu Formel (1). $j+\frac{1}{2}$ ist aber ganzzahlig, und deshalb wird:

$$\sin(j+\tfrac{1}{2})(\Phi+2\pi) = \sin(j+\tfrac{1}{2})\,\Phi\,,$$

während früher bei ganzzahligem l:

$$\sin(l+\tfrac{1}{2})(\Phi+2\pi) = -\sin(l+\tfrac{1}{2})\,\Phi$$

galt. Da aber für den Nenner gilt:

$$\sin\tfrac{1}{2}(\Phi+2\pi) = -\sin\tfrac{1}{2}\,\Phi\,,$$

so folgt für halbzahlige j:

$$\chi(2\pi_{(}\pm_{)}\Phi) = -\chi(\Phi)\,.$$

Jeder Charakter ändert bei Zufügung einer Drehung um 2π sein Vorzeichen, d. h. jeder Charakter ist zweideutig. Auch der Charakter der identischen Drehung kann entweder:

$$\chi(0) = 2l+1$$

oder:

$$\chi(2\pi) = -(2l+1)$$

betragen. Eindeudig ist nur der Charakter der Drehung um eine beliebige Achse um π,

$$\chi(\pi) = \chi(3\pi) = 0$$

und allgemein gilt:

$$\chi(\Phi) = \chi(4\pi-\Phi)\,.$$

Eine derartige zweideutige Darstellung der Kugeldrehgruppe kann natürlich nur zweideutige Darstellungen der Kristalldrehgruppen als irreduzible Bestandteile enthalten. Um diese zweideutigen Darstellungen zu erhalten, machen wir die Fiktion, der Kristall ginge bei Drehung um irgendeine Achse um 2π nicht in sich über, sondern erst bei Drehung um 4π. Wir definieren also ein neues Gruppenelement R, die Drehung um 2π (um irgendeine Achse) und ergänzen die Elemente der Kristalldrehgruppe durch diejenigen, welche durch Multiplikation mit R aus ihnen entstehen. Wir wollen die so entstehende Gruppe, welche doppelt so viele Elemente enthält wie die ursprüngliche, als *Kristalldoppelgruppe* bezeichnen und fragen nach ihren irreduziblen Darstellungen. Damit ist unser Ziel erreicht, jedes Element der einfachen Gruppe durch *zwei* Matrizen (eine zweideutige Matrix) darzustellen. Dieser Weg entspricht etwa der Konstruktion der Riemannschen Fläche zur Untersuchung mehrdeutiger Funktionen.

Die Doppelgruppe umfaßt mehr, aber nicht doppelt so viele Klassen wie die einfache. Denn die Drehung um Φ um eine bestimmte Achse gehört bei der Doppelgruppe im allgemeinen nicht in dieselbe Klasse wie die Drehung um die gleiche Achse um $2\pi \pm \Phi$; denn die Charaktere der beiden Drehungen sind ja in den reduziblen Darstellungen der Doppelgruppe[1]) verschieden. Ausgenommen hiervon ist der Fall $2\pi - \Phi = \Phi = \pi, \chi(\Phi) = 0$. Alle Drehungsklassen der einfachen Kristalldrehgruppe, welche Drehungen um π enthalten, entsprechen also je *einer*, alle anderen je *zwei* Klassen der Doppelgruppe. Dies werden wir bei Untersuchung der einzelnen „Doppelgruppen" bestätigt finden.

§ 8. Die tetragonale Doppelgruppe

Die *einfache* Gruppe der tetragonalen Drehungen läßt sich vollständig aufbauen aus den beiden Elementen

A = Drehung um die tetragonale Achse um $\pi/2$,

B = Drehung um irgendeine zweizählige Achse um π, wobei für die einfache Gruppe $A^4 = B^2 = E$ gilt.

Es umfaßt die Klasse

1) Die zweideutigen Darstellungen der Drehgruppe sind ja reduzible Darstellungen der Kristalldoppelgruppen.

E der einfachen Gruppe	das Element	$E = A^4$
C_2	„ „	A^2
C_3	die Elemente	$A,\ A^3$
C_4	„ „	$B,\ A^2 B,$
C_5	„ „	$AB,\ A^3 B,$
		$(AB)^2 = E.$

Wir erwarten, daß in der Doppelgruppe den Klassen E und C_3 der einfachen Gruppe je zwei, den übrigen je eine Klasse entspricht. Die Doppelgruppe erhalten wir, indem wir nunmehr

$$A^4 = B^2 = R, \quad R^2 = E$$

setzen. R ist vertauschbar mit allen übrigen Elementen der Gruppe. Man erhält folgende Klassen der Doppelgruppe:

E
R
$C_2 = A^2,\ A^6$
$C_3' = A,\ A^7$
$C_3'' = A^3,\ A^5$
$C_4 = B,\ A^2B,\ A^4B = RB = B^3,\ A^6B = (A^2B)^3$
$C_5 = AB,\ A^3B,\ A^5B,\ A^7B.$

Die 16 Elemente der Doppelgruppe ordnen sich also in 7 Klassen, zur Feststellung der Dimensionen der irreduziblen Darstellungen ist 16 in eine Summe von 7 Quadraten zu zerspalten:

$$16 = 2^2 + 2^2 + 2^2 + 1^2 + 1^2 + 1^2 + 1^2.$$

Aus den Relationen zwischen den Klassen:

$R^2 = E$	$C_2^2 = 2(E + R)$
$RC_2 = C_2$	$C_2C_3' = C_2C_3'' = C_3' + C_3''$
$RC_3' = C_3''$	$C_2C_4 = 2C_4$
$RC_3'' = C_3'$	$C_2C_5 = 2C_5$
$RC_4 = C_4$	$C_3'^2 = C_3''^2 = 2E + C_2$
$C_5 = C_5$	$C_3'C_3'' = 2R + C_2$
	$C_3'C_4 = C_3''C_4 = 2C_5$
	$C_3'C_5 = C_3''C_5 = 2C_4$
	$C_4^2 = C_5^2 = 4E + 4R + 4C_2$
	$C_4C_5 = 4C_3' + 4C_3''$

11*

folgt für die Charaktere der irreduziblen Darstellungen der tetragonalen Doppelgruppe das Schema der Tab. 7: Die ersten 5 Darstellungen sind die eindeutigen, die letzten beiden die zweideutigen Darstellungen der tetragonalen Drehgruppe.

Tabelle 7

Charaktere der tetragonalen Doppelgruppe

	E	R	C_2	C_3'	C_3''	C_4	C_5
Γ_1	1	1	1	1	1	1	1
Γ_2	1	1	1	1	1	-1	-1
Γ_3	1	1	1	-1	-1	1	-1
Γ_4	1	1	1	-1	-1	-1	1
Γ_5	2	2	-2	0	0	0	0
Γ_6	2	-2	0	$\sqrt{2}$	$-\sqrt{2}$	0	0
Γ_7	2	-2	0	$-\sqrt{2}$	$\sqrt{2}$	0	0

Wir haben noch die zweideutigen Darstellungen der Kugeldrehgruppe als Darstellungen der tetragonalen Doppelgruppe auszureduzieren. Die Charaktere der Klassen der tetragonalen Doppelgruppe in der $2j+1$-dimensionalen Darstellung der Kugeldrehgruppe sind nach Formel (4)

$$\chi = \frac{\sin (j+\frac{1}{2})\,\Phi}{\sin \frac{1}{2}\Phi} \qquad (j+\tfrac{1}{2} = \text{ganze Zahl}): \tag{2}$$

Drehwinkel	Charakter
$\Phi = 0$	$\chi_E = 2j+1$
$\Phi = 2\pi$	$\chi_R = -(2j+1)$
$\Phi = \pm\pi$	$\chi_2 = \chi_4 = \chi_5 = 0$
$\Phi = \pm\frac{\pi}{2}$	$\chi_3' = \dfrac{\sin\left(j+\frac{1}{2}\right)\frac{\pi}{2}}{\sin\frac{\pi}{4}} = \begin{cases} \sqrt{2}, & \text{wenn } j \equiv \frac{1}{2} \pmod 4 \\ 0, & \text{,, } j \equiv \frac{3}{2}, \frac{7}{2} \text{ (,, } 4) \\ -\sqrt{2}, & \text{,, } j \equiv \frac{5}{2} \text{ (,, } 4) \end{cases}$
$\Phi = \pm\frac{3\pi}{2}$	$\chi_3'' = -\chi_3'$

Wesentlich sind nur χ_E und χ_3'. Alle anderen Charaktere leiten sich eindeutig aus diesen beiden ab oder verschwinden. Man erhält im einzelnen:

Tabelle 8

j	Klassencharaktere in der reduziblen Darstellung, d. h. in der $(2j+1)$-dimensionalen Darstellung der Kugeldrehgruppe		Zerlegung der reduziblen Darstellung in irreduzible Bestandteile
	E	C_3'	
$1/2$	2	$\sqrt{2}$	Γ_6
$3/2$	4	0	$\Gamma_6 + \Gamma_7$
$5/2$	6	$-\sqrt{2}$	$\Gamma_6 + 2\Gamma_7$
$7/2$	8	0	$2\Gamma_6 + 2\Gamma_7$
$4\lambda + j'$	$8\lambda + 2j' + 1$	wie für j'	$2\lambda(\Gamma_6 + \Gamma_7)$ plus irreduzible Bestandteile für $j = j'$

§ 8. Zweideutige Darstellungen der hexagonalen und kubischen Symmetriegruppe

a) Von den 6 Klassen der *einfachen* hexagonalen Drehgruppe entsprechen drei einer Drehung um $\pm\pi$ und infolgedessen je *einer* Klasse der hexagonalen Doppelgruppe, die anderen drei (E, C_3, C_4) umfassen Drehungen um $0, \pm\frac{2\pi}{3}, \pm\frac{\pi}{3}$ und entsprechen daher je 2 Klassen der Doppelgruppe. Die Doppelgruppe enthält also 24 Elemente in 9 Klassen. Außer den 6 Darstellungen der einfachen Gruppe müssen also noch drei weitere Darstellungen der Doppelgruppe existieren; die Quadratsumme der Dimensionszahlen dieser Darstellungen muß gleich der Anzahl der gegenüber der einfachen Gruppe neu hinzukommenden Elemente sein, also gleich 12; demnach sind alle drei neuen Darstellungen (die zweideutigen Darstellungen der einfachen hexagonalen Drehgruppe) *zwei*dimensional.

Aus den leicht aufstellbaren Klassenrelationen erhält man die in der folgenden Tabelle aufgeführten Charaktere der einzelnen Klassen der hexagonalen Doppelgruppe in den drei neu hinzukommenden Darstellungen, welche als Darstellung der einfachen Gruppe zweideutig sind (die eindeutigen Darstellungen der einfachen Gruppe Γ_1 bis Γ_6 vgl. in Tab. 3).

Tabelle 9

Zweideutige Darstellungen der hexagonalen Drehgruppe

	E	R	C_2	C_3'	C_3''	C_4'	C_4''	C_5	C_6
Γ_7	2	−2	0	1	−1	$\sqrt{3}$	$-\sqrt{3}$	0	0
Γ_8	2	−2	0	1	−1	$-\sqrt{3}$	$\sqrt{3}$	0	0
Γ_9	2	−2	0	−2	2	0	0	0	0

In der $2j+1$-dimensionalen Darstellung der Kugeldrehgruppe kommen den Klassen der hexagonalen Doppelgruppe folgende Charaktere zu (Formel 2):

$$\chi_E = 2j + 1 \qquad \chi_R = -\chi_E$$

$$\chi_2 = \chi_5 = \chi_6 = 0$$

$$\chi_3' = \frac{\sin(2j+1)\frac{\pi}{3}}{\sin\frac{\pi}{3}} \begin{cases} 1, & \text{wenn } j \equiv \frac{1}{2} \pmod 3 \\ -1, & \quad " \quad j \equiv \frac{3}{2} \;(\;" \;\; 3) \\ 0, & \quad " \quad j \equiv \frac{5}{2} \;(\;" \;\; 3) \end{cases}$$

$$\chi_4' = \frac{\sin(2j+1)\frac{\pi}{6}}{\sin\frac{\pi}{6}} \begin{cases} \sqrt{3}, & \text{wenn } j \equiv \frac{1}{2} \text{ oder } \frac{3}{2} \pmod 6 \\ 0, & \quad " \quad j \equiv \frac{5}{2} \quad " \quad \frac{11}{2} \;(\;" \;\; 6) \\ -\sqrt{3}, & \quad " \quad j \equiv \frac{7}{2} \quad " \quad \frac{9}{2} \;(\;" \;\; 6) \end{cases}$$

$$\chi_3'' = -\chi_3' \qquad \chi_4'' = -\chi_4'.$$

Hieraus ergeben sich die Zerlegungen der zweideutigen Darstellungen der Kugeldrehgruppe in irreduzible zweideutige Darstellung der hexagonalen Drehgruppe wie folgt:

Tabelle 10

j	Zerlegung in irreduzible Bestandteile
$^1/_2$	Γ_7
$^3/_2$	$\Gamma_7 + \Gamma_9$
$^5/_2$	$\Gamma_7 + \Gamma_8 + \Gamma_9$
$^7/_2$	$\Gamma_7 + 2\Gamma_8 + \Gamma_9$
$^9/_2$	$\Gamma_7 + 2\Gamma_8 + 2\Gamma_9$
$^{11}/_2$	$2\Gamma_7 + 2\Gamma_8 + 2\Gamma_9$
$6\lambda + j'$	$2\lambda(\Gamma_7 + \Gamma_8 + \Gamma_9)$ + Zerlegung für j'

b) Die Doppeloktaedergruppe umfaßt 48 Elemente in 8 Klassen, da den Klassen E, C_3, C_5 der einfachen Oktaedergruppe je 2 Klassen der Doppelgruppe entsprechen müssen.

Die Oktaedergruppe besitzt also drei zweideutige Darstellungen, darunter eine vierdimensionale und zwei zweidimensionale ($48 = 24 + 4^2 + 2^2 + 2^2$, 24 gleich Quadratsumme der Dimensionszahlen der eindeutigen Darstellungen). Die Charaktere der einzelnen Klassen der Doppeloktaedergruppe in den zweideutigen irreduziblen Darstellungen der Oktaedergruppe sind:

Tabelle 11

	E	R	C_2	C_3'	C_3''	C_4	C_5'	C_5''
Γ_6	2	-2	0	$\sqrt{2}$	$-\sqrt{2}$	0	1	-1
Γ_7	2	-2	0	$-\sqrt{2}$	$\sqrt{2}$	0	1	-1
Γ_8	4	-4	0	0	0	0	-1	1

In der $2j+1$-dimensionalen Darstellung der Drehgruppe ist:

$$\chi_E = 2j+1 \qquad \chi_R = -\chi_E$$

$$\chi_2 = \chi_4 = 0$$

$$\chi_3' = \frac{\sin(2j+1)\frac{\pi}{4}}{\sin\frac{\pi}{4}} = \begin{cases} \sqrt{2}\,, & \text{wenn } j \equiv \frac{1}{2} \pmod 4 \\ 0 & \text{,, } j \equiv \frac{3}{2} \text{ oder } \frac{7}{2} \;(\text{,, } 4) \\ -\sqrt{2} & \text{,, } j \equiv \frac{5}{2} \;(\text{,, } 4) \end{cases}$$

$$\chi_5' = \frac{\sin(2j+1)\frac{\pi}{3}}{\sin\frac{\pi}{3}} = \begin{cases} 1 & \text{wenn } j \equiv \frac{1}{2} \pmod 3 \\ -1 & \text{,, } j \equiv \frac{3}{2} \;(\text{,, } 3) \\ 0 & \text{,, } j \equiv \frac{5}{2} \;(\text{,, } 3) \end{cases}$$

$$\chi_3'' = -\chi_3' \qquad \chi_5'' = -\chi_5'$$

Tabelle 12

Irreduzible Bestandteile der $2j+1$-dimensionalen zweideutigen Darstellung der Kugeldrehgruppe bei Ausreduktion als zweideutige Darstellung der Oktaedergruppe

j	Irreduzible Bestandteile	
$^1/_2$	Γ_6	
$^3/_2$	Γ_8	
$^5/_2$	$\Gamma_7 + \Gamma_8$	
$^7/_2$	$\Gamma_6 + \Gamma_7 + \Gamma_8$	
$^9/_2$	$\Gamma_6 \quad + 2\Gamma_8$	
$^{11}/_2$	$\Gamma_6 + \Gamma_7 + 2\Gamma_8$	
$6 + j'$	$\Gamma_6 + \Gamma_7 + 2\Gamma_8$	+ Bestandteile für $j = j'$ unter Vertauschung von Γ_6 mit Γ_7
$12\lambda + j'$	$2\lambda(\Gamma_6 + \Gamma_7 + 2\Gamma_8)$	+ Bestandteile für $j = j'$

Alle Drehimpulse bis $j = {}^3/_2$ stellen sich im kubischen Kristall beliebig ein, bei größerem Drehimpuls wird die Richtungsentartung teilweise aufgehoben.

c) Die rhombische Drehgruppe besitzt *eine* zweideutige irreduzible Darstellung: $\chi_E = 2$, $\chi_R = -2$, die Charaktere der Drehungen um die 3 Achsen um π sind 0. Die $2j + 1$-dimensionale zweideutige Darstellung der Kugeldrehgruppe enthält diese irreduzile Darstellung $j + {}^1/_2$ mal.

II. Die Eigenfunktionen nullter Näherung im Kristall

§ 9. Eigenfunktion eines Atoms mit mehreren Elektronen

Die Eigenfunktionen eines *freien* Atoms mit N Elektronen hängen außer von den Abständen der Elektronen voneinander und vom Kern auch von der Orientierung des Gesamtsystems im Raume ab; doch sind infolge der Kugelsymmetrie des Problems die $2l + 1$ Eigenfunktionen, die zu einem Term mit der azimutalen Quantenzahl l gehören, für *alle* Orientierungen des Systems im Raum bestimmt, wenn sie für *eine* Orientierung vorgegeben werden. Wir dürfen die Funktionen z. B. mit Wigner[1]) auf der $3N - 3$-dimensionalen „Hyperfläche" $x_1 = y_1 = x_2 = 0$ in Abhängigkeit von den übrigen Koordinaten der Elektronen in beliebiger Weise vorgeben, dann lassen sich mit Hilfe der $2l + 1$-dimensionalen Darstellung der Drehgruppe die Eigenfunktionen für jede beliebige andere Orientierung des Atoms im Raum (beliebige Koordinaten x_1, y_1, x_2) angeben. „Beliebig" ist die Vorgabe der Eigenfunktionen auf der Hyperfläche natürlich nur insofern, als es für das Verhalten der Eigenfunktionen gegenüber räumlichen Drehungen des Atoms auf ihre spezielle Wahl auf der Hyperfläche nicht ankommt; die wirklichen Eigenfunktionen müssen natürlich durch eine Lösung der Schrödingergleichung in extenso gewonnen werden.

Im Kristall wird nun das Verhalten der Eigenfunktionen durch die Symmetrie allein viel weniger bestimmt; man darf sie nicht nur auf einer $3N - 3$-dimensionalen Hyperfläche, sondern in einem ganzen Gebiet des $3N$-dimensionalen Raumes willkürlich vorgeben, ohne mit den durch die Symmetrie vorgeschriebenen Transformationseigenschaften in Konflikt zu kommen.

---

1) E. Wigner, Ztschr. f. Phys. **43**. S. 624, 640. 1927.

Läßt die Lage des Atomkernes im Kristall g Symmetrieoperationen zu, so dürfen für alle Lagen des ersten Elektrons innerhalb eines passend zu wählenden Sektors vom räumlichen Winkel $\frac{4\pi}{g}$, durch die kein Symmetrieelement geht, und beliebige Lagen der übrigen Elektronen die Eigenfunktionen des Atoms willkürlich vorgegeben werden. In den übrigen $g-1$ Gebieten gleicher Größe sind sie damit vermöge der für den gerade betrachteten Term charakteristischen gruppentheoretischen Darstellung der Symmetrieoperationen des Kristalles festgelegt. Die exakte Bestimmung der Eigenfunktionen erfordert also eine viel weitgehendere explizite Lösung der Schrödingergleichung als beim freien Atom.

Es liegt nun aber bekanntlich im Wesen der Störungstheorie — und wir wollen ja die *Störung* eines Atoms beim Einbau in den Kristall untersuchen — nur die Eigenfunktionen *nullter* Näherung anzugeben, d. h. aus den Eigenfunktionen des ungestörten (kugelsymmetrischen) Problems einfach die auszuwählen, an die sich die Eigenfunktionen des gestörten Problems stetig anschließen. Man tut also, als ob man auch im Kristall die Eigenfunktionen des Atoms dann schon vollständig kennte, wenn man sie auf der Hyperfläche von Wigner vorgibt und verwendet nur die Darstellung der Drehgruppe speziell in der als Darstellung der Kristallgruppe ausreduzierten Form. Dadurch transformieren sich bei Symmetrieoperationen des Kristalles nur solche Eigenfunktionen ineinander, die zum gleichen Term des Atoms im Kristall gehören. Doch dürfte die allgemeine Darstellung der Drehgruppe in ausreduzierter Form bezüglich einer Kristallsymmetriegruppe nicht einfach sein. Wir beschränken uns daher auf die Aufstellung der winkelabhängigen Eigenfunktionen nullter Näherung für ein einziges Elektron im Kristall, dessen Koppelung mit den übrigen Elektronen des Atoms wir aufgehoben denken. Mit solchen Eigenfunktionen wird man sich aber im allgemeinen auch zur Auswertung der Austauschintegrale im Kristall begnügen können[1]), indem man sie etwa mit den Hartreeschen radialabhängigen Eigenfunktionen kombiniert, welche ja gleichfalls für ein ent-

---

1) Die Suche nach solchen Eigenfunktionen bildete den Ausgangspunkt dieser Arbeit. Vgl. auch Ehlert, Ztschr. f. Phys., a. a. O.

koppeltes Elektron gelten und annimmt, daß die Einfügung des Atoms in den Kristall auf den Elektronenaustausch innerhalb eines Atoms keinen wesentlichen Einfluß hat.

§ 10. Eigenfunktionen nullter Näherung eines entkoppelten Elektrons im Kristall

Wir denken zunächst die Schrödingergleichung für ein von den übrigen entkoppelt gedachtes Elektron des *freien* Atoms in Polarkoordinaten separiert; zum Term mit der azimutalen Quantenzahl l gehören als winkelabhängige Eigenfunktionen die Kugelfunktionen $l$ter Ordnung. Dann suchen wir durch Linearkombinieren dieser Kugelfunktionen die richtigen Eigenfunktionen nullter Näherung zu gewinnen, die zu einem bestimmten Term des Elektrons *im Kristall* gehören, sich also bei Symmetrieoperationen des Kristallatoms entsprechend einer bestimmten irreduziblen Darstellung der Symmetriegruppe des Kristalls transformieren. Wir werden also alle diejenigen Kugelfunktionen zu einer „Gemeinschaft" zusammenordnen, die bei Symmetrieoperationen des Kristalls ineinander übergehen und dann durch geeignete Wahl von Linearkombinationen statt der ursprünglichen Kugelfunktionen diese Gemeinschaften möglichst klein zu machen suchen — das entspricht dem Aufsuchen der irreduziblen Darstellungen. Schließlich muß jedem k-dimensionalen irreduziblen Bestandteil der $2\,l+1$-dimensionalen Darstellung der Drehgruppe *eine* irreduzible Gemeinschaft von k Kugelfunktionen $l$ter Ordnung entsprechen. Eine so gefundene Gemeinschaft mit bestimmter Darstellungseigenschaft bei Symmetrieoperationen umfaßt dann die *richtigen* Eigenfunktionen nullter Näherung zu dem Kristallterm von eben dieser Darstellungseigenschaft, *wenn* kein weiterer Term der gleichen Darstellungseigenschaft aus dem gleichen Term des ungestörten Atoms entsteht, d. h. wenn die $2\,l+1$-dimensionale Darstellung der Drehgruppe die fragliche irreduzible Darstellung der Kristallgruppe nur einmal enthält. Denn die Eigenfunktionen nullter Näherung ψ_i sind zu bestimmen durch die Forderung der allgemeinen Störungstheorie:

$$(3)\quad \int V\psi_i\,\psi_k\,dx = 0, \quad i \neq k\;(i = -l,\cdots+l;\; k = -l,\cdots,+l)$$

V ist das Störungspotential, x steht für alle Koordinaten, ψ_i und ψ_k sollen sich bei Operationen der Gruppe des Kristallatoms entsprechend den Darstellungen (a_{ij}) und (b_{kl}) transformieren. In der bekannten Schlußweise von Wigner erhält man (g Anzahl der Gruppenelemente, R eine Substitution der Gruppe, bei der sich also V nicht ändert):

$$(4)\quad \left\{ \begin{aligned} \int V\, \psi_i(x)\, \psi_k(x)\, dx &= \frac{1}{g} \sum_R \int V(Rx)\, \psi_i(Rx)\, \psi_k(Rx)\, dx \\ &= \frac{1}{g} \sum_{j,l} \sum_R a_{ij}^R b_{kl}^R \int V\, \psi_j(x)\, \psi_l(x)\, dx \end{aligned} \right.$$

$= 0$, wenn $a \neq b$, d. h. bei Termen *verschiedener* Darstellung.
$= \delta_{ik}\, c_i$, wenn $a = b$.

Gehört also nur *ein Term* zu einer bestimmten Darstellung und ist ψ_i eine Eigenfunktion zu diesem Term, so verschwindet das Integral sowohl für solche ψ_k, die zu anderen Termen gehören, als auch für solche, die zum gleichen Term gehören; die Eigenfunktionen ψ_i erfüllen also die störungstheoretische Forderung (3).

Gehören dagegen zu einer irreduziblen Darstellung mehrere (n) Terme, so verschwindet das Integral (5) im allgemeinen nicht, wenn unter ψ_i und ψ_k zwei Eigenfunktionen verstanden werden, die zu zwei verschiedenen Termen gleicher Darstellung gehören. Die n „Gemeinschaften", die zur gleichen irreduziblen Darstellung der Kristallgruppe gehören, enthalten also offensichtlich i. allg. noch nicht die richtigen Eigenfunktionen des Problems und diese richtigen Eigenfunktionen lassen sich auf rein gruppentheoretischem Wege überhaupt nicht bestimmen, sondern erst unter Bezugnahme auf das speziell vorliegende Problem. Man muß, wie in der Störungstheorie üblich, aus den gegebenen $n\,k$ Kugelfunktionen der n Gemeinschaften durch Linearkombinieren $n\,k$ neue Eigenfunktionen Ψ_i bilden:

$$\Psi_i = \sum_{j=1}^{n\,k} b_{ij}\, \psi_j \qquad (i = 1, \ldots, n\,k)$$

und die Koeffizienten b_{ij} durch Lösung der üblichen Determinantengleichung vom Grade $n\,k$ bestimmen. Diese Gleichung liefert für den Termwert n k-fache Wurzeln, gleichzeitig erhält man die zu jedem gehörigen k Eigenfunktionen nullter Näherung. Diese Eigenfunktionen sind aber keine Linearaggregate von Kugelfunktionen mit *universellen* Koeffizienten, sondern die Koeffizienten hängen noch von den Einzelheiten des elektrischen Feldes des Kristalls ab. (Die relative Kompliziertheit dieses Falles, daß mehrere Terme zur gleichen Darstellung gehören, ist eng verknüpft mit dem allgemeinen Satz, daß Terme gleicher Darstellung sich nicht überschneiden [vgl. § 22].) (Ein Beispiel im nächsten Paragraphen.)

§ 11. Eigenfunktionen bei tetragonaler und hexagonaler Symmetrie

Um nun die Kugelfunktionen wirklich zu „Gemeinschaften" zusammenzuordnen, so daß sich bei beliebigen Symmetrieoperationen des Kristalls die Eigenfunktionen einer Gemeinschaft nur unter sich transformieren, fassen wir jede Symmetrieoperation auf als Produkt einer Vertauschung der positiven und negativen Richtung einer und derselben Achse, sowie der Achsen des Kristalls untereinander. Ferner schreiben wir die Kugelfunktionen in reeller Form $\sqrt{2}\, P_l^m(\cos\vartheta) \begin{smallmatrix}\cos\\ \sin\end{smallmatrix} m\varphi$; dann ändert bei Vertauschung von positiven und negativen Achsenrichtungen jede Kugelfunktion höchstens ihr Vorzeichen, geht aber nie in eine andere über. Wir brauchen uns also nur mehr um die Vertauschung *verschiedener* Achsen zu kümmern, und sehen sofort, daß im rhombischen Kristall jede reell geschriebene Kugelfunktion eine Gemeinschaft für sich bildet, weil Vertauschungen verschiedener Achsen nicht zur Gruppe gehören. Alle Eigenwerte sind also einfach, wie wir bereits früher gruppentheoretisch feststellten.

Bei tetragonal-holoedrischer Symmetrie der Atomlage sind die beiden zweizähligen Achsen X und Y vertauschbar. Bei dieser Vertauschung geht[1]) $\cos\varphi$ in $\pm\sin\varphi$ über und umgekehrt, ebenso $\cos m\varphi$ in $\pm\sin m\varphi$ für alle ungeraden m, während für gerade m $\cos m\varphi$ höchstens sein Vorzeichen ändert: Für ungerade m gehören $\cos m\varphi$ und $\sin m\varphi$ zur gleichen Gemeinschaft, für gerade m bildet jede Funktion eine eigene Gemeinschaft.

Analog gehören bei der hexagonal-holoedrischen Symmetrie (drei zweizählige Achsen senkrecht zur hexagonalen

---

1) Wir identifizieren die tetragonale Achse Z natürlich mit der Achse der Kugelfunktionen, außerdem benutzen wir stets Kugelfunktionen in der Normierung $\int (P_l^m)^2 \sin\vartheta\, d\vartheta = 1$.

Achse vertauschbar) $\cos m\varphi$ und $\sin m\varphi$ zum gleichen (zweifachen) Eigenwert, wenn m nicht durch 3 teilbar ist, dagegen zu zwei verschiedenen einfachen Eigenwerten, falls m ein Vielfaches von 3. In beiden Fällen erhält man die gruppentheoretisch bekannte Anzahl der Aufspaltungskomponenten (§ 5).

Wir dürfen aber in beiden Fällen nur dann behaupten, daß unsere Eigenfunktionen schon sämtlich die richtigen sind, wenn bei der Ausreduktion der $2l+1$-dimensionalen Darstellung der Kugeldrehgruppe nach den irreduziblen Darstellungen der Kristallgruppe keine irreduzible Darstellung mehrfach auftritt. Das ist bei tetragonaler Symmetrie für $l \leq 2$, bei hexagonaler für $l \leq 3$ der Fall. Für $l = 3$ dagegen gehören bei tetragonaler Symmetrie zwei Terme zur gleichen irreduziblen Darstellung der tetragonalen Gruppe, nämlich zur zweidimensionalen Darstellung Γ_5. Demnach sind hier nur diejenigen der von uns konstruierten Eigenfunktionen schon die endgültigen Eigenfunktionen nullter Näherung, welche einer eindimensionalen Darstellung der Kristallgruppe entsprechen, d. h. für sich allein eine Gemeinschaft bilden. Es sind dies P_3^0, $\sqrt{2}\, P_3^2 \cos 2\varphi$ und $\sqrt{2}\, P_3^2 \sin 2\varphi$. Die zu den Gemeinschaften P_3^1 und P_3^3 gehörigen Eigenfunktionen sind dagegen noch nicht endgültig, sondern es sind aus ihnen 4 Linearkombinationen ψ_i $(i = 1, 2, 3, 4)$ zu bilden, welche die Bedingung der Störungstheorie

$$\int V \psi_i \psi_k \, d\tau = \delta_{ik} \varepsilon_i$$

befriedigen. Wir setzen allgemein an

$$\psi_i = \psi_{nl}(r) \, (\alpha_{i1} \sqrt{2}\, P_3^1 \cos \varphi + \alpha_{i2} \sqrt{2}\, P_3^1 \sin \varphi + \alpha_{i3} \sqrt{2}\, P_3^3 \cos 3\varphi + \alpha_{i4} \sqrt{2}\, P_3^3 \sin 3\varphi)$$

und bekommen in bekannter Weise

$$\sum_{j=1}^{4} \alpha_{ij} (\varepsilon_{jk} - \delta_{jk} \varepsilon_i) = 0 \qquad k = 1, 2, 3, 4 ,$$

wobei

$$\varepsilon_{jk} = \int V \psi_j \psi_k \, d\tau$$

$$
(5)\quad\begin{cases}
\varepsilon_{11} = 2\int V\,\psi_{nl}^2(r)\,[P_3^1(\cos\vartheta)]^2\cos^2\varphi\,d\tau \\
\quad = 2\int V\,\psi_{nl}^2\,(P_3^1)^2\sin^2\varphi\,d\tau \\
\quad = \varepsilon_{22} = \int V\,\psi_{nl}^2\,(P_3^1)^2\,d\tau \\
\varepsilon_{33} = \int V\,\psi_{nl}^2\,(P_3^3)^2\,d\tau = \varepsilon_{44} \\
\varepsilon_{13} = \varepsilon_{31} = 2\int V\,\psi_{nl}^2\,P_3^1\,P_3^3\cos\varphi\cos 3\varphi \\
\quad = -2\int V\,\psi_{nl}^2\,P_3^1\,P_3^3\sin\varphi\,\sin 3\varphi \\
\quad = -\varepsilon_{24} = -\varepsilon_{42}\,.
\end{cases}
$$

Die Relationen ergeben sich daraus, daß V, das Kristallpotential, bei einer Drehung um die Z-Achse um $\frac{\pi}{2}$ in sich übergeht. Weiter folgt aus der Invarianz von V gegen Spiegelung an der XZ-Ebene (bzw. Drehung um X um π) stets, daß alle Integrale verschwinden, die cos und sin gemischt enthalten, z. B.

$$\varepsilon_{14} = 2\int V\,P_3^1\,P_3^3\,\psi_{nl}^2\cos\varphi\sin 3\varphi\,d\tau = 0\,.$$

Damit haben wir zur Bestimmung der Termverschiebung die Determinantengleichung

$$\begin{vmatrix} \varepsilon_{11}-\varepsilon & 0 & \varepsilon_{13} & 0 \\ 0 & \varepsilon_{11}-\varepsilon & 0 & -\varepsilon_{13} \\ \varepsilon_{13} & 0 & \varepsilon_{33}-\varepsilon & 0 \\ 0 & -\varepsilon_{13} & 0 & \varepsilon_{33}-\varepsilon \end{vmatrix} = [(\varepsilon_{11}-\varepsilon)(\varepsilon_{33}-\varepsilon)-\varepsilon_{13}^2]^2 = 0$$

Wir erhalten zwei einfache Eigenwerte:

$$\varepsilon', \varepsilon'' = \frac{\varepsilon_{11}+\varepsilon_{33}}{2} \pm \sqrt{\left(\frac{\varepsilon_{11}-\varepsilon_{33}}{2}\right)^2 + \varepsilon_{13}^2}$$

und als Eigenfunktionen zu ε':

$$\psi_1' = \sqrt{1-\delta}\,P_3^1\cos\varphi + \sqrt{1+\delta}\,P_3^3\cos 3\varphi\,,$$

$$\psi_2' = -\sqrt{1-\delta}\,P_3^1\sin\varphi + \sqrt{1+\delta}\,P_3^3\sin 3\varphi$$

mit $\delta = \frac{\varepsilon_{33}-\varepsilon_{11}}{\sqrt{(\varepsilon_{33}-\varepsilon_{11})^2 + 4\varepsilon_{13}^2}}$, ebenso als Eigenfunktionen zu ε'':

$$\psi_1'' = \sqrt{1+\delta}\, P_3^{\,1} \cos\varphi - \sqrt{1-\delta}\, P_3^{\,3} \cos 3\varphi\,,$$

$$\psi_2'' = \sqrt{1+\delta}\, P_3^{\,1} \sin\varphi + \sqrt{1-\delta}\, P_3^{\,3} \sin 3\varphi\,.$$

In § 22 werden wir feststellen, daß tatsächlich ε_{13} nicht verschwindet, daß also $\sqrt{2}\, P_3^{\,1} \begin{smallmatrix}\cos\\ \sin\end{smallmatrix} \varphi$ und $\sqrt{2}\, P_3^{\,3} \begin{smallmatrix}\cos\\ \sin\end{smallmatrix} 3\varphi$ noch nicht die richtigen Eigenfunktionen nullter Näherung sind, sowie ferner, daß ε_{13} und damit δ, also die Eigenfunktionen nullter Näherung selbst, stark abhängen von dem speziellen Problem, vor allem von „Bahnradius" des Elektrons und der Abweichung der Symmetrie von der kubischen.

Für $l > 3$ gehören nicht nur zum zweidimensionalen, sondern auch zu einigen eindimensionalen irreduziblen Bestandteilen der $2l+1$-dimensionalen Darstellung der Drehgruppe bei tetragonaler Symmetrie mehrere Terme, und man kann immer seltener die einfachen Funktionen $\sqrt{2}\, P_l^{\,m} \begin{smallmatrix}\cos\\ \sin\end{smallmatrix} m\varphi$ als

Tabelle 13

Eigenfunktionen zu den irreduziblen Darstellungen der tetragonalen und hexagonalen Drehgruppe[1])

Tetragonal			Hexagonal		
Darstellung bei		Eigenfunktionen	Darstellung bei		Eigenfunktionen
(Geraden λ)	(Ungeraden λ)		(Geraden λ)	(Ungeraden λ)	
Γ_1	Γ_2	$P_\lambda^0,\ \sqrt{2}\, P_\lambda^{4\mu} \cos 4\mu\varphi$	Γ_1	Γ_2	$P_\lambda^0,\ \sqrt{2}\, P_\lambda^{6\mu} \cos 6\mu\varphi$
Γ_2	Γ_1	$\sqrt{2}\, P_\lambda^{4\mu} \sin 4\mu\varphi$	Γ_2	Γ_1	$\sqrt{2}\, P_\lambda^{6\mu} \sin 6\mu\varphi$
Γ_3	Γ_4	$\sqrt{2}\, P_\lambda^{4\mu+2} \cos(4\mu+2)\varphi$	Γ_3	Γ_4	$\sqrt{2}\, P_\lambda^{6\mu+3} \cos(6\mu+3)\varphi$
Γ_4	Γ_3	$\sqrt{2}\, P_\lambda^{4\mu+2} \sin(4\mu+2)\varphi$	Γ_4	Γ_3	$\sqrt{2}\, P_\lambda^{6\mu+3} \sin(6\mu+3)\varphi$
Γ_5	Γ_5	$\sqrt{2}\, P_\lambda^{2\mu+1} \begin{smallmatrix}\cos\\ \sin\end{smallmatrix} (2\mu+1)\varphi$	Γ_5	Γ_5	$\sqrt{2}\, P_\lambda^{6\mu\pm2} \begin{smallmatrix}\cos\\ \sin\end{smallmatrix} (6\mu\pm2)\varphi$
			Γ_6	Γ_6	$\sqrt{2}\, P_\lambda^{6\mu\pm1} \begin{smallmatrix}\cos\\ \sin\end{smallmatrix} (6\mu\pm1)\varphi$

1) Die Eigenfunktionen sind natürlich noch nicht die „richtigen" Eigenfunktionen nullter Näherung.

Eigenfunktionen nullter Näherung im Kristall ansetzen. D. h. mit wachsender Azimutalquantenzahl wird die Aufspaltung im tetragonalen und ebenso im hexagonalen Kristall immer unähnlicher der gewöhnlichen Starkeffektaufspaltung im homogenen Feld, mit der sie für $l = 1$ (beim hexagonalen Kristall auch für $l = 2$) vollkommen übereinstimmt. Es macht sich immer mehr die Vierzähligkeit bzw. Sechszähligkeit der ausgezeichneten Achse geltend und verursacht Abweichungen von den Verhältnissen bei einer unendlich-zähligen Achse, welche die Symmetrieverhältnisse beim Starkeffekt im homogenen Feld wiedergibt.

§ 12. Eigenfunktionen bei kubisch-(holoedrischer) Symmetrie

Bei kubisch-holoedrischer Symmetrie sind alle drei Achsen X, Y, Z miteinander vertauschbar. Daraus folgt sofort, daß die drei Kugelfunktionen erster Ordnung: $P_1{}^0 = \sqrt{\frac{3}{2}} \cos \vartheta$ sowie $\sqrt{\frac{3}{2}} \sin \vartheta \frac{\cos}{\sin} \varphi$ durch Symmetrieoperationen der kubischen Symmetriegruppe ineinander übergeführt werden: Der P-Term spaltet nicht auf (vgl. § 4), die Wahl der Achse der Kugelfunktionen ist demnach beliebig, die Richtungsentartung des Drehimpulses bleibt bestehen.

Gehen wir zum D-Term über, so sieht man sofort, daß durch Vertauschungen der drei Kristallachsen X, Y, Z die zonale Kugelfunktion $P_2{}^0 = \sqrt{\frac{5}{2}} \left(\frac{3}{2} \cos^2 \vartheta - \frac{1}{2}\right)$ nur übergehen kann in eine Funktion, die $\cos \varphi$ (bzw. $\sin \varphi$) gar nicht oder in der zweiten Potenz enthält, das ist außer $P_2{}^0$ selbst nur $\sqrt{2}\, P_2{}^2 \cos 2\varphi = \sqrt{\frac{15}{8}} \sin^2 \vartheta \,(2 \cos^2 \varphi - 1)$. Wenn wir z. B. die Z-Achse durch die X-Achse ersetzen, geht $P_2{}^0$ über in:

$$P_2{}^0(x) = \sqrt{\frac{5}{2}} \left(\frac{3}{2} \sin^2 \vartheta \cos^2 \varphi - \frac{1}{2}\right)$$

$$= -\frac{1}{2} P_2{}^0 + \frac{\sqrt{3}}{2} \sqrt{2}\, P_2{}^2 \cos 2\varphi .$$

$P_2{}^0$ und $\sqrt{2}\, P_2{}^2 \cos 2\varphi$ gehören also zum gleichen, zweifach entarteten Eigenwert. Analog kann

$$\sqrt{2}\, P_2{}^2 \sin 2\varphi = \sqrt{\frac{15}{2}} \sin \vartheta \cos \varphi \sin \vartheta \sin \varphi$$

durch Achsenvertauschungen nur übergehen in solche Kugelfunktionen zweiter Ordnung, die $\cos\varphi$ oder $\sin\varphi$ (oder beide) je in der ersten Potenz enthalten, also außer in sich selbst in

$$\sqrt{2}\,P_2^1\cos\varphi \quad\text{und}\quad \sqrt{2}\,P_2^1\sin\varphi \quad \left(P_2^1=\sqrt{\tfrac{15}{4}}\sin\vartheta\cos\vartheta\right).$$

Wir bezeichnen die beiden Terme, in welche der D-Term bei kubischer Symmetrie im Kristall aufspaltet, mit D_γ (zweifacher Term, Darstellung Γ_3) und D_ε (dreifacher Term, Darstellung Γ_5). Die zugehörigen Eigenfunktionen stellen wir einerseits in ausgeschriebener Form und andererseits in der Schreibweise von Ehlert[1]) gegenüber. Dabei bedeutet:

$$(\alpha\beta\gamma) = r^{l+1}\frac{\partial^l}{\partial x^\alpha\,\partial y^\beta\,\partial z^\gamma}\left(\frac{1}{r}\right),\quad l=\alpha+\beta+\gamma.$$

Term	Darst.	Ausgeschriebene Eigenfunktion	Eigenfunkt. n. Ehlert
D_γ	Γ_3	$(2\gamma)_1 = \sqrt{\frac{5}{2}}\left(\frac{3}{2}\cos^2\vartheta-\frac{1}{2}\right)=P_2^0$	(002)
		$(2\gamma)_2 = \sqrt{2}\,P_2^2\cos 2\varphi = \sqrt{\frac{15}{8}}\sin^2\vartheta\cos 2\varphi$	(200)—(020)
D_ε	Γ_5	$(2\varepsilon)_1 = \sqrt{2}\,P_2^2\sin 2\varphi = \sqrt{\frac{15}{8}}\sin^2\vartheta\sin 2\varphi$	(110)
		$(2\varepsilon)_2 = \sqrt{2}\,P_2^1\cos\varphi = \sqrt{\frac{15}{2}}\sin\vartheta\cos\vartheta\cos\varphi$	(101)
		$(2\varepsilon)_3 = \sqrt{2}\,P_2^1\sin\varphi = \sqrt{\frac{15}{2}}\sin\vartheta\cos\vartheta\sin\varphi$	(011)

Analog erhalten wir die Eigenfunktionen zu den Aufspaltungsprodukten des F-Terms im Kristall, indem wir jeweils die Eigenfunktionen, die durch Vertauschung der Achsen auseinander hervorgehen, dem gleichen Term zuordnen.

1) W. Ehlert, a. a. O.

Term	Darst.	Ausgeschriebene Eigenfunktion	Eigenfunkt. n. Ehlert
F_β	Γ_2	$(3\beta) = \sqrt{2}\, P_3^2 \sin 2\varphi = \frac{\sqrt{210}}{4} \cos\vartheta \sin^2\vartheta \sin 2\varphi$	(111)
F_δ	Γ_4	$(3\delta)_1 = P_3^0 = \sqrt{\frac{7}{2}}\left(\frac{5}{2}\cos^3\vartheta - \frac{3}{2}\cos\vartheta\right)$	(003)
		$(3\delta)_2 = \sqrt{2}\sqrt{\frac{5}{8}}\, P_3^3 \cos 3\varphi - \sqrt{2}\sqrt{\frac{3}{8}}\, P_3^1 \cos\varphi$	(300)
		$(3\delta)_3 = \sqrt{2}\sqrt{\frac{5}{8}}\, P_3^3 \sin 3\varphi + \sqrt{2}\sqrt{\frac{3}{8}}\, P_3^1 \sin\varphi$	(030)
F_ε	Γ_5	$(3\varepsilon)_1 = \sqrt{2}\, P_3^2 \cos 2\varphi$	(201)—(021)
		$(3\varepsilon)_2 = \sqrt{2}\sqrt{\frac{3}{8}}\, P_3^3 \cos 3\varphi + \sqrt{2}\sqrt{\frac{5}{8}}\, P_3^1 \cos\varphi$	(120)—(102)
		$(3\varepsilon)_3 = \sqrt{2}\sqrt{\frac{3}{8}}\, P_3^3 \sin 3\varphi - \sqrt{2}\sqrt{\frac{5}{8}}\, P_3^1 \sin\varphi$	(012)—(210)

Dabei ist bekanntlich

$$P_3^1 = \sqrt{\frac{21}{2}}\left(\frac{5}{4}\cos^2\vartheta - \frac{1}{4}\right)\sin\vartheta,$$

$$P_3^3 = \frac{\sqrt{70}}{8}\sin^3\vartheta.$$

Ebenso für den G-Term:

Term	Darst.	Ausgeschriebene Eigenfunktion	Eigenfunktion nach Ehlert
G_α	Γ_1	$(4\alpha) = \sqrt{\frac{7}{12}}\, P_4^0 + \sqrt{2}\sqrt{\frac{5}{12}}\, P_4^4 \cos 4\varphi$	(400) + (040) + (004)
G_γ	Γ_3	$(4\gamma)_1 = \sqrt{2}\, P_4^2 \cos 2\varphi$	(400) − (040)
		$(4\gamma)_2 = \sqrt{\frac{5}{12}}\, P_4^0 - \sqrt{2}\sqrt{\frac{7}{12}}\, P_4^4 \cos 4\varphi$	2·(004) − (400) − (040)
G_δ	Γ_4	$(4\delta)_1 = \sqrt{2}\, P_4^4 \sin 4\varphi$	(310) − (130)
		$(4\delta)_2 = \sqrt{2}\sqrt{\frac{7}{8}}\, P_4^1 \cos\varphi - \sqrt{2}\sqrt{\frac{1}{8}}\, P_4^3 \cos 3\varphi$	(103) − (301)
		$(4\delta)_3 = \sqrt{2}\sqrt{\frac{7}{8}}\, P_4^1 \sin\varphi + \sqrt{2}\sqrt{\frac{1}{8}}\, P_4^3 \sin 3\varphi$	(031) − (013)

Term	Darst.	Ausgeschriebene Eigenfunktion	Eigenfunktion nach Ehlert
G_ε	Γ_ε	$(4\varepsilon)_1 = \sqrt{2}\, P_4^2 \sin 2\varphi$	(112)
		$(4\varepsilon)_2 = \sqrt{2}\sqrt{\frac{1}{8}}\, P_4^1 \cos\varphi + \sqrt{2}\sqrt{\frac{7}{8}}\, P_4^3 \cos 3\varphi$	(121)
		$(4\varepsilon)_3 = \sqrt{2}\sqrt{\frac{1}{8}}\, P_4^1 \sin\varphi - \sqrt{2}\sqrt{\frac{7}{8}}\, P_4^3 \sin 3\varphi$	(211)
		$P_4^0 = \sqrt{\frac{9}{2}}\left(\frac{35}{8}\cos^4\vartheta - \frac{15}{4}\cos^2\vartheta + \frac{3}{8}\right)$	
		$P_4^1 = \frac{3}{8}\sqrt{10}\,(7\cos^3\vartheta - 3\cos\vartheta)\sin\vartheta$	
		$P_4^2 = \frac{3}{8}\sqrt{5}\,(7\cos^2\vartheta - 1)\sin^2\vartheta$	
		$P_4^3 = \frac{3}{8}\sqrt{70}\,\cos\vartheta\sin^3\vartheta$	
		$P_4^4 = \frac{3}{16}\sqrt{35}\,\sin^4\vartheta$	

Alle unsere Eigenfunktionen sind die richtigen Eigenfunktionen nullter Näherung für das Elektron im Kristallatom; denn es gehören niemals zwei Terme zur gleichen Darstellung.[1]) Dies würde erst bei der Untersuchung der Aufspaltung des H-Terms der Fall sein.

§ 13. Beziehung der Termaufspaltung zur Bedeutung der Kugelfunktionen als Potential von Multipolen

Jede Kugelfunktion l-ter Ordnung stellt die Winkelabhängigkeit des Potentials eines Multipols derselben Ordnung dar, wie man dies am leichtesten aus der Maxwellschen Form der Kugelfunktionen

$$(\alpha\beta\gamma) = r^{l+1}\frac{\partial^l\left(\frac{1}{r}\right)}{\partial x^\alpha\,\partial y^\beta\,\partial z^\gamma}, \quad l = \alpha + \beta + \gamma$$

1) Für das Problem von Ehlert, die Schwingungseigenfunktionen von CH_4 zu bilden, fehlt noch eine Symmetrisierung, bei welcher aus den zum gleichen Term gehörigen Eigenfunktionen Linearaggregate mit bestimmten Symmetrieeigenschaften gegenüber Vertauschung der Ecken eines Tetraeders zu bilden sind.

erkennt. Jedem Term des Elektrons bei kubischer Symmetrie entspricht nun eine bestimmte Zerlegung („Partitio)" von l in drei Summanden α, β, γ, wie bereits Ehlert bemerkte. Bis $l = 3$ gilt auch die Umkehrung dieses Satzes, die eine sehr anschauliche Deutung der Termaufspaltung bei kubischer Symmetrie ermöglicht:

Für $l = 1$ erhält man drei linear unabhängige Eigenfunktionen (100), (010), (001), die den Potentialen eines in der X-, Y- bzw. Z-Achse eingestellten Dipols entsprechen. Diese Dipoleinstellungen sind selbstverständlich gleichberechtigt, die drei Eigenfunktionen gehören zum gleichen Eigenwert.

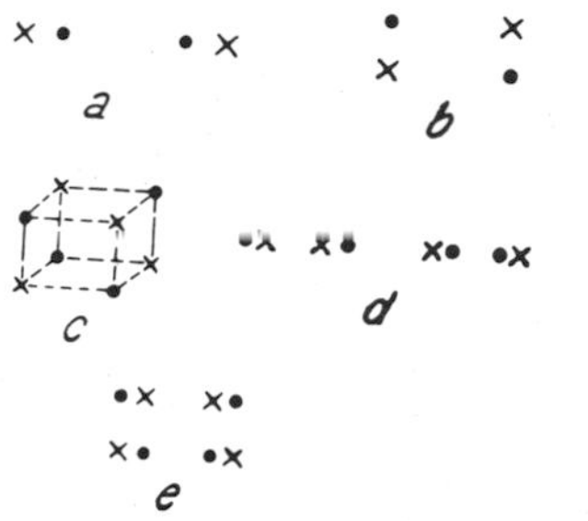

Quadrupol- und Oktopolklassen bei kubischer Symmetrie

Fig. 1

Für $l = 2$ gilt

$$(200) + (020) + (002) = r^3 \Delta \frac{1}{r} = 0.$$

Man erhält also:

a) die beiden linear unabhängigen Eigenfunktionen (002) und (200) — (020). Diese entsprechen den Potentialen von „gestreckten" Quadrupolen (Fig. 1a), d. h. solchen, die aus entgegengesetzt gleichen, *in* Richtung ihrer Achse gegeneinander verschobenen Dipolen bestehen. Da die Richtungen X, Y, Z gleichberechtigt sind, gehören beide Eigenfunktionen zum gleichen zweifachen Eigenwert D_γ.

b) Die drei linear unabhängigen Eigenfunktionen (011), (101), (110) stellen das Potential von flächenhaft ausgedehnten Quadrupolen dar, welche aus zwei *senkrecht* zu ihrer Achse verschobenen Dipolen bestehen (Fig. 1b). Die drei möglichen Orientierungen dieser Quadrupole in der YZ-, ZX- und XY-Ebene, welche den obigen Potentialen entsprechen, sind natürlich gleichberechtigt, die Eigenfunktionen gehören zum gleichen, dreifachen Eigenwert D_ε.

Für $l = 3$ erhält man:

a) Die Eigenfunktion (111), welche das Potential eines räumlichen Oktopols (Fig. 1c) darstellt, der aus zwei entgegengesetzt gleichen, senkrecht zu ihrer eigenen Ebene verschobenen flächenhaften Quadrupolen besteht. Die Eigenfunktion gehört zum einfachen Eigenwert F_β.

b) Die Eigenfunktionen (300), (030), (003), welche Potentiale von linear ausgedehnten Oktopolen (Fig. 1d) darstellen, gehören zum dreifachen Eigenwert F_δ.

c) Die Eigenfunktionen (120) — (102), (012) — (210), (201) — (021), Potentialfunktionen von flächenhaft ausgedehnten Oktopolen[1]), gehören zum dreifachen Eigenwert F_ε (Fig. 1e).

Bei höheren Multipolen ergeben sich kompliziertere Verhältnisse, vgl. Ehlert.

III. Das Atom unter dem Einfluß von Kristallfeldern verschiedener Größenordnung

Wir haben bisher allgemein das Verhalten eines mit einem bestimmten Drehimpuls begabten quantenmechanischen Systems in einem elektrischen Feld von vorgegebener Symmetrie behandelt. Nunmehr müssen wir die in der Einleitung aufgeführten drei Fälle diskutieren, welche sich durch die Größenordnung der Termaufspaltung im Kristall unterscheiden, und außerdem die Terme, welche man im starken, mittleren und schwachen Kristallfeld erhält, einander zuordnen.

§ 14. Die Winkelverteilung der Elektronendichte bei starkem Kristallfeld

Wir können im ersten der in der Einleitung unterschiedenen Fälle die Wechselwirkung der Elektronen des betrachteten Atoms durch gegenseitigen Austausch in erster Näherung vernachlässigen, und dem *einzelnen* Elektron einen Termwert und eine Eigenfunktion zuschreiben. Der wirkliche Termwert des Atoms ergibt sich durch Addition der einzelnen Elektronenterme, die Eigenfunktion durch Multiplikation der Elektroneneigenfunktionen und „Antisymmetrisieren" des Produkts.[2]) (Wir sprechen von „Eigenfunktion des einzelnen Elektrons"

1) Die Eigenfunktionen (210) + (012) = − (030) und die durch zyklische Vertauschung entstehenden sind identisch mit den unter b) genannten.

2) W. Heitler, Ztschr. f. Phys. **46**. S. 47. 1928.

sind uns aber dabei bewußt, daß vorgegeben nur die Form der Eigenfunktion, die Quantenzelle, ist, dagegen jedes der Elektronen des Atoms sich mit gleicher Wahrscheinlichkeit in dieser Quantenzelle befindet.)

Die Winkelabhängigkeit der Eigenfunktion des einzelnen Elektrons ist im *freien* Atom gegeben durch irgendeine Kugelfunktion l-ter Ordnung oder durch ein Linearaggregat solcher Funktionen, wenn l der Drehimpuls der „Bahn" des Elektrons ist (vgl. § 10). Alle $2(2l+1)$ Quantenzellen mit gleichem l (und gleicher Hauptquantenzahl n) gehören zum selben Elektronentermwert, und mangels Festlegung einer Achse durch ein äußeres Feld hat es keinen Sinn zu fragen, welcher der $2(2l+1)$ Zellen ein Elektron angehört: Seine Aufenthaltswahrscheinlichkeit (Dichte) hängt nur von der Entfernung vom Kern ab und bevorzugt keine Richtung im Raum, seine „Bahnebene" ist beliebig.

Im Kristall dagegen zerfällt die aus $2(2l+1)$-Quantenzellen bestehende Schale entsprechend den irreduziblen Darstellungen der Kristallsymmetriegruppe in mehrere Unterschalen, von denen jede einem anderen Elektronenterm entspricht. Ein einzelnes Elektron wird, wenn die durch die Kristallfelder bewirkte Aufspaltung genügend groß bzw. die Temperatur genügend tief ist, die zum tiefsten Term gehörige Unterschale aufsuchen, soweit diese noch nicht „besetzt" ist. Zu jeder Unterschale gehört aber eine ganz bestimmte winkelabhängige Eigenfunktion und die Aufenthaltswahrscheinlichkeit des einzelnen Elektrons wird deshalb in manchen Richtungen des Kristalls Maxima besitzen, in anderen Richtungen verschwinden, die Winkelverteilung der Elektronendichte ist geradezu ein Charakteristikum für jeden einzelnen Elektronenterm im Kristall.

Ein p-Elektron im tetragonalen Kristall z. B. wird sich je nach den speziellen elektrischen Feldern des Kristalls (vgl. § 22) entweder in die dem einfachen p_β-Term entsprechende Unterschale begeben oder in die dem zweifachen p_ε-Term entsprechende. Im ersten Falle ist die Winkelabhängigkeit seiner Dichte durch $\frac{3}{2}\cos^2\vartheta$ gegeben, die Dichte hat also in der tetragonalen Achse ($\vartheta = 0$) ein Maximum, in der Ebene der zweizähligen Achsen verschwindet sie. Im zweiten Falle ist die Dichte proportional $\sin^2\vartheta$, hat also den umgekehrten

Verlauf. Wenn etwa in der tetragonalen Achse nahe dem betrachteten Atom positive Ionen liegen, senkrecht dazu jedoch erst in erheblich größerer Entfernung, so wird der erste Fall einem Energieminimum entsprechen und vice versa.

Am interessantesten sind die Verhältnisse bei kubischer Symmetrie des Atoms im Kristall. Ein p-Elektron hat hier überhaupt keine bevorzugte Einstellung, alle „Plätze" der p-Schale sind energetisch gleichwertig und infolgedessen die Aufenthaltswahrscheinlichkeit eines p-Elektrons in nullter Näherung kugelsymmetrisch um den Kern verteilt. Ein d-Elektron dagegen kann entweder einen Platz in der vier Quantenzellen umfassenden d_γ-Unterschale einnehmen — das wird ein d-Elektron eines negativen Ions im NaCl-Typ z. B. stets tun — oder in der sechszelligen d_ε-Schale, je nachdem welcher der beiden Kristallterme tiefer liegt. Im ersten Fall wird seine Eigenfunktion entweder P_2^0 oder ebenso wahrscheinlich $\sqrt{2}\,P_2^2 \cos 2\varphi$ sein und seine Dichte bis auf einen vom Kernabstand abhängigen Faktor gegeben durch:

$$
(7) \quad \begin{cases} \varrho = (P_2^0)^2 + 2\,(P_2^2)^2 \cos^2 2\varphi = \frac{5}{4}\left(\frac{3}{2}\cos^2\vartheta - \frac{1}{2}\right)^2 \\ \qquad + \frac{5}{4}\cdot\frac{3}{4}\cdot\sin^4\vartheta \cos^2 2\varphi\,. \end{cases}
$$

(Die Wahrscheinlichkeiten, die sich aus beiden Eigenfunktionen ergeben, sind zu addieren.) Diese Dichte erreicht ein Maximum $\varrho = \frac{5}{4}$ in den drei vierzähligen Achsen

$$
\left(\vartheta = 0, \quad \vartheta = \frac{\pi}{2} \quad \varphi = 0, \quad \vartheta = \frac{\pi}{2} \quad \varphi = \frac{\pi}{2}\right)
$$

und ein Minimum $\varrho = 0$ in den dreizähligen Achsen

$$
\left(\cos\vartheta = \sqrt{\frac{1}{3}}, \quad \vartheta = 54^0\,44', \quad \varphi = \frac{\pi}{4}\right)\cdot
$$

Tabelle 14

Dichteverteilung eines d_γ-Elektrons bei kubischer Symmetrie

ϑ	$\varphi = 0$	10	20	30	40	45⁰
0	1,25	1,25	1,25	1,25	1,25	1,25
10	1,14	1,14	1,14	1,14	1,14	1,14
20	0,87	0,87	0,87	0,86	0,86	0,86
30	0,55	0,54	0,52	0,50	0,49	0,49
40	0,37	0,35	0,30	0,25	0,22	0,21
50	0,34	0,31	0,21	0,10	0,03	0,02
60	0,61	0,55	0,39	0,21	0,10	0,08
70	0,92	0,83	0,61	0,36	0,20	0,18
80	1,16	1,06	0,80	0,50	0,31	0,28
90	1,25	1,14	0,86	0,55	0,34	0,31

Fig. 2 stellt die Dichteverteilung graphisch dar.

Liegt andererseits der dreifache Term d_ε tiefer, so ist die Elektronendichte in der dreizähligen Achse (Würfeldiagonale) am größten.

Diese Bevorzugung einer Achse des Kristalls durch die Elektronendichte ist bereits ein Effekt nullter Näherung,

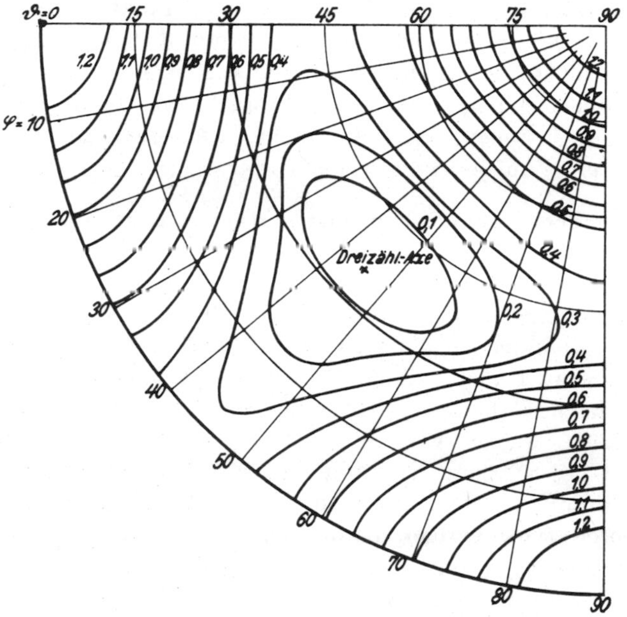

Elektronendichteverteilung des d_γ-Elektrons bei kubischer Symmetrie
Fig. 2

genau wie schon in nullter Näherung die Elektronen zweier H-Atome, die sich zum Molekül vereinigen, aufeinander zuzustreben scheinen.[1]) Als nullte Näherung ist dabei in üblicher Weise eine solche verstanden, die nur aus den Eigenfunktionen des ungestörten Problems die passenden herausgreift, ohne die Störung der Eigenfunktion zu berücksichtigen. Bei Durchführung weiterer Näherungen bleibt natürlich keine Dichtefunktion mehr kugelsymmetrisch, es ist eben dann die Zerlegung der Eigenfunktionen in ein Produkt eines radial-

1) F. London, Ztschr. f. Phys. **46.** S. 455. 1928.

abhängigen Faktors und einer Kugelfunktion überhaupt nicht mehr legitim. Es findet bei diesen weiteren Näherungen zwar keine Termaufspaltung mehr statt, aber immerhin noch eine stetige Verschiebung des Termwerts und Deformation der Eigenfunktion.

Auch die Verteilung der Gesamtladung einer abgeschlossenen Schale ist natürlich nicht mehr kugelsymmetrisch, wenn man die höheren Näherungen für die Eigenfunktion berechnet, d. h. die Deformation durch das Kristallfeld berücksichtigt. Dagegen bleibt in nullter Näherung die Kugelsymmetrie abgeschlossener Schalen erhalten, wie auch der Gesamttermwert der abgeschlossenen Schale in seiner Eigenschaft als ^1S-Term niemals eine Aufspaltung zeigen kann.

§ 15. Wechselwirkung der im Kristall orientierten Elektronen verschiedener Schalen

Wir haben nun die Aufgabe, die bei starkem und bei mittlerem Kristallfeld auftretenden Terme des Atoms im Kristall einander zuzuordnen (Fall 1 und 2 der Einleitung). Wir gehen in beiden Fällen aus von der Angabe der Quantenzahlen der außerhalb von abgeschlossenen Schalen befindlichen Elektronen, welche wir zunächst als entkoppelt annehmen. Im Fall des starken Kristallfeldes haben wir dann zunächst die Drehimpulse der einzelnen Elektronen zum Kristall zu orientieren (vgl. voriger Paragraph) und dann die Wechselwirkung der Elektronen unter Beibehaltung ihrer Orientierung zu untersuchen. Im zweiten Falle (mittleres Kristallfeld) haben wir die Wechselwirkung im freien Atom zu berücksichtigen, d. h. die Bahndrehimpulse der Elektronen zuerst zum Gesamtbahnimpuls l des Atoms zusammenzusetzen und dann die Einstellung *dieses* Drehimpulses im Kristall zu untersuchen. Bei *mittlerer* Kristallaufspaltung ist — ohne Anregung — zunächst der Grundzustand des freien Atoms fest vorgegeben, und im Kristall kann das Atom immer nur in einem Zustand sein, der aus dem Grundzustand durch Starkeffektaufspaltung hervorgeht. Dagegen sucht bei starkem Kristallfeld zunächst jedes Elektron für sich die energetisch tiefste Lage; durch Zusammenwirken der bereits orientierten Elektronen kann ein beliebiger Term des Atoms im Kristall entstehen, welcher bei Aufhebung des Kristallfeldes gar nicht immer in den Grundterm des freien Atoms überzugehen braucht.

Diesen Sachverhalt haben wir gruppentheoretisch zu untersuchen. Nehmen wir etwa zwei Elektronen mit den azimutalen Quantenzahlen l und λ an, so wird die Transformation ihrer Eigenfunktionen bei Drehungen des Atoms durch die Darstellungen d_l und d_λ der Drehgruppe gegeben. Im Fall 1 (starkes Kristallfeld) reduzieren wir d_l und d_λ zunächst einzeln als Darstellungen der Kristallsymmetriegruppe aus (Orientierung des einzelnen Elektrons im Kristall), es sei

$$d_l = \sum \alpha_{lk} \gamma_k,$$
$$d_\lambda = \sum \alpha_{\lambda\varkappa} \gamma_\varkappa;$$

die Energie des aus beiden Elektronen bestehenden Systems ist dann zunächst die Summe der Energien der einzelnen Elektronen, $E_k + E_\varkappa$; die Eigenfunktion ist als antisymmetrisiertes Produkt der Eigenfunktionen der einzelnen Elektronen anzusetzen und transformiert sich bei Symmetrieoperationen entsprechend der Darstellung $\gamma_k \cdot \gamma_\varkappa$ der Kristallgruppe. Nun berücksichtigen wir die Wechselwirkung der beiden Elektronen bei festgehaltener Orientierung, d. h. bei festgehaltenen Termen der einzelnen Elektronen. Dann spaltet der Term $E_k + E_\varkappa$ auf in mehrere Terme, welche den irreduziblen Bestandteilen von $\gamma_k \cdot \gamma_\varkappa$ entsprechen:

$$\gamma_k \cdot \gamma_\varkappa = \sum^i \beta^i_{k\varkappa} \Gamma_i .$$

Fall 2: Wir berücksichtigen zuerst die Elektronenwechselwirkung im freien Atom, indem wir $d_l \cdot d_\lambda$ als Darstellung der Drehgruppe ausreduzieren:[1])

$$d_l \cdot d_\lambda = \sum_{|l-\lambda|}^{l+\lambda} D_\varrho .$$

Dann setzen wir das fertige Atom in den Kristall ein, und haben D_ϱ als Darstellung der Kristallgruppe auszureduzieren:

$$D_\varrho = \sum \alpha_{\varrho i} \Gamma_i .$$

Die Reihenfolge der gruppentheoretischen Reduktionen korrespondiert der Reihenfolge der Störungsrechnungen, d. h. dem Größenverhältnis der Beiträge von Kristallfeld und Elektronenwechselwirkung zum Termwert. Die Anzahl der end-

1) E. Wigner u. J. v. Neumann, Ztschr. f. Phys. **49**. S. 73. E. Fues, ebenda **51**. S. 817. 1928.

gültigen Terme mit bestimmter Darstellungseigenschaft muß auf beiden Wegen gleich herauskommen. Die Zuordnung der einzelnen Terme bei starkem und mittlerem Kristallfeld zueinander kann durch die Forderung erfolgen, daß Kristallterme des Gesamtatoms, die zur gleichen Darstellung der Symmetriegruppe des Kristalls gehören, sich nicht überschneiden dürfen, wenn man sich das Kristallfeld langsam von mittlerer zu großer Stärke anwachsend denkt.

Beispiel: Um von Einschränkungen des Pauliprinzips frei zu sein, untersuchen wir die Terme eines Systems, bestehend aus zwei d-Elektronen mit *verschiedenen* Hauptquantenzahlen n_1 und n_2, im kubischen Kristall:

Fall 1: Große Aufspaltung im Kristall.

a) Ausreduktion der zur Eigenfunktion des Einzelelektrons gehörigen Darstellung der Drehgruppe als Darstellung der Kristallgruppe (vgl. Tab. 1):

$$d_2 = \gamma_3 + \gamma_5 .$$

Wir erhalten ohne Elektronenwechselwirkung vier Terme: Beide Elektronen können im Zustand γ_3, beide im Zustand γ_5 oder je eines in jedem Zustand sein, wobei es nicht gleichgültig ist, welches Elektron im Zustand γ_3 ist, da die beiden Elektronen durch die Hauptquantenzahl unterschieden sind. Die Terme ohne Wechselwirkung entsprechen den Darstellungen $\gamma_3 \cdot \gamma_3$, $\gamma_3 \cdot \gamma_5$, $\gamma_5 \cdot \gamma_3$ bzw. $\gamma_5 \cdot \gamma_5$ der Kristallgruppe.

b) Berücksichtigung der Elektronenwechselwirkung, Aufspaltung jedes Terms in a, entsprechend den irreduziblen Bestandteilen seiner Darstellung.[1])

$$\gamma_3 \cdot \gamma_3 = \Gamma_1 + \Gamma_2 + \Gamma_3,$$
$$\gamma_3 \cdot \gamma_5 = \Gamma_4 + \Gamma_5,$$
$$\gamma_5 \cdot \gamma_5 = \Gamma_1 + \Gamma_3 + \Gamma_4 + \Gamma_5 .$$

Fall 2: Mittlere Aufspaltung im Kristall.

a) Wechselwirkung der Elektronen im freien Atom:

$$d_2 \cdot d_2 = D_0 + D_1 + D_2 + D_3 + D_4 .$$

1) Wir bezeichnen die Darstellung, entsprechend der sich die Eigenfunktionen des einzelnen Elektrons transformieren, mit kleinen, diejenige der Eigenfunktion des Gesamtsystems mit großen griechischen Buchstaben.

Das freie Atom kann im *S*-, *P*-, *D*-, *F*-, *G*-Zustand sein; welches davon der tiefste Term ist, läßt sich ohne genaue Rechnung natürlich nicht entscheiden.

b) Einstellung des gesamten Bahndrehimpulses im Kristall, Ausreduktion *der* Darstellung der Drehgruppe, die dem Term

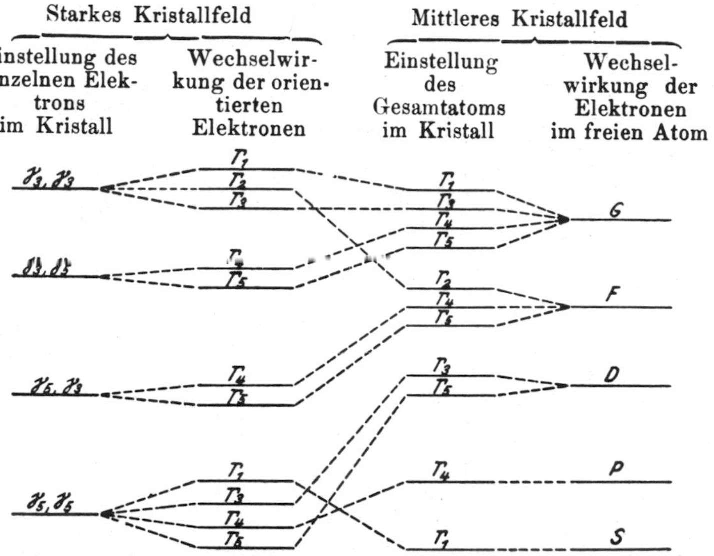

Terme eines Systems von zwei *d*-Elektronen verschiedener Hauptquantenzahl bei kubischer Symmetrie

Fig. 3

des Gesamtatoms entspricht, als Darstellung der Oktaedergruppe (Tab. 1):

$$D_0 = \Gamma_1,$$

$$D_1 = \Gamma_4,$$

$$D_2 = \Gamma_3 + \Gamma_5,$$

$$D_3 = \Gamma_2 + \Gamma_4 + \Gamma_5,$$

$$D_4 = \Gamma_1 + \Gamma_3 + \Gamma_4 + \Gamma_5.$$

Wir erhalten in Summa ersichtlich die gleichen Terme wie vorher. Die Zuordnung ist bei willkürlichen Annahmen über die gegenseitige Lage der Terme in Fig. 3 vorgenommen.

§ 16. Wechselwirkung von orientierten Elektronen der gleichen Schale

Um nur die nach dem Pauli-Prinzip erlaubten Terme eines Systems mehrerer Elektronen mit gleicher Haupt- und Azimutalquantenzahl zu erhalten, geht man am einfachsten analog vor wie beim entsprechenden Problem im freien Atom: Man hebt die noch bestehenden Termentartungen ganz oder teilweise auf, indem man etwa bei kubischer Symmetrie eine Achse etwas gedehnt annimmt und dadurch die Symmetrie auf tetragonale reduziert; jeder entartete Term des einzelnen Elektrons im kubischen System spaltet dabei in Terme niedrigerer Entartung im tetragonalen System auf, deren Wechselwirkung einfach zu behandeln ist. Man erhält auf diese Weise die Terme des Systems zweier Elektronen im Feld von tetragonaler Symmetrie, und hat diese dann wieder zu Termen im kubischen Feld zusammenzufügen. Die irreduziblen Darstellungen der Oktaedergruppe Γ_i enthalten nun, wenn man sie an Hand der Tabb. 1, 5, 7 und 11 als Darstellungen der tetragonalen Drehgruppe ausreduziert, folgende irreduzible Bestandteile[1]):

$$\Gamma_1 = G_1 \quad \Gamma_2 = G_3 \quad \Gamma_3 = G_1 + G_3 \quad \Gamma_4 = G_2 + G_5 \quad \Gamma_5 = G_4 + G_5$$
$$\Gamma_6 = G_6 \quad \Gamma_7 = G_6 \quad \Gamma_8 = G_6 + G_7$$

Es ist nützlich, auch die Entartung des zweifachen Terms G_5 der tetragonalen Gruppe noch aufzugeben, indem man eine der zweizähligen Achse der tetragonalen Symmetriegruppe nochmals vor der anderen auszeichnet, und dadurch die Symmetrie auf rhombische beschränkt. Die Ausreduktion ergibt:

$$G_1 = G_3 = \mathfrak{G}_1 \qquad G_2 = G_4 = \mathfrak{G}_2 \qquad G_5 = \mathfrak{G}_3 + \mathfrak{G}_4 .$$

Die Darstellungen der rhombischen Gruppe sind sämtlich eindimensional, das Produkt zweier Darstellungen ist also wieder eine irreduzible Darstellung:

$$\mathfrak{G}_1 \mathfrak{G}_i = \mathfrak{G}_i \quad \mathfrak{G}_i{}^2 = \mathfrak{G}_1 \quad \mathfrak{G}_2 \mathfrak{G}_3 = \mathfrak{G}_4 \quad \mathfrak{G}_3 \mathfrak{G}_4 = \mathfrak{G}_2 \quad \mathfrak{G}_4 \mathfrak{G}_2 = \mathfrak{G}_3 .$$

Wir behandeln als Beispiel die Wechselwirkung zweier d-Elektronen gleicher Hauptquantenzahl bei kubischer Symmetrie und starkem Kristallfeld. Die d-Schale spaltet zunächst auf

---

1) Wir schreiben in diesem Paragraphen Darstellungen der kubischen Gruppe mit griechischen, der tetragonalen mit lateinischen und der rhombischen mit deutschen Buchstaben.

in die $\gamma_3\,(d_\gamma)$-Unterschale mit 4 Plätzen (wir müssen bei Abzählung der Plätze den zwei möglichen Spinrichtungen Rechnung tragen) und die γ_5-Unterschale (Term d_ε) mit 6 Plätzen. Ist ein Elektron in jeder Unterschale, so entstehen durch Wechselwirkung sämtliche Terme des Zweielektronensystems, deren Darstellung irreduzible Bestandteile von

$$\gamma_3\,\gamma_5 = \Gamma_4 + \Gamma_5$$

sind, und zwar jeweils ein Tripletterm und ein Singuletterm, weil die Spinrichtung der beiden Elektronen wegen der verschiedenen Schwerpunktseigenfunktion noch beliebig bleibt. Das gesamte Quantengewicht dieser Terme ist $4 \cdot 6 = 24$.

Sind dagegen beide Elektronen in der γ_3-Unterschale, so sind von 16 Wechselwirkungsweisen nur $\frac{4 \cdot 3}{2} = 6$ erlaubt. Die Darstellungseigenschaften der Terme des Systems der zwei Elektronen sind wieder gegeben durch die irreduziblen Bestandteile von

$$\gamma_3 \cdot \gamma_3 = \Gamma_1 + \Gamma_2 + \Gamma_3\,,$$

aber die Vielfachheit der einzelnen Terme ist noch unbekannt. Wir denken uns die Symmetrie, wie oben beschrieben, auf tetragonale vermindert, dann spaltet die γ_3-Schale auf in die g_1 und g_3-Unterschale der tetragonalen Symmetrie mit je 2 Plätzen. Beide Elektronen in der $\mathfrak{G}_1$-Unterschale unterzubringen, ist nun nur auf eine Weise möglich, nämlich wenn sie entgegengesetzten Spin haben. Dagegen gibt es vier Möglichkeiten, je ein Elektron in g_1 und g_3 unterzubringen, weil dann der Spin beliebig ist. Man erhält:

$g_1 \cdot g_1 = G_1$	Quantengewicht	1	$m_s = 0$
$g_1 \cdot g_3 = G_3$	,,	4	$m_s = -1\;0\;0\;1$
$g_3 \cdot g_3 = G_1$	,,	1	$m_s = 0$

Das sind drei Singuletterme mit den Darstellungseigenschaften G_1, G_3, G_1 und ein Tripletterm G_3. Durch Vergleich mit (19) sieht man, daß der Term Γ_2 des Zweielektronensystems ein Triplett, Γ_1 und Γ_3 dagegen Singuletterme sind.

Ebenso behandeln wir die Wechselwirkung zweier Elektronen in der kubischen γ_5-Schale:

$$\gamma_5 \cdot \gamma_5 = \Gamma_1 + \Gamma_3 + \Gamma_4 + \Gamma_5 \quad \text{Quantengewicht } \frac{6 \cdot 5}{2} = 15\,.$$

Reduktion auf tetragonale Symmetrie:

$\gamma_5 = g_4 + g_5,$

$g_4 \cdot g_4 = G_1$ Quantengewicht $1\ m_s = 0,$

$g_4 \cdot g_5 = G_5$,, $8\ m_s = -1\,0\,0\,1,$

$g_5 \cdot g_5 = G_1 + G_2 + G_3 + G_4$,, $6\ m_s = \ ?$

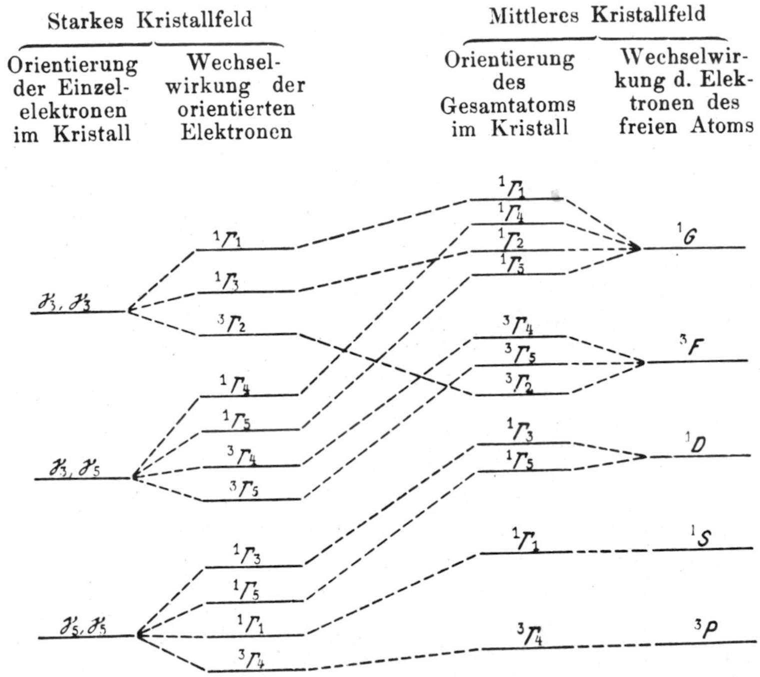

Terme eines Systems zweier d-Elektronen gleicher Hauptquantenzahl bei kubischer Symmetrie

Fig. 4

Weitere Reduktion auf rhombische Symmetrie:

$g_5 = \mathfrak{g}_3 + \mathfrak{g}_4,$

$\mathfrak{g}_3 \cdot \mathfrak{g}_3 = \mathfrak{G}_1$ Quantengewicht $1\ m_s = 0,$

$\mathfrak{g}_3 \cdot \mathfrak{g}_4 = \mathfrak{G}_2$,, $4\ m_s = -1\,0\,0\,1,$

$\mathfrak{g}_4 \cdot \mathfrak{g}_4 = \mathfrak{G}_1$,, $1\ m_s = 0.$

Zusammenfassung zu Zweielektronentermen bei tetragonaler Symmetrie:

$g_5\, g_5 = {}^1G_1 + {}^1G_2 + {}^1G_3 + {}^3G_4$ *oder* ${}^1G_1 + {}^3G_2 + {}^1G_3 + {}^1G_4\,.$

Zusammenfassung zu Termen bei kubischer Symmetrie:

$$\gamma_5\,\gamma_5 = {}^1\Gamma_1 + {}^1\Gamma_3 + {}^1\Gamma_4 + {}^3\Gamma_5 \quad \textit{oder} \quad {}^1\Gamma_1 + {}^1\Gamma_3 + {}^3\Gamma_4 + {}^1\Gamma_5\,.$$

Ob Γ_4 oder Γ_5 der Tripletterm ist, entscheidet sich je nach der Azimutalquantenzahl l der beiden γ_5-Elektronen, in unserem Fall $l = 2$ ist Γ_5 der Tripletterm.

Die Zuordnung der Terme des aus zwei d-Elektronen bestehenden quantenmechanischen Systems bei willkürlichen Annahmen über die Lage der Terme zueinander ersieht man aus Fig. 4.

In genau der gleichen Weise läßt sich die Wechselwirkung von mehr als zwei im Kristall orientierten Elektronen behandeln und so die Zuordnung von Termen im „starken“ und „mittleren“ Kristallfeld vollständig durchführen. Wir geben die Darstellungseigenschaften der Kristallterme, die bei Wechselwirkung von mehreren Elektronen der gleichen Unterschalung bei kubischer Symmetrie entstehen können:

Tabelle 15

Anzahl der Elektr.	Unterschale γ_3	Unterschale γ_4	Unterschale γ_5
1	${}^2\Gamma_3$	${}^2\Gamma_4$	${}^2\Gamma_5$
2	${}^1\Gamma_1 + {}^3\Gamma_2 + {}^1\Gamma_3$	$\left\{\begin{matrix} {}^1\Gamma_1 + {}^1\Gamma_3 + {}^1\Gamma_4 + {}^3\Gamma_5 \\ {}^1\Gamma_1 + {}^1\Gamma_3 + {}^3\Gamma_4 + {}^1\Gamma_5 \end{matrix}\right\}$	wie γ_4
3	${}^2\Gamma_3$	$\left\{\begin{matrix} {}^4\Gamma_1 + {}^2\Gamma_3 + {}^2\Gamma_4 + {}^2\Gamma_5 \\ {}^4\Gamma_2 + {}^2\Gamma_3 + {}^2\Gamma_4 + {}^2\Gamma_5 \end{matrix}\right\}$	${}^4\Gamma_1 + {}^2\Gamma_3 + {}^2\Gamma_4 + {}^2\Gamma_5$ ${}^4\Gamma_2 + {}^2\Gamma_3 + {}^2\Gamma_4 + {}^2\Gamma_5$
4	${}^1\Gamma_1$	wie bei 2 Elektronen in der Schale	
5	—	${}^2\Gamma_4$	${}^2\Gamma_5$
6	—	${}^1\Gamma_1$	${}^1\Gamma_1$

§ 17. Wechselwirkung zwischen Bahndrehimpuls und Spin

Ein Atom sei beschrieben durch seinen Bahndrehimpuls l und den Gesamtspin s (Termmultiplizität $2s + 1$). D. h. bei gemeinsamer Drehung aller Elektronenschwerpunkte um den Kern unter Beibehaltung der Spinrichtung transformiert sich die Hyperfunktion des Atoms entsprechend der Darstellung D_l der Drehgruppe; bei Drehung der Spinrichtung allein entsprechend der Darstellung D_s.[1])

1) Die erste Drehung entspricht der Transformation $P_{\mathfrak{R}}$ bei Wigner und J. v. Neumann, a. a. O., die zweite der Transformation $Q_{\mathfrak{R}}$.

Im Fall 2 der Einleitung (Starkeffektaufspaltung im Kristall *groß* gegen die Multiplettaufspaltung) stellt sich der Bahndrehimpuls l für sich zum Kristall ein; das *Gesamtmultiplett* ist als Ausgangszustand zu nehmen und spaltet im Kristall auf in Starkeffektkomponenten, welche den irreduziblen

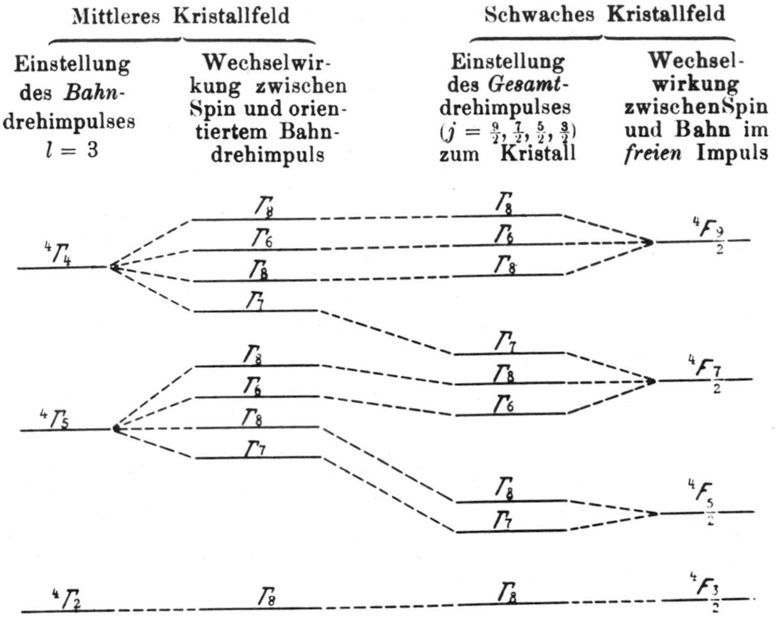

Einstellung eines Atoms im ^4F-Zustand im mittleren und schwachen Kristallfeld von kubischer Symmetrie

Fig. 5

Bestandteilen Γ_λ der Darstellung D_l der Kugeldrehgruppe entsprechen:

$$D_l = \sum \alpha_{l\lambda} \, \Gamma_\lambda \, .$$

Jede dieser Starkeffektkomponenten spaltet nun durch Wechselwirkung mit dem Spin weiter auf:

$$\Gamma_\lambda D_s = \sum \alpha_{\lambda\mu} \, \Gamma_\mu \, ,$$

dabei ist diese weitere Aufspaltung von der Größenordnung der Multiplettaufspaltung, doch ist die Anzahl der Komponenten eines Terms Γ_λ im allgemeinen nicht gleich $2s + 1$.

Im Fall 3 (Kristallaufspaltung *klein* gegen Multiplettaufspaltung) setzen sich Spin und Bahndrehimpuls zunächst zum Gesamtdrehimpuls j des Atoms zusammen:

$$D_l D_s = \sum_{|l-s|}^{l+s} D_j$$

und dieser orientiert sich zu den Kristallachsen:

$$D_j = \sum \alpha_{j\mu} \Gamma_\mu ,$$

d. h. jede Multiplettkomponente *einzeln* erleidet im Kristall eine weitere Aufspaltung von geringerer Größe als die Multiplettaufspaltung. Fig. 5 zeigt die Aufspaltung eines ^4F-Terms bei kubischer Symmetrie und mittlerem bzw. schwachem Kristallfeld.[1])

IV. Größe der Aufspaltung

§ 18. Allgemeine Formel für den Term eines Elektrons im Ionenkristall

Nachdem wir uns bisher qualitativ über den Starkeffekt im Kristall orientiert haben, wollen wir nun wenigstens für den Fall des starken Kristallfeldes (jedes Elektron sein eigener Termwert) auch die Größe der Aufspaltung berechnen. Dabei zeigt sich, daß ganz allgemein der Starkeffekt im inhomogenen Feld des Kristalls formal ein Effekt *erster* Ordnung ist, im Gegensatz zum Starkeffekt im homogenen Feld, bei dem die Störungsenergie erster Ordnung stets verschwindet. Von den Termen, die im Kristallfeld aus dem Term n, l des Elektrons im freien Atom entstehen, möge nur einer zur Darstellung Γ_λ der Kristallgruppe gehören. Dann beträgt die Lage dieses Terms λ relativ zu der des ungestörten Terms im freien Atom:

$$(8) \qquad E_\lambda = -\int e V \psi_{nl\lambda}^2 \, d\tau$$

$-Ve$ ist die potentielle Energie des Elektrons im Feld der fremden Atome des Kristalls, $\psi_{nl\lambda}$ eine beliebige von den zum Term (n, l, λ) gehörigen Eigenfunktionen nullter Näherung. $\psi_{nl\lambda}$ läßt sich schreiben als Produkt eines radialabhängigen Faktors und eines Linearaggregats von Kugelfunktionen:

$$(9) \qquad \psi_{nl\lambda} = \psi_{nl} \sum_{m=0}^{l}{}' c_{\lambda m} P_l^m (\cos\vartheta) \begin{matrix}\cos\\ \sin\end{matrix} m\varphi .$$

1) Praktisch kommen wohl meist Übergänge zwischen mittleren und starkem (bzw. mittlerem und schwachem) Kristallfeld vor.

Wir brauchen also Integrale der Form:

(10) $K'_{m\mu} = -\int V e\, \psi^2_{nl}(r) P_l^m(\cos\vartheta) P_l^\mu(\cos\vartheta) {}^{\cos}_{\sin} m\varphi\, {}^{\cos}_{\sin} \mu\varphi\, d\tau.$

Ersichtlich können wir mit Hilfe solcher Integrale auch dann sämtliche Termwerte bestimmen, wenn mehrere Terme zur gleichen Darstellung der Kristallgruppe gehören (§ 10/11) und man zur Berechnung der Eigenwerte ein spezielles Störungsproblem ansetzen muß, in welches die Matrixglieder:

$$\varepsilon_{ii} = -\int V e\, \psi^2_{nli}\, d\tau,$$

$$\varepsilon_{ik} = -\int V e\, \psi_{nli} \psi_{nlk}\, d\tau$$

eingehen.

Nun verschwinden bei mindestens rhombisch-holoedrischer Symmetrie von den Integralen $K_{m\mu}$ zunächst alle, welche $\cos m\varphi$ und $\sin \mu\varphi$ gemischt enthalten. Denn bei Spiegelung an der XZ-Ebene geht V in sich, dagegen $\cos m\varphi \sin\mu\varphi$ in den entgegengesetzt gleichen Wert über, das Integral $K_{m\mu}$ muß aber gegen derartige Änderungen der Integrationsvariablen invariant sein, also verschwinden. Wir schreiben jetzt:

$$K^{\cos}_{m\mu} = -\int V e\, \psi^2_{nl}(r) P_l^m(\cos\vartheta) P_l^\mu(\cos\vartheta) \cos m\varphi \cos\mu\varphi\, d\tau$$
$$= -\frac{1}{2}\int V e\, \psi^2_{nl} P_l^m P_l^\mu [\cos(m-\mu)\varphi + \cos(m+\mu)\varphi],$$

$$K^{\sin}_{m\mu} = -\frac{1}{2}\int V e\, \psi^2_{nl} P_l^m P_l^\mu [\cos(m-\mu)\varphi - \cos(m+\mu)\varphi].$$

Drehen wir das Koordinatensystem um die Z-Achse durch π, so geht V in sich über, bei ungeraden $m \pm \mu$ wechselt aber $\cos(m \pm \mu)\varphi$ sein Vorzeichen, so daß:

$$\int V e\, \psi^2_{nl} P_l^m P_l^\mu \cos(m \pm \mu)\varphi\, d\tau = 0, \quad m \pm \mu \text{ ungerade.}$$

Jetzt spezialisieren wir von rhombischer auf tetragonale Symmetrie. Dann geht V auch bei Drehungen um Z durch $\pi/2$ stets in sich über, während $\cos(m \pm \mu)\varphi$ bei dieser Drehung sein Vorzeichen wechselt, falls $\frac{m \pm \mu}{2}$ ungerade ist. Also bleiben bei

13*

tetragonaler und natürlich auch bei kubischer Symmetrie nur die Integrale:

$$(10a)\quad K_{m\mu} = -\int V e\,\psi_{nl}^2 P_l^m P_l^\mu \cos 4\sigma\varphi\, d\tau \qquad (4\sigma = m \pm \mu)$$

von Null verschieden, analog bei hexagonaler Symmetrie:

$$(10b)\quad K_{m\mu} = -\int V e\,\psi_{nl}^2 P_l^m P_l^\mu \cos 6\sigma\varphi\, d\tau \qquad (6\sigma = m \pm \mu).$$

Wir setzen nun voraus, daß wir einen Ionenkristall vor uns haben und schreiben das Potential V als Summe der von allen Ionen außer dem gerade betrachteten „Aufion" herrührenden Potentiale (erregendes Potential):

$$(11)\qquad V = \sum_i \frac{l_i}{r_i}.$$

Wir legen den Nullpunkt unseres Polarkoordinatensystems in den Kern des Aufions; die Koordinaten des $i$ten Ions (Kraftions) seien R_i, Θ, Φ, die des Aufpunktes r, ϑ, φ; der Winkel zwischen den vom Kern des Aufions nach diesen beiden Punkten gezogenen Radiivektoren α, r_i ist der Abstand Aufpunkt-Kraftion.

Wir entwickeln das vom $i$ten Ion auf den Aufpunkt ausgeübte Potential nach Kugelfunktionen, die wir in quantenmechanisch üblicher Weise normieren $(\int (P_l^m)^2 \sin\vartheta\, d\vartheta = 1)$:

$$\frac{1}{r_i} = \sum_{s=0}^{\infty} \frac{r^s}{R_i^{s+1}} \sqrt{\frac{2}{2s+1}}\, P_s^0(\cos\alpha).$$

Wegen der Definition von α folgt aus dem Additionstheorem der Kugelfunktionen:

$$\frac{1}{r_i} = \sum_{s=0}^{\infty} \frac{r^s}{R_i^{s+1}} \frac{2}{2s+1} \sum_{\varrho=-s}^{+s} P_s^\varrho(\cos\Theta)\, P_s^\varrho(\cos\vartheta)\, e^{i\varrho(\Phi-\varphi)}$$

und der Beitrag des $i$ten Ions zum Integral $K_{m\mu}$ wird:

$$K_{m\mu}^i = -ee_i \int \psi_{nl}^2(r) r^2\, dr \sum_{s=0}^{\infty} \frac{r^s}{R_i^{s+1}} \cdot \frac{2}{2s+1} \sum_{\varrho=-s}^{+s} P_s^\varrho(\cos\Theta)$$

$$\cdot \int_0 \sin\vartheta\, d\vartheta\, P_s^\varrho(\cos\vartheta)\, P_l^m(\cos\vartheta)\, P_l^\mu(\cos\vartheta)$$

$$\cdot \int_0^{2\pi} d\varphi \cos 4\sigma\varphi\,(\cos\varrho\varphi\cos\varrho\Phi + \sin\varrho\varphi\sin\varrho\Phi) = -ee_i$$

$$\cdot \int \psi_{nl}^2 r^2\, dr \sum_{s=0}^{\infty} \frac{r^s}{R_i^{s+1}} \sqrt{\frac{2}{2s+1}}\, P_s^{4\sigma}(\cos\Theta) \cos 4\sigma\Phi \cdot \alpha_{lm\mu}^{s\sigma}$$

wobei:

(12) $$\alpha^{s\sigma}_{l\,m\,\mu} = \sqrt{\frac{2}{2s+1}}\int P_s^{4\sigma}\, P_l^m\, P_l^\mu \sin\vartheta\, d\,\vartheta$$

der „Entwicklungskoeffizient" des Kugelfunktionsproduktes $P_l^m\, P_l^\mu$ nach der Kugelfunktion $P_s^{4\sigma}$ ist. Man kann α als ein Multipolmoment *s*ter Ordnung der „Übergangsfunktion" $P_l^m P_l^\mu$ bezeichnen, es ist ein Zahlenfaktor von der Größenordnung **1** (vgl. § 19).

Nun ist:

(10c) $$K^i_{m\,\mu} = -\, e\, e_i \sum_{s,\,\sigma} \frac{P_s^{4\sigma}(\cos\Theta)\cos 4\,\sigma\,\Phi}{R_i^{s+1}}\, \overline{r^s}\, \alpha^{s\sigma}_{l\,m\,\mu}\,,$$

wobei:

(13) $$\overline{r^s} = \int r^s\, \psi_{n\,l}^2(r)\, r^2\, d\, r$$

der Mittelwert der *s*ten Potenz des Abstandes des Elektrons vom Kern des Atoms ist, also eine Kleinigkeit mehr als die *s*te Potenz des „Bahnradius" des Elektrons. Der Beitrag des Momentes *s*ter Ordnung zu $K^i_{m\,\mu}$ ist also etwa proportional zu

$$\frac{e^2}{R_i}\cdot\frac{\overline{r^s}}{R_i^{s+1}} \approx \text{Gitterenergie}\cdot\left(\frac{\textbf{Bahnradius des Elektrons}}{\textbf{Abstand des Kraftions vom Aufion}}\right)^s$$

Da aber der Bahnradius kaum je größer als ein Viertel des Gitterabstandes sein wird, tragen die höheren Momente nur sehr wenig zu $K^i_{m\,\mu}$ bei, und wir können uns auf die Betrachtung kleiner s beschränken. Nun ist aber sicher $s \geqslant 4\,\sigma$, so daß allein $\sigma = 0$ oder 1 für die Termberechnung von Interesse sein werden.

A. $\sigma = 0$. Solche Integrale treten nach (10a) nur auf, wenn $m = \mu$ ist, haben also stets die Form $-\int V\, e\, \psi_{n\,l}^2\, (P_l^m)^2\, d\,\tau$. Die Betrachtung solcher Integrale *allein* genügt z. B. zur Berechnung der Größe der Terme, die bei kubischer Symmetrie aus einem *d*- oder *f*-Term entstehen. Man erhält zunächst:

$$K^i_{m\,m} = -\, e\, e_i \sum_{s=0}^{\infty} \frac{P_s^0(\cos\Theta)}{R_i^{s+1}}\, \overline{r^s}\cdot\alpha^{s\,0}_{l\,m\,m}\,.$$

Nach Definition der Kugelfunktionen ist:

$$\frac{P_s^0(\cos\Theta)}{R_i^{s+1}} = \frac{1}{s!}\left(\frac{\partial^s \frac{1}{R_i}}{\partial z^s}\right)_{x\,=\,y\,=\,z\,=\,0},$$

also wenn man über die Beiträge aller Ionen i summiert:

$$K_{mm} = -e\int V\,\psi_{nl}^2(r)\,[P_l^m(\cos\vartheta)]^2\,d\tau$$

$$= -e\sum_{s=0}^{\infty}\overline{r^s}\cdot\frac{1}{s!}\left(\frac{\partial^s V}{\partial z^s}\right)_0 \alpha_{lm}^s .$$

Dabei ist:

$$\text{(12a)}\qquad \alpha_{lm}^s = \alpha_{lmm}^{s0} = \int P_s^0 (P_l^m)^2 \sin\vartheta\, d\vartheta .$$

Für $s = 0$ wird $\alpha_{lm}^0 = 1$, $\overline{r^0} = 1$, also:

$$K_{mm} = -eV_0 - e\sum_{s=1}^{\infty}\overline{r^s}\cdot\frac{1}{s!}\left(\frac{\partial V}{\partial z^s}\right)_0\cdot\alpha_{lm}^s .$$

$-eV_0$ ist die potentielle Energie, welche das Elektron besitzen würde, wenn es im Kern des Aufions konzentriert wäre. Nun ist aber der *Schwerpunkt* aller Terme, die im Kristall aus *einem* Term (n, l) des freien Atoms entstehen, gegen diesen Term des freien Atoms gerade um den Betrag:

$$E_0 = -\frac{e}{2l+1}\int V\,\psi_{nl}^2(r)\sum_\lambda (P_{l\lambda})^2\,d\tau = -eV_0$$

verschoben; denn die Dichtefunktionen nullter Näherung $\psi_{nl}^2(r)\,P_{l\lambda}^2$, die zu den einzelnen Kristalltermen (n, l) gehören, setzen sich nach dem Additionstheorem der Kugelfunktionen zusammen zu der kugelsymmetrischen Elektronendichte der abgeschlossenen Schale $(2l+1)\,\psi_{nl}^2(r)$, die Integration von V über eine Kugelschale liefert sodann das Potential V_0 am Orte des Kerns.

Ein Term des Elektrons im Kristallatom mit der Eigenfunktion $\psi_{nl}(r)\,P_l^m(\cos\vartheta)\,e^{\pm im\varphi}$ hat also relativ zum Schwerpunkt aller Kristallterme mit der gleichen Hauptquantenzahl n und Azimutalquantenzahl l die Lage:

$$\text{(14)}\qquad E_m = K_{mm} - E_0 = -e\sum_{s=1}^{\infty}\overline{r^s}\cdot\frac{1}{s!}\left(\frac{\partial^s V}{\partial z^s}\right)_0 \alpha_{lm}^s .$$

Den Hauptbeitrag zu (14) liefern, wie gesagt, die ersten Reihenglieder. Jedes Glied ist ein Produkt dreier unabhängiger Faktoren:

1. Mittelwert der $s$ten Potenz des Abstandes des Elektrons vom Kern.

2. $s$te Ableitung des Gitterpotentials V am Ort des Kerns des Aufions.

3. Eigentlicher Aufspaltungsfaktor $\alpha^s_{l\,m}$ (Multipolmoment der Elektronendichte).

B. $\sigma = 1$. Summen der Form (10c) mit $\sigma = 1$ treten auf bei Berechnung folgender Integrale (vgl. 10a):

$$-e\int V\,\psi_{n\,l}^2\,[P_l^2\,(\cos\vartheta)]^2\,(\cos^2 2\,\varphi - \sin^2 2\,\varphi)\,d\,\tau \qquad (2+2=4),$$

$$-e\int V\,\psi_{n\,l}^2\,P_l^1\,(\cos\vartheta)\,P_l^3\,(\cos\vartheta)\cos\varphi\cos 3\,\varphi\,d\,\tau \qquad (1+3=4),$$

$$-e\int V\,\psi_{n\,l}^2\,P_l^{m+2}\,P_l^{m-2}\cos(m+2)\,\varphi\cos(m-2)\,\varphi\,d\,\tau \qquad \bigl((m+2)-(m-2)=4\bigr).$$

Der erste Integraltyp z. B. bestimmt den Abstand der beiden Terme mit den Eigenfunktionen $\sqrt{2}\,P_l^2\cos 2\,\varphi$ und $\sqrt{2}\,P_l^2\sin 2\,\varphi$ bei tetragonaler Symmetrie, die beiden anderen Typen treten vor allem dann auf, wenn mehrere Terme zur gleichen Darstellung gehören (vgl. das Matrixelement ε_{13} in § 11, sowie § 22).

Das einzig wichtige Summenglied in (10c) ist in diesem Falle das Glied $s = 4$ (wegen der Abnahme der Glieder mit wachsendem s). Für dieses gilt:

$$\sqrt{\frac{2}{9}}\,P_4^4 = \sqrt{\frac{35}{128}}\sin^4\vartheta$$

$$\frac{1}{R_i^5}\sqrt{\frac{35}{128}}\sin^4\Theta\cos 4\,\Phi = \frac{1}{16}\cdot\frac{1}{4\,!}\left(4\,\frac{\partial^4}{\partial x^4} + 4\,\frac{\partial^4}{\partial y^4} - 3\,\frac{\partial^4}{\partial z^4}\right)\frac{1}{R}\,.$$

Ferner ist mit Rücksicht auf die tetragonale Symmetrie:

$$\frac{\partial^4 V}{\partial x^4} = \frac{\partial^4 V}{\partial y^4},$$

also:

$$(15)\quad \begin{cases} K^{m\,\mu} = -e\cdot\dfrac{1}{4\,!}\cdot\dfrac{1}{16}\left(8\,\dfrac{\partial^4 V}{\partial x^4} - 3\,\dfrac{\partial^4 V}{\partial z^4}\right)\overline{r^4} \\ \cdot\displaystyle\int\sin^4\vartheta\,P_l^m\,P_l^\mu\sin\vartheta\,d\,\vartheta + \text{Glieder höherer Ordnung}. \end{cases}$$

Bei kubischer oder nahe kubischer Symmetrie ist auch:

$$\frac{\partial^4 V}{\partial x^4} = \frac{\partial^4 V}{\partial z^4}\,.$$

also:

$$(15\text{a})\quad \begin{cases} K_{m\mu} = -e \cdot \frac{1}{4!} \cdot \frac{5}{16} \frac{\partial^4 V}{\partial z^4} \overline{r^4} \\ \cdot \int \sin^4 \vartheta\, P_l^m P_l^\mu \sin \vartheta\, d\vartheta + \text{Glieder höherer Ordnung.} \end{cases}$$

Bei hexagonaler Symmetrie ist überall 6 an Stelle von 4 zu schreiben, zu den Integralen mit $\sigma = 0$ (vgl. A) treten also bestenfalls Beiträge der Multipolmomente 6. Ordnung.

§ 19. Entwicklung von Quadraten und Produkten von Kugelfunktionen nach Kugelfunktionen

Uns interessieren einmal die Entwicklungskoeffizienten von Kugelfunktionsquadraten nach zonalen Kugelfunktionen:

$$(12\text{a})\quad \alpha_{lm}^s = \sqrt{\frac{2}{2s+1}} \int P_s^0 (P_l^m)^2 \sin \vartheta\, d\vartheta,$$

sodann die Entwicklungskoeffizienten nach der Kugelfunktion P_4^4:

$$(16)\quad \begin{cases} \beta_{lm} = \int \sin^4 \vartheta\, (P_l^m)^2 \sin \vartheta\, d\vartheta & \text{für } m = 2 \\ \beta'_{lm} = \int \sin^4 \vartheta\, P_l^{m-1} P_l^{m+1} \sin \vartheta\, d\vartheta & \text{für } m = 2 \\ \beta''_{lm} = \int \sin^4 \vartheta\, P_l^{m-2} P_l^{m+2} \sin \vartheta\, d\vartheta & \text{für } m \geqslant 2, \end{cases}$$

welche in (14) bzw. (15) eingehen.

Wir beginnen mit α_{lm}^s: Da $[P_l^m(\cos\vartheta)]^2$ ein Polynom in $\cos\vartheta$ vom Grade $2l$ ist, läßt es sich nach den ersten $2l$ zonalen Kugelfunktionen *allein* entwickeln:

$$\alpha_{lm}^s = 0 \quad \text{für} \quad s > 2l.$$

D. h. die Elektronendichteverteilung hat keine höheren Multipole als solche der $2l$ten Ordnung, unsere Reihe (14) zur Berechnung der Terme im Kristall ist stets endlich. Außerdem verschwinden alle Entwicklungskoeffizienten mit ungeradem Index s, weil $(P_l^m)^2$ eine gerade Funktion in $\cos\vartheta$ ist. Es bleibt zu berechnen:

$$\alpha_{lm}^s \quad \text{für} \quad s = 2\sigma, \quad 0 \leq \sigma \leq l.$$

Hier interessieren vor allem die Entwicklungskoeffizienten mit kleinem s, welche wir nach der von Sommerfeld[1]) zur

1) A. Sommerfeld, Wellenmechanischer Ergänzungsband, S. 63.

Normierung der Kugelfunktionen benutzten Methode berechnen. Wir setzen also für einen Faktor $P_l^m(x)$ den Differentialausdruck:

$$P_l^m(x) = \sqrt{\frac{l-m!}{l+m!} \cdot \frac{2l+1}{2}} \cdot \frac{(1-x^2)^{m/2}}{2^l \cdot l!} \cdot \frac{d^{l+m}}{dx^{l+m}} \cdot (x^2-1)^l$$

den anderen schreiben wir als Polynom in x, wobei wir die beiden höchsten Potenzen benötigen (die andern verschwinden bei späteren Differentiationen)

$$P_l^m(x) = \sqrt{\frac{l-m!}{l+m!} \cdot \frac{2l+1}{2}} \cdot \frac{2l!}{2^l \cdot l!\; l-m!} \cdot (1-x^2)^{m/2}$$
$$\cdot \left(x^{l-m} - \frac{(l-m)(l-m-1)}{2(2l-1)} x^{l-m-2} + \ldots\right)$$

und erhalten durch partielle Integration:

$$a_{lm}^2 = \frac{2l+1!}{2^{2l+1} l!^2\, l+m!} \int_{-1}^{+1} dx \left(\frac{3}{2}x^2 - \frac{1}{2}\right)(1-x^2)^m$$
$$\cdot \left(x^{l-m} - \frac{(l-m)(l-m-1)}{2(2l-1)} x^{l-m-2} \pm \ldots\right) \cdot \frac{d^{l+m}}{dx^{l+m}} (x^2-1)^l$$
$$= (-)^{l+m} \frac{2l+1!}{2^{2l+2} l!^2\, l+m!} \int_{-1}^{+1} (x^2-1)^l \cdot \frac{d^{l+m}}{dx^{l+m}} \Big[(1-x^2)^m$$
$$\cdot (3x^2-1)\left(x^{l-m} - \frac{(l-m)(l-m-1)}{2(2l-1)} x^{l-m-2} \pm \ldots\right)\Big]$$
$$= (-)^l \cdot \frac{2l+1!}{2^{2l+2} \cdot l!^2} \int (x^2-1)^l \cdot \Big[\frac{3}{2}(l+m+2)(l+m+1)x^2$$
$$- \left(1 + 3m + 3\,\frac{(l-m)(l-m-1)}{2(2l-1)}\right)\Big]\, dx$$
$$= (-)^l \cdot \frac{2l+1!}{2^{2l+2} l!^2} \int \frac{3}{2} (l+m+2)(l+m+1)(x^2-1)^{l+1}$$
$$+ \left[\frac{3}{2}(l+m+2)(l+m+1) - \left(1 + 3m + 3\,\frac{(l-m)(l-m-1)}{2(2l-1)}\right)\right]$$
$$\cdot (x^2-1)^l\, dx$$
$$= - \frac{(l+1)^2}{(2l+2)(2l+3)} \cdot 3 \cdot (l+m+2)(l+m+1)$$
$$+ \frac{1}{2} \cdot \left(\frac{3}{2}(l+m+2)(l+m+1) - \left(1 + 3m + 3\,\frac{(l-m)(l-m-1)}{2(2l-1)}\right)\right).$$

Die letzte Zeile folgt sofort aus:

$$\int_{-1}^{+1} (x^2 - 1)^\lambda \, dx = (-)^\lambda \cdot \frac{2^{2\lambda} \cdot \lambda!^2}{2\lambda + 1!} \cdot 2 .$$

Durch elementare Umformung ergibt sich schließlich:

$$(17) \qquad \alpha_{lm}^2 = \int (P_l^m)^2 P_2^0 \sqrt{\frac{5}{2}} \sin \vartheta \, d\vartheta = \frac{l(l+1) - 3m^2}{(2l+3)(2l-1)} .$$

In ganz analoger Weise berechnet man α_{lm}^4, wobei aber die *drei* höchsten Potenzen in dem Polynom gebraucht werden:

$$(17a) \qquad \alpha_{lm}^4 = \frac{3}{4} \cdot \frac{3l^2(l+1)^2 - 30l(l+1)m^2 + 35m^4 - 6l(l+1) + 25m^2}{(2l+5)(2l+3)(2l-1)(2l-3)} .$$

Die Berechnung der weiteren Entwicklungskoeffizienten wird nun immer mühevoller, weil immer mehr Glieder der Reihe $P_l^m(x)$ herangezogen werden müssen, sie sind ja auch für unsere Zwecke von geringerer Wichtigkeit. Immerhin kann man für kleine azimutale Quantenzahlen ($l = 3$ und 4) auch die nächsten beiden Koeffizienten α_{lm}^6 und α_{lm}^8 leicht angeben, weil sich nämlich die Entwicklungskoeffizienten α_{lm}^{2l-2} und α_{lm}^{2l} wieder einfach berechnen lassen. Wir schreiben zur Berechnung von α_{lm}^{2l} die Kugelfunktion P_{2l}^0 als Differentialausdruck und $(P_l^m)^2$ als Polynom, wobei nur die höchste Potenz in x gebraucht wird:

$$(P_l^m)^2 = \binom{2l}{l-m} \frac{2l+1!}{2^{2l+1} \cdot l!^2} (-)^m (x^{2l} \pm \ldots)$$

$$\sqrt{\frac{2}{4l+1}} \, P_{2l}^0 = \frac{1}{2^{2l} \cdot 2l!} \frac{d^{2l}}{dx^{2l}} (x^2 - 1)^{2l} .$$

Durch partielle Integration ergibt sich aus:

$$\alpha_{lm}^{2l} = \binom{2l}{l-m} \frac{(-)^m (2l+1)}{2^{4l+1} \cdot l!^2} \cdot \int (x^{2l} \pm \ldots) \frac{d^{2l}}{dx^{4l}} (x^2 - 1)^{2l} ,$$

$$(17b) \qquad \alpha_{lm}^{2l} = (-)^m \cdot \binom{2l}{l-m} \cdot \frac{\binom{2l}{l}}{\binom{4l+1}{2l}} .$$

Analog erhält man:

$$(17c) \qquad \alpha_{lm}^{2l-2} = (-)^m \cdot \frac{1}{2} \cdot \binom{2l}{l-m} \cdot \frac{\binom{2l-2}{l-1}}{\binom{4l-1}{2l-2}} \cdot \left(1 - \frac{4l-1}{l^2} m^2\right) .$$

(vgl. Tab. 16).

Tabelle 16

Multipolmomente s ter Ordnung der proportional dem Kugelfunktionsquadrat $(P_l^m)^2$ verteilten Elektronendichte = Entwicklungskoeffizienten von $(P_l^m)^2$ nach zonalen Kugelfunktionen,

$$\alpha_{l\,m}^{s} = \sqrt{\frac{2}{2s+1}} \int_{-1}^{+1} (P_l^m)^2 P_s^0 \, dx .$$

$l =$	1		2			3			
$m =$	0	1	0	1	2	0	1	2	3
$s = 2$	$\frac{2}{5}$	$-\frac{1}{5}$	$\frac{2}{7}$	$\frac{1}{7}$	$-\frac{2}{7}$	$\frac{4}{15}$	$\frac{3}{15} = \frac{1}{5}$	0	$-\frac{5}{15} = -\frac{1}{3}$
$s = 4$	—	—	$\frac{6}{21} = \frac{2}{7}$	$-\frac{4}{21}$	$\frac{1}{21}$	$\frac{6}{33} = \frac{2}{11}$	$\frac{1}{33}$	$-\frac{7}{33}$	$\frac{3}{33} = \frac{1}{11}$
$s = 6$	—	—	—	—	—	$\frac{100}{429}$	$-\frac{75}{429} = -\frac{25}{143}$	$\frac{30}{429} = \frac{10}{143}$	$-\frac{5}{429}$

$l =$	4				
$m =$	0	1	2	3	4
$s = 2$	$\frac{20}{77}$	$\frac{17}{77}$	$\frac{8}{77}$	$-\frac{7}{77} = -\frac{1}{11}$	$-\frac{28}{77} = -\frac{4}{11}$
$s = 4$	$\frac{162}{1001}$	$\frac{81}{1001}$	$-\frac{99}{1001} = -\frac{9}{91}$	$-\frac{189}{1001} = -\frac{27}{143}$	$\frac{126}{1001} = \frac{18}{143}$
$s = 6$	$\frac{20}{143}$	$-\frac{1}{143}$	$-\frac{22}{143} = -\frac{2}{13}$	$\frac{17}{143}$	$-\frac{2}{143}$
$s = 8$	$\frac{490}{2431}$	$-\frac{392}{2431}$	$\frac{196}{2431}$	$-\frac{56}{2431}$	$\frac{7}{2431}$

Entwicklungskoeffizienten $\beta = \int P_l^m P_l^\mu \sin^4 \vartheta \sin \vartheta \, d\vartheta$

$l =$	2	3		4		
$m =$	2	2	3	2	3	4
$\mu =$	2	2	1	2	1	0
$\beta =$	$\frac{16}{21}$	$\frac{16}{33}$	$-\frac{16}{165}\sqrt{15}$	$\frac{432}{1001}$	$-\frac{144}{1001}\sqrt{7}$	$\frac{144}{5005}\sqrt{70}$

Nun berechnen wir die Konstanten (16). β_{lm} läßt sich auf die bisher behandelten Integrale reduzieren:

$$\beta_{lm} = \int (P_l^m)^2 \sin^4 \vartheta \sin \vartheta \, d\vartheta$$

$$= \int (P_l^m)^2 \cdot \left(\sqrt{\frac{2}{9}} \cdot \frac{8}{35} \cdot P_4^0 - \sqrt{\frac{2}{5}} \cdot \frac{16}{21} \cdot P_2^0 + \sqrt{2} \cdot \frac{8}{15} \cdot P_0^0 \right).$$

$$(18) \quad \beta_{lm} = 2 \cdot \frac{3l^2(l+1)^2 + 2l(l+1)m^2 + 3m^4 - 14l(l+1) - 15m^2 + 12}{(2l+5)(2l+3)(2l-1)(2l-3)}.$$

Für die beiden anderen Fälle bekommt man durch die analoge Rechnung wie früher:

$$(18') \quad \beta'_{lm} = -6 \cdot \sqrt{(l+m+1)(l+m)(l-m+1)(l-m)} \cdot \frac{l(l+1) + m^2 - 4}{(2l+5)(2l+3)(2l-1)(2l-3)}$$

$$(18'') \quad \beta''_{lm} = 6 \cdot \frac{\sqrt{(l+m+2)(l+m+1)(l+m)(l+m-1)(l-m+2)(l-m+1)(l-m)(l-m-1)}}{(2l+5)(2l+3)(2l-1)(2l-3)}.$$

Für $m = 2$ wird aus (18) und (18'):

$$(18\text{a}) \quad \beta_{l2} = 6 \cdot \frac{(l+2)(l+1)l(l-1)}{(2l+5)(2l+3)(2l-1)(2l-3)}.$$

$$(18'\text{a}) \quad \left\{ \beta'_{l2} = -6 \cdot \sqrt{(l+3)(l+2)(l-1)(l-2)} \cdot \frac{l(l+1)}{(2l+5)(2l+3)(2l-1)(2l-3)}. \right.$$

§ 20. Die Ableitungen des Gitterpotentials

Bei kubischer Symmetrie ist keine der drei Achsen ausgezeichnet, es folgt also aus:

$$\Delta V = 0$$

sofort:

$$(19) \quad \frac{\partial^2 V}{\partial x^2} = \frac{\partial^2 V}{\partial y^2} = \frac{\partial^2 V}{\partial z^2} = 0.$$

Das Kristallfeld ist ohne Wirkung auf das Quadrupolmoment der Elektronenverteilung, erst das Moment vierter Ordnung trägt zum Wert des Terms bei. Bei nicht kubischer Symmetrie ist dagegen $\frac{\partial^2 V}{\partial z^2}$ natürlich endlich. Die Termaufspaltung ist also bei nicht kubischer Symmetrie im allgemeinen erheblich größer $\left[\text{proportional} \left(\frac{\text{Bahnradius}}{\text{Gitterabstand}}\right)^2\right]$ als bei kubischer Symmetrie $\left[\text{proportional} \left(\frac{\text{Bahnradius}}{\text{Gitterabstand}}\right)^4\right]$.

Wir berechnen die $s$te Ableitung des Gitterpotentials nach einer Achsenrichtung, $\frac{\partial^s V}{\partial z^s}$, am Orte eines positiven Ions in einem Kristall von NaCl-Typ nach der Madelungschen Methode.[1]) Dazu zählen wir kartesische Koordinaten vom Kerne unseres Ions aus. Das erregende Potential $V(x, 0, 0)$ setzt sich zusammen aus einem Beitrag der Reihe, in der das Ion selbst liegt (X-Achse) einem Beitrag der übrigen Reihen der Ebene XY und einen Beitrag der anderen Ebenen $z = \text{const}$. Wir betrachten die drei Beiträge zu $\frac{1}{s!}\frac{\partial^s V}{\partial x^s}$ einzeln, den Abstand zweier nächster Nachbarn nennen wir a, die Ionenladung E.

1. Beitrag der Ionen der eigenen Reihe zum Potential:

$$V_1(x,0,0) = -\frac{E}{a}\left(\frac{1}{1-\frac{x}{a}} + \frac{1}{3-\frac{x}{a}} + \cdots + \frac{1}{1+\frac{x}{a}} + \frac{1}{3+\frac{x}{a}} + \cdots\right)$$

$$+\frac{E}{a}\left(\frac{1}{2-\frac{x}{a}} + \frac{1}{4-\frac{x}{a}} + \cdots + \frac{1}{2+\frac{x}{a}} + \frac{1}{4+\frac{x}{a}} + \cdots\right).$$

Also:

$$\frac{1}{s!}\left(\frac{\partial^s V_1(x,0,0)}{\partial x^s}\right)_{x=0} = 0$$

wenn s ungerade. Für gerade s wird:

$$\text{(20a)}\quad \begin{cases} \frac{1}{s!}\left(\frac{\partial^s V_1}{\partial x^s}\right)_0 = -\frac{2E}{a^{s+1}}\left(1^{-(s+1)} - 2^{-(s+1)} + 3^{-(s+1)} \pm \cdots\right) \\ \qquad = -\frac{2E}{a^{s+1}} r_{s+1}. \end{cases}$$

Es ist $r_3 = 0{,}9016$, $r_5 = 0{,}9722$, $r_7 = 0{,}9926$ usw.

2. Beitrag der Nachbarreihen. Die ϱte Nachbarreihe im Abstand $y = \varrho a$ enthält am Punkt $(0, \varrho a, 0)$ ein positives oder

1) E. Madelung, Physikal. Ztschr. 19. S. 524. 1918.

negatives Ion, je nachdem ob ϱ gerade oder ungerade ist, und erzeugt in der X-Achse das Potential[1]):

$$V_\varrho(x,0,0) = \frac{4E}{a}(-)^\varrho \sum_{l=1,3,5,\dots} K_0(\pi l \varrho) \cos\frac{\pi l x}{a}$$

$$\frac{1}{s!}\frac{\partial^s V_\varrho}{\partial x^s} = \frac{4E}{a^{s+1}}(-)^{\varrho+\frac{s}{2}}\frac{\pi^s}{s!}\sum_{l=1,3,5,\dots} K_0(\pi l \varrho)\, l^s \cos\frac{\pi l x}{a}\,.$$

für gerade s. Wir summieren über alle Reihen der $X\,Y$-Ebene (jede tritt doppelt auf, „rechts" und „links" von der Reihe unseres Ions) und setzen $x = 0$.

$$\text{(20b)}\quad \frac{1}{s!}\left(\frac{\partial^s V_2}{\partial x^s}\right)_0 = \frac{8E}{a^{s+1}}\cdot(-)^{\frac{s}{2}}\frac{\pi^s}{s!}\sum_{l=1,3,5,\dots} l^s \sum_{\varrho=1,2,3,\dots}(-)^\varrho K_0(\pi l \varrho).$$

3. Beitrag der Nachbarebene im Abstand $z = \varrho a$ [im Punkte $(0, 0, \varrho a)$ sitzt ein Ion mit der Ladung $(-)^\varrho E$]:

$$V'_\varrho = (-)^\varrho \frac{8E}{a}\sum_{l=1,3,5,\dots}\sum_{m=1,3,5,\dots}\frac{e^{-\pi\varrho\sqrt{l^2+m^2}}}{\sqrt{l^2+m^2}}\cos\frac{\pi l x}{a}$$

$$\frac{1}{s!}\frac{\partial^s V_\varrho'}{\partial x^s} = \frac{8E}{a^{s+1}}(-)^{\frac{s}{2}}\frac{\pi^s}{s!}\sum_{l=1,3,5,\dots}\sum_{m=1,3,5,\dots}\frac{e^{-\pi\varrho\sqrt{l^2+m^2}}}{\sqrt{l^2+m^2}}\, l^s\, (-)^\varrho \cos\frac{\pi l x}{a}\,. \qquad \text{(für gerade } s\text{)}$$

Beitrag aller Nachbarebenen:

$$\text{(20c)}\quad \begin{cases} \dfrac{1}{s!}\left(\dfrac{\partial^s V_3}{\partial x^s}\right)_0 = \dfrac{16E}{a^{s+1}}(-)^{\frac{s}{2}}\dfrac{\pi^s}{s!} \\ \qquad \cdot \displaystyle\sum_{l=1,3,5,\dots}\sum_{m=1,3,5,\dots}(-)^\varrho\frac{e^{-\pi\varrho\sqrt{l^2+m^2}}}{\sqrt{l^2+m^2}}\,. \end{cases}$$

Demnach ist die gesamte $s$te Ableitung des erregenden Potentials im Gitterpunkt:

$$\text{(20)}\quad \frac{1}{s!}\left(\frac{\partial^s V}{\partial x^s}\right)_0 = -\frac{2E}{a^{s+1}}r_{s+1} + \frac{8E}{a^{s+1}}(-)^{\frac{s}{2}}\frac{\pi^s}{s!}\sum_{l=1,3,5,\dots} l^s \beta_l\,,$$

---

1) $K_0(x) = H_0^{(1)}(ix)$ = **erste Hankelsche Funktion mit imaginärem Argument.**

wo:

(20d) $$\beta_l = \sum_{\varrho=1,2,3,\ldots} (-)^{\varrho} \left(K_0 (\pi l \varrho) + 2 \sum_{m=1,3,5,\ldots} \frac{e^{-\pi \varrho \sqrt{l^2+m^2}}}{\sqrt{l^2+m^2}} \right),$$

$$\beta_1 = -\ 0{,}0450 \qquad \beta_3 = -\ 0{,}650 \cdot 10^{-4}$$
$$\beta_5 = -\ 0{,}92 \cdot 10^{-7} \qquad \beta_7 = -\ 0{,}138 \cdot 10^{-9}.$$

Damit wird z. B.:

(20e) $$\left\{ \begin{aligned} \frac{1}{4!} \left(\frac{\partial^4 V}{\partial z^4} \right)_0 &= -\ 3{,}58 \frac{E}{a^5} \\ \frac{1}{6!} \left(\frac{\partial^6 V}{\partial z^6} \right)_0 &= -\ 0{,}82 \frac{E}{a^7}. \end{aligned} \right.$$

Nunmehr denken wir uns unseren NaCl-Kristall in Richtung einer Würfelkante Z etwas gedehnt, so daß wir einen Kristall von tetragonaler Symmetrie bekommen.[1]) Das Verhältnis der vierzähligen zu den zweizähligen Achsen sei $\frac{c}{a} = 1 + \varepsilon$. Dann verschwinden natürlich die zweiten Ableitungen des erregenden Potentials nach den Achsenrichtungen nicht mehr. Der Beitrag der Ionen der XY-Ebene zu $\frac{\partial^s V}{\partial x^s}$ bleibt zwar konstant, jedoch der Beitrag der Parallelebenen $z = \varrho\, a\, (1 + \varepsilon)$ wird kleiner, was physikalisch sofort einleuchtet und formal dadurch herauskommt, daß der Exponent jeder e-Potenz in (20c) den Faktor $\frac{c}{a} = 1 + \varepsilon$ erhält. Da nun die Parallelebenen einen positiven Beitrag zu $\frac{\partial^2 V}{\partial x^2}$ liefern[2]), so wird:

$$\frac{\partial^2 V}{\partial x^2} = \frac{\partial^2 V}{\partial y^2} < 0, \qquad \text{wenn } \varepsilon > 0$$

und wegen $\Delta V = 0$ also:

$$\frac{\partial^2 V}{\partial z^2} > 0.$$

---

1) Ein solcher Kristall ist in der Natur nicht beobachtet, es wäre aber die einfachst mögliche Struktur eines tetragonalen Ionenkristalls.

2) Denn $(-)^{\frac{s}{2}} = -\ 1$, den Hauptbeitrag liefert die erste Parallelebene auf jeder Seite ($\varrho = 1$), wegen des Faktors $(-)^{\varrho}$ ist daher $\frac{\partial^2 V_3}{\partial x^2}$ positiv.

Das heißt: in einem genügend gedehnten NaCl-Kristall liegen diejenigen Terme am tiefsten, bei denen das Moment zweiter Ordnung der Elektronendichte in Richtung der tetragonalen Achse positiv ist [vgl. (14)], also die Terme kleinster elektrischer Quantenzahl m (oberer Index der Kugelfunktion P_l^m). Voraussetzung dabei ist, daß die Dehnung groß genug ist, damit die Wirkung des Moments vierter Ordnung gegen das Moment zweiter Ordnung verschwindet (§ 22). Bei einem gestauchten Kristall vom NaCl-Typ liegen die Verhältnisse natürlich umgekehrt, ebenso kehren sich für die Elektronen des negativen Ions die Vorzeichen um.

§ 21. Das Aufspaltungsbild bei kubischer Symmetrie

Die Größe der Termaufspaltung wird bei kubischer Symmetrie im wesentlichen bestimmt durch die potentielle Energie der Multipolmomente vierter Ordnung der Elektronendichte im Gitterfeld. Der Term n, l, λ ist nach § 18 in seiner relativen Lage zum Schwerpunkt aller Terme mit gleichem n und l gegeben durch:

$$(21)\quad E_\lambda = -e\frac{\overline{r^4}}{4!}\frac{\partial^4 V}{\partial z^4}\int P_{l\lambda}^2\left(P_4^0+\frac{5}{8}\sin^4\vartheta\cos 4\varphi\right)\sin\vartheta\, d\vartheta\, d\varphi,$$

wo $P_{l\lambda}$ irgendeine der winkelabhängigen Eigenfunktionen ist, welche zum Term (n, l, λ) gehören, also ein Lineaggregat von Kugelfunktionen $l$ter Ordnung.[1]) Die absolute Größe der Aufspaltung in einem Kristall vom NaCl-Typ bei einwertigen Ionen ist nach (20e) bis auf einen Faktor der Größenordnung 1 gleich:

$$D = -e\frac{\overline{r^4}}{4!}\frac{\partial^4 V}{\partial z^4} = \frac{e^2}{a^5}\overline{r^4}\cdot 3{,}58 = \text{etwa } \frac{e^2}{5a_H}\cdot\frac{3{,}58}{5^4}$$

$$= 1{,}1\cdot 10^{-3}\cdot\frac{e^2}{a_H} = 2\cdot 1{,}1\cdot 10^{-3}\cdot\text{Rydbergkonst.} = \text{etwa } 250\,\text{cm}^{-1}.$$

1) Der Beitrag der vernachlässigten Glieder 6. Ordnung würde sich zu dem der Glieder 4. Ordnung etwa verhalten wie

$$\frac{\dfrac{\overline{r^6}}{6!}\dfrac{\partial^6 V}{\partial z^6}}{\dfrac{\overline{r^4}}{4!}\dfrac{\partial^4 V}{\partial z^4}} = \frac{0{,}92\dfrac{E}{a^7}}{3{,}58\dfrac{E}{a^5}}\frac{\overline{r^6}}{\overline{r^4}} = \text{etwa } \frac{1}{4}\left(\frac{\text{Bahnradius}}{\text{Gitterabstand}}\right)^2 = \text{etwa } 0{,}01,$$

d. h. die Vernachlässigung verursacht unter Zugrundelegung der Daten des NaCl-Typs nur etwa 1 Proz. Fehler.

wenn der Bahnradius des Elektrons gleich dem Wasserstoffradius a_H und der Gitterabstand $a = 5\,a_H$ angenommen wird. Die Aufspaltung ist also von der Größenordnung der Multiplettaufspaltung. Unsere Termberechnung ist demnach nicht exakt, da sie nicht nur die Wechselwirkung zwischen Spin und Bahn vernachlässigt, obwohl diese von gleicher Größenordnung ist wie der Einfluß des Kristallfeldes, sondern sogar die Wechselwirkung zwischen den einzelnen Elektronen außerhalb der abgeschlossenen Schalen. Trotzdem wollen wir sie zur Veranschaulichung der gruppentheoretisch bestimmten Aufspaltung weiter durchführen.

Betrachten wir nun z. B. ein positives Ion im NaCl-Typ, so ist die relative Lage der Terme bis auf den für alle Terme vom gleichen n, l konstanten *positiven* Faktor D gegeben durch:

$$\text{(21a)} \qquad \varepsilon_\lambda = \int P_{l\lambda}^2 \left(P_4^0 + \frac{5}{8} \sin^4 \vartheta \cos 4\varphi \right) \sin \vartheta \, d\vartheta \, d\varphi .$$

Für ein d-Elektron ist z. B. die Lage des zweifach entarteten Kristallterms d_γ relativ zum Schwerpunkt der beiden Terme d_γ und d_ε gegeben durch:

$$E_\gamma - E_0 = \frac{2}{7} D,$$

denn zu d_γ gehört die Eigenfunktion P_2^0 und nach Tab. 16 ist $\int (P_2^0)^2 P_4^0 = \frac{2}{7}$. Zum dreifachen Term d_ε gehört u. a. die Eigenfunktion $P_2^1 e^{i\varphi}$, es gilt nach Tab. 16 $\int P_2^1 e^{i\varphi} P_2^1 e^{-i\varphi} \cdot P_4^0 = -\frac{4}{21}$, also ist[1]):

$$E_\varepsilon - E_0 = -\frac{4}{21} D .$$

Der zweifache Term, bei dem die maximale Elektronendichte in den vierzähligen Achsen konzentriert ist (§ 14), liegt also beim positiven Ion im Kristall vom NaCl-Typ *höher* als der dreifache mit seiner Elektronenkonzentration in den dreizähligen Achsen. Das ist sehr einleuchtend: Die nächsten negativen Nachbarn des Ions liegen in der vierzähligen Achse und suchen selbstverständlich die Elektronen des positiven Ions möglichst von sich abzustoßen. Beim negativen Ion liegen natürlich die

1) Wie dies auch aus der Definition des Termschwerpunktes folgt, ist $2E_\gamma + 3E_\varepsilon = 5E_0$.

Verhältnisse umgekehrt, ebenso würden sie sich beim Übergang zum CsCl-Typ umkehren.

Ein Elektron im f-Zustand kann das Maximum seiner Aufenthaltswahrscheinlichkeit entweder in den Würfeldiagonalen (einfacher Term f_β) oder in den Würfelkanten (dreifacher Term f_δ) oder in den Flächendiagonalen (dreifacher Term f_ε) haben. Zum Term f_δ gehört u. a. die Eigenfunktion P_3^0, also ist:

$$E_\delta - E_0 = +\frac{2}{11} D .$$

Zu f_β gehört die Eigenfunktion $\sqrt{2}\, P_3^2 \sin 2\varphi$, zu f_ε unter anderem die Eigenfunktion $\sqrt{2}\, P_3^2 \cos 2\varphi$, also ist:

$$E_\beta + E_\varepsilon - 2E_0 = 2\int (P_3^2)^2 V \psi_{nl}^2(r)\, d\tau = -2 \cdot \frac{7}{33} D .$$

Unter Zuhilfenahme der Definition des Termschwerpunktes:

$$7E_0 = E_\beta + 3E_\delta + 3E_\varepsilon$$

bekommt man für die Lage der Terme:

$$E_\beta - E_0 = -\frac{4}{11} D$$

$$E_\varepsilon - E_0 = -\frac{2}{33} D$$

$$E_\delta - E_0 = +\frac{2}{11} D .$$

Beim positiven Ion des NaCl-Kristalls liegt der Term f_β am tiefsten, weil hier das Elektron am weitesten vom negativen Nachbarn entfernt in den Würfeldiagonalen sein Dichtemaximum hat, der Term f_δ am höchsten (Dichtemaximum in den Kanten), während f_ε mit dem Maximum in der Flächendiagonale des Würfels eine Mittelstellung einnimmt. Das Abstandsverhältnis der Terme ist:

$$E_\delta - E_\varepsilon : E_\varepsilon - E_\beta = 4 : 5$$

und zwar gilt dies *stets* bei kubischer Symmetrie, nicht nur beim NaCl-Typ.

Schließlich kommen wir zur Aufspaltung des g-Terms und finden unter Verwendung von Tab. 16:

$$E_\alpha - E_0 = 14 \cdot \frac{18}{1001} D$$

$$E_\delta - E_0 = 7 \cdot \frac{18}{1001} D$$

$$E_\gamma - E_0 = 2 \cdot \frac{18}{1001} D$$

$$E_\varepsilon - E_0 = -13 \cdot \frac{18}{1001} D .$$

g_α entspricht einem Dichtemaximum in den vierzähligen, g_ε in den dreizähligen Achsen.

§ 22. Das Aufspaltungsbild bei tetragonaler Symmetrie. Maß der „Tetragonalität“

Die Lage eines Terms mit den Quantenzahlen (n, l, λ) relativ zum Schwerpunkt sämtlicher Terme mit gleichem n, l ist:

$$(22) \quad \begin{cases} E_\lambda - E_0 = -e \frac{\overline{r^2}}{2!} \left(\frac{\partial^2 V}{\partial z^2}\right)_0 \alpha_{l\lambda} - e \frac{\overline{r^4}}{4!} \\ \cdot \left(\frac{\partial^4 V}{\partial z^4} \cdot \alpha'_{l\lambda} + \frac{1}{8}\left(8 \frac{\partial^4 V}{\partial x^4} - 3 \frac{\partial^4 V}{\partial z^4}\right) \beta_{l\lambda}\right), \end{cases}$$

wo (vgl. § 18):

$$(22\,a) \quad \begin{cases} \alpha_{l\lambda} = \int P_{l\lambda}^2 \, P_2^0 \qquad \alpha'_{l\lambda} = \int P_{l\lambda}^2 \, P_4^0 \\ \beta_{l\lambda} = 2 \int P_{l\lambda}^2 \sin^4 \vartheta \cos 4\varphi . \end{cases}$$

Voraussetzung ist dabei, daß zur Darstellung γ_λ der tetragonalen Gruppe nur der eine Term E_λ mit der Eigenfunktion $P_{l\lambda}$ gehört. Geht man von tetragonaler zu kubischer Symmetrie über, so verschwindet $\frac{\partial^2 V}{\partial z^2}$, während $\frac{\partial^4 V}{\partial z^4}$ und $\frac{\partial^4 V}{\partial x^4}$ sich nur relativ wenig ändern. Solange diese Glieder 4. Ordnung überhaupt eine Rolle spielen, d. h. eben bei nahezu kubischer Symmetrie, kann überdies $\frac{\partial^4 V}{\partial x^4} = \frac{\partial^4 V}{\partial z^4}$ gesetzt werden. Es liegt nun nahe, die Größe des Terms in Beziehung zu setzen zu:

$$D = -e \frac{\overline{r^4}}{4!} \frac{\partial^4 V}{\partial z^4},$$

um einen Vergleich mit den Verhältnissen bei kubischer Symmetrie zu ermöglichen:

$$(23)\qquad E_\lambda - E_0 = D\left(\varepsilon_\lambda + \frac{\overline{r^2}}{\overline{r^4}}\cdot\frac{\frac{1}{2!}\left(\frac{\partial^2 V}{\partial z^2}\right)_0}{\frac{1}{4!}\left(\frac{\partial^4 V}{\partial z^4}\right)_0}\alpha_{l\lambda}\right)$$

mit:

$$(21\,\mathrm{a})\qquad \varepsilon_\lambda = \alpha'_{l\lambda} + \frac{5}{16}\beta_{l\lambda} = \int P_{l\lambda}^2\left(P_4^0 + \frac{5}{8}\sin^4\vartheta\cos 4\varphi\right)d\tau .$$

Dabei hängt die Konstante D nur noch wenig davon ab, ob das Atom kubische oder tetragonale Symmetrie besitzt. Beim NaCl-Typ ist $D = 3{,}58\cdot\frac{e^2}{a^5}$. Uns interessiert vor allem die zweite Größe in der Klammer, welche bei kubischer Symmetrie verschwindet und bei tetragonaler Symmetrie den Unterschied zur kubischen ausmacht. Wir nennen sie die *wirksame Tetragonalität*.

$$(24)\qquad u = \frac{\overline{r^2}}{\overline{r^4}}\cdot\frac{\frac{1}{2!}\left(\frac{\partial^2 V}{\partial z^2}\right)_0}{\frac{1}{4!}\left(\frac{\partial^4 V}{\partial z^4}\right)_0}$$

und bekommen damit:

$$(23\,\mathrm{a})\qquad E_\lambda - E_0 = D(\varepsilon_\lambda + u\,\alpha_{l\lambda}).$$

Die wirksame „Tetragonalität" bestimmt die relative Lage der Starkeffektkomponenten eines Elektronenterms bei tetragonaler Symmetrie. Sie ist umgekehrt proportional dem Quadrat des „Bahnradius" des Elektrons. Um eine Konstante der Atomsymmetrie zu erhalten, definieren wir die *absolute Tetragonalität*:

$$(24\,\mathrm{a})\qquad U = \frac{\frac{1}{2!}\left(\frac{\partial^2 V}{\partial z^2}\right)_0}{\frac{1}{4!}\left(\frac{\partial^4 V}{\partial z^4}\right)_0}$$

als das Verhältnis der zweiten zur vierten Ableitung des erregenden Gitterpotentials nach der Richtung der tetragonalen Achse am Orte des Atomkerns. Bei kubischer Symmetrie ist $U = 0$, als nahezu kubisch werden wir eine Symmetrie bezeichnen, wenn die Tetragonalität U klein ist (etwa kleiner als 0,1), dagegen haben wir bei großem U eine ausgesprochen tetragonale Symmetrie vor uns.[1]) Die Einführung der Tetragonalität rechtfertigt sich dadurch, daß sie maßgebend ist für die relative Lage der Kristallterme, d. h. auch für die stabilste Elektronenanordnung in einem Atom von vorgegebener Lage im Kristall.

Nachgewiesen haben wir diese Bedeutung der Tetragonalität allerdings erst für den Fall, daß zu jeder irreduziblen Darstellung der tetragonalen Gruppe nur je ein Term gehört. Daß sie die gleiche Bedeutung (mit der gleichen Voraussetzung $\frac{\partial^4 V}{\partial x^4} \approx \frac{\partial^4 V}{\partial z^4}$!) auch im Falle mehrerer Terme mit gleicher Darstellung besitzt, zeigen wir an dem in § 11 begonnenen Beispiel (dies läßt sich leicht verallgemeinern). Die beiden Terme, die bei der azimutalen Quantenzahl $l = 3$ zur zweidimensionalen Darstellung gehören, sind gegeben durch:

$$E_5', E_5'' = \frac{\varepsilon_{11} + \varepsilon_{33}}{2} \pm \sqrt{\left(\frac{\varepsilon_{11} - \varepsilon_{33}}{2}\right)^2 + \varepsilon_{13}^2}\,. \tag{6}$$

Dabei ist nach (5) unter Heranziehung von (14), (15) und Tab. 16:

$$\varepsilon_{11} = - eV_0 - \frac{1}{5} e \frac{\overline{r^2}}{2!}\left(\frac{\partial^2 V}{\partial z^2}\right)_0 - \frac{1}{33} e \frac{\overline{r^4}}{4!}\left(\frac{\partial^4 V}{\partial z^4}\right)_0,$$

$$\varepsilon_{33} = - eV_0 + \frac{1}{3} e \frac{\overline{r^2}}{2!}\left(\frac{\partial^2 V}{\partial z^2}\right)_0 - \frac{1}{11} e \frac{\overline{r^4}}{4!}\left(\frac{\partial^4 V}{\partial z^4}\right)_0,$$

$$\varepsilon_{13} = \frac{16}{165}\sqrt{15}\, e \frac{\overline{r^4}}{4!} \cdot \frac{1}{16}\left(8 \frac{\partial^4 V}{\partial x^4} - 3 \frac{\partial^4 V}{\partial z^4}\right)_0 \approx \frac{\sqrt{15}}{33} \frac{\overline{r^4}}{4!}\left(\frac{\partial^4 V}{\partial z^4}\right)_0 .$$

Es wird:

$$\frac{\varepsilon_{11} + \varepsilon_{33}}{2} + eV_0 = D\left(\frac{2}{33} - \frac{u}{15}\right),$$

$$\frac{\varepsilon_{11} - \varepsilon_{33}}{2} = D\left(-\frac{1}{33} + \frac{4u}{15}\right),$$

$$\varepsilon_{13} = - D \cdot \frac{\sqrt{15}}{33}\,.$$

---

1) Bei einem gedehnten Kristall vom NaCl-Typ ist die Tetragonalität stets negativ (vgl. § 20) und für kleine Dehnungen gilt:

$$U \approx -1{,}71\, \varepsilon\,,$$

dementsprechend ist bei einem Bahnradius von etwa 1/4 des Gitterabstandes die wirksame Tetragonalität

$$u \approx -30\,\varepsilon = -30 \frac{c-a}{a}\,.$$

Bei Stauchung kehrt sich das Vorzeichen um, dagegen ist es unabhängig vom Vorzeichen der Ladung des betrachteten Ions.

$$(25)\quad \begin{cases} E_5' - E_0 = D\left(\frac{2}{33} - \frac{u}{15} + \frac{4}{3}\sqrt{\frac{u^2}{5^2} - \frac{u}{110} + \frac{1}{11^2}}\right) \\ E_5'' - E_0 = D\left(\frac{2}{33} - \frac{u}{15} - \frac{4}{3}\sqrt{\frac{u^2}{5^2} - \frac{u}{110} + \frac{1}{11^2}}\right). \end{cases}$$

Auch hier erweist sich die Lage der Terme als im wesentlichen abhängig von der wirksamen Tetragonalität u; der Proportionalitätsfaktor D ist derselbe wie in Formel (23a).

Bei kleiner wirksamer Tetragonalität (nahezu kubischer Symmetrie) erhält man aus (25):

$$(25\text{a})\quad \begin{cases} E_5' - E_0 = D\left(\frac{2}{11} - \frac{2u}{15} + \frac{11}{40}u^2 + \cdots\right) \\ E_5'' - E_0 = D\left(-\frac{2}{33} \qquad - \frac{11}{40}u^2 + \cdots\right). \end{cases}$$

D. h.: E_5' geht bei kubischer Symmetrie ($u = 0$) in den Term f_δ über, E_5'' in den Term f_ε. Bei geringer Abweichung von der kubischen Symmetrie ändert sich der Termwert stetig mit der Abweichung. Bei ausgesprochen tetragonaler Symmetrie ($u \gg l$) wird dagegen:

$$E_5' - E_0 = D\left(-\frac{u}{15} + \frac{2}{33} + \frac{4}{15}|u| - \frac{1}{33}\frac{u}{|u|} + \frac{1}{u}(\cdots)\right).$$

(Zu E_5' gehört das positive Vorzeichen der Quadratwurzel, daher muß u in Absolutstriche gesetzt werden.) Für positive große u bekommt man:

$$(25\text{b})\quad E_5' - E_0 = D\left(\frac{u}{5} + \frac{1}{33} + \frac{1}{u}\cdots\right),$$

dagegen für $u < 0$:

$$(25\text{c})\quad E_5' - E_0 = D\left(-\frac{u}{3} + \frac{1}{11} + \frac{1}{u}\cdots\right).$$

Ebenso:

$$(25\text{d})\quad \begin{cases} E_5' - E_0 = D\left(-\frac{u}{3} + \frac{1}{11} + \frac{1}{u}\cdots\right) & u > 0 \\ E_5'' - E_0 = D\left(\frac{u}{5} + \frac{1}{33} + \frac{1}{u}\cdots\right) & u < 0. \end{cases}$$

D. h.: Bei ausgesprochen tetragonaler Symmetrie ist es belanglos, daß E_5' und E_5'' zwei Terme mit der gleichen Darstellung der tetragonalen Gruppe sind. Man erhält (bis auf geringfügige Abweichungen von der 6. Ordnung) zwei Terme, welche einfach den beiden Einstellungen des Elektronendrehimpulses $l = 3$ mit den Komponenten $m = 1$ bzw. $m = 3$ in Richtung der tetragonalen Achse entsprechen:

$$E_3 - E_0 = \varepsilon_{33} - E_0 = D\left(-\frac{u}{3} + \frac{1}{11}\right) = \frac{e}{3} \cdot \frac{\overline{r^2}}{2!}\left(\frac{\partial^2 V}{\partial z^2}\right)_0 - \frac{e}{11}\frac{\overline{r^4}}{4!}\left(\frac{\partial^4 V}{\partial z^4}\right)_0,$$

$$E_1 - E_0 = \varepsilon_{11} - E_0 = D\left(\frac{u}{5} + \frac{1}{33}\right) = -\frac{e}{5} \cdot \frac{\overline{r^2}}{2!}\left(\frac{\partial^2 V}{\partial z^2}\right)_0 - \frac{e}{33}\frac{\overline{r^4}}{4!}\left(\frac{\partial^4 V}{\partial z^4}\right)_0.$$

Aber E_5' entspricht bei positiver Tetragonalität einer Einstellung des Drehimpulses mit der Komponente $m = 1$ in Richtung der tetragonalen Achse, dagegen bei negativer der Einstellung $m = 3$. E_5' bleibt *stets* — unabhängig von u — der *höhere* der beiden Terme, und man kann ihm deshalb nicht durchgehend für alle u die gleiche elektrische Quantenzahl, etwa $m = 3$, zuordnen, weil diese einmal dem tieferen, einmal dem höheren Term entspricht. Wenn wir, von positiver Tetragonalität kommend, den Kristall adiabatisch deformiert denken und uns der kubischen Symmetrie nähern, so hat es eben keinen Sinn mehr, von einer Komponente $m = 1$ des Drehimpulses in Richtung der tetragonalen Achse zu sprechen, sobald nämlich diese tetragonale Achse nicht mehr genügend vor den beiden anderen Achsen ausgezeichnet ist. Erst wenn wir mit unserer Deformation über die kubische Symmetrie so weit hinausgehen, daß die Symmetrie wieder ausgesprochen, aber nunmehr negativ tetragonal wird, können wir wieder eine Komponente des Drehimpulses der Elektronenbahn in Richtung der tetragonalen Achse definieren, diese ist aber nunmehr $m = 3$. Dies Beispiel kann als eine Illustration aufgefaßt werden zu dem allgemeinen Satz, daß Terme gleicher Darstellung sich nicht überschneiden dürfen. Wir könnten natürlich auch die Eigenfunktionen nullter Näherung nach § 11 (Gl. 7) in Abhängigkeit von u darstellen und würden dabei den gleichen Übergang von P_3^1 über die Eigenfunktion bei kubischer Symmetrie zu P_3^3 bekommen, wenn wir u von großen positiven zu großen negativen Werten übergehen lassen.

In den Figg. 6—8 ist die Abhängigkeit der Termwerte (relativ zum Termschwerpunkt gerechnet) von der wirksamen Tetragonalität u dargestellt; die Terme sind dabei relativ zu dem Faktor $D = -e\frac{\overline{r^4}}{4!}\left(\frac{\partial^4 V}{\partial z^4}\right)$ aufgetragen.[1]) Zur Orientierung ist als Abszisse neben der wirksamen Tetragonalität u auch noch das Achsenverhältnis c/a angegeben, welches ein gedehnter Kristall vom NaCl-Typ bei der Tetragonalität u ungefähr besitzen muß, wenn der Bahnradius des Elektrons zu $^1/_4$ des Gitterabstandes angenommen wird.

Fig. 6 zeigt das Aufspaltungsbild des d-Terms in Abhängigkeit von der Tetragonalität. Da zu keiner Darstellung zwei Terme gehören, läßt sich jedem Term ein m-Wert (elektrische Quantenzahl, Komponente des Drehimpulses in Richtung der tetragonalen Achse) und eine von u unabhängige winkelabhängige Eigenfunktion zuordnen, überdies hängt der Wert jedes Terms linear von u ab. Bei ausgesprochen tetragonaler Symmetrie bekommt man eine *weite* Aufspaltung proportional $\left(\frac{\text{Bahnradius}}{\text{Gitterabstand}}\right)^2$ zwischen Termen mit *verschiedenem* m, für welche ungefähr die übliche Intervallregel gilt (Abstand zwischen m und $m+1$ proportional $(m+1)^2 - m^2 = 2m+1$) und eine enge Aufspaltung proportional $\left(\frac{\text{Bahnradius}}{\text{Gitterabstand}}\right)^4$ zwischen den beiden Termen mit den Eigenfunktionen

$$\sqrt{2}\, P_2^2 \cos 2\varphi \quad \text{und} \quad \sqrt{2}\, P_2^2 \sin 2\varphi .$$

Fig. 7 zeigt die Aufspaltung des f-Terms in Abhängigkeit von u. Die Termkurven E_5' und E_5'' bilden Hyperbeln, deren Asymptoten die Geraden $\varepsilon_{11} - E_0$ und $\varepsilon_{33} - E_0$ sind. Diese Geraden würden die zu den elektrischen Quantenzahlen 1 und 3 gehörigen Terme darstellen, falls sich diese Quantenzahlen für alle u definieren ließen.

Fig. 8 endlich zeigt das Aufspaltungsbild des g-Terms, wobei vier Terme in Abhängigkeit von u sich als Hyperbeln darstellen. Bei großer Tetragonalität erhält man wieder eine Termordnung nach m; der Term $m = 2$ ist dabei nochmals aufgespalten (Eigenfunktionen $\sqrt{2}\, P_4^2 \cos 2\varphi$ und $\sqrt{2}\, P_4^2 \sin 2\varphi$),

1) In Wirklichkeit wird sich dieser Faktor selbst natürlich auch noch, wenn auch langsam, mit der Änderung der Tetragonalität ändern, doch wird diese Änderung abhängig sein von der speziellen Struktur des Kristalls, während die relative Lage und die relativen Abstände der Terme nur von der Größe der Tetragonalität abhängen.

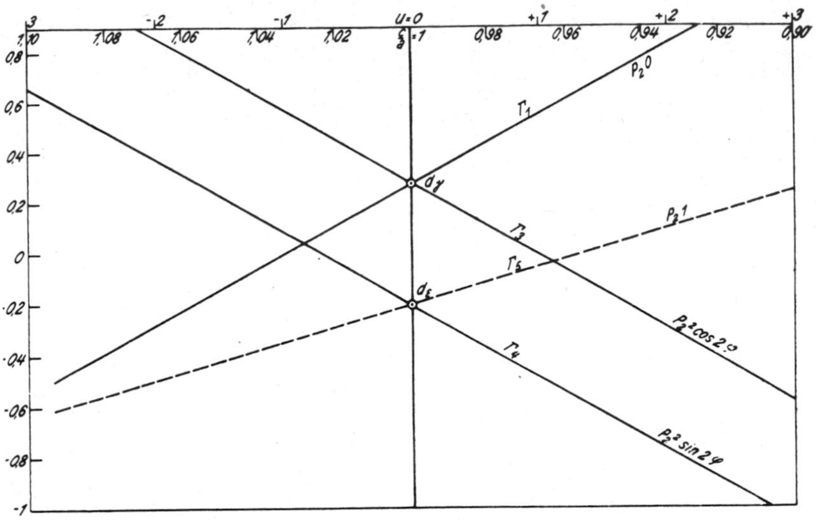

Aufspaltungsbild des *d*-Terms bei tetragonaler Symmetrie

Fig. 6

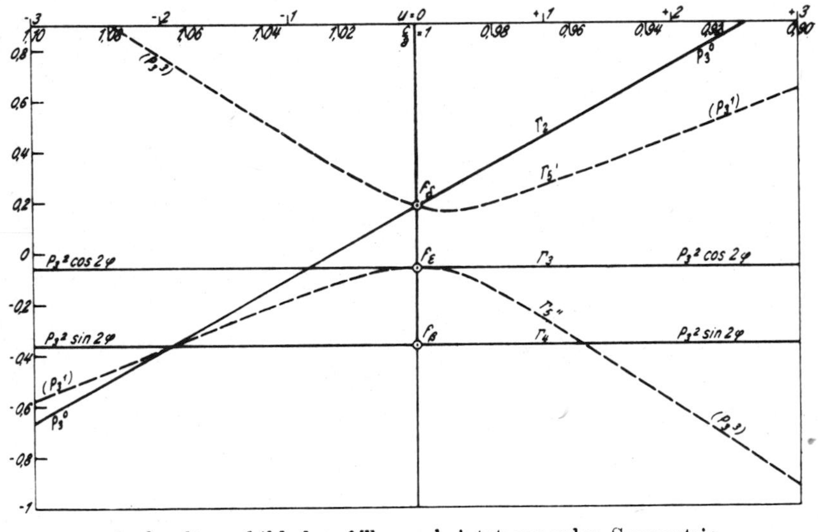

Aufspaltungsbild des *f*-Terms bei tetragonaler Symmetrie

– – – – zweifach entartete

——— einfache Terme

Fig. 7

die entsprechende Aufspaltung für $m = 4$ dagegen verschwindet praktisch, weil sie erst durch das Multipolmoment 8. Ordnung hervorgerufen wird.

Was die absolute Größe der Aufspaltung betrifft, so kann diese bei ausgesprochen tetragonaler Symmetrie erheblich größere Werte erreichen als bei kubischer. Nehmen wir etwa

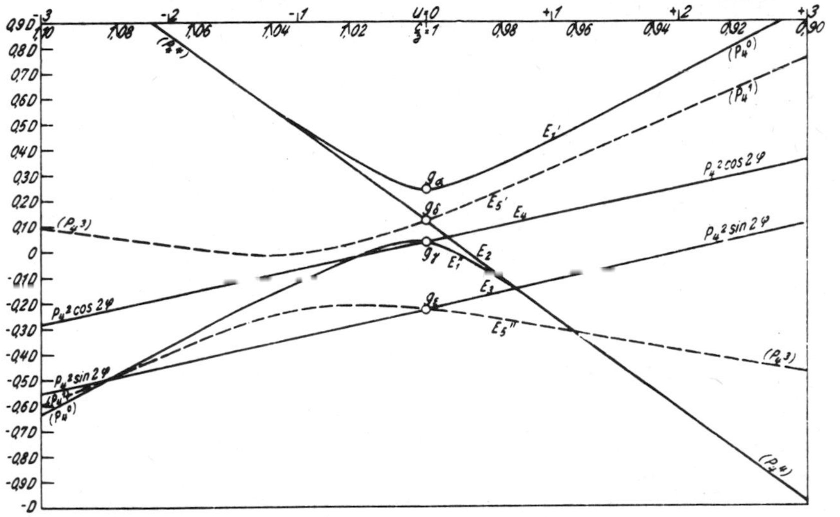

Aufspaltungsbild des *g*-Terms bei tetragonaler Symmetrie

Fig. 8

einen Kristall vom gestauchten NaCl-Typ an, so wird bei einem Achsenverhältnis

$$\frac{c}{a} = \frac{7}{8} \quad \text{etwa} \quad \frac{1}{2!}\left(\frac{\partial^2 V}{\partial z^2}\right)_0 = -1 \cdot \frac{E}{a^3},$$

also bei einem Bahnradius von $^1/_5$ des Gitterabstands und einfach geladenen Ionen der Termwert:

$$E_\lambda - E_0 \approx + \frac{\overline{r^2}}{a^2} \cdot \frac{e^2}{a} \cdot \alpha_{l\lambda} \approx \frac{1}{25}\,\frac{e^2}{a}\,\alpha_{l\lambda} \approx$$

$$0{,}008 \cdot 2 \cdot \text{Rydbergkonstante} \cdot \alpha_{l\lambda} = \text{etwa } 2000\ \text{cm}^{-1}.$$

Die Aufspaltung kann also leicht über 1000 cm^{-1} hinausgehen.

Indem die Tetragonalität die gegenseitige Lage der Kristallterme regelt, bestimmt sie, wie gesagt, auch die stabilste Elektronenanordnung im Kristall. Z. B. ist für ein *d*-Elektron eines positiven Ions im Falle stark negativer Tetragonalität (gedehnter NaCl-Kristall) die Einstellung des Elektrons in der tetragonalen Achse die stabilste (Eigenfunktion P_2^0), bei weniger starker Dehnung erhält man ein Energieminimum bei einer *schiefen* Einstellung, bei der das Dichtemaximum auf einem Kegel um die tetragonale Achse mit dem Öffnungswinkel $\pi/2$ gelegen ist (P_2^1). Bei *kubischer* Symmetrie liegt im stabilsten Zustand das Dichtemaximum in den dreizähligen Achsen, um schließlich bei positiver Tetragonalität (gestauchter NaCl-Kristall) in die Flächendiagonalen der zur tetragonalen Achse senkrechten Ebene überzugehen (Eigenfunktion $\sqrt{2}\; P_2^2 \sin 2\,\varphi$). Ebensolche Betrachtungen kann man an Hand der Figg. 6—8 für negative Ionen sowie für andere Elektronenterme durchführen. Die in den Figuren aufgezeichneten relativen Termabstände sind unabhängig von der speziellen Struktur des Kristalls, wie ja auch der ganze Begriff der Tetragonalität.

§ 23. Ausblick auf Anwendungen der Theorie

Die vorliegende Theorie gibt einen Überblick über den Einfluß elektrischer Felder von bestimmter Symmetrie auf ein Atom. Sie wird sich also auch auf die Behandlung von symmetrischen Molekülen mit Vorteil anwenden lassen.[1]) Eine direkte physikalische Bestätigung müßte durch Analyse des Spektrums von Kristallen zu erlangen sein; es scheint, als ob die Salze der seltenen Erden, welche allein scharfe Absorptionslinien geben, als Beleg für die Theorie dienen können.[2]) Es

---

1) Vgl. die Anmerkungen in § 12, 13 usw. bezüglich der Beziehungen unserer Untersuchung zu der von Ehlert über das CH_4, a. a. O.

2) Hrn. Dr. Schütz (Tübingen) bin ich für den Hinweis hierauf zu Dank verpflichtet.

Anm. bei der Korrektur: Die genannten Absorptionslinien zeigen im Magnetfeld eine Aufspaltung in nur wenige (häufig 2) Komponenten mit oft sehr großem Abstand voneinander. Wenn man bedenkt, daß das Kristallfeld eine weitgehende Termaufspaltung bedingt, ist dies leicht verständlich: Die ungestörten Kristallterme (ohne Magnetfeld) sind im allgemeinen nur zweifach oder gar nicht entartet, das Magnetfeld hebt diese letzte Richtungsentartung auf und spaltet den Term in nur ein bis zwei Komponenten. Der Abstand dieser Komponenten voneinander ist andererseits proportional der magnetischen Quantenzahl m, die wegen der großen Gesamtimpulsmomente (j bis etwa 10) der seltenen Erden sehr erhebliche Werte annehmen kann. Eine tiefergehende Analyse dieser Verhältnisse ist wegen des unübersichtlichen experimentellen Materials heute nicht möglich.

wäre ferner denkbar, daß die Theorie bei Vervollkommnung durch eine Theorie des Austauscheffekts die polymorphen Umwandlungen erklären könnte, doch lassen wir dies noch dahingestellt. Wichtig erscheint das negative Resultat, daß bei kubischer Symmetrie P-Terme nicht aufspalten (und allgemein S-Terme). Schließlich lassen sich die Abweichungen der Atomsymmetrie im Kristall von der Kugelsymmetrie übersehen; ein Vergleich der Größe der Termaufspaltung eines (n, l)-Elektrons mit seinem Termwert im freien Atom gestattet auch eine quantitative Abschätzung der Abweichung. Für den vierwertigen Kohlenstoff (Diamant) sollte man allerdings in nullter Näherung keine wesentliche Abweichung von der Kugelsymmetrie erwarten, da der Grundterm ein ^4S-Term ist, und demnach nur der Spin sich verschieden orientieren kann, nicht aber der Bahnimpuls. Vermutlich wird hier erst die *Deformation* der L-Schale die röntgenographisch beobachtete Abweichung von der Kugelsymmetrie verursachen.

(Eingegangen 20. Juli 1929)

1

Term Division in Crystals

H. BETHE

Translated from Annalen der Physik,
Ser. 5, Vol. 3, No. 2, pp. 133–138 (1929),
by the SLA Translation Center

The effect of an electric field of a given symmetry (crystal field) on an atom is treated in terms of wave mechanics. The terms of the atom split up in a manner which depends on the symmetry of the field and on the angular momentum l (or j) of the atom. S-terms do not divide in general; p-terms do not divide in fields of cubic symmetry. In the case where the single electrons of the atom can be treated separately (canceled interaction in the atom) the eigenfunction to zero approximation is given for every term in the crystal; from these, a grouping of the electron density characteristic for the term is obtained according to the axes of symmetry of the crystal. (The magnitude of the term division varies within the order of a few hundred cm^{-1}.) For tetragonal symmetry a quantitative measure can be defined for the deviation from cubic symmetry which determines clearly the most stable arrangement of the electrons in the crystal.

Introduction

The wave mechanics of crystals can be based on two starting points. We may consider the crystal as a complete whole, describe it by a spatial-periodical potential field and eigenfunctions of the same spatial periodicity, the modulation of which by the finer structure of the atoms contained in the crystal we shall consider only secondarily. This method has been used, by Block[1] in particular, to treat electrical conductivity and seems particularly convenient in the case of extensively "free" electrons. Or we may start from a free atom and consider its perturbation in the crystal in analogy with the method of London and Heitler[2]; this method can be used, in particular, to calculate lattice energy, lattice spacings, and the like.

[1]F. Block, *Z. Physik,* **52,** 555 (1929).
[2]F. London and W. Heitler, *Z. Physik,* **44,** 455 (1927).

A perturbation of the free atom when incorporated in a crystal must be expected for two reasons: On the one hand, the atom will enter into an electron exchange with the other atoms of the crystal; that is, its permutation group is changed. This exchange effect must be treated entirely in analogy with a molecule; at the most there is a purely quantitative difference in regard to the number of neighbors with which an exchange can take place. On the other hand, the atom in the crystal is influenced by an electric field of definite symmetry originating from the other atoms which cancels out the directional degeneration of the free atom. The qualitative difference between the molecule problem and the crystal problem lies in this different spatial symmetry. We shall discuss the specific dislocations of the free atom caused by the symmetry of the crystal without first considering the electron exchange phenomena between the atoms of the crystals; at the same time we shall create a basis for discussion of the latter.

An electric field of definite symmetry will cause a division of the terms of the undistorted atom which is analogous to the Stark effect division and will be characteristic even for the symmetry of the field, that is, for the symmetry of the position of the atom in the crystal. The number of components into which a term of the free atom is divided increases with decreasing symmetry. The amount of division can be of very different order, and three cases can be differentiated.

1. The Stark effect division in the crystal field is large compared with the distance between different multiplets. The effect upon the atom caused by the crystalline field is stronger than the interaction of the electrons in the atom and cancels their coupling to a first approximation. We use a model of the free atom as a basis for which only the population coefficients of the quantum cells characterized by main quantum number n_i and azimutal quantum number l_i (of the single electron) are given, but the term division is not considered because of exchange degeneration. (We can include the electrostatic mutual effects of the electrons in accordance with the Hartree method of the self-consistent field.)

We then consider to a first approximation the perturbation of the single electrons located outside closed shells by the field of the other atoms in the crystal, that is, the possible adjustments of the orbital angular momentum l_i of the single electron in regard to the axes of the crystal and the resulting Stark effect division of the terms of the atom. The electron exchange within the atom would have to be taken into account to a second approximation; this, in general, results in further term division, the order of magnitude of which naturally corresponds to the distance between various multiplets of the free atom. Finally, the interaction between orbital angular momentum and spin determines the multiplet division in the conventional manner.

2. The second case involves crystal cleavage of average value, that is, small by comparison with the distance between different multiplets, but large compared with the term differences of one and the same multiplet. In this case (probably the most likely one) we must start from the free atom, with consideration of the term division by exchange degeneration but without consideration of the interaction between spin and orbit. Then this atom would have to be inserted in the crystal and the adjustments of its overall orbital angular momentum l to the axes of the crystals and the corresponding term values investigated and finally the interaction between spin and orbit for constant orientation of the latter in the crystal.

3. The third case occurs when crystal cleavage is small compared with the term spacing within a multiplet. The completely "finished" atom, with consideration of the interaction between electron orbits and spin, is subjected to the dislocation by the electric field of the crystal. The result is total orbital angular momentum j of the atom with regard to the axes of the crystal, not orbital angular momentum as in cases 1 and 2 because this now remains tightly coupled with the spin; in cases 1 and 2 coupling is canceled by the crystal field.

We shall use the following quantum numbers to describe Stark effect division:

λ, azimuthal crystal quantum number that characterizes the orientation of the orbital angular momentum of the atom in the crystal
λ_i, azimuthal crystal quantum number of a single (ith) electron
l_i, conventional azimuthal quantum number of the single electron in a free atom
μ, inner crystal quantum number (orientation of the total angular momentum j of the atom in the crystal)

Since crystal quantum numbers do not concern "true" quantum numbers in general, which could be interpreted as angular momenta about an axis or the like, but only the differentiation between different group-theoretical properties of representation of single terms, we do not write them as numbers but as Greek letters.

In the three cases given above, then, an atom is characterized by the following quantum numbers:

1. Large crystal splitting: n_i, l_i, λ_i, λ, μ.
2. Average crystal splitting: n_i, l_i, l, λ, μ.
3. Small crystal splitting: n_i, l_i, l, j, μ.

We shall first consider the three possible cases: (1) adjustment of the orbital momentum of the single electron, (2) adjustment of the entire orbital momentum, and (3) adjustment of the overall angular momentum of the atom in the crystal, in general, and we shall calculate the number of division components of a term for given angular momentum (representation of the rotary group) and given symmetry of the position of the atom in the crystal in terms of group theory. The case of the adjustment of a half-integral total angular momentum (equivocal representation of the rotary group) must be treated separately. Then the angle-dependent factors of the eigenfunctions of zero approximation corresponding to the individual division components, as needed for exchange calculations, will be indicated. Next we shall discuss the three cases which are differentiated by the order of magnitude of the crystal splitting, with special consideration of the symmetry of the electron density distribution. Finally, we shall calculate the magnitude of the splitting for ionic crystals.

Solution in Terms of Group Theory

It is well known that Schrödinger's differential equation of the free atom is invariant with regard to arbitrary rotations of the system of coordinates as well as an inversion at the nucleus (and with regard to exchange of the electrons, which is not of interest to us a priori). Its substitution group[1] is also the group of rotation

[1]E. Wigner, *Z. Physik,* **43,** 624 (1927).

of the spheres, each one of which has an irreducible representation of the dimensions $2l+1$ ($l = 0\ 1\ 2\ \ldots$).

If we incorporate the atom in a crystal, the symmetry of the potential is reduced from spherical symmetry to the symmetry of the position assumed by the atom in the crystal, for instance, when incorporated in a crystal of the NaCl type, reduction to the cubic-holohedral symmetry, for the ZnS type, to cubic-hemihedral symmetry. The substitution group of Schrödinger's differential equation for the atom under consideration comprises only those symmetry operations which leave the position of the nucleus of the atom considered unchanged and at the same time transfer the entire crystal to itself (symmetry group of the crystal atom). The method of treating such a reduction of the symmetry has been indicated in a general way by Wigner. We have only to reduce completely the representation of the original substitution group corresponding to a given term of the unperturbed atom as representation of the new substitution group, which is a subgroup of the earlier one, and then obtain number and variety of the terms, into which the unperturbed term considered is divided for the reduction of the symmetry performed. It is entirely immaterial whether the perturbation is extensive or slight or whatever form it may have, and whether the perturbing energy is calculated only to a first order or in any higher order; determination of a definite symmetry of the perturbing potential is perfectly adequate.

In order to reduce the $(2l+1)$-dimensional representation of the rotary group practically as a representation of a definite crystal symmetry group (the octahedral group, for instance), we use the fundamental theorem of the group theory: every reducible representation of a group can be reduced to its irreducible constituents in one, and only one, way; the character of every group element in the reducible representation is equal to the sum of characters assigned to the element in the irreducible constituents. We must therefore know the character of every symmetry operation of the crystal which leaves the atomic nucleus in its position—on the one hand, in the irreducible representations of the substitution group of the crystal atom, and on the other hand, in the irreducible representations of the substitution group of the free atom.

Every symmetry operation of the crystal atom can be composed of a pure rotation and eventually an inversion at the nucleus, similar to every symmetry operation of the free atom. We need only to calculate the character of a given rotation in the $(2l+1)$-dimensional representation of the rotary group, however; if an inversion is added to the rotation, the character is simply multiplied by $+1$ or -1 according as a positive or a negative term is involved.[1] The most general rotation is that about an arbitrary axis by an angle of Φ. It is represented most simply by the transformation of the spherical functions, with subscript l referring to the axis of rotation:

$$f_\mu(x) = P_l^{u}(\cos\theta)e^{iu\phi}.$$

Our rotation R is to convert θ in itself, ϕ into $\phi+\Phi$, hence $f_\mu(x)$ into

$$f_\mu(Rx) = P_l^{u}(\cos\theta)e^{i\mu(\phi+\Phi)},$$

[1]See E. Wigner and J. von Neumann, *Z. Physik,* **49,** 73, 91 (1928).

i.e., our rotation is represented by the matrix

$$\begin{pmatrix} e^{-il\Phi} & 0 & \cdots\cdots\cdots & 0 \\ 0 & & e^{-(l-1)\Phi} \cdots\cdots\cdots & 0 \\ \cdots\cdots & \cdots\cdots & \cdots\cdots\cdots & \cdots \\ 0 & 0 & \cdots\cdots\cdots & e^{il\Phi} \end{pmatrix}$$

with the character

$$\chi(\Phi) = \frac{\sin(l + \frac{1}{2})\Phi}{\sin \frac{1}{2}\Phi} .$$

Reprinted from *Proc. Sect. Sci.*, **XXXIII**, Koninklifke Akad. van Wetenschappen, Amsterdam, 962–965 (1930)

2

Théorie générale de la rotation paramagnétique dans les cristaux*

H. A. KRAMERS

§ 2. *Théorème général sur les propriétés d'un système atomique dans un champ électrique.*

Considérons un système de noyaux et d'électrons placé dans un champ de force extérieur. L'expression classique pour l'énergie, en fonction des coördonnées et des impulsions, a été calculée par Darwin [1]) qui a tenu compte des corrections relativistes jusqu'aux termes contenant $\frac{1}{c^2}$. Cette expression consiste en deux parties $H_1 + H_2$; H_2 est la partie due aux forces extérieures, tandis que H_1 renferme les termes qui subsistent quand ces forces disparaissent. Les termes de H_1 sont de degré pair par rapport

[1]) C. G. DARWIN, Phil. Mag. **39**, 537, 1920.

*See *Group Theory and Solid State Physics: I,* P. H. Meijer, ed., International Science Review Series, Gordon and Breach, New York, 1964, for a translation into English.

aux impulsions; comme opérateur dans l'équation de SCHRÖDINGER, H_1 est donc une expression réelle. D'autre part H_2 contient des termes qui sont linéaires dans les impulsions, ce qui donne des termes imaginaires dans l'opérateur qui y correspond. Ces termes disparaissent pourtant si le champ est d'origine purement électrique. Dans ce cas, l'équation de SCHRÖDINGER, qui détermine les fonctions d'ondes dans les états stationnaires, est donc réelle. Il s'ensuit le théorème bien connu que, si φ est une solution, $\varphi^\star$ sera encore une solution pour la même valeur de l'énergie.

Comme condition que l'état considéré n'est pas dégénéré on aura, si a désigne une constante:

$$\varphi^\star = a\,\varphi = a\,a^\star\,\varphi^\star \quad . \quad . \quad . \quad . \quad . \quad . \quad . \quad . \quad (10)$$

c. à d.

$$|\,a\,|^2 = 1$$

Si la condition (10) est remplie, la fonction d'ondes pourra donc être mise sous forme réelle $a^{1/2}\varphi$.

Si l'on tient compte des spins des électrons, le théorème susdit permet une généralisation intéressante. Considérons d'abord l'expression de l'énergie quand on ne néglige plus les spins. Cette expression a été donnée par HEISENBERG dans son mémoire sur le spectre de l'hélium. [1]) Récemment M. BREIT [2]) a consacré un mémoire très important à cette question, dans lequel il examine l'interaction entre les particules sur la base de la théorie du spin due à DIRAC. Les expressions de BREIT et de HEISENBERG ne sont pas tout à fait identiques; cependant elles ont en commun que la partie H_1 de l'énergie qui ne dépend pas des forces extérieures, considérée comme fonction des composants des impulsions et des spins, ne contient que des termes de degré pair, dont les coëfficients sont réels. La contribution H_2 des forces extérieures contient des termes du degré 0, 1 et 2. Si le champ est purement électrique les coëfficients des termes du degrè 0 et 2 sont réels tandis que ceux du premier degré sont imaginaires.

Soit n le nombre des électrons. En négligeant les spins éventuels des noyaux, dont l'influence énergétique n'a pas d'importance en général, une solution de l'équation de SCHRÖDINGER est donnée par un système de 2^n fonctions d'ondes, que nous dèsignons par

$$\varphi_{s_1, \dots s_k, \dots, s_n} \quad . \quad . \quad . \quad . \quad . \quad . \quad . \quad . \quad . \quad (11)$$

où s_k signifie le moment d'impulsion (divisé par $h/2\pi$) que le spin du $k^{\text{ième}}$ électron possède dans une direction donnée ($s_k = + {}^1/_2$ ou $- {}^1/_2$).

Etant donné les propriétés spéciales dont jouit l'opérateur de l'énergie en l'absence de forces magnétiques extérieures et que nous venons de signaler, on pourra affirmer le théorème suivant: *quand* (11) *satisfait à*

---

[1]) W. HEISENBERG, Zs. für Phys. **39**, 514, 1926.

[2]) G. BREIT, Phys. Rev. **34**, 553, 1929. Voir en particulier sa formule (48).

l'équation de SCHRÖDINGER, *une autre solution, qui correspond à la même valeur caractéristique de l'énergie, sera donnée par:*

$$\varphi'_{s_1 \ldots s_n} = (-1)^{\sum_k s_k - n/2} \varphi^*_{-s_1, \ldots -s_n} \quad . \quad . \quad . \quad . \quad . \quad (12)$$

Nous démontrons cette proposition par un calcul direct. Désignons les composants du spin du k-ième électron, divisés par $h/4\pi$, par s^k_x, s^k_y, s^k_z. La manière dont ils opèrent sur une fonction (11) est décrite par les matrices de PAULI:

$$s^k_x \rightarrow \begin{vmatrix} 1 & 0 \\ 0 & -1 \end{vmatrix}, \; s^k_y \rightarrow \begin{vmatrix} 0 & 1 \\ 1 & 0 \end{vmatrix}, \; s^k_z \rightarrow \begin{vmatrix} 0 & -i \\ i & 0 \end{vmatrix}, \quad . \quad . \quad . \quad (13)$$

où x est la direction à laquelle se rapporte le symbole (11). On aura donc:

$$\left. \begin{aligned} s^k_x \varphi_{s_k} &= (-1)^{s_k - 1/2} \varphi_{s_k} \\ s^k_y \varphi_{s_k} &= \qquad\quad \varphi_{-s_k} \\ s^k_z \varphi_{s_k} &= -i(-1)^{s_k - 1/2} \varphi_{-s_k} \end{aligned} \right\} \quad . \quad . \quad . \quad . \quad . \quad . \quad . \quad (14)$$

Les indices $s_{k'}$ ($k' \neq k$), que nous avons omis, ne sont pas changés.

Nous écrivons l'équation de SCHRÖDINGER dans la forme:

$$H(s^k_x, s^k_y, s^k_z) \varphi_{s_k} = E\varphi_{s_k} \quad . \quad . \quad . \quad . \quad . \quad . \quad . \quad (15)$$

A part des composants des spins, H dépend encore des coördonnées et des impulsions des électrons et des noyaux. Prenons l'expression complexe conjuguée de (15), en nous rappelant que l'opérateur qui représente une impulsion est imaginaire et que les termes de degré impair dans les s^k sont multipliés par un nombre impair d'impulsions:

$$H(-s^{*k}_x, -s^{*k}_y, -s^{*k}_z) \varphi^*_{s_k} = E\varphi^*_{s_k}$$

En vue de (13) nous en déduisons:

$$H(-s^k_x, -s^k_y, s^k_z) \varphi^*_{s_k} = E\varphi^*_{s_k} \quad . \quad . \quad . \quad . \quad . \quad . \quad (16)$$

En posant:

$$\varphi''_{s_k} = \varphi^*_{-s_k}$$

l'équation (16) prend la forme:

$$H(s^k_x, \; -s^k_y, \; -s^k_z) \varphi''_{s_k} = E\varphi''_{s_k} \quad . \quad . \quad . \quad . \quad . \quad (17)$$

Introduisons maintenant la fonction φ' définie par (12):

$$\varphi'_{s_k} = (-1)^{\sum_k s_k - n/2} \varphi''_{s_k}$$

On voit aisément, à l'aide de (14), que s_x, s_y, s_z, opèrent sur φ' de la même façon que s_x, $-s_y$, $-s_z$ opèrent sur φ''. On conclut donc de (17):

$$H(s^k_x, \; s^k_y, \; s^k_z) \varphi'_{s_k} = E\varphi'_{s_k},$$

ce qui prouve la validité de notre théorème.

Examinons si l'état stationnaire auquel correspond (11) pourra être non-dégénéré. La condition nécessaire sera que les fonctions (12) ne diffèrent des fonctions (11) que par un facteur commun que nous désignons par a:

$$(-1)^{\Sigma s_k - n/2} \varphi^*_{-s_k} = a\, \varphi_{s_k}\,.$$

En prenant les valeurs conjuguées et en remplaçant s_k par $-s_k$:

$$(-1)^{-\Sigma s_k - n/2} \varphi_{s_k} = a^* \varphi^*_{-s_k} = a^* a\, (-1)^{-\Sigma s_k + n/2} \varphi_{s_k}\,.$$

Conséquemment

$$a^* a = (-1)^n\,.$$

Cette condition ne peut être remplie que si n est pair; dans ce cas on pourra mettre $a = 1$. Si n est impair, on a le théorème suivant: les états stationnaires d'un système atomique sont toujours dégénérés quand le système contient un nombre impair d'électrons, le degré de dégénération étant un nombre pair.

Si le nombre d'électrons est pair et si l'état stationnaire est non-dégénéré, l'énergie ne sera pas influencée en première approximation par un champ magnétique. En effet, l'énergie perturbatrice Ω qui correspond à un tel champ ne contient en première approximation que des termes linéaires dans les impulsions du type $f p f$ et des termes linéaires dans les spins du type $g s$ (f et g fonctions des coördonnées). Il suffira de considérer un terme $T = f p f = \frac{h}{2\pi i} f \frac{\partial}{\partial x_k} f$ et un terme $T = g\, s^k_x$. Pour calculer l'énergie de perturbation qui correspond à ces termes il faudra connaître les intégrales suivantes.

$$\left.\begin{aligned} \overline{T} = \sum_k \int \varphi^*_{s_k} T \varphi_{s_k} d\tau = \sum_k (-1)^{\Sigma s_k - n/2} \int \varphi_{-s_k} T \varphi_{s_k} d\tau = \\ = \Sigma (-1)^{-\Sigma s_k - n/2} \int \varphi_{s_k} T \varphi_{-s_k} d\tau \end{aligned}\right\} , \quad (18)$$

Pour les deux types de termes, que nous avons envisagés, on aura toujours:

$$\int \varphi_{s_k} T \varphi_{-s_k} d\tau = -\int \varphi_{-s_k} T \varphi_{s_k} d\tau \quad . \quad . \quad . \quad . \quad . \quad (19)$$

Si nous introduisons (19) dans le dernier membre de (18), celui-ci devient égal au négatif du second membre (observez que $(-1)^{-\Sigma s_k} = (-1)^{\Sigma s_k}$ si n est pair), donc $\overline{T}$ est nul. Les états stationnaires non-dégénérés sont donc toujours non-magnétiques.

Si le nombre d'électrons est impair, les niveaux seront magnétiques en général. Dans le paragraphe suivant nous examinerons ce problème de plus près.

62*

Reprinted from *Phys. Rev.*, **41**, 208–215 (1932)

3

Theory of the Variations in Paramagnetic Anisotropy Among Different Salts of the Iron Group

By J. H. VAN VLECK
University of Wisconsin

(Received June 6, 1932)

A theoretical explanation is given of why nickel salts are nearly isotropic magnetically, while those of cobalt exhibit over 25 percent anisotropy even though the Ni^{++} and Co^{++} ions are both in F states and are adjacent in the periodic table. The cause is an inversion of the levels in the crystalline Stark effect in passing from the configuration $d^8\,{}^3F$ (Ni^{++}) to $d^7\,{}^4F$ (Co^{++}). If the crystalline field has only rhombic symmetry, but with the deviations from cubic symmetry comparatively small, extension of the methods in Penney and Schlapp's preceding paper shows that a nearly isotropic level will be the ground level in Ni^{++}, but an anisotropic one in Co^{++}. It is to be particularly noted that the inversion exists purely in virtue of the difference between the configurations d^7 and d^8, and does not require different crystalline fields in Ni and Co compounds. The theory predicts that hydrated Ni salts conform closer to Curie's law than those of Co, and have a Curie constant more nearly equal to the "spin only" value $4NS(S+1)(he/4\pi mc)^2/3k$. This agrees with experiment. Other pairs of ions are cited in the iron group in which the inversion phenomenon is encountered, with attendant diversity in magnetic behavior. The nearly perfect magnetic isotropy of manganous salts is trivial, as the ground state of Mn^{++} is ${}^6S$; the slight anisotropy may be due to a small amount of incipient j-j coupling or to distortion of the constancy of orbital angular momentum by the crystalline field, so that the orbital magnetic moment does not vanish completely in S states.

NO ONE who has examined the measurements of the principal susceptibilities of salts of the iron group can fail to note how remarkably the amount of magnetic anisotropy varies with the nature of the cation. Typical determinations are, for instance, those by Rabi,[1] according to which the ammonosulphates of Mn, Co, Ni, and Cu exhibit anisotropies amounting respectively to 1, 30, $1\frac{1}{2}$, 20 percent.

The nearly perfect isotropy of the manganous salts is exactly what one expects inasmuch as the Mn^{++} ion has a ${}^6S$ ground state. Nearly complete symmetry should also be found in ferric salts, as Fe^{+++} and Mn^{++} have the same configuration, but adequate data on ferric compounds are wanting. The usual cause of magnetic anisotropy is the unsymmetrical partial freezing of orbital angular momentum by the lattice forces. The orbital magnetism is thus largely destroyed by these forces, but not completely. Particularly important is usually the coupling between the spin and the remains of the orbital angular momentum, which are not centro-symmetric and so destroy the equivalence of different spin orientations. In S states with perfect Russell-Saunders coupling there is no orbital angular momentum to congeal, and so the usual cause of anisotropy disappears in manganous salts. The ordinary explanation of the small residuum (about 1 percent) of anisotropy found in

[1] I. I. Rabi, Phys. Rev. **29**, 173 (1927).

manganous compounds is the magnetic spin-spin coupling, since the electrostatic exchange coupling is well known not to cause anisotropy. However, it seems doubtful whether the magnetic spin-spin forces could be great enough to cause even the 1 percent dissymmetry in manganous salts of high "magnetic dilution," such as e.g., hydrated sulphates, where the separation of the paramagnetic ions is great. Therefore we wish to suggest the possibility of a small amount of incipient j-j coupling, not enough to distort appreciably the g-factors from their Russell-Saunders values, but enough to impart a slight amount of orbital angular momentum to S states which would then behave anisotropically. Along with this is also, perhaps more important, the angular momentum which arises because the crystalline fields do not have complete central symmetry. In quantum mechanics the orbital angular momentum cannot be constant in time in noncentral fields[2] and so cannot vanish completely in the presence of the latter.

So much for the rather straightforward case of Mn. Much more puzzling is the great difference in isotropy between nickel and cobalt salts. The ions Co^{++} and Ni^{++} are adjacent in the periodic table, and both are in F states (respectively $d^7\,{}^4F$ and $d^8\,{}^3F$). Nevertheless Ni salts are nearly as isotropic as those of Mn, while Co salts are the least symmetrical of the whole group. Closer examination reveals that precisely this behavior is to be expected if the crystalline field possesses only rhombic symmetry, but with the departures from cubic symmetry relatively small. The development of the crystalline potential then takes the form

$$V = \sum_i [f(r_i) + Ax_i^2 + By_i^2 - (A + B)z_i^2 + D(x_i^4 + y_i^4 + z_i^4)], \quad (1)$$

provided we neglect the departures from cubic symmetry in the fourth but not the second order terms. The summation need be extended for our purposes only over the electrons not in closed shells, i.e., the d electrons in the ions of the iron group, inasmuch as completed shells exhibit no orientation effect and so contribute nothing interesting to (1). The reader is referred to the preceding paper by Penney and Schlapp and to reference 3 in case he desires further general exposition of the use of crystalline potentials in connection with magnetic problems.[3] That the rhombic or second order part of (1) is subordinate to the cubic or fourth order part is evidenced by the fact that paramagnetic salts are nearly isotropic in many cases, and especially by the success which has attended Penney and Schlapp's preceding calculation of the temperature dependence of the susceptibilities of rare earth salts involving Nd and Pr under the assumption that only the cubic portion of (4) need be considered. They have also tried calculations keeping only the rhombic part of (1), but the wrong temperature dependence is then obtained in the rare earths, as well as far too much anisotropy in Ni salts.

If the "cubic" or fourth-order part of (1) is the dominant noncentral term, then from Bethe's[4] group theory of levels in crystalline fields, one finds that

[2] Cf. J. H. Van Vleck, Phys. Rev. **31**, 600 (1928).
[3] J. H. Van Vleck, "The Theory of Electric and Magnetic Susceptibilities," section 73.
[4] H. Bethe, Ann. d. Physik **3**, 133 (1929); especially pp. 164–167, 196–199.

the Stark splitting is arranged either as in Fig. 1, or as in this figure turned upside down. The vertical lines represent the matrix elements of magnetic moment along some one principal direction, say x. The separation of the components of Γ_4 or of Γ_5 is due only to the rhombic terms, since Γ_4, Γ_5 (in Bethe's notation) are triply degenerate "Darstellungen" in strictly cubic fields.

If Fig. 1 is right side up, the orbital magnetic moment will be largely suppressed, and its "remains" will be nearly isotropic for the following reason. By perturbation theory,[5] the magnetic moment induced in the state a by application of a magnetic field H along x is given by the expression

$$2H \sum_k | \mu_x(ak) |^2 / h\nu(ka), \tag{2}$$

where the $\mu(ak)$ are the matrix elements of magnetic moment in the absence of H but in a system of representation which diagonalizes (1). The matrix μ

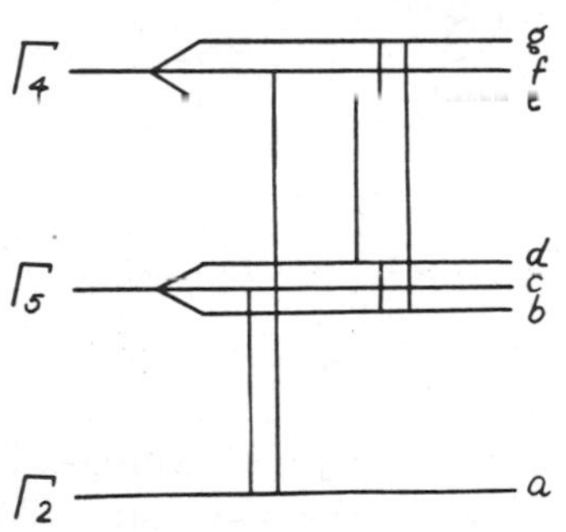

Fig. 1.
(upright in Ni^{++}, inverted in Co^{++})

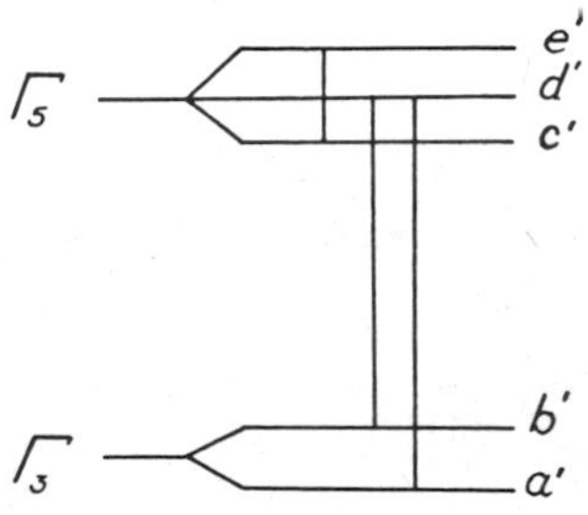

Fig. 2.
(upright in Cu^{++}, inverted in Fe^{++})

will contain no diagonal elements, because, as the writer has shown elsewhere,[3] crystalline fields of no more than rhombic symmetry suppress the diagonal elements of orbital angular momentum (neglecting perturbations by spin-orbit interaction, which restores some of the diagonal part in systems with an odd number of electrons). In the case shown in Fig. 1, the summation over k embraces only c and f, but if one changes the direction of H to another principal axis y or z, one encounters the matrix elements of μ_y, μ_z and these may join a to other components of Γ_4 and Γ_5 than c and f. One finds that the magnetic anisotropy of a exists only in virtue of the difference between the frequencies $\nu(ab)$, $\nu(ac)$, $\nu(ad)$ or between $\nu(ae)$, $\nu(af)$, $\nu(ag)$, and these percentage differences are small since the rhombic is small compared to the cubic separation.

Let us, however, turn Fig. 1 upside down, and suppose that the separations e-f-g, though small compared to a-e, etc., are nevertheless large compared to kT, making only g a normal state. The orbital magnetic moment will then be much less completely quenched than without the inversion, since

[5] See, for instance, p. 145 of reference 3.

the denominators $\nu(ge)$, $\nu(gf)$ will be involved in the formula analogous to (2). Furthermore there will be vastly more anisotropy than without the inversion. This will be true not merely because the percentage difference between $\nu(ge)$ and $\nu(gf)$ is greater than that between $\nu(ab)$ and $\nu(ac)$ or $\nu(ad)$ but also because the full symmetry of moment in the limiting case of perfectly cubic fields is achieved only with equal distribution of atoms among the three components of Γ_4 rather than concentration purely in one component g.

It is clear that one can understand the difference between Ni^{++} and Co^{++} if one can justify turning Fig. 1 upside down in passing from one of these ions to the other. Now it actually turns out that for a *fixed* sign of A, B, D in (1), precisely this behavior is to be expected theoretically. The argument is as follows. Let us consider matrix elements in the L, M_L system of representation rather than in the final system which diagonalizes (1). Then from the properties of the rotation group Penney and Schlapp find in the preceding paper that, regardless of the number n of electrons in the incomplete shell the matrix elements of $\sum(x_i^4+y_i^4+z_i^4)$ are of the form

$$\sum_i(x_i^4 + y_i^4 + z_i^4)(LM_L; LM_L) = Qf(L, M_L) + 3n\overline{r^4}/5, \tag{3}$$

where Q depends on L, S, n but not on M_L, and where

$$f(L, M_L) = M_L^2(7M_L^2 - 6L^2 - 6L + 5) + 3L(L^2 - 1)(L + 2)/5. \tag{4}$$

Besides (3) there are also elements of the form $\Delta M_L = \pm 4$ proportional to Q, as well as elements nondiagonal in L, but the explicit form of these does not concern us. Our interest is in the sign of the proportionality factor Q, which can be determined by the following adaptation of the Goudsmit-Slater method[6] of diagonal sums. For the configuration d^8, the only arrangement in m_l, m_s quantization consistent with $M_L=\sum m_l=3$, $M_S=\sum m_s=1$ is $m_l=2$, $m_l=1$, $m_l=0$ each twice, $m_l=-1$, $m_l=-2$ each once. The spur invariant for $M_L=3$, $M_S=1$ thus consists of but a single term, which yields a ^3F state. Hence, since f is even in M_L,

$$Q(d^8\,{}^3F)\cdot f(3, 3) = Q(d^2D)\,[3f(2, 2) + 3f(2, 1) + 2f(2, 0)]. \tag{5}$$

Here $Q(d^8\,{}^3F)$ means the value of Q appropriate to the configuration $d^8\,{}^3F$, while $Q(d\,{}^2D)$ of course relates to a single isolated d electron. The term in $\overline{r^4}$ has been omitted since it is invariant of the vector addition involved in constructing the Russell-Saunders coupling. For the case $M_L=3$, $M_S=3/2$ of d^7, one has the same arrangement of m_l's as before, except that $m_l=0$ is filled only once, and consequently

$$Q(d^7\,{}^4F)f(3, 3) = Q(d^2D)\,[3f(2, 2) + 3f(2, 1) + f(2, 0)]. \tag{6}$$

On comparing (5) and (6) and noting that

$$2f(2, 2) + 2f(2, 1) + f(2, 0) = 0, \tag{7}$$

one sees that $Q(d^8\,{}^3F)$ is the negative of $Q(d^7\,{}^4F)$. Of course the diagonal ele-

[6] S. Goudsmit, Phys. Rev. **31**, 945 (1928); J. C. Slater, *ibid.* **34**, 1293 (1929).

ments (3) are not the same as the characteristic values of $\sum(x_i^4+y_i^4+z_i^4)$, but except for the difference in Q, the matrix representing this expression in the L, M_L system is the same for $d^8\,{}^3F$ as $d^7\,{}^4F$ inasmuch as L is the same in both cases. Hence the characteristic values are obtained by the same canonical transformation and differ merely in the proportionality factor Q, which will thus cause inversion in the passage from Ni to Co. It is particularly gratifying that alteration of the sign of D in (1) is thus not necessary to invert the crystalline Stark effects. The crystals of Ni and Co hydrated sulphates, etc. have presumably so nearly the same structure and ionic arrangement that it would be illogical, to say the least, to postulate radically different fields in the two cases.

In the foregoing we have neglected entirely the spin in the interest of simplicity. Actually the spin is the main cause of the paramagnetism in Ni and Co, though not of its anisotropy. In point of fact, the spin-orbit interaction is comparable with the effect of the rhombic dissymmetry, but is subordinate to the "cubic" term, so that the inversion of cubic levels still comes into play. Inclusion of the spin-orbit interaction makes the remains of the orbital magnetic moment contribute to the part of the susceptibility which is inversely proportional to temperature, rather than merely a term such as (2) which is independent of temperature. When Fig. 1 is right side up, the spin interacts effectively with a smaller and more isotropic orbital moment than when Fig. 1 is inverted. The most potent elements of μ in the spin-orbit coupling, which is much looser than in the free gaseous state, can be shown[3] to be those whose associated frequencies $\nu(ij)$ are small. Hence the departures of the Curie constant C from the Stoner-Bose "spin only" value

$$C = 4NS(S+1)(he/4\pi mc)^2/3k \tag{8}$$

should be larger in Co than in Ni salts, and deviations from Curie's law in highly hydrated salts should appear at higher temperatures with a Co than with a Ni cation. (The hydration is to avoid exchange coupling between atoms.) This agrees with experiment. The deviation from (8) is about 30 percent in the nickel sulphates, as compared with 70 percent in those of cobalt. Further calculations by Schlapp and Penney, to appear shortly, predict that in hydrated nickel salts the deviations should become important only at temperatures near liquid helium. This is confirmed by Gorter, de Haas, and v. d. Handel's recent remeasurement[7] of $NiSO_4\cdot 7H_2O$ at low temperatures (down to liquid hydrogen) which has obliterated Jackson's previous large negative Δ and restored Curie's law almost perfectly. On the other hand, $CoSO_4\cdot 7H_2O$ apparently demands a Δ of about 14 in the Weiss-Curie formula $\chi = C/(T+\Delta)$, although it must be cautioned that here new measurements are not available.

By arguments very similar to the previous, one can show that the inversion phenomenon should also be found in the pair Fe^{++}, Cu^{++}, respectively $d^6\,{}^5D$ and $d^9\,{}^2D$, and also pairs of ions reciprocally related on the left and right sides of the group, such as (Cr^{+++}, Co^{++}) or (Ni^{++}, V^{+++}). We call ions re-

[7] C. J. Gorter, W. J. de Haas, and J. v. d. Handel, Leiden Comm. 218d.

ciprocally related if they have respectively x and $10\text{-}x$ d electrons, since then their ground states have the same L, S.

Adequate data are wanting on single crystals of Cr and V salts, but the Curie constant does have much more closely the Stoner-Bose value (8) for Cr^{+++} than for Co^{++}, as one should expect from the foregoing. In $Cr_2(SO_4)_2 K_2SO_4 \cdot 24H_2O$ the deviation from (8) is only about 2 percent.[8] The variation in the applicability of (8) is thus even more pronounced in the pair Cr^{+++}, Co^{++} (2 *vs.* 70 percent deviation) than in the pair Co^{++}, Ni^{++} (70 *vs.* 30). This is understandable on the ground that the multiplet structure and hence the spin-orbit interaction for the free ion are considerably smaller in Cr^{+++} than in Co^{++} or Ni^{++}. Of course the departures from (8) will be greater the larger this interaction, all other circumstances being equal. Consequently the alteration in free multiplet width and the inversion phenomenon reinforce each other in accentuating the deviations from (8) in Co^{++} as compared to Cr^{+++}.

The pair Fe^{++}, Cu^{++} require rather careful examination. Here we encounter D rather than F levels. Now Bethe[4] has proved that D terms split under the potential (1) in the fashion shown in Fig. 2, where the separation of the components of Γ_3 or of Γ_5 is due entirely to the deviations from cubic symmetry occasioned by the rhombic terms. The first thing to be remarked is that Fig. 2 is upright in Cu^{++} and inverted in Fe^{++} if Fig. 1 is upright in Ni^{++}. This follows from the fact that $Q(d^9\,{}^2D)$ has the opposite sign from $Q(d^8\,{}^3F)$ inasmuch as by the method of diagonal sums

$$Q(d^9\,{}^2D)f(2,2) = Q(d^2D)\,[3f(2,2) + 4f(2,1) + 2f(2,0)] = -\,Q(d^2D)f(2,2), \quad (9)$$

and by (4) the bracketed sum in (9) has the value $-12/5$ as compared with $36/5$ for the corresponding bracketed sum in (5). Bethe[4] has shown that for a fixed sign of Q, as in a one electron system, the D level requires inversion of Fig. 2 if the F level requires an upright Fig. 1. When the sign of Q changes, there are thus two cancelling inversions which together leave Fig. 2 upright.

To be consistent with our interpretation of Ni and Co, it thus appears[9]

[8] W. J. de Haas and C. J. Gorter, Leiden Comm. 208c.

[9] At this juncture it is perhaps well to remark that the reversal of the sign of Q is perhaps a more general and certain phenomenon for the pairs Ni^{++}, Co^{++} and Fe^{++}, Cu^{++} than for the pair Ni^{++}, Cu^{++}. The reversal for the former pair is contingent only upon (7) and hence takes place for any potential whose diagonal elements vanish when summed over M_L except for an additive constant, such as $3n\overline{r^4}/5$ in (3), which is independent of the type of vector addition. That the additive constant is indeed independent of the vector compounding is shown by the theorem (cf. Niessen, Phys. Rev. **35**, 274 ff., 1929) that $\Sigma_{M_L} F(n_1 n_2 \cdots M_L;\, n_1 n_2 \cdots M_L) = (2L+1) \times \overline{F}(n_1 n_2 \cdots;\, n_1 n_2 \cdots)$ where the matrix elements relate to any function $F(r, \theta, \phi)$ of polar coordinates and $4\pi\overline{F} = \iint F \sin\theta\, d\theta\, d\phi$. The function $\overline{F}$ is a purely radial one and consequently invariant of the vector addition involved in Russell-Saunders coupling etc. Hence the Ni^{++}, Co^{++} inversion is surely found for other forms of potential besides (1), and the great diversity in anisotropy in this pair requires only that the crystal symmetry be dominantly but not completely cubic. On the other hand after Eq. (9) explicit use was made of the numerical form (4) of the matrix elements of the fourth degree part of (1) in proving the reversal in the sign of Q for the pair Ni^{++}, Cu^{++}. It is hence conceivable, though improbable, that if the cubic potential requires large sixth, eighth, etc. degree terms for its representation in addition to the fourth order terms, the reversal in the sign of Q may exist for the Ni^{++}, Co^{++} and Cu^{++}, Fe^{++} pairs, but not for Ni^{++}, Cu^{++}. Fig. 2 would then be upright in Fe^{++} and inverted in Cu^{++}.

necessary that Fig. 2 be upright in Cu^{++} and inverted in Fe^{++}. Now Bethe[4] has demonstrated that the portion Γ_3 of Fig. 2 is "nonmagnetic," and so μ consequently has no matrix elements of the form $\mu(a'b')$, whereas the elements inside Γ_5 do not all vanish. The expression analogous to (2) will consequently be much larger in Fe^{++}, where e' is a ground state, than in Cu^{++}, because of the existence of small denominators such as $\nu(e'c')$. Hence the remains of the orbital angular momentum will be larger in Fe^{++} than Cu^{++}, and, all other things being equal, the distortion from (8) with attendant magnetic anisotropy should be much more accentuated in Fe^{++} than in Cu^{++}. Actually the anisotropy of cupric salts (e.g., 26 percent in $CuK_2(SO_4)_2 \cdot (6H_2O)$ is usually greater than for ferrous ones (e.g., 16 percent in $FeK_2(SO_4)_2 \cdot 6H_2O$), while (8) holds no more closely for Cu^{++} than for Fe^{++}. About 20 percent deviations from (8) are found in both cases. Because of the somewhat greater anisotropy in the cupric case, it thus appears at first sight that the inversion of Fig. 2 comes at the wrong place. However, "all other things" are not equal. The free multiplet structure is wider for Cu than Fe, so that if there were no inversion the spin-orbit distortion would be bigger in Cu than in Fe. Also the spin quantum number has the small value 1/2 in Cu^{++}, compared with 2 in Fe^{++}, so that residual orbital angular momentum of given magnitude is relatively more important in Cu^{++} than in Fe^{++}. These two effects may more than counteract the fact that the denominators in the formula for the induced orbital moment are smaller in Fe^{++} than in Cu^{++}. One thing to be particularly emphasized in connection with the question of anisotropy is the following. The orbital moment induced in a' in Fig. 2 exists, to be sure, solely in virtue of the matrix element $\mu(a'c')$, $\mu(a'd')$, $\mu(a'e')$ rather than $\mu(a'b')$, and hence tends to be small because the frequencies $\nu(a'c')$ etc. are large. However, unlike the case of a in Fig. 1, this moment in a' does not owe its anisotropy purely to departures of the ratios $\kappa = \nu(a'c')/\nu(a'd')$ etc. from unity. Instead the full cubic isotropy of the induced moment in Γ_3 is achieved only when the two components of Γ_3 are equally populated, whereas we suppose the separation $a' - b'$ large compared to kT, so that only a' is inhabited. Consequently, the anisotropy in the induced orbital moment of a', Fig. 2 is of the order of magnitude unity, like that of g in Fig. 1 or e' in Fig. 2, whereas that in a, Fig. 1 is only of the order $\kappa - 1$. By an anisotropy of the order unity we mean that the differences between the x, y, z components are comparable with the components themselves. Of course the anisotropy is in any case diminished because the residual orbital angular momentum is overshadowed by the spin, but we can say that for given deviations from (8), the anistropies should be of the same order of magnitude regardless of whether or not Fig. 2 is inverted, and inversion accentuates the anisotropy only in so far as the deviation from (8) is enhanced, whereas we have seen that in the Fe^{++}, Cu^{++} pair there are other counteracting tendencies which forestall this enhancement. The effect of inversion on anisotropy is thus a different story for Fig. 2 than Fig. 1. It must be further remembered that Fe, Cu are not adjacent in the periodic table, and their salts' crystalline fields need not resemble each other as closely as those of Co, Ni, thus obscuring the purity of the inversion phenomenon. The inversion becomes less important the more one increases the rhombic terms at the expense of the cubic ones.

In order to make Figs. 1, 2 upright at Ni^{++}, Cu^{++} respectively rather than Co^{++}, Fe^{++}, it is necessary that the constant D in (1) be positive. This follows since the quotient of the bracketed sum in Eq. (5) and f (3, 3) is positive, making Q have the same sign in Ni^{++} as for a one electron system, where Bethe[4] shows that Fig. 1 is upright if D is positive. The positive choice of D agrees with Penney and Schlapp's preceding calculations on the susceptibilities of the rare earths, but iron and rare earth salts are so widely different that it seems scarcely necessary that the sign of D be the same in both cases.

Detailed numerical calculations amplifying and testing quantitatively the foregoing ideas will be published by Schlapp and Penney and by Jordahl. The writer wishes to thank them for valuable discussions and comments.

Reprinted from *Phys. Rev.*, **42**, 666–686 (1932)

4

Influence of Crystalline Fields on the Susceptibilities of Salts of Paramagnetic Ions. II. The Iron Group, Especially Ni, Cr and Co

By ROBERT SCHLAPP AND WILLIAM G. PENNEY*
Department of Physics, University of Wisconsin

(Received October 13, 1932)

The present paper is concerned with the calculation of the paramagnetic susceptibility of highly hydrated crystals of the iron group elements Ni, Cr and Co. On the assumption that the metallic ion is subject to a crystalline electric field, predominantly cubic but also with a smaller rhombic term, the Hamiltonian function in a magnetic field $\boldsymbol{H}$ is given by

$$D(x^4+y^4+z^4)+Ax^2+By^2-(A+B)z^2+\lambda(\boldsymbol{L}\cdot\boldsymbol{S})+\beta\boldsymbol{H}\cdot(\boldsymbol{L}+2\boldsymbol{S})$$

the numerical value of λ being known from the work of Laporte but the other constants yet to be determined. It actually proves possible to formulate and solve approximately the resulting secular equations and so obtain the first and second order Zeeman effects and hence the susceptibility. For all three ions $L=3$, so that the orbital problem is the same for all. This problem is exactly soluble, the energy levels consisting of two triplets and a singlet, the singlet not lying between the triplets. The effect of the introduction of the spin and its coupling to the orbit then leads to a determinant of order 21 for Ni and of order 28 for Cr and Co. That for Ni factors into one of order 10 and one of order 11, while those for Cr and Co factor into two determinants, identical except for the sign of the coefficient of H. On the assumption of a cubic field of the same sign and of approximately the same magnitude for all three ions the orbit-spin, together with the rhombic field, is able to remove the degeneracy of the lowest level in Ni and Cr only in a high approximation, while with Co the degeneracy is removed in first approximation. This difference accounts for the isotropy of Ni and Cr compared with the anisotropy of Co. In order to obtain agreement with experiment it is necessary to assume that in Ni the singlet of the orbital problem lies lowest. It then follows from the work of Van Vleck that the singlet also lies lowest for Cr but that for Co the singlet lies highest. When the singlet lies lowest, the square of the magneton number is given by the "spin only" value $4S(S+1)$, together with a small orbital contribution of order λ/D, whose sign can be either positive or negative. Actually it is positive for Ni and negative for Cr. In order to fit the results on the principal susceptibilities of Ni, it is necessary to take $D=1260$ cm^{-1}, $A=176$ cm^{-1}, $B=352$ cm^{-1}, the magnitude of λ being -335 cm^{-1}. For Ni and Cr the theory requires that for the mean susceptibility $\chi=Q+P/T$, where P and Q are constants, Q being uniquely determined when P is fixed. Choosing P so that χT passes through the experimental point at 170°K we find that good agreement is obtained over the whole temperature range. For Cr $\lambda=87$ cm^{-1} and we find $D=3730$ cm^{-1}, but we cannot determine A or B since there are no data on the principal susceptibilities.

Computational difficulties prevent the accurate solution of the Co problem. The situation is complicated by the experimental data not being complete. It proves necessary to consider a sextet which is soluble only numerically in the general case but perturbation theory can be applied when either the orbit-spin is large compared with the rhombic field or *vice-versa*. We obtain fair agreement with experiment and our calculations indicate that good agreement would be obtained in an intermediate case.

* Commonwealth Fund Fellow.

Introduction

IN THE following paper the idea of crystal fields of definite symmetry, developed by Van Vleck and others,[1] and already used in a previous article by the authors,[2] is applied to calculate the susceptibilities of salts of the elements Ni, Co and Cr. There are two respects in which the present problem differs from that of susceptibilities in the rare earth group. In the first place, the incomplete shell which is responsible for the paramagnetism of the iron elements consists of $3d$ electrons, which are much more strongly affected by the crystal fields than the more sheltered $4f$ shell of the rare earths. In the second place the orbit-spin coupling, which determines the multiplet width, is usually smaller in the iron group than in the rare earths. For the latter it was allowable to suppose that each multiplet component underwent a "Stark effect" due to the crystal field, without distortion on account of the other multiplet components. In the iron group, however, the electric field of the crystal is able to break down the relatively weak coupling between orbit and spin, producing an electric Paschen-Back effect; the orbit-spin coupling may be treated hence as a perturbation on an unperturbed problem which neglects the spin. This unperturbed, or orbital problem, as we shall call it, is the same for all three ions Ni^{++} Co^{++} Cr^{++}, since they all have an F state ($L=3$) as ground state.

We assume that the crystal field has no more than rhombic symmetry.[3] The high degree of isotropy of Ni salts suggests that in this case the departure from cubic symmetry is small. Now it is known that a field which is nearly cubic decomposes the seven coincident levels of the F state (without spin) into a single level and two triplets, the single level lying outside the triplets and the triplet widths being small compared with the singlet-triplet or the triplet-triplet separations. If the spin and its coupling to the orbit be included, further decompositions of these levels occur. The general theory of susceptibilities shows that Curie's law will cease to be obeyed at low temperatures if kT becomes comparable with the separation of the lowest group of levels. The close conformity of Ni salts to Curie's law over a range of temperature from 300°K down to 14°K thus requires that a very narrow group of levels must lie considerably below all others. These conditions are satisfied if the single level of the orbital problem lies below the others, and on this assumption it is possible to account qualitatively for both the small anisotropy and the conformity to Curie's law. This arrangement of levels, however, appears to preclude an explanation on the same lines of the much greater anisotropy of the very similar and sometimes isomorphous salts of Co, and of the considerable departures from Curie's law which they exhibit. To ac-

[1] J. H. Van Vleck, *Theory of Electric and Magnetic Susceptibilities*, Oxford (1932).

[2] W. G. Penney and R. Schlapp, Phys. Rev. **41,** 194 (1932). Attention may be called here to a printers error in this paper. Minus signs were omitted in Eqs. (8) and (9) which should read $q=-I/10395$ for (8) and $q=-I/32670$ for (9). Moreover, in the secular determinant for Pr $a=\frac{1}{2}pD$ (not pD) and similarly for Nd $A=6ap(14)^{1/2}$.

[3] The assumption of a rhombic field not predominantly cubic was found to lead to very large asymmetry, in contradiction with experiment.

count for the behavior of Co salts it is necessary to suppose that the levels of the orbital problem in Co are inverted relatively to those in Ni. That such an inversion is actually to be expected in passing from Ni and Cr to Co has been neatly demonstrated by Van Vleck.[4]

Experimental Data

It is useful at this stage to review the experimental data available on the hydrated salts of Ni, Cr and Co. We restrict ourselves to salts of large magnetic dilution, so that exchange effects may be neglected. Determinations of the three principal susceptibilities of the double sulphates of Co with ammonium, potassium and rubidium, have been made by Rabi[5] at 300°K. Jackson[6] has measured the susceptibility of powdered $Ni(NH_4)_2(SO_4)_2 \cdot 6H_2O$ and Gorter, de Haas and v. d. Handel[7] that of powdered $NiSO_4 \cdot 7H_2O$ over a range of temperature between 14°K and 290°K. The graph of $1/\chi$ against T is approximately a straight line through the origin in both cases. Jackson[6] has measured the three principal susceptibilities of $Co(NH_4)_2(SO_4)_2 \cdot 6H_2O$ at various temperatures down to 14°K. His values of the susceptibility extrapolated to a temperature of 300°K differ considerably from Rabi's, and there is only one determination between 20° and 290°K. As far as one can judge, however, the graph of $1/\chi$ against T is a straight line for each of the three principal susceptibilities, down to a temperature of about 50°K, below which the curve bends downwards slightly, so that the susceptibilities are higher than those predicted by the relation $\chi = C/(T+\Delta)$ of Weiss.

Very recently determinations over a temperature range from 250°K to 360°K have been made by Bartlett[8] for crystalline cobalt ammonium sulphate and certain other crystals. They seem to be the most reliable measurements yet taken, being consistent and in agreement with Rabi's at the single temperature used by him. We are indebted to Dr. Bartlett for communicating these results to us in advance of publication.

The susceptibility of potassium chrome alum in powder form has been measured by de Haas and Gorter[9] at various temperatures between 290°K and 14°K. They find that the law $\chi = C/T$ is closely obeyed over the whole range. Chrome alum forms crystals in the cubic system so that it may be expected to be magnetically isotropic.[10]

[4] J. H. Van Vleck, Phys. Rev. **41**, 208 (1932).

[5] I. I. Rabi, Phys. Rev. **29**, 184 (1927).

[6] L. C. Jackson, Phil. Trans. Roy. Soc. London, **224**, 1 (1922), Leiden Com. 163.

[7] C. J. Gorter, W. J. de Haas and v. d. Handel, Proc. Amst. Acad. **34**, 1 (1931), Leiden Com. 218d.

[8] B. W. Bartlett, Phys. Rev. **41**, 818 (1932).

[9] W. J. de Haas and C. J. Gorter, Leiden Com. 208d.

[10] Measurements of the susceptibilities of the paramagnetic cubic crystal pyrite were made long ago by Voigt and Kinoshita (Ann. d. Physik **24**, 492 (1907)) who found it to be magnetically isotropic. There does not, however, seem to be any reason why magnetic dissymmetry should not exist in cubic crystals, as the electric field acting on the ion may have a lower symmetry than the lattice. See also reference **18**.

THE ION IN A PERFECTLY CUBIC FIELD

Before considering the secular determinant explicitly, it is instructive to look at the problem from a more general point of view. The analysis of Bethe[11] shows that the cubic field breaks up the F level into three, corresponding to the irreducible representations Γ_2, Γ_4, Γ_5, of the cubic group, in his notation. The level Γ_5 lies between Γ_2 and Γ_4; Γ_2 is single, while Γ_4 and Γ_5 are each triply degenerate. We shall see later that the intervals between Γ_2, Γ_4, Γ_5 are of the order 10^4 cm^{-1}. The reader is referred to Van Vleck's[4] paper for the demonstration of the fact that in Ni and Cr Γ_2 lies lowest and in Co Γ_4 lies lowest, for a given sign of D in the Hamiltonian. The level Γ_2 is non-magnetic; that is to say an atom in this state has no average orbital magnetic moment. The level Γ_4 is magnetic. Hence if Γ_2 is lowest the orbit is "quenched" i.e., contributes nothing to the susceptibility except a term independent of temperature. If, however, Γ_4 is lowest a certain portion survives.

We have now to consider the influence of the spin. Inclusion of the spin S ($=1$ for Ni and 3/2 for Co and Cr) without interaction with the orbit makes each level of the orbital problem have an additional $(2S+1)$-fold degeneracy, which is partially removed by the interaction. By the methods of Bethe's paper the decomposition of the levels is found by reducing the six direct products $\Gamma_i D_k$, ($i=2, 4, 5$; $k=1, 3/2$) to represent the cubic group. Here D_k is the representation group for the rotation of the spin k alone. The result is, in Bethe's notation,

$$\begin{aligned} \Gamma_2 D_1 &= \Gamma_5, & \Gamma_2 D_{3/2} &= \Gamma_8 \\ \Gamma_4 D_1 &= \Gamma_2 + \Gamma_3 + \Gamma_4 + \Gamma_5, & \Gamma_4 D_{3/2} &= \Gamma_6 + \Gamma_7 + 2\Gamma_8, \\ \Gamma_5 D_1 &= \Gamma_1 + \Gamma_3 + \Gamma_4 + \Gamma_5, & \Gamma_5 D_{3/2} &= \Gamma_6 + \Gamma_7 + 2\Gamma_8. \end{aligned} \tag{1}$$

Here Γ_6, Γ_7, Γ_8 are the "zweideutig" representations of the cubic group, of dimensions 2, 2, 4, respectively, which always arise with half-integral quantum numbers. These equations state that for Ni, Cr and Co the orbit-spin-interaction does not split the cubic level Γ_2 but splits each of the levels Γ_4, Γ_5 into four components.

Let us suppose that the level Γ_2 of the orbital problem lies lowest. The above reductions show that under the orbit-spin interaction this level does not break up, but remains triply degenerate (in Ni) or quadruply degenerate (in Co and Cr); no energy differences arise in consequence of different orientations of the spin, which therefore remains entirely free at all temperatures to orientate itself along the magnetic field. If the orbital contribution to the moment be neglected, the magneton number would be the Bose-Stoner or "spin only" value $[4S(S+1)]^{1/2}$. A further deduction is that the orbit-spin interaction causes the state Γ_2 to interact with components of Γ_4 and Γ_5 as is seen from the threefold occurrence of Γ_5 or Γ_8 on the right-hand side of (1). This produces a sharing of properties, and in particular gives rise to an orbital contribution to the magnetic moment in the state Γ_2 which is of order

[11] H. Bethe, Ann. d. Physik 3, 133 (1929).

λ/D. Thus the orbit-spin coupling produces, in an ion in a cubic field, departures from the Bose-Stoner value which may be either positive or negative according to the sign of λ/D.

The circumstances are not quite so simple if the state Γ_4 lies lowest. Here the orbit-spin coupling partially removes the degeneracy, so that different orientations of the spin have different energies, although, of course, it is not possible to associate a definite axial quantization of the spin with each of the levels. Thus the spin is only partially free and the orbital contribution will also be modified.

The Constant Δ

In a cubic field the quantity Δ of the Curie-Weiss formula $\chi = C/(T+\Delta)$ is given to a first approximation by the ratio of the coefficients of $1/T$ and $1/T^2$ in the expansion of the susceptibility in inverse powers of T. If we make the usual assumption that the magnetic moment in the absence of the magnetic field contains, besides low-frequency elements $M(n, n')$, only high-frequency elements, and none of intermediate frequency, it is easily shown[12] that

$$k\Delta = \left\{ \sum_{nn'} W_n \,|\, M(n, n') \,|^2 / \sum |\, M(n, n') \,|^2 \right\} - \overline{W}$$

the summation being over the group of levels connected by low-frequency elements, and $\overline{W}$ being their mean energy. When the level Γ_2 of the orbital problem, which is not split up by the orbit-spin interaction, lies lowest, the magnetic mean center and the energetic mean center, whose difference gives $k\Delta$ according to the last equation, necessarily coincide. Hence in this case the susceptibility is of the form $\chi = C/T$, correct to terms in $1/T^2$. To this approximation the ion in a cubic field behaves as if it were in the gaseous state.

If the level Γ_4 lies lowest, the magnetic mean center and the energetic mean center do not necessarily coincide, so that the susceptibility will in general have a term in $1/T^2$. Thus leaving aside the question of asymmetry produced by a rhombic field, which is considered in the next section, we should expect Ni and Cr to conform much more closely to Curie's law than Co, as is indeed found to be the case.

Asymmetry Due to a Rhombic Field

Let us for the moment neglect the spin. The effect of superposing a rhombic field on the cubic field is, as shown by Bethe,[11] to remove all the degeneracy in the orbital problem, the appropriate reduction being

$$\Gamma_2 = G_1, \quad \Gamma_4 = G_2 + G_3 + G_4, \quad \Gamma_5 = G_2 + G_3 + G_4, \tag{2}$$

where G_1, G_2, G_3, G_4 are the four one-dimensional representations of the rhombic group. Fig. 1 shows diagrammatically the decomposition of the levels under the various fields. The level Γ_2 is seen to be completely isolated

[12] C. J. Gorter, Arch. Musee Teyler, 7 (3), 183 (1932). This formula can readily be obtained from the equation for $k\Delta$ on page 197, reference 2, $\overline{W}$ in this case having been chosen to be zero.

from the others. The rhombic field alone, unlike the orbit-spin coupling, does not lead to a sharing of properties between Γ_2 and the other states. If Γ_2 is lowest the rhombic field does not give rise to any orbital contribution to the part of the susceptibility depending on the temperature. The part independent of the temperature is rendered slightly asymmetrical. Although no asymmetry is introduced directly by the rhombic field, the orbit-spin interaction, as we have seen, evokes an orbital contribution to the susceptibility, and this will be rendered anisotropic by the rhombic field. The anisotropy is thus a second order effect; the rhombic field may be comparatively large without producing much anisotropy. Neither the rhombic field alone nor the orbit-

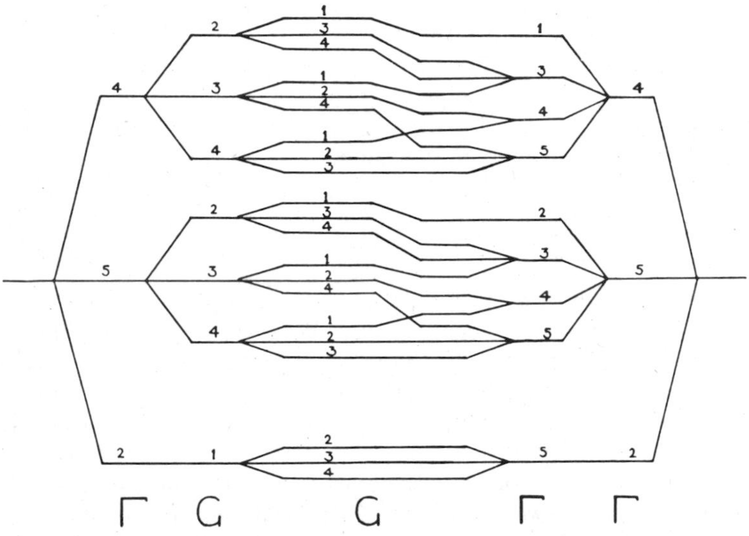

Fig. 1. Fig. 1 shows how the ^3F state in nickel is affected by the cubic field, the rhombic field and the orbit-spin coupling. By starting at the left with the free ion, the cubic field splits the single level into three, the numbers and symbols underneath denoting to which representation, in Bethe's notation, the levels belong. The application of the rhombic field splits Γ_4 and Γ_5 each into three, leaving Γ_2 single. The addition of the orbit-spin then removes all the remaining degeneracy. Making the rhombic field zero leaves only orbit-spin and cubic, the way the levels come together being shown. Removal of the orbit-spin leaves only the cubic field, and making this shrink to zero gives once again the free ion.

spin interaction alone can split the level Γ_2 of the cubic field. Both acting together will remove all degeneracy, but the separation produced will depend on a cross term and in addition is a second order effect in λ owing to the vanishing of the mean orbital angular momentum. If the level Γ_4 is lowest, these conclusions do not hold, both the anisotropy and Δ being first order effects of the rhombic field.

If Γ_2 lies lowest, as in Ni and Cr, it is possible to prove a result very similar to that which was shown to hold for the rare earths, namely that the expansion of χ for a crystal powder should contain no term in $1/T^2$. The Hamiltonian inclusive of the magnetic field is invariant for a half turn about the magnetic field. The group consisting of this operation and the identical operation has two one-dimensional (eindeutig) representations, associated respectively with the odd and even series of values of the quantum number M. (M is not a good quantum number when the electric field and $(L\cdot S)$ are diagonalized.) It is clear that of the three constituent levels ϵ_1, ϵ_2, ϵ_3 of Γ_2, two, (ϵ_1, ϵ_2) belong to one representation, and the third (ϵ_3) to the other. The third level has therefore no magnetic or other connection with the other two, which are linked to each other. In the presence of the magnetic field, the level ϵ_3 is unaltered, while ϵ_1, ϵ_2 undergo equal and opposite displacements away from each other. This neglects the high-frequency shift, which is practically the same for all three levels. The value of $k\Delta$ is therefore $(\epsilon_1+\epsilon_2+\epsilon_3)/3-(\epsilon_1+\epsilon_2)/2=(2\epsilon_3-\epsilon_1-\epsilon_2)/6$, an expression which is of course invariant of the origin of energy. On permuting the axes the "field free" levels ϵ_1, ϵ_2, ϵ_3 undergo a corresponding permutation; if we average the above expressions for $k\Delta$ over three cyclic permutations, which corresponds to finding the value of $k\Delta$ for a crystal powder, the result is seen to vanish.

We now turn our attention to the secular determinant of the problem in order to give the considerations of the foregoing sections a more quantitative form.

The Secular Determinant

The Hamiltonian

We assume a Hamiltonian function

$$\sum_i [D(x_i^4 + y_i^4 + z_i^4) + Ax_i^2 + By_i^2 - (A + B)z_i^2] + \lambda(\boldsymbol{L}\cdot\boldsymbol{S}) + \beta\boldsymbol{H}\cdot(\boldsymbol{L} + 2\boldsymbol{S}) \quad (3)$$

where A, B, D are constants specifying the crystal field, λ is the constant of the orbit-spin interaction, β is the Bohr magneton $eh/4\pi mc$ and $\boldsymbol{H}$ is the magnetic field.

In the Hamiltonian (3) the dominant term is the term in D. The most general field of rhombic symmetry which is nearly cubic would give a Hamiltonian containing other terms in addition to those written down. These would be of higher order and no greater generality would be obtained by their inclusion. The rhombic term and the orbit-spin coupling are of comparable magnitude. The magnetic energy may always be regarded as a small perturbation in calculating the susceptibility, even though the field is strong enough to produce a Paschen-Back effect, always provided that the magnetic separations do not become comparable with kT so that saturation effects occur. This follows from an application of the principle of spectroscopic stability due to Van Vleck.[13]

The matrix elements

The following matrix elements in the (M_L, M_S) system of quantization are required. They have been obtained by the method used in the previous paper.[2]

$$D\sum_i(x_i^4 + y_i^4 + z_i^4)(M_L, M_L)$$

$$= \text{const.} + q'DM_L^2[7M_L^2 + 5 - 6L(L + 1)],$$

$$D\sum(x_i^4 + y_i^4 + z_i^4)(M_L, M_L \pm 4)$$

$$= \tfrac{1}{2}q'D[(L \mp M_L)!(L \pm M_L + 4)!/(L \pm M_L)!(L \mp M_L - 4)!]^{1/2},$$

[13] Reference 1, page 231.

$$\sum[Ax_i^2 + By_i^2 - (A+B)z_i^2](M_L, M_L)$$
$$= \text{const.} + \tfrac{1}{2}a(A+B)[3M_L^2 - L(L+1)],$$
$$\sum[Ax_i^2 + By_i^2 - (A+B)z_i^2](M_L, M_L \pm 2)$$
$$= -a(A-B)[(L\mp M_L)!(L\pm M_L+2)!/(L\mp M_L-2)!(L\pm M_L)!]^{1/2}/4.$$

Here q and a are the ratios of the matrix elements calculated for a system of n electrons to those for a one-electron system; the sign of q has been discussed by Van Vleck.[4] The summation is over all the electrons of the incomplete group. The two additive constants, as well as q and a, are independent of M_L. These matrix elements are all diagonal in L. The elements non-diagonal in L are not required, since the crystal field is assumed not to destroy the Russell-Saunders coupling of the vectors $\boldsymbol{l}_i$, $\mathbf{s}_i$. Since the elements do not involve the spin, they may be regarded as diagonal in M_S. We also require

$$(\boldsymbol{L}\cdot\boldsymbol{S})(M_L, M_S; M_L, M_S) = M_L M_S,$$
$$(\boldsymbol{L}\cdot\boldsymbol{S})(M_L, M_S; M_L\pm 1, M_S\mp 1)$$
$$= \tfrac{1}{2}[(S\pm M_S)(S\mp M_S+1)(L\mp M_L)(L\pm M_L+1)]^{1/2}.$$

The secular determinant

The orbital problem, being common to all three ions Ni, Co, Cr, may be treated first. The orbital terms of the Hamiltonian are all of type $\Delta M_L = 0, \pm 2, \pm 4$, so that the secular determinant

$$\mathcal{H}(M_L, M_S; M_L', M_S') - \delta(M_L, M_S; M_L', M_S')W = 0,$$

breaks up into two factors, one of the fourth order involving $M_L = \pm 3, \pm 1$, and one of the third order involving $M_L = 0, \pm 2$. These factors are symmetrical about the principal and secondary diagonals, so that it is necessary to write down only the first row and central elements of the second row of each

$$\begin{vmatrix} -3Dq+15\sigma & -15^{1/2}\delta & 15^{1/2}Dq & 0 \\ & -5Dq-9\sigma & -6\delta & \end{vmatrix}, \quad \begin{vmatrix} -13Dq & -30^{1/2}\delta & 5Dq \\ & -12\sigma & \end{vmatrix}$$

Here σ, δ have been written for $a(A\pm B)/2$ and $q = 12q'$. The terms in D can be diagonalized[14] by means of unitary transformations SS^{-1}, TT^{-1} with

$$S = (\tfrac{1}{4})\begin{pmatrix} h & k & -k & -h \\ -k & h & -h & k \\ k & h & h & k \\ h & -k & -k & h \end{pmatrix} \quad \begin{matrix} h = 5^{1/2} \\ k = 3^{1/2} \end{matrix} \quad T = (\tfrac{1}{2})^{1/2}\begin{pmatrix} 1 & 0 & 1 \\ 0 & 2^{1/2} & 0 \\ -1 & 0 & 1 \end{pmatrix}$$

We may denote the matrices S, T by $S(N, M_L)$, $T(N, M_L)$. The columns are numbered by M_L having values $-3, -1, 1, 3$, for S and $-2, 0, 2$ for T. The

[14] The wave functions $S\psi$, $T\psi$ which the transformations S, T introduce are precisely those given by Bethe (reference 11, page 166). They diagonalize the cubic and rhombic fields except for matrix elements of the rhombic field between different cubic levels.

rows are numbered by a new "cubic" quantum number N which may be supposed to take on the same set of values as M_L. N has no obvious physical meaning, but served to identify the roots W_N in the cubic field according to the scheme

$$W_{-3} = W_0 = W_3 = 0, \; W_{-2} = W_{-1} = W_1 = -8Dq, \; W_2 = -18Dq.$$

The relations (2) show that the transformation which diagonalizes the terms in D will factorize the orbital problem into three quadratics and a singlet. These are

$$\begin{vmatrix} 6(\sigma-\delta) & -(15)^{1/2}(3\sigma+\delta) \\ -(15)^{1/2}(3\sigma+\delta) & -8Dq \end{vmatrix} \quad \begin{vmatrix} -8Dq & (15)^{1/2}(3\sigma-\delta) \\ (15)^{1/2}(3\sigma-\delta) & 6(\sigma+\delta) \end{vmatrix}$$

$$\begin{vmatrix} -8Dq & -2(15)^{1/2}\delta \\ -2(15)^{1/2}\delta & -12\sigma \end{vmatrix} \qquad -18Dq$$

We denote the roots of these by (r_{-3}, r_{-1}), (r_1, r_3), (r_{-2}, r_0) and r_2, where the suffix is a new "rhombic" quantum number taking on the same values as N or M_L. This set of roots is, of course, invariant if the coefficients of the rhombic field be permuted cyclically, $A \rightarrow B \rightarrow -(A+B)$ but they undergo the cyclic permutation $r_{-1} \rightarrow r_{-2} \rightarrow r_1$, $r_{-3} \rightarrow r_0 \rightarrow r_3$, while r_2 is invariant. Let the transformation matrices which diagonalize these quadratics for given values of D, A and B be

$$\begin{array}{c|cc} & -3 & -1 \\ \hline -3 & r & s \\ -1 & -s & r \end{array} \qquad \begin{array}{c|cc} & 1 & 3 \\ \hline 1 & t & u \\ 3 & -u & t \end{array} \qquad \begin{array}{c|cc} & -2 & 0 \\ \hline -2 & p & q \\ 0 & -q & p \end{array} \qquad \begin{array}{c|c} & 2 \\ \hline 2 & 1 \end{array} \tag{4}$$

The values of the elements can be calculated for any given values of D, A and B. The columns are numbered by the "cubic" quantum number N and the rows by the "rhombic" quantum number Q. We now introduce the spin and have to differentiate between Ni, Cr and Co.

Nickel

Mathematical theory

For nickel $S=1$, so that $M_S = -1, 0, 1$. The secular determinant, of order 21, breaks up into one of the tenth and one of the eleventh order, involving, respectively, even and odd values of $M = M_L + M_S$. This follows since the complete Hamiltonian contains only terms of the type $\Delta M = 0, \pm 2, \pm 4$. We are interested primarily in the root $-18Dq$ which lies below the others. It occurs once (with $M_S = 0$) in the eleventh order determinant, and twice ($M_S = \pm 1$) in the tenth order determinant. Transforming the eleventh order determinant to the (N, M_S) representation, we require the element

$$\mathcal{H}(2, 0; N', M_S') = T(2, 0; M_L, 0)\,\mathcal{H}(M_L, 0; M_L', M_S')R^{-1}(M_L', M_S'; N', M_S'),$$

where R stands for S if $M_S' = \pm 1$, and for T if $M_S' = 0$. We have included the quantum number M_S in T and R as though they were diagonal in M_S. They are indeed independent of M_S. In the N, M_S representation the only nonvanishing elements of the orbit-spin and magnetic energies are found to be

$$\mathcal{H}(2,0;1,1) = \mathcal{H}(2,0;-1,1) = \mathcal{H}(2,0;1,-1) = -\mathcal{H}(2,0;-1,-1) = (2)^{1/2}\lambda$$
$$\mathcal{H}(2,0;-2,0) = 2\omega.$$

We know from (1) that the diagonalization of the orbit-spin terms involves the solution of cubic equations, so that it is simpler to diagonalize rhombic field terms instead. This is accomplished by the matrices (4). The relevant matrix elements in the (Q, M_S) system of representation are

$$\mathcal{H}(2,0;-3,-1) = \mathcal{H}(2,0;-3,1) = (2)^{1/2}\lambda S,$$
$$\mathcal{H}(2,0;-1,-1) = \mathcal{H}(2,0;-1,1) = (2)^{1/2}\lambda r,$$
$$\mathcal{H}(2,0;1,-1) = \mathcal{H}(2,0;1,1) = (2)^{1/2}\lambda t,$$
$$\mathcal{H}(2,0;3,-1) = \mathcal{H}(2,0;3,1) = -(2)^{1/2}\lambda\omega,$$
$$\mathcal{H}(2,0;-2,0) = 2\omega p, \mathcal{H}(2,0;0,0) = -2\omega q,$$

from which the first approximation to the energy can be found. The tenth order determinant is not quite so simple, for the root $-18Dq$ occurs twice, with $M_S = \pm 1$, the degeneracy not being removed by the rhombic field. Suppose the Hamiltonian has been transformed to the (Q, M_S) system, i.e., to the form $\mathcal{H}_0 + \lambda_0\mathcal{H}_1 + \omega_0\mathcal{H}_2$, where $\mathcal{H}_0$ is diagonal and $\mathcal{H}_1$ has no diagonal terms. Apply the transformation

$$(1 + \lambda S)(\mathcal{H}_0 + \lambda\mathcal{H}_1 + \omega\mathcal{H}_2)(1 - \lambda S + \lambda^2 S^2 + \cdots)$$

and choose S so as to make the coefficient of λ vanish in this expression. Then the Hamiltonian becomes

$$\mathcal{H}(n,m) = \mathcal{H}_0(n,m) + \lambda^2 \sum \mathcal{H}_1(n,i)\mathcal{H}_1(i,m)/h\nu(n,i) + \omega\mathcal{H}_2(n,m)$$
$$+ \lambda\omega \sum [\mathcal{H}_1(n,i)\mathcal{H}_2(i,m)/h\nu(n,i) - \mathcal{H}_2(n,i)\mathcal{H}_1(i,m)/h\nu(i,m)]$$
$$+ \cdots.$$

There are now terms in λ^2 on the diagonal which remove the degeneracy[15] and the coefficient of ω is altered by a term of order $\lambda|D$. We can now set up the quadratic secular problem connected with the two coincident roots, and solve it on the assumption that the magnetic field is small. The two resulting values of W,[16] together with that obtained from the eleventh order determinant, are given below.

$$-18Dq + 4\lambda^2(\alpha_1 + \alpha_2) + \omega^2(1 + 8\lambda\alpha_1)/\lambda^2(\alpha_2 - \alpha_3) + 4\omega^2\alpha_1,$$
$$-18Dq + 4\lambda^2(\alpha_1 + \alpha_3) - \omega^2(1 + 8\lambda\alpha_1)/\lambda^2(\alpha_2 - \alpha_3) + 4\omega^2\alpha_1,$$
$$-18Dq + 4\lambda^2(\alpha_2 + \alpha_3) \qquad + 4\omega^2\alpha_1.$$

Here

$$\alpha_1 = p^2/(r_2 - r_{-2}) + q^2/(r_2 - r_0),$$
$$\alpha_2 = r^2/(r_2 - r_{-1}) + s^2/(r_2 - r_3),$$
$$\alpha_3 = t^2/(r_2 - r_1) + u^2/(r_2 - r_3).$$

15 J. H. Van Vleck, Phys. Rev. **33**, 467 (1929).

16 There are actually first order terms in the magnetic field of order (λ^2/cubic sepn.)^2 but the contribution of these to the susceptibility is so small that they can be completely neglected.

If the axes undergo a permutation represented by $A \to B \to -(A+B)$ it is readily verified that the roots r undergo a corresponding permutation, and $\alpha_1 \to \alpha_2 \to \alpha_3$. The susceptibility along the z axis is found from the general formula

$$\chi = -(N/H) \sum (\partial W/\partial H) e^{-W/kT} / \sum e^{-W/kT}.$$

On the assumption that the exponentials can be expanded, this gives

$$\chi_1 = (8N\beta^2/3kT)[1 + 8\lambda\alpha_1 + \theta_1/kT + \cdots] - 8N\beta^2\alpha_1, \tag{5A}$$

where terms in $1/T^3$ and above have been discarded, and

$$\theta_1 = 2\lambda^2(\alpha_2 + \alpha_3 - 2\alpha_1)/3.$$

The term independent of temperature arises as usual from the term in H^2 in the energy. The other two principal susceptibilities are obtained by permuting the axes cyclically, so that the mean of the three principal susceptibilities is

$$\bar{\chi} = (8N\beta^2/3kT)[1 + (8\lambda/3 - kT)(\alpha_1 + \alpha_2 + \alpha_3)], \tag{5B}$$

in which there is rigorously no term in $1/T^2$.

If the crystal field is assumed to have cubic symmetry, the lowest level has a first order effect, and the susceptibility is

$$\chi = (8N\beta^2/3kT)(1 - 4\lambda/5D) + 4N\beta^2/5D. \tag{5C}$$

Comparison with experiment

In the Hamiltonian the constant λ of the orbit-spin coupling is known, while Aa, Ba, Dq are to be determined from the observed susceptibilities. A measurement of the mean susceptibility at one temperature will enable us to determine the one parameter D if we assume as an approximation that Aa and Ba vanish. The assumption of a purely cubic field is a convenient approximation in estimating the order of magnitude of D. We shall consider later the effect of the rhombic field.

In Ni the multiplet is inverted; its over-all width is given by $|\Delta\nu| = \lambda S(2L+1)$. By using the value 2347 cm^{-1} given by Laporte[17] for $\Delta\nu$ we obtain $\lambda = -335$ cm^{-1}. The observed value 26.56×10^{-6} of the susceptibility at 170°K,[7] giving $\chi T = 45.15 \times 10^{-4}$, then leads to a value of Dq from Eq. (5) equal to 1485 cm^{-1}, which corresponds to an over-all separation due to the cubic field of the order of 3 volts. Thus according to (5) the graph of χT against T is a straight line which we have chosen to pass through the experimental point at 170°K, and which cuts the χT axis at $\chi T = 43.64 \times 10^{-4}$. If we had calculated Dq from experimental points at different temperatures, slightly different values would have been obtained. In Fig. 2 we have plotted the experimental values of χT obtained by Gorter, de Haas, and van den Handel,[7] using T as abscissa. It is seen that the experimental points, with the

[17] O. Laporte, Zeits. f. Physik **47**, 761 (1928).

exception of those at low temperatures, lie fairly close to the theoretical curve. The relation proposed by Gorter, de Haas, and van den Handel is $\chi(T+3) = \text{const.}$ In Fig. 2 we have also plotted the experimental values of $\Delta(T+3)$ as a function of T; the approximation of this function to a constant is seen to be very poor indeed. We conclude that the observations are represented much more closely by $\chi T = \text{const.}$ than by $\chi(T+3) = \text{const.}$ It should be remembered that this method of plotting the experimental data, which is equivalent to plotting the square of the effective magneton number as a function of T, is a much more severe test than plotting $1/\chi$ against T, as is usually done.

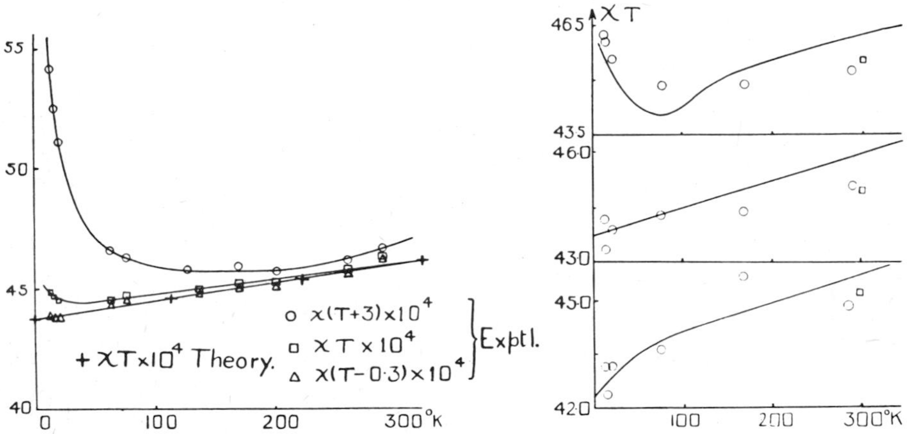

Fig. 2A. Fig. 2B.

Fig. 2. The curves in Fig. 2A serve a double purpose. In the first place they show that a formula of the type $\chi T = A + BT$ demanded by theory for Ni, represents the experimental points better than the curve $\chi(T+3\cdot 0) = \text{const.}$ given by Gorter, de Haas and v. d. Handel.[7] In the second place it is seen that the experimental values of $\chi(T-0\cdot 3) = \text{const.}$ fit the theoretical curve for χT exactly. The addition of -0.3 to T is a minor effect and perhaps represents the result of exchange forces not envisaged in the crystalline potential we have assumed. The one arbitrary parameter in the theory has been chosen to fit the observed magneton number at 170°K. The slope of the χT curve is then uniquely determined and its agreement with experiment is a good confirmation of the theory. Fig. 2B shows the theoretical principal susceptibilities χT of Ni plotted against T, and the experimental points of Jackson corrected by Gorter, de Haas and v. d. Handel,[7] which are marked in circles (see text).

There still remains the deviation of the three low temperature points from the theoretical curve. The present calculation takes no account of intermolecular actions, such as exchange effects, which are not describable by an electric field of definite symmetry. These effects are known to be capable of giving rise to a term Δ which will however be small on account of the high dilution of the salt. This correction will be important at low temperatures. If we plot $\chi(T-0.3)$ against T, the experimental points are brought to lie much more nearly on a straight line. It may be that the term $\Delta = 0.3°$ is a measure of inter-molecular actions other than the crystal fields here contemplated,

We must now consider the effect of the rhombic field in producing asymmetry in the principal susceptibilities. Using Rabi's[5] values for $NiSO_4(NH_4)_2SO_4 \cdot 6H_2O$ and equating them to the three expressions obtained by permuting cyclically the indices in (5A), we obtain three simultaneous equations for α_1, α_2, α_3, whose solution is

$$\alpha_1 = -7.94 \times 10^{-5}, \ \alpha_2 = -8.12 \times 10^{-5}, \ \alpha_3 = -8.29 \times 10^{-5}.$$

From these we have to determine the three parameters Dq, Aa and Ba which specify the crystal field. To set up and solve the algebraic equations connecting these parameters with the α's would be very lengthy, so we have to recourse to the method of trial and error. Thus we find that a crystal field having the Hamiltonian

$$1260(x^4 + y^4 + z^4) + 176(x^2 + 2y^2 - 3z^2)$$

gives

$$\alpha_1 = -7.96 \times 10^{-5}, \ \alpha_2 = -8.13 \times 10^{-5}, \ \alpha_3 = -8.24 \times 10^{-5}.$$

Better agreement could be obtained by using a slightly larger value of the constant of the rhombic field and by changing the ratios of the coefficients of the rhombic field, but it is not, perhaps, worth while pursuing numerical accuracy when the experimental precision is not very high. It is instructive, particularly for comparison with cobalt, to observe how little dissymmetry is produced by a comparatively large rhombic term. Thus the field given above produces an over-all separation of the cubic level Γ_4 amounting to about one-half the interval separating this cubic level from the level Γ_5. The separation produced in the level Γ_2 has an over-all width of 1.5 cm^{-1} so that the individual values of Δ for the three axes are almost negligible, and the expansions of the exponentials which we have used is legitimate even at liquid hydrogen temperatures. At extremely low temperatures Δ is relatively important and it is this fact that accounts for the different behavior of the principal χT for small values of T, shown in Fig. 2B.

Gorter, de Haas and v. d. Handel[7] have given values of the principal susceptibilities of $Ni(SO_4) \cdot 7H_2O$ using their own values of the mean susceptibility together with the differences in the principal susceptibilities found by Jackson.[6] The exactitude of these values is open to question, but to illustrate how they check with the theory we have plotted the experimental values (shown by circles) and the theoretical curves *using for the constants of the crystal field those values found for* $Ni(SO_4)_2(NH_4)_2 \cdot 6H_2O$. The agreement is very good, considering the sensitivity of the method of plotting the results, and the experimental results confirm the existence of a $1/T^2$ term for the individual axes although there is none in the mean. The values found by Rabi[5] on the ammonium salt are marked by squares. Since, as far as we can tell, the constants of the crystal field acting on the Ni ion are exactly equal in the two salts, it seems likely that the ions surrounding the Ni ions are the same and in the same relative positions in the two salts.

CHROMIUM

Chrome alum, whose susceptibility at temperatures down to that of liquid helium has been measured by de Haas and Gorter,[9] forms cubic crystals, so that no differences in the principal susceptibilities are to be expected.[18] It will be sufficient to suppose the crystal field to have cubic symmetry. Van Vleck[14] has shown that for Cr, whose ground state is ^4F, the matrix elements of $\sum_i(x_i^4+y_i^4+z_i^4)$ have the same sign as those of $(x^4+y^4+z^4)$ calculated for a single electron system; that is the coefficient q is positive. The root $-18Dq$ (Γ_2) is accordingly lowest. The secular problem inclusive of spin is of order 28; but on account of the selection rule $\Delta M=0, \pm 2, \pm 4$, obeyed by the Hamiltonian ($\Delta M=0, \pm 4$ if the rhombic field is absent) the secular determinant breaks up into the product of two, which are identical except as regards the sign of the terms in II. This is, of course, an example of the Kramers degeneracy.[19] Reference to the diagram in Bethe[11] shows that the orbit-spin interaction is incapable of removing the degeneracy of Γ_2 in any approximation, so that we need consider only the terms in II, which give rise to first and second order Zeeman effects. Fixing our attention on one of the two secular determinants of order 14, we observe first of all that the root Γ_2 occurs twice ($N=2$, with $M_S=-\frac{1}{2}, \frac{3}{2}$, say). In passing from the original M_L, M_S representation to that in which the cubic field is diagonal, the spin terms $2M_S$ will of course remain on the diagonal, so that the two occurrences of the root Γ_2 have first order moments 3ω and $-\omega$ from this determinant and -3ω and ω from the other. If this were all, the magneton number would have the "spin only" value $(15)^{1/2}$, verifying that the spin is free. But we have still to consider the off-diagonal terms involving ω, which represent the contribution of the orbit, and which are diagonal in M_S. We readily find for the elements satisfying this condition

$$\mathcal{H}(2, -\tfrac{1}{2}; -2, -\tfrac{1}{2}) = -\lambda + 2\omega$$

$$\mathcal{H}(2, 3/2; -2, 3/2) = 3\lambda + 2\omega$$

which gives for the levels in the presence of the field

$$\mp\omega(1-2\lambda/5Dq) - 2\omega^2/5Dq,$$

$$\pm 3\omega(1-2\lambda/5Dq) - 2\omega^2/5Dq.$$

Disregarding the high-frequency term for the moment, we obtain for the susceptibility

$$\chi = (15N\beta^2/3kT)(1-2\lambda/5Dq)^2.$$

[18] The chrome alum $KCr(SeO_4)_2 \cdot 12H_2O$, which forms cubic crystals has in its absorption spectrum a narrow doublet whose separation is roughly 4 cm^{-1} (cf. K. Schnetzler, Ann. d. Physik **10**, 373 (1931)). If this doublet is due to the doubling of the basic level Γ_2, the Hamiltonian must contain non-cubic terms, besides the predominant cubic terms. This follows since the orbit-spin coupling does not decompose Γ_2. If this is so, this alum should exhibit slight asymmetry in its principal susceptibilities.

[19] H. A. Kramers, Proc. Amst. Acad. **33**, 959 (1930).

Here λ is positive, so that the magneton number should be less than the "spin-only" value, which is actually found to be the case. From the value 912 cm^{-1} given by Laporte[17] for the over-all separation of the ^4F multiplet in chromium, we deduce $\lambda = 87$ cm^{-1} and taking de Haas and Gorter's[9] value 19.02 for the Weiss magneton number of Cr, we obtain $Dq = 3730$ cm^{-1}. This justifies our neglect of the high-frequency term, which is proportional to λ/D. The experimental results do not show any trace of high-frequency effects. It is not possible to place much reliance on the above estimate of the magnitude of the separation due to the cubic field, since a small change in the experimental magneton number would produce a very considerable change in the calculated value of Dq. It need scarcely be pointed out that for Cr, as for Ni, the introduction of even a large rhombic field will not appreciably affect the isotropy of the susceptibility. As yet, however, no measurements have been made on the principal susceptibilities of Cr salts.

Cobalt

Mathematical theory

It may be stated here that our calculations on Co are not as complete as those on Ni and Cr, but the difficulties are only in the numerical computation. There seems to be no doubt, however, that good agreement with experiment could be obtained by a more exhaustive trial-and-error procedure. The ground state of cobalt is ^4F, and the secular determinant is of order 28. On account of the Kramer's degeneracy, it breaks up into two determinants of order 14, identical except for the sign of the terms in the magnetic field. The orbital part of the problem is the same as for nickel, where the ground state was also an F state, so that in a cubic field the roots are 0, $-8Dq$, $-18Dq$. On account of the inversion discussed above, the level 0 (denoted Γ_4 above) is now lowest, and occurs six times in each secular determinant of order 14, namely, with $N = -3, 0, 3$ and $M_S = \frac{3}{2}, \frac{1}{2}$ or $-\frac{3}{2}, -\frac{1}{2}$. The portion of the determinant involving these roots, which coincide in the absence of a rhombic field and orbit-spin coupling, is

$$\begin{matrix} 6B - 3l\omega & -3m + 3\omega/2 & 0 & 0 & n & 0 \\ -3m + 3\omega/2 & 6A - 3l\omega & 0 & 0 & -n & 0 \\ 0 & 0 & 6B + l\omega & m + 3\omega/2 & 2m & n \\ 0 & 0 & m + 3\omega/2 & 6A + l\omega & 2m & -n \\ n & -n & 2m & 2m & 6C - \omega & 0 \\ 0 & 0 & n & -n & 0 & 6C + 3\omega \end{matrix} \qquad (6)$$

Here $l = 1 + 15\lambda/32D$, $m = 3\lambda/4$, $n = 3(3)^{1/2}\lambda/4$. Interaction between the levels $0, -8Dq, -18Dq$ has been taken account of with sufficient accuracy by the diagonal terms in $\omega\lambda/D$. This amounts to discarding the high-frequency part of the susceptibility; if we do this we may restrict ourselves to this sixth order determinant in calculating the levels. As the sextic secular equation is not soluble when the rhombic field is comparable with the orbit-spin coupling, we must assume that the former is much smaller than the latter or *vice-versa.* Our calculations indicate that the two influences are in fact of comparable magnitude, which makes close numerical agreement difficult to obtain without more elaborate computations. Here we deal with the two extreme cases only.

Orbit-spin greater than rhombic field

Consider first the case where the orbit-spin coupling is greater than the rhombic field. The orbit-spin interaction alone is capable of partially removing the degeneracy in (6). When we do not restrict ourselves to interactions within the sextet the degeneracy which survives the cubic field and the orbit-spin interaction is given by the resolution $\Gamma_4 D_{3/2} = \Gamma_6 + \Gamma_7 + 2\Gamma_8$, i.e., two singlets and a quadratic occurring twice. When we restrict ourselves to interactions within the sextet the degeneracy must at least be as great as this, which ensures that the sextet will have simple roots when σ, δ, ω all vanish. These are readily found to be $15\lambda/4$, $3\lambda/2$ (twice), $-9\lambda/4$ (three times), and the corresponding form of the sextet, with these roots on the diagonal, can easily be written down. The energy levels in the presence of the magnetic field have now to be found on the assumption that the rhombic field is small compared with the orbit-spin interaction. In cobalt $\lambda = -180$ cm^{-1} so that the triply degenerate level $-9\lambda/4$ will have such a small Boltzmann factor that its contribution to the susceptibility may be neglected even at room temperatures; this level does not affect the moment of the level $15\lambda/4$, and in calculating its influence on the moment of the levels $3\lambda/2$, which is relatively less important in any case, we may suppose it to remain undecomposed by the rhombic field. But in obtaining the moments of the two levels $3\lambda/2$, it is necessary to allow the rhombic field to remove this degeneracy. We have calculated the level $15\lambda/4$ correct to a third order approximation and the two levels $3\lambda/2$ to a second order approximation.[20] The calculation is straightforward but too elaborate to be given here. To illustrate the type of result obtained, we give the energy levels in the presence of the magnetic field only for the lowest level, correct to a second order perturbation calculation. For brevity the third order terms have been omitted. We find

$$W = 15\lambda/4 - \omega(13/6 + 5\lambda/8D) + [(36\sigma - 35\omega - 165\lambda\omega/16D)^2 + 432\delta^2]/405\lambda + \cdots.$$

The expressions for the other levels are of the same type.

At sufficiently low temperatures the square of the effective Bohr magneton number $n_B{}^2 = 3\chi kT/N\beta^2$ is given by three times the square of the coefficient of the term in H. Hence, if we extrapolate the experimental values of χT to $T=0$, we obtain three equations which theoretically enable us to determine σ, δ and D. The values obtained in this way are however so sensitive to variations in χT at $T=0$ within the range of possible error that it is

[20] The details of the inclusion of the third order terms in the perturbation problem will be considered by Mr. Jordahl in his paper on Cu.

preferable to proceed differently. The argument of Van Vleck[4] shows that if the cubic field acting on the metallic ion be the same in nickel ammonium sulphate as in cobalt ammonium sulphate the constant Dq has the same value numerically. We accordingly assume $Dq = -1200$ cm^{-1}, thereby giving up the possibility of obtaining values of σ and δ for arbitrarily given values of $(\chi T)_0$; instead we assume values for σ and δ and calculate the susceptibilities at various temperatures. The values which have been chosen for illustration are $\sigma = \delta = 20$, corresponding to a term $40(x^2 - z^2)$ in the Hamiltonian, which gives roughly the right degree of asymmetry. It is of course possible to choose different values for the individual coefficients of x^2, y^2, z^2, but the computations are laborious and do not give any new information. Two points deserve mention. In the first place, a much smaller rhombic field is required in cobalt salts to produce the observed asymmetry than is needed for nickel salts despite the much greater isotropy of the latter. In the second place the calculated mean susceptibility for the three orientations is consistently greater than that observed. Fig. 3A shows the calculated values of $n_B^2 = 3kT\chi/N\beta^2$ plotted against T for the three magnetic axes of $Co(NH_4)_2(SO_4)_2 \cdot 6H_2O$. The trend of these curves may readily be understood in a qualitative way. At low temperatures the only level contributing to the susceptibility is the lowest. Since this has both a first and a second order Zeeman effect $\chi T = a + bT$, a and b being constants. At higher temperatures the two states $3\lambda/2$ begin to contribute to the susceptibility, but this is counteracted by the depopulation of the lowest level; these higher levels have smaller Zeeman effects than the levels $15\lambda/4$ so that the curve of χT against T rises less steeply and tends to an almost constant value. At still higher temperatures the three levels $-9\lambda/4$ would also contribute to the susceptibility but these temperatures are not reached experimentally.

The value of n_B^2 is plotted also for the case where there is no rhombic field and this curve is shown dotted in Fig. 3A. It should be noticed that the effect of the rhombic field is to produce asymmetry and also to lower the mean value of the three principal susceptibilities.[21] Unfortunately, it is not within the limits of the present approximation to make the rhombic field sufficiently large to give agreement with experiment, as then the convergence would be poor. It is very reasonable, however, to suppose that a larger

[21] Let us imagine the magnitude of the rhombic field is varied from a very large value down to zero. The behavior of the three principal susceptibilities is as follows. Orientation (3) starts at the "spin only" value, decreases and then starts to increase again, finally ending on the curve for zero rhombic field shown in Fig. 3A. Orientation (2) starts slightly above the "spin only" value due to the introduction by the orbit spin coupling of small diagonal elements in the orbital angular momentum, representing the contribution of the higher cubic levels, and then decreases at low temperatures but increases at higher temperatures, ending finally with (3). Orientation (1) is rather complicated. It starts with (2) but the susceptibility increases rapidly with decreasing rhombic field, and develops a hump at low temperatures. This is because for this orientation the lowest level has a large first order Zeeman effect. The susceptibility then begins to fall again, passing through a representative curve shown in Fig. 3B. The hump fades out, the susceptibility decreasing at low and increasing at high temperatures, finally ending with (2) and (3).

rhombic field would give good agreement with experiment, except possibly at low temperatures. Since the only measurements at low temperatures are those of Jackson,[6] made as long ago as 1922, and which at other temperatures are known to be greatly in error, this discrepancy is not worth considering. When better experimental data are available another effort will be made to obtain a better solution of this troublesome sextet.

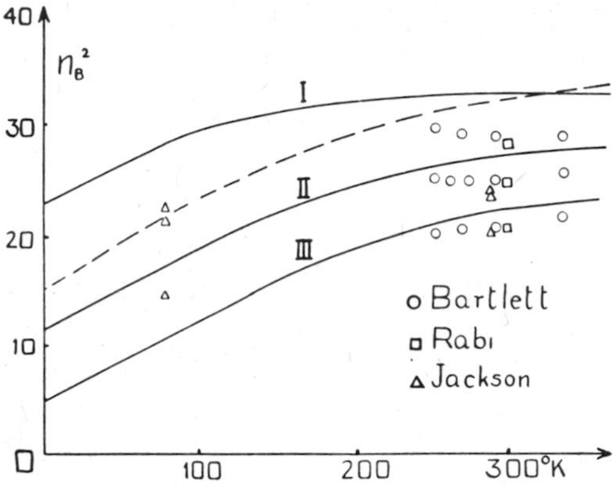

Fig. 3A.

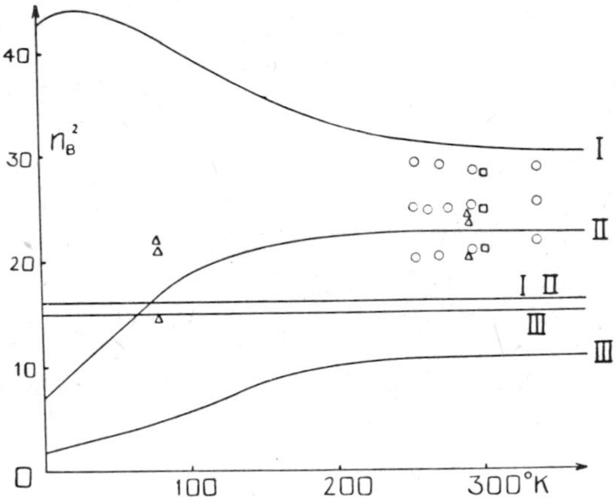

Fig. 3B.

Fig. 3. The heavy lines in Figs. 3 show the calculated values of n_B^2 for Co $(NH_4)_2(SO_4)_2 \cdot 6H_2O$ for the two rhombic fields (3A) $40(x^2-z^2)$ (3B) $200(x^2-z^2)$, while for comparison, the experimental points obtained by different observers are given. The numbers on the curves denote the axes to which the curves refer. The dotted lines in Fig. 3A is for zero rhombic field; the horizontal straight lines in Fig. 3B are for extremely large rhombic fields. It is reasonable to suppose that a rhombic field intermediate to the values we have taken would give agreement with experiment. It is significant that the results of Bartlett[8] along axis (1) do actually seem to be falling with increasing T, as would be the case for an intermediate field. The agreement at low temperatures is not good but the experimental values are those of Jackson[6] and may easily be in error.

Rhombic field greater than orbit-spin

In the hope that a perturbation calculation from the other limiting case would be more effective, calculations were also made from this end. Here the orbit spin is subsidiary to the rhombic field. The rhombic field can be seen to decompose the sextuply degenerate level into three doubly degenerate ones, with separations proportional to the coefficients of the rhombic field. The orbit-spin interaction does not remove the remaining degeneracy in the first approximation, but only in the second. Expressions for the energy levels and their first and second order Zeeman effects can be written down, and the susceptibility calculated for given values of the constants D, A, B, C of the crystal field, such that the rhombic field alone produces a separation which is small compared with that due to the orbit-spin alone. In the absence of the spin the energy levels are $6Aa$, $6Ba$, $-6a(A+B)$. We suppose that $-6a(A+B)$ lies below $6Aa$ and $6Ba$ and calculate the Zeeman effects of the two lowest roots correct to a third approximation, retaining terms in the magnetic field up to H^2, in the usual way. The Zeeman effects are different according as the magnetic field acts along the x, y, or z axes. The exact expressions are long, but for illustration we give the two lowest roots, when the magnetic field acts along the x, y and z axes, correct only up to a second order perturbation calculation. For brevity the third order terms have been omitted.

$$\left.\begin{aligned} W_{\pm}(x) &= W_{\pm} - \omega_{+}[1 \pm (\alpha + 2\beta)\theta] \pm 2\omega_{+}{}^2\alpha^2\theta^3/3\lambda^2, \\ W_{\pm}(y) &= W_{\pm} - \omega_{-}[1 \mp (2\alpha + \beta)\theta] \pm 2\omega_{-}{}^2\beta^2\theta^3/3\lambda^2, \\ W_{\pm}(z) &= W_{\pm} + \omega[1 \pm (\alpha - \beta)\theta] \pm 2\omega^2\theta^3(\alpha + \beta)^2/3\lambda^2 \end{aligned}\right\}$$

where

$$\begin{aligned} W_{\pm} &= -6a(A + B) + 45\lambda^2(\beta - \alpha)/16 \pm 9\lambda^2\theta/4, \\ \theta &= (\alpha^2 + \beta^2 + \alpha\beta), \quad \omega_{+} = \omega(1 + 9\lambda/128D + 9\lambda\beta/4), \\ \omega_{-} &= \omega(1 + 9\lambda/128D - 9\lambda\alpha/4). \end{aligned}$$

We have written $\alpha = 1/6a(B-C)$ and β, γ for the cyclic permutations. From these expressions the principal susceptibilities may be calculated. For illustration, a rhombic field represented by $200(x^2-z^2)$ in the Hamiltonian has been taken. The result is shown by the curves in Fig. 3B. It is readily verified that the curves correspond with those in Fig. 3A as shown by the numbering. The values of $n_B{}^2$ for a very large rhombic field are also shown by the horizontal straight lines. These are only limiting curves, however, since we have assumed the cubic field to predominate and therefore we cannot make the rhombic field as large as we please.

We have now calculated the susceptibilities (*i*) for a rhombic field $40(x^2-z^2)$ (*ii*) for a rhombic field $200(x^2-z^2)$, inclusive of orbit-spin coupling in both cases. The latter alone produces an over-all splitting of the level Γ_4 of the cubic field amounting to roughly 1000 cm^{-1}, while the separations produced by fields (*i*) and (*ii*) are respectively 480 cm^{-1} and 2400 cm^{-1}. The principal susceptibilities calculated on the basis of field (*i*) show roughly the right degree of asymmetry, but are too high. Interpolating between (*i*) and (*ii*), it seems that a rhombic field of magnitude intermediate between (*i*) and (*ii*), but much nearer (*i*), will give good agreement with experimental values, except possibly at low temperatures, where however, the experimental values are in considerable doubt. Neither of the two methods of approximation used above are applicable in this intermediate region and the calculation would consist in the numerical solution of the original sextic secular equation.

It is interesting to notice that there would have been no gain in generality if we had added to the Hamiltonian terms representing a field of tetragonal symmetry or terms of rhombic symmetry of higher degree, provided the field of cubic symmetry always predominates. In order to see this we need only observe that in the orbital problem the lowest level Γ_4 of the cubic field is split by the rhombic field into the three levels G_2, G_3, G_4, no two of which belong to the same representation of the rhombic group. Consequently it is not possibly to change the moments of these levels by changing the type of rhombic field nor can this be accomplished even by the superposition of a tetragonal field, since this is only a particular form of rhombic field. Because the moments are fixed, the susceptibility can be changed only through the relative position of the energy levels and as there are three levels, it needs only two parameters to specify them. The two parameters A and B are capable of doing this.

Conclusion

In the present paper no account has been taken of the variation with temperature of the constants of the crystal field. The very small changes in interatomic distances caused by thermal expansion may possibly affect these constants quite appreciably because the force between ions in a crystal is known to vary very rapidly with the distance. We have moreover assumed that the principal axes of the various types of crystalline fields all coincide. Perhaps a better approximation to the actual state of affairs would be to assume fields of different symmetry, whose principal axes were inclined to each other, the relative orientations depending on temperature in some complicated way. There would then arise the possibility of an explanation of the results of Bartlett,[8] who finds that the orientation of the principal susceptibilities relative to the crystallographic axes depend on temperature, the total variation being of the order of 5 degrees in a range of temperature 100°C.

From the considerations developed in this and the preceding paper, it should be evident that it is only rarely that the constant Δ of the experimenters has any theoretical interpretation. In order that it may have, it is necessary that the expansion $\chi = C/T - C\Delta/T^2 + \cdots$, should converge very rapidly. This condition may be expressed in another form, that the Stark separations of the levels contributing to the susceptibility should be small compared with kT. This is not satisfied in the rare earths nor with Co but our calculations have shown that it is satisfied with Ni and Cr. That the susceptibility of the rare earths can be represented by the Curie-Weiss law is merely fortuitous. Even here the value of Δ obtained depends on the temperature at which the measurements are made and in this sense Δ has no theoretical significance.

One of the most surprising facts revealed by our calculations of the susceptibilities in crystals is that a field of cubic symmetry should be capable of allowing such excellent agreement to be obtained with experiment. At first sight there seems to be no reason whatever for the field to possess cubic, or even nearly cubic, symmetry. In the case of Ni it was definitely established that a field predominantly rhombic and of the form $Ax^2+By^2-(A+B)z^2$, was incapable of giving the observed principal susceptibilities. The next assumption is naturally a field of cubic symmetry together with a much smaller rhombic term, an assumption which has proved completely successful with Ni, Cr and Co. For the rare earths, where measurements of the principal susceptibilities are lacking, and only the variation with temperature of the mean susceptibility has been observed, good agreement with experiment is obtained on the assumption of a cubic field alone. Without understanding why complicated crystals should have such simple crystalline fields, it must at least be conceded that the evidence in favor of a predominant field of cubic symmetry is strong. Whether or not there are other types of field which will give equally as good agreement with experiment remains to be seen.

In our calculations of paramagnetic susceptibilities, both of the rare earths and of the elements of the iron group, the sign of D in Eq. (3) has been consistently positive. In a Letter to the Editor[22] Gorter finds that this choice of the sign of D agrees with there being water molecules (or else oxygen ions) arranged at the corners of an octahedron around the paramagnetic ion.

The writers wish to place on record their thanks to Professor J. H. Van Vleck, to whose constructive and stimulating criticisms the present work owes a great deal.

[22] C J Gorter, Phys. Rev. **42**, 437 (1932).

Reprinted from *Phys. Rev.*, **42**, 437–438 (1932)

5

Note on the Electric Field in Paramagnetic Crystals

The work[1] of Kramers, Bethe, and especially of Van Vleck and his collaborators has created the possibility of drawing conclusions from magnetic data about the electric fields in paramagnetic crystals and hence about the spatial arrangement of atoms and molecules in the crystal.

In his recent paper Van Vleck[1] has shown that it is possible to account for the different magnetic behavior of crystals of hydrated Co and Ni compounds by assuming that to a first approximation the electric fields possess cubic symmetry around the magnetic ion, but that in the second approximation a rhombic term must be added. Quantitative calculations amplifying the theory are to be published shortly by Schlapp and Penney in this journal.

If the potential energy of an electron in the lattice can be developed as a power series in the displacement from the center of the magnetic ion, the terms which give rise to a decomposition of the energy levels are

$$\Phi = \Sigma_i\{Ax_i^2 + By_i^2 + cz_i^2 + D(x_i^4 + y_i^4 + z_i^4)\}.$$

Van Vleck concludes that D must be positive to account for the experimental data on hydrated salts of the iron group. The purpose of the present note is to consider what atomic groupings will lead to a positive D.

Since the contribution to D due to the different charges in the neighborhood of the central ion is proportional to R^{-5} (where R is the distance from the central ion), it is evident that only the immediate neighbors will give a noticeable contribution to the value of D.

If the metal ion is surrounded by 6 oxygen ions or water-dipoles[2] in an octahedral arrangement, D will be positive.[3] If on the contrary the ion is surrounded by 8 or 4 negative charges in a cubic or tetrahedral arrangement, D will be negative. This leads to the conclusion that in the hydrated salts of the iron group the metal ion is surrounded by six molecules of crystal water.

Dr. C. A. Beevers of the University of Liverpool kindly expressed to me his opinion that, from the result of x-ray researches, the arrangement indicated above may be regarded as probable, though not proved for the alums and the hepta- and hexa-hydrated sulphates. It seems probable to me that also in solutions the metal ions will be surrounded by six water molecules.

Penney and Schlapp have performed calculations on rare earth salts, and have shown that it is possible to explain the temperature variation of the susceptibilities of the octahydrates of Pr and Nd sulphates, assuming a cubic field again with a positive value of D. Consequently, here also, the octahedral grouping of the oxygen atoms is suggested. This demands six oxygen neighbors for the metal ion, but in the substances so far investigated there are only four water molecules to each such ion. Hence it is necessary to suppose either that oxygen atoms belonging to the SO_4-group figure among the immediate neighbors, or else that a water molecule may be shared by two metal ions. Both assumptions could give rise to deviations from cubic symmetry, which can perhaps account for any magnetic anisotropy which may be disclosed when measurements, on the principal susceptibilities of single crystals of rare earth salts become available.

A quantitative discussion of the arrangements proposed above will be able to decide whether the picture corresponds to reality.

C. J. Gorter

Natuurk. Laborat. v. Teylers Stichting,
Haarlem, Holland,
September 15, 1932.

[1] H. A. Kramers, Comm. Leiden **60**; Proc. Amsterdam Acad. **33**, 959 (1930); H. Bethe, Ann. d. Physik **3**, 133 (1929); Zeits. f. Physik **60**, 218 (1930); J. H. Van Vleck, *The Theory of Electric and Magnetic Susceptibilities*, Oxford University Press; Phys. Rev. **41**, 208 (1932); O. M. Jordahl, W. G. Penney and R. Schlapp, Phys. Rev. **40**, 637 (1932); W. G. Penney and R. Schlapp, Phys. Rev. **41**, 194 (1932); Schlapp and Penney, Phys. Rev. (in press).

[2] The dipoles will orient themselves in the field of the positive ion with the negative charge inside. With the octahedral arrangement the negative charges are at the face centers of a cube embracing the positive ion.

[3] Cf., for instance, H. Bethe, Ann. d. Physik **3**, 196 (1929).

Reprinted from *Phys. Rev.*, **46**, 79 (1934)

6

Information Concerning Crystal Structure from Data on Magnetic Susceptibilities

O. M. JORDAHL

Beevers and Lipson[1] have recently published their results on the crystal structure of the pentahydrate copper sulphate. It seems worth while to call attention to the general agreement between their experimental results and the conclusions which the author published[2] after an analysis of the magnetic susceptibility data for this salt.

The author has shown that the temperature variation of the principal magnetic susceptibilities of the hexahydrate double sulphates of copper is adequately explained on the assumption that the metal ion is located in a crystalline field of predominantly cubic symmetry. By analogy with the known crystal structure of other hexahydrate sulphates, it seemed reasonable to ascribe to the water molecules of these copper salts an octahedral arrangement which would provide the crystalline field of desired symmetry. The data which were available on the magnetic susceptibility of the pentahydrate copper sulphate[3] seemed to require a similar potential field for the copper ion in this salt and consequently a similar arrangement of the crystal groups. Although there are only five water molecules per copper atom in $CuSO_4:5H_2O$, the author suggested[2] that the octahedral structure, and the crystalline field of predominantly cubic symmetry, might be obtained by the sharing of two water groups between the two copper atoms of the unit crystal cell.

Beevers and Lipson actually find[1] that each copper atom is surrounded by an octahedron consisting of four water groups and two oxygen molecules. Thus the crystalline field should have predominantly cubic symmetry as was assumed. The departure from cubic symmetry is due to the effect of the other atoms or groups in the unit cell and to the fact that the field of an oxygen atom is different from that of a water dipole. The asymmetry of the crystalline field is secured in the calculations by superposing a smaller field of rhombic symmetry.

The important result of this investigation is that it is often possible to obtain some information about the crystal structure of salts from an analysis of the data on the principal susceptibilities. The information thus obtained should be valuable in the case of crystals of low symmetry where the x-ray analysis is difficult and the results possibly ambiguous. However, an accurate analysis requires precise data on the principal susceptibilities over a large range of temperatures extending as low as possible. Such data are at the present time very meager, and for many salts are entirely lacking.

O. M. JORDAHL

Knox College,
Galesburg, Illinois,
June 9, 1934.

[1] C. A. Beevers and H. Lipson, Nature **133**, 215 (1934).
[2] O. M. Jordahl, Phys. Rev. **45**, 87 (1934).
[3] W. J. de Haas and C. J. Gorter, Leiden Comm. 210d (1930).

Reprinted from *Proc. Roy. Soc. London,* **A161**, 220–235 (1937)

7

Stability of Polyatomic Molecules in Degenerate Electronic States
I—Orbital Degeneracy

By H. A. Jahn, *Davy-Faraday Laboratory, The Royal Institution*
and E. Teller, *George Washington University, Washington, D.C.**

(*Communicated by F. G. Donnan, F.R.S.—Received* 17 *February* 1937)

Introduction

In the following we investigate the conditions under which a polyatomic molecule can have a stable equilibrium configuration when its electronic state has orbital degeneracy, i.e. degeneracy not arising from the spin. We shall show that stability and degeneracy are not possible simultaneously unless the molecule is a linear one, i.e. unless all the nuclei in the equilibrium configuration lie on a straight line. We shall see also that the instability is only slight if the degeneracy is due solely to electrons having no great influence on the binding of the molecule.

* This research was carried out when the authors were working in the Sir William Ramsay Laboratories of Inorganic and Physical Chemistry, University College, London.

We first note that if accidental degeneracy (i.e. degeneracy not caused by symmetry) is disregarded then a degenerate electronic state necessarily entails a symmetrical nuclear configuration. Thus in order to cover all cases we may first consider each possible type of symmetry separately and discuss what nuclear configurations are consistent with each symmetry. A given molecule will possess a continuous set of configurations consistent with one definite type of symmetry, and among these configurations there may be one with a minimum electronic energy. This configuration is then stable with respect to all totally symmetrical nuclear displacements (i.e. displacements which do not disturb the symmetry). We shall have to investigate its stability with respect to all other nuclear displacements.

Now a nuclear configuration cannot be stable if the electronic energy for neighbouring configurations depends linearly upon any one of the nuclear displacements. There may be reasons of symmetry, however, which preclude such a linear dependence, and we illustrate the occurrence and non-occurrence of this respectively by the two following examples. In these examples there is the added complication that the nuclear displacements in question cause a splitting of the degenerate electronic state into states with different energies. This complication will be the rule rather than the exception, since the displacements reduce the symmetry of the original configuration. A linear dependence upon a nuclear displacement of the energy of any one of these states formed by the splitting is then sufficient to cause instability.

1—Two Examples

In our first example we consider the motion of a single electron in the field of three nuclei lying on a straight line. The states of the electron can then be classified as σ, π, δ, etc., states according as the component of the electron's orbital angular momentum along the nuclear axis is 0, ± 1, ± 2, etc. (in units of $h/2\pi$). The σ states are non-degenerate, whilst the π, δ and further states are each twofold degenerate, corresponding to the fact that the electron may move either clockwise or anticlockwise about the nuclear axis. If, now, one of the nuclei (say the middle one) is displaced through a distance d perpendicular to the axis, then the axial symmetry will be destroyed and the degeneracy removed. Each twofold degenerate state will split into two states, one symmetrical with respect to reflexions in the plane of the nuclei and the other antisymmetrical with respect to the same plane. These states will have different energies E_s and E_a. When the nuclear displacement is varied these states and their energies will change continuously, but their symmetry will remain and it is clear that when the displacement is $-d$ the states and energies will be the same as before. Thus the two energies E_s and E_a must be even functions of d, and all that is required for stability with respect to this nuclear displacement is that the function be positive in both cases (see fig. 1).

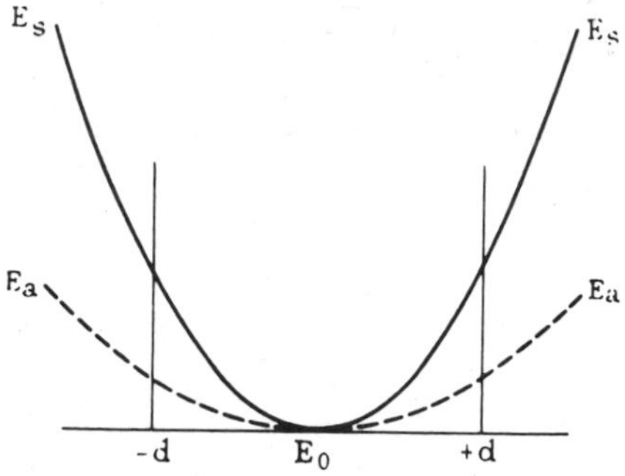

FIG. 1

In our second example we consider the motion of a single electron in the field of a plane square configuration of four identical nuclei. There is a similarity between the two examples in that, in the first example once the wave function on any half-plane through the axis is given, then it is determined by symmetry for all other such half-planes: a rotation through the angle θ multiplies the wave function by $e^{i\lambda\theta}$, where λ is the axial component of the orbital angular momentum. In the second example, on the other hand, the symmetry about the axis through the centre of the square, although considerably reduced, still determines the wave function completely on any four half-planes at right angles to each other, once the wave function has been given on any one of them. The function is again multiplied by $e^{i\lambda\theta}$, where here θ is restricted to the values $\pm\pi/2$, π and λ can have the values 0, $\pm$ 1. The state with $|\lambda| = 1$ is again twofold degenerate. We will discuss the stability of the square configuration for this degenerate electronic state. Consider the two displaced configurations I and II depicted in fig. 2. These

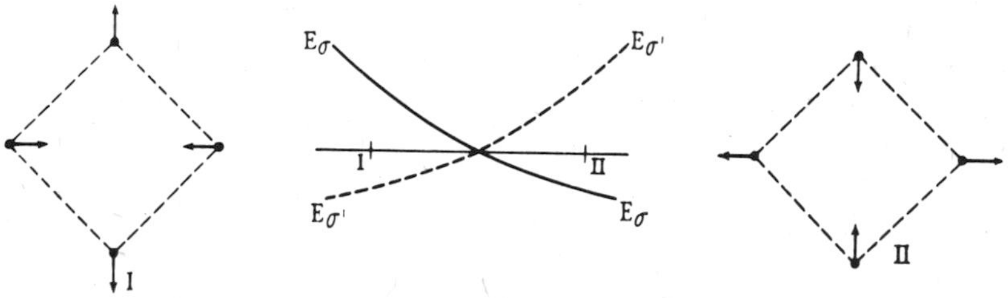

FIG. 2

may be regarded clearly as belonging to a positive and negative value of one and the same nuclear displacement. This nuclear displacement destroys the fourfold axial symmetry, replacing it by a twofold one. The degenerate state will split into two states ϕ_σ and $\phi_{\sigma'}$, the first having a node in the horizontal plane of symmetry σ and the second a node in the vertical plane σ'. Now it is clear that since the configurations I and II are geometrically congruent and the planes σ', σ in II correspond respectively to the planes σ, σ' in I, the energy of ϕ_σ for the configuration I must be the same as the energy of $\phi_{\sigma'}$ for the configuration II. Thus if E_σ, $E_{\sigma'}$ are the corresponding energies, we must have the relation

$$E_\sigma(\mathrm{I}) = E_{\sigma'}(\mathrm{II}).$$

Similarly

$$E_\sigma(\mathrm{II}) = E_{\sigma'}(\mathrm{I}).$$

Thus the two energy levels E_σ, $E_{\sigma'}$ have crossed over at the energy E_0 of the undisplaced configuration (see fig. 2). There is here thus no apparent reason of symmetry which precludes a linear dependence of the energy levels upon the nuclear displacement in the neighbourhood of E_0, and the square configuration therefore will not in general be a stable equilibrium configuration for the degenerate electronic state.

2—General Theorem

In the above two examples we have seen how the type of molecular symmetry can determine whether the energy of a degenerate electronic state should depend linearly or not upon nuclear displacements. It is our purpose to study this influence for all possible types of symmetry. We do this by an application of group theory to perturbation calculation. We restrict ourselves in this paper to orbital degeneracy and reserve the consideration of the special effects due to spin degeneracy for a second paper. The electronic energy levels of the displaced configuration are the characteristic values of a perturbation matrix, the elements of which may be expressed as a power series in the nuclear displacements. Unless those perturbation matrix elements which are linear in the nuclear displacements all vanish, then at least one perturbed energy level will depend linearly upon a nuclear displacement. Now the elements of these matrices are integrals which are necessarily invariant with regard to the group of symmetry of the equilibrium configuration. The integrands, on the other hand, are products whose transformation properties are determined by those of the degenerate electronic wave functions and the nuclear displacements. From an investigation of these transformation properties it may be deduced whether the invariance of the integrals and the transformation properties of the integrands are compatible. Unless this is the case the integrals vanish. If, however, they are compatible the matrix elements depending linearly upon the nuclear displacements will be in general different from zero. It will be found that all linear matrix elements will necessarily vanish only when the nuclear configuration shows complete axial symmetry, i.e. only when all the nuclei in the molecule lie on a straight line. *All non-linear nuclear configurations are therefore unstable for an orbitally degenerate electronic state.* Thus if we know of a polyatomic molecule that the nuclei in the equilibrium configuration do not all lie on a straight line, then we know at the same time that its ground electronic state does not possess orbital degeneracy. We should exclude from our consideration, however, orbitally degenerate electronic states in which the degenerate electrons do not contribute appreciably to the molecular binding and are not perturbed therefore by nuclear displacements. Such is the case for the inner degenerate electronic shells of the paramagnetic rare earth ionic salts. If, apart from such instances, a non-linear molecule is paramagnetic, then we may conclude that this paramagnetism must be due to the action of the spin alone, for it is well known that a non-degenerate orbital state can possess neither a mean orbital angular momentum nor a mean orbital magnetic moment (cf. van Vleck 1932).

3—Mathematical Formulation and Group-theoretical Considerations

We shall have to consider nuclear configurations Q which can be obtained from the given symmetrical configuration Q_0 by adding a linear combination of a complete set of displaced configurations Q_r which may be chosen orthogonal to each other, to the translations and rotations and also to all the totally symmetrical displacements:

$$Q = Q_0 + \sum_r Q_r \eta_r.$$

Here the η_r are infinitesimally small quantities which we will call the (non-totally symmetrical) normal displacements of the configuration Q_0. It is to be noted that the η_r can be so chosen that they transform according to irreducible representations of the symmetry group of Q_0 (cf. Wigner 1930).

The energy operator $H(Q)$ for the electronic motion in the field of the nuclei fixed in the configuration Q can be expanded as a power series in the η_r:

$$H = H_0 + \sum_r V_r(q)\,\eta_r + \sum_{rs} V_{rs}(q)\,\eta_r \eta_s + \dots,$$

where the V_r, V_{rs} are functions of the electronic coordinates q alone, since the kinetic energy of the electrons is included in the energy operator H_0.

Let E_0 be the energy level of the degenerate electronic state and ϕ_ρ a complete set of orthogonal wave functions such that

$$H_0\phi_\rho = E_0\phi_\rho.$$

In the configuration Q this energy level either may be split up into a number of different energy levels E_α or it may be merely displaced (i.e. the degeneracy may not be removed). The configuration Q_0 can be stable with respect to all nuclear displacements only if each of these energy levels E_α has a minimum for $Q = Q_0$. We denote the contribution to the perturbed energy E_α arising from the linear terms $\sum_r V_r\eta_r$ by E_α^I, putting

$$E_\alpha = E_0 + E_\alpha^I + \ldots.$$

Perturbation theory shows that the perturbation energies E_α^I are the characteristic values of a perturbation matrix having the following elements:

$$M_{\rho\sigma} = \Sigma\,\eta_r \int \phi_\rho^* V_r \phi_\sigma d\tau.$$

Now it is clear that if E is a characteristic value of this matrix for a given set of values of the η_r, then $-E$ is a characteristic value for the configuration obtained by changing the sign of all the η_r. Thus unless the characteristic values of this matrix all vanish (in which case the matrix itself must vanish) the configuration Q_0 cannot possibly be a stable one.

If the representation of the symmetry group of the configuration Q_0 which is subtended by the products $\phi_\rho^* V_r \phi_\sigma$ does not contain the identical representation, then the integrals $\int \phi_\rho^* V_r \phi_\sigma d\tau$ will vanish, for these integrals are necessarily invariant with respect to the symmetry operations. If this condition does not hold, then the integrals in general will not vanish. Now the normal displacements η_r may be chosen so that the V_r transform according to irreducible representations V of the group of symmetry. The degenerate wave functions ϕ_ρ will also subtend a representation Φ of the group of symmetry, which representation will be in general irreducible. An exception occurs here, as Wigner (1932) has shown, when the degeneracy arising from the invariance of the wave equation with respect to time reversal is considered; this may cause an "accidental" degeneracy so that the wave functions of the energy level subtend a representation which is the sum of two irreducible representations. For our purpose, however, this double representation Φ may be treated as though it were a single one. It may then be shown that if we restrict ourselves to orbital degeneracy (i.e. if we do not consider the spin wave functions) the representation Φ and consequently the wave functions ϕ_ρ can be chosen always real. The products $\phi_\rho V_r \phi_\sigma$ are linearly independent except that of course

$$\phi_\sigma V_r \phi_\rho = \phi_\rho V_r \phi_\sigma.$$

They will therefore transform according to the product representation $V[\Phi^2]$, where $[\Phi^2]$ denotes the representation of the symmetrical product of Φ with itself (cf. Weyl 1928). Thus if there exists a normal displacement (other than a totally symmetrical one) for which $V[\Phi^2]$ contains the identical representation, the symmetrical configuration Q_0 will not be stable.

We may illustrate these general considerations by applying them to the two examples discussed in the introduction. In the first example the group of symmetry for a general position of the nuclei on the axis is C_∞^v. The degenerate irreducible representations are all two-dimensional, and there is an infinite series of them, denoted by $E_1, E_2, \ldots, E_k, \ldots$. It can be shown (cf. Tisza 1933) that the representation of the symmetrical product of E_k is given by

$$[E_k^2] = A_1 + E_{2k},$$

where A_1 is the unit representation. Placzek (1934) has shown that for linear nuclear configurations all non-totally symmetrical vibrations are of the type E_1. One finds

$$E_1[E_k^2] = E_1(A_1 + E_{2k}) = E_1 + E_{2k+1} + E_{2k-1},$$

so that the product $V[\Phi^2]$ never contains the identical representation A_1 for any non-totally symmetrical vibration and any degenerate electronic state Φ. Thus, as has already been seen above, there is no reason of symmetry why the linear nuclear configuration should not be stable in any of the degenerate states. In the second example the reverse is the case. Here the group of symmetry is D_4^h, and there are only two degenerate representations, viz. E_g and E_u, g and u referring to even and odd representations with respect to inversion in the centre of symmetry. These representations are both two-dimensional. The considerations of Wigner (1930) show that the square configuration possesses non-totally symmetrical displacements of types B_{1g}, B_{1u}, B_{2g} and E_u. Since it can be shown that

$$[E_g^2] = [E_u^2] = A_{1g} + B_{1g} + B_{2g},$$

and since $B_{1g}^2 = B_{2g}^2 = A_{1g}$, we see that $B_{1g}[E_g^2]$, $B_{1g}[E_u^2]$, $B_{2g}[E_g^2]$ and $B_{2g}[E_u^2]$ all contain the identical representation A_{1g}. Hence normal displacements of the type B_{1g} or B_{2g} render the square configuration unstable for a degenerate electronic state of either type. One verifies easily that the nuclear displacement considered in the example is of the type B_{2g}.

4—Proof of General Theorem

Excluding the groups of complete axial symmetry, we have to show, for all configurations and any degenerate one-valued representation of the group of symmetry, that a symmetrical nuclear configuration necessarily possesses non-totally symmetrical normal displacements transforming according to irreducible representations V such that $V[\Phi^2]$ contains the identical representation. This is equivalent to showing that $[\Phi^2]$ in its reduced form contains at least one of the representations V.

Now Wigner (1930) has shown how to calculate for any given nuclear configuration how many normal displacements of each irreducible type occur. In order to apply this calculation of Wigner to all possible symmetrical molecules we note that any symmetrical configuration must contain sets of equivalent nuclei whose positions are transformed into one another by the various symmetry operations. For any given group of symmetry there are various possible kinds of such equivalent sets according as the points lie on no symmetry element or on one, two or more of them. We have applied the method of Wigner to all possible sets of equivalent points for the various groups of symmetry, and the results of our calculations are tabulated in Table I.

In this table the notation of Placzek (1934), Mulliken (1933), Lennard-Jones (1934) and Tisza (1933) is used for the groups and their irreducible representations. Column I gives the designation of the group, columns II and III give the reduction into irreducible constituents of the translations and rotations respectively, column IV gives a designation to the various kinds of equivalent points for each group, column V gives the number of equivalent points in each individual set, column VI gives the symmetry elements, if any, on which these points lie, whilst the last column VII gives the reduction into irreducible constituents for the whole set of nuclear displacements inclusive of translations, rotations and normal displacements for a "molecule" consisting of the equivalent points given in column IV. By subtracting the translations and rotations from this last column we find what types of normal displacements occur. This subtraction has not been carried out in the table, since for any molecule more than one set of equivalent points may occur (i.e. more than one row of the table is required, or the same row required more than once), whilst the subtraction of the translation and rotation has of course only to be carried out once. The table will be useful not

TABLE I

I	II	III	IV	V	VI	VII
C^v_∞	A_1+E_1	E_1	a	1	All	A_1+E_1
$C^{vi}_\infty = D^h_\infty$	$A_{1u}+E_{1u}$	E_{1g}	a	2	C_∞, σ_v	$A_{1g}+A_{1u}+E_{1g}+E_{1u}$
			b	1	All	$A_{1u}+E_{1u}$
C_{2p+1}	$A+E_1$	$A+E_1$	a	$2p+1$	None	$3(A+E_1+\ldots+E_p)$
			b	1	C_{2p+1}	$A+E_1$
C_{2p}	$A+E_1$	$A+E_1$	a	$2p$	None	$3(A+B+E_1+\ldots+E_{p-1})$
			b	1	C_{2p}	$A+E_1$
D_{2p+1}	A_2+E_1	A_2+E_1	a	$2(2p+1)$	None	$3(A_1+A_2)+6(E_1+E_2+\ldots+E_p)$
			b	$2p+1$	C_2	$A_1+2A_2+3(E_1+E_2+\ldots+E_p)$
			c	2	C_{2p+1}	$A_1+A_2+2E_1$
			d	1	All	A_2+E_1
D_{2p}	A_2+E_1	A_2+E_1	a	$4p$	None	$3(A_1+A_2+B_1+B_2)+6(E_1+E_2+\ldots+E_{p-1})$
			b	$2p$	C_2	$A_1+2A_2+B_1+2B_2+3(E_1+E_2+\ldots+E_{p-1})$
			c	$2p$	C_2'	$A_1+2A_2+2B_1+B_2+3(E_1+E_2+\ldots+E_{p-1})$
			d	2	C_{2p}	$A_1+A_2+2E_1$
			e	1	All	A_2+E_1
$C^i_{2p+1} = S_{2(2p+1)}$	A_u+E_{1u}	A_g+E_{1g}	a	$2(2p+1)$	None	$3(A_g+A_u+E_{1g}+E_{1u}+\ldots+E_{pg}+E_{pu})$
			b	2	C_{2p+1}	$A_g+A_u+E_{1g}+E_{1u}$
			c	1	All	A_u+E_{1u}
$C^i_{2p} = C^h_{2p}$	A_u+E_{1u}	A_g+E_{1g}	a	$4p$	None	$3(A_g+A_u+B_g+B_u+E_{1g}+E_{1u}+\ldots+E_{p-1,g}+E_{p-1,u})$
			b	$2p$	σ_h	$2A_g+A_u+\begin{cases}B_g+2B_u\ (p\text{ odd})\\2B_g+B_u\ (p\text{ even})\end{cases}+E_{1g}+2E_{1u}+2E_{2g}+E_{2u}$ $+\ldots+\begin{cases}2E_{p-1,g}+E_{p-1,u}\ (p\text{ odd})\\E_{p-1,g}+2E_{p-1,u}\ (p\text{ even})\end{cases}$
			c	2	C_{2p}	$A_g+A_u+E_{1g}+E_{1u}$
			d	1	All	A_u+E_{1u}

D^i_{2p+1} $= D^d_{2p+1}$	$A_{2u} + E_{1u}$	$A_{2g} + E_{1g}$	*a*	$4(2p+1)$	None	$3(A_{1g} + A_{1u} + A_{2g} + A_{2u}) + 6(E_{1g} + E_{1u} + \ldots + E_{pg} + E_{pu})$
			b	$2(2p+1)$	σ_d	$2A_{1g} + A_{1u} + A_{2g} + 2A_{2u} + 3(E_{1g} + E_{1u} + \ldots + E_{pg} + E_{pu})$
			c	$2(2p+1)$	C_2	$A_{1g} + A_{1u} + 2A_{2g} + 2A_{2u} + 3(E_{1g} + E_{1u} + \ldots + E_{pg} + E_{pu})$
			d	2	C_{2p+1}, σ_d	$A_{1g} + A_{2u} + E_{1g} + E_{1u}$
			e	1	All	$A_{2u} + E_{1u}$
$D^i_{2p} = D^h_{2p}$	$A_{2u} + E_{1u}$	$A_{2g} + E_{1g}$	*a*	$8p$	None	$3(A_{1g} + A_{1u} + A_{2g} + A_{2u} + B_{1g} + B_{1u} + B_{2g} + B_{2u})$ $+ 6(E_{1g} + E_{1u} + \ldots + E_{p-1,g} + E_{p-1,u})$
			b	$4p$	σ	$2A_{1g} + A_{1u} + A_{2g} + 2A_{2u} + \begin{cases} B_{1g} + 2B_{1u} + 2B_{2g} + B_{2u} \ (p \text{ odd}) \\ 2B_{1g} + B_{1u} + B_{2g} + 2B_{2u} \ (p \text{ even}) \end{cases}$ $+ 3(E_{1g} + E_{1u} + \ldots + E_{p-1,g} + E_{p-1,u})$
			c	$4p$	σ'	$2A_{1g} + A_{1u} + A_{2g} + 2A_{2u} + \begin{cases} 2B_{1g} + B_{1u} + B_{2g} + 2B_{2u} \ (p \text{ odd}) \\ B_{1g} + 2B_{1u} + 2B_{2g} + B_{2u} \ (p \text{ even}) \end{cases}$ $+ 3(E_{1g} + E_{1u} + \ldots + E_{p-1,g} + E_{p-1,u})$
			d	$4p$	σ_h	$2A_{1g} + A_{1u} + 2A_{2g} + A_{2u} + \begin{cases} B_{1g} + 2B_{1u} + B_{2g} + 2B_{2u} \ (p \text{ odd}) \\ 2B_{1g} + B_{1u} + 2B_{2g} + B_{2u} \ (p \text{ even}) \end{cases}$ $+ 2E_{1g} + 4E_{1u} + 4E_{2g} + 2E_{2u} + \ldots + \begin{cases} 4E_{p-1,g} + 2E_{p-1,u} \ (p \text{ odd}) \\ 2E_{p-1,g} + 4E_{p-1,u} \ (p \text{ even}) \end{cases}$
			e	$2p$	σ_h, C_2, σ	$A_{1g} + A_{2g} + A_{2u} + \begin{Bmatrix} B_{1g} \ (p \text{ odd}) \\ B_{1u} \ (p \text{ even}) \end{Bmatrix} + B_{2g} + B_{2u}$ $+ E_{1g} + 2E_{1u} + 2E_{2g} + E_{2u} + \ldots + \begin{cases} 2E_{p-1,g} + E_{p-1,u} \ (p \text{ odd}) \\ E_{p-1,g} + 2E_{p-1,u} \ (p \text{ even}) \end{cases}$
			f	$2p$	σ_h, C_2', σ'	$A_{1g} + A_{2g} + A_{2u} + B_{1g} - B_{1u} + \begin{cases} B_{2u} \ (p \text{ odd}) \\ B_{2g} \ (p \text{ even}) \end{cases}$ $+ E_{1g} + 2E_{1u} + 2E_{2g} + E_{2u} + \ldots + \begin{cases} 2E_{p-1,g} + E_{p-1,u} \ (p \text{ odd}) \\ E_{p-1,g} + 2E_{p-1,u} \ (p \text{ even}) \end{cases}$
			g	2	C_{2p}, σ, σ'	$A_{1g} + A_{2u} + E_{1g} + E_{1u}$
			h	1	All	$A_{2u} + E_{1u}$
C^h_{2p+1} $= S_{2p+1}$	$A'' + E_1'$	$A' + E_1''$	*a*	$2(2p+1)$	None	$3(A' + A'' + E_1' + E_1'' + \ldots + E_p' + E_p'')$
			b	$2p+1$	σ_h	$2A' + A'' + 2E_1' + E_1'' + 2E_2' + E_2'' + \ldots + 2E_p' + E_p''$
			c	2	C_{2p+1}	$A' + A'' + E_1' + E_1''$
			d	1	All	$A'' + E_1'$

TABLE I—(*continued*)

I	II	III	IV	V	VI	VII
$D^h_{2p+1} = D^v_{2p+1}$	$A''_2+E'_1$	$A'_2+E''_1$	*a*	$4(2p+1)$	None	$3(A'_1+A''_1+A'_2+A''_2)+6(E'_1+E''_1+\ldots+E'_p+E''_p)$
			b	$2(2p+1)$	σ_v	$2A'_1+A''_1+A'_2+2A''_2+3(E'_1+E''_1+\ldots+E'_p+E''_p)$
			c	$2(2p+1)$	σ_h	$2A'_1+A''_1+2A'_2+A''_2+4E'_1+2E''_1+\ldots+4E'_p+2E''_p$
			d	$2p+1$	σ_h, σ_v, C_2	$A'_1+A'_2+A''_2+2E'_1+E''_1+\ldots+2E'_p+E''_p$
			e	2	C_{2p+1}	$A'_1+A''_2+E'_1+E''_1$
			f	1	All	$A''_2+E'_1$
C^v_{2p+1}	A_1+E_1	A_2+E_1	*a*	$2(2p+1)$	None	$3(A_1+A_2)+6(E_1+\ldots+E_p)$
			b	$2p+1$	σ	$2A_1+A_2+3(E_1+\ldots+E_p)$
			c	1	All	A_1+E_1
C^v_{2p}	A_1+E_1	A_2+E_1	*a*	$4p$	None	$3(A_1+A_2+B_1+B_2)+6(E_1+\ldots+E_{p-1})$
			b	$2p$	σ	$2A_1+A_2+2B_1+B_2+3(E_1+\ldots+E_{p-1})$
			c	$2p$	σ'	$2A_1+A_2+B_1+2B_2+3(E_1+\ldots+E_{p-1})$
			d	1	All	A_1+E_1
S_{4p}	$B+E_1$	$A+E_{2p-1}$	*a*	$4p$	None	$3(A+B+E_1+\ldots+E_{2p-1})$
			b	2	C_{2p}	$A+B+E_1+E_{2p-1}$
			c	1	All	$B+E_1$
$S^v_{4p} = D^d_{2p}$	B_1+E_1	A_2+E_{2p-1}	*a*	$8p$	None	$3(A_1+A_2+B_1+B_2)+6(E_1+E_2+\ldots+E_{2p-1})$
			b	$4p$	σ_v	$2A_1+A_2+2B_1+B_2+3(E_1+E_2+\ldots+E_{2p-1})$
			c	$4p$	C_2	$A_1+2A_2+2B_1+B_2+3(E_1+E_2+\ldots+E_{2p-1})$
			d	2	C_{2p}	$A_1+B_1+E_1+E_{2p-1}$
			e	1	All	B_1+E_1
T	F	F	*a*	12	None	$3A+3E+9F$
			b	6	C_2	$A+E+5F$
			c	4	C_3	$A+E+3F$
			d	1	All	F
T_d	F_2	F_1	*a*	24	None	$3A_1+3A_2+6E+9F_1+9F_2$
			b	12	σ	$2A_1+A_2+3E+4F_1+5F_2$
			c	6	C_2, σ	$A_1+E+2F_1+3F_2$
			d	4	C_3, σ	$A_1+E+F_1+2F_2$
			e	1	All	F_2

$T_h = i \times T$	F_u	F_g	a	24	None	$3A_g + 3A_u + 3E_g + 3E_u + 9F_g + 9F_u$
			b	12	σ	$2A_g + A_u + 2E_g + E_u - 4F_g + 5F_u$
			c	8	C_3	$A_g + A_u + E_g + E_u + 3F_g + 3F_u$
			d	6	C_2, σ	$A_g + E_g + 2F_g + 3F_u$
			e	1	All	F_u
O	F_1	F_1	a	24	None	$3A_1 + 3A_2 + 6E + 9F_1 + 9F_2$
			b	12	$6C_2$	$A_1 + 2A_2 + 3E + 5F_1 + 4F_2$
			c	8	$3C_2$	$A_1 + A_2 + 2E + 3F_1 + 3F_2$
			d	6	$3C_2, C_4$	$A_1 + E + 3F_1 + 2F_2$
			e	1	All	F_1
$O_h = i \times O$	F_{1u}	F_{1g}	a	48	None	$3(A_{1g} + A_{1u} + A_{2g} + A_{2u}) + 6(E_g + E_u) + 9(F_{1g} + F_{1u} + F_{2g} + F_{2u})$
			b	24	σ_h	$2(A_{1g} + A_{2g}) + A_{1u} + A_{2u} + 4E_g + 2E_u + 4(F_{1g} + F_{2g}) + 5(F_{1u} + F_{2u})$
			c	24	σ_d	$2A_{1g} + A_{1u} + A_{2g} + 2A_{2u} + 3E_g + 3E_u + 4F_{1g} + 5F_{1u} + 5F_{2g} + 4F_{2u}$
			d	12	$C_2^d, \sigma_h, \sigma_d$	$A_{1g} + A_{2g} + A_{2u} + 2E_g + E_u + 2F_{1g} + 3F_{1u} + 2F_{2g} + 2F_{2u}$
			e	8	C_3, σ_d	$A_{1g} + A_{2u} + E_g + E_u + F_{1g} + 2F_{1u} + 2F_{2g} + F_{2u}$
			f	6	$C_2^v, C_4, \sigma_h, \sigma_d$	$A_{1g} + E_g + F_{1g} + 2F_{1u} + F_{2g} + F_{2u}$
			g	1	All	F_{1u}
I	F_1	F_1	a	60	None	$3A + 9(F_1 + F_2) + 12G + 15H$
			b	30	C_2	$A + 5(F_1 + F_2) + 6G + 7H$
			c	20	C_3	$A + 3(F_1 + F_2) + 4G + 5H$
			d	12	C_5	$A + 3F_1 + F_2 + 2G + 3H$
			e	1	All	F_1
$I_h = i \times I$	F_{1u}	F_{1g}	a	120	None	$3(A_g + A_u) + 9(F_{1g} + F_{1u} + F_{2g} + F_{2u}) + 12(G_g + G_u) + 15(H_g + H_u)$
			b	60	σ	$2A_g + A_u + 4F_{1g} + 5F_{1u} + 4F_{2g} + 5F_{2u} + 6(G_g + G_u) + 8H_g + 7H_u$
			c	30	C_2, σ	$A_g + 2F_{1g} + 3F_{1u} + 2F_{2g} + 3F_{2u} + 3(G_g + G_u) + 4H_g + 3H_u$
			d	20	C_3, σ	$A_g + F_{1g} + 2F_{1u} + F_{2g} + 2F_{2u} + 2(G_g + G_u) + 3H_g + 2H_u$
			e	12	C_5, σ	$A_g + F_{1g} + 2F_{1u} + F_{2u} + G_g + G_u + 2H_g + H_u$
			f	1	All	F_{1u}

merely in establishing our theorem but also in showing what normal displacements any given molecule possesses. Tables for this purpose restricted to the crystallographic point groups have already been published by Placzek (1934) and further tables for special molecules have been given by Bright Wilson (1934).

Now in order to establish our theorem for each group of symmetry we must investigate what is the minimum number of points from which a molecule of the corresponding symmetry can be composed. For instance, if we want to construct a molecule of the symmetry T it will not be sufficient to take six atoms and place them on the twofold axes (the points (b) in Table I), for this configuration would have the higher symmetry O_h. To this end we list in Table II the minimum number of the different kinds of points which are necessary and sufficient to produce each of the various groups of symmetry. In this table an entry such as $2(a)$ means that two complete sets of points of the type (a) have to be taken to produce the necessary symmetry. Types which are not listed can never produce the symmetry no matter how many complete sets of such a type are taken, e.g. the type (b) for the group T.

TABLE II

Group	Minimum points required to produce symmetry of group
C^v_∞	$2(a)$
C^{vi}_∞	(a)
C_{2p+1}	$2(a)$
C_{2p}	$2(a)$
D_{2p+1}	(a)
D_{2p}	(a)
C^i_{2p-1}	(a)
C^i_{2p}	$2(a)$, $2(b)$ or $(a)+(b)$
D^i_{2p-1}	(a) or (b)
D^i_{2p}	(a), (b), (c), (d), (e) or (f)
C^h_{2p-1}	$2(a)$, $2(b)$ or $(a)+(b)$
D^h_{2p-1}	(a), (b), (c) or (d)
C^v_{2p-1}	$2(a)$, $2(b)$, $(a)+(b)$, $(a)+(c)$ or $(b)+(c)$
C^v_{2p}	$2(a)$, $2(b)$, $2(c)$, $(a)+(b)$, $(a)+(c)$, $(a)+(d)$, $(b)+(c)$, $(b)+(d)$ or $(c)+(d)$
S_{4p}	$2(a)$
S^v_{4p}	(a) or (b)
T	(a)
T_d	(a), (b) or (d)
T_h	(a) or (b)
O	(a)
O_h	(a), (b), (c), (d), (e) or (f)
I	(a)
I_h	(a), (b), (c), (d) or (e)

TABLE III

Group	Symmetrical product of degenerate representations
C_{∞}^{v}	$[E_k^2] = A_1 + E_{2k} \quad (k = 1, 2, \ldots)$
C_{∞}^{vi}	$[E_{kg}^2] = [E_{ku}^2] = A_{1g} + E_{2k,g} \quad (k = 1, 2, \ldots)$
C_{2p+1}	$[E_k^2] = A + E_{2k}$ for $k \leqq p/2$
	$= A + E_{2p+1-2k}$ for $k > p/2 \quad (k = 1, 2, \ldots, p)$
C_{2p}	$[E_k^2] = A + E_{2k}$ for $k < p/2$
	$= A + 2B$ for $k = p/2$
	$= A + E_{2p-2k}$ for $k > p/2 \quad (k = 1, 2, \ldots, p-1)$
D_{2p+1} and C_{2p+1}^{v}	$[E_k^2] = A_1 + E_{2k}$ for $k \leqq p/2$
	$= A_1 + E_{2p+1-2k}$ for $k > p/2 \quad (k = 1, \ldots, p)$
D_{2p} and C_{2p}^{v}	$[E_k^2] = A_1 + E_{2k}$ for $k < p/2$
	$= A_1 + B_1 + B_2$ for $k = p/2$
	$= A_1 + E_{2p-2k}$ for $k > p/2 \quad (k = 1, \ldots, p-1)$
C_{2p+1}^{i}	$[E_{kg}^2] = [E_{ku}^2] = A_g + E_{2k,g}$ for $k \leqq p/2$
	$= A_g + E_{2p+1-2k,g}$ for $k > p/2 \quad (k = 1, \ldots, p)$
C_{2p}^{i}	$[E_{kg}^2] = [E_{ku}^2] = A_g + E_{2k,g}$ for $k < p/2$
	$= A_g + 2B_g$ for $k = p/2$
	$= A_g + E_{2p-2k,g}$ for $k > p/2 \quad (k = 1, \ldots, p-1)$
D_{2p+1}^{i}	$[E_{kg}^2] = [E_{ku}^2] = A_{1g} + E_{2k,g}$ for $k \leqq p/2$
	$= A_{1g} + E_{2p+1-2k,g}$ for $k > p/2 \quad (k = 1, \ldots, p)$
D_{2p}^{i}	$[E_{kg}^2] = [E_{ku}^2] = A_{1g} + E_{2k,g}$ for $k < p/2$
	$= A_{1g} + B_{1g} + B_{2g}$ for $k = p/2$
	$= A_{1g} + E_{2p-2k,g}$ for $k > p/2 \quad (k = 1, \ldots, p-1)$
C_{2p+1}^{h} and D_{2p+1}^{h}	$[E_k'^2] = [E_k''^2] = A' + E_{2k}'$ for $k \leqq p/2$
	$= A' + E_{2p+1-2k}'$ for $k > p/2 \quad (k = 1, \ldots, p)$
S_{4p}	$[E_k^2] = A + E_{2k}$ for $k < p$
	$= A + 2B$ for $k = p$
	$= A + E_{4p-2k}$ for $k > p \quad (k = 1, \ldots, 2p-1)$
S_{4p}^{v}	$[E_k^2] = A_1 + E_{2k}$ for $k < p$
	$= A_1 + B_1 + B_2$ for $k = p$
	$= A_1 + E_{4p-2k}$ for $k > p \quad (k = 1, \ldots, 2p-1)$
T	$[E^2] = A + E$
	$[F^2] = A + E + F$
T_d and O	$[E^2] = A_1 + E$
	$[F_1^2] = [F_2^2] = A_1 + E + F_2$
T_h	$[E_g^2] = [E_u^2] = A_g + E_g$
	$[F_g^2] = [F_u^2] = A_g + E_g + F_g$
O_h	$[E_g^2] = [E_u^2] = A_{1g} + E_g$
	$[F_{1g}^2] = [F_{1u}^2] = [F_{2g}^2] = [F_{2u}^2] = A_{1g} + E_g + F_{2g}$
I	$[F_1^2] = [F_2^2] = A + H$
	$[G^2] = A + G + H$
	$[H^2] = A + G + 2H$
I_h	$[F_{1g}^2] = [F_{1u}^2] = [F_{2g}^2] = [F_{2u}^2] = A_g + H_g$
	$[G_g^2] = [G_u^2] = A_g + G_g + H_g$
	$[H_g^2] = [H_u^2] = A_g + G_g + 2H_g$

We require, further, the symmetrical products of the various degenerate representations. These may be taken from the work of Tisza (1933), for they are identical with the representations subtended by the first overtones of the corresponding normal vibrations. For convenience we tabulate in Table III these symmetrical product representations for all degenerate representations.

Using Tables I, II and III, our general theorem will be verified easily. For instance, according to Table II the group of symmetry O_h can be produced by one complete set of equivalent points of any of the types (*a*), (*b*), (*c*), (*d*), (*e*) or (*f*). But from Table I we see that each of these sets of points has normal displacements of the type E_g, and according to Table III, $E_g[\Phi^2]$ always contains the identical representation for any degenerate representation Φ of the group O_h. We must, of course, always remember that displacements of the type A_{1g} (or A_1 or A or A_g) must not be made use of, since the molecular configurations considered are always taken to be stable with respect to all totally symmetrical displacements.

5—Conclusion

In concluding we would like to point out again that the forces which tend to destroy the symmetrical configuration, and which we have shown exist in cases of orbital degeneracy, will be important only if the degenerate electrons participate strongly in the binding of the molecule. The effect may be small both if the degenerate electrons are in inner atomic shells or if they are in highly excited states. Spin degeneracy may also produce similar effects, but these, too, will be small, since the coupling of spin and nuclear motion will depend upon the interaction of the spin with the orbital motion of the electrons, which interaction, at least for light elements, is small. Furthermore, a general theorem of Kramers (1930) and Wigner (1932) shows that for molecules containing an odd number of electrons there is a twofold spin degeneracy which cannot be split by any electrical forces. Such twofold spin degeneracy cannot therefore produce any instability of the molecular configuration. It can be shown, however, that with the exception of this one type of degeneracy all degenerate electronic states of non-linear molecules are unstable whether the degeneracy is due to electronic orbits or to spin. The proof of this statement, together with a more detailed discussion of the order of magnitude of splitting, will be given in a second paper.

The authors are indebted to Professor L. Landau for a discussion which led to the formulation of the theorem here established. The research was carried out in the Sir William Ramsay Laboratories of Inorganic and Physical Chemistry, University College, London. We are greatly indebted to the Imperial Chemical Industries, Ltd., for grants which enabled us to carry out the work, and above all to Professor F. G. Donnan, Director of the Laboratory, for his great hospitality and kind interest.

Summary

It is shown that orbital electronic degeneracy and stability of the nuclear configuration are incompatible unless all the atoms of a molecule lie on a straight line. The proof is based on group theory and is therefore valid only if accidental degeneracy is disregarded. If the electrons causing the degeneracy are not essential for molecular binding, only a slight instability will result. Table I, which is needed to prove the theorem, can also be used to obtain the number of proper vibrations of a given symmetry type for any polyatomic molecule.

References

Bright Wilson, E. 1934 *J. Chem. Phys.* **2**, 432.
Kramers, H. A. 1930 *Proc. Acad. Sci. Amst.* **33**, 959.
Lennard-Jones, J. E. 1934 *Trans. Faraday Soc.* **30**, 70.
Mulliken, R. S. 1933 *Phys. Rev.* **43**, 279.
Placzek, G. 1934 *Handbuch der Radiologie*, **6**, II, 205.
Tisza, L. 1933 *Z. Phys.* **82**, 48.
van Vleck 1932 "Electric and Magnetic Susceptibilities," p. 273. Oxford.
Weyl 1928 "Gruppentheorie und Quantenmechanik," p. 115. Leipzig.
Wigner, E. 1930 *Nachr. Ges. Wiss. Gottingen*, p. 133.
— 1932 *Nachr. Ges. Wiss. Gottingen*, p. 546.

The Covalent Bond–Valence Bond Theory

II

Editor's Comments on Papers 8 and 9

8 **Pauling:** *The Nature of the Chemical Bond. Application of Results Obtained from the Quantum Mechanics and from a Theory of Paramagnetic Susceptibility to the Structure of Molecules*

9 **Pauling:** *The Nature of the Chemical Bond. III. The Transition from One Extreme Bond Type to Another*

While physicists such as Bethe, Van Vleck, and others were applying group theoretical techniques to the quantum mechanical description of the electronic structure and magnetism of ionic transition metal compounds, chemist Linus Pauling and physicist J. C. Slater were attempting to modify the Lewis–Heitler–London electron pair bond concept [26, 27] to account for the magnetic moments and electron structures of complex ions. The success of Pauling and Slater in doing this is a monumental tribute to their genius, especially in view of the inadequacies we now know to be inherent in the valence bond theory. Symmetry is not explicitly used in the development of the theory; however, it underlies the concept of orbital hybridization so clearly expounded by Pauling, who received the Nobel Prize in chemistry for his work.

Paper 8 is the initial description by Pauling of the valence bond theory; Paper 9 applies the valence bond theory directly to the stereochemistry and bonding character of transition-metal coordination compounds. Slater was approaching the problem from a similar point of view. While both scientists grasped the fundamental concepts of orbital hybridization and electron pair bond formation, it was Pauling who captured the imagination and thought of the chemical community.

Reprinted from *J. Amer. Chem. Soc.*, **53**, 1367–1369, 1400 (1931)

8

[Contribution from Gates Chemical Laboratory, California Institute of Technology, No. 280]

THE NATURE OF THE CHEMICAL BOND. APPLICATION OF RESULTS OBTAINED FROM THE QUANTUM MECHANICS AND FROM A THEORY OF PARAMAGNETIC SUSCEPTIBILITY TO THE STRUCTURE OF MOLECULES

By Linus Pauling

Received February 17, 1931   Published April 6, 1931

During the last four years the problem of the nature of the chemical bond has been attacked by theoretical physicists, especially Heitler and London, by the application of the quantum mechanics. This work has led to an approximate theoretical calculation of the energy of formation and of other properties of very simple molecules, such as H_2, and has also provided a formal justification of the rules set up in 1916 by G. N. Lewis for his electron-pair bond. In the following paper it will be shown that many more results of chemical significance can be obtained from the quantum mechanical equations, permitting the formulation of an extensive and powerful set of rules for the electron-pair bond supplementing those of Lewis. These rules provide information regarding the relative strengths of bonds formed by different atoms, the angles between bonds, free rotation or lack of free rotation about bond axes, the relation between the quantum numbers of bonding electrons and the number and spatial arrangement of the bonds, etc. A complete theory of the magnetic moments of molecules and complex ions is also developed, and it is shown that for many compounds involving elements of the transition groups this theory together with the rules for electron-pair bonds leads to a unique assignment of electron structures as well as a definite determination of the type of bonds involved.[1]

I. The Electron-Pair Bond

The Interaction of Simple Atoms.—The discussion of the wave equation for the hydrogen molecule by Heitler and London,[2] Sugiura,[3] and Wang[4] showed that two normal hydrogen atoms can interact in either of two ways, one of which gives rise to repulsion with no molecule formation, the other

[1] A preliminary announcement of some of these results was made three years ago [Linus Pauling, *Proc. Nat. Acad. Sci.*, **14**, 359 (1928)]. Two of the results (90° bond angles for *p* eigenfunctions, and the existence, but not the stability, of tetrahedral eigenfunctions) have been independently discovered by Professor J. C. Slater and announced at meetings of the National Academy of Sciences (Washington, April, 1930) and the American Physical Society (Cleveland, December, 1930).

[2] W. Heitler and F. London, *Z. Physik*, **44**, 455 (1927).

[3] Y. Sugiura, *ibid.*, **45**, 484 (1927).

[4] S. C. Wang, *Phys. Rev.*, **31**, 579 (1928).

to attraction and the formation of a stable molecule. These two modes of interaction result from the identity of the two electrons. The characteristic resonance phenomenon of the quantum mechanics, which produces the stable bond in the hydrogen molecule, always occurs with two electrons, for even though the nuclei to which they are attached are different, the energy of the unperturbed system with one electron on one nucleus and the other on the other nucleus is the same as with the electrons interchanged. Hence we may expect to find electron-pair bonds turning up often.

But the interaction of atoms with more than one electron does not always lead to molecule formation. A normal helium atom and a normal hydrogen atom interact in only one way,[5] giving repulsion only, and two normal helium atoms repel each other except at large distances, where there is very weak attraction.[5,6] Two lithium atoms, on the other hand, can interact in two ways,[7] giving a repulsive potential and an attractive potential, the latter corresponding to formation of a stable molecule. In these cases it is seen that only when each of the two atoms initially possesses an unpaired electron is a stable molecule formed. The general conclusion that an electron-pair bond is formed by the interaction of an unpaired electron on each of two atoms has been obtained formally by Heitler[8] and London,[9] with the use of certain assumptions regarding the signs of integrals occurring in the theory. The energy of the bond is largely the resonance or interchange energy of two electrons. This energy depends mainly on electrostatic forces between electrons and nuclei, and is not due to magnetic interactions, although the electron spins determine whether attractive or repulsive potentials, or both, will occur.

Properties of the Electron-Pair Bond.—From the foregoing discussion we infer the following properties of the electron-pair bond.

1. *The electron-pair bond is formed through the interaction of an unpaired electron on each of two atoms.*

2. *The spins of the electrons are opposed when the bond is formed, so that they cannot contribute to the paramagnetic susceptibility of the substance.*

3. *Two electrons which form a shared pair cannot take part in forming additional pairs.*

In addition we postulate the following three rules, which are justified by the qualitative consideration of the factors influencing bond energies. An outline of the derivation of the rules from the wave equation is given below.

[5] G. Gentile, *Z. Physik.*, **63**, 795 (1930).

[6] J. C. Slater, *Phys. Rev.*, **32**, 349 (1927).

[7] M. Delbrück, *Ann. Physik*, **5**, 36 (1930).

[8] W. Heitler, *Z. Physik*, **46**, 47 (1927); **47**, 835 (1928); *Physik. Z.*, **31**, 185 (1930), etc.

[9] F. London, *Z. Physik*, **46**, 455 (1928); **50**, 24 (1928); "Sommerfeld Festschrift," p. 104; etc.

4. *The main resonance terms for a single electron-pair bond are those involving only one eigenfunction from each atom.*

5. *Of two eigenfunctions with the same dependence on r, the one with the larger value in the bond direction will give rise to the stronger bond, and for a given eigenfunction the bond will tend to be formed in the direction with the largest value of the eigenfunction.*

6. *Of two eigenfunctions with the same dependence on θ and φ, the one with the smaller mean value of r, that is, the one corresponding to the lower energy level for the atom, will give rise to the stronger bond.*

Here the eigenfunctions referred to are those for an electron in an atom, and r, θ and φ are polar coördinates of the electron, the nucleus being at the origin of the coördinate system.

It is not proposed to develop a complete proof of the above rules at this place, for even the formal justification of the electron-pair bond in the simplest cases (diatomic molecule, say) requires a formidable array of symbols and equations. The following sketch outlines the construction of an inclusive proof.

It can be shown[10] that if Ψ is an arbitrary function of the independent variables in a wave equation

$$(H - W)\psi = 0$$

then the integral

$$E = \int \Psi^* H \Psi d\tau$$

called the variation integral, is always larger than W_0, the lowest energy level for the system. A function Ψ containing several parameters provides the best approximation to the eigenfunction ψ_0 for the normal state of the system when the variational integral is minimized with respect to these parameters. Now let us consider two atoms A and B connected by an electron-pair bond, and for simplicity let all the other electrons in the system be paired, the pairs being either lone pairs or pairs shared between A or B and other atoms. Let us assume that there are available for bond formation by atom A several single-electron eigenfunctions of approximately the same energy, and that the change in energy of penetration into the core is negligible compared with bond energy. Then we may take as single-electron eigenfunctions

$$\psi_{Ai} = \Sigma_k a_{ik} \psi^0_{Ak}$$

in which the a_{ik}'s are numerical coefficients and the ψ^0_{Ak}'s are an arbitrary set of single-electron eigenfunctions, such as those obtained on separating the wave equation in polar coördinates. From the ψ_{Ai}'s there is built up a group composed of atom A and the atoms to which it is bonded except atom B, such that all electrons are paired except one, corresponding to the eigenfunction ψ_{Ai}, say. From atom B a similar group with one unpaired electron is built. The interaction energy of these two groups can then be calculated with the aid of the variational equation through the substitution of an eigenfunction for the molecule built of those for the two groups in such a way that it has the correct symmetry character. The construction of this eingenfunction and evaluation of the integral would be very laborious; it will be noticed, however, that this problem is formally similar to Born's treatment[11] of the interaction of two atoms in S states, based on Slater's treatment of atomic eigenfunctions, and the value of E is found to be

$$E = W_A + W_B + J_E + J_X - \Sigma_Y J_Y - 2\Sigma_Z J_Z$$

---

[10] A clear discussion is given by C. Eckart, *Phys. Rev.*, **36**, 878 (1930).

[11] M. Born, *Z. Physik*, **64**, 729 (1930).

Summary

With the aid of the quantum mechanics there is formulated a set of rules regarding electron-pair bonds, dealing particularly with the strength of bonds in relation to the nature of the single-electron eigenfunctions involved. It is shown that one single-electron eigenfunction on each of two atoms determines essentially the nature of the electron-pair bond formed between them; this effect is accentuated by the phenomenon of concentration of the bond eigenfunctions.

The type of bond formed by an atom is dependent on the ratio of bond energy to energy of penetration of the core (*s*–*p* separation). When this ratio is small, the bond eigenfunctions are *p* eigenfunctions, giving rise to bonds at right angles to one another; but when it is large, new eigenfunctions especially adapted to bond formation can be constructed. From *s* and *p* eigenfunctions the best bond eigenfunctions which can be made are four equivalent tetrahedral eigenfunctions, giving bonds directed toward the corners of a regular tetrahedron. These account for the chemist's tetrahedral atom, and lead directly to free rotation about a single bond but not about a double bond and to other tetrahedral properties. A single *d* eigenfunction with *s* and *p* gives rise to four strong bonds lying in a plane and directed toward the corners of a square. These are formed by bivalent nickel, palladium, and platinum. Two *d* eigenfunctions with *s* and *p* give six octahedral eigenfunctions, occurring in many complexes formed by transition-group elements.

It is then shown that (excepting the rare-earth ions) the magnetic moment of a non-linear molecule or complex ion is determined by the number of unpaired electrons, being equal to $\mu_S = 2\sqrt{S(S+1)}$, in which S is half that number. This makes it possible to determine from magnetic data which eigenfunctions are involved in bond formation, and so to decide between electron-pair bonds and ionic or ion-dipole bonds for various complexes. It is found that the transition-group elements almost without exception form electron-pair bonds with CN, ionic bonds with F, and ion-dipole bonds with H_2O; with other groups the bond type varies.

Examples of deductions regarding atomic arrangement, bond angles and other properties of molecules and complex ions from magnetic data, with the aid of calculations involving bond eigenfunctions, are given.

PASADENA, CALIFORNIA

Reprinted from *J. Amer. Chem. Soc.*, **54**, 988–995 (1932)

9

[CONTRIBUTION FROM GATES CHEMICAL LABORATORY, CALIFORNIA INSTITUTE OF TECHNOLOGY, NO. 292]

THE NATURE OF THE CHEMICAL BOND. III. THE TRANSITION FROM ONE EXTREME BOND TYPE TO ANOTHER[1]

BY LINUS PAULING

RECEIVED NOVEMBER 9, 1931 PUBLISHED MARCH 5, 1932

A question which has been keenly argued for a number of years is the following: if it were possible continuously to vary one or more of the parameters determining the nature of a system such as a molecule or a crystal, say the effective nuclear charges, then would the transition from one extreme bond type to another take place continuously, or would it show discontinuities? For example, are there possible all intermediate bond types between the pure ionic bond and the pure electron-pair bond? With the development of our knowledge of the nature of the chemical bond it has become evident that this question and others like it cannot be answered categorically. It is necessary to define the terms used and to indicate the point of view adopted; and then it may turn out, as with this question, that no statement of universal application can be made.

In the following sections, after a discussion of the properties of ionic compounds and compounds containing electron-pair bonds, the transition from one extreme to the other is considered. It is concluded that in some cases the transition could take place continuously, whereas in others an effective discontinuity would appear.

Bond Type and Atomic Arrangement

The properties of a compound depend on two main factors, the nature of the bonds between the atoms, and the nature of the atomic arrangement. It is convenient to consider that actual bonds approach more or less closely one or another of certain postulated extreme bond types (ionic, electron-pair, ion-dipole, one-electron, three-electron, metallic, etc.), or

[1] A part of the material of this paper was presented to the American Chemical Society at Buffalo, New York, September 2, 1931, under the title "The Structure of Crystals and the Nature of the Chemical Bond."

are intermediate between one extreme type and another, or involve two or more co-existent types. The satisfactory description of the atomic arrangement in a crystal or molecule necessitates the complete determination of the position of the atoms relative to one another. Sometimes individual details of the atomic arrangement are of interest; for example, in case the atoms are held together in part by strong bonds and in part by very weak bonds, it is instructive to mention the type of atomic aggregate formed by the strong bonds (finite molecules or complex ions, or molecules or complex ions of very great extent in one, two or three dimensions). In other cases the number of nearest neighbors of each atom (its coördination number) and their relative positions are items of interest.

There is, of course, a close relation between atomic arrangement and bond type. Thus the four single bonds of a carbon atom are directed toward the corners of a tetrahedron. But tetrahedral and octahedral configurations are also assumed in ionic compounds, so that it is by no means always possible to deduce the bond type from a knowledge of the atomic arrangement.

An abrupt change in properties in a series of compounds, as in the melting points or boiling points of halides, is often taken as indicating an abrupt change in bond type. Thus of the fluorides

	NaF	MgF_2	AlF_3	SiF_4	PF_5	SF_6
M. p.	980°	1400°	1040°	−77°	−83°	−55°

those of high melting points have been described as salts, and the others as covalent compounds. Actually the Al–F bond is no doubt closely similar to the Si–F bond. The abrupt change in properties between AlF_3 and SiF_4 is due to a change in atomic arrangement. In NaF, MgF_2 and AlF_3 the coördination number of the metal (six) is greater than the stoichiometric ratio of non-metal to metal atoms, so that each non-metal atom or ion is held jointly by two or more metal atoms or ions, resulting in high melting and boiling points. In SiF_4, PF_5 and SF_6 the coördination number is just equal to this ratio. The discrete molecules of composition given by the formulas are held only loosely together, and the substances melt and boil easily. As pointed out long ago by Kossel, this ease of fusion and volatilization would be expected for symmetrical ionic molecules, and is not sound evidence for the presence of electron-pair bonds. Volatility does not depend mainly on bond type, but on the atomic arrangement and the distribution of bonds.

The Ionic Bond

The theoretical treatment of the properties of ionic crystals and molecules has been carried farther than that of other types of atomic aggregates. The Born theory of crystal energy permits the calculation to within a few per cent. of the energy of dissociation of an ionic crystal into gaseous ions when the structure of the crystal and the equilibrium interionic distances are known. Moreover, it is now possible to predict interionic distances to within 2 or 3%, with the use either of Goldschmidt's set[2] of ionic radii, derived from the experimentally observed interionic distances in crystals of simple structure, many of which were prepared for the first time and studied with x-rays in his Mineralogical Institute in Oslo, or of a closely similar set obtained[3] by a method of treatment based on the theory of the electron distribution in atoms and ions. The screening constants[4] used in the derivation of this set were derived in part theoretically (for light ions) and in part from the experimental values of the mole refraction of atoms and ions. It has been recently found[5] that screening constants can be obtained purely from observed ionization potentials

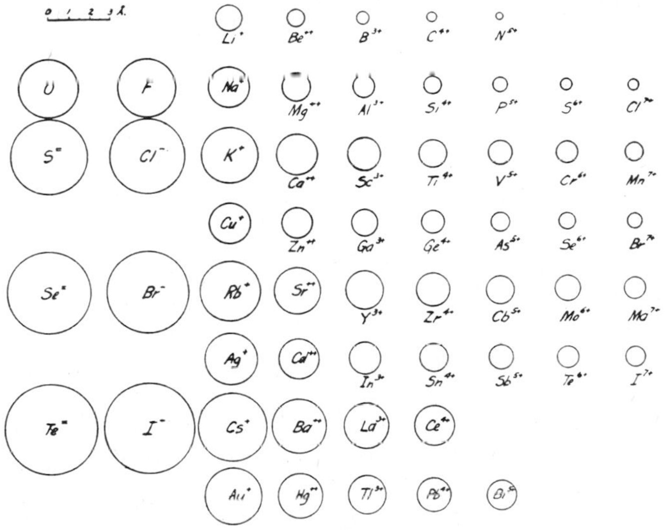

Fig. 1.—The sizes of spherically-symmetrical ions. The radii of the spheres are taken equal to the crystal radii of the ions.

and x-ray term values. The new set of screening constants is in nearly complete agreement with the old one, which gives a feeling of confidence in this method of treating the properties of many-electron atoms and monatomic ions.

The fundamental character of the understanding of ionic crystals which has been obtained is seen from the nature of the dependence of ionic radii on atomic number, shown in Fig. 3 of Ref. 3. The univalent radii lie on

[2] V. M. Goldschmidt, "Verteilungsgesetze der Elemente," Oslo, 1926, Vol. VII.

[3] Linus Pauling, THIS JOURNAL, **49**, 765 (1927).

[4] Linus Pauling, *Proc. Roy. Soc.* (London), **A114**, 181 (1927).

[5] Linus Pauling and J. Sherman, *Z. Krist.*, **81**, 1 (1932).

a smooth curve for each isoelectronic sequence; these are corrected in a systematic way for the effect of increasing Coulomb attraction in pulling highly charged ions more closely together to obtain the crystal radii. As is seen in Fig. 1, anions are very large, larger than all cations except those of the alkali metals. Moreover, the anions occur in pairs of nearly the same radius, $O^{=}$ and F^-, etc. This is the reason that hydroxyl ion and fluorine ion so often replace each other isomorphously in minerals, such as topaz, $Al_2SiO_4(OH,F)_2$. (The proton has little effect on the crystal radius.)

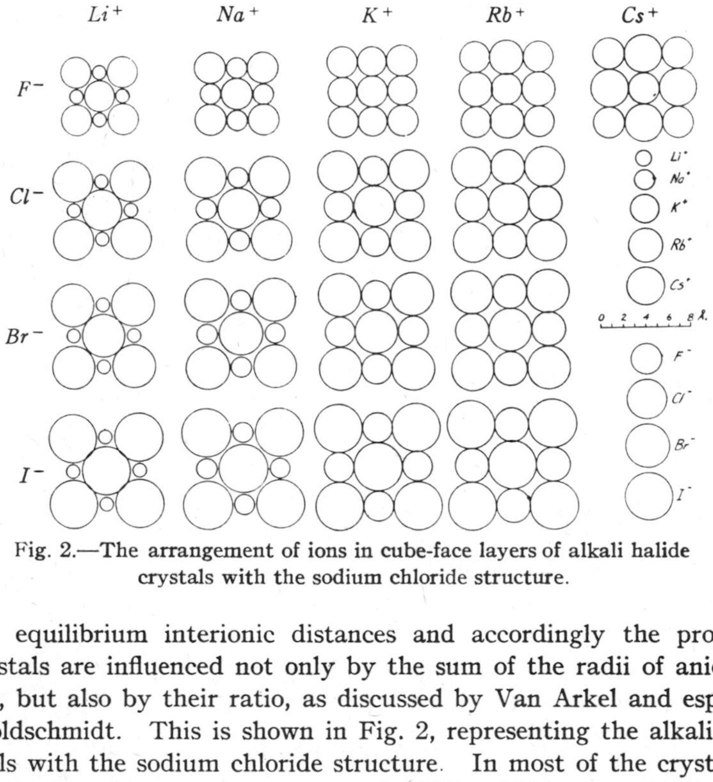

Fig. 2.—The arrangement of ions in cube-face layers of alkali halide crystals with the sodium chloride structure.

The equilibrium interionic distances and accordingly the properties of crystals are influenced not only by the sum of the radii of anion and cation, but also by their ratio, as discussed by Van Arkel and especially by Goldschmidt. This is shown in Fig. 2, representing the alkali halide crystals with the sodium chloride structure. In most of the crystals the anions and cations are in contact. In LiCl, LiBr and LiI, however, the anions are so large relative to the cations that they come into mutual contact, and the size of the unit of structure is determined by the anion radius alone, the cations being left to rattle around in the interstices. This determination of the dimensions of the units and the arrangement of the anions by the anions alone has been especially discussed by W. L. Bragg and co-workers,[6] who have utilized it extensively in studying the structures of silicate minerals. In LiF, NaCl, NaBr and NaI the phe-

[6] W. L. Bragg and J. West, *Proc. Roy. Soc.* (London), **A114**, 450 (1927).

nomenon of *double repulsion* (anion–cation as well as anion–anion repulsion operative) causes the unit to be larger than would be expected from either the anion–cation or anion–anion radius sum. Anion contact and double repulsion have a striking effect on the course of the properties of the alkali halides, particularly the melting points and boiling points.[7]

The conception of simple ionic crystals as coördinated structures, developed by Ewald, Goldschmidt and others, has been shown to be applicable to complex crystals by W. L. Bragg's determination of the structure of a number of the silicate minerals,[8] and through the study of brookite[9] and topaz[10] led ultimately to the formulation of a set of principles governing the structure of complex ionic crystals.[11] These rules may be illustrated by the application of one of them to the question of the isomorphous replacement of OH^- by F^-. The electrostatic valence rule states that in a stable crystal each anion tends to have its charge balanced by the electrostatic bonds of adjacent cations, the strength of a bond from a cation being taken as the ratio of its charge to its coördination number; in other words, the valence of the cation is considered as evenly distributed among the anions coördinated about it. Thus Si^{4+} with coördination number 4 has bonds of strength 1, Al^{3+} with coördination number 6, bonds of strength $^1/_2$. In topaz, $Al_2SiO_4(OH,F)_2$; chondrodite, $Mg_5Si_2O_8(OH,F)_2$; etc., the OH^- ions are in positions of total strength of bonds from cations other than hydrogen equal to 1, and so they may be replaced by F^- ions. But in staurolite, $H_2FeAl_4Si_2O_{12}$, there are not two oxygens in positions of total bond strength 1, but rather four with total bond strength 3/2, suggesting strongly the presence of two $[OHO]^=$ groups. In consequence fluorine should not be expected to occur in this mineral, in agreement with observation.

The Electron-Pair Bond

It was shown by G. N. Lewis and Irving Langmuir that unique electronic structures involving electron-pair bonds can be assigned with considerable certainty to a great many molecules and crystals. For many others, however, decision between two or more alternative structures has been difficult; moreover, the theory has not led to predictions regarding atomic arrangements other than the postulated tetrahedral arrangement of four electron pairs. Considerable further progress has now been made possible through the deduction from the quantum mechanics of a set of rules regarding electron-pair bonds, and in particular the discovery of a simple semi-quantitative treatment of bond eigenfunctions which

[7] Linus Pauling, This Journal, **50,** 1036 (1928).
[8] W. L. Bragg, *Z. Krist.*, **74,** 237 (1930).
[9] Linus Pauling and J. H. Sturdivant, *ibid.*, **68,** 239 (1928).
[10] Linus Pauling, *Proc. Nat. Acad. Sci.*, **14,** 603 (1928).
[11] Linus Pauling, This Journal, **51,** 1010 (1929).

gives much information regarding the strength and mutual orientation of the bonds which can be formed by various atoms.[12]

It has been found that the strength and direction of an electron-pair bond formed by an atom are determined essentially by one electronic eigenfunction. The bond tends to be formed in the direction in which

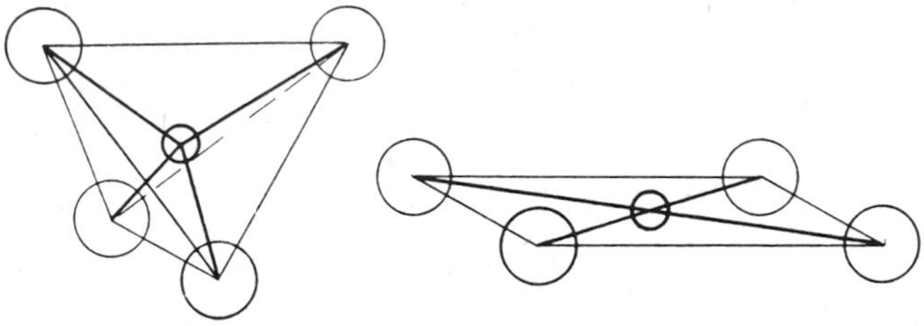

Tetrahedral, sp^3 (2.000). C in diamond and compounds. Zn in $[Zn(CN)_4]^{=}$, etc.

Square, dsp^2 (2.694). Ni in $[Ni(CN)_4]^{=}$, Pd in $[PdCl_4]^{=}$, Pt in $[PtCl_4]^{=}$, Au in $[AuCl_4]^{-}$, etc.

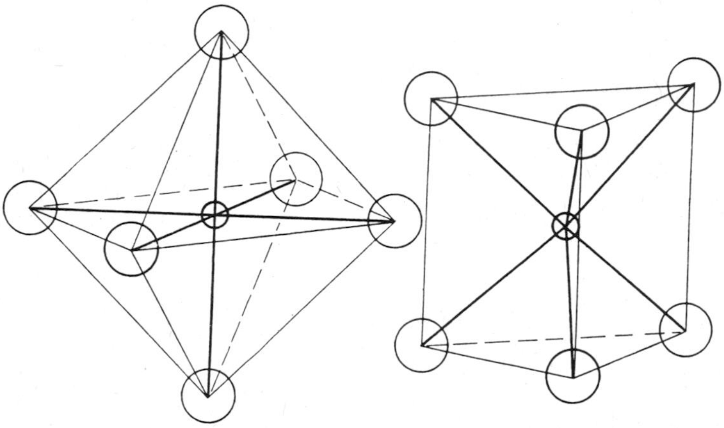

Octahedral, d^2sp^3 (2.923). Co in $[Co(CN)_6]^{\equiv}$, Pd in $[PdCl_6]^{=}$, Pt in $[PtCl_6]^{=}$, etc.

Trigonal–prismatic, (2.983). Mo in MoS_2, and perhaps in $[Mo(CN)_6]^{=}$.

Fig. 3.—The relative orientations of various electron-pair bonds.

the eigenfunction has its maximum value, and the greater the concentration of the eigenfunction in the bond direction the stronger the bond will be. The spherically-symmetrical *s* eigenfunction can form a bond in any direction of strength 1 according to the semi-quantitative treatment, and a *p* eigenfunction a bond of strength 1.732 in either of two opposite directions. But in most atoms which form four or more bonds

[12] (a) Linus Pauling, *Proc. Nat. Acad. Sci.*, **14**, 359 (1928); (b) THIS JOURNAL, **53**, 1367 (1931); (c) *Phys. Rev.*, **37**, 1185 (1931); (d) J. C. Slater, *ibid.*, **37**, 481 (1931); **38**, 1109 (1931).

the *s* and *p* eigenfunctions do not retain their identity, being instead combined to form new eigenfunctions, better suited to bond formation. The best bond eigenfunction which can be formed from the one *s* and three *p* eigenfunctions in a given shell has the strength 2.000. Moreover, three other equivalent bond eigenfunctions can also be formed, and the four bonds are directed toward the corners of a regular tetrahedron (Fig. 3). This result immediately gives the quantum-mechanical justification of the chemist's tetrahedral carbon atom, with all its properties, such as free rotation about a single bond (except when restricted by steric effects) and lack of it about a double bond, and shows that many other atoms direct their bonds toward tetrahedron corners.

The nature of possible bond eigenfunctions involving *d* eigenfunctions depends on the number of *d* eigenfunctions available. Bivalent nickel contains eight unshared 3*d* electrons, which require at least four of the five 3*d* eigenfunctions, leaving only one for bond formation. When this is combined with the 4*s* and the three 4*p* eigenfunctions of nearly the same energy, it is found that not five but only four equivalent bond eigenfunctions, of strength 2.694, can be formed, and that these are directed toward the corners of a square. Thus the $[Ni(CN)_4]^{=}$ ion should have a square rather than a tetrahedral configuration. This was suggested for complexes containing bivalent palladium and platinum by Werner, and was verified by Dickinson[13] by the x-ray study of K_2PdCl_4 and K_2PtCl_4. We expect a similar configuration for complexes of trivalent copper, silver and gold.

When two *d* eigenfunctions are available, as in trivalent cobalt, quadrivalent palladium and platinum, etc., six equivalent bond eigenfunctions of strength 2.923 and directed toward the corners of a regular octahedron can be formed. These form the bonds in a great many octahedral complexes.

It is interesting to note, as pointed out to me by Mr. J. L. Hoard, that these considerations lead to an explanation of the stability of trivalent cobalt in electron-pair bond complexes as compared to ionic compounds. The formation of complexes does not change the equilibrium between bivalent and trivalent iron very much, as is seen from the electrode potentials, while a great change is produced in the equilibrium between bivalent and trivalent cobalt.

$$\left.\begin{array}{lr} Fe^{++} = Fe^{+++} + E^- & -0.74\ v. \\ [Fe(CN)_6]^{4-} = [Fe(CN)_6]^{3-} + E^- & -0.49\ v. \end{array}\right\rangle -0.25\ v.$$

$$\left.\begin{array}{lr} Co^{++} = Co^{+++} + E^- & -1.8\ v. \\ [Co(CN)_6]^{4-} = [Co(CN)_6]^{3-} + E^- & +0.3\ v. \end{array}\right\rangle -2.1\ v.$$

The effect is so pronounced that covalent compounds of bivalent cobalt are difficult to prepare, decomposing water with the liberation of hydro-

[13] R. G. Dickinson, This Journal, **44**, 2404 (1922).

gen. The explanation is contained in Fig. 4. In the ions Co^{++}, Co^{+++}, Fe^{++}, and Fe^{+++} there is room for all unshared electrons in the $3d$ subshell or inner subshells. When octahedral bonds are formed only three $3d$ eigenfunctions are left for occupancy by unshared electrons. These are enough for bivalent and trivalent iron and for trivalent cobalt, but they hold only six of the seven electrons of bivalent cobalt. The seventh electron must accordingly occupy an outer unstable orbit, causing the complex to become unstable.

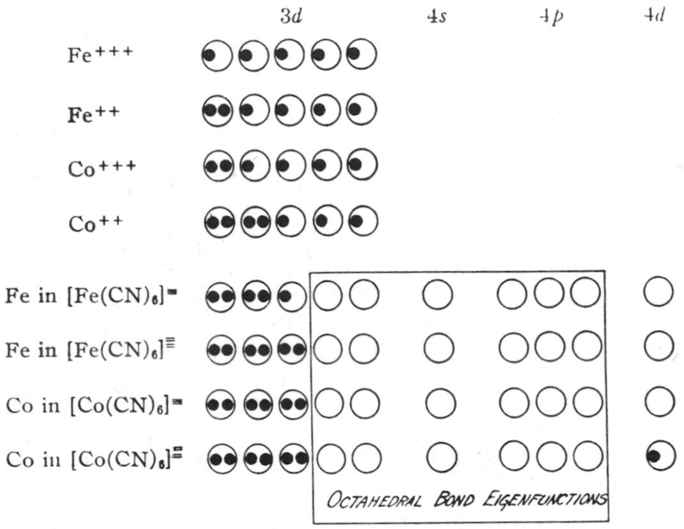

Fig. 4.—A diagram representing the electronic configurations of certain complex ions. Each circle represents a single-electron positional eigenfunction, each dot an electron.

The structure[14] of the mineral molybdenite, MoS_2, in which each molybdenum atom is surrounded by six sulfur atoms at the corners of a triangular prism, has long been puzzling, for if the crystal were composed of ions an octahedral arrangement would be more stable, and the large number of octahedral complexes containing electron-pair bonds would suggest an octahedral electron-pair bond structure also. But the molybdenum atom contains only two unshared d electrons, and it has been recently found by Mr. Ralph Hultgren that in this case six equivalent bonds of strength 2.983 (considerably greater than that of octahedral bond eigenfunctions) directed toward the corners of a trigonal prism may be formed. The theoretical ratio of altitude to base of the prism is 1000, while the experimental value is 1.007 ± 0.038. It would be interesting to determine whether a similar configuration holds for $[Mo(CN)_6]^{=}$; a di-substituted complex such as $[Mo(CN)_4Cl_2]^{=}$ should then exist in three isomeric forms.

[14] R. G. Dickinson and Linus Pauling, This Journal, **45**, 1466 (1923).

Molecular Orbital Theory

III

Editor's Comments on Papers 10, 11, and 12

10 Mulliken: *Electronic Structures of Polyatomic Molecules and Valence*

11 Mulliken: *Electronic Structures of Polyatomic Molecules and Valence. IV. Electronic States, Quantum Theory of the Double Bond*

12 Van Vleck: *The Group Relation Between the Mulliken and Slater–Pauling Theories of Valence*

The third bonding theory to gain prominence in the early 1930s has come to be known as the "molecular orbital theory." Its development stems directly from the creative mind of Robert S. Mulliken. Mulliken, unlike Pauling and Slater, took advantage of the group theoretical results of Bethe to formulate a one-electron bonding theory which gave results in essential agreement with those of Pauling and Slater but went considerably further "in its ability to deal with spectroscopic terms and ionization potentials."

It was very difficult to judge which, if any, of Mulliken's papers on the electronic structure of molecules could be excluded from this volume. His first paper on the subject (Paper 10) clearly spelled out the fact that molecular orbital wavefunctions correspond to representations of the symmetry group of the molecule in question. The applicability of the group theory of Bethe and Wigner to the symmetry classification of chemical bonds is apparent. Mulliken, however, clearly presents a viewpoint different from the ionic model of Bethe and Van Vleck. Furthermore, he avoids the artificiality of the Pauling–Slater theory with its discrete number of electron pairs or bonding electrons. The second Mulliken paper [28] presents "to a large extent a preliminary critical review" of the theory of one-electron orbital wavefunctions, the basis of the molecular orbital (MO) theory. The third paper in the series by Mulliken [29] deals primarily with ethylene and explained the photochemically induced *cis–trans* isomerizations about a double bond.

The fourth paper (Paper 11), reproduced in its entirety, completes the MO theory with a discussion and illustration of the symmetry adapted linear combination of atomic orbitals to form molecular orbitals, the SALCAO–MO technique, still so very useful in describing the bonding in chemical systems. Mulliken here displays a familiarity with group theory and quantum mechanics far beyond that of most scientists in the 1930s who were concerned with chemical structure. In addition to clearly applying the linear combination of atomic orbitals–molecular orbitals to a number of molecules with different symmetries, he extended the group theoretical work of Bethe and Wigner. Furthermore, he synthesized the cumbersome group theoretical nomenclature of these authors with the more convenient spectroscopic symbolism of G. Placzek. Indeed, the nomenclature employed is that commonly used in the chemical community today and is referred to as "Mulliken symbols." As a bonus, this paper by Mulliken extends the work of Bethe and Wigner. The irreducible representations of the 32 crystallographic point groups are presented. These symmetry concepts are applied to selection rules for transitions between electronic states of molecules.

A particularly interesting footnote, from a historical point of view, is the sixth one in this article. The influence of John H. Van Vleck on the Mulliken theory is apparent. Van Vleck's contributions to the entire field are overwhelming and in the opinion of this editor have not been adequately recognized. The symmetry relationships which connect the valence bond and molecular orbital theories are delineated in Paper 12 by Van Vleck. He shows how the directional characteristics of hybridized orbitals ingeniously constructed by Pauling relate to the symmetry properties of the Mulliken molecular orbitals. This article paves the way for the synthesis by Van Vleck of both theories with the Bethe crystal field theory.

Reprinted from *Phys. Rev.*, **40**, 55–62 (1932)

10

ELECTRONIC STRUCTURES OF POLYATOMIC MOLECULES AND VALENCE

By R. S. Mulliken
Ryerson Physical Laboratory, University of Chicago

(Received, February 23, 1932)

Abstract

The electronic structures of a number of molecules and ions (H_2O, NH_3, CH_4, CF_4, CI_4, ClO_4^-, $SO_4^=$, ClO_3^-, $SO_3^=$, CO_2, and others) are briefly described in terms of one-electron wave functions; other molecular types are easily described in a similar way, and will be discussed in later detailed papers. Many valence phenomena can be understood using these wave functions and a simple rule based on the Pauli principle.

THE electronic structures of polyatomic molecules can probably best be understood by expressing them in terms of one-electron wave functions. The forms of these are conditioned by the symmetry of the molecule, which is that given by the arrangement of the nuclei.

Following are the electron configurations of a number of simple hydrides with ten electrons: HF, OH^-, $1s^2\ 2s^2\ 2p\sigma^2\ 2p\pi^4$; H_2O, NH_2^-, $1s^2\ 2s^2\ 2pa^2\ 2pb^2\ 2pc^2$; H_3O^+, NH_3, $1s^2\ 2s^2\ 2p[\pi]^4\ 2p[\sigma]^2$; NH_4^+, CH_4, $1s^2\ 2s^2\ 2p^6$. The order in which the symbols are written is that of decreasing firmness of binding. The $2s$ and $2p$ types, more or less modified, of course, always remain distinct; the $2p$ type does not split up in the case of regular tetrahedral symmetry (CH_4), splits into two types σ and π in linear molecules (HF) or into a different but somewhat related pair of types $[\pi]$ and $[\sigma]$ in molecules having the symmetry of a trigonal pyramid (NH_3), and splits into three types when the symmetry is that of an isosceles triangle (H_2O). The amount of splitting corresponding to each kind of molecular symmetry can be easily determined by group theory methods.[1] The results just stated are based on the kinds of symmetry empirically known for the nuclear configurations in H_2O, NH_3, and CH_4.

Empirical data on the energy of formation of each of the molecules HF, H_2O, NH_3, CH_4 from their atoms indicate that the tetrahedral arrangement of the hydrogen nuclei, combined with the one-electron wave functions adapted thereto, gives a relatively high stability: the energies of formation *per hydrogen atom*, in kilocalories, are about 148, 110, 87, and 91 for HF, H_2O, NH_3, CH_4. The value for CH_4 is higher than one would get by an extrapolation from the other values. A theoretical reason for this may probably be found in the form of the wave functions. In CH_4 the four H nuclei are embedded in the wave functions of the six equivalent modified-$2p$ wave functions and to a somewhat less extent in those of the two $2s$ wave functions. All these wave functions, of course, correspond to "representations of the tetrahedral group", i.e., have symmetries of a tetrahedral type. The present

[1] Cf. H. Bethe, Ann. d. Physik **[5]**, **3**, 133 (1929).

explanation of the structure and stability of CH_4 in terms of one-electron wave functions seems at first sight not closely related to that given by Pauling[2] and Slater.[3] The zeroth order wave functions which they use do indeed indicate that tetrahedral symmetry should give high stability, but it seems doubtful whether their wave functions are the most appropriate ones (cf. note added in proof, below).

With the triangular pyramid arrangement as in NH_3, the three H nuclei are embedded in the four $2p[\pi]$ and to a lesser degree in the two $2s$ wave functions. But the two $2p[\sigma]$ electrons in NH_3 avoid the region of the H nuclei, and as a result are relatively loosely bound and reduce the energy of formation of the molecule. Direct evidence of these electrons in NH_3 is given by the low ionization potential of this molecule, which is 11.1 volts. CH_4, on the other hand, has an ionization potential of 14.5 volts. One may predict roughly that ionization of a $2p[\pi]$ electron of NH_3 would take 15 volts, and of a $2s$ electron 23 volts in CH_4 and 25 volts in NH_3. Many features of the chemical behavior of NH_3, e.g., the formation of NH_4^+, $BCl_3 \cdot NH_3$, $Cu^{++}(NH_3)_4$, are reasonably interpreted as conditioned by the stabilization of the two loosely bound $2p[\sigma]$ electrons of NH_3 under the influence of an additional nucleus, giving relations somewhat closely related to those for tetrahedral symmetry.

For the ionization potentials of H_2O the following rough estimates may be given: $2pc$, 13.2 volts (*observed*); $2pb$, $2pa$, 16 and 17 volts; $2s$, 30 volts. The $2pc$ type in H_2O is relatively loosely bound like the $2p[\sigma]$ in NH_3.

Among the molecules or "radicals" related to the foregoing, but containing fewer electrons, there is interesting experimental evidence concerning CH_4^+, CH_3, CH_3^+. The energy of dissociation of CH_4 into CH_3+H is perhaps roughly 120 kilocalories, or at any rate much more than the average value 91 kcal per H atom for $C+4H \rightarrow CH_4$. Probably the CH_3 radical is pyramidal like NH_3, with an electron configuration $1s^2\ 2s^2\ 2p[\pi]^4\ 2p[\sigma]$, the $2p[\sigma]$ electron making it relatively unstable. Taking 5.2 volts (120 kcal) for $CH_4 \rightarrow CH_3+H$, and using the result of Hogness and Kvalnes[4] that CH_3^+ ions are produced from CH_4 at 15.5 volts ($CH_4 \rightarrow CH_3^+ + H + \epsilon$), while CH_4^+ ions are produced at 14.5 volts, one gets 10.3 volts for the ionization potential of CH_3, a result in harmony with the above interpretation. When CH_3 molecules unite with other atoms or radicals, as in C_2H_6 or CH_3Cl, the loosely bound $2p[\sigma]$ electron becomes firmly bound by being shared with another nucleus which at the same time shares an electron with the CH_3 radical. Thus there is formed an "electron-pair bond". The stabilization of the $2p[\sigma]$ electron here is similar in nature to that effected when NH_3 forms NH_4^+ or unites with BCl_3, although in terms of conventional valence theory the two latter processes are quite different from the present one.

More complicated molecules can be treated by extending the methods used above. The following additional *valence rule*, with which the preceding

[2] L. Pauling, J. Am. Chem. Soc. **53,** 1367, 3225 (1931).
[3] J. C. Slater, Phys. Rev. **37,** 481; **38,** 325, 1109 (1931).
[4] Phys. Rev. **32,** 942 (1928).

results also are in harmony, is needed: Every nucleus in a molecule tends to be surrounded by an electron density distribution corresponding to some stable electron configuration having a total charge approximately equal to or somewhat exceeding the charge of the nucleus; the electron density distribution as a whole, and the individual wave functions, have symmetries adapted to the configuration of nuclei surrounding the given nucleus. By "stable configuration" is meant a set of wave functions completely occupied by electrons (i.e., a set of closed shells) and of such type that further electrons could go only into wave functions of distinctly higher energy,—usually of higher quantum number, from the point of view of the central nucleus. The truth of the first part of the rule follows at once from the fact that for the electrons in the neighborhood of every nucleus in a molecule the Pauli exclusion principle makes effectively nearly the same requirements as for electrons in an isolated atom.

This simple rule, together with the energy-decrease which is normal when a bonding electronic wave function is brought near an incomplete atom, suffices to give a qualitative explanation of molecule-formation and of the ordinary numerical aspects of valence (combining ratios), and covers polar and non-polar valence under a single viewpoint. The first part of the rule is in most respects merely a somewhat generalized re-statement in present-day quantum language of the familiar rules of Lewis and Langmuir (cf. especially the "octet"-forming tendency emphasized by these authors). Essentially this same principle has been used earlier by Dunkel[5] and others and still earlier but rather less definitely by Knorr.[6] Dunkel's results on electron configurations in polyatomic molecules are more or less similar to and in some cases the same as those obtained here. Knorr's results on valence phenomena are also in many respects similar to the present.

The principal novelty in the present viewpoint in regard to molecular structure, aside from the consideration of the effect of the symmetry of the molecule on the wave functions, consists in the following: in general no attempt is made to treat the molecule as *consisting of* atoms or ions. Attempts to regard a molecule as consisting of specific atomic or ionic units held together by discrete numbers of bonding electrons or electron-pairs are considered as more or less meaningless, except as an approximation in special cases, or as a method of calculation. It is believed that the main physical content of the assumption that a molecule consists of specified atoms or ions, or even of a quantum-mechanical hybrid of several different sets of these, could better be expressed in terms of electric moments. A molecule is here regarded as a set of *nuclei*, around each of which is grouped an electron configuration closely similar to that of a free atom in an external field, except that the outer parts of the electron configurations surrounding each nucleus usually belong, in part, jointly to two or more nuclei. The electron configuration surrounding each nucleus can be described to a good approximation *with reference to that nucleus* in terms of its own set of electronic quantum

[5] Dunkel, Zeits. f. phys. Chem. [*B*] **7,** 81; **10,** 434 (1930).

[6] Knorr, Zeits. f. anorgan. Allgem. Chem. **129,** 109 (1923).

numbers or one-electron wave functions. In the case of shared electrons, the same wave functions may be given quite different approximate formulations or descriptions from the points of view of different nuclei. The procedure just outlined is useful for a qualitative understanding of many features of chemical valence, especially combining ratios and valence saturation. A more accurate and detailed method of dealing with shared electrons is to make use of one-electron wave functions whose effective domains embrace more than one atom.

The molecules CF_4 and CI_4 will serve as illustrations of both methods. Presumably the arrangement of the nuclei is tetrahedral as in methane. First let us consider CF_4. The electrons immediately surrounding the carbon nucleus in all probability constitute, from the latter's point of view, a set of closed shells $1s^2\ 2s^2\ 2p^6$, the one-electron wave functions being much like those in methane. Other electrons in the molecule, namely, those specially attached to the fluorine atoms, perhaps function from the point of view of the carbon nucleus as 3- or 4- quantum electrons. Their connection with the carbon nucleus is, at any rate, very remote.

Now considering the electron configuration from the point of view of a fluorine nucleus, the electrons immediately surrounding the latter undoubtedly constitute a set of closed shells $1s^2\ 2s^2\ 2p^6$, while other electrons are too distant to be of much importance. From the point of view of the electrons near one F nucleus, the rest of the molecule (approximately CF_3) sets up a field of force which is nearly symmetrical around the F-C axis, although strictly speaking it has symmetry like that in NH_3 or better, as in CH_3F. This field causes a splitting of the $2p^6$ group of the F nucleus in question into $2p[\sigma]^2\ 2p[\pi]^4$, of which $2p[\sigma]^2$ is shared with the C nucleus. In fact the four pairs $2p[\sigma]^2$ belonging to the four fluorine nuclei are all shared by the carbon nucleus, which regards these same electrons as constituting its own outer shell $2s^2\ 2p^6$.

The complete molecule may now be approximately described as consisting of $1s^2$ close to the C nucleus, $1s^2 2s^2 2p[\pi]^4$ near each F nucleus, and eight shared wave functions. (Strictly speaking, the fluorine types $1s$, $2s$, $2p[\pi]$ each split up as a result of "resonance" interactions, but for most purposes this effect can be neglected.) *Each* of the shared wave functions surrounds the C nucleus and reaches out to and around each F nucleus. Two of them correspond approximately to a linear combination consisting to the extent of perhaps about 50 percent of a tetrahedralized but uncombined $2s$ carbon wave function plus $12\frac{1}{2}$ percent each of four uncombined $2p[\sigma]$ fluorine wave functions. The other six consist of similar combinations of tetrahedralized uncombined carbon $2p$ plus uncombined $2p[\sigma]$ of the four F atoms, probably with the latter predominant, corresponding to the strong tendency of the F to be F^-. The resulting wave functions are largely concentrated in the regions between the C and the F nuclei and so are acceptable to the various nuclei as parts of their outer shells, at the same time being very effective in producing chemical binding; only a small fraction of the density is behind the F nuclei or (in the $2s$ type) too close to the C nucleus to be effective. Although eight electrons, of two types, take part in the binding of each F atom, only about two of these on the average are in action at any one instant between the C and any one F. It is these two electrons which, in spite of the fact that they belong to *two types* different in energy and one contributing about $\frac{1}{2}$ electron and the other $1\frac{1}{2}$ electrons on the average, are regarded by the F nucleus as fairly satisfactorily representing its shell of two $2p[\sigma]$ electrons.

In the molecule CI_4, the much weaker electron affinity of the I atom very likely has the following results: (1) two of the shared wave functions are such as might be formed from say possibly 88 percent of carbon $2s$ and only 3 percent each of the iodine $5p\{\sigma\}$'s, i.e., the carbon nucleus shares its $2s$ electrons very little; (2) the other six shared wave functions are such as might be formed from say possibly 50 percent of carbon $2p$. Thus the four halogen atoms in CI_4 are held practically by about six electrons, instead of by about eight as in CF_4. This causes CI_4 to be chemically relatively unstable.

The electronic structure of such groups as BF_4^-, $SO_4^=$, ClO_4^- is doubtless of the same type as that of CF_4 and CI_4. Each O nucleus in $SO_4^=$ and ClO_4^- shares two electrons which from its point of view are $2p[\sigma]$, these eight electrons functioning at the same time as (tetrahedralized) $3s^23p^6$ for the S or Cl atom. In the groups $SO_3^=$ and ClO_3^- only six electrons of the expected shell $3s^23p^6$ of the central nucleus can be shared, otherwise the 2-quantum oxygen shells would be more than filled. Now $SO_3^=$ and ClO_3^- have a pyramidal structure,[7] hence the group $3p^6$ must be subdivided into $3p[\pi]^43p[\sigma]^2$, with the $[\sigma]$ type less firmly bound than the $[\pi]$ (cf. discussion of NH_3 above). Doubtless the two $3p[\sigma]$ electrons, whose wave functions avoid the vicinity of the O nuclei, are the unshared ones, while $3s^23p[\pi]^4$ describes the shared electrons from the viewpoint of the central nucleus. From the viewpoint of the O nuclei, these same six electrons function as three pairs of $2p[\sigma]$ electrons, one pair for each O nucleus. The existence of the unshared $3p[\sigma]$ electrons is in harmony with conclusions of Zachariasen.[7]

The electronic structures of many other types of molecules can be described with some confidence by proceeding according to the methods illustrated above. These will be discussed in forthcoming more detailed publications. A beginning has been made for some of these in a recent article.[8] Other authors, especially Hund and Herzberg, have also been attacking the problem from more or less similar viewpoints. One more example, that of the CO_2 molecule, will be given here. This illustrates particularly well the uselessness of trying to decide whether molecules are composed of atoms or ions and in what states. The examples already given also illustrate the same point, but are more complicated.

It is now generally accepted that CO_2 is a linear molecule. Hence its electrons can be classified, like those of a diatomic molecule, as σ, π, δ . . . . Since CO_2 is diamagnetic, its normal state in all probability has a configuration of closed shells. Consideration of available evidence makes it probable that there are *two* closed shells (π^4) of π electrons. Hence the complete electron

[7] Cf. W. H. Zachariasen, J. Am. Chem. Soc. **53,** 2123 (1931).
[8] R. S. Mulliken, Chem. Reviews **9,** 347 (1931).

configuration is of the type $\sigma^2\sigma^2\sigma^2\sigma^2\sigma^2\sigma^2\sigma^2\pi^4\pi^4$.Consideration of the available evidence, and application of the valence rule stated above, according to which each nucleus tends to be surrounded by what from its point of view is a closed shell, leads to a rather definite description of the wave functions. They may be approximately described by writing the electron configuration as follows:

$$(1s)^2_{O}(1s)^2_{O}(1s)^2_{C}(2s)^2_{O}(2s)^2_{O}(\sigma)^2_{C-O}(\sigma)^2_{C-O}(\pi 2p)^4_{C-O}(\pi^* 2p)^4_{O}.$$

The first six electrons are essentially $1s$ electrons of the O and C atoms, the next four,—which might also be written $(\sigma 2s)^2(\sigma^* 2s)^2$,—are slightly modified oxygen $2s$ electrons. All these are non-bonding electrons or nearly so (strictly, $\sigma 2s$ are somewhat bonding, $\sigma^* 2s$ somewhat anti-bonding). Next come eight bonding electrons (four σ and four π), enough to give two bonding pairs between the C and each O, in agreement with conventional valence theory. Finally there are four non-bonding (or slightly anti-bonding) π electrons which remain near the O nuclei. The two pairs of σ bonding electrons have wave functions which are concentrated mainly around the C nucleus and, especially, between the C and the O nuclei, while the π bonding electrons are concentrated mainly between the C and the two O's, half on the side of each O. From the point of view of the carbon nucleus the two σ^2 pairs represent $2s^2 2p\sigma^2$ while the π^4 group represents $2p\pi^4$, so that the carbon nucleus has a complete L shell. From the point of view of either O nucleus, about *half* of the electron density represented by the bonding electrons $\sigma^2\sigma^2\pi^4$ is near enough to it to count as belonging to its L shell, and counts as $2p\sigma^2 2p\pi^2$. The additional $2p\pi^2$ necessary to complete the L shell is represented by that half of the electron density corresponding to$(\pi^* 2p)^4$ which is near the nucleus in question. This picture of the wave functions is exactly what one would get by pushing two $O^{=}$ ions up against a C^{++++} ion, but it is equally what one would expect from $2O^{-}+C^{++}$, or $2O^{+}+C^{=}$, or $2O+C$. Nevertheless it is foolish to try to think of the molecule as consisting of any one of these sets of ions or atoms.

The writer is indebted to Professors C. Eckart and J. H. Van Vleck for helpful suggestions.

Added in Proof. Further consideration indicates that the present method gives results which are usually essentially in agreement with those of Pauling and Slater, but that it goes considerably farther. Hund also has developed the method of molecular one-electron wave functions and has obtained important results.[9] He has not, however, as yet paid much attention to the matter of finding wave functions having the proper symmetry with respect to the nuclear configurations. Recently[10] he has expressed the conclusion that the Pauling and Slater method of electron-pair "localized bonding" is a poorer approximation than a method like the present one, but nevertheless a very valuable approximation corresponding rather well to the conventional electron-paid bond theory of homopolar valence. In the writer's opinion, the present method gives the possibility of going behind conventional valence theory

[9] F. Hund, Zeits. f. Physik **73,** 1 (1931).
[10] F. Hund, Zeits. f. Physik **73,** 565 (1932).

and understanding both the rules and the exceptions together, and furthermore is well adapted to intermediate cases, e.g., cases on the border-line between homopolar and heteropolar valence. The present method also goes beyond that of Slater and Pauling in its ability to deal with spectroscopic terms and ionization potentials.

In the case of molecules like NH_3 and ClO_3^- having the symmetry of a trigonal pyramid, the wave functions above called s and $p[\sigma]$ both belong to the same representation of the symmetry-group. [This particular symmetry-group has not been considered by Bethe.] A consequence, overlooked in the discussion given above, is that the true molecular wave functions must correspond to (perturbed) linear combinations of s and $p[\sigma]$. If one should take equal parts of s and $p[\sigma]$, the resulting zeroth approximation wave functions may be written as $(s-p[\sigma])$, concentrated largely on the opposite side of the N or Cl nucleus from the three H or O nuclei, and $(s+p[\sigma])$, concentrated near the plane of the three H or O nuclei. The $(s+p[\sigma])$ type is well suited to share with the $p[\pi]$ type in binding the three H or O nuclei, while the $(s-p[\sigma])$ type avoids the neighborhood of all the nuclei, and agrees well with Zachariasen's conclusions as to the unshared pair of electrons in ClO_3^- and similar molecules.[7] This unshared type is, however, extremely well suited to the formation of a new bond with an additional atom to form a structure of approximate or exact tetrahedral symmetry, e.g., ClO_4^- from ClO_3^- or CH_3Cl from CH_3.

Actually, of course, because of the difference in energy between s and p in the unperturbed central atom, one must expect instead of $(s+p[\sigma])$ and $(s-p[\sigma])$, types intermediate between the former and s, and between the latter and $p[\sigma]$. The actual wave functions in order of energy, in say NH_3 or CH_3, can perhaps be fairly well described as (*a*) $1s$; (*b*) $2s[\sigma]$ obtained as a linear combination of $2s$ of the central atom, $1s$ of the three H atoms, and a lesser proportion of $2p[\sigma]$ of the central atom, with the $1s$ wave functions so introduced as to conform to the proper representation of the symmetry group; (*c*) $2p[\pi]$ composed of $2p\pi$ of the central atom mixed with $1s$ of the H atoms; (*d*) $2p[\sigma]$ consisting mainly of $2p\sigma$ of the central atom, a smaller proportion of $2s$, and a small proportion of $1s$ of the H atoms.

In molecules like NO_3^-, $CO_3^=$, and presumably SO_3, with a plane arrangement of the nuclei,[7] the wave functions correspond to representations of the trigonal symmetry-group, which are the same as those of the hexagonal group.[1] Here, from the standpoint of the central atom, the types s, $p[\pi]$ and $p[\sigma]$ are possible. Of these $p[\pi]$ is well suited to bonding and s is moderately so, while $p[\sigma]$ avoids the plane of the nuclei. In this case s and $p[\sigma]$ belong to different representations and so cannot hybridize as in NH_3 and $SO_3^=$ to form a stronger bonding type. Assuming two shared electrons per O atom, the central atom has the incomplete (but diamagnetic) shell $s^2p[\pi]^4$, the $p[\sigma]$ wave function being unoccupied. From the point of view of an O nucleus, the two electrons which it shares are $p[\sigma]$,—this must not be confused with the central-atom-viewpoint kind of $p[\sigma]$,—while its unshared electrons are $1s^22s^22p[\pi]^4$ (cf. discussion of CF_4, above). Possibly the twelve

$p[\pi]$ electrons of the three O atoms interact in such a way as to be available for partial sharing with the central atom, so that in effect two of their number, on the average, function as $p[\sigma]$ electrons of the central atom and so after all complete the latter's otherwise incomplete p shell.

Many other interesting results can be obtained. In the case of atoms with d electrons, these closely parallel the results of Pauling,[2] even to the explanation of diamagnetic and paramagnetic susceptibilities. For example, the square arrangement and diamagnetism of $PtCl_4^{=}$ and the octahedral arrangement and diamagnetism of $PtCl_4^{=}$ can be easily predicted or understood. These results are based on Bethe's work[1] showing that d wave functions split into two types d_ϵ (3-fold orbital degeneracy) and d_γ (2-fold degeneracy) in a field of octahedral or tetrahedral symmetry; d_ϵ and d_γ are of forms which appear to be respectively specially adapted to bond-formation with tetrahedral and octahedral symmetry. The tetrahedral ions $CrO_4^{=}$ and MnO_4^{-} are tentatively interpreted as having shared central-atom configurations $3d_\epsilon^6 4s^2$, thus being closely analogous to $SO_4^{=}$ and ClO_4^{-} with $3s^2 3p^6$. This interpretation appears more plausible than Pauling's, which involves a hybrid of $3d$, $4s$, and $4p$ electrons. SF_6 is interpreted as being octahedral with $3s^2 3p^6 3d_\gamma^4$ around the S core. Further details will be given later.

Reprinted from *Phys. Rev.*, **43**, 279–301 (1933)

11

Electronic Structures of Polyatomic Molecules and Valence. IV. Electronic States, Quantum Theory of the Double Bond

ROBERT S. MULLIKEN,* *Ryerson Physical Laboratory, University of Chicago*

(Received November 29, 1932)

The possible types of electronic states of polyatomic molecules (assuming fixed nuclei and neglecting spin fine structure) are discussed and tabulated (Table I) with the help of simple group theory methods, applying results of Bethe and Wigner. A notation for electronic states (ψ's) and molecular orbitals (ϕ's) for molecules having any type of symmetry to be found among the 32 crystal classes, is adopted; this is essentially the same as that used by Placzek for designating the vibrational states of molecules. It is shown how the possible ψ's corresponding to any given electron configuration (set of ϕ's) can be determined for any type of symmetry; for the more complicated cases, the results are tabulated (Table V). It is shown how all the *selection rules* for transitions between electronic states of molecules can be easily determined. Limitations resulting here from the application of the Franck-Condon principle are discussed. Extending work of Bethe, tables are given (Tables II–IV) showing how the various types of electronic states of atoms and of diatomic and polyatomic molecules (S, P, Σ^+, ΔA, etc.) go over into various other types of states if the symmetry of the original system is decreased. Examples are given showing how electronic wave functions (ψ's) of molecules can be constructed which conform to the possible types (Table I) allowed by the symmetry of the nuclear skeleton, and which at the same time, with Slater's method, are antisymmetrical in the electrons (cf. section 2 and Eqs. (9–12)). It is shown that for molecules having all their electrons in closed shells or electron-pair bonds, zeroth approximation ψ's which conform to the identical representation of the molecule's symmetry group (analogous to ^1S of atoms and $^1\Sigma^+$ or $^1\Sigma_g^+$ of diatomic molecules) can be built up either by using electron-pair bonds or by using molecular orbitals. The approximate construction of molecular orbitals as linear combinations of atomic orbitals, in such a way that they conform to the types allowed by the symmetry of the molecule, is discussed and illustrated (cf. Eqs. (3, 8)). Several statements made in a previous paper (III) of this series, on the quantum theory of the double bond, are here justified by the methods mentioned above, thereby also providing examples of the application of the latter. Some additional details concerning the nature of the double bond are given. Finally, it is shown that the model of the double bond given in III should according to the theory be altered somewhat for the perp. form of the molecule, in a way which offers the possibility of improved agreement with experiment.

INTRODUCTION

1. Symmetry of electronic wave functions

In I, II and III of this series[1] the use of molecular orbitals for shared electrons in describing and interpreting the electronic states of polyatomic molecules has been discussed and illustrated.[2] Discussion of further examples will be prefaced by some general considerations.—Through an oversight, the writer failed to mention in II that Hückel also[3] has been following a similar program in his work on the structure of the benzene ring, its derivatives, and certain other organic compounds. In this connection Hückel also has made a comparison of the methods of molecular orbitals and of electron-pair bonds.

For a molecule with fixed nuclei, the complete electronic wave function ψ is restricted to one of certain types which depend on the symmetry of the nuclear skeleton. In the language of group theory, ψ must conform to an irreducible representation of the symmetry group of the corresponding Schrödinger equation,—which contains a potential energy whose symmetry is that of the nuclear skeleton. Or more briefly, one may say that every ψ must belong to an irreducible

* Fellow of the John Simon Guggenheim Memorial Foundation.

[1] R. S. Mulliken, Phys. Rev. **40**, 55; **41**, 49, 751 (1932). Hereafter designated as I, II, III.

[2] Cf. II, also J. C. Slater, Phys. Rev. **41**, 255 (1932), for a comparison of the method of electron-pair bonds with that of molecular orbitals.

[3] E. Hückel, Zeits. f. Physik **70**, 204; **72**, 310 (1931); **76**, 628 (1932). Also it should be mentioned that the interpretation of the structure of C_6H_6 attributed in II (p. 56) exclusively to Hund had already appeared in Hückel's paper on this molecule.

representation of the symmetry group of the nuclear skeleton. Corresponding statements apply to every molecular orbital φ. In nature ψ is of course further restricted, in accordance with the Pauli principle, to forms antisymmetrical in the electrons.

In general a knowledge, for any nuclear configuration, of the different irreducible representations of its symmetry group, since these determine the forms or types to which ψ and the φ's may belong, is important in determining the number, spacing, and degree of degeneracy of molecular electronic states, and the selection rules for transitions between them, also for determining the possible states of the dissociation products of a molecule.

Like the electronic φ's and ψ's, the possible states of vibration of a (rotationless) molecule also conform to the irreducible representations of the symmetry group of the molecular skeleton (in its equilibrium or in some more symmetrical configuration).[4, 5] Finally, the *total* electronic $\times$vibrational *state*, excluding rotation, must belong to such a representation. This is strictly true even when the electronic and vibrational parts of the wave function cannot be even approximately separated.

The process of finding the irreducible representations for any given type of molecular symmetry is accomplished in an easy and instructive way (cf. section 4), by using a little group theory. The problem has already been solved, for all or most of the kinds of symmetry likely to occur in actual molecules, by Bethe[6] and Wigner.[4] The results have been applied by Wigner and by Placzek[5] to the case of nuclear vibrations, of importance especially for Raman and infrared spectra.

An important secondary problem is that of notation. It has seemed best here to adopt for classifying electronic φ's and ψ's (Table I below) essentially the same notation Placzek has used for describing vibrational states. The same notation could well be used also for the electronic φ's and ψ's of atoms in crystals (Bethe's problem[6]), and the same or a similar notation for describing the electronic$\times$vibrational states of molecules. Placzek's notation has marked advantages over the Bethe notation used in I–III of this series, in being more descriptive.

These problems will be taken up again in section 3. In section 2 the matter of building up good approximate ψ's which are antisymmetrical in the electrons will be considered.

In classifying electronic states of polyatomic molecules, complications often arise because of the existence of more than one fairly stable arrangement of the same set of nuclei. Such different arrangements as are chemically stable (chemical isomers) can when in their normal states most conveniently be treated as distinct individuals. In excited states of such molecules, the relative stabilities of different arrangements are in general altered. It may then often be advisable to regard a variety of nuclear arrangements as belonging to a single molecular species. This is of course always necessary to a greater or less degree when one considers excited molecules in which strong vibrations or internal rotations are occurring.

Even for unexcited molecules belonging to a single chemical species it is not always true that there is just a single very stable type of nuclear configuration. In C_2H_6, for example, only very weak forces[7] oppose a relative rotation of the two CH_3 groups around the C–C axis. Hence in discussing the electronic structure of C_2H_6, it is perhaps best to assume only such symmetry as is common to the various forms differing by arbitrary rotations of this kind. Another less extreme example is found in NH_3, where a plane

[4] E. Wigner, Göttinger Nachr., Math.-Phys. Klasse, p. 133 (1930). Wigner extended Bethe's results (see reference 6) to include the symmetry groups of all the 32 crystal classes.

[5] G. Placzek, article on Raman and infrared spectra, to appear soon in *Marx's Handbuch der Radiologie*. The writer is indebted to Dr. Placzek for the use of his tables before publication and for valuable discussions, also for calling his attention to reference 4. Placzek gives his tables without direct use of group theory, but they are essentially the same as the group theory results (cf. Table I below).

[6] H. Bethe (Ann. d. Physik [5], **3**, 133 (1929)) used group theory in determining the irreducible representations to which the ψ and φ's of an atom may belong when in a field of force corresponding to that produced by its neighbors in a crystal. The writer is greatly indebted to Professor J. H. Van Vleck for calling his attention to the applicability of Bethe's results to molecular electron wave functions.

[7] Cf. H. Eyring, J. Am. Chem. Soc. **54**, 3191 (1932).

form is nearly as stable as the pyramidal equilibrium forms.

2. Use of atomic and molecular orbitals to build up ψ

The complete electronic wave function ψ of a molecule can be conveniently approximated by an antisymmetrical linear combination of products of atomic or molecular one-electron wave functions, each a product of an orbital factor φ and a spin factor σ. This can be done in various ways. In any case one may start with a determinant form as follows (N is a normalizing factor):

$$U = N\begin{vmatrix} \varphi_1\sigma_1(1) & \varphi_1\sigma_1(2) & \cdots \\ \varphi_2\sigma_2(1) & \varphi_2\sigma_2(2) & \cdots \\ \cdots & \cdots & \cdots \end{vmatrix} \quad (1)$$

This is antisymmetrical in the electrons 1, 2, 3, . . . Using a set of such U's as unperturbed functions, one can obtain the desired wave functions ψ as linear combinations of them (*general method* of Slater[8, 9]). In so doing, one includes in any linear combination only such U's as have the same M_S (resultant spin magnetic quantum number).[8] [In the case of atoms, one includes only such U's as are alike also in M_L.]

Various cases may arise. In general one has in the unperturbed system (no interactions between electrons) several U's of equal energy corresponding to a given set of φ's but various arrangements of σ's giving the same M_S (spin degeneracy), and to a variety of sets of equivalent φ's for any given spin arrangement (orbital degeneracy). In some cases (e.g., an atom or molecule built of closed shells of atomic or molecular orbitals respectively), there is no such degeneracy and a single U suffices.—In the case of any molecule, every linear combination of U's must be so chosen as to conform to an irreducible representation of the molecule's symmetry group (cf. section 1).

To get the best practical approximation (Slater's most general practical method,[10] (see also reference 9, page 1111)) one includes not only such U's as are really degenerate in the unperturbed system, but also other U's of equal M_S, etc., whose energies lie in the same neighborhood. It should be noted that the U's belonging to a really degenerate set always belong to a definite electron configuration, i.e., to a definite set of φ's (counting degenerate φ's as belonging to a single type), distributed, if there is more than one atom, with a definite set of φ's for each atom. The use of an approximation built up only from a really degenerate set of U's may therefore appropriately be referred to as the *method of the pure electron configuration*.

As applied to molecules, the method of the pure electron configuration may be specialized or approximated in various ways. The present *method of molecular orbitals* is a special form in which, following Lennard-Jones, atomic φ's (orbitals) are used for inner or unshared electrons (usually in atomic closed shells), molecular φ's for outer, shared or valence, electrons. The form used by Slater[8, 9] which for convenience will be called the "*method of atomic orbitals*," is one in which atomic φ's exclusively are used. A special case of the method of atomic orbitals, sometimes identical with it but in general representing a simpler but cruder approximation, is the Slater-Pauling *method of electron-pair bonds*. This last is applicable only to chemically saturated molecules in their normal states, i.e., to a restricted but particularly important class of molecular states.

Following a method first used by Bloch for metals[11] and later used by Hückel, Hund, and others, molecular orbitals will as a matter of convenience usually be approximated here by linear combinations of atomic orbitals, although eventually we may hope to obtain forms which are better approximations. When atomic orbitals are used in constructing molecular orbitals, the resulting ψ is in the final analysis expressed entirely in terms of atomic orbitals, but is nevertheless not in general identical with that obtained with the "method of atomic orbitals" as defined above. For excited states, to be sure, the approximate ψ's given by the two methods are very often identical: examples, ${}^3\Sigma_u{}^+$ and ${}^1\Sigma_u{}^+$ states of H_2 built up, respectively, in the atomic orbital method, from 2 H(1s) and from $H^+ + H^-(1s^2)$. For saturated molecules in their normal states, however, and whenever there is

[8] J. C. Slater, Phys. Rev. **34**, 1293 (1929).
[9] J. C. Slater, Phys. Rev. **38**, 1109 (1931).
[10] For examples of this case cf. E. U. Condon, Phys. Rev. **36**, 1121 (1930), and other papers.

[11] F. Bloch, Zeits. f. Physik **52**, 555 (1928).

at least one pair of electrons which form a valence bond, the two methods are never identical. Nevertheless the approximations given by the two methods can of course always be made identical if we generalize (and complicate) each, dropping their common "pure electron configuration" limitation, and form linear combinations with U's belonging to other configurations. It is still true, however, that the point of view and method of approach are different in the method of molecular orbitals than in that of atomic orbitals or of electron-pair bonds. Slater[2] has pointed out the usefulness of considering problems from both points of view.

In building up ψ's for complicated molecules according to the present method, it will often be useful to proceed in two or more stages, first assigning electron configurations composed of (atomic and) molecular orbitals for separate parts of the molecule (*radicals*), then combining these. Two courses are then open for the construction of ψ for the complete molecule. (1) One may proceed in accordance with the method of molecular orbitals, using molecular orbitals of the total molecule for those electrons which may reasonably be considered as shared by the two or more radicals, but keeping *radical orbitals* for those electrons which are shared within, but not between radicals (and of course using atomic orbitals for electrons which belong to particular atoms and are not shared at all). (2) Or one may proceed in analogy with the method of atomic orbitals, building up the final ψ entirely from orbitals of the various radicals (and atomic orbitals for the completely unshared electrons). One may form electron-pair bonds from radical orbitals if the latter are known to give bonding.

Good examples of molecules which can be built up out of radicals are C_2H_4 and its derivatives. Both procedures (1) and (2), but especially (1), will be used in section 8*d* below for building up ψ's of C_2H_4 using CH_2 orbitals. A symmetrical molecule like C_2H_4 when treated in this way is analogous to a homopolar diatomic molecule. In plane C_2H_4 the analogy is close, in perp. C_2H_4 less so in some respects. In practice (cf. the [z] and [x] orbitals of CH_2 used in forming C_2H_4) those radical orbitals which act as valence orbitals of a radical, forming bonding electron pairs in the complete molecule, are really often to a fairly good approximation just atomic orbitals of certain atoms between which binding chiefly occurs. Cf. the [z] and [x] orbitals of CH_2 used in forming the C=C double bond in C_2H_4; these are not far different from C atom orbitals. Hence one need not fear that the use of the method of molecular orbitals in complex molecules necessarily means using orbitals which are spread over a large number of atoms. Even in large molecules, it will be found that one arrives at molecular orbitals which usually fade out after bridging the gap between any atom and one or more of its immediate neighbors. Especially in hydrides, however, many details of chemistry may prove to be better understandable by admitting molecular orbitals which do extend with appreciable density somewhat farther than this (cf. section 8*e* for an example; a still better example is probably B_2H_6).

2a. Method of electron-pair bonds. It will be instructive first to say something about the electron-pair bond method. In this, the ψ of a molecule with definite bonds is approximated[9] by a linear combination of the type

$$\psi = N\begin{vmatrix} \varphi_1\alpha(1) & \varphi_1\alpha(2) & \cdots \\ \varphi_2\beta(1) & \varphi_2\beta(2) & \cdots \\ \varphi_3\alpha(1) & \varphi_3\alpha(2) & \cdots \\ \varphi_4\beta(1) & \varphi_4\beta(2) & \cdots \\ \text{etc.} & \cdots & \cdots \end{vmatrix} - N\begin{vmatrix} \varphi_1\beta & \cdots \\ \varphi_2\alpha & \cdots \\ \varphi_3\alpha & \cdots \\ \varphi_4\beta & \cdots \\ \cdots & \cdots \end{vmatrix} + N\begin{vmatrix} \varphi_1\beta & \cdots \\ \varphi_2\alpha & \cdots \\ \varphi_3\beta & \cdots \\ \varphi_4\alpha & \cdots \\ \cdots & \cdots \end{vmatrix} - N\begin{vmatrix} \varphi_1\alpha & \cdots \\ \varphi_2\beta & \cdots \\ \varphi_3\beta & \cdots \\ \varphi_4\alpha & \cdots \\ \cdots & \cdots \end{vmatrix} + \text{etc.} \qquad (2)$$

The number of U's in this expression depends on the number of bonds. For H_2, with one bond, there are just two U's, each in the form of a determinant with two rows and columns, and with φ_1 and φ_2 denoting H atom 1*s* orbitals, one for atom A, the other for atom B.[9] For a molecule He_2, with no bonds, there would be only one U, with $\varphi_1 = \varphi_2$, and $\varphi_3 = \varphi_4$, φ_1 and φ_3 referring to a a 1*s* orbital on He atom A or B respectively.

For H_2O, with two bonds and approximately a 90° angle between them, ψ would consist of four U's, with φ_1 and φ_3 representing, say, a $2p_y$

and a $2p_z$ oxygen orbital, φ_2 and φ_4 representing $1s$ orbitals of the two hydrogen atoms; each of the four determinants would also contain terms corresponding to six more electrons and to wave functions $\varphi_4\alpha$, $\varphi_5\beta$, $\varphi_6\alpha$, $\varphi_6\beta$, $\varphi_7\alpha$, and $\varphi_7\beta$, where φ_5, φ_6, φ_7 refer to oxygen $1s$, $2s$, and $2p_x$ orbitals, which do not form bonds. Similarly the Slater-Pauling model of NH_3 is approximated by using three bonds between nitrogen p_x, p_y, and p_z and hydrogen $1s$ orbitals. In every case, orbitals which are to be used in forming bonds belong to different atoms and appear with opposite spins (α and β) on the two atoms (cf. reference 9, top page 1128, for details).

In dealing with compounds of N, O, F, and their homologues, Slater and Pauling assume as a good approximation that it is only the p electrons which form the bonds. They generally use as suitable zeroth approximations atomic orbitals of the types p_x, p_y, and p_z (i.e., $f(r)\sin\theta\cos\varphi$, $f(r)\sin\theta\sin\varphi$, and $f(r)\cos\theta$). For the univalent atoms such as H, Na, they, of course, use s orbitals.

For the carbon atom, Slater and Pauling use "tetrahedral" orbitals. These comprise four energetically and geometrically equivalent linear combinations[12] of $2s$ and $2p$ orbitals pointing toward the corners of a tetrahedron. Strongest binding of other atoms is then obtained if the latter are at the corners of a tetrahedron. For CH_4, the expression for ψ written down according to Eq. (2) contains sixteen U's. Each U has $\varphi_1=\varphi_2=$ carbon $1s$, $\varphi_3=\frac{1}{2}(s+p_x+p_y+p_z)$ of carbon, $\varphi_4=$ the H $1s$ which overlaps φ_3, and so on to φ_{10}.

For any saturated molecule (unshared electrons all in atomic closed shells and shared electrons all in electron-pair bonds), ψ, as well as every one of the U's, if chosen as in Eq. (2) and with atomic orbitals properly adapted to the symmetry of the molecule, can be shown to belong always to the "identical representation" of the symmetry group of the molecule, with zero spin (^1A or $^1A_1$ or $^1A_{1g}$, etc.).

The proof is as follows (cf. sections 3, 4 and Table I for necessary group theory and discussion of symmetry types). First we note that for a saturated molecule, every properly-constructed approximate U and ψ of the type found in Eq. (2) must contain for every atom one wave function for every bond which the atom forms, and for any atom which forms more than one bond of the same kind, the zeroth approximation wave functions used for these bonds must be equivalent, i.e., must transform one into another under the operations of the symmetry group (cf. e.g., the Pauling-Slater tetrahedral orbitals). [If this last condition is not met, one must in general use a linear combination of several expressions of the form of ψ in Eq. (2) in order to get a final ψ which is a representation of the symmetry group. For molecules which are not saturated, it is always necessary to form such linear combinations, by using the general form of the method of atomic orbitals, since the more special method of electron-pair bonds is not applicable.] If these conditions are fulfilled, then the effect on any U of any symmetry operation belonging to the symmetry group defined by the nuclear configuration is readily seen to be merely to permute some of the rows in its determinant (Eq. (1)), but never to eliminate any row nor to introduce any new kind of row. Every symmetry operation either leaves all φ's unchanged, or replaces some of them by other equivalent ones; it does not affect the spins (α, β). The totality of permutations produced by any operation can always be expressed in terms of a certain number of specified transpositions (exchanges) of rows. Now the value of a determinant is multiplied by -1 if an odd number, by $+1$ if an even number of transpositions of its rows is made. One now sees (1): for the totality of electrons which are in atomic closed shells, every symmetry operation produces an even number of transpositions of rows, simply because there is an even number of wave functions in the closed shells of each atom; equivalent atoms of course have equivalent closed shells; (2): the wave functions of any two electrons (one on each of two atoms) which form a bond undergo parallel transformations under any symmetry operation, and from this it is easily seen to follow that the total number of transpositions of rows resulting from the action of any symmetry operation on the bonding electron wave functions is even. Hence for a saturated molecule whose ψ is constructed as above specified, the total number of

[12] L. Pauling, J. Am. Chem. Soc. 53, 1378 (1931).

transpositions produced in each U by any symmetry operation is even, so that each U, and ψ, is multiplied by $+1$ for every operation of its symmetry group. This behavior is that which characterizes the *identical representation* of any symmetry group; this representation has the same symmetry as the nuclear skeleton itself.

It should be noted that the result just proved holds even if the actual molecule, e.g., H_2O, does not have the ideal valence angles (90°, etc.) of Pauling and Slater, provided only that the axes for the atomic orbitals used are chosen with proper regard for the actual symmetry of the molecule. The reader can readily verify the correctness of the result of the last paragraph for the cases of H_2, He_2, H_2O, NH_3, CH_4, whose ψ's were given above, by testing what happens to each U or ψ when subjected to each operation (cf. Table I) of the appropriate symmetry group.

2b. Method of molecular orbitals. We shall assume that the symmetry of the molecule is known, empirically or perhaps from the Pauling-Slater electron-pair bond rules. In order to obtain ψ's, one inserts the proper orbitals, some atomic and molecular, into Eq. (1). A suitable linear combination of the resulting U's formed in exactly the same way as in the case[8] of an atom.

The main problem is usually that of finding suitable molecular orbitals. In the first place, these must always be representations of the symmetry group defined by the arrangement of the nuclei. For molecules in their normal states, one must furthermore usually select only *bonding* molecular orbitals. These are characterized by giving a relatively high probability density between the nuclei which they bind together. The fulfillment of this condition is assured, if one constructs the molecular orbitals approximately by taking linear combinations of atomic orbitals, by using only combinations which add between nuclei.

One can also construct a variety of antibonding and of partially bonding molecular orbitals, which should be useful mostly in describing excited or repulsive states of molecules. (Examples of antibonding orbitals: σ^*2p in N_2, cf. II; C–C antibonding orbitals $[x-x]$ in C_2H_4, —cf. III, also section 8*b* (below).)

In case degenerate (atomic or) molecular orbitals, not all in closed shells, are present in the U's of Eq. (1), one must take care to form the ψ's as linear combinations of the U's in such a way that each is a representation of the symmetry group of the molecule. The method is similar to that used for atoms[8] or diatomic molecules when degenerate orbitals are present. [Usually in the present method atomic orbitals, corresponding to unshared electrons, occur only in closed shells and so cause no trouble even if degenerate.] If only nondegenerate molecular orbitals are present, and all atomic orbitals are in closed shells, linear combinations of U's need to be taken only to remove spin degeneracy,[8] and each resultant ψ is then always a representation of the symmetry group *if the molecular orbitals are.* If all orbitals occur only in closed groups (cf. II, page 51), whether atomic or molecular, one has always the identical representation. The truth of these statements can be proved by reasoning similar to that applied to a similar matter in section 2*a*, but rather simpler. The examples to be found in section 8*d* (cf. Eqs. (9–12) should also be instructive.

2c. Construction of molecular orbitals as linear combinations of atomic orbitals. The manner in which molecular orbitals can be constructed from atomic orbitals so as to conform to definite representations of the molecular group can be seen from some examples (cf. also I, II, III, etc.). Thus, molecular orbitals of H_2^+ can be formed as sums (bonding, types σ_g, π_u, etc.) or differences (antibonding, types σ_u, π_g, etc.) of H atom orbitals, namely *const.* $(\varphi_A+\varphi_B)$ or *const.* $(\varphi_A-\varphi_B)$, where φ_A and φ_B refer to equivalent orbitals of the two H nuclei.

As a simple example of a polyatomic case, a certain bonding orbital of CH_2 (or H_2O), belonging to the identical representation a_1 of point group $\mathbf{C_{2v}}$ (cf. Tables I, Ia), and related to the carbon (or oxygen) atomic orbital $2p_z$, can be approximated as follows (cf. Eq. (8) in section 8*a* for other CH_2 or H_2O orbitals):

$$[z]=a(2p_z)+b(\alpha+\beta)+c(2s)+\cdots. \qquad (3)$$

Here α and β refer to $1s$ orbitals of two H atoms placed at equal distances from the C or O atom, while $2s$ and $2p_z$ refer to orbitals of the latter. Equality of the coefficients of α and β is necessary here in order to have the molecular orbital [z] conform to a_1 of $\mathbf{C_{2v}}$, while the relative signs of a and b must be such that, in the regions where α and β strongly overlap $2p_z$, they have the same sign as the latter. In Eq. (3), c is relatively small in the case of H_2O, but larger in that of CH_2. The ratios b/a and c/a are undetermined coefficients. An antibonding orbital would be obtained if the coefficient b were taken with reversed sign; at the same time the relative magnitudes of a, b, c would be more or less changed.—It may be remarked that the symmetry type $\mathbf{C_{2v}}$ is so simple that everything can be easily seen and worked out without using group theory. For molecules of higher symmetry, however, the group theory treatment is very convenient.

In regard to the possibility of improving the approximation given by Eq. (3), by adding further terms such as $d(3s)$, $e(3p_z)$, $f(3d_z)$, the reader is referred to an illuminating discussion by Slater which is applicable here (reference 9, page 1111). In general, it is profitable only to include atomic orbitals which are fairly much alike in energy. (In the case of H_2O, even the term $c(2s)$ in Eq. (3) could probably better be dropped.) Exact equality of energy of all the orbitals, such as one has in H_2^+, is not necessary.

In Eq. (3) each of the terms $a(2p_z)$, $c(2s)$, and $b(\alpha+\beta)$, separately conforms like [z] itself to representation a_1 of $\mathbf{C_{2v}}$. Such a relation is usual in cases like Eq. (3) where a molecular orbital is built up around an orbital (here $2p_z$) of a central or dominant atom. Further examples (NH_3 and CH_4 types) will be found in V. Only such orbitals of the dominant atom as conform to the final desired representation can be used, for example $2p_z$ and $2s$, but not $2p_x$ or $2p_y$, in Eq. (3).

Often the inclusion of more than one orbital of the dominant or central atom yields a hybrid orbital which gives increased overlapping with the orbitals of the other atoms and so gives stronger bonding. This fact can often be used as a guide in forming a qualitative estimate of the relative magnitudes of two coefficients such as c and a in Eq. (3). One could indeed systematically seek out the "best bonding" hybrid orbitals of the central atom, i.e., orbitals giving maximum overlapping with those of the outer atoms to be bound, as Pauling has done[12] in connection with electron-pair bonds.[13]

One has, however, in general no right to assign to two coefficients like a and c in Eq. (3) such relative values as would correspond to a BBH (best bonding hybrid). In general, if $a_B\varphi_1+c_B\varphi_2$ represents a BBH, then to get a best approximation in Eq. (3), it is correct to use the ratio c_B/a_B only if φ_1 and φ_2 are actually degenerate. If φ_1 and φ_2 are only approximately degenerate, the ratio must be less (or greater), and must approach zero (or infinity) as φ_1 and φ_2 become more unequal in energy. Clearly, however, the larger energy decrease obtainable from stronger bonding weights the scales in favor of hybrids which approximate BBH's.

In seeking approximations corresponding to valence theory, Pauling's procedure for electron-pair bonds is to use alternatively either a BBH or a simple atomic orbital according as it appears probable from chemical and other evidence that the former or the latter gives the better approximation; he does not use intermediate types. This procedure is indeed unavoidable if a *simple* approximation is to be obtained in terms of electron-pair bonds, but it is evidently at a disadvantage in this respect as compared with the present more flexible and (when desirable) more noncommittal method; and it is more subject to possible errors of judgment. The present method has even a certain advantage in having less that it must predict or decide, for this makes it better able to be guided by empirical, including spectroscopic, data. The problem of the structure of the double bond (cf. III, and section 8 below), and of the structure of NH_3, to be discussed in V, are examples of this.

Application of Group Theory

3. Objects and notation

First we might seek to determine the irreducible representations (cf. section 1) for each kind of symmetry that is possible for a polyatomic molecule. Most, at least, of the symmetry types likely to be found in actual molecules, as well,

[13] When a hybrid $a\psi_1+c\psi_2$ of two orbitals is formed, there is of course also another one orthogonal to it, $c\psi_1-a\psi_2$. Of these two hybrid orbitals, one may be well adapted to binding in one direction, the other in another.

apparently, as a number not likely so to occur, are included among those of the 32 crystal classes. Symmetry types with n-fold axes, with $n=5$, 7, 8, or more, are not included. Table I in section 4 gives the representations for all the symmetry types of the 32 crystal classes. If the representations for other types should be needed (e.g., the ring molecule C_5H_{10} may have a five-fold axis), they can be obtained by the same methods (cf. section 4) used for the 32 crystal classes.

For each crystal class there is a symmetry group, composed of all the operations to which the crystal or molecule could be subjected without changing its appearance or aspect viewed from any fixed position.[14] Such symmetry groups are often called point groups.[14] The 32 point groups, it may be noted, correspond to a considerably smaller number of abstract groups, since many of the point groups are, when considered abstractly, identical with others.

The results, as already obtained by Bethe and Wigner (cf. section 1) but arranged somewhat differently to suit the present application, are given in Table I. The different representations of each group are designated by symbols usually the same as those used by Placzek for molecular vibrational states. As noted in section 1, this has advantages over Bethe's simple listing of representations as Γ_1, Γ_2, $\cdots$, γ_1, γ_2, $\cdots$.

Although for the symmetry group of an atom or of a diatomic or linear molecule the number of representations is infinite (s or S, P or p, D, $\cdots$; Σ^+ or σ, Σ^-, Π or (π, Δ, $\cdots$), it is finite and rather small for all the 32 point groups. For designating the representations of these, capital letters are used here for resultant electronic states (ψ's), small letters for orbitals (φ's), just as for atoms and diatomic molecules. The multiplicity of resultant states is denoted by a left-hand superscript as for atoms and diatomic molecules. No attempt will be made for the present to develop a notation for multiplet components or spin fine structure; it should be noticed that the numerical subscripts $_1$, $_2$, and $_3$ used here for certain representations belong to the *orbital* description. [Cf. Bethe[6] for a treatment of the spin structures.]

In Table I the same symbols are often used for different point groups, but each symbol has, at least in part, a rather definite significance. Thus A, B, a, b refer to nondegenerate, E, e to twofoldly degenerate, T, t to threefoldly degenerate, states or orbitals. A or a means symmetrical, B or b antisymmetrical, for a rotation of $2\pi/n$ around the (or an) n-fold principal axis. (If there is no axis, A is used.) The subscripts $_1$ and $_2$ have varying meanings; superscripts $'$ and $''$ mean symmetrical and antisymmetrical for reflection in a plane (σ_h) perpendicular to the principal axis.

Besides using general symbols such as A_1, a_2, $b_2 \cdots$ analogous to S, p, π, etc., we shall of course feel free to add more specific symbols to describe particular states or orbitals, just as we use symbols like $3p$, $2p\pi$, etc., for atoms and diatomic molecules. In particular we shall use, as already in I, II and III, a variety of symbols such as $[s]$, $[2p_z]$, $[\sigma]$, $[\pi]$, $3d_\beta$, and so on, which indicate that the molecular orbital in question is derived from, or related to, some particular type of atomic or diatomic orbital.

The objects of the following sections, insofar as group theory problems are concerned, may be summarized as follows: (1) to give all the types of electronic states (corresponding to the irreducible representations) for the 32 point groups; (2) to determine the selection rules for these; (3) to determine the nature of the resultant electronic states corresponding to any given electron configuration (e.g., $a_1^2a_2b_1e^3$); (4) to determine what happens to the orbitals and states of a molecule of given symmetry when the symmetry is altered; (5) to find what relations exist between the electronic states of a molecule and those of its (atomic or molecular) dissociation products, for various modes of dissociation. The results for (1), (2), (3), and (4) are given below; those for (5) will be considered later. For atoms and diatomic molecules the corresponding results are already well known.

4. Representations of the 32 point groups

Table I can be completely derived with the help of a limited knowledge of the theory of finite groups, not difficult to acquire in connection with a study of some of the examples

[14] For details, cf. e.g., P. P. Ewald, *Handbuch der Physik*, Vol. 24. J. Springer, Berlin, 1927; or R. W. G. Wyckoff, *The Structure of Crystals:* The Chemical Catalog Company, New York, 1931.

given in the table. The meaning of Table I will now be explained briefly, so that it can be used without further knowledge of group theory, also the method of its derivation will be given. Table Ia shows how the symmetries of a number of molecules are distributed among various point groups.

Given a point group whose irreducible representations are desired, one first divides the symmetry elements (i.e., operations), whose total number may be called g, into a number, let us call it r, of group-theory *symmetry classes*[15] (not to be confused with crystal classes) each containing, let us say, h equivalent elements, where h varies from one class to another.[16] Every group includes the element E (identity), which always forms a class by itself, with $h=1$. An important theorem is: *the number of irreducible representations of any group equals r.* One can construct a square table of representations ($\Gamma_1, \Gamma_2, \cdots \Gamma_r$), classes ($E \equiv C_1, C_2, \cdots C_r$), and *characters* ($\chi$) as shown (also cf. Table I for specific examples).

	E	C_2	..	C_r
Γ_1	$\chi_1^{(1)}$	$\chi_2^{(1)}$	..	$\chi_r^{(1)}$
Γ_2	$\chi_1^{(2)}$	$\chi_2^{(2)}$	..	$\chi_r^{(2)}$
..	..		..	..
Γ_r	$\chi_1^{(r)}$	$\chi_2^{(r)}$	..	$\chi_r^{(r)}$

Any character $\chi_m^{(j)}$ describes the effect on the representation Γ_j of *any* operation of the symmetry class C_m (the notation $\Gamma_1, \cdots C_2, \cdots$ will be replaced by more specific designations in Table I). In our case the representations define the possible types of electronic wave functions ψ or φ. The representations of the 32 point groups are all either 1-, 2-, or 3-dimensional, which for φ's and ψ's means nondegenerate, twofoldly or threefoldly degenerate.

For a nondegenerate φ or ψ, belonging say to Γ_j, $\chi_m^{(j)}$ is merely a factor by which φ or ψ is multiplied when subjected to a symmetry operation of the class C_m. For a twofoldly degenerate φ or ψ, one has of course a set of two mutually orthogonal wave functions, say φ_1, φ_2. When these are subjected to any symmetry operation of the molecule's point group, each is in general transformed in such a way that one gets a new mutually orthogonal set φ_1', φ_2', where $\varphi_1' = a_{11}\varphi_1 + a_{12}\varphi_2$ and $\varphi_2' = a_{21}\varphi_1 + a_{22}\varphi_2$. The character χ_m for such a φ(or ψ) is merely the sum $a_{11}+a_{22}$ (i.e., the spur of the matrix of coefficients a_{ij}), taken for any operation of the class C_m. The extension to threefoldly degenerate φ's and ψ's is obvious. Important is the fact that $\chi = a_{11} + a_{22}(+a_{33})$ is independent of how one selects the original two (or three) mutually orthogonal φ's or ψ's.

For the class E, $a_{ij}=0$ except that $a_{11}=a_{22}$ $(=a_{33})=1$, always, so that the character $\chi_1^{(j)}$ for any representation Γ_j is always equal to the number of dimensions of the latter. The following relation[16] then suffices to determine, for any group, how many of its r representations there are of each number of dimensions:

$$(\chi_1^{(1)})^2 + (\chi_1^{(2)})^2 + \cdots + (\chi_1^{(r)})^2 = g. \quad (4)$$

Every symmetry group has a one-dimensional representation, called the *identical representation*,[15] always symbolized by A, A_1, or A_{1g} in Table I, for which every χ is $+1$, so that any φ or ψ belonging to it has the same symmetry as the nuclear skeleton itself, as is not true of any other representation.

For most of the point groups the χ's are all integers (cf. Table I), but for a few, some of them are complex numbers. In general, φ's or ψ's belonging to different representations cannot have equal energy except in isolated special cases, and n-foldly degenerate φ's or ψ's ordinarily appear only for n-dimensional representations. But when the χ's are complex, one finds that some of the representations occur in pairs, such that the χ's of one member of a pair are conjugate complex to those of the other (cf. e.g., the groups **C**$_3$, **C**$_4$ in Table I). One can then readily show that the two representations of such a pair are conjugate complex to each other (φ or $\psi = A \pm Bi$); and from this it follows[4] by insertion in the Schrödinger equation (if possible magnetic interactions are neglected) that A and B, and so $A \pm Bi$, etc., are equal in energy and so should be considered as belonging to a single degenerate state.

[15] For a convenient survey of the application of group theory to atomic and diatomic problems, cf. B. L. van der Waerden, *Die gruppentheoretische Methode in der Quantenmechanik*, J. Springer, Berlin, 1932. Also Wigner's book and Eckart's article in Rev. Mod. Phys. for atomic problems.

[16] Cf. pages 11, 76, 77 of reference 15. Placzek[5] uses a similar method without group theory.

TABLE I. ***Irreducible representations of the 32 point groups.***

Triclinic

$\mathbf{C_1}$	E
A	1

Also $\mathbf{C_i} = \mathbf{C_1} \times i$

Monoclinic

$\mathbf{C_2}$; *Co*		E	C_2
	$\mathbf{C_s}$	E	$iC_2=\sigma$
A; z	A'	1	1
B; x or y	A''	1	-1

Also $\mathbf{C_{2h}} = \mathbf{C_2} \times i$

Orthorhombic

$\mathbf{V} \equiv \mathbf{D_2}$; *Co*		E	$C_2(z)$	$C_2(y)$	$C_2(x)$
	$\mathbf{C_{2v}}$	E	$C_2(z)$	$iC_2(y)=\sigma_v$	$iC_2(x)=\sigma_v$
A_1	A_1	1	1	1	1
B_1; z	A_2	1	1	-1	-1
B_2; y	B_1	1	-1	1	-1
B_3; x	B_2	1	-1	-1	1

Also $\mathbf{V_h} \equiv \mathbf{D_{2h}} = \mathbf{V} \times i$

Tetragonal

$\mathbf{D_4}$; *Co*	E	C_2	$2C_4$	$2C_2$	$2C_2'$
$\mathbf{C_{4v}}$	E	C_2	$2C_4$	$2iC_2=\sigma_v$	$2iC_2'=\sigma_d$
$\mathbf{V_d} \equiv \mathbf{D_{2d}}$	E	C_2	$2iC_4=2S_4$	$2C_2$	$2iC_2'=\sigma_d$
A_1	1	1	1	1	1
A_2; z	1	1	1	-1	-1
B_1	1	1	-1	1	-1
B_2	1	1	-1	-1	1
E; $x \pm iy$	2	-2	0	0	0

$\mathbf{C_4}$; *Co*	E	C_2	C_4	C_4^3
$\mathbf{S_4}$	E	C_2	S_4	S_4^3
A; z	1	1	1	1
B	1	1	-1	-1
E; $x \pm iy$	$\{1$	-1	$-i$	$i\}$
	$\{1$	-1	i	$-i\}$

Also $\mathbf{D_{4h}} = \mathbf{D_4} \times i$, and $\mathbf{C_{4h}} = \mathbf{C_4} \times i$

Hexagonal

$\mathbf{D_6}$; *Co*			E	C_2	$2C_3$	$2C_6$	$3C_2$	$3C_2'$
	$\mathbf{C_{6v}}$		E	C_2	$2C_3$	$2C_6$	$3iC_2=3\sigma_d$	$3iC_2'=3\sigma_v$
		$\mathbf{D_{3h}}$	E	$iC_2=\sigma_h$	$2C_3$	$2iC_6=2S_6$	$3C_2$	$3iC_2'=3\sigma_v$
A_1	A_1	A_1'	1	1	1	1	1	1
A_2; z	A_2	A_2'	1	1	1	1	-1	-1
B_1	B_2	A_1''	1	-1	1	-1	1	-1
B_2	B_1	A_2''	1	-1	1	-1	-1	1
E^*	E^*	E'	2	2	-1	-1	0	0
E^*_*; $x \pm iy$	E^*_*	E''	2	-2	-1	1	0	0

Also $\mathbf{D_{6h}} = \mathbf{D_6} \times i$

$\mathbf{C_6}$; *Co*	E	C_6	C_3	C_2	C_3^2	C_6^5
A	1	1	1	1	1	1
B	1	-1	1	-1	1	-1
E^*	$\{1$	ω^2	$-\omega$	1	ω^2	$-\omega\}$
	$\{1$	$-\omega$	ω^2	1	$-\omega$	$\omega^2\}$
E^*_*	$\{1$	ω	ω^2	-1	$-\omega$	$-\omega^2\}$
	$\{1$	$-\omega^2$	$-\omega$	-1	ω^2	$\omega\}$

Note: $\omega = e^{2\pi i/6} = -\omega^4$

Also $\mathbf{C_{6h}} = \mathbf{C_6} \times i$; also $\mathbf{C_{3h}} = \mathbf{C_3} \times \sigma_h$, with states A', A'', E', E''

TABLE I. (*Continued*).

Rhombohedral

D3; *Co*	E	$2C_3$	$3C_2'$
C3v	E	$2C_3$	$3iC_2'=3\sigma_v$
A_1	1	1	1
A_2; z	1	1	−1
E; $x\pm iy$	2	−1	0

C3	E	C_3	C_3^2
A	1	1	1
E	$\left\{\begin{matrix}1\\1\end{matrix}\right.$	$\begin{matrix}\omega\\\omega^2\end{matrix}$	$\left.\begin{matrix}\omega^2\\\omega\end{matrix}\right\}$

Note: $\omega=e^{2\pi i/3}$

Also $\mathbf{D_{3d}}=\mathbf{D_3}\times i$; $\mathbf{C_{3i}}\equiv\mathbf{S_6}=\mathbf{C_3}\times i$

Cubic

O; *Co*	E	$3C_2$ (x, y, z)	$6C_4$	$6C_2$	$8C_3$
Td	E	$3C_2$	$6iC_4=S_4$	$6iC_2=6\sigma_d$	$8C_3$
A_1	1	1	1	1	1
A_2	1	1	−1	−1	1
E	2	2	0	0	−1
T_1; (x, y, z)	3	−1	1	−1	0
T_2	3	−1	−1	1	0

T	E	$3C_2$	$4C_3$	$4C_3'$
A	1	1	1	1
E	$\left\{\begin{matrix}1\\1\end{matrix}\right.$	$\begin{matrix}1\\1\end{matrix}$	$\begin{matrix}\omega\\\omega^2\end{matrix}$	$\left.\begin{matrix}\omega^2\\\omega\end{matrix}\right\}$
T	3	−1	0	0

Note: $\omega=e^{2\pi i/3}$

Also $\mathbf{O_h}=\mathbf{O}\times i$; $\mathbf{T_h}=\mathbf{T}\times i$

Besides Eq. (4), two further relations[17] can be given by means of which the characters for all the possible representations of any finite group can be completely determined. The results given in Table I have been obtained[4, 6] in this way. For the various characters of any one representation (cf. the small table above on page 287)

$$\sum_{i=1}^{r} h_i\chi_i\bar{\chi}_i=g, \tag{5}$$

where $\bar{\chi}_i$ is the complex conjugate of χ_i; and also

$$h_ih_k\chi_i\chi_k=\chi_1\sum_{l=1}^{r} c_{(ik)l}h_l\chi_l, \tag{6}$$

where the coefficients $c_{(ik)l}$ are those which appear when one takes the "product of two classes" C_i and C_k (cf. reference 15, p. 170):

$$C_iC_k=C_kC_i=\sum_{l=1}^{r} c_{(ik)l}C_l. \tag{6a}$$

In order to use Eq. (6*a*) it is of course necessary to have a multiplication table of products of classes C_iC_k (including the case $i=k$). Such tables can be constructed by obvious methods when needed, for any symmetry group (cf. reference 6).

Explanation of tables. At the top of each table are given at the left (**bold-faced type**) the point groups to which it applies, at the right (*italic type* the elements of symmetry belonging to each point group. The notations for the point groups and for the symmetry elements are essentially those of Schönflies,[14] except that in the case of certain elements (reflections and rotary-reflections) an additional designation of the form iC_n (cf. Bethe[6]), showing how the element could be obtained as the product of a pure rotation C_n and the inversion i, is given for convenience. [In some cases also a ′ or an x, y, or z has been added to help make clear the exact operation which is meant; Ewald[14] gives complete details as to the various operations. In every case the (or a) principal axis of symmetry is called the z axis,—even in the monoclinic system, where the axis is usually called y.] The symmetry elements are arranged in symmetry classes according to group theory[17] the number preceding the symbol for any symmetry element being the *number h* of equivalent elements of this kind forming the class (if no number is given, $h=1$). The *total number g* of symmetry elements belonging to any point group can be obtained by adding the h's, e.g., $g=2$ for $\mathbf{C_2}$ and $\mathbf{C_s}$, 8 for $\mathbf{D_4}$. *In the main* and lower part of each table are given at

[17] Cf. A. Speiser, *Die Theorie der Gruppen von endlicher Ordnung*, second edition (J. Springer, Berlin, 1927), especially Chaps. 1, 2, 11, 12; cf. p. 28 for definition of a class, pp. 174–6 for Eqs. (4)–(6*a*). For an explanation of the relations of group theory to the present problem, reference 6 is valuable (cf. also reference 4).

the left the symbols, and at the right the characters χ, of all the irreducible representations of the point groups designated at the top of the table.

Outside each table are indicated one or two point groups not included in the table, e.g., $\mathbf{D_{6h}}$ in the case of the hexagonal table. Each of these contains the same symmetry classes as for some group in the table, plus an equal number of others each generated by multiplying one of the original set by the operation i (inversion), or in one case by σ_h. For example $\mathbf{D_{6h}} = \mathbf{D_6} \times i$ has all the six classes E, C_2, $2C_3$, etc. of $\mathbf{D_6}$ plus the six classes i, iC_2, $2iC_3$, etc. Each point group $\mathbf{G} \times i$ has two irreducible representations for each one of $\mathbf{G}$. For example, corresponding to A_1, A_2, etc. of $\mathbf{D_6}$ one has for $\mathbf{D_{6h}}$ the representations A_{1g}, A_{1u}, A_{2g}, A_{2u}, etc. A_{1g} and A_{1u} have the same characters as A_1 for the symmetry classes E, C_2, $\cdots$ common to $\mathbf{D_6}$ and $\mathbf{D_{6h}}$, while the characters of A_{1g} and A_{1u} for the classes i, iC_2, etc. are respectively $+1$ and -1 times the characters of A_1 for the classes E, C_2, etc. Similar relations hold for A_{2g}, A_{2u}, and so on.

For use in connection with the determination of selection rules, the behavior of the coordinates x, y, z under the *rotational operations* of each crystal system (monoclinic, triclinic, etc.) is given in the table or tables for that system (*heading Co*, and *symmetry classes as given in first line of table*). The behavior of each coordinate, or of a pair $(x \pm iy)$ or set of three (x, y, z in the cubic system only) of coordinates which are equivalent, is given by a set of characters whose correctness the reader can easily verify. Every such set of characters is found to agree with that of some representation in the table (cf. tables). For the operation i *the character for a coordinate or set of these* is always just the negative of that for operation E, and *for any symmetry class* niC_m *it is always just the negative of that for* nC_m.

TABLE Ia. *Some probable examples of molecules having symmetries belonging to various point groups.*

Cs: NOCl, C_2H_4 derivatives like C_2H_3Cl, C_2H_2ClBr, $C_2HClBrI$. **C2**: perp. $C_2H_2Cl_2$. **C2h**: plane *trans*-$C_2H_2Cl_2$. **C2v**: plane *cis*-$C_2H_2Cl_2$, $(Cl_2C)CH_2$, CH_2Cl_2, H_2O_2, H_2O, NO_2, SO_2, NH_2Cl, HCHO. **V**: partly rotated C_2H_4. **Vh**: plane C_2H_4. **Vd**: perp. C_2H_4. **D4h**: $PtCl_4^-$. **C4v**: distorted $PtCl_4^-$ (Pt out of plane). **D6h**: C_6H_6. **C6v**: C_6H_6 with planes of C and H displaced. **D3h**: NO_3^-, O_3 (if triangular), C_2H_6 when trigonal. **D3**, **D3d**: rotated forms of C_2H_6. **C3v**: NH_3, ClO_3^-, CH_3F, PCl_5(?) **Td**: CH_4, MnO_4^-. **Oh**: SF_6, $PtCl_6^-$.

5. Selection rules

One can easily obtain the polarization and selection rules which limit transitions between the electronic states of a molecule conforming to any point group in Table I. To do this, one makes use of the fact that the coordinates x, y, z themselves,—or combinations of equivalent coordinates, e.g., (x, y) or $(x \pm iy)$, which are analogous to degenerate φ's or ψ's,—always belong like the φ's and ψ's to definite representations of the point groups, and can be characterized by sets of χ's (cf. Table I, and last paragraph of "Explanation" following it). To find out with what states a given state with wave function ψ_i can combine, for an electric moment Q, one makes use of the expansion[16]

$$Q\psi_i = \Sigma a_{ij}\psi_j. \tag{7}$$

Here Q is proportional in the case of one electron, for dipole transitions, to x, y, or z, or to a combination $(x+iy, x-iy)$ or (x, y, z); or if there are several electrons one replaces for example x by a sum of x's, and so on. Every transition $\psi_i \leftrightarrow \psi_j$ is allowed for which a Q can be found giving $a_{ij} \neq 0$ in Eq. (7), but it is forbidden if $a_{ij} = 0$ for all Q's. If $a_{ij} = 0$ for all but one Q, the transition is polarized accordingly. Selection rules for quadrupole and other transitions can be obtained by using suitable expressions for Q. Selection rules obtained as above, of course, apply also to vibrational[5] and electronic$\times$vibrational transitions.

In order to determine selection rules, Eq. (7) is applied in the following way. For any ψ_i belonging to any irreducible representation of a given point group, and for a given Q, one first determines for the product $Q\psi_i$ a set of characters, by multiplying the character of ψ_i for each symmetry class by the character of Q for the same class.

Very often the resulting set of characters of $Q\psi_i$ is at once identified as belonging to one of the irreducible representations of the group; this happens whenever Q or ψ_i, or both, belong to one-dimensional representations. This means that only ψ_j's which belong to the representation thus identified have $a_{ij} \neq 0$ in Eq. (7), and only such ψ_j's can combine with ψ_i with a moment of the type Q.

Example: let Q be z and let ψ_i belong to representation B_1 of point group $\mathbf{C_{4v}}$; cf. Table I, tetragonal system. Multiplying the character-systems of z and B_1, the resulting character-system is seen to be that of B_1, since in this case z belongs to the identical representation (characters 1, 1, 1, 1, 1).[To get the correct result, one must take care to notice that the signs of the characters of z, for the operations iC_2 and iC_2', must be reversed as compared with those given in the table, since the latter apply to the operations C_2 and C_2' (cf. last paragraph of explanation under Table I). A similar precaution must be taken in other cases too.] Hence one concludes that for a z moment, states (φ's or ψ's) belonging to B_1 can combine only with other states belonging to the same representation. Similarly one finds that for electric moments $x+iy$, states of type B_1 can combine only with those of type E. One sees thus that for dipole transitions the B_1 type combines only with types B_1 and E. The reader can easily determine the selection rules for A_1, A_2, and B_2 states by the same method.

Whenever Q and ψ_i both belong to representations which are more-than-one-dimensional, the system of characters obtained for Q_i is not that of an irreducible representation. In this case one has to resolve each character of $Q\psi_i$ into a sum of characters, in such a way that the resulting (two or more) sets of characters are those of irreducible representations. Such a resolution is always possible, and the result is unique.

Example: let Q be $x\pm iy$ and let ψ_i belong to E of $\mathbf{C_{4v}}$. The character-system (4, 4, 0, 0, 0) of the product $(x\pm iy)\psi_E$ is seen to be the sum of those of the representations A_1, A_2, B_1, and B_2 of $\mathbf{C_{4v}}$. We conclude that states belonging to E can combine, for a moment $x\pm iy$, with all these four types. From the characters of $z\psi_E$ one concludes further that for a z moment, states of type E can combine only with other states of type E.

For any point group $\mathbf{G}\times i$, the types of states are the same as for group $\mathbf{G}$, except that there is one $_g$ and one $_u$ representation for each representation of group $\mathbf{G}$ (cf. "Explanation of tables" under Table I for details). It is easily shown by making use of the fact that every Q of type x, y, z or $x\pm iy$ changes sign under the operation i, that for dipole transitions one has for any group $\mathbf{G}\times i$ just the same selection rules as for the corresponding group $\mathbf{G}$, *plus the rule* $g\leftrightarrow u$ ($g\leftrightarrow g$, $u\leftrightarrow u$ forbidden). This is the same rule which holds for all systems having a center of inversion (atoms, electronic ψ of homopolar diatomic molecules, total ψ of any molecule, etc.).

5a. Franck-Condon principle. The electronic selection-polarization rules derived in section 5 hold strictly only for the case that the symmetry of the molecule belongs to the same point group in the initial and final states. (Changes of dimensions not causing a change in the point group do not, however, affect the selection rules.) The most probable transitions, according to the Franck principle, are those in which the nuclear configuration does not change its dimensions or velocities. Since even moderately large deviations from the Franck principle, which occur only with low probability, would rarely if ever cause more than a moderate break-down in the electronic selection rules, we conclude that, for any given initial state, in emission or absorption, transitions which violate the electronic selection rules belonging to the point group of this initial state should occur only with very low, usually negligible, intensity.

In the case of electronic transitions which do *not* violate the selection rules of the initial point group, the extent of deviations from the predictions of the Franck-Condon rule strictly applied, should presumably be similar to that in diatomic molecules. For an initial state without nuclear vibration or internal rotation, the application of the Franck-Condon principle shows that in many cases an electronic change may result in high-amplitude vibrations or internal rotations (cf. III for an example). These may give rise to large changes in the arrangement of the nuclei, but this makes little difference for the electronic selection rules, since according to Franck these depend mainly on what happens at the instant the light quantum is absorbed or emitted, before the nuclei have had time to move much.

If one has an initial state of high-amplitude vibration or internal rotation such that the nuclei are continually passing through a variety of configurations, one must determine what the selection rules are for each different point group whose symmetry the nuclei take on. Only such electronic selection rules as are common to all the different configurations are strict. Furthermore, if as usual there are preferred configurations in which the nuclei spend most of their time, and if these, but not intermediate, configurations, demand certain selection rules, then these rules must hold approximately.—In dealing with all such cases, one needs of course to know the rules for correlating electronic states of different point groups (cf. section 6, paragraph just before Table II, but also section 8*c*).

6. Resolution of representations of symmetry groups (atomic and molecular) into those of groups of lower symmetry

An important problem is the resolution of a reducible representation into irreducible ones.[15, 17] An example where this problem occurs, and the very simple method of solving it, have been given in section 5. Other important examples occur when, starting with a physical system having relatively high symmetry, the symmetry is altered in such a way that the new symmetry group is a subgroup of the old one, i.e., is a group containing just a part of the symmetry elements of the old group, but no new ones. Examples are: (*a*) an atom subjected to a perturbing electric field, as in a crystal[6] or as in the formation of a molecule; or imagined modified by the splitting of its nucleus to give a diatomic or polyatomic molecule; (*b*) a molecule of high symmetry distorted to one of lower symmetry, for example tetrahedral CH_4 deformed by pulling one H atom out of position.

In many cases, degenerate representations split up partly or wholly when the symmetry of a molecular system is decreased, but the total number of different irreducible representations generally decreases. The new representations can be obtained by the following method. One tests the effect of each class of symmetry elements of the *new group* on any desired representation (or specifically on a particular φ or ψ) of the old group, and writes down the resulting set of characters. If the original representation was nondegenerate, this set of characters is at once identified as belonging to a definite representation of the new group. Otherwise, one has a reducible representation which can be resolved into a sum of irreducible representations of the new group by the method stated in section 5 (second from last paragraph). The result is always unique.[15, 17] Examples are given by Bethe,[6] and in the following Tables II–IV.

In the course of the present work, it will be convenient for various purposes to have reduction tables showing how atomic→polyatomic, diatomic→polyatomic, and polyatomic→polyatomic representations of lower symmetry. Tables II–IV are not complete, but cover a number of the cases most likely to be needed. Other results can be worked out easily when needed (cf. Table II, note *c* and Table III, notes *a*, *d*, *e*). The tables with their notes are self-explanatory.

In using any table, care should be taken to be sure that the relations between the axes of the two groups with which one is concerned are the same as in the table. Otherwise the table is not applicable. In the tables, the only cases given are those where the *z* axes of the more and the less symmetrical group with which any table deals are coincident. Such cases are the most common in practice.

Sometimes one has two symmetry groups with some elements in common, others peculiar to each, and wishes to know in what way the representations of the one group would go over into those of the other if the symmetry were altered from the one to the other case. An *example* is the correlation of energy levels between the states of plane and "perpendicular" C_2H_4 (cf. III, and section 8*c* below). In such cases, just those symmetry elements which are common to the two groups are also possessed by the symmetry group corresponding to an arrangement of nuclei intermediate between those of the two original cases. One then reduces the representations of each of the latter in terms of those of the intermediate group, which is a sort of greatest common factor. Then one applies the usual rule for adiabatic correlations, namely that these are so made that, on the energy-level diagram, no two lines cross which denote states belonging to the same representation of the intermediate symmetry group.

Explanation of Table IV. The meaning of the tables should be clear from the following detailed interpretation of the fifth small table, headed $\mathbf{D_{4h}}|\mathbf{V_h}\ \mathbf{V_d}\ \mathbf{C_{4v}}$. This table shows what each of the ten representations of $\mathbf{D_{4h}}$ goes over into if the symmetry is reduced to $\mathbf{V_h}$, or $\mathbf{V_d}$, or $\mathbf{C_{4v}}$. For example, A_{1g}, A_{1u}, A_{2g}, etc., of $\mathbf{D_{4h}}$ go over respectively into A_{1g}, A_{1u}, B_{1g}, etc. of $\mathbf{V_h}$, or into A_1, B_1, A_2, etc. of $\mathbf{V_d}$, or into A_1, A_2, A_2, etc. of $\mathbf{C_{4v}}$. Also obviously (not in the tables) A_{1g} and A_{1u} of $\mathbf{D_{4h}}$ would go into A_1 of $\mathbf{D_4}$, and so on.

Table IV is by no means complete; other reductions when needed can easily be obtained, as were those given, by methods described in sections 6, 5. In all cases the z axes of the two groups considered are taken as coincident; otherwise different results would in general be obtained. In some cases ($\mathbf{C_{4v}}\rightarrow\mathbf{C_{2v}}$, $\mathbf{C_{6v}}\rightarrow\mathbf{C_{3v}}$) the result depends also on which of two sets of vertical planes of the larger group is identified

TABLE II. *Resolution of atomic representations into irreducible representations of molecular symmetry groups.*

Atom	$\mathbf{O_h}$	$\mathbf{T_d}$	$\mathbf{D_{6h}}$	$\mathbf{C_{3v}}$	$\mathbf{D_{4h}}$	$\mathbf{C_{2v}}$
	(all g, u)		(all g, u)		(all g, u)	
S_g, S_u	A_1	A_1, A_2	A_1	A_1, A_2	A_1	A_1, A_2
P_g, P_u	T_1	T_1, T_2	$A_2+E_*^*$	A_2+E, A_1+E	A_2+E	$A_2+B_1+B_2$, $A_1+B_1+B_2$
D_g, D_u	$E+T_2$	$E+T_2$, $E+T_1$	A_1+E^* $+E_*^*$	A_1+2E, A_2+2E	A_1+B_1 $+B_2+E$	$2A_1+A_2+B_1+B_2$, $A_1+2A_2+B_1+B_2$

Notes: (*a*) S_g, S_u, P_g, P_u, D_g, D_u are usually written S, S^0, P, P^0, D, D^0. (*b*) The table holds also for orbitals as follows: s behaves like S_g, p like P_u, d like D_g. (*c*) The results for $\mathbf{O_h}$, $\mathbf{D_{6h}}$, and $\mathbf{D_{4h}}$ are from reference 6, while the remaining results, and any desired results for other point groups, can be obtained from those for $\mathbf{O_h}$, $\mathbf{D_{6h}}$, and $\mathbf{D_{4h}}$ by using Tables I and IV. (*d*) Bethe[6] gives additional results, for the relations between F, G, . . . atomic states and the representations of $\mathbf{O_h}$, $\mathbf{D_{6h}}$, and $\mathbf{D_{4h}}$. (*e*) For the relations between atomic and diatomic (or linear-molecule) representations, cf. Wigner and Witmer, Zeits. f. Physik **51**, 859 (1928); or reference 15 or Mulliken, Rev. Mod. Phys. **4**, 1932 (bottom page 20).

TABLE III. *Resolution of representations of diatomic (or linear-molecule) groups* $\mathbf{C_{\infty v}}$ *and* $\mathbf{D_{\infty h}}$ *into those of some other molecular symmetry groups.*

$\mathbf{D_{\infty h}}$	$\mathbf{D_{6h}}$	$\mathbf{C_{6v}}$	$\mathbf{C_{3v}}$	$\mathbf{D_{4h}}$	$\mathbf{C_{4v}}$	$\mathbf{C_{2v}}$
$\Sigma^+_{g,u}$	A_{1g}, A_{2u}	A_1	A_1	A_{1g}, A_{2u}	A_1	A_1
$\Sigma^-_{g,u}$	A_{2g}, A_{1u}	A_2	A_2	A_{2g}, A_{1u}	A_2	A_2
$\Pi_{g,u}$	$E_{*g,u}^*$	E_*^*	E	$E_{g,u}$	E	B_1+B_2
$\Delta_{g,u}$	$E^*_{g,u}$	E^*	E	$B_{1g,u}+B_{2g,u}$	B_1+B_2	A_1+A_2
$\Phi_{g,u}$	$B_{1g,u}+B_{2g,u}$	B_2+B_1	A_1+A_2	$E_{g,u}$	E	B_1+B_2
$\Gamma_{g,u}$	$E^*_{g,u}$	E^*	E	$A_{1g,u}+A_{2g,u}$	A_1+A_2	A_1+A_2

Notes: (*a*) Table III holds only if the symmetry (z) axis of the diatomic case coincides with the z axis of the other case. Other results are obtained for other relations between the axes of the two cases. (*b*) Table III of course applies equally well for the corresponding small letters, e.g., $\sigma_{g,u}$ behave like $\Sigma^+_{g,u}$ and give a_{1g}, a_{2u}, and so on. (*c*) Table III is applicable also (dropping g's and u's) for the resolution of the representations Σ^+, Σ^-, Π, Δ, . . . of the group $\mathbf{C_{\infty v}}$ ($\mathbf{D_{\infty h}}=\mathbf{C_{\infty v}}\times i$) into representations of $\mathbf{C_{6v}}$, $\mathbf{C_{4v}}$, $\mathbf{C_{3v}}$, $\mathbf{C_{2v}}$, etc., but the representations of $\mathbf{C_{\infty v}}$ cannot, of course, be resolved into those of $\mathbf{D_{6h}}$, $\mathbf{D_{4h}}$ or other groups of type $\mathbf{G}\times i$. (*d*) Resolution, when possible, into representations of other point groups can be effected by using first Table III, then Table I. Resolution of representations of $\mathbf{C_{\infty v}}$ and $\mathbf{D_{\infty h}}$ into those of point groups belonging to the cubic system is not possible. Even and odd (g and u) representations of $\mathbf{D_{\infty h}}$ must in general be treated separately. (*e*) In constructing Table III, one needs first an auxiliary table of characters showing how the representations of $\mathbf{C_{\infty v}}$ and $\mathbf{D_{\infty h}}$ behave under various symmetry operations of the point groups. Such a table can easily be constructed by the reader by using the following relations. For the operation E, $\chi=1$ for Σ, 2 for all other states. For a rotation through any angle φ around the z axis, $\chi=1$ for all Σ states, $\chi=e^{i\Lambda\varphi}+e^{-i\Lambda\varphi}=2\cos\Lambda\varphi$ for states with $\Lambda>0$ (Π, Δ, . . . states). For reflection in any plane through the z axis, $\chi=+1$ for Σ^+, -1 for Σ^-, 0 for all other states; σ behaves like Σ^+. (Cf. Wigner and Witmer, Zeits. f. Physik **51**, 862, 1928, or reference 15, p. 40.) These results apply as well to $\mathbf{C_{\infty v}}$ as to $\mathbf{D_{\infty h}}$. For $\mathbf{D_{\infty h}}$ one needs also the following: for operation i, χ is ±1 times its value for E, according as the diatomic state is g or u; for reflection in the xy plane, which is equivalent to i times a rotation of π around the z axis, χ is equal to ±1 (depending on the sign of χ for i) times the χ for the rotation; for rotation by π around any axis perpendicular to the z axis, which is equivalent to i times reflection in any plane passing through the z axis, χ is equal to ±1 (depending on the sign of χ for i) times χ for the reflection, and so, as one easily finds, has the value $+1$ for Σ^+_g and Σ^-_u, -1 for Σ^-_g and Σ^+_u, 0 for all other states.

TABLE IV. *Resolution of polyatomic representations into those of groups of lower symmetry.*

C_{2h}	C_i	C_s
$A_{g,u}$	$A_{g,u}$	A', A''
$B_{g,u}$	$A_{g,u}$	A'', A'

V_h	C_{2h}	C_{2v}
$A_{1g,u}$	$A_{g,u}$	A_1, A_2
$B_{1g,u}$	$A_{g,u}$	A_2, A_1
$B_{2g,u}$	$B_{g,u}$	B_1, B_2
$B_{3g,u}$	$B_{g,u}$	B_2, B_1

V	C_2
A_1	A
B_1	A
B_2	B
B_3	B

C_{2v}	C_2
A_1	A
A_2	A
B_1	B
B_2	B

D_{4h}	V_h	V_d	C_{4v}
(all g, u)	(all g, u)		
A_1	A_1	A_1, B_1	A_1, A_2
A_2	B_1	A_2, B_2	A_2, A_1
B_1	A_1	B_1, A_1	B_1, B_2
B_2	B_1	B_2, A_2	B_2, B_1
E	B_2+B_3	E	E

V_d	V	C_{2v}
A_1	A_1	A_1
A_2	B_1	A_2
B_1	A_1	A_2
B_2	B_1	A_1
E	B_2+B_3	B_1+B_2

C_{4v}	C_{2v}
A_1	A_1
A_2	A_2
B_1	A_1 or A_2
B_2	A_2 or A_1
E	B_1+B_2

D_{6h}	D_{3d}	D_{3h}	C_{6v}
(all g, u)	(all g, u)		
A_1	A_1	A_1', A_1''	A_1, A_2
A_2	A_2	A_2', A_2''	A_2, A_1
B_1	A_2	A_1'', A_1'	B_2, B_1
B_2	A_1	A_2'', A_2'	B_1, B_2
E^*	E	E', E''	E^*
E^*_*	E	E'', E'	E^*_*

D_{3h}	C_{3v}
A_1'	A_1
A_2'	A_2
A_1''	A_2
A_2''	A_1
E'	E
E''	E

C_{6v}	C_{3v}
A_1	A_1
A_2	A_2
B_2	A_2 or A_1
B_1	A_1 or A_2
E^*	E
E^*_*	E

D_{3d}	C_{2h}	C_{3v}
(all g, u)	(all g, u)	
A_1	A	A_1, A_2
A_2	B	A_2, A_1
E	$A+B$	E

D_3	C_2
A_1	A
A_2	B
E	$A+B$

C_{3v}	C_s
A_1	A'
A_2	A''
E	$A'+A''$

O_h	D_{4h}	T_d
(all g, u)	(all g, u)	
A_1	A_1	A_1, A_2
A_2	B_1	A_2, A_1
E	A_1+B_1	E
T_1	A_2+E	T_1, T_2
T_2	B_2+E	T_2, T_1

T_d	V_d	C_{3v}
A_1	A_1	A_1
A_2	B_1	A_2
E	A_1+B_1	E
T_1	A_2+E	A_2+E
T_2	B_2+E	A_1+E

with a set of vertical planes of the smaller group; this accounts for the alternatives given in the tables.

7. Determination of possible resultant electron states for various electron configurations

An important problem is that of determining, for a given electron configuration, what are the possible electronic states. For example with a configuration $a_1^2b_2e^3$ of a molecule having the symmetry $\mathbf{C_{4v}}$, these states would be just ^3E and ^1E.

First we may consider the case that the molecule has its electrons all in different, molecular, orbitals. One forms the product, for each symmetry operation, of the characters χ for the various orbitals which are occupied, and thus gets a definite system of characters for the resultant product representation.[6, 15] This either is immediately identified as an irreducible representation of the group, defining the type of the resultant electronic state, or else it can be resolved into a sum of such representations, corresponding to several possible resultant states (cf. sections 5, 6 for the method). The resultant spin has of course all the possible values one would get for the same number of non-equivalent electrons in an atom or diatomic molecule.

Examples: (a) given the configuration a_2b_2 of $\mathbf{C_{4v}}$, the product representation is immediately identified as B_1, and the states are $^3B_1$ and $^1B_1$. (b) Given $a_2b_1e^*$ of $\mathbf{D_6}$, one gets $^4E^*_*$, $^2E^*_*$, $^2E^*_*$. (c) Given $e^*e^*_*$ of $\mathbf{D_6}$, one gets the product character-system 4, -4, 1, -1, 0, 0, and the states $^3B_1$, $^3B_2$, $^3E^*_*$, $^1B_1$, $^1B_2$, $^1E^*_*$.

The same method can be used if one thinks of a molecule as composed of a core plus one or more outer, perhaps valence, electrons. For example, with point group $\mathbf{C_{4v}}$, a core of type ${}^3B_2$ plus an outer electron of type e give states ${}^4E$ and ${}^2E$.—The same point of view can also often be applied in determining the possible states of a united-system (molecule) in relation to the states of two part-systems (atoms or radicals) which come together. Discussion of such dissociation problems will, however, be postponed.

In a polyatomic molecule a group of $2n$ electrons occupying any n-foldly degenerate molecular orbital functions is a closed shell; it is required by the Pauli principle to have zero resultant spin, and it belongs to the identical representation of the molecule's point group, i.e., all the χ's are $+1$ (cf. fifth following paragraph for proof). The totality of electrons in closed shells composed of molecular orbitals, plus the totality of unshared electrons assigned to closed shells of atomic orbitals (assuming that no atomic closed shell has been removed *in toto* from the molecule), can always be regarded as a core, whose state always belongs to the identical representation with zero spin (cf. fifth following paragraph for proof). From the rule for getting the χ's for a product representation, it is now evident that such a core, since all its χ's are $+1$, can be disregarded in finding the nature of the resultant state of the whole molecule, in the same way that closed shells can be disregarded in the case of atoms or diatomic molecules.—It may also be noted here that, except for their spins, electrons in molecular orbitals belonging to the identical representation (a_1, a_{1g}, etc.) can also be disregarded in determining the resultant state.

Next we must consider the case where, aside from closed shells, two or more equivalent electrons are present in degenerate orbitals. (Two electrons in a nondegenerate orbital of course form a closed shell.) In case both equivalent and non-equivalent electrons are present, one of course first finds the resultant states of each of these separately, then treating the one set of electrons as a core, finds the final resultant states.

Bethe has attacked the problem by a method similar to that customary for the analogous atomic problem. He assumes the molecule, belonging to a specified point group, to be subjected to a perturbing electric field of sufficiently low symmetry so that the degenerate representations of the original group are split up. The Pauli principle can then be applied, and the nature of the allowed states for the desired case determined by resynthesizing the perturbed representations into those of the original group (cf. reference 6, pages 177–180). This method (hereafter called *method A*) is, however, not in general adequate, since the correlations by means of which one goes backward from the perturbed to the unperturbed representations do not always give unambiguous results (cf. e.g., the two alternative sets of Γ's for Bethe's $g_5{}^2$ of the tetragonal holohedral and $\gamma_5{}^2$ of the cubic holohedral group), although it is clear that for every set of equivalent electrons there must exist a unique set of resultant electronic states. Unique results can, however, be obtained by using method A in combination with another method (B) described below.

In the monoclinic, triclinic and orthorhombic systems of point groups, all the representations are nondegenerate, so that the present problem does not arise. In the tetragonal, hexagonal, and rhombohedral systems, we must consider cases like e^2, e^3, where e is twofoldly degenerate. In the cubic system we have e^2, e^3, also t^2, t^3, t^4, t^5, where t is threefoldly degenerate. It will be sufficient to consider just the three holohedral point groups $\mathbf{D_{4h}}$, $\mathbf{D_{6h}}$, and $\mathbf{O_h}$, for as can easily be seen, any desired results for other point groups having degenerate representations can be obtained by a process of resolution and comparison, by using Tables IV, I. (An example is given in the third following paragraph.)

Beginning with $\mathbf{D_{4h}}$, *method B* proceeds as in the following example. From Table III we note that π_u, π_g of the diatomic group $\mathbf{D_{\infty h}}$ go over, if the symmetry is reduced to $\mathbf{D_{4h}}$, into e_u, e_g. Likewise $\pi_u{}^2$ and $\pi_g{}^2$ of $\mathbf{D_{\infty h}}$ must go over, for the imaginary case of no coupling between the two electrons, into $(e_u)^2$ and $(e_g)^2$ of $\mathbf{D_{4h}}$. If now one allows some coupling between the electrons, one gets for $\mathbf{D_{\infty h}}$ the states ${}^3\Sigma^-{}_g$, ${}^1\Delta_g$, ${}^1\Sigma^+{}_g$ (the same for $\pi_u{}^2$ as for $\pi_g{}^2$). Since these results are independent of the strength of the coupling, these states of $\mathbf{D_{\infty h}}$ must be correlated with the states obtainable from $(e_u)^2$ and $(e_g)^2$ of $\mathbf{D_{4h}}$. Reference to Table III shows that ${}^3\Sigma^-{}_g$, ${}^1\Delta_g$, ${}^1\Sigma^+{}_g$ of $\mathbf{D_{\infty h}}$ go over into ${}^3A_{2g}$, ${}^1B_{1g}$, ${}^1B_{2g}$, and ${}^1A_{1g}$ of $\mathbf{D_{4h}}$, which are, then, the desired possible resultant states of the configuration $(e_u)^2$, and of $(e_g)^2$. In a similar way, since π_u^3 gives $^2\Pi_u$ and π_g^3 gives $^2\Pi_g$, one concludes that $(e_u)^3$ of $\mathbf{D_{4h}}$ gives ^2E_u and $(e_g)^3$ gives ^2E_g.

In a similar way it follows, from the fact that π_g^4 or π_u^4 gives the identical representation $^1\Sigma^+{}_g$, that $(e_u)^4$ or $(e_g)^4$ gives $^1A_{1g}$. In an analogous manner, one can easily show, for any point group outside the cubic system, that a molecular closed shell always gives the identical representation of the group. For point groups of the cubic system, a different method, like that used below for such groups, gives a corresponding result.—If one regards as a core the unshared electrons which in the present method are assigned to closed shells of atomic orbitals, it follows from (1) of the proof given in section 2*a* that this core belongs to the identical representation of the molecule's point group, provided equivalent atoms have the same sets of closed shells (as they of course have in saturated molecules). Hence the totality of electrons in such atomic closed shells, together with those in closed shells of molecular orbitals, can be treated as a core which belongs to the identical representation.

If we are interested not in $\mathbf{D_{4h}}$ as above, but for example in $\mathbf{V_d}$, to which perp. C_2H_4 belongs (cf. section 8*b*), one easily finds from Table IV, by resolution of representations of $\mathbf{D_{4h}}$ into those of $\mathbf{V_d}$, that the set of states corresponding to e^2 of $\mathbf{V_d}$ is $^3A_2$, $^1B_1$, $^1B_2$, $^1A_1$, while that corresponding to e^3 is ^2E.

Next considering $\mathbf{D_{6h}}$, we find from Table III that $\pi_{g,u}$ go over into $e^*_{*g,u}$, and $\delta_{g,u}$ into $e^*{}_{g,u}$. Matching the states of e_*^{*2},—either g or u,—against those of π^2, and those of e^{*2} against those of δ^2 (which are $^3\Sigma^-{}_g$, $^1\Gamma_g$, $^1\Sigma^+{}_g$), one finds that both e^{*2} and e_*^{*2} give rise to the set $^3A_{2g}$, $^1E^*{}_g$, $^1A_{1g}$. Similarly one finds that $e^*{}_g{}^3$ gives $^2E^*{}_g$, $e^*_{*g}{}^3$ gives $^2E^*_{*g}$, $e^*{}_u{}^3$ gives $^2E^*{}_u$, $e^*_{*u}{}^3$ gives $^2E^*_{*u}$.

For the group $\mathbf{O_h}$, method *A* gives the desired results if it is applied twice, reducing one time from $\mathbf{O_h}$ to $\mathbf{D_{4h}}$, then again from $\mathbf{O_h}$, through $\mathbf{T_d}$, to $\mathbf{C_{3v}}$. Reduction to $\mathbf{D_{4h}}$ alone, or to $\mathbf{C_{3v}}$ alone, gives ambiguous results, but when the various sets of states obtained from the two reductions are compared, it is found that there is always one and only one set of states which is common to both.

First we may consider e_g^2 and e_u^2 of $\mathbf{O_h}$. Reducing to $\mathbf{D_{4h}}$, e_g or e_u gives $a_{1g}+b_{1g}$ or $a_{1u}+b_{1u}$ (cf. Table IV). Hence 3_g^2 of $\mathbf{O_h}$ corresponds to $(a_{1g}+b_{1g})^2=a_{1g}^2+a_{1g}b_{1g}+b_{1g}^2$ of $\mathbf{D_{4h}}$. The groups a_{1g}^2 and b_{1g}^2 of $\mathbf{D_{4h}}$ are closed shells (cf. Table I), so each gives $^1A_{1g}$, while $a_{1g}b_{1g}$ gives $^3B_{1g}$ and $^1B_{1g}$; so altogether we have $2^1A_{1g}+^3B_{1g}+^1B_{1g}$ of $\mathbf{D_{4h}}$. As is easily verified, e_u^2 of $\mathbf{O_h}$ gives the same result. Now going backward to states of $\mathbf{O_h}$, using Table IV, we find two possibilities: (*a*) $^3A_{2g}+^1E_g+^1A_{1g}$; (*b*) $^3A_{2g}+^1A_{2g}+2^1A_{1g}$. To decide between these, we must use the reduction to $\mathbf{C_{3v}}$.

Reducing to $\mathbf{C_{3v}}$, one first finds (Table IV) that both e_g and e_u of $\mathbf{O_h}$ go into e of $\mathbf{T_d}$ and thence into e of $\mathbf{C_{3v}}$. Hence to find either e_g^2 or e_u^2 of $\mathbf{O_h}$, we must find e^2 of $\mathbf{C_{3v}}$. Using the result obtained in a previous paragraph for $\mathbf{D_{6h}}$, that $e^*{}_g{}^2=e^*{}_u{}^2=e^*_{*g}{}^2=e^*_{*u}{}^2=^3A_{2g}+^1E^*{}_g+^1A_{1g}$, and reducing to $\mathbf{C_{3v}}$ with the help of Table IV, we find that e^2 of $\mathbf{C_{3v}}=^3A_2+^1E+^1A_1$. Going backward from $\mathbf{C_{3v}}$ to $\mathbf{T_d}$ and thence to $\mathbf{O_h}$ by Table 4, we find as possibilities for e_g^2 and e_u^2 of $\mathbf{O_h}$ the following: $^3A_{2g}+^1T_{2g}$; $^3A_{2g}+^1E_g+^1A_{1g}$, and six other sets which are obviously out of the question because each contains one or more odd (u) states.

One sees that the correct result is $^3A_{2g}+^1E_g+^1A_{1g}$, since this and only this is common to the possibilities offered by the reductions to $\mathbf{D_{4h}}$ and to $\mathbf{C_{3v}}$. The result is the same here as that given by Bethe. By similar methods one can show that e_g^3 gives ^2E_g, e_u^3 gives ^2E_u, for $\mathbf{O_h}$.

The results given above, and others which have been obtained for $\mathbf{O_h}$ by the method just used, are summarized in Table V. The results for $\mathbf{O_h}$ differ in part from those given by Bethe (cf. his Table XV), in that a decision is made between some alternatives left open by him. (Bethe considered that both alternatives were possible, depending on circumstances, but this seems not to be correct.)

TABLE V. *Resultant states for various numbers of equivalent electrons in molecular orbitals.*

No. Els.	e_g, e_u of $\mathbf{D_{4h}}$	e^*_g, e^*_u of $\mathbf{D_{6h}}$	e^*_{*g}, e^*_{*u} of $\mathbf{D_{6h}}$	e of $\mathbf{C_{3v}}$
1	$^2E_{g,\,u}$	$^2E^*_{g,\,u}$	$^2E^*_{*g,\,u}$	^2E
2	$^3A_{2g}+{}^1B_{1g}+{}^1B_{2g}+{}^1A_{1g}$	$^3A_{2g}+{}^1E^*_g+{}^1A_{1g}$	$^3A_{2g}+{}^1E^*_g+{}^1A_{1g}$	$^3A_2+{}^1E+{}^1A_1$
3	$^2E_{g,\,u}$	$^2E^*_{g,\,u}$	$^2E^*_{*g,\,u}$	^2E
4	$^1A_{1g}$	$^1A_{1g}$	$^1A_{1g}$	$^1A_1$

	e_g, e_u of $\mathbf{O_h}$	t_{1g}, t_{1u} of $\mathbf{O_h}$	t_{2g}, t_{2u} of $\mathbf{O_h}$
1	$^2E_{g,\,u}$	$^2T_{1g,\,u}$	$^2T_{2g,\,u}$
2	$^3A_{2g}+{}^1E_g+{}^1A_{1g}$	$^3T_{1g}+{}^1A_{1g}+{}^1E_g+{}^1T_{2g}$	$^3T_{1g}+{}^1A_{1g}+{}^1E_g+{}^1T_{2g}$
3	$^2E_{g,\,u}$	$^4A_{1g,\,u}+{}^2E_{g,\,u}+{}^2T_{1g,\,u}+{}^2T_{2g,\,u}$	$^4A_{2g,\,u}+{}^2E_{g,\,u}+{}^2T_{1g,\,u}+{}^2T_{2g,\,u}$
4	$^1A_{1g}$	Same as for two electrons	
5		Same as for one electron	
6		$^1A_{1g}$	$^1A_{1g}$

APPLICATION TO CH_2 AND C_2H_4

8. The molecules CH_2 and C_2H_4

In III of this series, the formation of C_2H_4 from $2CH_2$ was discussed. Several statements were made there, without proof, whose justification depends on the results of the present paper. In order to give this justification and also to illustrate the methods described above, the electronic structures of CH_2 and C_2H_4 will be considered again now. The following discussion and that in III supplement each other, and should be read together.

Applications to other molecules will be given in later papers. The basis for a number of conclusions stated in I of this series can, however, now easily be found by the reader, if he wishes, in the Tables I–IV given above.

8a. The molecule CH_2. The CH_2 molecule, assuming it to have the form of an isosceles triangle (cf. III), has the symmetry of the point group $\mathbf{C_{2v}}$ (cf. Table I). Electronic configurations of CH_2 were given in III in terms of molecular orbitals called $[s]$, $[x]$, $[y]$, $[z]$, and respectively capable of being approximated by linear combinations of $2s$, $2p_x$, $2p_y$, $2p_z$ orbitals of the carbon atom with hydrogen $1s$ orbitals, hereafter called α and β. With the choice of x, y, z axes described in III it will be found, on testing their behavior under the symmetry operations of $\mathbf{C_{2v}}$, that carbon $2s$, $2p_x$, $2p_y$, $2p_z$ respectively belong to the representations a_1, b_1, b_2, a_1 of $\mathbf{C_{2v}}$. The linear combinations with α and β, formed in such a way that they still belong to these same representations, are

$$\left.\begin{aligned} &[s]=a(2s)+b(\alpha+\beta)+c(2p_z); \quad [x]=2p_x.\\ &[y]=a'(2p_y)+b'(\alpha-\beta).\\ &[z]=a''(2p_z)+b''(\alpha+\beta)+c''(2s). \end{aligned}\right\} \quad (8)$$

The fact that $2s$ and $2p_z$ of carbon belong to the same representation allows and requires them to hybridize somewhat with each other when CH_2 is formed. Since α and β by themselves are not representations of $\mathbf{C_{2v}}$, their hybridization with the carbon orbitals, necessary to obtain bonding orbitals in CH_2, can take place freely provided the coefficients of α and β are so related that the resulting hybrids belong to representations of $\mathbf{C_{2v}}$. In the case of $[x]$, these coefficients are zero because $2p_x$ belongs to a representation b_1 which demands that the plane of the three nuclei shall be a nodal plane.

The fact that the electron configuration $1s^2[s]^2[y]^2[z]^2$ of Eq. (1) of III gives a $^1A_1$ state (called $^1\Gamma_1$ in the notation of III) is an illustration of the rule that a set of closed shells always gives the identical representation (cf. section 7). On applying the rules of section 7, one finds that $\cdots[z][x]$ in Eq. (1) of III, which is of the type $\cdots a_1b_1$, gives a $^3B_1$ and a $^1B_1$ state (called $^3\Gamma_3$ and $^1\Gamma_3$ in III), of which we may reasonably expect the $^3B_1$ to have the lower energy.

8b. Formation of C_2H_4. The normal plane form of C_2H_4 (rotation angle 0° or 180° in Fig. 2 of III) has the symmetry $\mathbf{V_h}$, the perp. form (rotation angle 90° or 270°) the symmetry $\mathbf{V_d}$, while all intermediate forms belong to $\mathbf{V}$: cf. Table I. In order to make the notation used in III conform to the revised notation of Table I, the following changes must be made: *for plane*

C_2H_4 in III (Eqs. (2, 3, 5) and elsewhere), *read* ${}^1A_{1g}$, ${}^3B_{1u}$, ${}^1B_{1u}$ *everywhere instead of* ${}^1\Gamma_{1g}$, ${}^3\Gamma_{4u}$, ${}^1\Gamma_{4u}$; ***for perp.*** C_2H_4 in III (Eqs. (6, 7), Fig. 1, and elsewhere), *read* ${}^3A_2$, ${}^1B_1$, ${}^1B_2$, ${}^1A_1$ *instead of* ${}^3\Gamma_3$, ${}^1\Gamma_2$, ${}^1\Gamma_4$, and ${}^1\Gamma_1$.

The orbitals $[x+x]$, and so on, of plane C_2H_4 (cf. III) have been so constructed from $[x]$, and so on, of CH_2 that they conform to representations of $\mathbf{V_h}$. As the reader can easily verify by testing the behavior of each under the symmetry operations of $\mathbf{V_h}$ (cf. Table I), the orbitals $[x+x]$, $[x-x]$, $[y+y]$, $[y-y]$, $[z+z]$ of plane C_2H_4 belong respectively to the representations b_{3u}, b_{2g}, b_{2u}, b_{3g}, a_{1g}. Any electron configuration consisting of closed shells, e.g., that in Eq. (2) or (3) of III, gives a ${}^1A_{1g}$ state. By applying the rule given in section 7, it is found that $\cdots[x+x][x-x]$, which is of the type $\cdots b_{3u}b_{2g}$, gives a ${}^3B_{1u}$ and a ${}^1B_{1u}$ state (${}^3\Gamma_{4u}$ and ${}^1\Gamma_{4u}$ in the notation of III, cf. Eq. (5)).

Next we may consider perp. C_2H_4 (symmetry $\mathbf{V_d}$), and its formation from $2CH_2$, each in the state $\cdots[z][x]$, ${}^3B_1$ with their planes at right angles. The axes appropriate to $\mathbf{V_d}$ then are an x and a y axis whose directions make 45° angles with the x and y directions of the two CH_2, and a z axis coincident with the z axes of both. To get stable forms of perp. C_2H_4, one forms the bonding pair $[z+z]^2$ from the two $[z]$ electrons of $2CH_2$. The orbital $[z+z]$ of perp. C_2H_4, although constructed in zeroth approximation from two $[z]$ each as in Eq. (8), is not identical with $[z+z]$ of plane C_2H_4. It belongs to representation a_1 of $\mathbf{V_d}$.

Let us now denote by $[x]_A$ and $[x]_B$ the $[x]$ orbitals of the two CH_2, whose x axes, it should be remembered, are at right angles. On testing the effect of the symmetry operations of $\mathbf{V_d}$ on $[x]_A$ and $[x]_B$, one finds changes of the type

$$\begin{cases}[x]_A\rightarrow c_{AA}[x]_A+c_{AB}[x]_B\\ [x]_B\rightarrow c_{BA}[x]_A+c_{BB}[x]_B.\end{cases}$$

By writing down the sums $c_{AA}+c_{BB}$ and regarding them as characters (cf. section 4), one finds that the ***pair*** $[x]_A$, $[x]_B$ gives exactly the set of characters belonging to the two-dimensional representation e of $\mathbf{V_d}$. This shows that the CH_2-radical orbitals $[x]_A$, $[x]_B$, degenerate for $2CH_2$, remain so if perp. C_2H_4 is formed, and constitute suitable zeroth approximations for molecular orbitals of perp. C_2H_4.

Exactly the same situation holds for the pair of CH_2-radical orbitals $[y]_A$, $[y]_B$, and therefore the four electrons $[y]_A^2[y]_B^2$ of $2CH_2$, in spite of the fact that each of the two pairs already forms a closed shell of CH_2, all belong in perp. C_2H_4 to a single degenerate type of C_2H_4-molecule orbitals belonging to representation e, and together form a larger closed shell of type e^4. This is true even though (or even if) these electrons are not shared in any real sense by the two CH_2 radicals.

In III (cf. especially Eqs. (6, 7) and reference 6) the types $e\{[y]_B, [y]_A\}$ and $e\{[x]_A, [x]_B\}$ were called $[\pi]_y$ and $[\pi]_x$ because the relation between $[y]_B$ and $[y]_A$, or between $[x]_A$ and $[x]_B$, is rather similar to that between the two orbitals which belong to a representation π of a diatomic molecule. (One should not attach too much significance to this notation, however.) The fact that both the types $\{[x]_A, [x]_B\}$ and $\{[y]_B, [y]_A\}$ belong to the same representation e shows that there must be more or less hybridization between them. Possible consequences of some importance are discussed in section *8e*. Until then they will for simplicity be treated as independent.

In the same way as for any degenerate pair of orbitals, one may replace $[x]_A$, $[x]_B$ of perp. C_2H_4 by any two mutually orthogonal linear combinations, for example by *const.* $\{[x]_A+[x]_B\}$ and *const.* $\{[x]_A-[x]_B\}$. (Similarly with $[y]_B$, $[y]_A$.) This is instructive when used in making comparisons with $[x+x]$ and $[x-x]$ of plane C_2H_4. (One should recall that $[x+x]$ is just an abbreviation for *const.* $\{[x]_A+[x]_B\}$.)

In plane C_2H_4, $[x+x]$ and $[x-x]$ are far apart in energy and are respectively strongly bonding and strongly antibonding, while in perp. C_2H_4 they belong to a single degenerate representation and are therefore rather obviously essentially nonbonding, as are also of course $[y+y]$ and $[y-y]$ of perp. C_2H_4 (cf. also III, beginning of paragraph containing Eq. (6)). Between plane and perp. C_2H_4 lies a continuous set of intermediate cases. Everywhere except for perp. C_2H_4, it is necessary to use orbitals of the types $[x+x]$ and $[x-x]$, which then belong to different nondegenerate representations of the appropriate point group. Just for the angles 90° and 270°, $[x+x]$ and $[x-x]$ become degenerate, and can if desired be replaced by $[x]_A$, $[x]_B$.

It is of interest to note here how the unshared-electron notation $[x][x]$ automatically becomes appropriate as the rotation angle approaches 90°, corresponding to the gradual breaking of the second bond of the double bond by twisting it. [In the case of $[y]^2[y]^2$ (cf. Eqs. (2, 6, 7) of III), the unshared-electron (CH_2-radical) notation is used for all angles, but for a different reason, namely that we have arbitrarily agreed to use it for electrons which are essentially unshared.] For perp. C_2H_4, both $[x]$ and $[y]$ orbitals may be considered as CH_2-radical or as C_2H_4-molecule orbitals with equal appropriateness.

8c. Electronic states, correlations, selection rules for C_2H_4. The problem of determining the possible electron states corresponding to an electron configuration e^2 (in particular, $[\pi]_x^2$ of $\mathbf{V_d}$) has been solved in section 7. The states are $^3A_2$, $^1B_1$, $^1B_2$, $^1A_1$ as stated (except for changed notation) in III (cf. Fig. 1).

Next it may be well to justify the correlations shown in Fig. 1 between the states of plane and perp. C_2H_4. This is readily done by using Table IV to see how the representations of $\mathbf{V_d}$ (perp. C_2H_4) and of $\mathbf{V_h}$ (plane C_2H_4) go over into those of $\mathbf{V}$ (intermediate angles). The results are: A_1 or B_1 of $\mathbf{V_d}$ can go (by way of A_1 of $\mathbf{V}$) into either A_{1g} or A_{1u} of $\mathbf{V_h}$, A_2 or B_2 of $\mathbf{V_d}$ (by way of B_1 of $\mathbf{V}$) into B_{1g} or B_{1u} of $\mathbf{V_h}$, while E of $\mathbf{V_d}$ splits (into B_2+B_3 of $\mathbf{V}$, which go) into B_{2g} or B_{2u} plus B_{3g} or B_{3u} of $\mathbf{V_h}$. Also, singlet→singlet, triplet→triplet. After changing the notation, Fig. 1 of III will be found consistent with these rules. One might, however, raise the question whether an adiabatic correlation scheme is appropriate to the problem considered in III (absorption of ultraviolet light followed by spontaneous relative rotation of the two halves of C_2H_4), since for such rotations the wave function cannot very well be separated into an electronic and a rotational part. Lack of separability might change the restrictions given above, except the singlet→singlet . . . rule, but, as it happens, could not effectively alter the correlations shown in Fig. 1 and so would not lead to any change in the conclusions reached in III.

The fact that a transition $^1A_{1g}\rightarrow{}^1B_{1u}$ of $\mathbf{V_h}$, identified in III (except for change of notation) with the ultraviolet absorption of C_2H_4, is allowed by the selection rules can now be verified by the method of Eq. (7); and it is seen that the electric moment is parallel to the z axis.

8d. Wave functions (ψ) and dissociation of C_2H_4. The various states of C_2H_4 have so far been described in terms of electron configurations, i.e., sets of (atomic and) molecular orbitals, but expressions have not been given for the wave functions of the molecule (cf. sections 2, 2a, 3b). It is of interest to see how approximate ψ's can be constructed (a) with C_2H_4-molecule orbitals for the valence or shared electrons, (b) with CH_2-radical orbitals as one would use atomic orbitals in the method of atomic orbitals. (The reader should refer at this point to the last two paragraphs of section 2.)

Using C_2H_4-molecule orbitals, one has for the normal state of plane C_2H_4

$$\psi=N\begin{vmatrix} [x+x]\alpha(1) & [x+x]\alpha(2) & \cdots(3) & \cdots(4) & \\ [x+x]\beta(1) & \cdots\beta(2) & (3) & (4) & \\ [z+z]\alpha(1) & (2) & (3) & (4) & \text{etc.} \\ [z+z]\beta(1) & (2) & (3) & (4) & \\ & \text{etc.} & & & \end{vmatrix} : {}^1A_{1g} \quad (9)$$

Here "etc." refers to electrons 5 to 16, which are all in CH_2-radical or C-atom closed shells (cf. discussion of H_2O following Eq. (2)). The two excited states $^3B_{1u}$ and $^1B_{1u}$ are given, for $M_S=0$, by

$$\psi=N\begin{vmatrix} [x+x]\alpha(1) & \cdots(2) & \\ [x-x]\beta(1) & \cdots(2) & \text{etc.} \\ \text{etc. as before} & & \end{vmatrix} \pm N\begin{vmatrix} [x+x]\beta(1) & \cdots(2) & \\ [x-x]\alpha(1) & \cdots(2) & \text{etc.} \\ \text{etc. as before} & & \end{vmatrix} : \begin{cases} {}^3B_{1u} \\ {}^1B_{1u} \end{cases} \quad (10)$$

Another excited state is the following:

$$\psi = N\begin{vmatrix} [x-x]\alpha(1) & \cdots(2) \\ [x-x]\beta(1) & \cdots(2) \quad \text{etc.} \\ \text{etc.} & \end{vmatrix} : {}^1A_{1g} \qquad (9')$$

The approximations (9) and (9′), which belong to the same representation, could both be improved by forming linear combinations whereby a little of (9′) is admixed with (9) and *vice versa*. That the expressions given for ψ in (9), (10), and (9′) actually belong to the representations A_{1g} and B_{1u} of $\mathbf{V_h}$ can be verified by testing the behavior of each under the symmetry operations of $\mathbf{V_h}$. The spin character of each can also be verified easily.

Expressions (9), (10), and (9′) apply also to forms of C_2H_4 intermediate between the plane and perp. forms, the states then being ${}^1A_1$ of $\mathbf{V}$ for (9) and (9′), ${}^3B_1$, ${}^1B_1$ for (10). Expression (10) still holds even for perp. C_2H_4, and the two states then prove to have exactly the symmetry properties of ${}^3A_2$ and ${}^1B_2$ of $\mathbf{V_d}$. Although expressions (9) and (9′) both conform to ${}^1A_1$ of $\mathbf{V}$, neither conforms to any representation of $\mathbf{V_d}$. Instead one must form the two linear combinations

$$(\text{Perp. } C_2H_4) \quad \psi = N\begin{vmatrix} [x+x]\alpha(1) & \cdots(2) \\ [x+x]\beta(1) & \cdots(2) \quad \text{etc.} \\ \text{etc.} & \end{vmatrix} \mp N\begin{vmatrix} [x-x]\alpha(1) & \cdots(2) \\ [x-x]\beta(1) & \cdots(2) \quad \text{etc.} \\ \text{etc.} & \end{vmatrix} : \begin{cases} {}^1B_1 \\ {}^1A_1 \end{cases} \qquad (9'')$$

Evidently the two types (9) and (9′) which differ greatly in energy and are slightly admixed for plane C_2H_4 must admix more and more as one goes from plane to perp. C_2H_4, until in the latter they are mixed in equal proportions (Eq. (9″)). This is connected with the fact, noted in an earlier paragraph, that the orbitals $[x+x]$ and $[x-x]$, although differing greatly in energy for plane C_2H_4, become degenerate for perp. C_2H_4, so that (9) and (9′) converge toward the same energy as one approaches perp. C_2H_4. The two energy curves starting approximately from (9) and (9′) of plane C_2H_4 avoid coming together, however, by interacting strongly to give the two states ${}^1A_1$ and ${}^1B_1$ of (9″). These states, together with ${}^3A_2$ and ${}^1B_2$, whose ψ's are given by Eq. (10), are of just the four types which, as we have seen in an earlier paragraph, are expected according to the group theory method when we have e^2 of $\mathbf{V_d}$, the type e here being represented by the two forms $[x+x]$, $[x-x]$.

Even more interesting results for perp. C_2H_4 are obtained by building up the ψ's using the pair of perp. C_2H_4 orbitals $[x]_A$, $[x]_B$ instead of the equivalent forms $[x+x]$, $[x-x]$. In terms of $[x]_A$, $[x]_B$,—which, it may be recalled, can also be regarded equally well as orbitals of CH_2,—the four states of perp. C_2H_4 just discussed appear as

$$\psi = N\begin{vmatrix} [x]_A\alpha(1) & \cdots(2) \\ [x]_B\beta(1) & \cdots(2) \quad \text{etc.} \\ \text{etc.} & \end{vmatrix} \pm N\begin{vmatrix} [x]_A\beta(1) & \cdots(2) \\ [x]_B\alpha(1) & \cdots(2) \quad \text{etc.} \\ \text{etc.} & \end{vmatrix} : \begin{cases} {}^3A_2 \\ {}^1B_1 \end{cases} \qquad (11)$$

$$\psi = N\begin{vmatrix} [x]_A\alpha(1) & \cdots(2) \\ [x]_A\beta(1) & \cdots(2) \quad \text{etc.} \\ \text{etc.} & \end{vmatrix} \mp N\begin{vmatrix} [x]_B\alpha(1) & \cdots(2) \\ [x]_B\beta(1) & \cdots(2) \quad \text{etc.} \\ \text{etc.} & \end{vmatrix} : \begin{cases} {}^1B_2 \\ {}^1A_1 \end{cases} \qquad (12)$$

By multiplying out each of the four cases given by Eqs. (11, 12), one finds that each is identical with one of those obtained by multiplying out the expressions given by Eqs. (9″, 10). (Nothing of interest is lost if all the "etc." parts are dropped before multiplying.)

The forms of Eqs. (11) and (12) show that ${}^3A_2$ and ${}^1B_1$ tend to dissociate so as to leave one $[x]$ electron on each CH_2, but ${}^1B_2$ and ${}^1A_1$ so as to leave both on one CH_2, corresponding to $CH_2^+ + CH_2^-$. Of course the actual adiabatic dissociation processes would be mostly different.

The most probable adiabatic correlations are shown in Fig. 1 of III.

Consideration of Eqs. (9″–12) and of the integrals representing interactions between electrons in $[x+x]$ and $[x-x]$ orbitals indicates that the four states $^3A_2$, $^1B_1$, $^1B_2$, $^1A_1$ of perp. C_2H_4 lie within a moderate energy range. In this connection it should not be forgotten that the two parts of the molecule are always being held together strongly by the $[z+z]^2$ bond. Definite predictions can, however, hardly be made without careful study. The arrangement given in Fig. 1 of III seems plausible.

It will now be instructive to consider the formation of plane and perp. C_2H_4 by the method of atomic orbitals treating each CH_2 like an atom (cf. end of section 2). For simplicity we may disregard the $[z]$ electrons of CH_2, since their behavior is essentially the same as that of the $1s$ hydrogen electrons in the formation of H_2. The latter is to a considerable extent also true of the $[x]$ electrons of CH_2. In fact Eqs. (11) and (12) above, if we regard $[x]_A$ and $[x]_B$ as CH_2 and not as C_2H_4 orbitals, correspond exactly in form to the ψ's for the four states of H_2 ($^1\Sigma_g^+$, $^3\Sigma_u^+$, $^1\Sigma_u^+$, $^1\Sigma_g^+$) derivable from $2H(1s)$ and from $H+H(1s^2)$. Eqs. (11) and (12) really apply not only for perp. C_2H_4 but also for plane and intermediate nuclear configurations, where they are the correct forms for the atomic orbital method. The forms which belong to $^3A_2$, $^1B_2$, $^3B_2$, $^1A_1$ for perp. C_2H_4 in Eqs. (11, 12) belong respectively to $^3B_1$, $^1A_1$, $^1B_1$, $^1A_1$ for intermediate and to $^3B_{1u}$, $^1A_{1g}$, $^1B_{1u}$, $^1A_{1g}$ for plane C_2H_4.

Comparing Eqs. (9, 10, 9′) with (11, 12), for plane C_2H_4, the relations and differences are exactly analogous to those between the methods of molecular and atomic orbitals as applied to the four states of H_2 mentioned above. On multiplying out the various expressions, those for the $^3B_{1u}$ and $^1B_{1u}$ states as given by the two methods are identical, while those for the two $^1A_{1g}$ states differ characteristically, but can be brought into agreement by abandoning pure electron configurations (cf. section 2) and taking suitable admixtures of the two $^1A_{1g}$ forms in each case (cf. II, section 13, after dropping the spins from the present equations). For perp. C_2H_4, the two methods become identical in all respects. All these relations are true, however, only provided we omit all but the $[x]$ electrons of CH_2 from consideration.

8e. Partial persistence of second bond in perp. C_2H_4. In an earlier paragraph it was mentioned that there must be more or less hybridization between the two e orbital types $[\pi]_x=\{[x]_A, [x]_B\}$ and $[\pi]_y=\{[y]_B, [y]_A\}$. Of these two types, it should be noted $[\pi]_y$ is presumably decidedly the lower in energy. The two resulting hybrid e types would be of the forms

$$q=\{a[y]_A+b[x]_B\},\ \{a[y]_B+b[x]_A\},\quad\text{and}$$

$$r=\{a[x]_B-b[y]_A\},\ \{a[x]_A-b[y]_B\}. \tag{13}$$

Complete hybridization would make $a=b$, and would make the two types closely similar to the types π and π^* of O_2 (described as $(\pi+\pi)$ and $(\pi-\pi)$ in III). They would differ in zeroth approximation from π and π^* only because the $[y]$ orbitals contain contributions from hydrogen $1s$ and are C–H bonding (cf. Eq. (8) above).

Actually $a>b$ must hold, but it is reasonable to suppose that $a\gg b$ is not true. Then, as is obvious from its form, the lower-energy type q, which is the more closely related to $[\pi]_y$, has more or less C–C bonding power while type r has more or less C–C anti-bonding power. Since, corresponding to $[\pi]_y{}^4[\pi]_x{}^2$ for the case of no hybridization, one has q^4r^2, the result is that hybridization tends to produce a net C–C bonding effect like the O–O bonding effect of $\pi^4\pi^{*2}$ in O_2. It appears, then, that the double bond in perp. C_2H_4 is intermediate in character between the model given in III and the double bond ($\sigma^2\pi^4\pi^{*2}$) of O_2.

We now see that a 90° rotation of the two parts of a C_2H_4 molecule after all does not entirely destroy the second bond of the double bond, so that the energy difference between the normal state of plane C_2H_4 and the lowest states of perp. C_2H_4 should be less than in Fig. 1 of III. This would make the energy differences between the excited plane states and the perp. states correspondingly greater, and increase the probability of the correctness of the interpretation given in III of certain photochemical experiments.

The change in Fig. 1 just mentioned also brings it into better agreement with calculations from chemical data (cf. III, last sentence before

Reprinted from *J. Chem. Phys.*, **3**, 803–806 (1935)

The Group Relation Between the Mulliken and Slater-Pauling Theories of Valence

12

J. H. VAN VLECK, *Harvard University*

(Received October 7, 1935)

By means of the group theory of characters, it is shown that there is an intimate relation between Mulliken's molecular orbitals and the Slater-Pauling directed wave functions. One can pass from the former to the latter by making a simple transformation from an irreducible to a reducible representation. Consequently the same formal valence rules are usually given by either method, and one can understand generally why wave functions of the central atom which are nonbonding in Mulliken's procedure are likewise never employed in constructing Pauling's "hybridized" linear combinations.

TWO distinct viewpoints have been particularly developed in applying quantum mechanics to problems of valence,[1] namely, the Heitler-London-Pauling-Slater method of the electron pair bond, and the method of molecular orbitals used by Hund, Lennard-Jones, Mulliken, and others. The two procedures represent different approximations to the solution of a complicated secular equation. The method of molecular orbitals permits factorization into one-electron problems, but at the expense of adequate cognizance of the terms due to electron repulsion, which are too fully recognized in the H-L-P-S procedure. A characteristic feature of the latter is the "hybridization," whereby linear combinations of states of different azimuthal quantum number for the central atom are necessary in fields of, for example, tetrahedral symmetry. It was shown by the writer that in the case of carbon compounds the two theories, though superficially different, predicted similar results on geometrical arrangement.[2] It is the purpose of the present paper to show that the equivalence is general in the sense that one formulation will give the same formal stereochemical valence principles as the other. Thus it is futile to discuss whether the Mulliken or Pauling theory will give better working rules in compounds formally amenable to electron pair treatment. Both, for instance, suggest the now classic Pauling square configuration for $Ni(CN)_4^{--}$ in view of the diamagnetism of this ion. (A tetrahedral model would give paramagnetism with either method.[3]) One can only inquire which procedure involves the more reasonable hypotheses. It seems to us that in the case of the transition elements, one must probably decide in favor of the Mulliken formation as a simple qualitative description, though perhaps a poor quantitative approximation. It is difficult to believe, for instance, that the $Fe(CN)_6^{4-}$ radical has the Pauling structure $Fe^{4-}(CN)_6$, since the Fe ion certainly is unwilling to swallow four extra electrons. The conventional ionic model $Fe^{++}(CN^-)_6$, on the other hand, probably goes too far in the other direction. The Mulliken viewpoint has here the advantage of allowing an arbitrary distribution of charge between Fe and $(CN)_6$, depending on how one weights the various atomic orbitals in forming a molecular orbital as a linear combination of them. Only a limited significance should, however, be given to any purported preference between the two methods, as each represents a solution of the secular equation only under certain extreme conditions. The true wave function is in reality a combination of H-LJ-M and H-L-P-S functions, along with many ingredients intermediate between these two extremes, and so either theory is bound to have some semblance of truth. The latter functions, for instance, can be amplified

[1] L. Pauling, J. Am. Chem. Soc. **53**, 1367 (1931); J. C. Slater, Phys. Rev. **38**, 1109 (1931); R. S. Mulliken, Phys. Rev. **40**, 55; **41**, 49, 751; **43**, 279 (1932–3); J. Chem. Phys. **1**, 492 (1933); **3**, 375, 506 (1935). For other references, or more detailed introduction on the methods which we compare, see J. H. Van Vleck and A. Sherman, Rev. Mod. Phys. **7**, 167 (1935). Other workers besides Mulliken have contributed to the molecular orbital procedure, but we sometimes refer to the latter as Mulliken's method, since we are concerned with the application to polyatomic molecules in the light of symmetry groups, an aspect considered primarily by Mulliken.

[2] J. H. Van Vleck, J. Chem. Phys. **1**, 219 (1933).

[3] Cf. Pauling, reference 1, and J. H. Van Vleck and A. Sherman, Rev. Mod. Phys. **7**, 206, 221 (1935).

by taking linear combinations with terms representing different stages of charge transfer until finally just the right polarity is obtained.

Molecular Orbitals

We shall confine our discussion to the case where a central atom attaches n atoms arranged in some symmetrical fashion (tetrahedron, square, etc.) characteristic of a crystallographic point group.[4] Let $\psi(\Gamma)$ be a wave function of the central atom which has the proper symmetry, i.e., whose transformation scheme under the covering operations of the group is that characteristic of some irreducible representation Γ. Let ψ_i be a wave function of attached atom i. We shall assume that only one orbital state need be considered for each attached atom, and that this state is either an s state or else is symmetric (as in a $2p\sigma$ bond) about the line joining the attached to the central atom. The method of molecular orbitals in its simplest form[5] seeks to construct solutions of the form

$$\psi=\psi(\Gamma)+\Sigma_i a_i\psi_i. \tag{1}$$

The coefficients a_i must be so chosen that $\Sigma_i a_i\psi_i$ transforms in the fashion appropriate to the irreducible representation Γ. Now the important point is that bases for only certain irreducible representations can be constructed out of linear combinations of the ψ_i. To determine which, one ascertains the group characters associated with the transformation scheme, usually reducible, of the original attached wave functions ψ_i before linear combinations are taken. This step is easy, as the character χ_D for a covering operation D is simply equal to q, where q is the number of atoms left invariant by D. This result is true inasmuch as D leaves q of the atoms alone, and completely rearranges the others, so that the diagonal sum involved in the character will contain unity q times, and will have zeros for the other entries. The scheme for evaluating the characters is reminiscent of that in the group theory of molecular vibrations employed by Wigner and by E. B. Wilson, Jr.,[6] but is simpler since we are not interested in displacements of atoms from equilibrium, and in consequence all nonvanishing entries are unity rather than some root of unity. After the characters have been found, the determination of the constituent irreducible representations proceeds in the usual way by means of the theorem that the unresolved characters must equal the sum of the primitive characters contained therein.

As an illustration, we may consider a complex containing six atoms octahedrally arranged, i.e., located at the centers of the six cube faces. Then the symmetry group is the cubic one O_h, and the characters associated with the arrangement of attached atoms are

$$\begin{array}{lcccccccccc} & E & C_2 & C_4 & C_2' & C_3 & I & IC_2 & IC_4 & IC_2' & IC_3 \\ \chi= & 6 & 2 & 2 & 0 & 0 & 0 & 4 & 0 & 2 & 0. \end{array} \tag{2}$$

Here $\chi(C_4)$, for instance, means the character for the covering operation consisting of rotation about one of the fourfold or principal cubic axes (normals to cube faces) by $2\pi/4$. Any rotation about such an axis leaves two atoms invariant, and hence $\chi(C_2)=\chi(C_4)=2$. On the other hand, $\chi(C_2')=\chi(C_3)=0$ since no atoms are left invariant under rotations about the twofold or secondary cubic axes (surface diagonals) or about the threefold axes (body diagonals). Inversion in the center of symmetry is denoted by I. By using tables of characters for the group O_h, one finds that the irreducible representations contained in the character scheme (2) are, in Mulliken's notation,[4]

$$A_{1g},\ E_g,\ T_{1u}. \tag{3}$$

The irreducible representations corresponding to various kinds of central orbitals are shown below:

$$\begin{array}{llllll} \text{orbit} & s & p & d\gamma & d\epsilon & \overbrace{\qquad\qquad}^{f} \\ \text{rep.} & A_{1g} & T_{1u} & E_g & T_{2u} & A_{2u},\ T_{1u},\ T_{2u}. \end{array} \tag{4}$$

The notation for the various kinds of d wave functions is that of Bethe,[7] *viz.*,

[4] See R. S. Mulliken, Phys. Rev. **43**, 279 (1933) or references 6 and 7 if further background is desired on the aspects of group theory and crystallographic symmetry which we use.

[5] Called by Mulliken the LCAO ("linear combination of atomic orbitals") form. For a critique of this type of approximation see R. S. Mulliken, J. Chem. Phys. **3**, 375 (1935). It appears to have been first suggested by Lennard-Jones.

[6] E. Wigner, Gött. Nachr., p. 133 (1930); E. B. Wilson, Jr., J. Chem. Phys. **2**, 432 (1934); Phys. Rev. **45**, 706 (1934).

[7] H. Bethe, Ann. d. Physik **3**, 165 (1929).

TABLE I.

Symmetry	Central orbitals s	p	d	f	Attached orbitals No.	Representations
Tetrahedral (T_d)	A_1	T_2	$E(d\gamma)$, $T_2(d\epsilon)$	A_1, T_1, T_2	4	A_1, T_2
Trigonal (D_{3h})	A_1'	A_2'', E'	A_1', E', E''	$A_1', A_2', A_2'', E', E''$	3	A_1', E'
					6	A_1', A_2'', E', E''
Tetragonal (D_{4h})	A_{1g}	A_{2u}, E_u	$A_{1g}, B_{1g}, B_{2g}, E_g$	$A_{2u}, B_{1u}, B_{2u}, 2E_u$	4	A_{1g}, B_{1g}, E_u
					8	$A_{1g}, A_{2u}, B_{1g}, B_{2u}, E_u, E_g$

$$\psi(d\gamma_1)=(1/12)^{\frac{1}{2}}f(r)(3z^2-r^2),\quad \psi(d\gamma_2)=\tfrac{1}{2}f(r)(x^2-y^2), \tag{5}$$

$$\psi(d\epsilon_1)=f(r)xy,\quad \psi(d\epsilon_2)=f(r)xz,\quad \psi(d\epsilon_3)=f(r)yz. \tag{6}$$

By comparison of (3) and (4) we see that is it impossible for $d\epsilon$ orbitals to form any partnerships with attached orbitals, in agreement with Mulliken's conclusion[8] that $d\gamma$ rather than $d\epsilon$ is particularly adapted to forming octahedral bonds. Mulliken's arguments were mainly of a qualitative nature. The preceding considerations enable us to formulate the situation more succinctly, as they show that the $d\epsilon$ orbitals are entirely nonbonding.

In Table I we give the irreducible representations, in Mulliken's notation, contained in the central and attached orbitals for compounds of other types of symmetry. When 3 or 4 atoms are trigonally or tetragonally attached, we have supposed that the plane of these atoms is a plane of symmetry, as in $(NO_3)^-$ or $Ni(CN)_4^{--}$. When there is no such symmetry plane, as in NH_3, the distinctions between u and g, or between primes and double primes, are to be abolished,[9] and the symmetries degenerate to C_{3v}, C_{4v} instead of D_{3h}, D_{4h}. When 6 atoms are attached in the scheme D_{3h}, or 8 in D_{4h}, they are arranged respectively at the corners of a trigonal and a square prism.

The results given in Table I are obtained by the same method as in the octahedral example. The explicit forms of the linear combinations of the attached orbitals which transform irreducibly, or in other words the values of the coefficients a_i in (1) have been tabulated by Van Vleck and Sherman[10] in many instances, and so need not be repeated here. The a_i for the octahedral case are also given in Eqs. (2)–(7) of the following paper.

Note particularly that in the tetrahedral complexes, the $d\epsilon$ orbitals of the central atom are bonding, as there are attached orbitals of similar group properties with which they can combine, while the $d\gamma$ orbitals are nonbonding. The reverse was true of octahedral compounds—a result at first a little surprising in view of the isomorphism of the groups T_d and O_h. This reversal was also deduced by Mulliken[8] from the geometrical study of the way the central wave functions "overlap."

It will be observed that if eight atoms are attached, their full bonding power is not utilized unless one includes f wave functions for the central atom, since the representation B_{2u} is not included in s, p, or d. Now f wave functions usually have too high energy to be normally available, or else are so sequestered in the interior of the atom as to be of no value for bonding because of small overlapping. Even if eight atoms are attached at the corners of a cube, central f wave functions must be included in order to realize all possible bonding partnerships, for results always true of tetragonal symmetry surely apply to cubic symmetry, which is a special case of the latter. On the other hand, no f functions are needed for six atoms attached either octahedrally or at the corners of a trigonal prism. We thus have an indication of why it is that coordination numbers of six are common in nature, while those of eight are rare.

METHOD OF DIRECTED ELECTRON PAIRS

We now turn to the method of Pauling and Slater. Here the procedure is to use hybridized

[8] R. S. Mulliken, Phys. Rev. **40**, 55 (1932).

[9] In adapting Table I to the case C_{3v}, the following irregularity, however, is to be noted: one must replace A_1', A_1'', A_2', A_2'', respectively, by A_1, A_2, A_2, A_1 rather than by A_1, A_1, A_2, A_2 as one would guess.

[10] J. H. Van Vleck and A. Sherman, Rev. Mod. Phys. **7**, 219 (1935).

central orbitals, i.e., linear combinations of orbitals of different azimuthal quantum number, in such a way that the resulting central wave function projects out especially in some one direction in space, and so is adapted to form an electron pair with one particular attached atom. Thus for tetrahedral compounds Pauling and Slater[1] use wave functions which are linear combinations of s and p, or alternatively as Pauling[1] shows, of s and $d\epsilon$ wave functions. For octahedral compounds, Pauling finds that $sp^3d\gamma^2$ combinations are appropriate, and $sp^2d\gamma$ for tetragonal. Hultgren[11] proves that eight atoms cannot be attached (at least symmetrically) by means of unidirectional electron pair bonds formed from s, p, and d wave functions. Incidentally, the present paper shows that sp^3d^3f functions are needed to hold eight atoms.[12] It will be noted that the wave functions involved in the Pauling unidirectional linear combinations are precisely those which are bonding in the method of molecular orbitals. For example, Pauling, like Mulliken, makes no use of $d\gamma$ orbitals for tetrahedral compounds, or of $d\epsilon$ for octahedral. Such coincidences have hitherto appeared something of a mystery, but as immediate explanation, as follows, is furnished by group theory.

In the Pauling-Slater theory, one desires the central wave functions to possess unilateral directional properties so as to be correlated with one particular attached atom. Hence the P-S central functions must have the same transformation properties as do those ψ_i of the attached atoms before linear combinations of the latter are taken. Thus the problem of finding the linear combinations of the central orbitals which exhibit the proper directional properties is simply the reverse of finding the proper linear combinations of the attached orbitals in the Mulliken procedure. The difference is only that in the P-S theory, the linear combinations are in the central rather than attached portion, and their construction corresponds to transformation from an irreducible representation to a reducible one, of structure similar to that belonging to the original ψ_i's, rather than to the inverse transformation.[13] Clearly, the same irreducible representations are needed in the construction of a given reducible representation as those contained in the resolution of the latter into its irreducible parts. Pauling has obviously shown considerable ingenuity in constructing his wave functions without using the status in terms of group theory.

Generality of the Results

The argument underlying Table I, etc., ostensibly assumed that the molecular orbital be expressible as a linear combination of atomic orbitals, but is readily seen to be still applicable provided only that the charge cloud of any attached orbital be symmetric about the line joining the given attached atom to the central one. Hence the atomic orbitals can be of what James calls the flexible type, i.e., contain parameters which can be varied in the Ritz method, and which allow for the fact that chemical combination distorts the atomic orbitals from what they would be in the free condition. This admission of flexibility is fortunate, for it is well known that it is a bad quantitative approximation[5] to express a molecular orbital as a linear combination of undistorted atomic orbitals. The Ritz variational problem is, of course, to be of the 1 rather than n electron type, so that the generality in our analysis by means of molecular orbitals is roughly comparable with that in the Hartree method.

One thing which the preceding analysis does not do is to tell us what is the best arrangement of atoms in case the symmetry group does not uniquely determine this arrangement. For instance, by examining the overlapping of wave functions, Hultgren[11] finds that when six atoms are attached at the corners of a trigonal prism, the binding is firmest if the sides of the prism are square. This fact cannot, however, be inferred from our group theory considerations, as there are no additional elements of symmetry when the sides are square rather than rectangular.

[11] R. Hultgren, Phys. Rev. **40**, 891 (1932).

[12] Similar conclusions on the type of bonds necessary to attach eight atoms have also been obtained in unpublished work of R. S. Mulliken.

[13] This transformation has been explicitly given by the writer in the case of methane (J. Chem. Phys. **1**, 177 (1933)), but he did not discuss its group-theoretical significance.

The Synthesis—Ligand Field Theory

IV

Editor's Comments on Papers 13, 14, and 15

13 Van Vleck: *Valence Strength and the Magnetism of Complex Salts*

14 Kimball: *Directed Valence*

15 Moffitt and Ballhausen: *Quantum Theory*

Four papers have been regarded as critical to the ultimate synthesis of crystal field theory with the valence bond and molecular orbital theories. The first two of these are the contributions of John H. Van Vleck. He, more than any other individual of the period, recognized the common features of all three theories, understood the particular significance of each of the theories, compared them (Paper 12 in Part III), and used them to help understand magnetism (Paper 13).

In 1935 Van Vleck displayed the common features of the three theories used at that time to describe electronic structure in molecules. He pointed out the weaknesses in each and their individual strengths. A particularly significant result, presented in Paper 13, is the conclusion that octahedral and tetrahedral complexes of nickel(II) should display a total spin of $S = 1$ independent of the theory used to predict the result.

George E. Kimball, another physicist, put together the orbital hybridization theory and symmetry in a way very appropriate to chemical thought. He published his *Directed Valence* (Paper 14) just as World War II was developing in intensity in Europe. After the war this article had a significant impact on the chemical world because it synthesized, simply and elegantly, the very popular valence bond theory with the concepts of symmetry. Few chemists recognized that most of the theory described in Kimball's paper could be found in the earlier work of Van Vleck.

Kimball's 1940 paper became popular reading for chemists in the 1950s, especially as the power of group theory as applied to quantum mechanical problems began to be understood. His table (XXIV) of bond arrangements has found wide employment by chemists attempting to understand the structure and bonding of molecules of widely varying symmetries. The publication of the important book *Quantum Chemistry* by Eyring, Walter, and Kimball also helped to popularize this article.

The final paper in the set on synthesis is the very important review article by W. Moffitt and C. J. Ballhausen (Paper 15). It is written by chemists and appeared at approximately the same time that many of the first important papers applying ligand field theory to chemical problems were being written. Five years later most of the material in the Moffitt–Ballhausen paper had become included in important texts in physical–inorganic chemistry.

A review article such as Paper 15 often paves the way for full acceptance of a particular scientific point of view. Moffitt, a young thermodynamics expert and squash enthusiast who tragically died on the courts at Harvard only a few months after publication of this review, teamed with Ballhausen to produce this highly significant review with its emphasis on the crystal field theory, a theory generally neglected by chemists until that time. Here, for the first time, significant distinctions were given the terms "weak field," "strong field," and "intermediate field," which are of such substantial importance to the understanding of most coordination compounds. The beginnings of crystal field spectroscopy are also reviewed and placed in context with the results of magnetic studies. By comparing valence bond, molecular orbital, and crystal field theories, the authors suggested that certain specific modifications should be made in crystal field theory, such as: "when e_g orbitals are regarded as occupied in excited states, these should be viewed as antibonding counterparts of bonding orbitals already filled by electrons originally on the ligands." Some double-bonding with t_{2g} orbitals is possible. Charge-transfer bands in the spectra of transition-metal complexes fall outside the realm of crystal field theory. Ligand field theory had come of age!

Reprinted from *J. Chem. Phys.*, **3**, 807–813 (1935)

Valence Strength and the Magnetism of Complex Salts

13

J. H. VAN VLECK, *Harvard University*

(Received October 7, 1935)

Certain complex salts, notably ferro- and ferricyanides, have susceptibilities much lower than those predicted by the Bose-Stoner "spin only" formula. The first interpretation was that given by Pauling on the basis of (I) directed wave functions. In the present paper it is shown that alternative explanations are possible with (II) the crystalline potential model of Schlapp and Penney, or with (III) Mulliken's method of molecular orbitals. In any of the theories, the interatomic forces, if sufficiently large, will disrupt the Russell-Saunders coupling, and make the deepest state have a smaller spin, and hence smaller susceptibility, than that given by the Hund rule. This situation is not to be confused with that in normal paramagnetic salts, such as sulphates or fluorides, where only the spin-orbit coupling is destroyed. The similarity of the predictions with all three theories is comforting, since any one method in valence usually involves rather questionable approximations. Because of this similarity, a preference between the theories cannot be established merely from ability to interpret the anomalously low magnetism of the cyanides. Covalent bonds, as in cyanides, seem to be more effective in suppressing magnetism than are ionic ones, as in fluorides, but so far the evidence to this effect is empirical rather than theoretical.

IN most salts of the iron group, the susceptibility has approximately the value

$$\chi = 4N\beta^2 S(S+1)/3kT, \qquad (1)$$

where β is the Bohr magneton $he/4\pi mc$, and where the spin S has the value given by the Hund rule that the ground state is that of maximum multiplicity compatible with the Pauli principle. Formula (1) was suggested by Bose and Stoner. The theoretical explanation has been known for some time,[1] and is that the crystalline field is so strong as to destroy the coupling of the spin angular momentum S and orbital angular momentum L to a quantized resultant J. It can be shown that with such an "electric Paschen-Back effect," the orbital magnetism is largely quenched, and the susceptibility has nearly the value (1).

There are, however, certain compounds of elements of the iron group, notably iron cyanides and the various cobaltammines, for which the susceptibility has a value very much lower than that given by (1). In fact, these compounds are diamagnetic if they involve a complex ion containing an even number of electrons (e.g., $Fe(CN)_6^{4-}$), or have a susceptibility of an order of magnitude corresponding to one free spin, i.e., to $S=\frac{1}{2}$ in (1), if this number is odd (e.g., $Fe(CN)_6^{3-}$). The first specific model accounting for this behavior was given by Pauling,[2] but the explanation may be couched in more general language as follows: the interatomic forces are so very large as to destroy not merely the spin-orbit but also the Russell-Saunders coupling, and make the deepest state that of lowest possible spin rather than of maximum spin as given by the Hund rule[3] Large spin would be an advantage as far as a free atom is concerned, but the point which we wish to make is that the interatomic energy is decreased by lowering the total spin regardless of whether one makes the calculation in any of three ways:

(I) the directed electron pair bond, sometimes called the Heitler-London-Pauling-Slater approximation,

(II) the method of crystalline fields, particularly adapted to ideally ionic compounds, and

(III) Mulliken's method of molecular orbitals.[4]

There is thus a formal similarity between the results of various approaches, in many respects paralleling that on the subject of directional valence stressed in the preceding paper. The cyanide anomalies have been previously interpreted only on the basis of I, although Mulliken[4] did state that magnetic behavior might often be equally intelligible with the mechanism III.

The particular examples which we shall consider are compounds in which the central or

[1] J. H. Van Vleck, *The Theory of Electric and Magnetic Susceptibilities*, Chap. XI.

[2] L. Pauling, J. Am. Chem. Soc. **53**, 1367 (1931); L. Pauling and M. L. Huggins, Zeits. f. Krist. **87**, 205 (1934).

[3] For this rule see, for instance, Pauling and Goudsmit, *The Structure of Line Spectra*, p. 165.

[4] R. S. Mulliken, Phys. Rev. **40**, 55 (1932).

paramagnetic atom has a coordination number of six resulting from six neighboring atoms or ions octahedrally arranged. The clusters $Fe(CN)_6^{4-}$, $Fe(CN)_6^{3-}$, FeF_6^{3-}, in $K_4Fe(CN)_6$, $K_3Fe(CN)_6$, $(NH_4)_3FeF_6$, respectively, are structures of this character. It is not necessary that the crystal as a whole have cubic symmetry. For instance, $K_3Fe(CN)_6$ is monoclinic. From these examples we trust it will be sufficiently clear how the similarity between the various methods is demonstrated in other cases.

The deepest state of the Fe^{+++} ion is a ^6S term, and indeed the observed susceptibility[5] of $(NH_4)_3FeF_6$ approximately equals that given by (1) with $S=5/2$. The susceptibility[6] of $K_3Fe(CN)_6$, however, is much nearer the value given by (1) with $S=\frac{1}{2}$. The interatomic forces should thus be relatively more important in the cyanides than in the fluorides, and it is, in fact, known that the former are firmer compounds than the latter.

I. The Method of Directed Electron Pair Bonds

The magnetic behavior of $Fe(CN)_6^{3-}$ is explained by Pauling on the basis of the structure $Fe^{3-}(CN)_6$. This model need not be literally true, but is taken to be sufficiently typical, i.e., a sufficiently common phase in the resonance through various stages of polarity, to serve for purposes of discussion. According to Pauling,[2] the Fe^{3-} ion attaches its six atoms of coordination by means of six $sp^3d\gamma^2$ electron pair bonds. These six bonds consume the $4s$, $4p$, and $d\gamma$ states,[7] and house six electrons of Fe^{3-}. (Only one electron can be assigned to a state of the central atom which is used for bonding purposes; for if a state houses two electrons, their spins compensate each other, and they are not free to form electron pair bonds with outside atoms.) The Fe^{3-} ion has 11 electrons in all to house (apart from completed inner shells). The five electrons not absorbed by the electron pair bonds can only be accommodated in the three $3d\epsilon$ states. Thus two of these states will be filled twice. The electrons which are thus "doubled up" must have their spins antiparallel because of the Pauli principle, and hence make no contribution to the susceptibility. The fifth electron, however, has a free spin, so that the susceptibility should be given approximately by (1) with $S=\frac{1}{2}$ provided one can overlook orbital contributions to the susceptibility. Actually they are not negligible, as Howard shows in the next paper. In $Fe(CN)_6^{4-}$, there is an additional electron to house, so that the fifth electron loses its private state, and there can only be diamagnetism, in agreement with experiment.

II. The Method of the Crystalline Potential

The simplest model for explaining the varying magnetic properties is that used with success by Penney and Schlapp[8] primarily on hydrated sulphates of the rare earth and iron groups. The basic assumption is that the effect of neighboring atoms can be represented by means of a static potential. One may distinguish between three cases depending on the strength of the crystalline field: (a) the field is so weak that the inner quantum number J has a meaning (b) it is so large as to prevent J, but not L or S from being a good quantum number, and (c) it is still more powerful, and able to destroy Russell-Saunders coupling, i.e., the compounding of the individual l's to a resultant L, so that L loses its validity as a quantum number. Cases (a) and (b) are those studied by Penney and Schlapp, (a) being characteristic of the rare earths, and (b) of the hydrated sulphates of the iron group. We shall now show that (c) furnishes an explanation of the behavior of the cyanides and other salts having abnormally low susceptibilities.

In the compounds in which we are at present interested, the crystalline field is of dominantly cubic symmetry. In a field of this symmetry type, it can be shown by means of group theory,[9] or otherwise, that a d electron has the Stark pattern shown in Fig. 1. The separation of the

[5] E. Cotton-Feytis, Ann. Chim. **4**, 9 (1925).
[6] Cf. for instance, L. C. Jackson, Proc. Roy. Soc. **A140**, 695 (1933).
[7] For explanation of the notation $d\gamma$, $d\epsilon$ see Eqs. (5)–(6) of the preceding paper in this issue.
[8] W. G. Penney and R. Schlapp, Phys. Rev. **41**, 194 (1932); R. Schlapp and W. G. Penney, ibid. **42**, 666 (1932); O. Jordahl, ibid. **45**, 87 (1934); R. Janes, ibid. **48**, 78 (1935).
[9] See, for instance, H. Bethe, Ann. d. Physik **3**, 143 (1929).

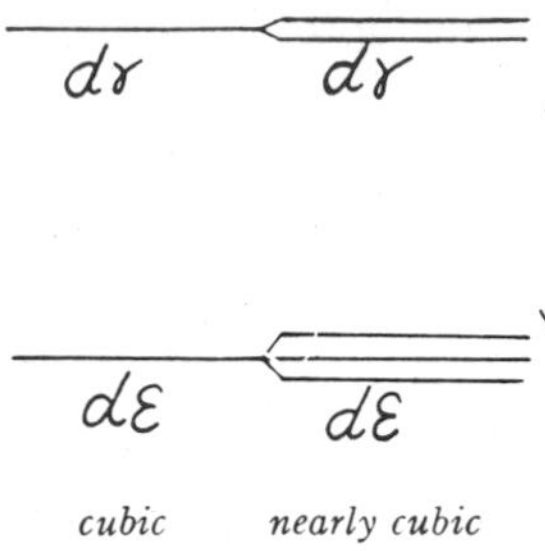

Splitting in method of crystalline potentials

FIG. 1.

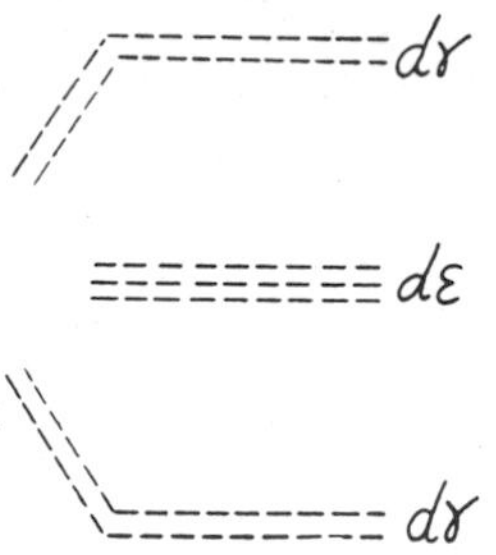

Splitting in method of molecular orbitals

FIG. 2.

levels which branch out from a common origin on the left is, by hypothesis, comparatively small, and owes its existence to the deviations of the field from perfect cubic symmetry. Do not confuse Fig. 1, which is for one electron, and the closely related crystalline Stark diagrams given by Penney, Schlapp, and Van Vleck for whole atoms in which the electrons are space quantized collectively rather than individually. Fig. 1 is right side up, as Gorter[10] shows, so long as the negative ions surrounding the paramagnetic cation are octahedrally arranged. Fig. 1 should be inverted in case the coordination number is four rather than six, i.e., the symmetry tetrahedral rather than octahedral. Now if we assume the conventional polar structure $Fe^{+++}(CN^-)_6$, the Fe^{+++} ion has five electrons outside closed shells, and there are five electrons to be housed in the various states of Fig. 1. Clearly the deepest energy is obtained by assigning the five electrons to the three components of $d\epsilon$. Then all the electrons but one double up, and there is only one free spin, while in $Fe^{++}(CN^-)_6$ there would be no free spin. Thus with a sufficiently large field to make individual space quantization a good approximation, the deepest state is quite different from that ^6S given by the Hund rule.

It is interesting to note the significance of the state $d^5\,{}^6S$ in terms of Fig. 1. Usually Russell-Saunders coupling involves collective quantization in such a way as not to permit any simple interpretation in terms of a one-electron diagram such as Fig. 1. However, when the spin quantum number equals half the number of electrons, as in $d^5\,{}^6S$, the spins are all mutually parallel. Such an alignment is possible only if each electron has its own private orbital. Hence a state of the configuration d^5 with $S=5/2$ involves one electron in each of the five levels of Fig. 1. This distribution has an invariant significance (one to a state) regardless of how the axis of space quantization is chosen. Hence this state must be an S state, as is also proved by other means and as is reflected in the high magnetic isotropy of manganous and ferric salts of the sulphate variety. When, however, one has the cyanide case, there are three different ways in which the five electrons may be distributed among $d\epsilon$, since the private orbital may be the lower, middle, or upper component of $d\epsilon$. True cubic symmetry is achieved only if the three components are populated equally. Hence $K_3Fe(CN)_6$ can exhibit a high degree of magnetic anisotropy. This point is studied quantitatively by J. Howard in the next paper.

As we have seen, the energy due to the crystalline field is lowest when the Hund rule is broken down. However, the internal energy of the Fe^{+++} ion is lower with $S=5/2$ than with $S=\frac{1}{2}$. (This is the meaning of the Hund rule.) Thus whether the total internal plus external energy is lower or higher for $S=5/2$ than for $S=\frac{1}{2}$ is a question purely of the size of the crystalline potential relative to the strength of the Russell-Saunders coupling. The fact that both cases are actually realized suggests that the two energies are of the same order of magnitude, and such indeed appears to be the case. The crystalline fields deduced by Schlapp, Penney, Jordahl, and Janes[8] for the hydrated sulphates usually amount to about 3 volts.[11]

[10] C. J. Gorter, Phys. Rev. **42**, 437 (1932).

[11] In Cr^{+++} an over-all splitting of about 8 volts is obtained by Schlapp and Penney, a value which seems unduly high and out of line with their other results. As they intimate, the explanation of this discrepancy is probably that in Cr^{+++}, the computation of the splitting is unusually sensitive to small experimental errors in the determination of the absolute value of the susceptibility because the latter has very nearly the "spin-only" value in this particular ion. Schlapp and Penney employed the Leiden data on chrome alum. It is interesting to note that a splitting (3 volts) of about the usual size is yielded by Janes's[8] recent measurements on $K_3Cr(SCN)_64H_2O$. Incidentally, the separation 17,200 cm^{-1} which Janes obtains in cupric salts relates to the over-all splitting, rather to the constant D as stated in his article, and so is not unreasonable.

The strength of the Russell-Saunders coupling is measured by the separation of the various levels arising from the configuration d^5 in the free ion. Spectroscopic data indicate that this separation amounts to only about 4 or 5 volts.[12] Hence the crystalline field would not have to be much larger than in the examples studied by Schlapp and Penney in order to overpower the Russell-Saunders coupling,[13] and so it is plausible that this situation is realized in the cyanides.

Even in the case studied by Schlapp and Penney, labeled (b) above, the separation between the various states of the configuration $3d^x$ is not larger than the crystalline potential. Consequently one naturally wonders whether they were justified in neglecting matrix elements between states of different L but similar S belonging to the same configuration. Neglect of matrix elements between states differing both in L and S is legitimate if the intervals between such states are large compared with the spin-orbit interaction, i.e., with multiplet widths. This condition is practically always fulfilled. In Fe^{+++} or Mn^{++} there can be no question of rigor, since the configuration d^5 has only one sextet state. Also in Cr^{++} or Fe^{++} (d^4 or d^6) there is only one quintet, while in Cu^{++} (d^9) there is only one state of any character, a doublet. In Cr^{+++} or Co^{++}, however, the states of highest multiplicity are ^4F and ^4P, while in Ni^{++}, they are ^3F and ^3P. The separation between these two states is only about 2 volts[14] in either Co^{++} or Ni^{++}. Spedding[15] criticizes the work of Schlapp and Penney on the ground that they do not consider the perturbing effect of ^3P in their calculations of the ^3F state of Ni^{++}, or of ^4P on the ^4F of Cr^{+++}, Co^{++}. However, Spedding has overlooked the fact that in Cr^{+++} or Ni^{++} the Stark component of the F level which is important for the magnetic calculations is that which belongs to the representation A_2[16] of the cubic group. On the other hand, ^3P or ^4P belongs to the representation T_1, and so is incapable of perturbing A_2, if the field is really cubic. This statement is not quite true when one allows for the fact that the spin-orbit distortion makes the symmetry different from that characteristic of orbit alone. In other words, one ought rigorously to consider the irreducible transformation properties of the spin-orbit rather than just orbital wave functions. However, the spin-orbit distortion is subordinate, and so the perturbation due to this cause is negligible. Furthermore the Stark component A_2 of ^3F or ^4F involves different rhombic representations than do ^3P or ^4P, and so cannot be perturbed by the latter even when one considers the deviations from cubic symmetry, as long as rhombic symmetry is preserved. Hence Schlapp and Penney's conclusions on the nearly perfect magnetic isotropy of Ni^{++}, Cr^{++} and their close conformity to (1), are unaffected. In the case of Co^{++}, the ground level is the Stark component T_1 of $d^7\ ^4F$, and can be perturbed by $d^7\ ^4P$. We hope to study this effect later more fully. In Co^{++}, anyway, the agreement which Schlapp and Penney obtained with experiment was qualitative rather than quantitative. Rough preliminary examination suggests that it will probably be improved by considering the perturbation by ^4P. Hence, we conclude that all the vital features of Schlapp and Penney's calculations are unaffected by considering the incipient breakdown of Russell-Saunders coupling, and that in Co^{++} the quantitative agreement may actually be bettered.

III. The Method of Molecular Orbitals

This method differs from the crystalline field procedure II in that the structural unit for the wave function is the whole complex ion (e.g., $Fe(CN)_6^{3-}$) rather than the single central atom. Both II and III utilize one-electron wave functions, and so group theory based on symmetry properties can still be used to obtain information about the character of the levels. However, there is the difference that if a central orbital is bonding, a given representation appears more often than in method II because of the fact that, in the language of the preceding paper, there are both central and attached orbitals belonging to the same irreducible representation. The proper

[12] Spectroscopic data are not available on the Fe^{+++} ion. However, the separation of the various terms belonging to the configuration d^5 in Fe^{+++} should be somewhat greater than (probably a little less than double) the corresponding separations in the homologous ion Cr^+, and the interval $d^5\ ^6S-d^5\ ^4G$, for instance, is known to be 2.5 volts in Cr^+.

[13] Unfortunately it does not appear possible to estimate directly from theoretical considerations the crystalline splitting to be expected even with an ideal ionic structure and assumed interatomic separations obtained from Pauling's atomic radii. So one can only deduce the splittings empirically from the magnetic data. The difficulty is that one does not know well enough the effective charge Z to be used in computing the d wave functions. Clearly Z should be somewhat greater than 4, the value corresponding to perfect screening, and somewhat less than the effective charge Z_{ip} deduced from ionization potentials, as the value of Z to be used in computing $\overline{r^4}$ etc. is less than that involved in $\overline{1/r}$. Extrapolated spectroscopic data indicate that Z_{ip} is about 6. In unpublished calculations, Howard finds that the interval $d\gamma-d\epsilon$ amounts to 4 volts if $Z=4$ and to 1 volt if $Z=6$, provided the further assumptions are made that the distance Fe–F is 1.91A, and that the F^- ions act like point charges.

[14] The intervals $d^7\ ^4F-d^7\ ^4P$ of Co^{++} and $d^8\ ^3F-d^8\ ^3P$ of Ni^{++} have not been observed directly, but probably do not differ greatly from the intervals $d^7(^4F)4s^2-d^7(^4P)4s^2$ of Co I, and $d^8(^3F)4s^2-d^8(^3P)4s^2$ of Ni I, respectively. These latter intervals are known and are both 1.9 volts. That the addition of the $4s$ electrons does not change too materially the separations of the core states is indicated, for instance, by the fact that the frequency difference $d^4(^3F)4s\ ^4F-d^4(^3G)4s\ ^4G$ in Cr II deviates only 4 percent from the difference $d^4\ ^3F-d^4\ ^3G$ in Cr III (see Bacher and Goudsmit's tables).

[15] F. H. Spedding and G. C. Nutting, J. Chem. Phys. **2**, 421 (1935).

[16] For explanation of the notation for the group representations see R. S. Mulliken, Phys. Rev. **43**, 279 (1933).

linear combination of such central and attached orbitals is found by solving a secular equation. In the case of the octahedral group, the wave functions belonging to the irreducible representation E_g are

$$\psi(E_g)=\alpha\psi(3d\gamma_1)+(1-\alpha^2)^{\frac{1}{2}}(1/12)^{\frac{1}{2}}(2\psi_3+2\psi_6-\psi_1-\psi_4-\psi_2-\psi_5), \quad (2a)$$

$$\psi''(E_g)=(1-\alpha^2)^{\frac{1}{2}}\psi(3d\gamma_1)-\alpha(1/12)^{\frac{1}{2}}(2\psi_3+2\psi_6-\psi_1-\psi_4-\psi_2-\psi_5), \quad (2b)$$

$$\psi'(E_g)=\alpha\psi(3d\gamma_2)+(1-\alpha^2)^{\frac{1}{2}}\tfrac{1}{2}(\psi_1+\psi_4-\psi_2-\psi_5), \quad (3a)$$

$$\psi'''(E_g)=(1-\alpha^2)^{\frac{1}{2}}\psi(3d\gamma_2)-\alpha\tfrac{1}{2}(\psi_1+\psi_4-\psi_2-\psi_5), \quad (3b)$$

where the wave function of attached atom i is denoted by ψ_i, and the explicit forms of the central wave functions $\psi(d\gamma)$ are as given in Eqs. (5)–(6) of the preceding paper. Attached atoms 1, 4 are supposed located on the x axis; 2, 5 on the y; and 3, 6 on the z. The value of α is determined, at least in principle by solving a secular equation. Similarly, the wave functions corresponding to A_{1g} and T_{1u} are

$$\psi(A_{1g})=\beta\psi(4s)+(1-\beta^2)^{\frac{1}{2}}(\tfrac{1}{6})^{\frac{1}{2}}(\psi_1+\psi_2+\psi_3+\psi_4+\psi_5+\psi_6), \quad (4a)$$

$$\psi(T_{1u})=\gamma\psi(4p\sigma_x)+(1-\gamma^2)^{\frac{1}{2}}(\tfrac{1}{2})^{\frac{1}{2}}(\psi_1-\psi_4), \quad (5a)$$

$$\psi'(T_{1u})=\gamma\psi(4p\sigma_y)+(1-\gamma^2)^{\frac{1}{2}}(\tfrac{1}{2})^{\frac{1}{2}}(\psi_2-\psi_5), \quad (6a)$$

$$\psi''(T_{1u})=\gamma\psi(4p\sigma_z)+(1-\gamma^2)^{\frac{1}{2}}(\tfrac{1}{2})^{\frac{1}{2}}(\psi_3-\psi_6), \quad (7a)$$

together with four other wave functions (4b)–(7b) which are similar except that β, $(1-\beta^2)^{\frac{1}{2}}$ are replaced, respectively, by $(1-\beta^2)^{\frac{1}{2}}$, $-\beta$, etc. The notation $p\sigma_x$ means a p state with $m_l=0$ when the quantization is relative to the x axis. The basic theoretical principles underlying the construction of wave functions such as (2)–(7) are outlined in the preceding article.[17]

The lower roots of the quadratic secular equations associated with (2)–(7) represent lower energy than for the free atom, i.e., are bonding levels, while the upper roots give higher energy than for the free atoms, i.e., are antibonding. The $d\epsilon$ wave functions are nonbonding, as shown in the previous paper. The situation is illustrated in Fig. 2. All these statements on the positions of the levels are surely true as long as the important term for bonding is the so-called Hund resonance integral, which is the off-diagonal matrix element connecting the central and attached portions. Since diagonal sums are unaffected by the addition of off-diagonal elements, it follows that the sum of the two roots of the quadratic must equal an energy level of the free central atom, plus one of the attached atom. Hence if one root is lowered in virtue of the resonance integral, the other must be raised.

Let us now discuss the case of $Fe(CN)_6^{3-}$. We can consider each CN radical as contributing one electron (*viz.*, the electron which corresponds to a $2p\sigma$ wave function directed towards the central atom). There are thus 17 electrons in all to house (apart from completed inner shells). Twelve electrons can be housed in the lower roots of the secular equations associated with (2)–(7), since the Pauli principle allows two electrons to an orbital state. The remaining five electrons are then to be accommodated in the three nonbonding, i.e., purely central $d\epsilon$ orbitals, which are preferable to antibonding states. One, and only one, of these five electrons can have a private orbital, and so only one spin contributes to the susceptibility. With $Fe(CN)_6^{4-}$, the last six electrons are snugly accommodated in the three $d\epsilon$ levels, and so there is diamagnetism. Of course, if the bonding action is weak enough, several electrons may prefer to occupy private orbitals, even though antibonding, in order to secure large spin and low internal energy for the central atom. This is a refinement not included in the crude forms of the method of molecular orbitals, for it is the interelectronic exchange energy, not incorporated in the one-electron problem, which is lowered by making the spins parallel. Thus we see that, as in the other methods, the susceptibility will have the low values characteristic of the cyanides only if the bonding action is larger than the Russell-Saunders structure of the central atom.

Comparison of the Various Methods

We have now seen that all three descriptions are capable of explaining the anomalously low

[17] Cf. also J. H. Van Vleck and A. Sherman, Rev. Mod. Phys. 7, 218–222 (1935).

magnetism of the cyanides. The question naturally arises as to which corresponds the closest to reality. It seems to us that in the cyanides one must decide in favor of III or possibly I. One reason is that method III is the most general, since it is noncommittal on the amount of polarity, which can be anything between $Fe^{3+}(CN^-)_6$ and $Fe^{9-}(CN^+)_6$, depending on the size of the coefficients α, β, γ. In fact, II can be regarded as a special case of III, wherein these coefficients are zero and the CN orbitals have lower energies than those of the central atom, so that the twelve electrons digested by (2–7) are located on the CN radicals. Method I is also to a certain extent a special case of III, corresponding to

$$\alpha=\beta=\gamma=\sqrt{\tfrac{1}{2}}, \qquad (8)$$

but we have seen in the preceding paper that in any event I differs from III in assuming localized or unidirectional bonds. Eq. (8) may be regarded as the condition for ideal covalency, as when it is satisfied, the twelve bonding electrons spend equal amounts of time on the $(CN)_6$ group and on the central atom. The actual behavior is not as extreme as in (8), for the structure $Fe^{3-}(CN)_6$ given by (8) requires too much negative charge on the Fe ion, as mentioned at the beginning of the preceding paper.

A decision between II and III is furnished by the fact that $(NH_4)_3FeF_6$ obeys the ordinary formula (1), and so has a much higher magnetism than $K_3Fe(CN)_6$. Now it is known from other evidence[18] that $(NH_4)_3FeF_6$ is much more polar in its structure, i.e., more nearly $(NH_4^+)_3Fe^{3+}(F^-)_6$, than is $K_3Fe(CN)_6$. With the method of the crystalline potential, the fields responsible for the Stark splitting are those due to the negative charges surrounding the paramagnetic cation. Hence if method II were always applicable, the quenching of magnetism would be more complete in the fluoride than in the cyanide, contrary to experiment. On the other hand, in the method III of molecular orbitals, the strength of the bond or separation of the levels stands in no immediate relation to the amount of polarity, and it is perfectly possible to have a covalent structure quench magnetism more completely than an ionic one. In fact, if the resonance integrals are the determining factor in chemical bonds, one comes nearer to (8) the greater the interatomic forces available to suppress the magnetism.

Pauling has argued that the suppression of magnetism is evidence in favor of the electron pair mechanism I in $Fe(CN)_6^{3-}$. However, we have seen that II or III can also be effective in quenching magnetism, so that it does not seem possible directly to infer the bond type from magnetic behavior. We agree with Pauling that covalent bonds are more effective in destroying magnetism than ionic, but the reasoning by which we reach this conclusion is somewhat different and is based on the empirical comparison of $(NH_4)_3FeF_6$ and $K_3Fe(CN)_6$ made above. Once it has been established empirically in a few cases that covalent bonds destroy magnetism more than ionic, one is probably justified in extrapolating to other cases, and supposing the bonds to be covalent when most of the magnetism has been destroyed.

If method III is to be preferred to II, the calculations of Schlapp and Penney,[8] based ostensibly on II, must be given a somewhat different interpretation than previously. They by no means lose applicability, for they are based largely on the symmetry group properties, and so retain practically as much significance with III as with II, except that the size of the crystalline potential is not to be taken too literally. The splittings of a few volts may now relate to the magnitude of the Hund resonance integrals rather than of the crystalline potential, but remain comparable to the Russell-Saunders structure, making it still reasonable that one should sometimes have the cyanide behavior and sometimes the sulphate.

One feature which is common to methods I, II, III is that nickel salts should have a susceptibility corresponding to $S=1$ rather than to $S=0$ in (1) even if the cubic splitting is large compared with the Russell-Saunders structure. This statement is true if the grouping about the nickel ion is either tetrahedral or octahedral. The demonstration has been given by Pauling[2] with mechanism I and by Van Vleck and Sherman[17] with III. If II is used, one has only to note that Ni^{++} has two more electrons than Fe^{++}, so

[18] L. Pauling, J. Am. Chem. Soc. **54**, 988 (1932); Pauling and Huggins, reference 2.

that there are eight rather than six electrons to be housed in the five levels of Fig. 1. Hence regardless of whether Fig. 1 is upright or inverted, two electrons can have private orbitals without materially diminishing the energy, provided the rhombic splitting in Fig. 1 is small. To explain the observed diamagnetism of $K_2Ni(CN)_4$, one is thus led to predict a square (tetragonal) rather than tetrahedral configuration for the $Ni(CN)_4^{--}$ ion. This prediction is confirmed by x-ray measurements[19] on $BaNi(CN)_4 4H_2O$. On the other hand, $Ni(NH_3)_4SO_4$ is paramagnetic, and this fact suggests that there may be a tetrahedral structure for the group $(NH_3)_4$.

Our arguments to explain the diamagnetism of $K_4Fe(CN)_6$ apply equally well to diamagnetic cobaltic compounds such as the cobaltammines, since Co^{+++} is isoelectronic with Fe^{++}. The writer has shown elsewhere[20] that if the coordination number is six, and if (1) is valid (the sulphate rather than cyanide case), cobaltous compounds should show marked anisotropy and deviations from Curie's law, while the reverse should be true of nickel ones. If the coordination number is four, the roles of cobalt and nickel should be interchanged. Blue and pink cobaltous salts are supposed to involve coordination numbers of four and six, respectively.[21] It would therefore be interesting to see if one could prepare a blue cobaltous salt conforming closely to Curie's law. Attempts of R. B. Janes to do this have so far proved unsuccessful. The explanation may be either that the magnetic dilution is insufficient, so that the exchange forces between paramagnetic ions complicate the temperature dependence, or else that one is in the critical transition region wherein the mechanism for suppressing the magnetism below (1) just comes into play. In the latter event both states with $S=\frac{1}{2}$ and $S=3/2$ are populated, and the temperature variation will not be simple.

In conclusion, we may say that it should be regarded as reassuring and significant that all three mechanisms I, II, III permit a rational understanding of magnetism lower than that given by (1) and the Hund rule. Since none of the methods of approximation can be regarded as satisfactory from a quantitative standpoint, a property which is common to all three must be regarded as on a much firmer basis than if it is characteristic only of one approach.

The writer wishes to thank Mr. John Howard for interesting and valuable discussions.

[19] H. Brasseur, A. de Rassenfoss and J. Piérard, Zeits. f. Krist. **88**, 210 (1934).

[20] J. H. Van Vleck, Phys. Rev. **41**, 208 (1932). In this connection see reference 10 regarding the relation between coordination number and the sign of the crystalline field.

[21] R. Hill and O. R. Howell, Phil. Mag. **48**, 833 (1924).

Reprinted from *J. Chem. Phys.*, **8**, 188–198 (1940)

Directed Valence*

14

George E. Kimball
Department of Chemistry, Columbia University, New York, New York

(Received October 19, 1939)

The problem of directed valence is treated from a group theory point of view. A method is developed by which the possibility of formation of covalent bonds in any spatial arrangement from a given electron configuration can be tested. The same method also determines the possibilities of double and triple bond formation. Previous results in the field of directed valence are extended to cover all possible configurations from two to eight *s*, *p*, or *d* electrons, and the possibilities of double bond formation in each case. A number of examples are discussed.

Introduction

Problems of directed valence, like most problems of molecular structure, can be attacked by either of two methods: the method of localized electron pairs (Heitler-London) or the method of molecular orbitals (Hund-Mulliken). While it is now realized[1] that these methods are but different starting approximations to the same final solution, each has its advantages in obtaining qualitative results. Theories of directed valence based on the methods of localized pairs have been developed by Slater[2] and Pauling[3] and extended by Hultgren.[4] The method of molecular orbitals has been developed principally by Hund[5] and Mulliken.[6] These methods have been compared extensively by Van Vleck and Sherman.[1]

In the previous papers no attempt has been made to discover *all* the possible stable electron groups which lead to directed valence bonds, nor have the possibilities of double bond formation been completely explored. In the present paper both of these deficiencies in the theory have been removed.

Method

Pauling's method consists of finding linear combinations of *s*, *p*, and *d* orbitals which differ from each other only in direction. Thus he has found combinations which are directed toward the corners of a tetrahedron, others directed toward the corners of a square, and others directed toward the corners of an octohedron. Electrons occupying these new orbitals can then resonate with unpaired electrons occupying orbitals of other atoms lying in the directions of these new orbitals and so form covalent bonds with these atoms. The further these orbitals project in the direction of the surrounding atoms, the stronger should be the resulting bonds.

In order to construct such sets of orbitals, it is most convenient to make use of group theory. Each set of equivalent directed valence orbitals has a characteristic symmetry group. If the operations of this group are performed on the orbitals, a representation, which is usually reducible, is generated. By means of the character table of the group[7] this representation, which we shall call the σ representation, can be reduced to its component irreducible representations. The *s*, *p*, and *d* orbitals of the atom also form representations of the group, and can also be divided into sets which form irreducible representations.[8]

Let us refer to the set of equivalent valence orbitals as the set $\mathfrak{S}$. If the transformation which reduces this set is T, then the set $T\mathfrak{S}$ can be broken up into subsets, each of which forms a basis for one of the irreducible representations of the symmetry group of $\mathfrak{S}$.

$$T\mathfrak{S}=\sum a_i\mathfrak{S}_i, \tag{1}$$

* Presented at the Boston meeting of the American Chemical Society, September 15, 1939. Publication assisted by the Ernest Kempton Adams Fund for Physical Research of Columbia University.

1 Van Vleck and Sherman, Rev. Mod. Phys. **7**, 167 (1935).

2 Slater, Phys. Rev. **37**, 841 (1931).

3 Pauling, J. Am. Chem. Soc. **53**, 1367, 3225 (1931).

4 Hultgren, Phys. Rev. **40**, 891 (1932).

5 Hund, Zeits. f. Physik **73**, 1 (1931); **73**, 565 (1931); **74**, 429 (1932).

6 Mulliken, Phys. Rev. **40**, 55 (1932); **41**, 49 (1932); **41**, 751 (1932); **43**, 279 (1933).

7 For the group theoretical methods used here see Wigner, *Gruppentheorie* (Vieweg, Braunschweig, 1931); Weyl, *Theory of Groups and Quantum Mechanics* (tr. Robertson) (Methuen, London, 1931); Van der Waerden, *Gruppentheoretische Methoden in der Quantenmechanik* (Springer, Berlin, 1932).

8 Bethe, Ann. d. Physik (5) **3**, 133 (1929).

where $\mathfrak{S}_i$ belongs to the ith irreducible representation. If we let $\mathfrak{R}$ be the set of available orbitals of the atom, these may always be chosen so that they fall into subsets each of which is a basis for one of the irreducible representations:

$$\mathfrak{R}=\sum b_i\mathfrak{R}_i. \quad (2)$$

If each of the coefficients b_i in (2) is equal to or greater than the corresponding a_i in (1), we may form the new subset $\mathfrak{R}'$, given by

$$\mathfrak{R}'=\sum a_i\mathfrak{R}_i, \quad (3)$$

in which each orbital transforms in exactly the same way as the corresponding orbital in (1). If we now apply the inverse transformation T^{-1} to this set we obtain a set of orbitals $T^{-1}\mathfrak{R}'$ which must have exactly the symmetry properties of the desired set $\mathfrak{S}$.

To illustrate this process, consider the set $\mathfrak{S}=\sigma_1+\sigma_2+\sigma_3$ consisting of three valence orbitals lying in a plane and making equal angles of 120° with each other. If the atom in question has available s, p, and d orbitals, the set $\mathfrak{R}$ consists of the nine orbitals s, p_x, p_y, p_z, d_z, d_{xz}, d_{yz}, d_{xy}, d_{x+y}. The symmetry group of $\mathfrak{S}$ is D_{3h}, and the set $\mathfrak{S}$ may be reduced by the transformation

$$\begin{aligned}\sigma_1'&=1/\sqrt{3}(\sigma_1+\sigma_2+\sigma_3)\\ \sigma_2'&=1/\sqrt{6}(2\sigma_1-\sigma_2-\sigma_3)\\ \sigma_3'&=1/\sqrt{2}(\sigma_2-\sigma_3).\end{aligned} \quad (3)$$

Of these orbitals, σ_1' belongs to the representation[9] A_1' while σ_2' and σ_3' belong to E'. Hence we find for (1)

$$\mathfrak{S}=A_1'+E'. \quad (4)$$

TABLE I. *Character table for trigonal orbitals.*

D_{3h}	E	σh	$2C_3$	$2S_3$	$3C_2$	$3\sigma v$
A_1'	1	1	1	1	1	1
A_2'	1	1	1	1	−1	−1
A_1''	1	−1	1	−1	1	−1
A_2''	1	−1	1	−1	−1	1
E'	2	2	−1	−1	0	0
E''	2	−2	−1	1	0	0
s	1	1	1	1	1	1
p	3	1	0	−2	−1	1
d	5	1	−1	1	1	1
σ	3	3	0	0	1	1
π	6	0	0	0	−2	0

[9] The notation used here is that of Mulliken, Phys. Rev. **43**, 279 (1933).

The set $\mathfrak{R}$ is already reduced, for s and d_z belong to A_1', p_z to A_2'', the pairs p_x, p_y and d_{xy}, $-d_{x+y}$ to E', and the pair d_{xz}, d_{yz} to E''. Hence (2) becomes

$$\mathfrak{R}=2A_1'+A_2''+E'+E''. \quad (5)$$

Since the coefficients in (5) are each not less than the corresponding coefficient in (4), the desired directed orbitals are possible. In fact, since we have a choice between s and d_z for the orbital

TABLE II. *Reduction table for trigonal orbitals.*

D_{3h}	A_1'	A_1''	A_2'	A_2''	E'	E''
s	1	0	0	0	0	0
p	0	0	0	1	1	0
d	1	0	0	0	1	1
σ	1	0	0	0	1	0
π	0	0	1	1	1	1

belonging to A_1', we may construct two different sets of directed orbitals. If we choose for $\mathfrak{R}'$ the set

$$\mathfrak{R}'=s+p_x+p_y \quad (6)$$

and apply the inverse transformation

$$\begin{aligned}\sigma_1&=\frac{1}{\sqrt{3}}\sigma_1'+\frac{2}{\sqrt{6}}\sigma_2',\\ \sigma_2&=\frac{1}{\sqrt{3}}\sigma_1'-\frac{1}{\sqrt{6}}\sigma_2'+\frac{1}{\sqrt{2}}\sigma_3',\\ \sigma_3&=\frac{1}{\sqrt{3}}\sigma_1'-\frac{1}{\sqrt{6}}\sigma_2'-\frac{1}{\sqrt{2}}\sigma_3',\end{aligned} \quad (7)$$

we obtain Pauling's trigonal orbitals

$$\begin{aligned}\sigma_1&=\frac{1}{\sqrt{3}}s+\frac{2}{\sqrt{6}}p_x,\\ \sigma_2&=\frac{1}{\sqrt{3}}s-\frac{1}{\sqrt{6}}p_x+\frac{1}{\sqrt{2}}p_y,\\ \sigma_3&=\frac{1}{\sqrt{3}}s-\frac{1}{\sqrt{6}}p_x-\frac{1}{\sqrt{2}}p_y.\end{aligned} \quad (8)$$

On the other hand, this theory shows that we can everywhere replace s by d_z, or in fact by any linear combination $\alpha s+\beta d_z$ provided $\alpha^2+\beta^2=1$. Similarly p_x and p_y can be replaced by d_{xy} and d_{x+y}.

It is not necessary, however, to carry out the actual determination of T and the reduction of $\mathfrak{S}$ and $\mathfrak{R}$ to decide whether or not a given set of directed orbitals is obtainable from given atomic orbitals, for the coefficients a_i and b_i can be obtained very simply from the character tables of $\mathfrak{S}$ and $\mathfrak{R}$. The transformation matrices of the σ representation can be written down and the trace of each gives the character for the corresponding element of the group. The characters for $\mathfrak{R}$ are easily found by the method of Bethe.[8] We can thus construct a character table for the representations based on $\mathfrak{S}$ and $\mathfrak{R}$. We also can enter in the same table the characters for the irreducible representations, as given for example by Mulliken.[9] Such a character table for the plane trigonal case we have been considering is shown in Table I. By means of the orthogonality theorems for group characters, the coefficients a_i and b_i can easily be found. These are entered in a second table, to which I shall refer as the reduction table. The reduction table for our trigonal case is shown in Table II. Comparing the coefficients a_i (given in the row marked σ) with the coefficients b_i (given in the rows marked s, p, and d) we see immediately that the trigonal orbitals require one orbital which may be either s or d, and two p orbitals (remembering that the representation E' is of degree two) or two d orbitals. Hence the possible valence configurations are sp^2, dp^2, sd^2, and d^3.

It is interesting to note that the method of molecular orbitals leads to identical results, but by a rather different route. In this method we consider first the set of orbitals on the atoms surrounding the central atom. If this set consists of orbitals symmetrical about the line joining each external atom to the central atom, then these external orbitals form a basis for a representation of the symmetry group which is identical with the σ representation. The reduction of this representation then corresponds to the resonance of these external orbitals among themselves. The formation of molecular orbitals then takes place by the interaction between these reduced external orbitals and the orbitals of the central atom. This interaction can only take place, however, between orbitals belonging to the same representation. Hence, to obtain a set of molecular orbitals equal in number to the number of external atoms, it is necessary that each of the reduced external orbitals be matched with an orbital from the set $\mathfrak{R}$ which belongs to the same representation. The condition for this is again that $b_i \geqslant a_i$, so that the same result is reached as before.

The possibilities of double or triple bond formation are most easily discussed in terms of molecular orbitals. The principal type of multiple bond consists of two parts: first, a pair of electrons in an orbital symmetrical about the axis of the bond; and second, one or more pairs of electrons in orbitals which are not symmetrical about the axis. The first pair of electrons form a bond which differs in no way from the ordinary single, or σ, bond. The other pairs are ordinarily in orbitals which are antisymmetric with respect to a plane passed through the axis. They may be regarded as formed by the interaction of two p orbitals, one on each atom, with axes parallel to each other and perpendicular to the axis of the bond. We shall refer to orbitals of this type as π orbitals.

In a polyatomic molecule, consisting of a central atom and a number of external atoms bound to it, bonds of this type may also be formed. As far as the external atoms are concerned, the condition for the formation of π bonds is the presence of p orbitals at right angles to the bond axes. These p orbitals, however, will resonate among themselves to form new orbitals which are bases of irreducible representations of the symmetry group of the molecule. This reduction can be carried out in the same way as the reduction of the σ representation. We first determine the representation generated by the p orbitals of the external atoms. Since there are two such p orbitals per external atom, this representation, which will be referred to as the π representation, will have a degree twice that of the σ representation. The characters of this representation are then computed, and entered in the character table. This has been done for the case of a plane trigonal molecule in Table I. The component irreducible representations are found as before, and entered in the reduction table.

The condition for the formation of a π bond is now that there be an orbital of the central atom belonging to the same representation as one of

TABLE III. *Resolution table for linear bonds.*

$D_{\infty h}$	Σ_g^+	Σ_u^+	Π_g	Π_u	Δ_g	Δ_u
s	1	0	0	0	0	0
p	0	1	0	1	0	0
d	1	0	1	0	1	0
σ	1	1	0	0	0	0
π	0	0	1	1	0	0

TABLE IV. *Resolution table for angular bonds.*

C_{2v}	A_1	A_2	B_1	B_2
s	1	0	0	0
p	1	0	1	1
d	2	1	1	1
σ	1	0	0	1
π	1	1	1	1

the irreducible components of the π representation. Since this molecule is already supposed to be held together by σ bonds, it is not necessary that all of the irreducible components of the π representation be matched by orbitals of the central atom. However, unless at least half of the irreducible components of the π representation are so matched, it will be impossible to localize the π bonds, and the resulting molecule will be of the type ordinarily written with resonating double bonds.

Thus in the plane trigonal case, Table II shows that π bond formation is possible through the p orbital belonging to $A_2''(p_z)$, the d orbitals belonging to $E''(d_{xz},d_{yz})$ or through the two orbitals belonging to E' which are not used in forming the original σ bonds. Since, however, the σ bonds are probably formed by a mixture of both the p and the d E' orbitals, these last two π bonds are probably weaker than the others. In general we shall divide the π bonds into two classes, calling them "strong" if they belong to representations not used in σ bond formation, and "weak" if they belong to those representations already used in σ bond formation.

Because of the "resonating" character of π bonds, it is usually difficult to form a mental picture of them. In this trigonal case which we have been discussing, we may imagine the p_z orbital of the central atom to interact in turn with the p_z orbitals of the external atoms. If the orbitals d_{xz} and d_{yz} are available, these may be combined with p_z, in the same way that p_x and p_y can be combined with s, to form three directed π orbitals. No such simple picture, however, seems to be available for the π bonds formed by d_{xy} and d_{x+y}.

It should be noted that this method does not predict directly the type of bond arrangement formed by any given electron configuration. Instead it merely tells whether or not a given arrangement is possible. In many cases it is found that several arrangements are possible for a single configuration of electrons. In these cases the relative stability of the various arrangements must be decided by other methods, such as Pauling's "strength" criterion, or consideration of the repulsions between nonbonded atoms.

RESULTS

The results of these calculations are most conveniently arranged according to the coordination number of the central atom. For the sake of completeness all of the results, including those previously obtained by Pauling, Hultgren and others, are contained in the following summary.

Coordination number 2

If the central atom forms bonds with two external atoms only two arrangements of the bonds are possible: a linear arrangement (group $D_{\infty h}$) and an angular one (group C_{2v}). The resolution tables for these are given in Tables III and IV. The configurations sp and dp can lead to either arrangement. The linear arrangement, however, is favored by both the repulsive forces and the possibilities for double bond formation and is therefore the stable arrangement for these configurations. The configurations ds, d^2 and p^2 on the other hand must be angular.

In the linear arrangement two p and two d orbitals are available for double bond formation, while in the angular arrangement only two strong π bonds are possible, one of which must be through a d orbital, the other of which may be formed by either a d or a p orbital. The other two possible π bonds are weakened by the fact that their orbitals belong to representations already used by the σ bonds.

The angular nature of the p^2 bonds in such molecules as H_2O and H_2S is well known, and need not be discussed further. As examples of double bond formation in molecules having this

TABLE V. *Resolution table for trigonal pyramid bonds.*

C_{3v}	A_1	A_2	E
s	1	0	0
p	1	0	1
d	1	0	2
σ	1	0	1
π	1	1	2

TABLE VI. *Resolution table for three unsymmetrical bonds in a plane.*

C_{2v}	A_1	B_2	B_1	B_2
s	1	0	0	0
p	1	0	1	1
d	2	1	1	1
σ	2	0	1	0
π	1	1	2	2

primary valence structure the nitroso compounds are typical. The prototype of these compounds is the nitrite ion NO_2^-. In this ion each oxygen atom is joined to the nitrogen by a σ bond involving one of the p orbitals of the nitrogen. In addition to these bonds, however, there is a π bond which belongs to the representation B_1 and cannot be localized. Its component orbitals are the p_x orbital of the nitrogen atom and the normalized sum of the p_x orbitals of the two oxygen atoms. The remaining parts of the oxygen p orbitals are occupied by unshared pairs of electrons. Because of the distributed character of the π bond no single valence bond picture can be drawn for this molecule, but only the resonating pair

$$O^- - N \qquad\qquad O = N$$
$$\diagdown\kern-0.6em\diagdown \qquad\qquad\qquad \diagdown$$
$$O \qquad\qquad\qquad O^-.$$

In the true nitroso compounds RNO, and also in the nitrosyl halides ClNO, etc., the situation is very similar. The lack of complete symmetry does not change any of the essential features of the structure of these molecules. In these cases, however, the difference in electro-negativity between oxygen and the atom or group R may cause ionic structures to play an important role. Pauling has suggested for example that in ClNO the important structures are

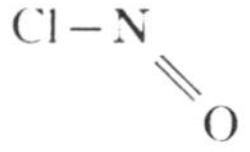

$$Cl - N \qquad \text{and} \qquad Cl^- \; N$$
$$\diagdown\kern-0.6em\diagdown \qquad\qquad\qquad \diagdown\kern-0.6em\diagdown$$
$$O \qquad\qquad\qquad\qquad O^+.$$

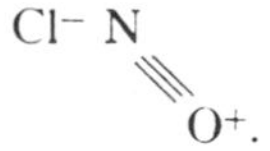

The bond formation in both these structures is in accord with the present theory.

Examples of the linear configuration are found in the ions of the $Ag(NH_3)_2^+$ type and those of the I_3^- type. In $Ag(NH_3)_2^+$ the configuration is certainly sp. In I_3^- and its relatives it is not certain whether the promotion of one of the $5p$ electrons of the central atom is to the $5d$ or $6s$ orbital, but in either case the resulting configuration (ps or pd) should give the observed linear arrangement.

The effect of double bond formation is shown clearly by CO_2 and SO_2. In CO_2 the valence configuration of the carbon is sp^3. The primary structure of σ bonds is sp, which requires the molecule to be linear. The other two p orbitals form the two π bonds. In SO_2, on the other hand, the primary bonds are certainly formed by two p orbitals, thus producing angular arrangement. The customary way of writing the structure of this molecule,

$$O = S^+$$
$$\diagdown$$
$$O^-$$

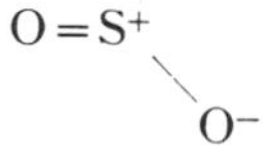

(analogous to NO_2^-) is quite possible, since the π bond can be formed by the third p orbital of the sulfur, but the structure

$$O = S$$
$$\diagdown\kern-0.6em\diagdown$$
$$O$$

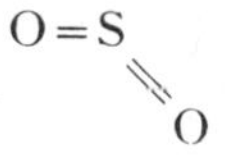

is also possible and probably as important as the first. In this structure one of the d orbitals of the sulfur acts as an acceptor for one of the electron pairs of the oxygen. It should be noted that the formation of two double bonds by an atom does not require the linear arrangement.

Coordination number 3

For this number of σ bonds three arrangements are important: the plane trigonal arrangement (group D_{3h}) already discussed, the trigonal pyramid (group C_{3v}), and an unsymmetrical plane arrangement with two of the bond angles equal, but not equal to the third (group C_{2v}). The resolution tables for the last two are given in Tables V and VI. As has already been shown, the plane arrangement is stable for the configurations sp^2, sd^2, dp^2, and d^3. (The pyramidal configuration is also possible, but less stable because of repulsive forces. Such cases will simply be omitted in the future.) The configurations p^3 and d^2p are therefore the only ones leading to a pyramidal structure.[10] The configuration *dsp* leads to the unsymmetrical plane arrangement.

The possibilities of double bond formation in the plane trigonal arrangement have already been discussed. In the pyramidal arrangement no strong double bonds are possible, although weak π bonds are possible with *d* orbitals. Thus the sulfite ion, for example, is restricted to the structure

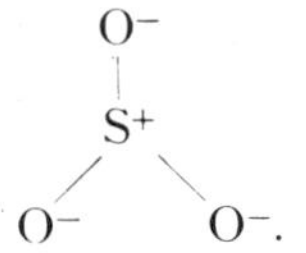

It is interesting to contrast this structure with that of SO_3. The most easily removed pair of electrons in $SO_3^=$ is the pair of 3*s* electrons in the sulfur atom. If these are removed, however, one of the 3*p* pairs forming a σ bond falls to the 3*s* level, thus making the valence configuration sp^2 and the molecule plane. The vacated 3*p* orbital is then filled by an unshared pair from the oxygen atoms to give a π bond. Two more unshared pairs of electrons from the oxygen atoms may interact with the 3*d* orbitals thus forming two more π bonds, which are somewhat weaker than the first. The SO_3 molecule has therefore three comparatively strong double bonds, while the $SO_3^=$ ion has only weak double bonds, if any at all. It is this lack of stabilizing double bonds which gives $SO_3^=$ its relative instability.

The remaining configuration, *dsp*, should give rise to three bonds in a plane forming two right angles. No examples of molecules of this configuration are known. This is hardly surprising, however, in view of the instability of this arrangement compared to the others. If molecules of this arrangement are to be found at all they would be complex ions of the transition elements. For example, if the ion $Ni(NH_3)_3^{++}$ existed it would be of this structure.

Coordination number 4

With a coordination number of four there are three arrangements which need consideration. The reduction tables for these are given in Tables VII, VIII, IX and X. Table VII is for the regular tetrahedral arrangement of the bonds. It is easily seen that the configurations sp^3 and d^3s lead to this arrangement. For double bond formation there remain two *d* orbitals which can form strong π bonds and two other *d* orbitals (if the σ bonds are sp^3) or two *p* orbitals (if the σ bonds are d^3s) which can form weaker π bonds. These possibilities of double bond formation are in accord with Pauling's suggestion that the structure of ions of the type XO_4 should be written with double bonds. Thus for the sulfate ion the primary σ structure arises from the configuration sp^3, but two of the unshared pairs of the O^- ions can be donated to the vacant *d*

TABLE VII. *Reduction table for tetrahedral bonds.*

T_d	A_1	A_2	E	T_1	T_2
s	1	0	0	0	0
p	0	0	0	0	1
d	0	0	1	0	1
σ	1	0	0	0	1
π	0	0	1	1	1

TABLE VIII. *Reduction table for tetragonal plane bonds.*

D_{4h}	A_{1g}	A_{1u}	A_{2g}	A_{2u}	B_{1g}	B_{1u}	B_{2g}	B_{2u}	E_g	E_u
s	1	0	0	0	0	0	0	0	0	0
p	0	0	0	1	0	0	0	0	0	1
d	1	0	0	0	1	0	1	0	1	0
σ	1	0	0	0	0	0	1	0	0	1
π	0	0	1	1	1	1	0	0	1	1

TABLE IX. *Reduction table for tetragonal pyramidal bonds.*

C_{4v}	A_1	A_2	B_1	B_2	E
s	1	0	0	0	0
p	1	0	0	0	1
d	1	0	1	1	1
σ	1	0	0	1	1
π	1	1	1	1	2

TABLE X. *Reduction table for irregular tetrahedral bonds.*

C_{3v}	A_1	A_2	E
s	1	0	0
p	1	0	1
d	1	0	2
σ	2	0	1
π	1	1	3

[10] The bond angles of 90° in this structure are not predicted by the reduction table alone, but can be easily found by the process of forming the actual orbitals as shown in the second section.

orbitals of the sulfur atom, thus giving the structure

$$O^- \diagdown \diagup O$$
$$S$$
$$O^- \diagup \diagdown O$$

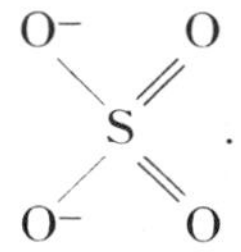

It must, however, be doubtful that two further π bonds are formed to give the structure

$$O \diagdown \diagup O$$
$$S^=$$
$$O \diagup \diagdown O$$

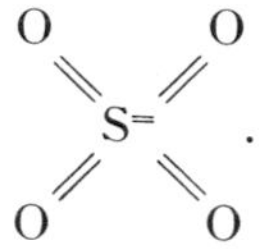

The situation in such molecules as $SiCl_4$ is similar.

From Table VIII it is seen that tetragonal planar (square) bonds are possible with configurations dsp^2 and d^2p^2. Beside the primary bonds, four strong π bonds are possible, using one p and three d orbitals. In $Ni(CN)_4^=$, for example, the primary valence configuration of the Ni is dsp^2, so that the structure

$$N$$
$$|||$$
$$C$$
$$|$$
$$N \equiv C - Ni^= - C \equiv N$$
$$|$$
$$C$$
$$|||$$
$$N$$

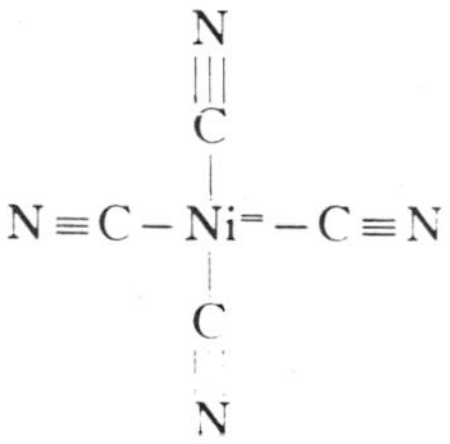

is one possible structure for this ion. By a shift of one of the π pairs forming the $C - N$ triple bond to the nitrogen, however, the carbon atom is left with an empty orbital, which can accept an electron pair donated by the nickel atom. In this way three π bonds can be formed between the nickel and carbon. At the same time one of the triple bond π pairs can shift in the opposite direction and be donated by the carbon to the empty p orbital of the nickel. These shifts would lead to the structure

$$N^-$$
$$C$$
$$N^- = C = Ni = C = N^-.$$
$$C$$
$$N^+$$

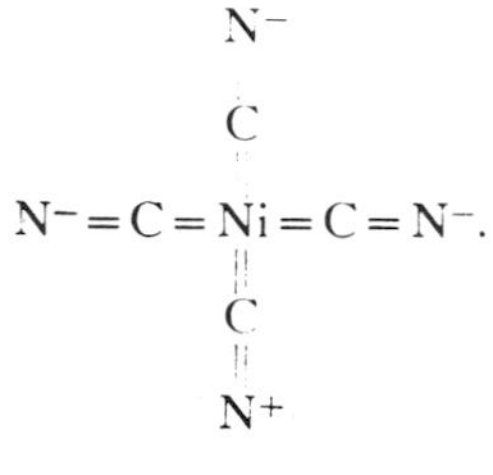

This structure affords an interesting contrast with that of $Ni(CO)_4$, in which one more pair of electrons must be accommodated. This pair occupies the d orbital of the nickel, thus pushing the valence configuration up to sp^3 and making the molecule tetrahedral. There are then only two π bonds possible, and the structure is

$$O^+$$
$$|||$$
$$C$$
$$|$$
$$O = C = Ni^= = C = O.$$
$$|$$
$$C$$
$$|||$$
$$O^+$$

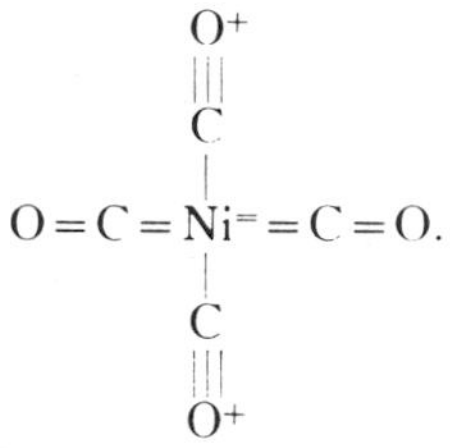

The configuration d^2p^2 is found in such ions as ICl_4^-, which are known to be plane.

From Table IX it appears that the configurations dp^3 and d^3p can lead to a tetragonal pyramid structure, while from Table X it appears that the "irregular tetrahedron" is also possible for these same configurations and also for the configuration d^2sp. By "irregular tetrahedron" is meant a structure in which three of the bonds are directed to the corners of an equilateral triangle and the fourth along the line perpendicular to the triangle at its center. The central atom is not necessarily located at the center of the triangle, but it is probably somewhat above it in the direction of the fourth bond. The choice between these two structures in the cases of the configurations dp^3 and d^3p is very close, with the advantage somewhat on the side of the irregular tetrahedron. The configuration d^4 can only have the pyramidal structure.

Examples of these configurations are rare. In the tetrahalides of the sulfur family we have the configuration p^3d produced by the promotion of one of the p electrons of the central atom to a d orbital. Unfortunately the spatial arrangements of these molecules are unknown. The other two configurations seem to be unstable. Such ions as $Fe(NH_3)_4^{++}$ and $Fe(CN)_4^=$ would have the configuration d^2sp, but do not seem to exist, which is perhaps some indication of this instability.

Coordination number 5

For this coordination number four bond arrangements are possible. In Table XI is given the

TABLE XI. *Reduction table for trigonal bipyramidal bonds.*

D_{3h}	A_1'	A_1''	A_2'	A_2''	E'	E''
s	1	0	0	0	0	0
p	0	0	0	1	1	0
d	1	0	0	0	1	1
σ	2	0	0	1	1	0
π	0	1	0	1	2	2

TABLE XII. *Reduction table for five tetragonal pyramidal bonds.*

C_{4v}	A_1	A_2	B_1	B_2	E
s	1	0	0	0	0
p	1	0	0	0	1
d	1	0	1	1	1
σ	2	0	0	1	1
π	1	1	1	1	3

reduction table for a trigonal bipyramid, which is seen to be stable for the configurations dsp^3 and d^3sp. In Table XII the bonds are directed from the center to the corners of a square pyramid. This is the arrangement expected for the configurations d^2sp^2, d^4s, d^2p^3, and d^4p. Table XIII is for the case of five bonds in a plane, directed toward the corners of a regular pentagon, the arrangement for the configuration d^3p^2, and Table XIV is for the case of five bonds directed along the slant edges of a pentagonal pyramid, the arrangement for the configuration d^5.

In PCl$_5$ and other molecules of this type, the pentavalent state is formed by the promotion of an s electron to the vacant d shell. The valence configuration is therefore dsp^3 and the bipyramidal structure is to be expected, in agreement with the results of electron diffraction studies. Table XI shows that two of the remaining d orbitals are capable of forming strong π bonds; the other two, two weak π bonds by accepting pairs of electrons. Neglecting weak bonds, the structure of PCl$_5$ is best written

$$\begin{array}{ccc} \text{Cl} & \text{Cl}^+ & \\ \diagdown & \| & \\ & \text{P} = -\text{Cl}. \\ \diagup & \| & \\ \text{Cl} & \text{Cl}^+ & \end{array}$$

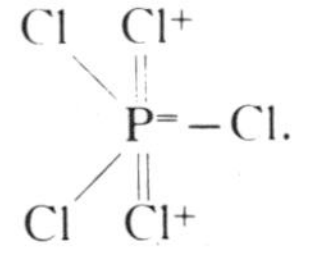

In the molecule Fe(CO)$_5$ the valence configuration is again dsp^3, and the bipyramidal structure is to be expected. Here π bonds can be formed by donation of pairs of d electrons from the iron through the carbon to the oxygens, as in Ni(CO)$_4$, giving the structure

$$\begin{array}{ccc} \text{O}^+ & \text{O} & \\ \diagdown\!\!\!\diagdown & \| & \\ \text{C} & \text{C} & \\ \diagdown & \| & \\ & \text{Fe} = -\text{C} \equiv \text{O}^+. \\ \diagup & \| & \\ \text{C} & \text{C} & \\ \diagup\!\!\!\diagup & \| & \\ \text{O}^+ & \text{O} & \end{array}$$

TABLE XIII. *Reduction table for pentagonal plane bonds.*

D_{5h}	A_1'	A_1''	A_2'	A_2''	E_1'	E_1''	E_2'	E_2''
s	1	0	0	0	0	0	0	0
p	0	1	0	0	1	0	0	0
d	1	0	0	0	0	1	1	0
σ	1	0	0	0	1	0	1	0
π	0	1	1	0	1	1	1	1

TABLE XIV. *Reduction table for pentagonal pyramidal bonds.*

C_{5v}	A_1	A_2	E_1	E_2
s	1	0	0	0
p	1	0	1	0
d	1	0	1	0
σ	1	0	1	1
π	1	1	2	2

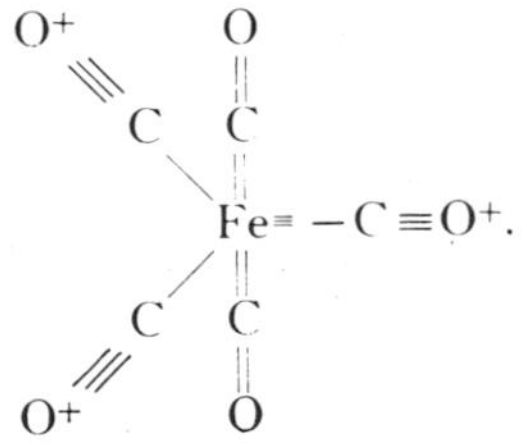

In IF$_5$, however, the valence configuration is p^3d^2, and the molecule should have the square pyramid structure. Here again two strong π bonds can be formed by donations of pairs of electrons from the F to the I, so that the structure may be

$$\begin{array}{ccc} & \text{F} & \\ \text{F}^+ & \diagdown & \\ \diagdown\!\!\!\diagdown & \diagdown & \\ & & \text{I} = = \text{F}^+, \\ \diagup & \diagup & \\ \text{F} & & \\ & \text{F} & \end{array}$$

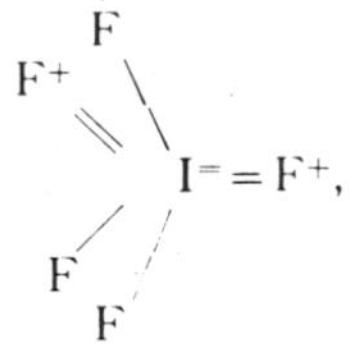

but the high electro-negativity of fluorine makes the existence of the double bonds doubtful. The spatial configuration of this molecule has not yet been determined experimentally.

The pentagonal configurations crowd the atoms so much that they must be unstable.

Coordination number 6

Six bonds may be arranged symmetrically in space in three ways: to the corners of a regular octahedron, to those of a trigonal prism, or to those of a trigonal antiprism (an octahedron stretched or compressed along one of the tri-

TABLE XV. *Reduction table for octahedral bonds.*

O_h	A_{1g}	A_{1u}	A_{2g}	A_{2u}	E_g	E_u	T_{1g}	T_{1u}	T_{2g}	T_{2u}
s	1	0	0	0	0	0	0	0	0	0
p	0	0	0	0	0	0	0	1	0	0
d	0	0	0	0	1	0	0	0	1	0
σ	1	0	0	0	1	0	0	1	0	0
π	0	0	0	0	0	0	1	1	1	1

TABLE XVI. *Reduction table for trigonal prismatic bonds.*

D_{3h}	A_1'	A_1''	A_2'	A_2''	E'	E''
s	1	0	0	0	0	0
p	0	0	0	1	1	0
d	1	0	0	0	1	1
σ	1	0	0	1	1	1
π	1	1	1	1	2	2

TABLE XVII. *Reduction table for trigonal antiprismatic bonds.*

D_{3d}	A_{1g}	A_{1u}	A_{2g}	A_{2u}	E_g	E_u
s	1	0	0	0	0	0
p	0	0	0	1	0	1
d	1	0	0	0	2	0
σ	1	0	0	1	1	1
π	1	1	1	1	2	2

gonal axes). The reduction tables for these possibilities are given in Tables XV, XVI, and XVII.

Table XV shows that octahedral bonds are formed by the configuration d^2sp^3 and no other. The commonness of this arrangement is therefore due to the fact that the configuration d^2sp^3 is the usual configuration of six valence electrons, rather than any particular virtue of the octahedral arrangement. This configuration arises both in the 6-coordinated ions of the transition elements, and in the molecules of the SF_6 type in which one s and one p electron are promoted to the next higher d level. Table XV also shows that all three remaining d orbitals are capable of forming strong π bonds. In the ferrocyanide ion $Fe(CN)_6^{-4}$, these d orbitals can donate their pairs of electrons to the nitrogen atoms, giving Pauling's structure

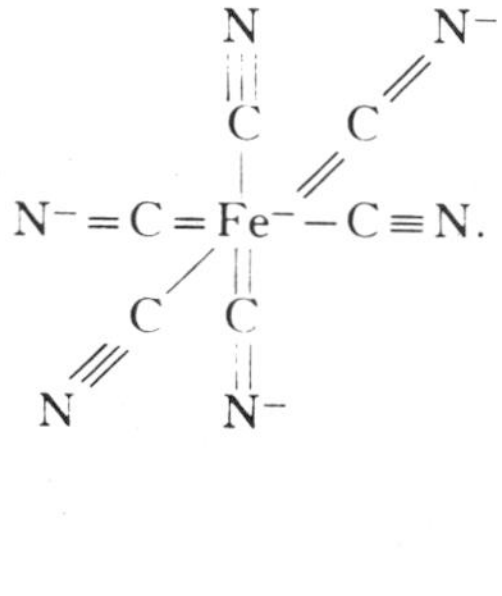

If SF_6 the remaining d orbitals are empty, and can accept pairs of electron from the fluorine atoms, giving the structure

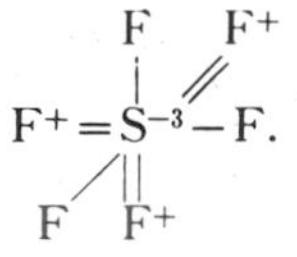

It is this structure which accounts for the great resistance of SF_6 to hydrolysis. It has been suggested by Sidgwick[11] that the hydrolysis of halides takes place either by their accepting a pair of electrons from the oxygen of a water molecule, with subsequent decomposition, or by donation of an unshared pair on the central atom to a hydrogen atom, with loss of hypohalous acid. With the usual single bonded structure of SF_6, the first mechanism is possible, and one would expect the hydrolysis to take place easily. With the double bonded structure, however, all of the electrons in the outer shell of the sulfur are taking part in bond formation, so that one should expect the same inertness as that displayed by CCl_4. It is interesting to note that SeF_6 is like SF_6, but TeF_6 is easily hydrolyzed. This must be due to the possibility of the Te atom's accepting a pair of oxygen electrons in the vacant $4f$ orbital.

In MoS_2 and WS_2 in the crystalline form the valence configuration is d^4sp and the arrangement is prismatic, as Hultgren[4] has pointed out. Table XVI indicates that the two empty p orbitals of the metal atom cannot form strong π bonds to the sulfur atoms. The configuration d^5p should also lead to the prismatic arrangement, but no examples are known.

The cases of the ions $SeBr_6^{-2}$ and $SbBr_6^{-3}$ are interesting in that the configuration in each of these should be p^3d^3. This configuration should lead to the antiprismatic arrangement, i.e., the octahedral symmetry should not be perfect. If, however, the unshared pair of s electrons is promoted to a d orbital, and one of the valence pairs slips into its place, the configuration will be the octahedral d^2sp^3. The deciding factor here may be the possibility of double bond formation offered by the octahedral but not by the antiprismatic arrangement.

[11] Sidgwick, *The Electronic Theory of Valency* (Oxford, 1927), p. 157.

The configurations d^3sp^2, d^5s and d^4p^2 do not permit any arrangement in which all of the bonds are equivalent. While the configuration d^3sp^2 might form bonds of a mixed type, e.g., the tetrahedral d^3s+the angular p^2, the bonds will be weak, and the configuration unstable. Molecules which might be expected to have this arrangement, e.g., $Fe(CN)_6^{-3}$, rearrange the electrons to obtain the more stable d^2sp^3 configuration. In $Fe(CN)_6^{-3}$ for example, the odd electron, which normally should occupy a $4p$ orbital, instead goes to a $3d$ orbital making the third $4p$ orbital available for the formation of octahedral bonds.

Coordination number 7

The coordination number 7 is extremely rare. The fact that it appears only in the heavier atoms, such as Zr, Cb, I, and Ta, leads one to suspect that for stability f electrons are necessary, although it is possible that the determining factor is ion size. Two arrangements of seven bonds have been observed:[12] the ZrF_7^{-3} structure, which may be obtained from the octahedron by adding an atom at the center of one face; and the TaF_7^{-2} structure, which may be obtained from the trigonal prism by adding an atom at the center of one of the square faces. The reduction tables for these two arrangements are given in Tables XVIII and XIX.

From Table XVIII we see that if no f orbitals are involved, the ZrF_7^{-3} type bonds can arise from the configurations d^3sp^3 and d^5sp. The configuration expected for ZrF_7^{-3} is d^5sp, so that it is not necessary to appeal to the f orbitals to form these bonds. This, however, does not preclude the possibility that f orbitals are used to strengthen the bonds.

Table XIX shows that the TaF_7^{-2} structure is possible with the configurations d^3sp^3, d^4sp^2, d^5sp, d^4p^3 and d^5p^2. The ion TaF_7^{-2} itself is isoelectronic with ZrF_7^{-3}, so that it is difficult to understand why it should prefer its structure to that of ZrF_7^{-3}. It is quite possible that the bonds in all the 7-coordinated molecules are so ionic that the

TABLE XVIII. *Reduction table for ZrF_7^{-3} type bonds.*

C_{3v}	A_1	A_2	E
s	1	0	0
p	1	0	1
d	1	0	2
f	2	1	2
σ	3	0	2
π	2	2	5

TABLE XIX. *Reduction table for TaF_7^{-2} type bonds.*

C_{2v}	A_1	A_2	B_1	B_2
s	1	0	0	0
p	1	0	1	1
d	2	1	1	1
f	2	1	2	2
σ	3	1	2	1
π	3	3	4	4

TABLE XX. *Reduction table for cubic bonds.*

O_h	A_{1g}	A_{1u}	A_{2g}	A_{2u}	E_g	E_u	T_{1g}	T_{1u}	T_{2g}	T_{2u}
s	1	0	0	0	0	0	0	0	0	0
p	0	0	0	0	0	0	0	1	0	0
d	0	0	0	0	1	0	0	0	1	0
f	0	0	0	1	0	0	0	1	0	1
σ	1	0	0	1	0	0	0	1	1	0
π	0	0	0	0	0	0	1	1	1	1

TABLE XXI. *Reduction table for tetragonal antiprismatic bonds.*

D_{4d}	A_{1g}	A_{1u}	A_{2g}	A_{2u}	E_1	E_2	E_3
s	1	0	0	0	0	0	0
p	0	0	0	1	1	0	0
d	1	0	0	0	0	1	1
σ	1	0	0	1	1	1	1
π	1	1	1	1	2	2	2

TABLE XXII. *Reduction table for dodecahedral bonds.*

V_d	A_1	A_2	B_1	B_2	E
s	1	0	0	0	0
p	0	0	0	1	0
d	1	0	1	1	1
σ	2	0	0	2	2
π	2	2	2	2	4

TABLE XXIII. *Reduction table for face-centered prismatic bonds.*

C_{2v}	A_1	A_2	B_1	B_2
s	1	0	0	0
p	1	0	1	1
d	2	1	1	1
σ	3	0	2	1
π	4	4	4	4

[12] Hampson and Pauling, J. Am. Chem. Soc. **60**, 2702 (1938); Hoard, J. Am. Chem. Soc. **61**, 1252 (1939).

TABLE XXIV. *Summary of stable bond arrangements and multiple bond possibilities.*

Coordination Number	Configuration	Arrangement	Strong π Orbitals	Weak π Orbitals	Table
2	sp	linear	p^2d^2	—	III
	dp	linear	p^2d^2	—	III
	p^2	angular	$d(pd)$	$d(sd)$	IV
	ds	angular	$d(pd)$	$p(pd)$	IV
	d^2	angular	$d(pd)$	$p(spd)$	IV
3	sp^2	trigonal plane	pd^2	d^2	II
	dp^2	trigonal plane	pd^2	d^2	II
	d^2s	trigonal plane	pd^2	p^2	II
	d^3	trigonal plane	pd^2	p^2	II
	dsp	unsymmetrical plane	pd^2	$(pd)d$	VI
	p^3	trigonal pyramid	—	$(sd)d^4$	V
	d^2p	trigonal pyramid	—	$(sd)p^2d^2$	V
4	sp^3	tetrahedral	d^2	d^3	VII
	d^3s	tetrahedral	d^2	p^3	VII
	dsp^2	tetragonal plane	d^3p	—	VIII
	d^2p^2	tetragonal plane	d^3p	—	VIII
	d^2sp	irregular tetrahedron	—	d	X
	dp^3	irregular tetrahedron	—	s	X
	d^3p	irregular tetrahedron	—	s	X
	d^4	tetragonal pyramid	d	$(sp)p$	IX
5	dsp^3	bipyramid	d^2	d^2	XI
	d^3sp	bipyramid	d^2	p^2	XI
	d^2sp^2	tetragonal pyramid	d	pd^2	XII
	d^4s	tetragonal pyramid	d	p^3	XII
	d^2p^3	tetragonal pyramid	d	sd^2	XII
	d^4p	tetragonal pyramid	d	sp^2	XII
	d^3p^2	pentagonal plane	pd^2	—	XIII
	d^5	pentagonal pyramid	—	$(sp)p^2$	XIV
6	d^2sp^3	octahedron	d^3	—	XV
	d^4sp	trigonal prism	—	p^2d	XVI
	d^5p	trigonal prism	—	p^2s	XVI
	d^3p^3	trigonal antiprism	—	sd	XVII
	d^3sp^2	mixed			
	d^5s	mixed			
	d^4p^2	mixed			
7	d^3sp^3	ZrF_7^{-3}	—	d^2	XVIII
	d^5sp	ZrF_7^{-3}	—	p^2	XVIII
	d^4sp^2	TaF_7^{-2}	—	dp	XIX
	d^4p^3	TaF_7^{-2}	—	ds	XIX
	d^5p^2	TaF_7^{-2}	—	ps	XIX
8	d^4sp^3	dodecahedron	d	—	XXII
	d^5p^3	antiprism	—	s	XXI
	d^5sp^2	face-centered prism	p	—	XXIII

directed nature of covalent bonds has little importance in determining the atomic arrangement.

Coordination number 8

Until very recently the spatial arrangement of no molecule involving a coordinate number of 8 had been determined experimentally. It had been supposed that the arrangements which were most probable were the cubic and the tetragonal antiprismatic. Hoard and Nordsieck[13] have now determined the arrangement of the ion $Mo(CN)_8^{-4}$ and have found that the structure is neither cubic nor antiprismatic, but instead that of a dodecahedron with triangular faces and symmetry $V_d(D_2^d)$. The reduction tables for these three cases are given in Tables XX, XXI, and XXII. Table XXIII is the reduction table for an arrangement which is derivable from a trigonal prism by placing two atoms at the centers of two of the rectangular faces. It is seen that the cubic requires the use of f orbitals, and even then arises only from the configurations d^3fsp^3 and d^3f^4s. The cubic arrangement should therefore not be ordinarily found. The antiprismatic arrangement is stable for the configurations d^4sp^3 and d^5p^3, and the dodecahedral for only d^4sp^3. The observed arrangement for $Mo(CN)_8^{-4}$, which has the configuration d^4sp^3, indicates that the dodecahedral arrangement has greater stability than the antiprismatic. The structure of such ions as TaF_8^{-3} and the molecule OsF_8 which should have the configuration d^5sp^2 should be the face-centered prismatic structure of Table XXIII.

Table XXIV summarizes these results. In it are given the most stable bond arrangements for each configuration of two to eight electrons in s, p or d orbitals. The orbitals available for strong π bonds and weak π bonds are also given. In those cases where there is a choice of two or more orbitals when only one can be chosen, the orbitals are enclosed in parentheses.

[13] Hoard and Nordsieck, J. Am. Chem. Soc. **61**, 2853 (1939).

Reprinted from *Ann. Rev. Phys. Chem.*, **7**, 107–136 (1956)

15

QUANTUM THEORY

BY W. MOFFITT AND C. J. BALLHAUSEN[1]

Department of Chemistry, Harvard University, Cambridge, Massachusetts

Previous reviews with the same or similar advertisement have concerned themselves with the theory of the electronic structure of molecules. No mention has been made, however, of perhaps its most strikingly successful variant, namely the crystal field theory. It seems unnecessary, therefore, to offer excuses for any lack of balance imposed on the series if we confine ourselves to transition-metal complexes in the present issue.

The crystal field theory has a long history. It was first developed by Becquerel (13), Bethe (15, 16), Kramers (62, 63, 64) and Van Vleck (97, 98). By 1932, with the work also of Gorter (37), Penney and Schlapp (82, 86), its most important features were already recognized and had explained many of the magnetic properties of salts of the transition metals and of the rare earths.[2] Essentially the same theory is used to-day in order to describe a much wider variety of effects, ranging from trends in the heats of hydration and the colours of aqueous ions to the hyperfine structure of their paramagnetic resonance spectra. Indeed, the only nontrivial many-electron systems which are more completely understood are free atoms and ions.

Our aim has been to provide both a review and a bibliography of the theory, but we separate these goals in our treatment. In a chapter of this size, it is impossible to develop the whole theory, so we have tried to cover those aspects of it which are of most concern to chemists. Where to be general is to risk being obscure, we have chosen simple examples to illustrate the argument; it should be stressed that these are chosen for no other reason. In the same way, the experimental values quoted are not to be regarded as in any way representative of an enormous number of publications which we have not attempted to compile. We have not tried to serve history, nor have we been at pains to accord priorities: many of the ideas that have proved to be most fruitful have been propounded in the course of interpreting particular experiments; it would be a major undertaking in a form of scholarship with which we have little familiarity to find their true originators. Finally, we are chemists writing for chemists. Our emphasis and notation have therefore differed at times from those of the physicists primarily responsible for the theory. Our survey of the pertinent literature was completed in December, 1955.

CRYSTAL FIELD THEORY

The crystal field theory concerns itself with the behaviour of rare earth

[1] On leave of absence from Chemical Laboratory A, Technical University of Denmark, Copenhagen.

[2] At the same time, Pauling (80) was developing a superficially very different approach which has become a standard part of chemical thinking and symbolism (vide infra).

and transition-metal salts, their solutions and complexes. It treats the metal ions as if they were subjected to electrostatic fields of force arising from the groups by which they are surrounded. Thus the anions, the waters of crystallization or of solvation, and the other ligands with which they are most intimately associated are assigned rather passive roles. These may, to be sure, be polarized by the presence of the cation they stabilize, but the motions of their electrons are assumed to remain unaffected by such factors as the optical excitation of the cation: the ligands are supposed to provide a potential field for the central ion which is the same for its ground as for its lower excited states.

The quantal treatment for this model is therefore easily formulated. The Hamiltonian for the cation consists of two terms:

$$\mathcal{H} = \mathcal{H}_F + \mathcal{V},$$

where $\mathcal{H}_F$ is the Hamiltonian for the free ion and $\mathcal{V}$ is the potential provided by its ligands. It is supposed that the eigenfunctions and eigenvalues of $\mathcal{H}_F$ are known. Accordingly, the potential $\mathcal{V}$ is regarded as a perturbation which determines the electronic motions and term values of the cation in the crystal, solution or complex.

Since the unperturbed functions are solutions of problems with full spherical symmetry, it is convenient to expand $\mathcal{V}$ in a series of normalized harmonics:

$$\mathcal{V} = \sum_i \sum_n \sum_m \mathcal{Y}_{nm}(\theta_i, \phi_i)\mathcal{R}_n(r_i),$$

the first summation being over the several electrons of the cation.[3] By far the most important term in this development is the spherically symmetric term with $n=0$:

$$\mathcal{V}_R = \sum_i (4\pi)^{-1/2}\mathcal{R}_0(r_i).$$

Thus $\mathcal{V}_R$ is responsible for the greater part of the lattice energy or heat of solution of a given cation. On the other hand, it has little effect on the optical and magnetic properties of crystals since, to a first approximation, it will only give rise to a uniform shift of all levels with the same number of d or of f electrons.[4] Of the remaining terms in the expansion, many may be

[3] The ligands are sometimes replaced by supposedly equivalent point charges, dipoles and so forth, in which case $\mathcal{R}_n(r_i)$ takes the form $a_n r_i^n + b_n r_i^{-(n+1)}$. On the other hand, if $\mathcal{V}$ is more nearly a self-consistent field, say, it will no longer satisfy Laplace's equation and $\mathcal{R}_n$ is a more general function of r_i. In most applications, the specific form of $\mathcal{R}_n$ is in any case of little importance, parameters containing it being determined semi-empirically.

[4] If it is supposed that the radius of the free cation is the same in its ground and lower excited states—that is, for example, that the radial parts of the d orbitals are the same both for the ^4F and ^4P levels of Cr^{+++}—, then the influence of $\mathcal{V}_R$ will only be felt as the result of second-order interactions with highly excited configurations in which, say, a $3d$ electron is promoted to a $4d$ level.

neglected. For example, if the effects to be described arise from degeneracies in incompleted d or f shells, all harmonics of odd order are ineffective; again, when only d electrons are involved, all potentials after the fourth-order terms exert no influence.[5] And finally, the potential $\mathcal{V}$ must have the same symmetry as the cationic site: if this is octahedral, $\mathcal{V}$ must remain invariant under all operations of O_h. It should be remarked that to good approximation $\mathcal{V}_R$ is the only component of $\mathcal{V}$ that is operative on S states in general—and therefore on closed shells and half-filled shells of maximum multiplicity in particular. Rather crudely, one may regard $\mathcal{V}_R$ as providing a spherical hole of constant potential, outside which it offers an infinite potential barrier to the cationic electrons; this view has the merit that it emphasizes the dependence of $\mathcal{V}_R$ on the effective ionic radius.

Octahedral Fields

Many of the principal features of the transition-metal ions emerge from the following simple model. The ions, whose lower electronic states all arise from the configurations (argon) $(3d)^n$, are situated in fields of perfect octahedral symmetry. Since the representation of O_h spanned by the harmonics of order less than five contains the identical representation only twice ($n=0$, 4), it follows that

$$\mathcal{V} = \mathcal{V}_R + \mathcal{V}_O,$$

$$\mathcal{V}_O = \sum_i \left\{ \mathcal{Y}_{40}(\theta_i, \phi_i) + \sqrt{\frac{5}{14}} \left[\mathcal{Y}_{44}(\theta_i, \phi_i) + \mathcal{Y}_{4-4}(\theta_i, \phi_i)\right] \mathcal{R}_4(r_i) \right\}.$$

The polar axis here, as elsewhere, is chosen to coincide with a particular fourfold axis of the octahedron; also, the $\mathcal{Y}_{nm}$ are normalized spherical harmonics whose phases are prescribed by Condon & Shortley (28). The crystalline forces are supposed to be much stronger than those attributable to spin-orbit coupling in the free ions. As has been remarked, $\mathcal{V}_R$ is very large and corresponds to the replacement of six ligands by a spherically symmetric field. Thus, if $\mathcal{V}$ represents six waters of crystallization, then we should expect the variation resulting from $\mathcal{V}_R$ in absolute term values to behave monotonically on going from Ti^{+++} to Co^{+++}, or from Ti^{++} to Zn^{++}, reflecting a more or less continuous decrease in the radii of the free ions. The potential $\mathcal{V}_O$, on the other hand, is more specific stereochemically, and acknowledges the existence of six negative groups at the respective vertices of the octahedron. It will remove the orbital degeneracy of the d electrons, since these will prefer to avoid regions where the electronic density of the ligands is greatest. However, the extent to which the orbital alignments are determined by $\mathcal{V}_O$ will depend on the number of electrons present and their correlative interactions with one another. Accordingly the influence of $\mathcal{V}_O$ on the ionic

[5] Any matrix representation of an odd order potential in a basis of orbitals with the same parity vanishes identically. Again, the direct product of two d orbital sets spans no representations of the rotation group of order higher than four.

term values will vary as a function of atomic number in a much more complicated manner than that of $\mathcal{V}_R$.

Single d electron.—The effect of the octahedral field is most clearly shown by considering the Ti^{+++} ion, which has a single $3d$ electron outside an argon shell. If higher configurations are ignored, $\mathcal{V}_O$ has no effect on the closed shell but simply splits the fivefold orbital degeneracy of the ion. The qualitative behaviour is easily determined by group theory, which shows that the reducible representation spanned by the five d orbitals contains the irreducible representations E_g and T_{2g} of the octahedral group O_h, the former being doubly degenerate and the latter threefold degenerate. The ^2D term therefore splits into ^2E_g and $^2T_{2g}$ levels.[6] More specifically, it is easily shown by symmetry that

$$t_{2g}{}^0 = \frac{1}{\sqrt{2}}(d\delta_+ - d\delta_-), \qquad t_{2g}{}^+ = d\pi_+, \qquad t_{2g}{}^- = d\pi_-;$$

$$e_g{}^a = d\sigma, \qquad e_g{}^b = \frac{1}{\sqrt{2}}(d\delta_+ + d\delta_-),$$

where σ, $\pi_\perp$, $\delta_\perp$ refer to $m_l = 0, \pm 1, \pm 2$ respectively. We see that the densities of the e_g electrons are directed towards the vertices of the octahedron, and therefore towards the ligands, whereas those of the t_{2g} electrons avoid these regions. Since the ligands are either anions or molecules, the negative ends of whose dipoles are directed towards the cation, it is evident that the t_{2g} orbitals are more stable than the e_g orbitals. Explicitly, their energies may be written

$$\epsilon(e_g) = \epsilon_0 + 6Dq, \qquad \epsilon(t_{2g}) = \epsilon_0 - 4Dq,$$

$$-4Dq = \int \phi^*(t_{2g}{}^0)\mathcal{V}_O\phi(t_{2g}{}^0)dv.$$

By convention, the overall splitting of a single d electron under $\mathcal{V}_O$ is set equal to $10Dq$, where $Dq > 0$ (86).[7]

The electronic spectrum of the $[Ti(H_2O)_6]^{+++}$ ion would be described as arising from the transition $(t_{2g})^{-1}(e_g)$, and be assigned an excitation energy of $10Dq$. If the symmetry is close to octahedral, we should expect a single weak band, since no change in parity occurs. Such a band is indeed found for the ion, with a maximum at 4900Å, corresponding to $Dq = 2040$ cm^{-1}. It is very broad, however, and shaded to the blue (39), suggesting that the degeneracies are partially resolved by fields of lower symmetry. Nevertheless, one might expect to understand, for example, the heat of hydration of the Ti^{+++} ion on the basis of $\mathcal{V}_R$ and an additional stabilization, $4Dq = 8160$ cm^{-1}, due to $\mathcal{V}_O$.

[6] When referring to the orbitals themselves, we call them e_g, t_{2g} respectively; the distinction is trivial for a single d electron in the same way that the terms of the hydrogen atom may be called ns, np, · · · , or n^2S, n^2P, · · · .

[7] The coefficients +6 and −4 may most easily be determined on noting that $\mathcal{V}_O$ makes no contribution to S states—such as that found on filling all d orbitals.

There is a well-known reciprocity in atomic spectroscopy between electrons and "holes." Thus the terms arising from $n(\leq 5)$ electrons outside an argon shell are the same as those arising from $(10-n)$ electrons outside the same shell or, equivalently, of n holes in the completed d^{10} shell. Since the interaction between two holes is the same repulsive coulombic term as between two electrons, these terms are in the same order for both d^n and d^{10-n} configurations. However, since the magnetic moment due to the circulation of a positron is the opposite of that due to an electron, the multiplet structure is inverted. A similar reciprocity exists in crystal field theory, say, between Ti^{+++} and Cu^{++}, whose configuration is d^9. It is clear that the ^2D state of Cu^{++} also splits into ^2E_g and $^2T_{2g}$ terms. But the order of these levels is now inverted because the electrostatic field due to the ligands has an opposite effect on holes as on electrons.

In order to explain the spectrum of the Cu^{++} aquo complex, a very broad and weak system with a maximum around 8000Å, it is necessary to invoke a considerable contribution from fields of lower symmetry: the hydrated ion is a distorted octahedron (17).

Weak fields.—In dealing with more than one electron or one hole, it is convenient to distinguish between two limiting cases. In both, the splitting due to the octahedral field is much larger than that due to spin-orbit coupling. But in the one case, which we pursue now, the crystal field is small with respect to the separation of different groups of multiplets, such as the ^4F and ^4P levels of Cr^{+++}. And in the other, it is large with respect to these intervals. They are, respectively, the weak and strong field limits. The most general, that is, the intermediate case is also the most common and will be considered later. As a rough, but by no means rigorous criterion, it is found that Pauling's "ionic" complexes correspond to weak fields and his "covalent" complexes to strong fields (81).

TABLE I

STATES AND ENERGIES IN WEAK OCTAHEDRAL FIELDS

Ground State of Free Ion	Octahedral States and Energies (Dq units)
$d:{}^2D$	$^2T_{2g}(-4)$, $^2E_g(+6)$
$d^2:{}^3F$	$^3T_{1g}(-6)$, $^3T_{2g}(+2)$, $^3A_{2g}(+12)$
$d^3:{}^4F$	$^4A_{2g}(-12)$, $^4T_{2g}(-2)$, $^4T_{1g}(+6)$
$d^4:{}^5D$	$^5E_g(-6)$, $^5T_{2g}(+4)$
$d^5:{}^6S$	$^6A_{1g}(0)$
$d^6:{}^5D$	$^5T_{2g}(-4)$, $^5E_g(+6)$
$d^7:{}^4F$	$^4T_{1g}(-6)$, $^4T_{2g}(+2)$, $^4A_{2g}(+12)$
$d^8:{}^3F$	$^3A_{2g}(-12)$, $^3T_{2g}(-2)$, $^3T_{1g}(+6)$
$d^9:{}^2D$	$^2E_g(-6)$, $^2T_{2g}(+4)$

$A_{1g}=\Gamma_1$; $A_{2g}=\Gamma_2$; $E_g=\Gamma_3$; $T_{1g}=\Gamma_4$; $T_{2g}=\Gamma_5$

In the case of weak fields, therefore, the quantum numbers L, S of the central ion retain their validity, and we have only to determine how a given L, S level is split by the octahedral field $\mathcal{V}_O$, without considering the possibility that different levels may interact under it. To a first approximation, the ground states of the aquo complexes of the ions in the first transition period may be treated in this way. The qualitative features are again most simply obtained by group theory. The symmetries of the states into which the ground level of the free ion splits are listed in Table I, at the bottom of which Bethc's notation for representations of the octahedral group is also given, to facilitate comparisons (15). These states are listed in increasing order of excitation. Thus the ground state of the Cr^{+++} ion in the field is the orbitally nondegenerate ${}^4A_{2g}$ state. Following such term symbols, numbers are given parenthetically which assess their energies due to $\mathcal{V}_O$ in terms of the parameter Dq for a single electron: the Cr^{+++} ion, for example, is stabilized by some $12Dq$ in the limit of weak fields.[8]

It is interesting to find how many electrons, on the average, occupy the t_{2g} and how many the e_g orbitals in the ground states of the crystalline ions. In general, these numbers will not be integers since, in weak fields, the tendency for maximum field stabilization—namely, the preferential filling of the t_{2g} shell—is subordinate to the forces of correlation between the electronic motions. The expectation values for these occupation numbers are most simply determined from Table I, by inspection: in order that the two electrons

[8] There are many different ways of obtaining these energies. The most general and also perhaps the most simple method for the weak field case is to observe that—within a manifold spanned by the $(2L+1)$ orbital states of given L, S, M_S—the matrix representation of $\mathcal{V}_O$ is directly proportional to that of the operator.

$$L_x^4 + L_y^4 + L_z^4 - \frac{1}{5} L(L+1)[3L(L+1) - 1]$$

$$= \frac{1}{20}[35 L_z^4 - 30L(L+1)L_z^2 + 25L_z^2 - 6L(L+1) + 3L^2(L+1)^2]$$

$$+ \frac{1}{8}[(L_x + iL_y)^4 + (L_x - iL_y)^4].$$

(There appears to be a misprint in the table of operator equivalents given by Bleaney & Stevens (22), where the factor $\frac{1}{8}$ preceding the last expression is given as $\frac{1}{5}$.) It is obtained on replacing x_i, y_i, z_i by L_x, L_y, L_z in the symmetric, homogeneous form of writing the angular dependence of a typical term of $\mathcal{V}_O$: $(x_i^4+y_i^4+z_i^4-\frac{3}{5}r_i^4)$, taking account of the fact that whereas x_i, y_i commute, L_x, L_y do not. This, and analogous results for different coupling schemes and different potentials, may be proved using Wigner coefficients (91) or the results of section 9^3 in Condon & Shortley (28). The proportionality constant is now determined by evaluating a single element, chosen for its simplicity, in both matrices. For example, the eigenfunction for the $L_z=3$, $S_z=3/2$ component of d^3, ${}^4F$ may be written $|d\sigma d\pi_+ d\delta_+|$, where $|\cdots|$ denotes a normalized and antisymmetrized product, and the absence of bars over the orbital symbols means that these are to be taken with α spin. The diagonal element for this row in the matrix $\mathcal{V}_O$ is just $+3Dq$, and the corresponding element in its operator equivalent is $+9$.

of ${}^3F(T_{1g})$ should distribute themselves between these two levels and have energy $-6Dq$ under $\mathcal{V}_O$, it is clear that on the average 9/5 must be found in the stable t_{2g} shell and only $\frac{1}{5}$ in the unstable e_g shell. For three, four, $\cdots$, eight electrons, these pairs of occupation numbers are (3, 0), (3, 1), (3, 2), (4, 2), (24/5, 11/5), (6, 2) respectively. In the excited states, of course, electron density is transferred from the t_{2g} to the e_g levels. Indeed, such numbers are entirely equivalent to prescriptions of energies.

[The general theory is described by Bethe (15), Van Vleck (98), Schlapp & Penney (86), Jordahl (52). Techniques are given by Wigner (108), Kramers (65), Condon & Shortley (28), Von der Lage & Bethe (107), Hellwege (43, 44), Stevens (91), Bell (14) and Judd (57).]

Applications of weak field theory.—The earliest applications of crystal field theory, in its weak field limit, were to the magnetic properties of salts. These are discussed later. At present we consider topics of greater chemical interest.

The lower electronic transitions may tentatively be assigned to excitations from the ground crystalline component to the higher ones. Thus the 5800Å system of the $[V(H_2O)_6]^{+++}$ ion (40) would be ascribed to the process ${}^3T_{1g} \rightarrow {}^3T_{2g}$. And similarly for the other ions. (In general, however, it is better to use intermediate field theory for such assignments.) These identifications enable one to determine the appropriate Dq in each case; for the V^{+++} aquo complex (d^2), we have $8\ Dq \approx 17{,}200\ \text{cm}^{-1}$.

Values of Dq, determined in this manner, have been used in order to rationalize the behaviour of the heats of hydration of the divalent and trivalent ions of the iron group. On the basis of the spherically symmetric potential $\mathcal{V}_R$ alone, one would expect these heats to increase steadily on going from left to right across the periodic table—for the radii of the free ions, before solvation, decrease uniformly in this order. However, from this standpoint, the values for $Mn^{++}({}^6S)$ and also for $Fe^{+++}({}^6S)$ are anomalously low. This is no longer surprising when the additional stabilization conferred on the aquo complexes by the octahedral component $\mathcal{V}_O$ of the field is taken into account. As an inspection of Table I will show, it is only the d^5 and d^{10} ions which do not benefit from this. On calculating the energies of stabilization from spectral data, the irregularities in the observed heats are almost quantitatively removed [Orgel (73)].

A related explanation has been given for the behaviour of the effective crystal radii of these ions. Empirically, it seems clear that these do not decrease monotonically as a function of atomic number, but show maxima at configurations d^4, d^5 and possibly again at d^9, d^{10}. Now it is clear that, for a given number of electrons, the more that are in the t_{2g} levels, the more closely may the ligands approach and, therefore, the smaller the crystal radius of the ion; occupation of the unstable e_g orbitals will imply a localization of electron density in the region of the approaching ligands which are accordingly repelled. The lowest crystal levels of the d^3, d^4 and d^5 configurations correspond successively to the occupation of the e_g orbitals by 0, 1 and 2 electrons, and a ready explanation for the crystal radii is obtained [Santen & Wieringen (84)].[9]

In general, the longer wave-length bands, and in particular those which are attributed to transitions between states arising from the same free ion level, are very broad. The factors contributing to these widths may be described at this point, though they will often be applied in contexts where neither the weak field approximation, nor the assumption of perfect octahedral symmetry, are valid. The appropriate generalization of the argument is not difficult and given elsewhere (77). If we ignore the broadening attributable to thermal fluctuations, the main part of the spectrum will consist of a symmetric vibrational progression—superposed, probably, on the 1-0 band of an asymmetric vibration (vide infra)—and will in many respects resemble that of a diatomic molecule. Clearly if the vibrational constants (by which we mean to include the equilibrium internuclear distances) in the ground and excited states are the same, a single sharp set of lines would be observed. However, the greater the difference in these constants, the broader will the overall spectrum appear. Now the factor responsible for differences in the interactions of a given ion with its ligands is just the relative number of its e_g and t_{2g} electrons or, equivalently, the stabilization of the ion by the field in units of Dq. Bands for which both upper and lower states have the same number of t_{2g} electrons are therefore expected to be sharp, but those for which this number changes may be very broad.

For example, the lowest quartet-quartet transition of the $[Cr(H_2O)_6]^{+++}$ ion, which corresponds to $^4A_{2g} \rightarrow {}^4T_{2g}$ lies at 5800Å and has a half width of some 1625 cm^{-1} (53). The difference in crystal field stabilization between the two states is $10Dq$, which corresponds to the excitation $(t_{2g})^3 \rightarrow (t_{2g})^2(e_g)$. By contrast, certain doublet states, which arise from essentially the same ground $(t_{2g})^3$ configuration, are also seen in the spectrum. The bands, or rather lines, corresponding to these as the upper states of spin-forbidden systems in crystals are so sharp that Zeeman splittings of only 1 cm^{-1} or so are readily observed (88, 89). [These factors were recognized by Freed (34) and Van Vleck (102), but first correctly formalized by Orgel (77).]

Strong fields.—We now consider the case where the crystal field splitting parameter Dq is large enough so that the torques exerted on the orbital motions of the electrons are determined by the lattice site symmetry, rather than by the correlative interactions between them. The forces tending to repel the electrons of the central ion from the ligands cause a complete "Paschen-Back" effect which destroys the L, S coupling at the ion. In place of these quantum numbers, we now have the occupation numbers of the t_{2g} and e_g orbitals. The lowest group of states arises by filling first the t_{2g} shell

[9] The argument may also be given in terms of octahedral stabilization to which it is equivalent. One can see why the heats of hydration will not follow the Dq analysis exactly: for the different effective crystal radii will also imply different contributions from $\mathcal{V}_R$ to the stability of the aquo complexes. The qualitative explanation is unaffected, however.

and only then, when six electrons have been accommodated, the less stable e_g levels. Such an assignment or crystal configuration will subsume a number of different states in cases of degeneracy, whose splitting is due to the coulomb repulsions of the electrons. These same repulsions are, of course, responsible for the separations of the different groups of multiplets in the spectrum of the free ion. In Table II, we list the lowest configurations and the crystal states into which they are split by the correlative interactions.

TABLE II

STATES AND ENERGIES IN STRONG OCTAHEDRAL FIELDS

Atomic Configuration	Crystal Configuration	Symmetry of Ground State	Octahedral Stabilization	Promotional Energy	Symmetries of Other States
d	(t_{2g})	$^2T_{2g}$	$4Dq$	0	—
d^2	$(t_{2g})^2$	$^3T_{1g}$	$8Dq$	$3F_2-15F_4$	$^1A_{1g}$, ^1E_g, $^1T_{2g}$
d^3	$(t_{2g})^3$	$^4A_{2g}$	$12Dq$	0	^2E_g, $^2T_{1g}$, $^2T_{2g}$
d^4	$(t_{2g})^4$	$^3T_{1g}$	$16Dq$	$6F_2+145F_4$	$^1A_{1g}$, ^1E_g, $^1T_{2g}$
d^5	$(t_{2g})^5$	$^2T_{2g}$	$20Dq$	$15F_2+275F_4$	—
d^6	$(t_{2g})^6$	$^1A_{1g}$	$24Dq$	$5F_2+255F_4$	—
d^7	$(t_{2g})^6(e_g)$	^2E_g	$18Dq$	$7F_2+105F_4$	—
d^8	$(t_{2g})^6(e_g)^2$	$^3A_{2g}$	$12Dq$	0	$^1A_{1g}$, ^1E_g
d^9	$(t_{2g})^6(e_g)^3$	^2E_g	$6Dq$	0	—

It will be noticed that, in every case, there is only one state of maximum multiplicity which, by Hund's principle, is therefore also the ground state. The multiplicities and symmetries predicted for these ground states are the same, both in the weak and strong field limits, for all cases except d^4, d^5, d^6 and d^7. This is due to the fact that, accidentally so to speak, the occupation numbers in the former limit differ appreciably from those in the latter only for these numbers of electrons. The total spin in the strong field limit is also uniformly lower—just as the octahedral stabilization is higher—in these cases. Thus the strong field limit refers to those complexes which, on the basis of his magnetic criterion, Pauling calls covalent (81).

The spectra of complexes whose ligands, such as the cyanide ion, give rise to strong fields are easily classified. In the d^2, d^3, d^4 and d^8 cases, there will be a number of spin-forbidden bands involving no change in crystal configuration; these bands should therefore be sharp, as is observed in complexes of Cr^{+++} (54). However, in the other cases, namely d^5, d^6, and d^7, the upper states of different multiplicity all involve at least one excitation $(t_{2g})^{-1}(e_g)$. Their spin-forbidden bands should therefore be broad; such a system has been described for diamagnetic Co^{+++} (54). Bands involving no spin change must all be accompanied by changes in crystal configuration and these are therefore uniformly broad.

The evaluation of the appropriate energies in the limit of strong fields is quite straight-forward.[10] These are listed for the ground states of the ions in the crystal, referring them to the ground states of the free ions. In general, the crystalline states in the strong field case are not stationary states of the free ion, but rather linear combinations of these, called valence states; their energies, apart from the crystal stabilization, are called promotional energies. Values for the parameters F_2, F_4 are obtained from the spectra of the free ions (28). Values of Dq, which depend both upon the nature of the ligands and the ions with which they complex, are generally determined semi-empirically. From Table II, quantitative criteria may be given to distinguish the weak from the strong field cases: for example, in the former limit, the field stabilization for d^6 ions is only $4Dq$; in the latter, however, it is as much as $24Dq$, but this has been achieved at the expense of some $5(F_2+51F_4)$ in promotional energy. When the last quantity is much greater than $20Dq$, the complex is paramagnetic and conversely when it is much less the complex is diamagnetic. Very frequently, the two crucial quantities are of comparable magnitude and these situations are treated in terms of intermediate coupling [Van Vleck (101), Howard (17), Kotani (60) and particularly Orgel (73, 76) who lists values of the parameters F_2, F_4, also Jørgensen (55) and Ballhausen (8). Valence states are reviewed by Moffitt (70)].

Fields of intermediate strength.—Although, as has been indicated, there may be discontinuities in the symmetries and multiplicities of the ground states of ions on going from the weak to the strong field case, the crystal field theory treats the transition domain in terms of the same physical picture—without explicitly invoking any discontinuous change in bond type. We shall illustrate the appropriate analysis, which deals with the intermediate coupling, for a particularly simple case, namely that of two d electrons, or of two holes in the completed d^{10} shell. For simplicity, we confine our attention to triplet states.

The free ions with configurations d^2 or d^8 have two triplet states, ^3F and ^3P, of which the former is more stable than the latter by some $15(F_2-5F_4)$ in both cases. Under the influence of a weak field, ^3F yields $^3A_{2g}$, $^3T_{2g}$ and $^3T_{1g}$ components, whereas the ^3P state is not split and transforms under $^3T_{1g}$. Since these are the only triplets that occur, we see that even in the strong field limit, the $^3A_{2g}$ and $^3T_{2g}$ states are correctly represented by their pure ^3F functions: these may also be written $(e_g)^2$ or $(t_{2g})^6(e_g)^2$ and $(t_{2g})(e_g)$ or $(t_{2g})^5(e_g)^3$ respectively. However, $^3F(T_{1g})$ and $^3P(T_{1g})$ may interact in stronger fields. That they will, in fact, do so is clear because the crystal field energy of the ground $^3T_{1g}$ state of d^2 is different in the two limits. A continuous change, with increasing Dq, will occur during which the $^3F(T_{1g})$ and $^3P(T_{1g})$ functions

[10] The octahedral field contributes terms which are just obtained by counting the numbers of electrons in the t_{2g} and e_g orbitals. The intra-atomic energies may be given in terms of Slater's F_2, F_4 parameters by the same methods as are used in atomic spectroscopy (28)—for the set of orbitals t_{2g}, e_g is just another listing of the usual atomic d orbitals.

are progressively contaminated by each other until finally, in the strong field limit, the eigenfunctions arise instead from the pure configurations $(t_{2g})^2$ or $(t_{2g})^4(e_g)^4$ and (t_{2g}) (e_g) or $(t_{2g})^5(e_g)^3$.[11]

We see that the transition between the strong and weak field cases—and therefore the compromise between correlative interactions, which are the basis of Hund's principle, and the tendency to minimize the crystal field energy—may be handled by the theory. And that a singlet state of d^6 should cross over several triplets and then the quintuplet ^5D state, the $^5T_{2g}$ component of which it usurps as the ground state of, for example, most Co^{+++} complexes, is no longer surprising and requires no special postulates other than a strong crystalline potential $\mathcal{V}_O$.[12]

In order to illustrate the degree of accuracy that may be obtained in this way, let us consider the spectrum of the complex $[Ni(NH_3)_6]^{++}$ ion. This has three bands, whose maxima lie at 10,700, 17,500 and 28,100 cm^{-1} respectively (56). The only reasonable assignment for the first is $^3A_{2g}\rightarrow{}^3T_{2g}$, which has an excitation energy of $10Dq$ and does not depend on the separation of the ^3F and ^3P states of the free ion—some 15,800 cm^{-1} (72). We take $Dq = 1070$ cm^{-1}, therefore, together with the last figure and calculate the positions of the other two bands, which are both $^3A_{2g}\rightarrow{}^3T_{1g}$ transitions. In this way,[11] we find 17,600 cm^{-1} for the second band and 30,300 cm^{-1} for the third. The agreement with the experimental values seems to flatter somewhat the basic premises of the theory.

[Finkelstein & Van Vleck (33), Abragam & Pryce (4), Orgel (73, 75), Tanabe & Sugano (93, 94). For a very striking confirmation of the theory see Jørgensen (56), but also Owen (79).]

The spectrochemical series.—Many regularities in the spectra of octahedral complexes had been noted some time before the advent of crystal field theory. It was observed, for example, that the successive replacement

[11] The diagonal elements of the $^3T_{1g}$ part of the energy matrix, whose rows and columns refer to the weak field case, are clearly $[E^0 \mp 6\ Dq]$ and $[E^0+15(F_2-5F_4)]$, where E^0 is just the energy of the free ^3F ion in the spherically symmetric field $\mathcal{V}_R$. The off-diagonal elements are found by noting that, as F_2, F_4 tend to zero, the eigenvalues must become those of the strong field limit, namely $\mp\ 8Dq$ and $\pm\ 2Dq$ respectively. With an appropriate phase choice, they are therefore $4Dq$ in both cases. Since $Dq>0$ by our convention, the upper signs refer to the d^2 configuration and the lower to d^8.

[12] The analysis we have given could equally well be developed by starting with the configurational functions and making the transition in the other direction; the two methods are, of course, entirely equivalent. It is interesting to note that the average numbers n_t, $n_e = n - n_t$ of electrons in the t_{2g}, e_g orbitals are easily obtained for a particular state K of d^n by plotting its energy E_K as a function of Dq. Since E_K, for a given Dq, assumes a stationary value with respect to the coefficients of, say, the various configurational functions, we have

$$dE_K/d[Dq] = (\partial E_K/\partial[Dq])_{n_t} = -4n_t + 6n_e.$$

This formula is useful in discussing band widths [Orgel (76)].

of I^- by Br^-, Cl^- and H_2O as the ligands surrounding a particular cation shifted the maxima of prominent band systems to progressively shorter wave lengths. Similar effects were noticed with other ligands, and it appeared possible to classify these on the basis of their spectral effects. This led to the construction of the spectrochemical series [Fajans (32), Tsuchida (95, 96)]:

$$I^- < Br^- < Cl^- \lesssim OH^- < F^- \lesssim H_2O < \text{pyridine} < NH_3 < \text{ethylene diamine} < CN^-.$$

Minor variants arose, depending mainly on the particular cation in conjunction with which they were developed. But the general trend of the ligands appeared to be roughly independent of the central ion. The spectra concerned, it should be remarked, were those diffuse systems which we now classify as spin-allowed. With certain conspicuous exceptions, the qualitative features of a given cation could be systematized very satisfactorily in this way. Those ligand replacements which did not fall into line were accompanied by radical changes in stereochemistry or a change in multiplicity of the central ion.

By applying the spectrochemical series to, for example, the d^9 (Cu^{++}, say) case, we see that we may rationalize the ligand sequence as one of increasing Dq. The excitation energy is $10Dq$ here, and the hypsochromic shifts must therefore correspond to increases in this parameter. More generally, let us suppose that the series represents a succession of increasing octahedral fields $\mathcal{V}_O$ when the relevant ligands are associated with a given central ion, whatever this may be. Let us further agree to consider only spin-allowed band systems. Then it may be proved that unless the multiplicity of the ground state changes, the spectrum of a given octahedral ion is shifted to shorter wave lengths on replacing its ligands by others lying above it in the spectrochemical series.

This "existence theorem" is easily established for magnetically normal complexes whose symmetry and spin for the ground states are those of the weak field limit.[13] In those cases, namely d^4, d^5, d^6 and d^7, where the ground state multiplicity is different in the strong and weak field limits, two quite distinct types of spectral behaviour may occur. For although the energy levels all move continuously as functions of Dq, as soon as, for example, the

[13] For the configurations d^4, d^5 and d^6, as well as for d and d^9, there is no distinction between the weak and strong field cases, since the free ions have only one, namely their ground state of maximum multiplicity. Their lower and upper states both arise from this term, and their separation is linear in Dq. The remaining ions, d^2, d^3, d^7 and d^8, have two states of maximum multiplicity, of which ^3F or ^4F is the ground state and ^3P or ^4P the excited state. In the weak field limit, the P states are not split but yield T_{1g} states. Their interactions with the ground states of d^2, d^7 for intermediate or strong fields can only depress the latter, so that the intervals, which increase linearly in both limits, increase more rapidly with Dq in the intermediate domain. And finally, for d^3 and d^8, whose ground states are the same in fields of all sizes, we have only to note that the excitation energies increase linearly in both limiting cases.

$^1A_{1g}$ state of Co^{+++} supplants its $^5T_{2g}$ state as the ground state, the selection rules change. No longer are the prominent bands in the spectrum due to quintuplet-quintuplet transitions, but singlet-singlets take their place. It is an entirely different set of states that is now "visible." A simplifying factor in this magnetically "anomalous" case arises from the fact that, in the strong field limit, the ground state is the only state of its total spin which arises from the lowest configuration. In this limit, or near it—and for present purposes all these "anomalous" cases fall into this category—, it is clear that an increase in Dq removes the ground state yet further from all other states of the same multiplicity, which must arise from excited crystal configurations: the existence of the spectrochemical series has been demonstrated for such complexes also.

It may be observed that the spectral shifts of different bands are governed by the same factor which accounts for their breadths. Analytically, both depend upon the quantity $d(\Delta E)/d(Dq)$.[12] Thus the absence of narrow spin-allowed bands and the existence of the spectrochemical series are closely related. Indeed, it is clear that the broader the band, the more it should be shifted on replacing the ligands by others near them in the spectrochemical series. However, it is only for the spin-allowed bands that the direction of the displacement is always the same. $d(\Delta E)/d(Dq)$ may be zero, or even negative and reverse this, for the spin-forbidden bands.

[General theories of colour have been given mainly by Orgel (73, 75, 77, 78) and by Tanabe & Sugano (93, 94). Other references are classified by the ions they consider most specifically: Ti^{+++}: Ilse & Hartmann (48); V^{+++}: Ilse & Hartmann (49); Cr^{+++}: Hartmann & Kruse (42); Mn^{++}: Schläfer (87); Co^{+++}: Basolo, Ballhausen & Bjerrum (12); Co^{++}: Abragam & Pryce (4), Ballhausen & Jørgensen (7); Ni^{++}: Hartmann & Fischer (41), Ballhausen (6, 10), Jørgensen (56); Cu^{++}: Abragam & Pryce (3), Bjerrum, Ballhausen & Jørgensen (17), Ballhausen (5).]

Spin-orbit coupling.—Systems with perfect octahedral symmetry are rarely, if ever, discovered. However, in order to understand magnetic effects in solids, it is most illuminating to see how the spin-orbit coupling changes on going from the free ion, first to a site subject to strong fields of this ideal kind.

The interaction of the spins of a system of d electrons with their own orbital motions in a central force field may be written

$$\mathcal{H}_\zeta = \zeta_d \sum_i \mathbf{l}_i \cdot \mathbf{s}_i,$$

where ζ_d is a characteristic of the radial part of the d orbitals and of the central field; l_i, s_i are supposed, throughout, to be expressed in units of $\hbar$. The same form for the spin-orbit interaction operator will also be used for the crystalline ions, though this involves an additional assumption. When L, S are good quantum numbers—e.g., for the free ion in Russell-Saunders coupling and for the weak field limit—, it is easy to show that $\mathcal{H}_\zeta = \lambda \mathbf{L} \cdot \mathbf{S}$, where λ, whose value depends on the particular L, S term considered, is a simple fraction of ζ_d; ζ_d is always positive, though λ need not be so. For the free ion, $\boldsymbol{S}$ and $\boldsymbol{L}$ couple to form a resultant J, of course, where the different values of $J=|L-S|, \cdots, L+S$ represent the individual multiplets. Their energies due to the spin-orbit coupling are given by the Landé rule, namely, $\frac{1}{2}\lambda[J(J+1)-L(L+1)-S(S+1)]$. Values of λ are generally around ± 100 cm^{-1}, though they are appreciably larger for Ni^{++} (-335 cm^{-1}) and for Cu^{++}(-828 cm^{-1}). Moreover, for ground states, it changes sign on going from a less to a more than half-completed shell: "holes" have positive charges.

It is easy to see by symmetry arguments which crystal states will show first-order multiplet separations. The orbital angular momentum operator transforms under T_{1g} of the group O_h. Accordingly, a term whose orbital symmetry is Γ, say, will exhibit properties of angular momentum and therefore propensities for spin-orbit coupling only if the direct product representation $\Gamma \times T_{1g} \times \Gamma$ contains the identity A_{1g}. It follows that the multiplet structure is expected to be pronounced only for those octahedrally coordinated ions with either T_{1g} or T_{2g} ground states.[14]

In order to illustrate the very striking changes which may occur on first tying $\boldsymbol{L}$ to the octahedral lattice and only afterwards allowing it to couple with $\boldsymbol{S}$, we take the case of a single d electron. The free ion has states ${}^2D_{3/2}$, ${}^2D_{5/2}$ with spin-orbit energies of $-(3/2)\zeta_d$ and ζ_d respectively. However, on placing this in the strong field, we obtain the sixfold ${}^2T_{2g}$ and fourfold ${}^2E_g$ levels. We label the possible functions $t_{2g}{}^0$, $\bar{t}_{2g}{}^0$, $e_g{}^a$ and so on, the bars over orbital symbols denoting that they are to be taken with β spin, and their absence showing α spin.

Remembering that $m=m_s+m_l$ is a good quantum number for the free atom, we see that the only interactions which occur under $\mathcal{H}_\zeta$, within the ${}^2T_{2g}$ manifold, are of $t_{2g}{}^0$ with $\bar{t}_{2g}{}^-$ and of $\bar{t}_{2g}{}^0$ with $t_{2g}{}^+$. The four ${}^2E_g$'s do not interact with each other at all. As a result, no splitting and no contribution to the energy of the ${}^2E_g$ state occurs—as had been anticipated. On the other hand, it is found that the ${}^2T_{2g}$ state is split into a doubly-degenerate level, ${}^2T_{2g}{}^7$ say, and a fourfold level ${}^2T_{2g}{}^8$. Their energies are ζ_d and $-\frac{1}{2}\zeta_d$ respec-

[14] The qualitative aspects of the multiplet structure are most easily seen by group theory also. In order to account for systems with an odd number of electrons, the octahedral group is replaced by the corresponding "double group," which contains the element R—a rotation through 2π about an arbitrary axis. In this way the representations of the simple group are augmented by certain double-valued representations, under which systems of odd spin transform. The symbol ${}^2T_{2g}$ now signifies a direct product representation of the double group, whose elements will act simultaneously on both space and spin variables; its irreducible components are called Γ_7 and Γ_8 by Bethe (15). Thus the degenerate ${}^2T_{2g}$ term is split into two levels by $\mathcal{H}_\zeta$, one of which is doubly-degenerate, namely ${}^2T_{2g}{}^7$, the other being the fourfold ${}^2T_{2g}{}^8$ state. On the other hand, the fourfold degeneracy of the ${}^2E_g$ term is not removed, its Γ_8 character being specified by the symbol ${}^2E_g{}^8$. It may be noticed that ${}^2E_g{}^8$ may interact, though it is not split, in second-order of $\mathcal{H}_\zeta$, with ${}^2T_{2g}{}^8$, but not with ${}^2T_{2g}{}^7$.

tively, so that ${}^2T_{2g}{}^8$ is the lower multiplet; its eigenfunctions are

$$t_k{}^8 = \frac{1}{\sqrt{3}}(\sqrt{2}t_{2g}{}^0 + \bar{t}_{2g}{}^-), \qquad t_l{}^8 = \frac{1}{\sqrt{3}}(-\sqrt{2}\bar{t}_{2g}{}^0 + t_{2g}{}^+),$$
$$t_m{}^8 = \bar{t}_{2g}{}^+, \qquad t_n{}^8 = t_{2g}{}^-.$$

Those for the upper ${}^2T_{2g}{}^7$ state are just

$$t_a{}^7 = \frac{1}{\sqrt{3}}(t_{2g}{}^0 - \sqrt{2}\bar{t}_{2g}{}^-), \qquad t_b{}^7 = \frac{1}{\sqrt{3}}(\bar{t}_{2g}{}^0 + \sqrt{2}t_{2g}{}^+).$$

The splittings are shown in Figure 1.

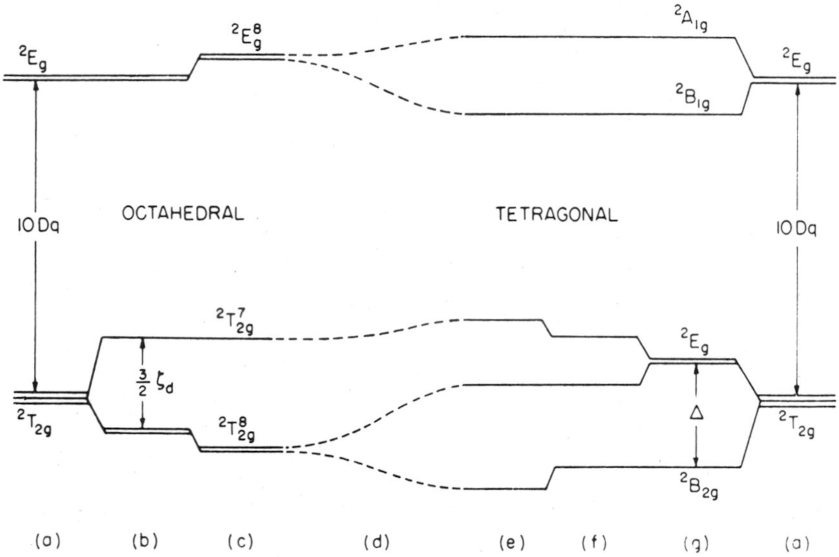

FIG. 1. Term splittings for a single *d* electron. (Each line represents a spin doublet.)
(a) Levels in a strong octahedral field,
(b) ...... with weak spin-orbit coupling,
(c) ...... with stronger spin-orbit coupling.
(g) Levels in strong octahedral and weak tetragonal fields,
(f) ...... with weaker spin-orbit coupling,
(e) ...... with spin-orbit coupling and tetragonal fields of comparable size.
(d) ...... Intermediate domain.

To cover the intermediate range, where spin-orbit coupling becomes stronger and eventually supplants the octahedral field in directing L, it may be noted that each of the ${}^2T_{2g}$ states interacts with one of the ${}^2E_g$ states under $\mathcal{H}_\zeta$—and these are the only further interactions that occur. Since the four off-diagonal elements concerned are all equal to $\sqrt{(3/2)}\cdot\zeta_d$, no splitting, but only a uniform displacement of all the four states occurs until, finally, when ζ_d overwhelms Dq, the free ion functions are recovered and J is a good quantum number again. To the first order in the parameter $(\zeta_d/10Dq) = \xi$, say, the perturbed functions for the lowest multiplet are typified by

$$t_m^8 - \sqrt{(3/2)} \cdot \xi e_g^a.$$

The second-order energy correction for this level is just $-(3/2)\zeta_d^2/10Dq$.

The most interesting features of these results appear when we consider the Zeeman splittings. In a magnetic field $\boldsymbol{H}$, the additional term

$$\mathcal{H}_M = \beta \boldsymbol{H} \cdot \sum_i (l_i + 2S_i) = \beta \boldsymbol{H} \cdot (\boldsymbol{L} + 2\boldsymbol{S})$$

must be added to the Hamiltonian. The separations within a given multiplet due to the field along some axis κ, say, may be expressed in the form $g_\kappa \beta H_\kappa$, where β is the Bohr magneton and g_κ the so-called spectroscopic splitting factor. For free ions, it is well known that

$$g = 1 + \frac{J(J+1) - L(L+1) + S(S+1)}{2J(J+1)},$$

whenever Russel-Saunders coupling holds. The g-factors are also isotropic—that is independent of the orientation of κ—for ions in octahedral sites. However, their values differ widely from those given by the Landé formula.

Evaluating $(L_z + 2S_z)$ using the unperturbed eigenfunctions for ${}^2T_{2g}{}^7$, we see that its levels are separated by just $2\beta H_z$ and therefore that $g = 2$. Similarly, the ${}^2E_g$ levels are split into two doublets of energy $\pm\beta H_z$, magnetic dipole transitions being allowed only between and not within them; once again, $g = 2$. The "spin-only" values are obtained in both cases. However, the ${}^2T_{2g}{}^8$ quartet is not split by the field at all, using zeroth-order functions.[15] Its g-factor differs from zero only by virtue of its "high frequency" terms, coupling it to the ${}^2E_g$ state. Under these conditions, it gives an undisplaced doublet, within which magnetic dipole transitions are forbidden, and two singlets to which they may make transitions. Since the latter have excitation energies $\pm 4\xi\beta H_z$, we see that $g = 4\xi$ for ${}^2T_{2g}{}^8$; because ζ_d is of order 100 cm^{-1}, but $10Dq$ is around 20,000 cm^{-1}, the g-factor is in any case rather small. By comparison, the g-factors are 4/5 and 6/5 for the ${}^2D_{3/2}$ and ${}^2D_{5/2}$ levels of the free ion.

It is easy to treat the complementary case of a single hole in a completed d^{10} shell. Here the free ${}^2D$ ion has an inverted multiplet structure, with ${}^2D_{5/2}$ $(-\zeta_d,\ g = 6/5)$ and ${}^2D_{3/2}([3/2]\zeta_d,\ g = 4/5)$. By definition, ζ_d and Dq remain positive, as does ξ therefore. Now ${}^2E_g$ is the ground state of the crystal and, in a perfectly octahedral field, it is four-fold degenerate. If we include cor-

[15] The way in which the precessional motion of $\boldsymbol{L}$ is anchored to the crystal lattice in these complexes has an illuminating parallel in diatomic spectroscopy. In the NO molecule, for example, the orbital angular momentum is partially quenched by the strong axial fields, so that it has no perpendicular components in, say, its ${}^2\Pi$ ground state. The lower ${}^2\Pi_{1/2}$ component also has a vanishing g-factor.

rections for its interactions with the $^2T_{2g}{}^8$ state, its eigenfunctions are typified by

$$e_g{}^b + \sqrt{(3/2)} \cdot \xi t_k{}^8.$$

In the presence of a magnetic field, the degeneracy is completely removed; the displacements are $\pm\beta H_z$ and $\pm\beta(1+4\xi)H_z$ respectively. By the selection rule $\Delta m = \pm 1$, we see that the corresponding g-factors are 2 and $2(1+4\xi)$.

To summarize, it is only the ions whose ground states in the crystal have orbital symmetry T_{1g} or T_{2g} which show large multiplet separations. The appropriate g-factors, and therefore also the bulk magnetic properties, will differ widely from those of the free ion. Other states, whose orbital degeneracy, if any, is only of the twofold E_g variety, show at best only second-order splittings. Their g-factors are close to the "spin-only" value of 2, or rather 2.0023 as given by quantum electrodynamics.

Fields of Lower Symmetry

There are, of course, many complexes whose symmetry is not octahedral. Indeed, the more sensitive one's measurements, the farther must one go to find a perfectly octahedral system. Even when the six ligands are identical, there are many factors tending to destroy their most regular arrangement about the cation. Crystals show, to a greater or lesser degree, both optical and magnetic anisotropies. This makes it very important to discuss the effects of departures from octahedral symmetry.

Tetragonal fields.—Consider, first, the case of a complex containing four identical ligands arranged in a square, at whose centre the cation is located. Immediately above and below it, two other ligands are placed: $[MX_4Y_2]$. The symmetry is now that of a distorted octahedron, namely D_{4h}. Examples of this are *trans*-$[Co(NH_3)_4Cl_2]^+$, or just $[H_2O \cdots Cu(H_2O)_4 \cdots OH_2]^{++}$ where two waters of crystallization are less closely coordinated to the central ion than are the other four. The most general potential is easily seen to be $(\mathcal{V}_R+\mathcal{V}_O+\mathcal{V}_T)$, where

$$\mathcal{V}_T = \sum_i \{\mathcal{Y}_{20}(\theta_i, \phi_i)\mathcal{R}_2(r_i) + \mathcal{Y}_{40}(\theta_i, \phi_i)\mathcal{R}_4'(r_i)\}.$$

We shall study the complex as if it were octahedral, say $[MX_6]$, and then add the tetragonal perturbation $\mathcal{V}_T$, whose effect is generally small with respect to that of $\mathcal{V}_O$. $\mathcal{V}_T$ will have the sense either of attracting or of repelling electrons in the neighbourhood of the ligands Y. If Y lies below X in the spectrochemical series, we should expect this to be an attraction.

Now suppose we have just one electron in this field. It is clear that the unstable e_g orbitals are split by $\mathcal{V}_T$ and span a_{1g} and b_{1g} of D_{4h}; their energies may be written

$$\epsilon[e_g{}^a(a_{1g})] = \epsilon_0 + 6Dq - 2Ds - 6Dt,$$
$$\epsilon[e_g{}^b(b_{1g})] = \epsilon_0 + 6Dq + 2Ds - Dt;$$

Ds, Dt are splitting parameters for $\mathcal{Y}_{20}$, $\mathcal{Y}_{40}$ respectively, in the same way that

Dq is for $\mathcal{V}_O$. Similarly, the lower three levels split into a doubly-degenerate e_g pair and a single b_{2g} level of D_{4h}:

$$\epsilon[t_{2g}^{\pm}(e_g)] = \epsilon_0 - 4Dq - Ds + 4Dt$$
$$\epsilon[t_{2g}^{0}(b_{2g})] = \epsilon_0 - 4Dq + 2Ds - Dt.$$

For the moment, we suppose Ds, Dt to be large with respect ot ζ_d.[16]

To see how these splittings affect a polyelectronic system, let us consider the spectra of diamagnetic Co^{+++} complexes, which may be treated with reasonable accuracy using the strong field approximation. The ground $^1A_{1g}$, $(t_{2g})^6$ state is non-degenerate, but the excited $^1T_{1g}$ and $^1T_{2g}$ states arising from the configuration $(t_{2g})^5(e_g)$ are split by the field: formally, $^1T_{1g}$ becomes $^1A_{2g}+{}^1E_g$ and $^1T_{2g}$ becomes $^1B_{2g}+{}^1E_g$ on going from O_h to D_{4h}. Owing to the electronic repulsions, $^1T_{2g}$ lies some $16(F_2-5F_4)$ above the lower excited $^1T_{1g}$ state.[17] Using the individual orbital levels which we have already evaluated, it may then be shown that the separation of the levels arising from the $^1T_{1g}$ state is

$$E[^1T_{1g}(A_{2g})] - E[^1T_{1g}(E_g)] = (35/4)Dt,$$

whereas the similar splitting of the $^1T_{2g}$ levels follows

$$E[^1T_{2g}(E_g)] - E[^1T_{2g}(B_{2g})] = 6Ds - (5/4)Dt.$$

Since the longer wavelength band system, $^1A_{1g}\rightarrow{}^1T_{1g}$, is of some further interest, we note that if Y, say Cl^-, is lower down in the spectrochemical series than X, say NH_3, then $Dt>0$, for $\mathcal{V}_T$ will stabilize the a_{1g} orbital. In this case, we should expect the band to be split, the shorter wave-length component being due to an upper state of symmetry A_{2g} in D_{4h}.

Spectroscopic splitting factors.—In order to illustrate the magnetic behaviour of ions in fields of distorted octahedral symmetry, we refer again to the case of a single d electron and to a single hole. To avoid further notational difficulties, we assume the symmetry to be tetragonal (D_{4h}) and throughout label orbitals according to their species symbols in this lower group.

We first examine the Ti^{+++} ion in such fields, taking $(5Dt-3Ds)=\Delta$, say, to be positive. Of the three orbital levels arising from the octahedral $^2T_{2g}$ term, b_{2g} is the lowest, the doubly-degenerate e_g levels lying some Δ above it. If ζ_d is very small, it is clear that spin-orbit coupling is only ap-

[16] It may be noticed that since the effect of all odd-order potentials vanishes in a representation containing only d electrons, certain curious features emerge. Without including higher configurations, it will never be possible to localize electrons on one side of an axis containing the cation, since the functions concerned are all of even parity. Thus the replacement of only one of the ligands of $[MX_6]$ by Y would have to be described by means of the same potential $\mathcal{V}_T$—or rather, one half of it—that determines the perturbation on forming *trans*-$[MX_4Y_2]$. A molecule of symmetry C_{4v}, like $[MX_5Y]$, would appear to have the higher symmetry D_{4h}.

[17] This is shown most easily by evaluating the energies of the two unique configurations $(e_g)^4(b_{2g})(b_{1g})$, $^1T_{1g}(A_{2g})$ and $(e_g)^4(b_{2g})(a_{1g})$, $^1T_{2g}(B_{2g})$. The concomitant splittings under $\mathcal{V}_T$ are obtained by an obvious extension of this.

preciable in the upper two levels (see figure). However, it will more often be the case that ζ_d and Δ are comparable (around 100 cm^{-1}), though both are small with respect to $10Dq$. We now find that b_{2g} mixes with $\bar{e}_g^-$, the energy difference between the two unperturbed levels being $(\Delta+\frac{1}{2}\zeta_d)$; their off-diagonal element in $\mathcal{H}_\zeta$ is $-(1/\sqrt{2})\zeta_d$. Similarly, $\bar{b}_{2g}$ and e_g^+ mix. The resulting eigenfunctions for the perturbed ground state, a Kramers spin doublet, take the forms $(Cb_{2g}+D\bar{e}_g^-)$, $(C\bar{b}_{2g}-De_g^+)$, where C and D are simple functions of ζ_d, Δ. Finding expectation values for (L_z+2S_z) and for (L_x+2S_x), their Zeeman splittings may be written $g_{\parallel}\beta H_z$ and $g_{\perp}\beta H_x$ respectively, for magnetic fields directed along, or normal to the four-fold axis. Explicitly,

$$g_{\parallel} = -1 + 3(\Delta + \tfrac{1}{2}\zeta_d)/\sqrt{(\Delta + \tfrac{1}{2}\zeta_d)^2 + 2\zeta_d^2},$$

$$g_{\perp} = 1 + (\Delta - \tfrac{3}{2}\zeta_d)/\sqrt{(\Delta + \tfrac{1}{2}\zeta_d)^2 + 2\zeta_d^2}.$$

These reduce to 0, of course, when $\zeta_d \gg \Delta$, for the tetragonal perturbation is then suppressed and the effective symmetry is octahedral again: the levels reduce to components of $^2T_{2g}{}^8$. On the other hand, if $\Delta \gg \zeta_d$, $g_{\parallel}$ and $g_{\perp}$ both become 2, and the orbital angular momentum is completely quenched in this state. The alum, $CsTi(SO_4)_2 \cdot 12H_2O$, which has been studied by paramagnetic resonance (23), has trigonal, rather than tetragonal symmetry; but the qualitative splitting of the octahedral $^2T_{2g}$ state is exactly the same, as are the theoretical expressions for $g_{\parallel}$, $g_{\perp}$ to this approximation (18). Experimentally, it is found that $g_{\parallel}=1.25$ and $g_{\perp}=1.14$. The low figures are to be expected on the basis of the simple formulae, but their actual values require a more detailed discussion, in which second-order corrections arising from the excited octahedral ^2E_g state are taken into account.

Let us now take a single d hole, say, a Cu^{++} ion in a field corresponding to an octahedron which is elongated along the four-fold axis. Such fields occur in the Tutton salts and probably also in the hydrated ion. They stabilize electrons in the a_{1g} orbital, particularly, and accordingly repel holes there. We shall talk of orbital levels as occupied by holes. The lowest octahedral ^2E_g state is therefore split, b_{1g} being more stable than a_{1g}.[18] Unlike the case considered above, however, both b_{1g} and a_{1g} levels are magnetically inert so far as their orbital parts are concerned, and they do not interact under $\mathcal{H}_\zeta$. As in the octahedral case, these levels only interact in second-order of $\xi=\zeta_d/10Dq$, namely with the octahedrally excited $^2T_{2g}{}^8$ state (since both λ and $\mathcal{V}_O$ change sign when applying to holes, ξ remains positive). The perturbed eigenfunctions turn out to be

$$b_{1g}' = b_{1g} + \xi\left(b_{2g} + \frac{1}{\sqrt{2}}\bar{e}_g^-\right), \qquad \bar{b}_{1g}' = \bar{b}_{1g} - \xi\left(\bar{b}_{2g} - \frac{1}{\sqrt{2}}e_g^+\right)$$

for the lowest doublet; the e_g, b_{2g} orbitals prefaced by the parameter ξ refer

[18] The possibility that a_{1g} lies below b_{1g}—which seems physically unreasonable—is excluded by these measurements for, on this basis, we should have found $g_{\parallel}=2$, $g_{\perp}=2(1+3\xi)$.

to tetragonal levels of the excited octahedral $^2T_{2g}$ state. Correspondingly, the g-factors are

$$g_{\|} = 2(1 + 4\xi), \qquad g_{\perp} = 2(1 + \xi).$$

From low temperature resonance spectra (20), their values are found to be 2.4, 2.1 respectively (on correcting for slight rhombic distortions), in admirable accord with the theory if $\xi = 0.05$. On a quantitative basis, and this is rather typical, the agreement is less satisfactory: taking $\lambda = -828$ cm^{-1}, one would predict the maximum of the optical spectrum to lie around 16,600 cm^{-1} (3). In fact, the broad system observed in the spectrum of the aqueous ion has a maximum at 12,500 cm^{-1} (17), and shows no bands in the expected region. It appears, therefore, that in order to account for the g-factors one must assume that the spin-orbit coupling constant ζ_d ($= -\lambda$, here) is diminished from its free ion value to only about 625 cm^{-1} for the crystalline ion.

Paramagnetic resonance work serves to confirm the simple generalizations with which we began this particular discussion. Salts with $^{4,3}A_{2g}$ ground states, like Cr^{+++} and Ni^{++}, have g-factors reasonably close to the spin-only value ($g = 2 \pm 36\lambda/25Dq$)—closer than Cu^{++} for which ζ_d is remarkably large. Owing to the change from normal to inverted multiplet structure for the free ions, g is less than 2.0023 for Cr^{+++} and V^{++}, but larger for Ni^{++}. Moreover, since the orbital ground states are non-degenerate, unlike the ^2E_g state of Cu^{++}, the g-factors are very nearly isotropic. Contrastingly, Co^{++} salts whose crystal states arise from $^4T_{1g}$, show their orbital angular momenta very strongly. Similarly, since they are susceptible to, for example, tetragonal distortions, their g-factors are highly anisotropic ($g_{\|} = 5.5$ to 7, $g_{\perp} = 2.5$ to 3.5).

Studies of these microwave spectra yield very detailed information about the lowest energy levels. Higher order spin-orbit interactions together with intraionic spin-spin coupling terms—factors which had been predicted on the basis of low temperature susceptibilities and relaxations, for example, by Van Vleck and Penney (99)—are seen very strikingly. They help in the interpretation of the specific heats at very low temperatures. The resonance spectra also often show well defined hyperfine structures. Quite apart from the useful information about nuclear moments that these provide, illuminating if occasionally puzzling facts are disclosed about the more intimate details of the electronic eigenfunctions. These are of equal interest both for valence theory and for the theory of atomic spectra.

[Sources for magnetic theory are Van Vleck (97, 106) and Casimir (26); for paramagnetic relaxation Gorter (38) and Cooke (29); for paramagnetic resonance Abragam & Pryce (2), Bleaney & Stevens (22) and Bowers & Owen (24).]

Spectral intensities.—By many standards, the electronic absorption bands responsible for the colours of the transition-metal complexes would be regarded as very weak: the broad, spin-allowed systems which we consider in these paragraphs have molar extinction coefficients which rarely exceed 100. On the other hand, many of these complexes appear to have centres of symmetry and, since no change in parity accompanies the excitations, these should be strongly forbidden, orbitally, in electric dipole radiation fields. It is perhaps even surprising that the oscillator strengths f of the main bands in the hexammines of Cr^{+++} and Co^{+++}, for example, should be as high as the 10^{-3} or so which are observed. The source of this intensity has been discussed several times, and two sets of explanations have been put forward. On the one hand, it is accepted that the symmetry has a centre of inversion (holohedral field), in which case the absorption probabilities are attributed either to magnetic dipole or to electric quadrupole radiation. Alternatively, it is supposed that a predominantly holohedral field is subject to hemihedral distortions arising either from the intimate properties of the static crystal field or from the vibrations of the ligands.

The former set are easily rejected in the present case, though they may be important in the spectra of rare earth salts. We observe that the orbital magnetic moments transform under T_{1g} of O_h, whereas the electric quadrupole tensor spans irreducible representations E_g and T_{2g} of this group. Now, for example, the 4750 Å ($^1A_{1g} \rightarrow {}^1T_{1g}$) and 3400 Å ($^1A_{1g} \rightarrow {}^1T_{2g}$) systems of $[Co(NH_3)_6]^{+++}$ are almost equally strongly absorbing (67). $^1A_{1g} \rightarrow {}^1T_{1g}$ is allowed for magnetic dipole, but forbidden in electric quadrupole radiation fields; the opposite holds for the second system: neither effect may separately account for the intensity of both band systems. Both are also too weak. Thus, on the basis of the magnetic dipole hypothesis, we should predict oscillator strengths of $(h\nu/6m_ec)\Sigma|L|^2$, where m_e is the electronic mass and $\Sigma|L|^2$ refers to the sum of the squares of the changes in angular momentum (in units of $\hbar^2$), on going from $^1A_{1g}$ to each of the $^1T_{1g}$ components. For diamagnetic cobaltic complexes of this kind, $\Sigma|L|^2 = 24$ in the strong field approximation, so that the computed f is about 2×10^{-5}. The observed 10^{-3} is quite inexplicable on such a basis and, as it turns out, even less so for the $^1A_{1g} \rightarrow {}^1T_{2g}$ system assuming electric quadrupole radiation.

The considerably stronger absorption of bands in, for example, *cis*-$[Co(en)_2Cl_2]^+$ than in their *trans*-analogue (12) shows that static hemihedral fields are a major factor in promoting the intensities of bands. But this seems unlikely to be the only factor. The oscillator strengths of the halogenopentammines of Cr^{+++} and Co^{+++} are larger, but not inordinately so than those of the corresponding hexammines (67, 68): it is a little difficult to believe that slight but permanent hemihedral distortions of the crystal fields in the latter should have as profound an effect as the substitution of an NH_3 by a Cl^- ligand.[21]

It appears, therefore, that the influence of asymmetric vibrational modes must account for the observed intensities of symmetric ions like the hexammines. The mechanics may be seen as follows. Suppose Q represents a normal coordinate for a particular asymmetric vibrational mode of odd parity. The instantaneous crystal field has the form $(\mathcal{V}_R + \mathcal{V}_0 + Q\mathcal{V}_Q)$, where $\mathcal{V}_Q$ is made up of odd-order harmonics, and represents the momentary distortion of what we have supposed to be a perfect octahedron. Since the electronic motions respond almost immediately to the nuclear motions, we see that the eigenfunctions for the ionic electrons will be functions of Q which take the form

$$\psi(Q) = \phi_g + cQ\phi_u.$$

ϕ_g represents the electronic eigenfunction in the undistorted field, its parity being g. In the asymmetric nuclear configuration, $\mathcal{V}_Q$ causes ϕ_g to interact with excited states ϕ_u of odd parity—particularly those corresponding to one-electron jumps $(3d)^{-1}(4p)$. For simplicity, let us suppose Q to represent the only vibrational mode, the corresponding harmonic oscillator functions being χ_p, χ_q for quantum numbers p, q respectively. Then the electronic and vibrational (therefore, "vibronic") state of the octahedral complex is represented by a function $\Psi_p = \psi\chi_p$. If the force constants for this vibration are the same both in the electronic ground state N and in an excited state, V say, then the probability of a transition from Np to Vq is proportional to

$$\left| \boldsymbol{R}_{Np,Vq} = \iint \Psi_{Np}{}^* \boldsymbol{R} \Psi_{Vq} dv dQ \right|^2,$$

where $\boldsymbol{R}$ represents the electric dipole operator. We find that

$$\boldsymbol{R}_{Np,Vq} = \int (c_N{}^*\phi_{Nu}{}^*\boldsymbol{R}\phi_{Vg} + c_V\phi_{Ng}{}^*\boldsymbol{R}\phi_{Vu})dv \int \chi_{Np}{}^*Q\chi_{Vq}dQ,$$

from which we see that transitions are allowed only if the vibrational states p, q differ by one quantum. At low temperatures, therefore, the spectrum will correspond to the 1-0 band, the 0-0 band being absent. It may be noted that the vibrational integral is proportional to the square root of p or q, whichever is the greater. At higher temperatures, therefore, the stronger 2-1, 3-2, · · · bands appear, corresponding to larger amplitudes of vibration and greater hemihedral distortions, and the total intensity of these electronically forbidden systems increases.[19]

Of the various vibrations the complex may undergo, we should expect the asymmetric stretching modes to be more effective in breaking down the electronic selection rules than the weaker bending modes. There is only one such mode for an octahedral $[MX_6]$ complex, namely τ_{1u} (small Greek letters are used to denote vibrational species which transform under the representation whose symbol, like T_{1u}, is obtained from them by using Roman capitals). The selection rules are found by applying symmetry arguments to the vibronic functions Ψ, rather than to the ϕ's alone. $\chi_0(\gamma)$ belongs to the A_{1g} species and $\chi_1(\gamma)$ to the Γ species. Thus $\phi(^1A_{1g})\chi_0(\tau_{1u})$ combines with $\phi(^1T_{1g})\chi_1(\tau_{1u})$ in electric dipole radiation fields (T_{1u}) because $(A_{1g} \times A_{1g})$

[19] In general, the force constants for Q will not, of course, be the same in both states, nor may we neglect other modes; however, a more exhaustive treatment leads to very similar conclusions and encounters difficulties of notation alone.

$\times T_{1u}\times(T_{1g}\times T_{1u})$ contains the identity representation. Unfortunately, for purposes of assignments, the former also combines with $\phi(^1T_{2g})\chi_1(\tau_{1u})$, which means that one cannot in this way identify the upper state symmetries unambiguously.

Let us, as a specific illustration of these principles, consider the diamagnetic complex ions *trans*-$[Co(en)_2Cl_2]^+$ and *trans*-$[Co(en)_2Br_2]^+$. These are essentially tetragonal and almost identical, spectroscopically, with their tetrammine analogues (12). The perchlorates crystallize in the monoclinic system, whose *a*-axes are almost coincident with the fourfold axis of the complex ions, and whose *b*-axes are almost normal to this axis. As we have seen, the octahedral $^1A_{1g}\rightarrow{}^1T_{1g}$ system should be split into two bands, that to the red having an upper ^1E_g state, and that to the blue having $^1A_{2g}$. The crystals have been studied using polarized light, and two bands are in fact observed in the requisite regions (e.g., at 6200 Å and 4300 Å in the one case). The longer wave-length system absorbs more strongly with the electric vector along the *a*-axis, though it also absorbs appreciably when this vector lies along the *b*-axis. The band to the blue, on the other hand, disappears almost entirely when the electric vector lies along the *a*-axis, but is reasonably strong for the other orientation (109). The electric vector along the *a*-axis transforms like A_{2u}, whereas when it is normal to this it transforms under E_u of D_{4h}. Applying our analysis, we see that $^1A_{2g}$ can be seen in absorption only with the simultaneous excitation of α_{1u} (*a*-axis) or of ϵ_u (*b*-axis) vibrational modes. Transitions to the ^1E_g state, however, may occur with the aid of ϵ_u (*a*-axis) or of α_{1u}, α_{2u}, β_{1u} or β_{2u} (*b*-axis) excitations. As it happens, there are no vibrations of species α_{1u}, neither stretching nor bending modes, but there are stretching modes of species α_{2u} and ϵ_u. Thus $^1A_{1g}\rightarrow{}^1A_{2g}$ should appear for *b*-axis polarization alone, whereas $^1A_{1g}\rightarrow{}^1E_g$ should be active for both orientations (all upper states being derived from the same octahedral $^1T_{1g}$ term, of course). It follows that the upper state of the shorter wave-length system must be $^1A_{2g}$, in complete agreement with predictions based on the spectrochemical series.

[Herzberg & Teller (46), Van Vleck (102), Broer, Gorter & Hoogschagen (25), Ballhausen (9), Ballhausen & Moffitt (11).]

The Jahn-Teller effect.—As a general rule, it is found that those transition-metal ions whose ground states in octahedral fields are orbitally degenerate, form complexes whose symmetry is lower, say tetragonal or rhombic. This may be due to many causes: crystal packing considerations for the complex within a lattice, repulsions between neighbouring ligands and so forth. However, it is also to be expected on very general grounds which seem to be of some importance to their chemistry. According to a theorem of Jahn and Teller, when the orbital state of an ion is degenerate for symmetry reasons, its ligands will experience forces distorting the octahedron until they assume a configuration, both of lower symmetry and of lower energy, thereby resolving the degeneracy.

The physical background for this may be seen very readily. Consider once more the case of a single d electron, or of a single hole, in a perfectly octahedral field. Let the two ligands on the z-axis move an equal distance S away from each other along their line of centres. At the same time, each of the other four ligands may be supposed to approach the central ion a distance $\frac{1}{2}S$, so that the net displacement is orthogonal to a totally symmetric expansion of the octahedron. The electrons of the ion are now in a crystal field which, to a first approximation, may be written $(\mathcal{V}_R+\mathcal{V}_O+S\mathcal{V}_T)$, where $\mathcal{V}_T$ describes the effect of the tetragonal distortion, per unit distance, and is independent of S. We have already seen that the degeneracy is at least partially removed in such fields. The d, ${}^2T_{2g}$ state splits into an orbital singlet and doublet, whereas d^9, ${}^2E_g$ splits into two orbital singlets; moreover, the splitting is linear in S. The total energy of the ground state therefore also contains terms linear in S and cannot represent a stable configuration, not being minimal with respect to this parameter: the ligands may be regarded as experiencing a set of non-totally symmetric forces which destroy their octahedral arrangement. Ultimately, an equilibrium configuration of lower symmetry is attained, whose geometry is determined by including quadratic and higher powers of the displacement coordinates in the energy expression.[20]

A rather curious but characteristic effect arises occasionally, not only for complexes of this kind but also in quite different molecular systems such as the cyclopentadienyl radical. It may show itself when the displacement S is a member of a degenerate set of symmetry coordinates for the undisplaced configuration. The case we have chosen illustrates this since S, together with another displacement S' span E_g of the octahedral group. In the latter, the two ligands on the y-axis approach, and the two on the x-axis move away from the central ion equal distances $\sqrt{3}\cdot S'$. Let us write $S=R\cos\chi$, $S'=R\sin\chi$. Then, by considering quadratic and higher terms in the energy, it is sometimes found that whereas R is strongly determined by the minimum energy requirement, the "phase" χ is only weakly determined—that is, it is almost a cyclic coordinate which does not appear in the potential energy expression. There exists a continuum of configurations, to each of which there

[20] It is easily seen that if the symmetry of an orbitally degenerate state is Γ in any particular point group, it will contain in its energy terms linear in all those displacements with symmetry given by $[\Gamma^2]$, which is the representation spanned by the symmetrical product of Γ with itself. This will always contain the identity representation, and therefore a totally symmetric displacement; however, the corresponding term in the energy vanishes since we suppose the system to have equilibrated with respect to such displacements. It is not obvious that to each electronically degenerate species Γ, there actually exists at least one non-totally symmetric displacement of the molecule or complex whose species is contained in $[\Gamma^2]$. Nevertheless, Jahn and Teller show this to be the case for all possible realizations of all possible point groups, unless all the atom are colinear. The theorem is extended by Jahn to cover spin-orbit coupling and the two-valued representation theory appropriate to systems of odd spin. It is only the Kramers doublets, of course, that cannot be split by such ligand displacements.

corresponds a single value of χ, all with much the same energy. We may regard the ligands as undergoing a but feebly hindered "internal rotation." What is curious is not, of course, that there is a number of different configurations of the same minimum energy—clearly there are three equivalent tetragonal distortions of an octahedron—but that these should interconvert so readily. The effect has been adduced to account for certain paramagnetic relaxation times as well as for the isotropic g-factors found for copper fluosilicate (19, 21).

[Jahn & Teller (50), Jahn (51). Magnetic applications by Van Vleck (103, 104, 105) and by Abragam & Pryce (1). Hydrocarbons by Liehr & Moffitt (66).]

Optical rotatory dispersion.—One of the methods that have been employed most usefully in studying the inorganic chemistry of complexes is that of optical rotation. In many cases, particularly polydentate ligands such as ethylene diamine are sufficiently firmly attached to the central ion that they are displaced only slowly in solutions of their complexes. Different optical isomers may therefore be separated and often racemize only rather slowly. Their existence offers an attractive method of assigning certain electronic transitions.

The crystal field for a dihedral complex ion like d-$[Co(en)_3]^{+++}$ may be written in the form $(\mathcal{V}_R+\mathcal{V}_O+\mathcal{V}_D)$, where $\mathcal{V}_D$ is a small hemihedral potential. On reflection or inversion, the field becomes $(\mathcal{V}_R+\mathcal{V}_O-\mathcal{V}_D)$, which refers to the l-enantiomorph. Let us assume that the fluctuating part of $\mathcal{V}_D$, due to the vibrations, may be ignored in assessing rotational propensities; this seems reasonable, since it can affect these properties only as a result of electrical or mechanical anharmonicities. $\mathcal{V}_D$ therefore represents the static perturbation on the octahedral field for the d-isomer. Its most important function in the present context is to add to the state representative ϕ_g, of even parity, small admixtures from higher states ϕ_u, which are odd and correspond to one-electron excitations like $(3d)^{-1}(4p)$. As a result, there is a non-vanishing electric dipole moment $\boldsymbol{R}_{NV}$ associated with, for example, the transition $N\rightarrow V$ even for the static, non-vibrating complex. However, this is small—of order v say—, since it arises entirely from the hemihedral perturbation. By contrast, the magnetic dipole moment $\beta \boldsymbol{L}_{NV}=\beta(\boldsymbol{L}_{VN})^*$ for the same transition depends almost entirely on the octahedral part of the field alone; according to the selection rules for magnetic dipole radiation fields, it is of order β, or much smaller, say $\iota\beta$.

Now the rotational strength of the transition $N\rightarrow V$ is proportional to the imaginary part of the quantity $\boldsymbol{R}_{NV}\cdot \boldsymbol{L}_{VN}$ (27). It refers to the optical activity of assemblies of randomly oriented molecules—say, in solution. Being a pseudo-scalar, it remains invariant under rotations but changes sign on inversion: it depends for its value on the specific d- or l-juxtaposition of the hemihedral and holohedral crystal fields. We see immediately that $\boldsymbol{R}_{NV}\cdot \boldsymbol{L}_{VN}$ is either of order βv or of the lower order $\beta\iota v$. For the two transitions ${}^1A_{1g}\rightarrow{}^1T_{1g}$ (magnetically allowed) and ${}^1A_{1g}\rightarrow{}^1T_{2g}$ (magnetically forbidden) of $[Co(en)_3]^{+++}$, it will be βv in the former case and only $\beta \iota v$ in the latter. The one should therefore show appreciably stronger optical rotatory powers than the other. This difference offers a ready method for distinguishing between the two transitions, and others like them. It seems difficult, in fact, to find any other experimental evidence which has as direct a bearing on such assignments[21] [Moffitt (71)].

Rare earths.—Many of the historical developments in crystal field theory are due to studies of the rare earths. However, from a chemical point of view, they are less interesting than the transition metals: their properties are a more or less continuous reflection of the lanthanide contraction, with the anticipated singularities at or near the xenon structure, the half-completed and the completed $4f$ shell. The $4f$ electrons are shielded from their ligands by the completed $5s$ and $5p$ shells, so that the crystal splittings of given ionic L, S, J terms are only about 100 cm^{-1}. Accordingly the coarser features of their generally rather sharp electronic spectra are characteristic more of the free ions than of their environment. Broader systems are observed, but these are not well characterized and probably arise from transitions of $4f$ electrons to $5d$ or other states. The crystal field, which generally has rather low symmetry (C_{3h} or even C_{3v}), is of much less importance than spin-orbit coupling, since the multiplet separations are around 1000 cm^{-1}.

Spectral studies of their salts in the infra-red, visible and ultra-violet are of interest since they yield, effectively, the atomic line spectra of the ions in a specific oxidation state at concentrations equivalent to enormous path lengths in the vapour phase. Their paramagnetic resonances are also important in physics, since the associated hyperfine structures enable one to determine nuclear spins and moments. Chemically, perhaps, the main interest of these studies would lie in getting information about local site symmetries in crystals.

[Becquerel (13), Bethe (16), Penney & Schlapp (82), Kynch (61), Spedding (90), Penney & Kynch (83), Giesekus (35, 36), Hellwege & Hellwege (45), Bleaney & Stevens (22), Satten (85), Judd (58).]

Discussion and Conclusion

Before discussing the relation of crystal field theory to other methods of approaching molecular problems, we might mention that several attempts have been made to calculate the octahedral splitting factor. On the basis of assumed models for the ligands, $\mathcal{V}_O$ is explicitly constructed and the necessary quadratures are performed to obtain Dq. If each ligand is replaced by an appropriately situated point dipole, Dq may generally be fitted using reason-

[21] One might expect the hemihedral fields in the tris-ethylene diamine complex of Co^{+++} to give an estimate of the probable static distortions of the cobaltic hexammine, which have been postulated as a possible source of the intensity of band systems in the latter complex. The optical rotatory dispersion data enable one to estimate $|\boldsymbol{R}_{NV}(\text{static})|^2$ for the dihedral complex (71). It turns out be to too small by a factor of fifty or so to account for the observed oscillator strengths in the hexammine.

able values of the parameters [Ilse & Hartmann (48), Bjerrum, Ballhausen & Jørgensen (17)]. However, if it is acknowledged that the electrons of the ligands are continuously distributed in space, these will overlap the electronic densities due to the central cation and orthogonality problems arise [Kleiner (59)]. Indeed, each successive refinement of such models appears to raise more questions than it answers. At present, we feel that any successes achieved by means of one model are no more encouraging than the failures of another are disheartening.

The two approaches most commonly applied to chemical problems of electronic structure are those of electron-pairing and of molecular orbitals. Both are, at first glance, very different from crystal field theory: in dealing with particular systems, even concerning transition-metal complexes, one or other of them is often used in preference to it; the metal carbonyls [Pauling (81)] and the recently developed bis-cyclopentadienyl compounds [Dunitz & Orgel (30, 31), Moffitt (69)] are examples that spring to mind. But in some ways, all three methods are remarkably similar, as was first pointed out by Van Vleck (100, 101). In the conventional valence theories, the e_g orbitals of the central cation have a tendency to accept electrons from the ligands, acting as donors—in Pauling's theory, it is these d orbitals that are used in forming his d^2sp^3 octahedral hybrids and also his square planar dsp^2 orbitals. When the tendency is strong (covalent complexes), the ground state of the ion must leave the relevant e_g orbitals vacant for bond formation. It is easy to see that this corresponds, so far as the cationic electrons are concerned, to the strong field limit of the theory we have been describing. Alternatively, when the tendency is weak (ionic complexes), the correlative interactions between the electrons overwhelm it and the electron assignments are made first according to the Hund principle of maximum multiplicity; this is equivalent to the limit of weak crystal fields [see also, Orgel (73, 76)].

The comparison of the different methods suggests certain modifications in crystal field theory. Thus, when e_g orbitals are regarded as occupied in excited states, these should be viewed as the antibonding counterparts of bonding orbitals already filled by electrons originally on the ligands. In so far as the latter electrons may be considered to migrate onto the central cation, so may electrons in the former be regarded as in part migrating to the ligands. Owen (79) has adopted this picture to rationalize the fact that spin-orbit coupling constants for free ions are rather larger than the values they apparently assume in crystal fields. It is also formally possible for the t_{2g} orbitals to participate in some double-bonding [Pauling (81)]. Stevens (92) has attributed the chlorine hyperfine structure observed in the paramagnetic resonance spectrum of the $[IrCl_6]^{=}$, (d^5, $^2T_{2g}$) complex to just such an effect. More strikingly, perhaps, it is well known that complexes absorb very strongly at shorter wavelengths in the ultraviolet, occasionally obscuring the relatively weak bands due to transitions within the d shell, whose properties we have described [see a review by Orgel (74), for example]. Neither the free ions nor the ligands separately show strong absorptions in this region, so that the concomitant transitions must arise from their special relation to each other in complexes. These so-called charge-transfer spectra indicate very clearly that ligands are not content with the passive roles ascribed to them by crystal field theory. It is remarkable, however, that the presence of the excited states of charge-transfer bands do not exert a greater influence on the lower frequency behaviour of transition-metal complexes. Indeed, the extraordinary success of crystal field theory suggests that their associated motions are separable, to high approximation, from those motions primarily responsible for the magnetic and optical behaviour of the complexes with which it deals.

It will be a long time before a method is developed to surpass in simplicity, elegance and power that of crystal field theory. Within its extensive domain, it has provided at very least a deep qualitative insight into the behaviour of a many-electron system. No other molecular theory, to our knowledge, has provided so many useful numbers which are so nearly correct. And none has a better immediate prospect of extending its chemical applications.

LITERATURE CITED

1. Abragam, A., and Pryce M. H. L., *Proc. Phys. Soc. (London)*, [A]**63**, 409 (1950)
2. Abragam, A., and Pryce M. H. L., *Proc. Roy. Soc. (London)*, [A]**205**, 135 (1951)
3. Abragam, A., and Pryce M. H. L., *Proc. Roy. Soc. (London)*, [A]**206**, 164 (1951)
4. Abragam, A., and Pryce, M. H. L., *Proc. Roy. Soc. (London)*, [A]**206**, 173 (1951)
5. Ballhausen, C. J., *Kgl. Danske Videnskab. Selsk. Mat. Fys. Medd.*, **29**, No. 4 (1954)
6. Ballhausen, C. J., *Kgl. Danske Videnskab. Selsk. Mat. Fys. Medd.*, **29**, No. 8 (1955)
7. Ballhausen, C. J., and Jørgensen, C. Klixbüll, *Acta Chem. Scand.* **9**, 397 (1955)
8. Ballhausen, C. J., and Jørgensen, C. Klixbüll, *Dan. Mat. fys. Medd.*, **29**, No. 14 (1955)
9. Ballhausen, C. J., *Acta Chem. Scand.*, **9**, 821 (1955)
10. Ballhausen, C. J., *Rec. trav. Chim.*, **75** (1956)
11. Ballhausen, C. J., and Moffitt, W., *J. Chem. Phys.*,
12. Basolo, F., Ballhausen, C. J., and Bjerrum, J., *Acta Chem. Scand.*, **9**, 810 (1955)
13. Becquerel, J., *Z. Physik*, **58**, 205 (1929)
14. Bell, D. G., *Rev. Mod. Phys.*, **26**, 311 (1954)
15. Bethe, H., *Ann. Physik*, [5]**3**, 133 (1929)
16. Bethe, H., *Z. Physik*, **60**, 218 (1930)
17. Bjerrum, J., Ballhausen, C. J., and Jørgensen, C. Klixbüll, *Acta Chem. Scand.*, **8**, 1275 (1954)
18. Bleaney, B., *Proc. Phys. Soc. (London)*, [A]**63**, 407 (1950)
19. Bleaney, B., and Ingram, D. J. E., *Proc. Phys. Soc. (London)*, [A]**63**, 408 (1950)
20. Bleaney, B., Bowers, K. D., and Ingram, D. J. E., *Proc. Phys. Soc. (London)*, [A]**64**, 758 (1951)
21. Bleaney, B., and Bowers, K. D., *Proc. Phys. Soc. (London)*, [A]**65**, 667 (1952)
22. Bleaney, B., and Stevens, K. W. H., *Repts. Progr. Phys.*, **16**, 108 (1953)
23. Bleaney, B., Bogle, G. S., Cooke, A. H., Duffus, R. J., O'Brien, M. C. M., and Stevens, K. W. H., *Proc. Phys. Soc. (London)*, [A]**68**, 57 (1955)

24. Bowers, K. D., and Owen, J., *Repts. Progr. Phys.*, **18**, 304 (1955)
25. Broer, L. J. F., Gorter, C. J., and Hoogschagen J., *Physica*, **11**, 231 (1945)
26. Casimir, H. B. G., *Magnetism* (Cambridge University Press, Cambridge, England, 93 pp. 1940)
27. Condon, E. U., *Rev. Mod. Phys.*, **9**, 432 (1937)
28. Condon, E. U., and Shortley, G. H., *Theory of Atomic Spectra* (Cambridge University Press, Cambridge, England, 441 pp., 1935)
29. Cooke, A. H., *Repts. Progr. Phys.*, **13**, 276 (1950)
30. Dunitz, J. D., and Orgel, L. E., *Nature*, **171**, 121 (1953)
31. Dunitz, J. D., and Orgel, L. E., *J. Chem. Phys.*, **23**, 954 (1955)
32. Fajans, K., *Naturwissenschaften*, **11**, 165 (1923)
33. Finkelstein, R., and Van Vleck, J. H., *J. Chem. Phys.*, **8**, 790 (1940)
34. Freed, S., *Phys. Rev.*, **38**, 2122 (1931)
35. Giesekus, H., *Ann. Physik* [6]**8**, 350 (1951)
36. Giesekus, H., *Ann. Physik* [6]**8**, 373 (1951)
37. Gorter, C. J., *Phys. Rev.*, **42**, 437 (1932)
38. Gorter, C. J., *Paramagnetic Relaxation* (Elsevier, Amsterdam, 127 pp., 1947)
39. Hartmann, H., and Schläfer, H. L., *Z. phys. Chem.*, **197**, 116 (1951)
40. Hartmann, H., and Schläfer, H., L., *Z. Naturforsch.*, **6a**, 754 (1951)
41. Hartmann, H., and Fischer-Wasels, H., *Z. phys. Chem.*, **4**, 297 (1955)
42. Hartmann, H., and Kruse, H., H., *Z. phys. Chem.*, **5**, 9 (1955)
43. Hellwege, K. H., *Ann. Physik*, [6]**4**, 127 (1948)
44. Hellwege, K. H., *Ann. Physik*, [6]**4**, 357 (1948)
45. Hellwege, A. M., and Hellwege, K. H., *Z. Physik*, **130**, 549 (1951)
46. Herzberg, G., and Teller, E., *Z. phys. Chem.*, **21B**, 410 (1933)
47. Howard, J. B., *J. Chem. Phys.*, **3**, 813 (1935)
48. Ilse, F. E., and Hartmann, H., *Z. phys. Chem.*, **197**, 239 (1951)
49. Ilse, F. E., and Hartmann, H., *Z. Naturforsch.*, **6a**, 751 (1951)
50. Jahn, H. A., and Teller, E., *Proc. Roy. Soc. (London)*, [A]**161**, 220 (1937)
51. Jahn, H. A., *Proc. Roy. Soc. (London)*, [A]**164**, 117 (1938)
52. Jordahl, O. M., *Phys. Rev.*, **45**, 87 (1934)
53. Jørgensen, C. Klixbüll, *Acta Chem. Scand.*, **8**, 1495 (1954)
54. Jørgensen, C. Klixbüll, *Acta Chem. Scand.*, **8**, 1502 (1954)
55. Jørgensen, C. Klixbüll, *Acta Chem. Scand.*, **9**, 116 (1955)
56. Jørgensen, C. Klixbüll, *Acta Chem. Scand.*, **9**, 1362 (1955)
57. Judd, B. R., *Proc. Roy. Soc. (London)* [A]**227**, 552 (1955)
58. Judd, B. R., *Proc. Roy. Soc. (London)* [A]**232**, 458 (1955)
59. Kleiner, W. H., *J. Chem. Phys.*, **20**, 1784 (1952)
60. Kotani, M., *J. Phys. Soc. (Japan)*, **4**, 293 (1949)
61. Kynch, G. J., *Trans. Faraday Soc.*, **33**, 1402 (1937)
62. Kramers, H. A., *Proc. Acad. Sci. (Amsterdam)*, **32**, 1176 (1929)
63. Kramers, H. A., *Compt. rend.*, **191**, 784 (1930)
64. Kramers, H. A., *Proc. Acad. Sci. (Amsterdam)*, **33**, 959 (1930)
65. Kramers, H. A., *Proc. Acad. Sci. (Amsterdam)*, **33**, 953 (1930)
66. Liehr, A. D., and Moffitt, W., *J. Chem. Phys.*, (In press)
67. Linhard, M., *Z. Elektrochem.*, **50**, 224 (1944)
68. Linhard, M., and Weigel, M., *Z. anorg. u. allgem., Chem.*, **266**, 49 (1951)
69. Moffitt, W., *J. Am. Chem. Soc.*, **76**, 3386 (1954)
70. Moffitt, W., *Rep. Prog. Phys.*, **17**, 173 (1954)

71. Moffitt, W., *J. Chem. Phys.*, (In press)
72. Moore, C., *Circular of the National Bureau of Standards*, 467, Vol. I (1949), II (1952)
73. Orgel, L. E., *J. Chem. Soc.*, 4756 (1952)
74. Orgel, L. E., *Quart. Rev.*, **8**, 422 (1954)
75. Orgel, L. E., *J. Chem. Phys.*, **23**, 1004 (1955)
76. Orgel, L. E., *J. Chem. Phys.*, **23**, 1819 (1955)
77. Orgel, L. E., *J. Chem. Phys.*, **23**, 1824 (1955)
78. Orgel, L. E., *J. Chem. Phys.*, **23**, 1958 (1955)
79. Owen, J., *Proc. Roy. Soc. (London)*, [A]**227**, 183 (1955)
80. Pauling, L., *J. Am. Chem. Soc.*, **53**, 1367 (1931)
81. Pauling, L., *Nature of the Chemical Bond* (Cornell University Press, Ithaca, N. Y., 450 pp., 1939)
82. Penney, W. G., and Schlapp, R., *Phys. Rev.*, **41**, 194 (1932)
83. Penney, W. G., and Kynch, G. J., *Proc. Roy. Soc. (London)*, [A]**170**, 112 (1939)
84. Santen, J. H., and Wieringen, J. S., *Rec. trav. chim.*, **71**, 420 (1952)
85. Satten, R. A., *J. Chem. Phys.*, **21**, 637 (1953)
86. Schlapp, R., and Penney, W. G., *Phys. Rev.*, **42**, 666 (1932)
87. Schläfer, H. L., *Z. phys. Chem.*, **4**, 116 (1955)
88. Spedding, F. H., and Nutting, G. C., *J. Chem. Phys.*, **2**, 421 (1934)
89. Spedding, F. H., and Nutting, G. C., *J. Chem. Phys.*, **3**, 369 (1935)
90. Spedding, F. H., *J. Chem. Phys.*, **5**, 316 (1937)
91. Stevens, K. W. H., *Proc. Phys. Soc. (London)*, [A]**65**, 209 (1952)
92. Stevens, K. W. H., *Proc. Roy. Soc. (London)*, [A]**219**, 542 (1953)
93. Tanabe, Y., and Sugano, S., *J. Phys. Soc. (Japan)*, **9**, 753 (1954)
94. Tanabe, Y., and Sugano, S., *J. Phys. Soc. (Japan)*, **9**, 766 (1954)
95. Tsuchida, R., *Bull. Chem. Soc. (Japan)*, **13**, 388 (1938)
96. Tsuchida, R., *Bull. Chem. Soc. (Japan)*, **13**, 436 (1938)
97. Van Vleck, J. H., *Electric and Magnetic Susceptibilities* (Oxford University Press, London, England, 384 pp., 1932)
98. Van Vleck, J. H., *Phys. Rev.*, **41**, 208 (1932)
99. Van Vleck, J. H., and Penney, W. G., *Phil. Mag.*, [S]**7**, 17, 961 (1934)
100. Van Vleck, J. H., *J. Chem. Phys.*, **3**, 803 (1935)
101. Van Vleck, J. H., *J. Chem. Phys.*, **3**, 807 (1935)
102. Van Vleck, J. H., *J. Phys. Chem.* **41**, 67 (1937)
103. Van Vleck, J. H., *J. Chem. Phys.*, **7**, 61 (1939)
104. Van Vleck, J. H., *J. Chem. Phys.*, **7**, 72 (1939)
105. Van Vleck, J. H., *Phys. Rev.*, **57**, 426 (1940)
106. Van Vleck, J. H., *Ann. Inst. Henri Poincaré*, **10**, 57 (1948)
107. Von der Lage, F. C., and Bethe, H., *Phys. Rev.*, **71**, 612 (1947)
108. Wigner, E., *Gruppentheorie* (Braunschweig, 332 pp., 1931)
109. Yamada, S., Nakakara, A., Shimura, Y., and Tsuchida, R., *Bull. Chem. Soc. (Japan)*, **28**, 222 (1955)

Development and Application of the Bonding Theories

V

Editor's Comments on Papers 16 Through 30

16 **Penney:** *A Note on the Bonding Powers of Groups of* d *Electrons*

17 **Tanabe and Sugano:** *On the Absorption Spectra of Complex Ions. I.*

18 **Tanabe and Sugano:** *On the Absorption Spectra of Complex Ions. II.*

19 **Orgel:** *Spectra of Transition-Metal Complexes*

20 **Owen:** *The Colours and Magnetic Properties of Hydrated Iron Group Salts, and Evidence for Covalent Bonding*

21 **Griffith:** *On the Stabilities of Transition Metal Complexes. I.*

22 **Holmes and McClure:** *Optical Spectra of Hydrated Ions of the Transition Metals*

23 **Schäffer and Jørgensen:** *The Nephelauxetic Series of Ligands Corresponding to Increasing Tendency of Partly Covalent Bonding*

24 **Ballhausen and Liehr:** *Intensities in Inorganic Complexes. Part II. Tetrahedral Complexes*

25 **Liehr and Ballhausen:** *Complete Theory of Ni(II) and V(III) in Cubic Crystalline Fields*

26 **Gill, Nyholm, and Pauling:** *Stereochemistry of Complex Halides of the Transition Metals*

27 **Parsons and Drickamer:** *Effect of Pressure on the Spectra of Certain Transition Metal Complexes*

28 **Cotton and Goodgame:** *New Tetrahedral Complexes of Nickel(II)*

29 **Fackler and Cotton:** *Electronic Structure and Molecular Association of Some Bis-(β-diketone) nickel(II) Complexes*

30 **Cotton and Harris:** *Molecular Orbital Calculations for Complexes of Heavier Transition Elements. III. The Metal–Metal Bonding and Electronic Structure of* $Re_2Cl_8^{2-}$

Papers 16 through 30 have been chosen to illustrate how completely and imaginatively chemists (primarily) adopted the ligand field and molecular orbital theories of the physicist and applied them to important chemical problems.

The chemical world was not quite ready (or perhaps interested enough) in 1940 to verify the relationships between the symmetry of coordinated transition-metal ions and bonding energies as suggested by Lord Penney in Paper 16. However, a renaissance in inorganic chemistry began to take place in the 1950s as the power of the symmetry-related theories became apparent. The Tanabe and Sugano papers (Papers 17 and 18) and the Orgel paper (Paper 19) dealt concisely with the colors and spectra of transition-metal ions. The former papers contain the famous diagrams so useful to the description of electronic *d-d* spectra.

The remaining papers in this section are concerned with "crystal field stabilization energies," Papers 20–22; *d-d* electronic properties, Papers 22–25 and 27; and stereochemistries, Papers 26, 28, and 29. The final paper in this group, Paper 30, records a particularly novel application of symmetry to the description of the stereochemistry and metal–metal bonding in $Re_2Cl_8^{2-}$, the first metallic compound recognized to contain a stereochemically important multiple (quadrupole) bond.

W. G. Penney "discovered" that the Bethe–Van Vleck crystal field theory explained some of the very interesting energy relationships that appear in the ionization potentials and heats of the first-row transition-metal hydrates. In this paper the discovery of crystal field stabilization energy is announced, although it had to be rediscovered by L. Orgel [*J. Chem. Soc.*, 4756 1952] twelve years later to have an impact on the chemical world. This entire paper by Lord Penney, unfortunately, was overlooked by chemists until the mid-1950s.

The two papers by Tanabe and Sugano, along with the following one (Paper 19) by Orgel, theoretically and experimentally established the basis of the adjusted crystal field theory—the beginnings of ligand field theory. The diagrams presented in the second of the papers by Tanabe and Sugano have been widely reproduced. The technique of adjusting D_q and B remains a highly satisfactory way to quantitatively evaluate the optical spectra of cubic transition-metal complexes.

Leslie Orgel captured the attention of British chemists and a few Americans in the mid1950s with his papers on spectra of transition-metal complexes. Prior to turning toward biological chemistry in the early 1960s, Orgel was one of the leading proponents of the Van Vleck theoretical techniques for explaining electronic spectra and properties of transition-metal complexes. His little book *An Introduction to Transition-Metal Chemistry Ligand Field Theory* (Methuen, London, 1960, 1966) is itself a classic. It introduced the theory to the chemistry undergraduate.

While the ligand field theory was being expounded at Cambridge by L. Orgel and H. C. Longuet-Higgins, Oxford was not quiet. Paper 20 by Owen substantially expanded the magnetic studies of Van Vleck, Penney, and others, paving the way for the outstanding work by J. Lewis, B. Figgis, and others that followed [see B. N. Figgis and J. Lewis, *Progr. Inorganic Chem.*, **6**, 37–239 (1964)].

John Griffith, like L. Orgel, F. A. Cotton, C. J. Ballhausen, C. K. Jørgensen, and others, represented the modern young chemist of the 1950s who thoroughly grasped the concepts of symmetry and quantum mechanics. Paper 21 by Griffith accounts theoretically for the apparent experimental absence of octahedral or tetrahedral transition-metal complexes with a spin degeneracy intermediate to the Bethe "ionic" and Mulliken "covalent" cases.

John Griffith left us an excellent treatise, *The Theory of Transition-Metal Ions* (Cambridge University Press, New York, 1961). The chemical community lost an intellectual giant in 1972 with his untimely death.

Few electronic spectroscopists of the 1950s showed the complete grasp of ligand field theory that is displayed by D. S. McClure and O. G. Holmes in their 1957 paper (Paper 22). Polarized absorption spectra are used to resolve *d-d* transitions and confirm assignments.

The idea that there exists a ligand series analogous to the spectrochemical series [R. Tsuchida, *Chemistry (Japan)*, **6**, 4 (1951)] which is based on the experimental reduction in the B parameter of G. Racah [*Phys. Rev.*, **63**, 367 (1943)] is expounded in Paper 23, presented at the Rome Conference on Coordination Chemistry. The nephelauxetic series is described in somewhat greater detail by C. K. Jørgensen in *Progr. Inorganic Chemistry*, **4**, 73 (1962). Both Schäffer and Jørgensen have independently contributed substantially to the understanding of electronic structures in coordination compounds.

Carl Ballhausen and Andrew Liehr cooperated to produce some highly significant papers describing the electronic spectra of coordination compounds. Two of their papers are reproduced here. The first (Paper 24) is concerned with the electronic spectra of tetrahedral complex ions and the second (Paper 25) presents the spin-orbit coupling augmented ligand field theory for cubic d^2 and d^8 ion complexes.

Liehr developed the theory for cubic symmetry complexes of each of the *d*-electron configurations and contributed substantially to the present-day understanding of the Jahn–Teller behavior in complexes [*Progr. Inorganic Chem.*, **3,** 281 (1962); **4,** 455 (1962); **5,** 385 (1963)]. Unfortunately, he ceased active work in chemistry during the middle 1960s. Much unpublished material by Liehr which deals with symmetry concepts in chemistry can be found. However, it is generally difficult to read because of Liehr's rather unconventional literary style.

Carl Ballhausen has continued to contribute much significant work, largely of an experimental nature, dealing with the electronic spectra and structures of coordination compounds. His spectroscopic studies, along with those of H. L. Schläfer and others, have contributed substantially to our understanding of coordination compounds. Both Schläfer and Ballhausen have given us excellent texts, each entitled *Ligand Field Theory* (C. J. Ballhausen, McGraw-Hill, New York, 1962; H. L. Schläfer and G. Gliemann, Wiley-Interscience, New York, 1969).

Sir Ronald S. Nyholm, an Australian from a small mining town whose streets are named for the chemical elements, contributed to nearly every aspect of the reawakening of inorganic chemistry in the 1950s and 1960s. An experimentalist of first rank, he and his students made University College, London, the mecca of coordination chemistry from 1956 to 1965.

Paper 26 establishes how it was possible, using ligand field theory, to determine, before their x-ray crystal structures were known, that $NiCl_4^{2-}$ and many other tetrahalides of the bivalent transition-metal ions are essentially tetrahedral in geometry. Sir Ronald's contributions to inorganic stereochemistry were numerous [see especially R. J. Gillespie and R. S. Nyholm, *Quart. Rev. Chem. Soc.*, **11,** 339 (1957)]. He died tragically in an automobile accident in 1971 after nearly twenty years of leadership in the field of coordination chemistry.

Peter Pauling may have developed his interest in structural chemistry from his illustrious father. A number of his crystallographic contributions to coordination chemistry are notable for their timeliness and quality.

Three papers have been chosen from the work of F. A. Cotton and his students at the Massachusetts Institute of Technology. They are interesting examples of the way in which ligand field theory and molecular orbital theory have been applied to inorganic systems in order to discuss symmetry related properties of the materials. The first (Paper 28) established the electronic properties to be expected for tetrahedral nickel(II) complexes, based on the theory of Liehr and Ballhausen (Paper 24). The second paper (Paper 29) established the oligomeric association in solution of nickel(II) acetylacetonate and other related β-diketone complexes. This work relied heavily on the spectral differences expected for octahedral, tetrahedral, and planar complexes of nickel(II). Paper 30 developed the criteria for multiple metal–metal bond formation in $[Re_2Cl_8]^{2-}$, a species initially thought to be low-spin, tetrahedral $[ReCl_4]^{1-}$.

Reprinted from *Trans. Faraday Soc.*, **36**, 627–633 (1940)

16

A NOTE ON THE BONDING POWERS OF GROUPS OF *d* ELECTRONS.

BY W. G. PENNEY.

Received 9th February, 1940.

The outstanding features of the bonding properties of *d* electrons have been discovered by Pauling,[1] using the simple criterion of maximum overlapping of wave functions in conjunction with the directional properties of atomic orbitals. Van Vleck [2] has explained how Pauling's results can be derived equally well from other approximations, such as those of molecular orbitals and the crystalline field potential. We shall now discuss a point which has previously been overlooked.

The ions Ca^{++}, Mn^{++}, and Zn^{++} in aqueous solution, or in hydrated crystals, are in *S* states, while the other ions intermediate to them in the periodic table, are in *D* or *F* states. Now *S* states are not split up by crystalline forces, but *D* and *F* states are, the order of magnitude of the splittings being 20,000 cm.$^{-1}$. As a result, the ions in *S* states are not so closely bound to their water clusters as the other ions, the energy differences being about 30 kcal./mol.

To estimate the differences in the binding energies of the various ions to their water clusters we envisage the following steps, shown diagrammatically in Fig. 1.

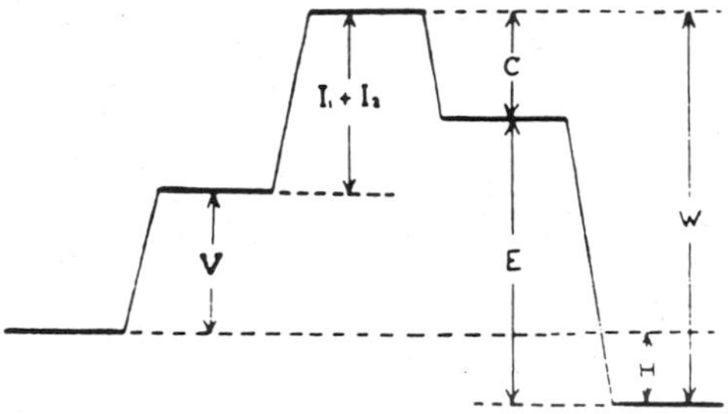

FIG. 1.

The initial system is the metal M in the solid state, together with a dilute acid solution, the amount of acid being just enough to form a divalent salt with the metal. Vaporise the metal, the energy of vaporisation being *V* kcal./mol. Now form the ion M^{++}, the first and second ionisation potentials being I_1 and I_2 kcal./mol. respectively. The resulting two electrons per atom may be regarded as absorbed by two H^+ ions in solution; the resulting normal atoms form a molecule and escape from the solution. Let the energy gained in this process be *C* kcal./mol. The final state of the system is a dilute aqueous solution of the metallic salt, and from thermochemical measurements, let us say, is *H* kcal./mol. more stable than the initial state. The heat of formation of the complex $M^{++} . 6H_2O$ from M^{++} and six water molecules is *E*, where

$$E = V + I_1 + I_2 - C + H.$$

The energy of the complex is $- E$, if the widely separated state is taken to be of zero energy.

[1] Pauling, *J. Amer. Chem. Soc.*, 1931, **53**, 1367.
[2] Van Vleck, *J. Chem. Physics*, 1935, **3**, 803.

627

Since we are concerned only in variations in E from metal to metal, we may disregard C completely, and consider the energy $-W$, where

$$-W = V + I_1 + I_2 + H.$$

The quantity H may be taken either from thermochemical data on the heats of formation of dilute solutions of various salts, or more simply is given directly as the heat of formation of aqueous ions, the heat of formation of the aqueous hydrogen ion H^+ (aq.) being taken arbitrarily as zero.

Our contention is that a plot of W against atomic number through the iron group sequence will show irregularities, Ca, Mn and Zn being about 30 kcal./mol less stable than the others. To test this, we clearly need the experimental values of V, I_1, I_2 and H.

The Heats of Vaporisation.

Probably the most accurate method of obtaining the heats of vaporisation of the iron group metals is from the slope of the log $p - T$ curve, where p is the vapour pressure at temperature T. Apart from Zn, the vapour pressure is not appreciable at temperatures ordinarily available. Apparently, no measurements of the vapour pressures of these metals have been made since 1914, and it is therefore not surprising that the values quoted in the literature for the heats of vaporisation vary widely. Grimm and Wolf [3] have made a critical examination of the published data. Sherman [4] and Bichowsky and Rossini [5] have recalculated most of the heats from the original data ; and the latter authors have selected what, in their opinion, are the best values. Curiously enough, Bichowsky and Rossini do not mention the article of Grimm and Wolf. One may therefore compare the two sets of values and thus obtain an estimate of the probable errors. Table I gives in the first two rows the vaporisation energies of various iron group metals according to Grimm and Wolf and to Bichowsky and Rossini respectively. The units are kcal./mol.

TABLE I.

	Ca.	Cr.	Mn.	Fe.	Co.	Ni.	Cu.	Zn.
V (G and W)	39	83	63	108	105	101	76	32
V (B and R)	48	88	74	94	85	85	81	27
I_1 . .	140	155	171	188	180	175	177	215
I_2 . .	272	382	361	371	397	418	465	412
H . .	130	42	49	21	17	15	−15	36

Even for a metal as volatile as zinc there is a discrepancy of 5 kcal./mol., while for Ni and Co the discrepancy is as large as 20 kcal./mol. Measurements on Ca, quoted by Bichowsky and Rossini, were last made

[3] Grimm and Wolf, *Handbuch der Physik*, 1934, **24**-2, 1073.
[4] Sherman, *Chem. Revs.*, 1932, **11**, 94.
[5] Bichowsky and Rossini, *Thermochemistry of Chemical Substances*, Reinhold, 1936.

in 1929, but even so, there is still a difference of opinion to the extent of 10 kcal./mol. There are no data on Ti, Sc and V.

If a plot of the above heats of sublimation is made, an irregular curve is obtained. The values jump up and down, and there can be no doubt that the true values do not lie on a simple smooth curve.

The Ionisation Potentials.

Table I also gives the first and second ionisation potentials, I_1 and I_2, taken from Bacher and Goudsmit.[6] In most cases, the values are accurate to about 1 kcal./mol. There is, however, some doubt concerning the value of I_1 for Co, since it is obtained by a wide extrapolation of meagre spectroscopic data. Bacher and Goudsmit give "about 8·5 e.v.," but this value seems at least 0·5 e.v. too high, when compared

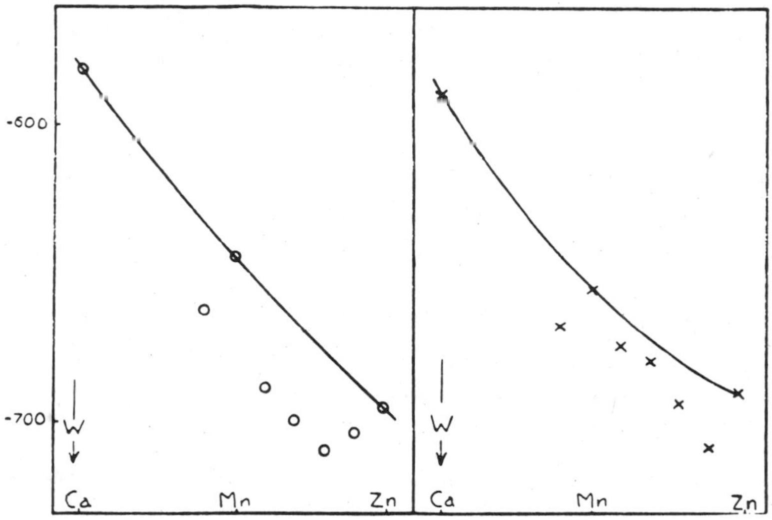

FIG. 2*a* and 2*b*.

with the other ionisation potentials. Bichowský and Rossini have re-examined the data and give 7·8 e.v. This is the value we take.

The Energies *H*.

The last row of Table I gives the heats of formation of the aqueous ions, taken from the book of Bichowsky and Rossini.

The Energies *W*.

Figs. 2*a* and 2*b* show plots of the energies *W* defined by equation (2) above, against atomic number, the former being based on the heats of vaporisation given by Grimm and Wolf, and the latter on those of Bichowsky and Rossini. The two curves are parabolas drawn through the Ca, Mn and Zn points. In both diagrams the points for the other atoms lie well below the parabolas.

[6] Bacher and Goudsmit, *Atomic Energy States*, McGraw Hill, 1932.

We now investigate whether it is possible to apply corrections to the points representing the other atoms in such a way that they should now lie on the true parabola. Any substantial errors in the heats of vaporisation then show up immediately.

Theory.

The stability of the cluster $X^{++}.6H_2O$ is very largely governed by the attractive force of the positive double charge on X to the negatively charged oxygen atoms of the water molecules, and electrostatic Coulombic and exchange forces between the electrons of the various atoms in the cluster. The only electrons of X which need be considered are those in the shells $3s$, $3p$ and $3d$. The first two of these are full in every case; certainly their effective size varies, but only in a regular way through the sequence of ions X. We may therefore assume that the exchange and Coulomb interactions with the shells $(3s)^2$ and $(3p)^6$ varies steadily through the sequence. Similarly, the interactions between the water molecules vary but little, and even then, in a regular way.

Let us imagine that the *d* electrons in every X are averaged over a sphere. All of the above statements now apply to these *d* distributions. Actually, of course, the *d* electrons do not adopt an arrangement of spherical symmetry. The *d* electrons in certain of the X ions become oriented in such a way that their interaction energy with the water molecules is minimised, thus accounting for the extra stability of certain clusters, as shown in Figs. 2*a* and 2*b*. Only in three of the ions are the *d* electrons compelled to adopt a state of spherical symmetry, namely Ca^{++} which has no *d* electrons, Mn^{++} which has half a complete shell, and Zn^{++} which has a complete shell.

The variation of the interaction energies of various $X^{++}.6H_2O$ clusters with the water molecules of the solution may be disregarded, since it is first small, and second regular.

From what has been said above, it is clear that we have now to calculate the differences in the energies of the most stable orientations of various groups of electrons d^0-d^{10} in the field of the water molecules, and the same configurations averaged over a sphere. The calculation is straightforward, and follows closely that developed by Penney and Schlapp.[7]

The Crystalline Potential.

Theoretical studies of the magnetic properties of hydrated salts of iron group elements [8] have led to the conclusion that the motion of the *d* electrons are not seriously affected by the crystalline forces. Consequently, the effect of the crystalline forces on the *d* electrons may be regarded as a perturbation, most conveniently represented as a Taylor's series in the electronic co-ordinates x_i, y_i, z_i, referred to the centre of the ion as origin. Since the potential has cubic symmetry, it may be expressed in the form

$$V(x_i, y_i, z_i) = f(r_i) + D(x_i^4 + y_i^4 + z_i^4 - 3r_i^4/5) + E(x_i^6 + \ldots) + \ldots,$$

or in other equivalent forms.

The energy levels of the free ion are well described by the Russell-Saunders scheme. The effect of V is to disrupt the magnetic coupling of L and S, but is not sufficient to spoil the significance of L and S. In

[7] Penney and Schlapp, *Physic. Rev.*, 1932, **41**, 194.
[8] Van Vleck, *Theory of Electric and Magnetic Susceptibilities*, 1932.

other words, the exchange and Coulomb interactions of the d electrons of the ion with the surrounding water molecules are of subsidiary importance to the exchange and Coulomb interactions between the d electrons themselves. If, therefore, the energy diagram of a free ion is drawn, neglecting the orbit-spin forces, then the effect of V may be shown as a splitting of the L levels, and the splittings of the lower L levels are small compared with the intervals between the L levels.

The potential V through the sequence Ca^{++} — Zn^{++} will vary, but not very much. We assume that it is constant. The effect of V on a d electron varies with $\overline{r^4}$, taken over the orbit. We neglect any variation throughout the sequence.

The term $f(r)$ in V has no orienting effect on d electrons; the sixth-order and higher terms have no effect at all. The only term remaining is the fourth-order one.

TABLE II.

Ion.	State.	Γ_1.	Γ_3.	Γ_4.	Γ_5.	Δ.
Ca^{++} .	$d^0\ ^1S$	0	—	—	—	0
Sc^{++} .	$d'\ ^2D$	—	12	—	−8	8
Ti^{++} .	$d^2\ ^3F$	24	—	−12	4	12
V^{++} .	$d^3\ ^4F$	−24	—	12	−4	24
Cr^{++} .	$d^4\ ^5D$	—	−12	—	8	12
Mn^{++} .	$d^5\ ^6S$	0	—	—	—	0
Fe^{++} .	$d^6\ ^5D$	—	12	—	−8	8
Co^{++} .	$d^7\ ^4F$	24	—	−12	4	12
Ni^{++} .	$d^8\ ^3F$	−24	—	12	−4	24
Cu^{++} .	$d^9\ ^2D$	—	−12	—	8	12
Zn^{++} .	$d^{10}\ ^1S$	0	—	—	—	0

Table II gives the effect of the potential $D(x^4 + y^4 + z^4 - 3r^4/5)$ on the various ions Ca^{++} — Zn^{++}. The first column gives the ion; the next its electronic state when free. The next group of columns gives the decomposition of the lowest L state under the action of the field, the notation being that of Bethe.[9] Levels belonging to symmetry types Γ_1 and Γ_2 are singlets, to Γ_3 are doublets, and to Γ_4 and Γ_5 are triplets (actually Γ_2 does not appear). The units of energy λ is given in terms of the constants of the ion and the crystalline field by the equation

$$\lambda = \overline{r^4}D/105.$$

The last column in Table II is the most interesting one from our present point of view. It gives Δ, the energy interval between the lowest level and the average position of all the levels. Since we have, in fact, chosen the origin of energy of the levels in each case to be the mean position, Δ is simply the negative of the lowest energy level.

[9] Bethe, *Ann. Physik*, 1929, **3**, 133.

A numerical value for λ may be had from work on the paramagnetic properties of the ions. Only in two cases are data available. According to Schlapp and Penney,[10] the overall separation of the level ^3F of Ni^{++}, due to the crystalline field of an octahedron of water molecules, is about 26,000 cm.$^{-1}$. Krishnan and Mookherji,[11] using the equations of Schlapp and Penney, find a smaller value 23,000 cm.$^{-1}$, for a number of hydrated crystals of Ni^{++}. Becquerel and Opechowski,[12] from calculations on the paramagnetic rotation of nickel salts, obtain a splitting 19,200 cm.$^{-1}$. Since, in every case, the crystalline potential is determined from the deviations of the magnetic properties from " spin only " values, and these are quite small, the agreement is fairly satisfactory.

Jordahl [13] estimates 19,000 cm.$^{-1}$ as the overall separation of the ^2D level of Cu^{++} due to an octahedron of water molecules. The calculations are particularly difficult in this case because a non-degenerate orbital level is lowest.

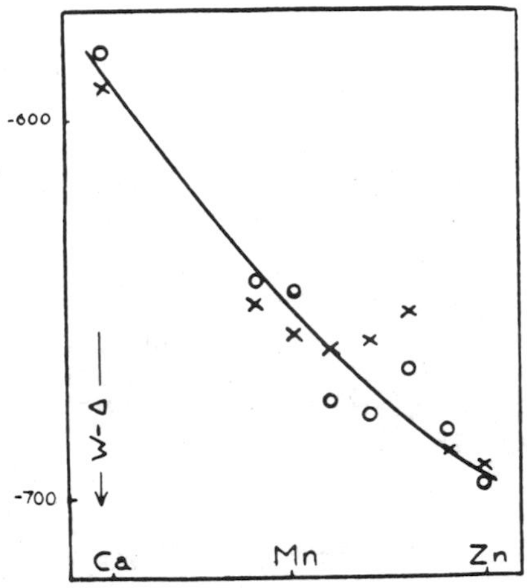

FIG. 3.

If the assumptions on which Table II are based are valid, the ratio of the overall separations in Ni^{++} and Cu^{++} is 9 : 5. The separations given above do not fit this closely; but if we take the separation for Cu^{++} as 14,000 cm.$^{-1}$, and for Ni^{++} as 21,000 cm.$^{-1}$, the discrepancies are not large. Unfortunately the present thermo-chemical data are so discordant that they are of little use for refinements of these values. The corresponding value of λ is 700 cm.$^{-1}$, or 1·98 kcal./mol.

Evaluating the various quantities Δ with this value of λ, and adding them to the energies $-W$ plotted in Figs. 2*a* and 2*b*, we have obtained results shown in Fig. 3. The circles are values of $-W + \Delta$ obtained from Grimm and Wolf's figures for the heats of sublimation of the metals, and the crosses from the corresponding figures of Bichowsky and Rossini. It will be seen that the representative points, although very scattered, do lie approximately on a smooth curve of little curvature. We have sketched in such a curve; clearly the data are not worthy of anything more elaborate.

An interesting feature of the curve is the way in which circles and crosses are distributed on both sides. From a study of Fig. 3, and a consideration of the probable errors in the theory given here, we venture on the following values of the heats of vaporisation of the metals (kcal./mol.).

M	Cr	Mn	Fe	Co	Ni	Cu
V	80	70	95	100	110	80

[10] Schlapp and Penney, *Physic. Rev.*, 1932, **42**, 666.
[11] Krishnan and Mookherji, *Phil. Trans.*, *A*, 1938, **237**, 135.
[12] Becquerel and Opechowski, *Physica*, 1939, **6**, 1039.
[13] Jordahl, *Physic Rev.*, 1934, **45**, 87.

Conclusion.

The theory developed above is clearly on the right lines. When more accurate data become available, the theory can be improved, for example, by allowing for the variation in the ionic radius throughout the series, and by including interaction with higher levels of the ions.

The writer wishes to express his thanks to Dr. Purcell for helpful discussion.

Imperial College of Science.

Reprinted from *J. Phys. Soc. Japan*, **9**, 753–766 (1954)

17

On the Absorption Spectra of Complex Ions. I

By Yukito TANABE
Department of Physics, University of Tokyo
and Satoru SUGANO
Institute of Broadcasting, N. H. K.

(Recieved June 11, 1954)

In order to explain the origin of both absorption bands and lines of the octahedral normal complex ions in which the central metal ions belong to iron group elements, the crystalline field approximation interpreted in somewhat generalized sense is adopted. For this purpose the calculation of the energy matrix elements for d^n ($n=1, 2, \ldots, 9$) electron configuration in cubic field is performed by means of Racah's method. Comparison of the results with experiments will be made in the following paper.

§ 1. Introduction

Much work, both theoretical and experimental, has been done about the ground states of the normal complex ions of the type $[XY_6]^{3+}$, $[XY_6]^{2+}$, where X represents an iron group element and Y represents a ligand such as H_2O, NH_3, ethylenediamine, etc.. Especially owing to recent investigations by the method of the paramagnetic resonance absorption, rather detailed knowledge of them has been obtained. As to the excited states, however, little is known of these normal complex ions, and, it seems, there remain many problems to be investigated further from the theoretical point of view.

The optical absorption of both crystals and solutions (aqueous in many cases) containing these complex ions have been studied since the days of Werner by many physicists and chemists, and it is known that the absorption spectra of solutions are quite similar (although the width and the intensity of bands are slightly different) to those of crystals containing the same complex ion as the solutions except for an intense edge absorption ($\log \mathcal{E}_{max} \sim 4$) in the ultra violet region which does not appear in the latter, so that they are considered to be due mainly to the absorption of the complex ion as a whole. The absorption spectra of these complex ions are characterized by the presence of several (one to three) bands in the near infrared, visible and ultra violet region (8000 cm^{-1} to 40000 cm^{-1}). Some of them show several sharp absorption lines besides these broad bands (e.g. Chrome Alum).

There have been several explanations of the origin of these spectra, but many of them are only of qualitative nature and it is not yet certain which is the correct one. Only two of them shall be quoted here, because they bear close relation to our calculation. Finkelstein and Van Vleck[1] treated the doublet lines of Chrome Alum at 6700 Å and obtained quantitative agreement between theory and experiment using the crystalline field approximation, but they left the origin of the broad bands unanswered. Recently, Hartmann and Schlaefer et al.[2] treated the problem of broad bands also from the same standing point. They maintained that these absorption bands were due to the transitions between the Stark components of the ground multiplet of the central metal ion, and showed that the number of broad bands (which are observed in the region up to 33000 cm^{-1}) could be reasonably explained on this assumption at least in case the central metal ion has no more than five d-electrons. But they made no mention of the line spectra at all. Though these explanations seem to be resonable, it is not clear why some of the transitions correspond to the broad bands and the others to the sharp lines. It seems necessary to examine if such an assignment should give the order of magnitude of the observed intensity and the band width theoretically, and if the calculation could predict the position of the absortion maxima satisfactorily.

In order to study to what degree the crystalline field approximation succeeds in explaining the absorption spectra of these complex ions including both the lines and bands, the energy matrix elements for the electron configuration d^n in cubic field are calculated in this paper.

§ 2. The Crystalline Field Approximation

These complex ions are known to be of the octahedral shape, and the central ion X is considered to be exposed to the cubic electric field (crystalline field) due to the octahedrally coordinated ligands Y.*

In the case of the rare earth salts, the crystalline field is relatively weak because of the screening effect of the electrons in the outer shells, so that, when one wants to study the electronic structure of the rare earth ions it is generally allowed to treat each *SL* multiplet separately and to consider the splittings of these multiplets due to the crystalline field. In our case of the iron group complex ions, however, the following treatment** will be more adequate, because here d-electrons are exposed directly to the strong crystalline field. In Cubic field the d-level splits into two levels, i.e., triply degenerate $d\varepsilon$ and doubly degenerate $d\gamma$ ($d\varepsilon$ is lower than the original d-level by $-4Dq$, and $d\gamma$ higher by $6Dq$, where Dq is the crystalline field parameter usually used.) and the electrons are accomodated in these two levels. For those complexes where the free $X^{3,2+}$ ion has the electrons configuration d^N, we now take the configurations $d\varepsilon^n d\gamma^{N-n}$ ($n=0$ to N) as the starting point.

If the crystalline field is sufficiently strong, we may neglect the interaction between these configurations. Then the problem is to find the positions of the $S\Gamma$ multiplets arising from each configuration mentioned above, but unfortunately the crystalline field is not so strong as to permit us to neglect the configuration interaction entirely and we must take it into account for some of the levels.

As is easily seen, in the calculation under the scheme which we mentioned to be suited for the weak field case, if the interaction between the Stark components of different *SL* multiplets is fully taken into account as in Finkelstein and Van Vleck's one, the results are of course the same as those calculated in our way, when we construct $d\varepsilon$, $d\gamma$ wave function there by proper linear combinations of the d-wave function.

It is here necessary to consider the meaning of the "crystalline field." Recently Kleiner[4] has calculated the strength of the crystalline field Dq of Chrome Alum semi-classically taking the overlapping of the charge cloud of the Cr^{3+} ion and that of H_2O into account, and found that the calculated value was by far smaller (the sign was wrong) than the value estimated from the magnetic susceptibility data, though his model seemed more reasonable than the point-dipole or -charge model[5],[6] which can give the resonable values of Dq in many cases. This is unfortunate, but this may not be so serious. It seems to us that, as far as Dq is considered as an empirical parameter, the validity of the crystalline field approximation is not so limited as one might suppose from his results, since this approximation can be interpreted in a generalized sense as follows.

That is, we consider $[XY_6]^{3,2+}$ complex ion as a single molecule whose filled orbitals with lower energies correspond to the inner shells or closed shells of the central ion and ligands in the free states which may be deformed in the complex ion, and whose next higher orbitals are f_2 and e (which correspond to $d\varepsilon$ and $d\gamma$ respectively) belonging to the irreducible representations F_{2g} and E_g of O_h group. This is just the model which corresponds to the crystalline field approximation. f_2 and e orbitals will not have the pure d-character in general, and may be approximated by the linear combinations of d-orbitals and the ligand orbitals. For $d\gamma$ this departure from the atomic d-character will be large, but also for $d\varepsilon$ such deformation would exist as Stevens has recently pointed out[7]. Thus in this generalized crystalline field formalism, Dq is now defined through the relation $\epsilon_e - \epsilon_{f_2} = 10Dq$ where ϵ_e and ϵ_{f_2} denote the orbital energies of e and f_2 orbitals. Futhermore in this molecular treatment, the integrals appearing in the calculation may be different from those in the usual crystalline field approximation, where they are expressed in terms of the Slater integrals for the free ions.

If we go further in this way there is no reason why only two levels f_2 and e should

* Throughout this and the following paper the Jahn-Teller effect is neglected.

** Similar calculations along this line has already been done by Kotani for the case of d^2 and by Kambe and Usui for d^3 [3].

be taken into account. But when we include further configurations the calculation would become somewhat tedious and the empirical treatment such as adopted in the following paper would become impossible. Thus we assume with a hope that the generalized crystalline field model will well approximate the reality, and carry out the calculation under this assumption.

In the next section, however, we calculate the energy matrix elements using the atomic $d\varepsilon$ and $d\gamma$, and express them in terms of A, B, C and Dq (A, B, C are the linear combinations of Slater integrals introduced by Racah). Thus, the effect of deformation appears only through the change of B, C from those for the free ions. This was done only to avoid the complexity caused by many unknown parameters which appear when e and f_2 are considered not to be pure $d\varepsilon$ and $d\gamma$, although the method of calculation is also applicable when the crystalline field approximation is taken in a generalized sence.

§ 3. The Calculation of Energy Matrices and the Results

When we adopt the scheme stated in § 2 and confine ourselves to the states which come from the configurations $d\varepsilon^n d\gamma^{N-n}(n=0,\ldots,N)$, the orthonormal basic functions to construct the energy matrix of symmetry species $S\Gamma$ are $\Psi(d\varepsilon^n(S_1\Gamma_1)d\gamma^{N-n}(S_2\Gamma_2)S\Gamma)$. The matrix elements of the scalar (cubic scalar in this case) operator $G_N=\sum_{i>j}^{N} g_{ij}$ between these states can be calculated by means of the following recurrence formula, which is a generalization of the formula (1) of R IV[8]:

$$
\begin{aligned}
&(\gamma_1^n(S_1\Gamma_1)\gamma_2^{N-n}(S_2\Gamma_2)S\Gamma|G_N|\gamma_1^{n'}(S_3\Gamma_3)\gamma_2^{N-n'}(S_4\Gamma_4)S\Gamma)\\
&=\sqrt{nn'}/(N-2)\sum_{S'S''\bar{S}\Gamma'\Gamma''\bar{\Gamma}}(\gamma_1^n S_1\Gamma_1\{|\gamma_1,\gamma_1^{n-1}(S'\Gamma')S_1\Gamma_1)(\gamma_1 S'\Gamma'(S_1\Gamma_1)S_2\Gamma_2 S\Gamma|\\
&\qquad\gamma_1,S'\Gamma'S_2\Gamma_2(\bar{S}\bar{\Gamma})S\Gamma)(\gamma_1^{n-1}(S'\Gamma')\gamma_2^{N-n}(S_2\Gamma_2)\bar{S}\bar{\Gamma}|G_{N-1}|\gamma_1^{n'-1}(S''\Gamma'')\gamma_2^{N-n'}(S_4\Gamma_4)\bar{S}\bar{\Gamma})\\
&\qquad\times(\gamma_1,S''\Gamma''S_4\Gamma_4(\bar{S}\bar{\Gamma})S\Gamma|\gamma_1 S''\Gamma''(S_3\Gamma_3)S_4\Gamma_4 S\Gamma)(\gamma_1\gamma_1^{n'-1}(S''\Gamma'')S_3\Gamma_3|\}\gamma_1^{n'}S_3\Gamma_3)\\
&+\sqrt{(N-n)(N-n')}/(N-2)\sum_{S'S''\bar{S}\Gamma'\Gamma''\bar{\Gamma}}(\gamma_2^{N-n}S_2\Gamma_2\{|\gamma_2^{N-n-1}(S'\Gamma')\gamma_2 S_2\Gamma_2)\\
&\qquad\times(S_1\Gamma_1,S'\Gamma'\gamma_2(S_2\Gamma_2)S\Gamma|S_1\Gamma_1 S'\Gamma'(\bar{S}\bar{\Gamma})\gamma_2 S\Gamma)\\
&\qquad\times(\gamma_1^n(S_1\Gamma_1)\gamma_2^{N-n-1}(S'\Gamma')\bar{S}\bar{\Gamma}|G_{N-1}|\gamma_1^{n'}(S_3\Gamma_3)\gamma_2^{N-n'-1}(S''\Gamma'')\bar{S}\bar{\Gamma})\\
&\qquad\times(S_3\Gamma_3 S''\Gamma''(\bar{S}\bar{\Gamma})\gamma_2 S\Gamma|S_3\Gamma_3,S''\Gamma''\gamma_2(S_4\Gamma_4)S\Gamma)\\
&\qquad\times(\gamma_2^{N-n'-1}(S''\Gamma'')\gamma_2 S_4\Gamma_4|\}\gamma_2^{N-n'}S_4\Gamma_4)\,.
\end{aligned}
\tag{3.1}
$$

Although γ_1, γ_2 and Γ's here are the symbols for the irreducible representations of the octahedral group (γ_1, γ_2 being E and F_2 respectively), this formula is generally valid for any other group.

In order to calculate this matrix element for the N electron system, it is necessary to know the transformation matrices $(S'\Gamma'S''\Gamma''(\bar{S}\bar{\Gamma}),S'''\Gamma'''S\Gamma|S'\Gamma',S''\Gamma''S'''\Gamma'''(\underline{S\Gamma})S\Gamma)$ and the coefficients of the fractional parentage $(\gamma^n S\Gamma\{|\gamma^{n-1}(S'\Gamma')\gamma S\Gamma)$ (c.f.p.) besides the matrix elements for the $N-1$ electron system.

The transformation matrices were simply calculated according to the definition (3) of R III[8], and the Clebsch-Gordan coefficients thereby used are given in Table I.

The c.f.p. for n-electrons are obtained solving the following linear equations which are (11) of R III, assuming that we know those for $n-1$ electrons and the transformation matrices of the type $(S''\Gamma'',\gamma\gamma(S'''\Gamma''')S\Gamma|S''\Gamma''\gamma(S'\Gamma')\gamma S\Gamma)$:

$$
\begin{aligned}
&\sum_{S'\Gamma'}(S''\Gamma'',\gamma\gamma(S'''\Gamma''')S\Gamma|S''\Gamma''\gamma(S'\Gamma')\gamma S\Gamma)\\
&\quad\times(\gamma^{n-2}(S''\Gamma'')\gamma S'\Gamma'|\}\gamma^{n-1}S'\Gamma')\\
&\quad\times(\gamma^{n-1}(S'\Gamma')\gamma S\Gamma|\}\gamma^n S\Gamma)=0\,.
\end{aligned}
\tag{3.2}
$$

Here $S'''\Gamma'''$'s denote the "forbidden states" for the configuration γ^2. For instance, in the case of $\gamma=d\varepsilon$ they are ${}^3A_1$, ${}^3E$, ${}^1F_1$ and ${}^3F_2$. c.f.p. calculated in this way are given at the end of this paper (Table II), together with the c.f.p. of the type $(\gamma^{n-2}(S'\Gamma')\gamma^2(S''\Gamma'')S\Gamma|\}\gamma^n S\Gamma)$ which were calculated according to (32) of R III and useful in some cases (see below).

Further relation analogous to Eq. (29) of R III was also used:

$$(\gamma, \gamma^{n-1}(S'\Gamma')S\Gamma|\}\gamma^n S\Gamma) = (-)^{S'+1/2-S+f(\gamma\Gamma'\Gamma)+n+1} \times(\gamma^{n-1}(S'\Gamma')\gamma S\Gamma|\}\gamma^n S\Gamma), \quad (3.3)$$

where

$$f(\gamma\Gamma'\Gamma)=1 \quad \text{for} \quad (\gamma\Gamma'\Gamma)=(EEA_2), (F_1F_1F_1), (F_2F_2F_1),$$
$$=0 \quad \text{otherwise.}$$

The formula (3.1) is rather complicated and the calculation becomes somewhat tedious when one uses this formula as it is. But it is often not necessary to cary out both summations and the labor is greatly reduced especially when one of the two sums becomes simple owing to the restriction on the summation parameters.

This fact applies to the calculation of the non-diagonal elements.

Without loss of generality we may assume $n'+3>n\geq n'$. When $n-n'=2$, applying (3.1) successively we see that this type of matrix elements has a single ancestor of the type $(\gamma_1^2S\Gamma|g|\gamma_2^2S\Gamma)$, and in this case, it can be shown (Appendix) that the ratio of two sums is a constant which depends only N, n, n', and therefore it is sufficient to calculate the simpler sum. When $n-n'=1$, we have two ancestors for this type of elements, i.e., $(\gamma_1\gamma_2S\Gamma|g|\gamma_2^2S\Gamma)$ and $(\gamma_1^2S\Gamma|g|\gamma_1\gamma_2S\Gamma)$ so that no simplification can be obtained. But in our case fortunately there is no ancestor of the type $(d\varepsilon d\gamma S\Gamma|g|d\gamma^2S\Gamma)$, because configurations $d\varepsilon d\gamma$ and $d\gamma^2$ have no common state $S\Gamma$, and we can simplify the formula (3.1) in a similar way as was made in the case $n-n'=2$. In the last case, i.e., when $n=n'$, there are three ancestors, namely $(\gamma_1^2S\Gamma|g|\gamma_1^2S\Gamma)$, $(\gamma_2^2S\Gamma|g|\gamma_2^2S\Gamma)$ and $(\gamma_1\gamma_2S\Gamma|g|\gamma_1\gamma_2S\Gamma)$. The non-diagonal elements in this case are, however, seen to originate only from the last type of the three ancestors mentioned above, i.e., $(\gamma_1\gamma_2S\Gamma|g|\gamma_1\gamma_2S\Gamma)$, so also in this case the same simplification is achieved in the calculation of these elements.

In some of the last cases the matrix elements of G_N are calculated more easily from those of G_{N-2} or from an ancestor directly by making use of the c.f.p. of the type

$$(\gamma^{n-2}(S'\Gamma')\gamma^2(S''\Gamma'')S\Gamma|\}\gamma^n S\Gamma)$$

than from those of G_{N-1}. The recurrence formulae of this type are easily obtained in a way analogous to (3.1) and some of these are already given in (33) of R III.

The matrix elements of the electrostatic interaction for two electron system are easily calculated and the results are given below. Here (u,v) and (ξ,η,ζ) denote the real orbitals which belong to the irreducible representations E_g and F_{2g} respectively of the octahedral group.

$$\left.\begin{array}{ll}
n-n'=0 & (d\varepsilon^2S\Gamma|g|d\varepsilon^2S\Gamma) \\
& S\Gamma={}^1A_1:\ (\xi\xi|g|\xi\xi)+2K(\zeta\eta)=(A+10B+5C), \\
& {}^1E:\ (\xi\xi|g|\xi\xi)-K(\zeta\eta)=(A+B+2C), \\
& {}^3F_1:\ J(\zeta\eta)-K(\zeta\eta)=(A-5B), \\
& {}^1F_2:\ J(\zeta\eta)+K(\zeta\eta)=(A+B+2C), \\
& (d\varepsilon d\gamma S\Gamma|g|d\varepsilon d\gamma S\Gamma) \\
& S\Gamma={}^1F_1:\ J(\zeta v)+K(\zeta v)=(A+4B+2C), \\
& {}^3F_1:\ J(\zeta v)-K(\zeta v)=(A+4B), \\
& {}^1F_2:\ J(\zeta u)+K(\zeta u)=(A+2C), \\
& {}^3F_2:\ J(\zeta u)-K(\zeta u)=(A-8B), \\
& (d\gamma^2S\Gamma|g|d\gamma^2S\Gamma) \\
& S\Gamma={}^1A_1: 2(uu|g|uu)-J(uv)-K(uv)=(A+8B+4C), \\
& {}^3A_2:\ J(uv)-K(uv)=(A-8B), \\
& {}^1E:\ J(uv)+K(uv)=(A+2C), \\
n-n'=1 & (d\varepsilon^2S\Gamma|g|d\varepsilon d\gamma S\Gamma) \\
& S\Gamma={}^3F_1: 2(\eta\xi|g|v\zeta)=(6B), \\
& {}^1F_2: 2(\eta\xi|g|u\zeta)=(2\sqrt{3}B), \\
n-n'=2 & (d\varepsilon^2S\Gamma|g|d\gamma^2S\Gamma) \\
& S\Gamma={}^1A_1: \sqrt{3/2}\{K(\xi u)+K(\xi v)\}=(\sqrt{6}(2B+C)),
\end{array}\right\} \quad (3.4)$$

$$ {}^1E: \sqrt{3}\ \{K(\xi u)-K(\xi v)\} = (-2\sqrt{3}\,B)\,, $$

where

$$ J(ab)=(ab|g|ab)\,, \qquad K(ab)=(ab|g|ba)\,. $$

To obtain these results the following relations were used, which come from the transformation properties of the basic orbitals;

$$
\left.\begin{aligned}
&(\xi\xi|g|\xi\xi)=(\eta\eta|g|\eta\eta)=(\zeta\zeta|g|\zeta\zeta)\,,\\
&(\xi\xi|g|\eta\eta)=(\eta\eta|g|\zeta\zeta)=(\zeta\zeta|g|\xi\xi)\,,\\
&(uu|g|uu)=(vv|g|vv)\,,\\
&(uu|g|vv)=(uu|g|uu)-J(uv)-K(uv)\,,\\
&(\eta\xi|g|v\zeta)=-(\eta\xi|g|\zeta v)\,,\\
&(\eta\xi|g|u\zeta)=(\eta\xi|g|\zeta u)\,,\\
&(\xi\xi|g|uu)=(\eta\eta|g|uu)\,,\\
&(\xi\xi|g|vv)=(\eta\eta|g|vv)\,,\\
&(\xi\xi|g|uu)-(\xi\xi|g|vv)=2/\sqrt{3}\ (\xi\xi|g|uv)\,,\\
&(\zeta\zeta|g|uu)=-1/2(\eta\eta|g|uu)+3/2(\eta\eta|g|vv)\,,\\
&(\zeta\zeta|g|vv)=3/2(\eta\eta|g|uu)-1/2(\eta\eta|g|vv)\,,
\end{aligned}\right\}
\tag{3.5}
$$

and from their reality;

$$
\left.\begin{aligned}
&(\zeta\zeta|g|\eta\eta)=K(\zeta\eta)\,, &&(uu|g|vv)=K(uv)\,,\\
&(\zeta\zeta|g|uu)=K(\zeta u)\,, &&(\zeta\zeta|g|vv)=K(\zeta v)\,,\\
&(\xi\xi|g|uu)=K(\xi u)\,, &&(\xi\xi|g|vv)=K(\xi v)\,.
\end{aligned}\right\}
\tag{3.6}
$$

Using these matrix elements of $G_2=1/r_{12}$, we can now proceed to calculate the matrix elements of $G_N=\sum_{i>j}^{N} 1/r_{ij}$. The complete energy matrix is obtained adding the orbital (crystalline field) energies to its diagonal elements.

As was mentioned at the end of § 2, we now return to the crystalline field approximation in normal sense, and assume the orbitals (u, v) and (ξ, η, ζ) are pure $d\gamma$ and $d\varepsilon$ respectively, namely, of the form;

$$
\left.\begin{aligned}
u&=R(r)Y_{20}(\theta\varphi),\\
v&=1/\sqrt{2}\ \ R(r)\{Y_{22}(\theta\varphi)+Y_{2-2}(\theta\varphi)\}\,,\\
\xi&=-i/\sqrt{2}\ R(r)\{Y_{21}(\theta\varphi)+Y_{2-1}(\theta\varphi)\}\,,\\
\eta&=-1/\sqrt{2}\ R(r)\{Y_{21}(\theta\varphi)-Y_{2-1}(\theta\varphi)\}\,,\\
\zeta&=\ i/\sqrt{2}\ \ R(r)\{Y_{22}(\theta\varphi)-Y_{2-2}(\theta\varphi)\}\,.
\end{aligned}\right\}
\tag{3.7}
$$

The matrix elements (3.4) are accordingly expressed in terms of Racah's parameters A, B, C as are shown at the right hand side of (3.4) in parenthesis.

The orbital energy for the configuration $d\varepsilon^n d\gamma^{N-n}$ is given by $(6N-10n)Dq$.

The results thus obtained are given in Table III, where the term $N(N-1)/2\cdot A$ in the diagonal elements is omitted. They were checked by examining whether they gave the energies of the free state when Dq was set equal to zero, or not.

Conjugation is simple in this case, that is, we can obtain the energy matrices for d^{10-N} from those for d^N merely changing the sign of Dq[1]. Conjugate states are given at the top of the columns of each matrix in Table III.

Table I. Tables of the clebsch-gordan coefficients. $(\Gamma_1\gamma_1\Gamma_2\gamma_2|\Gamma_1\Gamma_2\Gamma\gamma)$

$A_2\times A_2$

		Γ	A_1
		γ	e_1
γ_1	γ_2		
e_2	e_2		-1

$A_2\times E$

		Γ	E	
		γ	u	v
γ_1	γ_2			
e_2	u			-1
	v		1	

$A_2\times F_1$

		Γ	F_2		
		γ	ξ	η	ζ
γ_1	γ_2				
	α		1		
e_2	β			1	
	γ				1

$A_2\times F_2$

		Γ	F_1		
		γ	α	β	γ
γ_1	γ_2				
	ξ		-1		
e_2	η			-1	
	ζ				-1

$E\times E$

		Γ	A_1	A_2	E	
		γ	e_1	e_2	u	v
γ_1	γ_2					
u	u		$1/\sqrt{2}$		$-1/\sqrt{2}$	
	v			$1/\sqrt{2}$		$1/\sqrt{2}$
v	u			$-1/\sqrt{2}$		$1/\sqrt{2}$
	v		$1/\sqrt{2}$		$1/\sqrt{2}$	

Table I. (continued)

$E \times F_1$

γ_1	γ_2 \ Γ / γ	F_1 α	F_1 β	F_1 γ	F_2 ξ	F_2 η	F_2 ζ
	α	$-1/2$			$\sqrt{3}/2$		
u	β		$-1/2$			$-\sqrt{3}/2$	
	γ			1			
	α	$\sqrt{3}/2$			$1/2$		
v	β		$-\sqrt{3}/2$			$1/2$	
	γ						-1

$E \times F_2$

γ_1	γ_2 \ Γ / γ	F_1 α	F_1 β	F_1 γ	F_2 ξ	F_2 η	F_2 ζ
	ξ	$-\sqrt{3}/2$			$-1/2$		
u	η		$\sqrt{3}/2$			$-1/2$	
	ζ						1
	ξ	$-1/2$			$\sqrt{3}/2$		
v	η		$-1/2$			$-\sqrt{3}/2$	
	ζ			1			

$F_1 \times F_1$

γ_1	γ_2 \ Γ / γ	A_1 e_1	E u	E v	F_1 α	F_1 β	F_1 γ	F_2 ξ	F_2 η	F_2 ζ
	α	$-1/\sqrt{3}$	$1/\sqrt{6}$	$-1/\sqrt{2}$						
α	β						$-1/\sqrt{2}$			$-1/\sqrt{2}$
	γ					$1/\sqrt{2}$			$-1/\sqrt{2}$	
	α						$1/\sqrt{2}$			$-1/\sqrt{2}$
β	β	$-1/\sqrt{3}$	$1/\sqrt{6}$	$1/\sqrt{1}$						
	γ				$-1/\sqrt{2}$			$-1/\sqrt{2}$		
	α					$-1/\sqrt{2}$			$-1/\sqrt{2}$	
γ	β				$1/\sqrt{2}$			$-1/\sqrt{2}$		
	γ	$-1/\sqrt{3}$	$-2/\sqrt{6}$							

$F_1 \times F_2$

γ_1	γ_2 \ Γ / γ	A_2 e_2	E u	E v	F_1 α	F_1 β	F_1 γ	F_2 ξ	F_2 η	F_2 ζ
	ξ	$-1/\sqrt{3}$	$-1/\sqrt{2}$	$-1/\sqrt{6}$						
α	η						$1/\sqrt{2}$			$-1/\sqrt{2}$
	ζ					$1/\sqrt{2}$			$1/\sqrt{2}$	
	ξ						$1/\sqrt{2}$			$1/\sqrt{2}$
β	η	$-1/\sqrt{3}$	$1/\sqrt{2}$	$-1/\sqrt{6}$						
	ζ				$1/\sqrt{2}$			$-1/\sqrt{2}$		
	ξ					$1/\sqrt{2}$			$-1/\sqrt{2}$	
γ	η				$1/\sqrt{2}$			$1/\sqrt{2}$		
	ζ	$-1/\sqrt{3}$		$2/\sqrt{6}$						

Table I. (continued)

$F_2 \times F_2$

γ_1	γ_2 \ Γ / γ	A_1: e_1	E: u	E: v	F_1: α	F_1: β	F_1: γ	F_2: ξ	F_2: η	F_2: ζ
	ξ	$1/\sqrt{3}$	$-1/\sqrt{6}$	$1/\sqrt{2}$						
ξ	η						$1/\sqrt{2}$			$1/\sqrt{2}$
	ζ					$-1/\sqrt{2}$			$1/\sqrt{2}$	
	ξ						$-1/\sqrt{2}$			$1/\sqrt{2}$
η	η	$1/\sqrt{3}$	$-1/\sqrt{6}$	$-1/\sqrt{2}$						
	ζ				$1/\sqrt{2}$			$1/\sqrt{2}$		
	ξ					$1/\sqrt{2}$			$1/\sqrt{2}$	
ζ	η				$-1/\sqrt{2}$			$1/\sqrt{2}$		
	ζ	$1/\sqrt{3}$	$2/\sqrt{6}$							

Table II.

Tables of the C. F. P. $(\gamma^{n-1}(S'\Gamma')\gamma\, S\Gamma|\}\gamma^n S\Gamma)$.

$(d\varepsilon^{n-1}(S'\Gamma')d\varepsilon\, S\Gamma|\}d\varepsilon^n S\Gamma)$

$n=3$

$S\Gamma$ \ $S'\Gamma'$	$^1A_1$	$^3F_1$	^1E	$^1F_2$
$^4A_2$		1		
$^2F_1$		$1/\sqrt{2}$	$1/\sqrt{3}$	$1/\sqrt{6}$
^2E		$1/\sqrt{2}$		$1/\sqrt{2}$
$^2F_2$	$\sqrt{2}/3$	$1/\sqrt{2}$	$-1/3$	$-1/\sqrt{6}$

$n=4$

$S\Gamma$ \ $S'\Gamma'$	$^4A_2$	$^2F_1$	^2E	$^2F_2$
$^1A_1$				1
$^3F_1$	$1/\sqrt{3}$	$-1/2$	$-1/\sqrt{6}$	$1/2$
^1E		$\sqrt{3}/2$		$1/2$
$^1F_2$		$1/2$	$1/\sqrt{2}$	$1/2$

$n=5$

$S\Gamma$ \ $S'\Gamma'$	$^1A_1$	$^3F_1$	^1E	$^1F_2$
$^2F_2$	$1/\sqrt{15}$	$\sqrt{3}/\sqrt{5}$	$-\sqrt{2}/\sqrt{15}$	$-1/\sqrt{5}$

$(d\gamma^2(S'\Gamma')d\gamma\, {}^2E|\}d\gamma^3\, {}^2E)$

$S\Gamma$ \ $S'\Gamma'$	$^1A_1$	^1E	$^3A_2$
^2E	$1/\sqrt{6}$	$-1/\sqrt{3}$	$1/\sqrt{2}$

Tables of the C. F. P. $(\gamma^{n-2}(S_1\Gamma_1)\gamma^2(S_2\Gamma_2)S\Gamma|\}\gamma^n S\Gamma)$

$(d\varepsilon^{n-2}(S_1\Gamma_1)d\varepsilon^2(S_2\Gamma_2)S\Gamma|\}d\varepsilon^n S\Gamma)$

$n=4$

$S\Gamma = {}^1A_1$

$S_1\Gamma_1$ \ $S_2\Gamma_2$	$^1A_1$	$^3F_1$	^1E	$^1F_2$
$^1A_1$	$\sqrt{2}/3$			
$^3F_1$		$1/\sqrt{2}$		
^1E			$-1/3$	
$^1F_2$				$-1/\sqrt{6}$

$S\Gamma = {}^3F_1$

$S_1\Gamma_1$ \ $S_2\Gamma_2$	$^1A_1$	$^3F_1$	^1E	$^1F_2$
$^1A_1$		$1/3\sqrt{2}$		
$^3F_1$	$1/3\sqrt{2}$	$1/\sqrt{3}$	$1/3$	$1/\sqrt{6}$
^1E		$1/3$		
$^1F_2$		$1/\sqrt{6}$		

$S\Gamma = {}^1E$

$S_1\Gamma_1$ \ $S_2\Gamma_2$	$^1A_1$	$^3F_1$	^1E	$^1F_2$
$^1A_1$			$1/3\sqrt{2}$	
$^3F_1$		$-1/\sqrt{2}$		
^1E	$1/3\sqrt{2}$		$\sqrt{2}/3$	
$^1F_2$				$1/\sqrt{6}$

$S\Gamma = {}^1F_2$

$S_1\Gamma_1$ \ $S_2\Gamma_2$	$^1A_1$	$^3F_1$	^1E	$^1F_2$
$^1A_1$				$1/3\sqrt{2}$
$^3F_1$		$-1/\sqrt{2}$		
^1E				$1/3$
$^1F_2$	$1/3\sqrt{2}$		$1/3$	$-1/\sqrt{6}$

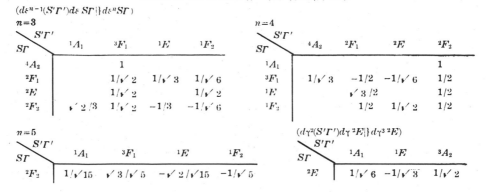

Table II. (continued)

$n=5$

$S\Gamma = {}^2F_2$

$S_1\Gamma_1$ \ $S_2\Gamma_2$	${}^1A_1$	${}^3F_1$	${}^1E$	${}^1F_2$
${}^4A_2$		$1/\sqrt{5}$		
${}^2F_1$		$\sqrt{3}/2\sqrt{5}$	$1/\sqrt{10}$	$1/2\sqrt{5}$
${}^2E$		$1/\sqrt{10}$		$-1/\sqrt{10}$
${}^2F_2$	$1/\sqrt{15}$	$-\sqrt{3}/2\sqrt{5}$	$-1/\sqrt{30}$	$-1/2\sqrt{5}$

$(d\gamma^2(S_1\Gamma_1)d\gamma^2(S_2\Gamma_2)\,{}^1A_1|\}d\gamma^4\,{}^1A_1)$

$S_1\Gamma_1$ \ $S_2\Gamma_2$	${}^1A_1$	${}^1E$	${}^3A_2$
${}^1A_1$	$1/\sqrt{6}$		
${}^1E$		$-1/\sqrt{3}$	
${}^3A_2$			$1/\sqrt{2}$

Table III.

Energy Matrices

d^3

${}^2F_2$ $(a\,{}^2D,\ b\,{}^2D,\ {}^2F,\ {}^2G,\ {}^2H)$

	$d\varepsilon^3({}^2F_2)d\gamma^4$	$d\varepsilon^4({}^3F_1)d\gamma^3$	$d\varepsilon^4({}^1F_2)d\gamma^3$	$d\varepsilon^5 d\gamma^2({}^1A_1)$	$d\varepsilon^5 d\gamma^2({}^1E)$
$d\varepsilon^3$	$-12Dq+5C$	$-3\sqrt{3}B$	$-5\sqrt{3}B$	$4B+2C$	$2B$
$d\varepsilon^2({}^3F_1)d\gamma$		$-2Dq-6B+3C$	$3B$	$-3\sqrt{3}B$	$-3\sqrt{3}B$
$d\varepsilon^2({}^1F_2)d\gamma$			$-2Dq+4B+3C$	$-\sqrt{3}B$	$\sqrt{3}B$
$d\varepsilon\, d\gamma^2({}^1A_1)$				$8Dq+6B+5C$	$10B$
$d\varepsilon\, d\gamma^2({}^1E)$					$8Dq-2B+3C$

${}^2F_1$ $({}^2P,\ {}^2F,\ {}^2G,\ {}^2H)$

	$d\varepsilon^3({}^2F_1)d\gamma^4$	$d\varepsilon^4({}^3F_1)d\gamma^3$	$d\varepsilon^4({}^1F_2)d\gamma^3$	$d\varepsilon^5 d\gamma^2({}^3A_2)$	$d\varepsilon^5 d\gamma^2({}^1E)$
$d\varepsilon^3$	$-12Dq-6B+3C$	$-3B$	$3B$	0	$-2\sqrt{3}B$
$d\varepsilon^2({}^3F_1)d\gamma$		$-2Dq+3C$	$-3B$	$3B$	$3\sqrt{3}B$
$d\varepsilon^2({}^1F_2)d\gamma$			$-2Dq-6B+3C$	$-3B$	$-\sqrt{3}B$
$d\varepsilon d\gamma^2({}^3A_2)$				$8Dq-6B+3C$	$2\sqrt{3}B$
$d\varepsilon d\gamma^2({}^1E)$					$8Dq-2B+3C$

${}^2E$ $(a\,{}^2D,\ b\,{}^2D,\ {}^2G,\ {}^2H)$

	$d\varepsilon^3({}^2E)d\gamma^4$	$d\varepsilon^4({}^1A_1)d\gamma^3$	$d\varepsilon^4({}^1E)d\gamma^3$	$d\varepsilon^0 d\gamma$
$d\varepsilon^3$	$-12Dq-6B+3B$	$-6\sqrt{2}B$	$-3\sqrt{2}B$	0
$d\varepsilon^2({}^1A_1)d\gamma$		$-2Dq+8B+6C$	$10B$	$\sqrt{3}(2B+C)$
$d\varepsilon^2({}^1E)d\gamma$			$-2Dq-B+3C$	$2\sqrt{3}B$
$d\gamma^3$				$18Dq-8B+4C$

Table III. (continued)

$^4F_1$ (^4P, ^4F)

	$d\varepsilon^4(^3F_1)d\gamma^3$	$d\varepsilon^5 d\gamma^2(^3A_2)$
$d\varepsilon^2(^3F_1)d\gamma$	$-2Dq$ $-3B$	$6B$
$d\varepsilon\, d\gamma^2(^3A_2)$		$8Dq$ $-12B$

$^4A_2$ (^4F)	$d\varepsilon^3$	$-12Dq-15B$	$d\varepsilon^3(^4A_2)d\gamma^4$
$^4F_2$ (^4F)	$d\varepsilon^2(^3F_1)d\gamma$	$-2Dq-15B$	$d\varepsilon^4(^3F_1)d\gamma^3$
$^2A_1$ (^2G)	$d\varepsilon^2(^1E)d\gamma^3$	$-2Dq-11B+3C$	$d\varepsilon^4(^1E)d\gamma^3$
$^2A_2$ (^2F)	$d\varepsilon^2(^1E)d\gamma$	$-2Dq+9B+3C$	$d\varepsilon^4(^1E)d\gamma^3$

d^4

$^3F_1$ ($a\,^3P$, $b\,^3P$, $a\,^3F$, $b\,^3F$, ^3G, ^3H)

	$d\varepsilon^2(^3F_1)d\gamma^4$	$d\varepsilon^3(^2F_1)d\gamma^3$	$d\varepsilon^3(^2F_2)d\gamma^3$	$d\varepsilon^4(^3F_1)d\gamma^2(^1A_1)$	$d\varepsilon^4(^3F_1)d\gamma^2(^1E)$	$d\varepsilon^4(^1F_2)d\gamma^2(^3A_2)$	$d\varepsilon^5 d\gamma$
$d\varepsilon^4$	$-16Dq$ $-15B+5C$	$-\sqrt{6}B$	$-3\sqrt{2}B$	$\sqrt{2}(2B+C)$	$-2\sqrt{2}B$	0	0
$d\varepsilon^3(^2F_1)d\gamma$		$-6Dq$ $-11B+4C$	$5\sqrt{3}B$	$\sqrt{3}B$	$-\sqrt{3}B$	$3B$	$\sqrt{6}B$
$d\varepsilon^3(^2F_2)d\gamma$			$-6Dq$ $-3B+6C$	$-3B$	$-3B$	$5\sqrt{3}B$	$\sqrt{2}(B+C)$
$d\varepsilon^2(^3F_1)d\gamma^2(^1A_1)$				$4Dq$ $-B+6C$	$-10B$	0	$3\sqrt{2}B$
$d\varepsilon^2(^3F_1)d\gamma^2(^1E)$					$4Dq$ $-9B+4C$	$-2\sqrt{3}B$	$-3\sqrt{2}B$
$d\varepsilon^2(^1F_2)d\gamma^2(^3A_2)$						$4Dq$ $-11B+4C$	$\sqrt{6}B$
$d\varepsilon\, d\gamma^3$							$14Dq$ $-16B+5C$

$^1F_2$ ($a\,^1D$, $b\,^1D$, $a\,^1G$, $b\,^1G$, ^1F, ^1I)

	$d\varepsilon^2(^1F_2)d\gamma^4$	$d\varepsilon^3(^2F_1)d\gamma^3$	$d\varepsilon^3(^2F_2)d\gamma^3$	$d\varepsilon^4(^3F_1)d\gamma^2(^3A_2)$	$d\varepsilon^4(^1F_2)d\gamma^2(^1E)$	$d\varepsilon^4(^1F_2)d\gamma^2(^1A_1)$	$d\varepsilon^5 d\gamma$
$d\varepsilon^4$	$-16Dq$ $-9B+7C$	$3\sqrt{2}B$	$-5\sqrt{6}B$	0	$-2\sqrt{2}B$	$\sqrt{2}(2B+C)$	0
$d\varepsilon^3(^2F_1)d\gamma$		$-6Dq$ $-9B+6C$	$-5\sqrt{3}B$	$3B$	$-3B$	$-3B$	$-\sqrt{6}B$
$d\varepsilon^3(^2F_2)d\gamma$			$-6Dq$ $+3B+8C$	$-3\sqrt{3}B$	$5\sqrt{3}B$	$-5\sqrt{3}B$	$\sqrt{2}(3B+C)$
$d\varepsilon^2(^3F_1)d\gamma^2(^3A_2)$				$4Dq$ $-9B+6C$	$-6B$	0	$-3\sqrt{6}B$
$d\varepsilon^2(^1F_2)d\gamma^2(^1E)$					$4Dq$ $-3B+6C$	$-10B$	$\sqrt{6}B$
$d\varepsilon^2(^1F_2)d\gamma^2(^1A_1)$						$4Dq$ $+5B+8C$	$\sqrt{6}B$
$d\varepsilon\, d\gamma^3$							$14Dq$ $+7C$

$^1A_1$ ($a\,^1S$, $b\,^1S$, $a\,^1G$, $b\,^1G$, ^1I)

	$d\varepsilon^2(^1A_1)d\gamma^4$	$d\varepsilon^3(^2E)d\gamma^3$	$d\varepsilon^4(^1A_1)d\gamma^2(^1A_1)$	$d\varepsilon^4(^1E)d\gamma^2(^1E)$	$d\varepsilon^6$
$d\varepsilon^4$	$-16Dq$ $+10C$	$-12\sqrt{2}B$	$\sqrt{2}(4B+2C)$	$2\sqrt{2}B$	0
$d\varepsilon^3(^2E)d\gamma$		$-6Dq$ $+6C$	$-12B$	$-6B$	0
$d\varepsilon^2(^1A_1)d\gamma^2(^1A_1)$			$4Dq$ $+14B+11C$	$20B$	$\sqrt{6}(2B+C)$
$d\varepsilon^2(^1E)d\gamma^2(^1E)$				$4Dq$ $-3B+6C$	$2\sqrt{6}B$
$d\gamma^4$					$24Dq$ $-16B+8C$

Table III. (continued)

d^4

^1E ($a\,^1D$, $b\,^1D$, $a\,^1G$, $b\,^1G$, ^1I)

	$d\varepsilon^2(^1E)d\gamma^4$	$d\varepsilon^3(^2E)d\gamma^3$	$d\varepsilon^4(^1E)d\gamma^2(^1A_1)$	$d\varepsilon^4(^1A_1)d\gamma^2(^1E)$	$d\varepsilon^4(^1E)d\gamma^2(^1E)$
$d\varepsilon^4$	$-16Dq$ $-9B+7C$	$6B$	$\sqrt{2}(2B+C)$	$-2B$	$-4B$
$d\varepsilon^3(^2E)d\gamma$		$-6Dq$ $-6B+6C$	$-3\sqrt{2}B$	$-12B$	0
$d\varepsilon^2(^1E)d\gamma^2(^1A_1)$			$4Dq$ $+5B+8C$	$10\sqrt{2}B$	$-10\sqrt{2}B$
$d\varepsilon^2(^1A_1)d\gamma^2(^1E)$				$4Dq$ $+6B+9C$	0
$d\varepsilon^2(^1E)d\gamma^2(^1E)$					$4Dq$ $-3B+6C$

$^3F_2$ (^3D, $a\,^3F$, $b\,^3F$, ^3G, ^3H)

	$d\varepsilon^3(^2F_1)d\gamma^3$	$d\varepsilon^3(^2F_2)d\gamma^3$	$d\varepsilon^4(^3F_1)d\gamma^2(^3A_2)$	$d\varepsilon^4(^3F_1)d\gamma^2(^1E$	$d\varepsilon^5d\gamma$
$d\varepsilon^3(^2F_1)d\gamma$	$-6Dq$ $-9B+4C$	$-5\sqrt{3}B$	$\sqrt{6}B$	$\sqrt{3}B$	$-\sqrt{6}B$
$d\varepsilon^3(^2F_2)d\gamma$		$-6Dq$ $-5B+6C$	$-3\sqrt{2}B$	$3B$	$\sqrt{2}(3B+C)$
$d\varepsilon^2(^3F_1)d\gamma^2(^3A_2)$			$4Dq$ $-13B+4C$	$-2\sqrt{2}B$	$-6B$
$d\varepsilon^2(^3F_1)d\gamma^2(^1E)$				$4Dq$ $-9B+4C$	$3\sqrt{2}B$
$d\varepsilon d\gamma^3$					$14Dq$ $-8B+5C$

$^1F_1$ (^1F, $a\,^1G$, $b\,^1G$, ^1I)

	$d\varepsilon^3(^2F_1)d\gamma^3$	$d\varepsilon^3(^2F_2)d\gamma^3$	$d\varepsilon^4(^1F_2)d\gamma^2(^1E)$	$d\varepsilon^5d\gamma$
$d\varepsilon^3(^2F_1)d\gamma$	$-6Dq$ $-3B+6C$	$5\sqrt{3}B$	$3B$	$\sqrt{6}B$
$d\varepsilon^3(^2F_2)d\gamma$		$-6Dq$ $-3B+8C$	$-5\sqrt{3}B$	$\sqrt{2}(B+C)$
$d\varepsilon^2(^1F_2)d\gamma^2(^1E)$			$4Dq$ $-3B+6C$	$-\sqrt{6}B$
$d\varepsilon d\gamma^3$				$14Dq$ $-16B+7C$

^3E (^3D, ^3G, ^3H)

	$d\varepsilon^3(^4A_2)d\gamma^3$	$d\varepsilon^3(^2E)d\gamma^3$	$d\varepsilon^4(^1E)d\gamma^2(^3A_2)$
$d\varepsilon^3(^4A_2)d\gamma$	$-6Dq$ $-13B+4C$	$-4B$	0
$d\varepsilon^3(^2E)d\gamma$		$-6Dq$ $-10B+4C$	$-3\sqrt{2}B$
$d\varepsilon^2(^1E)d\gamma^2(^3A_2)$			$4Dq$ $-11B+4C$

$^3A_2$ ($a\,^3F$, $b\,^3F$)

	$d\varepsilon^3(^2E)d\gamma^3$	$d\varepsilon^4(^1A_1)d\gamma^2(^3A_2)$
$d\varepsilon^3(^2E)d\gamma$	$-6Dq$ $-8B+4C$	$-12B$
$d\varepsilon^2(^1A_1)d\gamma^2(^3A_2)$		$4Dq$ $-2B+7C$

$^1A_2$ (^1F, ^1I)

	$d\varepsilon^3(^2E)d\gamma^3$	$d\varepsilon^4(^1E)d\gamma^2(^1E)$
$d\varepsilon^3(^2E)d\gamma$	$-6Dq$ $-12B+6C$	$6B$
$d\varepsilon^2(^1E)d\gamma^2(^1E)$		$4Dq$ $-3B+6C$

^5E (^5D)	$d\varepsilon^3(^4A_2)d\gamma$	$-6Dq-21B$	$d\varepsilon^3(^4A_2)d\gamma^3$
$^5F_2$ (^5D)	$d\varepsilon^2(^3F_1)d\gamma^2(^3A_2)$	$4Dq-21B$	$d\varepsilon^4(^3F_1)d\gamma^2(^3A_2)$
$^3A_1$ (^3G)	$d\varepsilon^3(^2E)d\gamma$	$-6Dq-12B+4C$	$d\varepsilon^3(^2E)d\gamma^3$

Table III. (continued)

d^5

$^2F_2$ ($a\,^2F$, $b\,^2F$, $a\,^2G$, $b\,^2G$, ^2H, ^2I, $a\,^2D$, $b\,^2D$, $c\,^2D$)

$d\varepsilon^5$	$-20Dq$ $-20B+10C$	$3\sqrt{6}B$	$\sqrt{6}B$	0	$-2\sqrt{3}B$	$4B+2C$	$2B$	0	0	0
$d\varepsilon^4(^3F_1)d\gamma$		$-10Dq$ $-8B+9C$	$3B$	$\sqrt{6}/2B$	$-3\sqrt{2}/2B$	$3\sqrt{6}/2B$	$3\sqrt{6}/2B$	0	$4B+C$	0
$d\varepsilon^4(^1F_2)d\gamma$			$-10Dq$ $-18B+9C$	$3\sqrt{6}/2B$	$-3\sqrt{2}/2B$	$5\sqrt{6}/2B$	$-5\sqrt{6}/2B$	C	0	0
$d\varepsilon^3(^2F_1)d\gamma^2(^3A_2)$				$-16B+8C$	$2\sqrt{3}B$	0	0	$-3\sqrt{6}/2B$	$-\sqrt{6}/2B$	0
$d\varepsilon^3(^2F_1)d\gamma^2(^1E)$					$-12B+8C$	$-10\sqrt{3}B$	0	$3\sqrt{2}/2B$	$3\sqrt{2}/2B$	$-2\sqrt{3}B$
$d\varepsilon^3(^2F_2)d\gamma^2(^1A_1)$						$2B+12C$	0	$-5\sqrt{6}/2B$	$-3\sqrt{6}/2B$	$4B+2C$
$d\varepsilon^3(^2F_2)d\gamma^2(^1E)$							$-6B+10C$	$-5\sqrt{6}/2B$	$3\sqrt{6}/2B$	$-2B$
$d\varepsilon^2(^1F_2)d\gamma^3(^2E)$								$10Dq$ $-18B+9C$	$3B$	$-\sqrt{6}B$
$d\varepsilon^2(^3F_1)d\gamma^3(^2E)$									$10Dq$ $-8B+9C$	$-3\sqrt{6}B$
$d\varepsilon\, d\gamma^4$										$20Dq$ $-20B+10C$

$^2F_1$ (^2P, $a\,^2F$, $b\,^2F$, $a\,^2G$, $b\,^2G$, ^2H, ^2I)

$d\varepsilon^4(^3F_1)d\gamma$	$-10Dq$ $-22B+9C$	$-3B$	$-3\sqrt{2}/2B$	$3\sqrt{2}/2B$	$-3\sqrt{2}/2B$	$-3\sqrt{6}/2B$	0	C
$d\varepsilon^4(^1F_2)d\gamma$		$-10Dq$ $-8B+9C$	$3\sqrt{2}/2B$	$3\sqrt{2}/2B$	$15\sqrt{2}/2B$	$5\sqrt{6}/2B$	$4B+C$	0
$d\varepsilon^3(^2F_1)d\gamma^2(^1A_1)$			$-4B+10C$	0	0	$10\sqrt{3}B$	$3\sqrt{2}/2B$	$-3\sqrt{2}/2B$
$d\varepsilon^3(^2F_1)d\gamma^2(^1E)$				$-12B+8C$	0	0	$-3\sqrt{2}/2B$	$-3\sqrt{2}/2B$
$d\varepsilon^3(^2F_2)d\gamma^2(^3A_2)$					$-10B+10C$	$2\sqrt{3}B$	$15\sqrt{2}/2B$	$-3\sqrt{2}/2B$
$d\varepsilon^3(^2F_2)d\gamma^2(^1E)$						$-6B+10C$	$5\sqrt{6}/2B$	$-3\sqrt{6}/2B$
$d\varepsilon^2(^1F_2)d\gamma^3$							$10Dq$ $-8B+9C$	$-3B$
$d\varepsilon^2(^3F_1)d\gamma^3$								$10Dq$ $-22B+9C$

Table III. (continued)

d^5

^2E ($a\,^2D$, $b\,^2D$, $c\,^2D$, $a\,^2G$, $b\,^2G$, ^2H, ^2I)

$d\varepsilon^4(^1A_1)d\gamma$	$-10Dq-4B+12C$	$10B$	$6B$	$6\sqrt{3}B$	$6\sqrt{2}B$	$-2B$	$4B+2C$
$d\varepsilon^4(^1E)d\gamma$		$-10Dq-13B+9C$	$-3B$	$3\sqrt{3}B$	0	$2B+C$	$2B$
$d\varepsilon^3(^2E)d\gamma^2(^1A_1)$			$-4B+10C$	0	0	$-3B$	$-6B$
$d\varepsilon^3(^2E)d\gamma^2(^3A_2)$				$-16B+8C$	$2\sqrt{6}B$	$-3\sqrt{3}B$	$6\sqrt{3}B$
$d\varepsilon^3(^2E)d\gamma^2(^1E)$					$-12B+8C$	0	$6\sqrt{2}B$
$d\varepsilon^2(^1E)d\gamma^3$						$10Dq-13B+9C$	$-10B$
$d\varepsilon^2(^1A_1)d\gamma^3$							$10Dq-4B+12C$

$^2A_1$ (^2S, $a\,^2G$, $b\,^2G$, ^2I)

$d\varepsilon^4(^1E)d\gamma$	$-10Dq-3B+9C$	$-3\sqrt{2}B$	0	$6B+C$
$d\varepsilon^3(^2E)d\gamma^2(^1E)$		$-12B+8C$	$-4\sqrt{3}B$	$3\sqrt{2}B$
$d\varepsilon^3(^4A_2)d\gamma^2(^3A_2)$			$-19B+8C$	0
$d\varepsilon^2(^1E)d\gamma^3$				$10Dq-3B+9C$

$^2A_2$ ($a\,^2F$, $b\,^2F$, ^2I)

$d\varepsilon^4(^1E)d\gamma$	$-10Dq-23B+9C$	$3\sqrt{2}B$	$-2B+C$
$d\varepsilon^3(^2E)d\gamma^2(^1E)$		$-12B+8C$	$-3\sqrt{2}B$
$d\varepsilon^2(^1E)d\gamma^3$			$10Dq-23B+9C$

$^4F_1$ (^4P, ^4F, ^4G)

$d\varepsilon^4(^3F_1)d\gamma$	$-10Dq-25B+6C$	$-3\sqrt{2}B$	C
$d\varepsilon^3(^2F_2)d\gamma^2(^3A_2)$		$-16B+7C$	$-3\sqrt{2}B$
$d\varepsilon^2(^3F_1)d\gamma^3$			$10Dq-25B+6C$

$^4F_2$ (^4F, ^4G, ^4D)

$d\varepsilon^4(^3F_1)d\gamma$	$-10Dq-17B+6C$	$\sqrt{6}B$	$4B+C$
$d\varepsilon^3(^2F_1)d\gamma^2(^3A_2)$		$-22B+5C$	$-\sqrt{6}B$
$d\varepsilon^2(^3F_1)d\gamma^3$			$10Dq-17B+6C$

^4E (^4D, ^4G)

$d\varepsilon^3(^2E)d\gamma^2(^3A_2)$	$-22B+5C$	$-2\sqrt{3}B$
$d\varepsilon^3(^4A_2)d\gamma^2(^1E)$		$-21B+5C$

$^6A_1$	(^6S)	$d\varepsilon^3(^4A_2)d\gamma^2(^3A_2)$	$-35B$
$^4A_1$	(^4G)	$d\varepsilon^3(^4A_2)d\gamma^2(^3A_2)$	$-25B+5C$
$^4A_2$	(^4F)	$d\varepsilon^3(^4A_2)d\gamma^2(^1A_1)$	$-13B+7C$

Appendix

Here we will show by induction that the following relation holds between the first and second sum in (3.1) in certain cases:

$$\left(\begin{array}{c}\text{the first sum in (3.1)}\\ \text{for } N \text{ electron system}\end{array}\right) = A(N{:}nn')\left(\begin{array}{c}\text{the second sum in (3.1)}\\ \text{for } N \text{ electron system}\end{array}\right), \qquad (A.1)$$

with

$$A(N:nn')=\{1+A(N-1:n-1,n'-1)\}/\{1+A^{-1}(N-1:n,n')\}\,. \qquad \text{(A.2)}$$

$A(N:nn')$ is a constant (including zero) depending only on N, n and n'. (When the sum on the right hand side of (A.1) is zero, we consider A^{-1} to be zero.)

For this purpose we assume similar relation for $N-1$ electron system, namely

$$\begin{pmatrix}\text{the first sum in (3.1)}\\ \text{for } N-1 \text{ electron system}\end{pmatrix} = A(N-1:nn')\begin{pmatrix}\text{the second sum in (3.1)}\\ \text{for } N-1 \text{ electron system}\end{pmatrix}. \qquad \text{(A.3)}$$

To verify the relation (A.1), we express

$$(\gamma_1^{n-1}(S'\Gamma')\gamma_2^{N-n}(S_2\Gamma_2)\overline{S}\,\overline{\Gamma}|G_{N-1}|\gamma_1^{n'-1}(S''\Gamma'')\gamma_2^{N-n'}(S_1\Gamma_4)\overline{S}\,\overline{\Gamma})$$

and

$$(\gamma_1^{n}(S_1\Gamma_1)\gamma_2^{N-n-1}(S'\Gamma')\overline{S}\,\overline{\Gamma}|G_{N-1}|\gamma_1^{n'}(S_3\Gamma_3)\gamma_2^{N-n'-1}(S''\Gamma'')\overline{S}\,\overline{\Gamma})$$

in (A.1) by

$$(\gamma_1^{n-1}(S'\Gamma')\gamma_2^{N-n-1}(S_0'\Gamma_0')\overline{S_0}\overline{\Gamma_0}|G_{N-2}|\gamma_1^{n'-1}(S''\Gamma'')\gamma_2^{N-n'-1}(S_0''\Gamma_0'')\overline{S_0}\overline{\Gamma_0})$$

using both the recurrence formula and the assumed relation (A.3), and apply the following relation for the transformation matrices;

$$\begin{aligned}&\sum_{\overline{\Gamma}}(\Gamma^1,\Gamma_3\Gamma_4(\Gamma^2)\Gamma|\Gamma^1\Gamma_3(\overline{\Gamma}),\Gamma_4\Gamma)\\&\quad(\Gamma_1\Gamma_2(\Gamma^1),\Gamma_3\overline{\Gamma}|\Gamma_1,\Gamma_2\Gamma_3(\Gamma'')\overline{\Gamma})\\&\quad(\Gamma_1,\overline{\Gamma}_2\overline{\Gamma}_3(\Gamma'')\overline{\Gamma}|\Gamma_1\overline{\Gamma}_2(\Gamma^3),\overline{\Gamma}_3\overline{\Gamma})\\&\quad(\Gamma^3\overline{\Gamma}_3(\overline{\Gamma}),\Gamma_4\Gamma|\Gamma^3,\overline{\Gamma}_3\Gamma_4(\Gamma^4)\Gamma)\end{aligned}
=
\begin{aligned}&\sum_{\overline{\Gamma}}(\Gamma_1\Gamma_2(\Gamma^1),\Gamma^2\Gamma|\Gamma_1,\Gamma_2\Gamma^2(\overline{\Gamma})\Gamma)\\&\quad(\Gamma_2,\Gamma_3\Gamma_4(\Gamma^2)\overline{\Gamma}|\Gamma_2\Gamma_3(\Gamma''),\Gamma_4\overline{\Gamma})\\&\quad(\overline{\Gamma}_2\overline{\Gamma}_3(\Gamma''),\Gamma_4\overline{\Gamma}|\overline{\Gamma}_2,\overline{\Gamma}_3\Gamma_4(\Gamma^4)\overline{\Gamma})\\&\quad(\Gamma_1,\overline{\Gamma}_2\Gamma^4(\overline{\Gamma})\Gamma|\Gamma_1\overline{\Gamma}_2(\Gamma^3),\Gamma^4\Gamma)\,.\end{aligned} \qquad \text{(A.4)}$$

(A.4) can be proved in the following way:

the left hand side

$$\begin{aligned}&=\sum_{\overline{\Gamma}}(\Gamma_1\Gamma_2(\Gamma^1),\Gamma_3\Gamma_4(\Gamma^2)\Gamma|\Gamma_1,\Gamma_2\Gamma_3(\Gamma'')(\overline{\Gamma})\Gamma_4\Gamma)\\&\qquad(\Gamma_1,\overline{\Gamma}_2\overline{\Gamma}_3(\Gamma'')(\overline{\Gamma})\Gamma_4\Gamma|\Gamma_1\overline{\Gamma}_2(\Gamma^3)\overline{\Gamma}_3\Gamma_4(\Gamma^4)\Gamma)\\&=\sum_{\overline{\Gamma}}(\Gamma_1\Gamma_2(\Gamma^1),\Gamma_3\Gamma_4(\Gamma^2)\Gamma|\Gamma_1,\Gamma_2\Gamma_3(\Gamma'')\Gamma_4(\overline{\Gamma})\Gamma)\\&\qquad(\Gamma_1,\overline{\Gamma}_2\overline{\Gamma}_3(\Gamma'')\Gamma_4(\overline{\Gamma})\Gamma|\Gamma_1\overline{\Gamma}_2(\Gamma^3)\overline{\Gamma}_3\Gamma_4(\Gamma^4)\Gamma)\end{aligned}$$

=the right hand side,

because

$$\sum_{\underline{\Gamma}}(\Gamma_1\Gamma''(\overline{\Gamma})\Gamma_4\Gamma|\Gamma_1,\Gamma''\Gamma_4(\underline{\Gamma})\Gamma)(\Gamma_1,\Gamma''\Gamma_4(\underline{\Gamma})\Gamma|\Gamma_1\Gamma''(\overline{\Gamma}')\Gamma_4\Gamma)=\delta(\overline{\Gamma},\overline{\Gamma}')\,.$$

To complete the proof for the simplification of the recurrence formula for non-diagonal elements described in § 3, it is necessary to show that the relation (A.3) holds for 4 electron system. This is possible only when a single ancestor exists and so the value of $A(3:nn')$ or $A^{-1}(3:nn')$ is equal to zero.

For the case of $n-n'=2$, where the ancestor of the type $(\gamma_1^2S\Gamma|g|\gamma_2^2S\Gamma)$ and the constants $A(3:2,0)$ and $A(3:3,1)$ are zero, we obtain the value of $A(4:3,1)$ as unity and so on, using the relation (A.2). $A(N:N,N-2)$ and $A(N:2,0)$ are always zero, because we have only one sum in the recurrence formula for these configurations. The general expression for $A(N:nn')$ in this case is

$$A(N:n,n-2)=(n-2)/(N-n)\,. \qquad \text{(A.7)}$$

As was mentioned in § 3, there is only one ancestor of the type $(\gamma_1^2S\Gamma|g|\gamma_1\gamma_2S\Gamma)$ for the case of $n-n'=1$ in our special problem of the octahedral group, and the constant A is given by

$$A(N:n,n-1)=(n-2)/(N-n)\,. \qquad \text{(A.8)}$$

For the non-diagonal elements in the case of $n-n'=0$, we see that A takes the following values:

$$A(N:n,n)=(n-1)/(N-n-1)\,. \qquad \text{(A.9)}$$

References

1) R. Finkelstein and J. H. Van Vleck: J. Chem. Phys. **8** (1940) 790.
2) F. E. Ilse and H. Hartmann: Zs. f. phys. Chem. **197** (1951) 239.
H. Hartmann and H. L. Schlaefer: *ibid.* **197** (1951) 115.
H. Hartmann and H. L. Schlaefer: Zs. f. Naturforschg. **6a** (1951) 751, 754, 760.
3) M. Kotani: unpublished.
K. Kambe and T. Usui: unpublished.
4) W. H. Kleiner: J. Chem. Phys. **20** (1952) 1784.
5) D. Polder: Physica **IX** (1942) 709.
6) J. H. Van Vleck: J. Chem. Phys. **7** (1939) 72.
7) K. W. H. Stevens: Proc. Roy. Soc. **A219** (1953) 542.
8) G. Racah: Phys. Rev. **63** (1943) 367; Phys. Rev. **76** (1949) 1352, cited as R III and R IV respectively.

ERRATA

On the Absorption Spectra in Complex Ions. I
by Y. Tanabe and S. Sugano

J. Phys. Soc. Japan 9 (1954) 753

1) On p. 758, in the table of Clebsch-Gordan coefficient $(F_1\beta F_1\beta|F_1F_1Ev)$, "$1/\sqrt{1}$" should be read as "$1/\sqrt{2}$".
2) On p. 760, in the energy matrix element for ^2E (Table III). "$-12Dq-6B+3B$" should be read as "$-12Dq-6B+3C$".
3) On p. 761, a two dimensional energy matrix is for "$^4F_1$".
4) On p. 761, on the third line from the top, "$^2A_1(^2G)\ d\varepsilon^2(^1E)d\gamma^3$ should be read as "$^2A_1(^2G)\ d\varepsilon^2(^1E)d\gamma$".
5) On p. 762, conjugate states for $^1F_1$, $d\varepsilon^3(^2F_1)d\gamma^3$, $d\varepsilon^3(^2F_2)d\gamma^3$, $d\varepsilon^4(^1F_2)d\gamma^2(^1E)$, and $d\varepsilon^5 d\gamma$ should be at the top of the first, second, third, and fourth columns, respectively.

Reprinted from *J. Phys. Soc. Japan*, **9**, 766–779 (1954)

18

JOURNAL OF THE PHYSICAL SOCIETY OF JAPAN Vol. 9, No. 5, SEP.—OCT., 1954

On the Absorption Spectra of Complex Ions II

By Yukito TANABE
Physics Department, Tokyo University
and Satoru SUGANO
Institute of Broadcating N.H.K.

(Received June 11, 1954)

The secular determinants obtained in the previous paper are solved for the energy levels which are important in the absorption spectra of the normal complex ions, leaving the crystalline field strength as a parameter. The values of B and C (Racah's parameters) there needed are determined from the observed spectra of free ions or in some cases by extrapolation.

The f-values of the transitions which connect the energy levels calculated are estimated and compared with the observed intensities. The difference of the spectral width among absorption bands and lines is also considered using the energy diagram obtained. Following the assignments determined by the above considerations, the calculated positions of lines and bands are rather in good agreement with the experimental data in divalent ions $[MX_6]^{2+}$ (M=Cr, Mn, Fe, Co, Ni), when we adjust the crystalline field parameter Dq suitably. In trivalent ions $[MX_6]^{3+}$ (M=Ti, V, Cr, Mn, Fe), it is necessary besides the adjustment of Dq to use smaller values of B and C than those of the free ions to obtain better agreement with experiments. The values of Dq thus determined are of reasonable magnitude close to those obtained in other ways. The decrease of B and C compared with those of the free ions might be connected with the recent Stevens' suggestion. It is interesting that, though the agreement is poor about $[Co(H_2O)_6]^{3+}$ and $[Co(NH_3)_6]^{3+}$ whose bindings are usually considered covalent, the qualitative explanation of their spectra is found possible in the crystalline field approximation.

§ Introduction

It has already been mentioned in the previous paper (**1**) that the complex ions under consideration show several forms of the absorption spectra, namely the relatively intense band, the weak band and the sharp line. No quantitative consideration, however, has been made to explain the origin of these differences.

We shall therefore investigate the nature

of the transitions due to the optical absorption in these complexes in § 3. Then the comparison with the experimental data will be given in § 4 using the energy level diagram obtained in § 2.

§ 2. Determination of *B*, *C* and the Resulting Energy Level Diagram

Because most of the spectra of the free ions of the iron group elements are not yet completely known, in order to make the extrapolation possible, the values of B and C for the doubly or triply ionized states were estimated in the following way.

Since the term values of the elements in the d^2-isoelectronic sequence, i.e., Ti III, V IV, Cr V, Mn VI and Fe VII are experimentally known[1)], we have determined the values of B and C so that they will give the observed values for ^3F, ^3P and ^1G. When we plot these values of B and C against the atomic number Z[2)], they approximately fall on a straight line: $B=145\ (Z-s)\ \text{cm}^{-1}$. $C=705\ (Z-s)\ \text{cm}^{-1}$ (compare with $B=129.5\ (Z-s)\ \text{cm}^{-1}$, $C=514.5\ (Z-s)\ \text{cm}^{-1}$ which one obtains using the Slater type wave function with $n^*=3$).

To obtain B and C in the d^3-isoelectronic sequence, we calculate $\Delta B=B(d^3)-B(d^2)$, $\Delta C=C(d^3)-C(d^2)$ using the observed value of the parent ion after Meshkov[3)] (for the term value of $^1S(d^2)$ we adopt the calculated value, since the observed values are lacking), so that they give term values of ^4F, ^4P and ^2H correctly. In Table I, the column (diff. (I)) gives the differences between the values calculated in this way and the observed ones, and the column (diff. (II)) gives differences between the observed values and the values calculated in the usual way using the values of B and C determined so as to give ^4F, ^4P and ^2H correctly. The values of $B(d^3)$, $C(d^3)$ thus obtained also fall on a straight line parallel to the line for d^2 determined previously.

For d^n-isoelectronic sequences ($n\geq 4$) where we have few observed term values, this method cannot be applied, and the values of B and C were determined roughly from the observed three levels suitably chosen. Then they were adjusted so that they fall on straight lines which are parallel to the lines of d^2- and d^3-isoelectronic sequences and equally spaced with the distance between the d^2- and d^3-lines.

In Fig. 1 and 2 the points marked by × indicate the values thus determined for d^2 and d^3. But for d^4 B-values are not plotted,

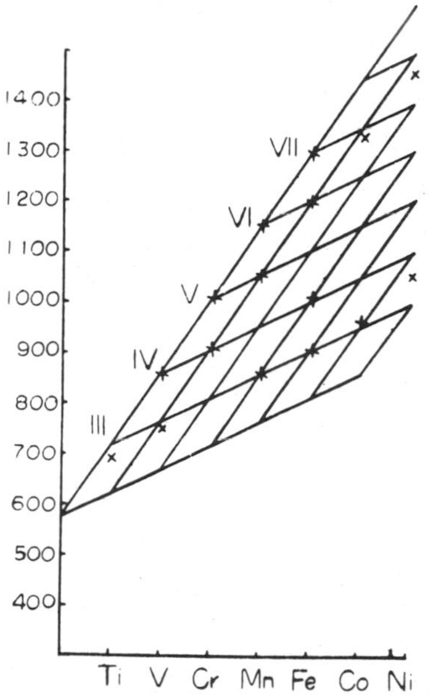

Fig. 1. B-values for various Z and ionicities.

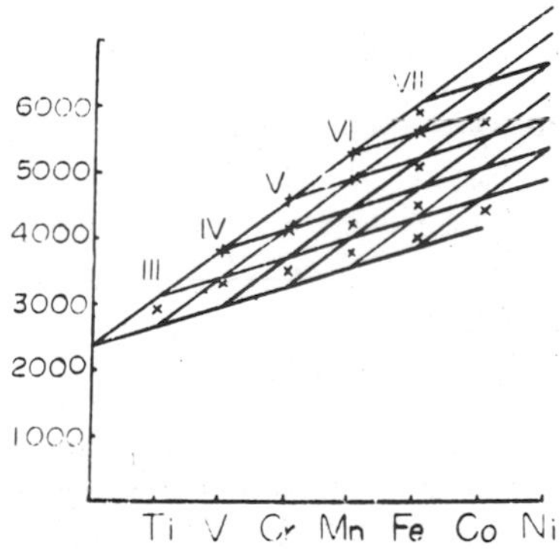

Fig. 2. C-values for various Z and ionicities.

since if they were determined from ^5D, ^3H, ^3G we have far smaller values than those expected (smaller about by 130 cm^{-1}). C-values for d^4 determined from ^5D, ^{3}H, ^3G fall on the points expected from the graph. For d^5 B, C-values are the average of the two determined from ^6S, ^4G, ^4D and ^4P, ^4F respectively. For d^6 they were determined from ^5D,

Table I. The deviations of the calculated term values from those of the observed. (obs.—calc.) (Values are given in cm^{-1})

d^2

V IV		Diff.	Cr V		Diff.	Mn VI		Diff.	Fe VII		Diff.
^3F	419		^3F	662		^3F	964		^3F	1346	
^1D	10960	−1397	^1D	13200	−1764	^1D	15336	−2065	^1D	17475	−2335
^3P	13344		^3P	15868		^3P	18344		^3P	20855	
^1G	18389		^1G	22060		^1G	25511		^1G	28915	
$B=862$			$B=1014$			$B=1159$			$B=1301$		
$C=3815$			$C=4617$			$C=5322$			$C=5981$		

d^3

V III		Diff. (I)	Diff. (II)	Cr IV		Diff. (I)	Diff. (II)	Mn V		Diff. (I)	Diff. (II)	Fe VI		Diff. (I)	Diff. (II)
^4F	336			^4F	554			^4F	813			^4F	813		
^2P	11327	−3484	−5579	^4P	14324			^4P	16781			^4P	19235		
^4P	11668			^2P	14229	−4341	−6985	^2G	18158	−195	−1670	^2G	20996	−292	−1961
^2G	12089	−43	−1040	^2G	15258	−106	−1366	^2P	22961	908	−2190	^2P	26297	820	−2685
^2D	16317	−590	−2443	^2D	20487	−788	−3652	^2D	24626	−661	−3172	^2D	28524	−686	−3516
^2H	16907			^2H	21214			^2H	25150			^2H	28981		
$B=755$				^2F	36735	2506	1751	$B=1065$				$B=1205$			
$C=3257$				$B=918$				$C=4919$				$C=5659$			
				$C=4133$											

Table II. The values of B, C* and γ determined in the way described in § 2. (cm^{-1})

	B	C	C/B		B	C	C/B
Ti III	694.6	2910.4	4.190				
V	755.4	3257.2	4.312	V IV	861.6	3814.9	4.428
Cr	810.	3565.	4.401	Cr	918.	4132.7	4.502
Mn	860.	3850.	4.477	Mn	965.	4450.	4.611
Fe	917.	4040.	4.406	Fe	1015.	4800.	4.729
Co	971.	4497.	4.633	Co	1065.	5120.	4.808
Ni	1030.	4850.	4.709	Ni	1115.	5450.	4.888

* These B-values are not greatly different from those values recommended by Catalan et al.[5] after our calculation has been performed, and the value determined by Meshkov[4] for V III.

^3H, ^3G, and for d^7 B's were determined from ^4F, ^4P, ^2G and C's are the average of the two determined from ^4F, ^4P, ^2G and ^4P, ^2G, ^2H respectively. For d^8 B and C were determined from ^3F, ^3P.

The calculated term values using these B, C do not show so good agreement with the observed ones which were not used in the determination of these values ($\Delta E \simeq 2000$–$3000\ cm^{-1}$ except some particular terms for which the error is even larger), so that we may expect that error of the same order would occur for the element, for example Co IV, for which spectral terms are not known yet. These errors are much larger than the errors that occur when one uses the values B, C determined by the least square method. But as our calculations are for molecular ions, we consider that the accuracy of such degree is sufficient for our purpose, judging from the nature of the crystalline field approximation. Values we adopt here are given in Table II. $\gamma = C/B$ is nearly constant for all elements and is about 4~5, while, if we adopt the Slater type wave function, $\gamma = 3.97$ and is independent of the effective nuclear charge.

With these values we solved the secular determinants including the interaction between configurations, and obtained the lowest eigenvalues among those belonging to the same $S\Gamma$. The results are shown in the energy diagrams in full line (Fig. 3–9). The other interesting levels were also solved neglecting the interaction between different configurations and are given in broken lines in the range $\Delta = Dq/B \geq 2$. Both the level energy (ordinate) which are measured always from the ground level and the strength of the crystalline field (abscissa) are scaled by B as unit. The values of γ and B used in the calculation of the level energy are also given. The observed term values and the lowest level of $d^{N-1}s$ configuration in the free state of the ion are indicated there.

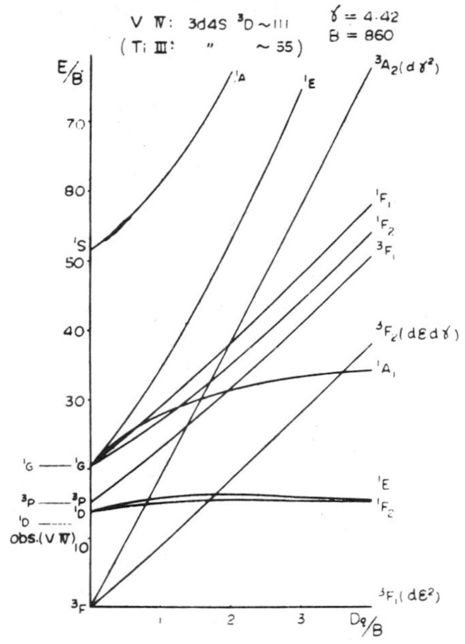

Fig. 3. The energy diagram for d².

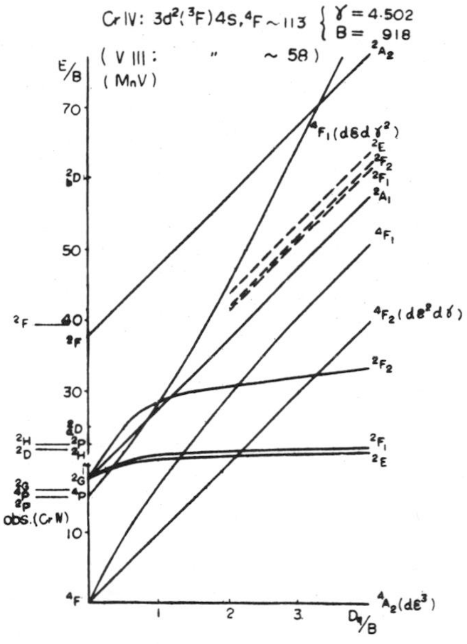

Fig. 5. The energy diagram for d³.

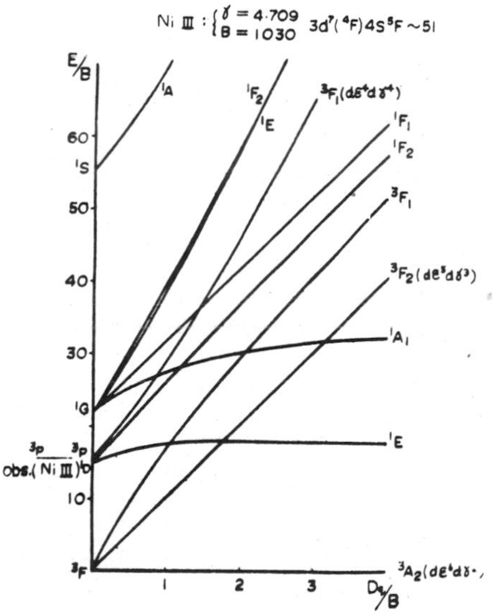

Fig. 4. The energy diagram for d⁸.

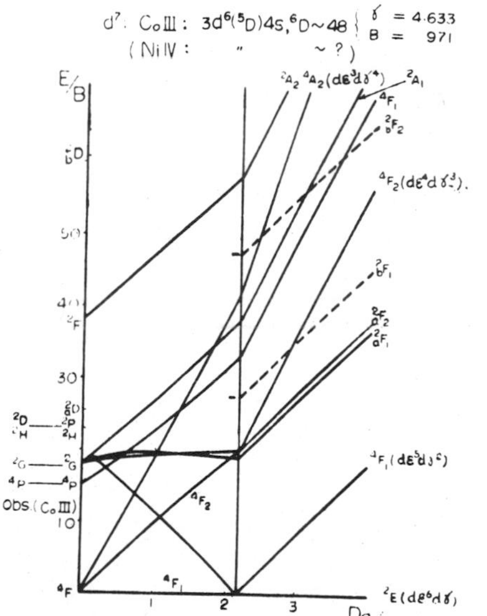

Fig. 6. The energy diagram for d⁷.

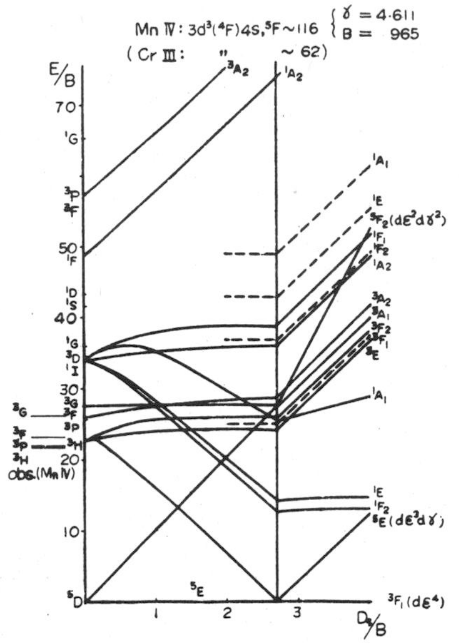

Fig. 7. The energy diagram for d^4.

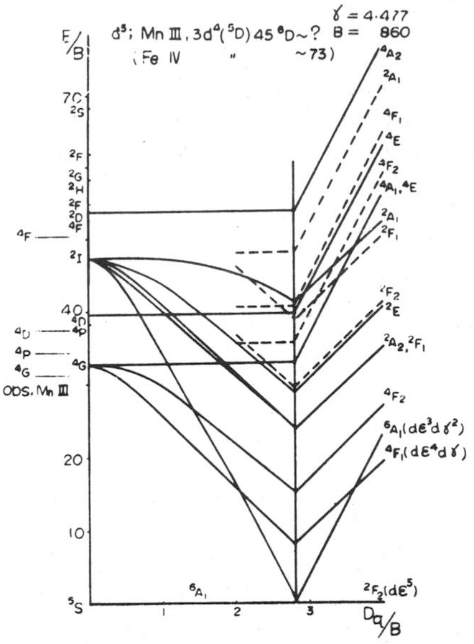

Fig. 9. The energy diagram for d^5.

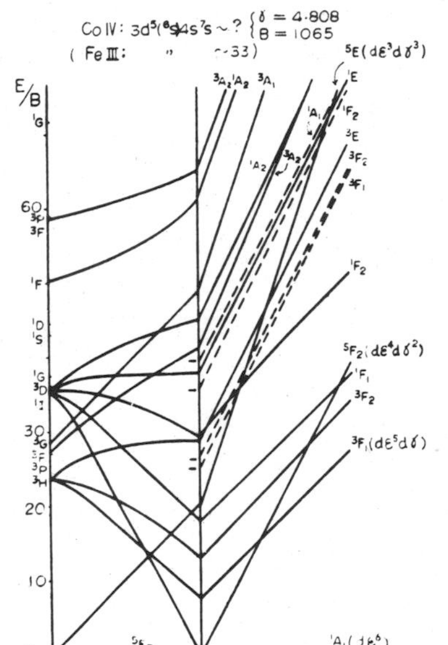

Fig. 8. The energy diagram for d^6.

§ 3. Nature of the Transitions

In this section we shall consider the intensity of transitions, selection rules and the problem of the width of the observed absorption, in order to obtain some theoretical foundations for the interpretation of the observed spectra.

The relation between the absorption coefficient $k(\nu)$ and the oscillator strength f is given by the relation

$$f=\frac{mc}{\pi Ne^2}\int k(\nu)d\nu\ , \qquad (3.1)$$

where N is the number of the absorption centres per cubic centimeter, and ν the frequency measured in sec^{-1}. For solutions, the intensity of the absorption is ordinarily given in terms of the extinction coefficient $\mathcal{E}$ defined through the relation $I=I_0 10^{-\mathcal{E}c_0 x}$ where x is the thickness of the solution in cm and c_0 is the concentration in mol per litre, so that correspondingly to (3.1) we have

$$f=\frac{2.3\times10^3}{N_{\rm Avog}}\cdot\frac{mc}{\pi Ne^2}\int \mathcal{E}(\nu)d\nu\ , \qquad (3.2)$$

where $N_{\rm Avog}$ is the Avogadro number. For a rough estimation of f, we can replace the integral in (3.1) or (3.2) with $2.5\times k_{\max}\Delta\nu$ or $2.5\times\mathcal{E}_{\max}\Delta\nu$ in which $\Delta\nu$ represents the half width, assuming the absorption curve to be gaussian.

Experimentally it is known that in solutions all the bands in question have the values of $\log\mathcal{E}\simeq1\sim2$, $\Delta\nu\simeq10^{14}\,sec^{-1}$ ($3000\,cm^{-1}$) except for the case of $[Fe(H_2O)_6]^{3+}$, whose bands observed at 14300 cm^{-1} and 18200 cm^{-1} have rather small intensity $\log \mathcal{E} \simeq -1$ and the width $\Delta\nu \simeq 2\times10^{14}$ sec^{-1}. Therefore, for the broad bands observed in solutions f-value can be estimated to be of the order of 2×10^{-4} if we take $\log\mathcal{E}=1$ (according to Sauer's measurement[6]) for the bands of chrome alum, f is of the order of 10^{-3}), but for the weak bands of $[Fe(H_2O)_6]^{3+}$ it is about 4×10^{-6}.

On the other hand, the doublet lines of the chrome alums have the oscillator strength $f\simeq3\times10^{-8}$ according to Sauer's results. The line (though it is rather broad, the reason why we regard it as a line will be given later in § 4.) observed at 24600 cm^{-1} at room temperature has the values $\log\mathcal{E}\simeq-1$, $\Delta\nu\simeq 4\times10^{13}$ sec^{-1}, so that its f-value becomes to be of the order of 10^{-6}.

All the observed absorption in question are thus considered to be due to forbidden transitions. This fact supports the interpretation that they are the transitions between the levels arising from the d^n-configuration, since all the levels then have the same parity so that the transitions between them are forbidden.

Among the forbidden transitions between the states with the same parity, the following transitions are able to have non-vanishing intensity; 1) the electric-dipole transition coupled with the vibration, 2) the electric-quadrupole transition and 3) the magnetic-dipole transition. Electric-dipole radiation due to naturally unsymmetric fields can be treated in a similar way to 1) and its transition probability may be of the same order of magnitude as those of 1), but we do not thuch it here because of its uncertainly. (See Van Vleck[7][8].)

We now estimate the f-values of those transitions in the same way as Van Vleck has done on the spectra of the rare earth salts.

For the intrasystem transition which connects the two states with the same multiplicity, the transition probability of the type 1) is given by

$$\sigma_I \simeq \sigma_{allow}(V_{hem}/h\nu')^2\,, \qquad (3.3)$$

where V_{hem} is a matrix element of the hemihedral part of the crystalline field, and ν' is the frequency interval separating the state in question from the excited states of the opposite parity, which are not limited to the excited states of the central metal ion but include those of the complex as a whole.

To obtain V_{hem} we expand the crystalline field $V(x,Q)$ in power series of Q where x denotes the electron coordinate and Q is the normal coordinates of the vibrational displacements

$$V(x,Q)=V_0(x)+\sum V_q^{(k)}Q_q^{(k)} + \tfrac{1}{2}\sum V_{qq'}^{(kk')}Q_q^{(k)}Q_{q'}^{(k')}\,, \qquad (3.4)$$

where $V_0(x)$ represents the crystalline field at the equilibrium position of the nuclei. We see that $V_q^{(k)}$ and $Q_q^{(k)}$ or $V_{qq'}^{(kk')}$ and $Q_q^{(k)}Q_{q'}^{(k)}$ have the same transformation properties, since the crystalline field has the properties

$$V(x,Q)=V(Rx,RQ)\,, \qquad (3.5)$$

where R denotes the symmetry operation. Now V_{hem} originates from the vibrational displacement with odd inversion symmetry F_{1u} and F_{2u}, that is, the main contribution to V_{hem} comes from $V_q(F_{1u})Q_q(F_{1u})$ and $V_q(F_{2u})Q_q(F_{2u})$ of the second term in (3.4). The order of magnitude of the second term is seen to be $(Q/r)V_0$. so that we can estimate V_{hem} using the relation

$$V_{hem}\simeq(\delta R/\bar{r})V_0\,, \qquad (3.6)$$

where δR is the zero-point amplitude of the vibration and $\bar{r}$ is the average of r. The amplitude δR is estimated from

$$2\pi^2\mu\nu_0^2(\delta R)^2=\tfrac{1}{2}h\nu_0\,,$$

where ν_0 is the fundamental frequency of the vibration and μ is the effective mass for the vibrating complex. For the numerical value of ν_0, we adopt 10^{13} sec^{-1} which may be compared with the infrared absorption at 2.4×10^{13} sec^{-1} and 4×10^{13} sec^{-1} of the cobalt complex recently found by Hill and Rosenberg[9]. For the value of μ, we take the mass of the central ion, $\mu\simeq4\times10^{-23}$ gr., as μ does not seem to differ from the mass of the central ion so much. Then we obtain the zero point amplitude to be of the order of 10^{-9} cm. Assuming the value $\bar{r}\simeq10^{-8}$ cm, $V_0\simeq10^4$ cm^{-1}, we have $V_{hem}\simeq10^3$ cm^{-1}. With this and $\nu'\simeq10^5$ cm^{-1}, we finally obtain

$$\sigma_I\simeq10^{-4}\cdot\sigma_{allow}\,. \qquad (3.7)$$

The f-value of the allowed transition can be estimated from the relation

$$f\simeq\frac{8\pi^2 m}{3h}\nu\bar{r}^2\,, \qquad (3.8)$$

where m is the electron mass and the frequency ν is assumed to be $6\times10^{14}\ \mathrm{sec}^{-1}$, so that the f-value of the allowed transition may be considered to be of the order of unity. The f-value of the case 1) thus would be of the order of

$$f_{\mathrm{I}}\simeq10^{-4}\,. \tag{3.9}$$

The transition probabilities of the case 2) and 3) for the intrasystem combinations are estimated from the relation

$$\text{for 2)}\qquad \sigma_{\mathrm{II}}=32\pi^6\nu^5e^2Q^2/5hc^5\,, \tag{3.10}$$

$$\text{for 3)}\qquad \sigma_{\mathrm{III}}=64\pi^4\nu^3M^2/3hc^3\,. \tag{3.11}$$

Here Q is the "quadrupole amplitude" which can be estimated from the relation $Q\simeq\bar{r}^2$, and M is the matrix element of the magnetic moment connecting the states in question and considered of the order of the Bohr magneton. Then the corresponding f-values are

$$\text{for 2)}\qquad f_{\mathrm{II}}\simeq2\times10^{-7}\,, \tag{3.12}$$

$$\text{for 3)}\qquad f_{\mathrm{III}}\simeq4\times10^{-6}\,. \tag{3.13}$$

So that, for the intrasystem combinations, it seems to us that the electric-dipole transition coupled with the vibration predominates over the other two cases in our problem, though the intensity of the magnetic-dipole transition has the possibility to become comparable with that of the case 1) as our estimates may be in error by a factor of 10 or 100.

For the intersystem combinations we must take the spin-orbit interaction into account. We thus consider the contamination of the state with different multiplicity due to the spin-orbit interaction as follows[10]

$$\Psi(\alpha_0\Gamma_T)=a\Psi(\alpha S\Gamma,\Gamma_T)+b\Psi(\alpha' S'\Gamma',\Gamma_T)\,, \tag{3.14}$$

where Γ_T is the irreducible representation of $S\times\Gamma$ and contamination occurs only among the states with the same Γ_T. Here it may generally be assumed that the coefficient a is nearly equal to unity, and b is of the order of magnitude $\Delta\nu/\nu$ where $\Delta\nu$ is the multiplet width and ν is the interval between different multiplets. Then the f-values of 1), 2) and 3) for the intersystem combinations are estimated from (3.9), (3.12) and (3.13) multiplying the factor b^2 respectively. Assuming the value of b as 3×10^{-2} which is estimated to be a little smaller than that of the rare earth salts, f-values thus obtained are

$$\left.\begin{array}{ll}\text{for 1)} & \bar{f}_{\mathrm{I}}\simeq10^{-7}\,,\\ \text{for 2)} & \bar{f}_{\mathrm{II}}\simeq2\times10^{-10}\,,\\ \text{for 3)} & \bar{f}_{\mathrm{III}}\simeq4\times10^{-9}\,.\end{array}\right\} \tag{3.15}$$

From these considerations we now have some theoretical ground to say that, to the relatively strong absorption bands, the intrasystem transitions may be assigned in which electric-dipole transitions coupled with vibrations probably predominates, whose estimated f-value 10^{-4} should be compared with the observed ones $10^{-3}\sim10^{-4}$, and that, to the weak bands of $[Fe(H_2O)_6]^{3+}$ or to the lines, the intersystem transitions may correspond, where estimated f-value $\simeq10^{-7}$ is not so different from those observed $10^{-6}\sim10^{-8}$.

We shall briefly consider the fact that the absorption bands of the triply ionized complex show always a larger intensity than those of the doubly ionized one by a factor 10 in ε. This difference of the absorption intensity between the doubly and the triply ionized complexes is thought to come from the difference in the strength of the crystalline field and from that in ν', because the factor σ_{allow} does not differ greatly in both cases. The crystalline field is generally stronger in the triply ionized complex ions as will be seen later. Further ν' in the latter may be smaller than that of the former since the excited states with the odd parity likely exist at relatively lower positions in the more tightly bound triply ionized complexes owing to their departure from the atomic character. This supposition will be only justified when it is tested by a more thorough treatment of the complex as a molecule.

The selection rules were examined on the individual cases 1), 2) and 3) in each complexes. All assignment adopted in the following section do not break such selection rules. In our problem of the perfect cubic field, difference of the absorption due to the polarization of the incident light does not, of course, occur, but if we introduce the field of the lower symmetry that may really exist in crystals, or if some of the ligands are replaced with other atoms or molecules, i.e., in $[M\,X_n\,Y_{6-n}]^{2+,3+}$ complexes, this occurs and further investigation on the selection rules in these cases will give more precise information as to the nature of the absorption.

The coexistence of broad bands and lines in iron group complexes is an interesting phenomenon. In the energy diagrams (Fig. 3→Fig. 9), we can see that the energy intervals between the levels belonging to the same configuration $d\varepsilon^n d\gamma^{N-n}$ becomes to be almost independent of the strength of the crystalline field in the region where Dq is $\gtrsim 2B$ (this is really what happens in the iron group complexes), in other words, in this region interactions between configurations are negligible. But those belonging to different configurations are strongly dependent of the field strength, so that we are led to assign the transitions which connect the levels belonging to the same configurations to the lines and those belonging to different configurations to broad bands. Thus, we consider that the strong crystalline field in the iron group complexes is the cause of such a phenomenon. In rare earth salts where the crystalline field is much weaker than that of the iron group salts by a factor of 10^2, the broad bands has not been observed in the visible region as far as we know. (The broad bands observed in the ultraviolet region in Ce^{3+} [11] or Yb^{3+} [12] salts seem to be somewhat different from those of the iron group complex in their origin.)

The observed half width $3000\,cm^{-1}$ of the broad bands would be explained from this view-point, since the crystalline field could fluctuate with the amplitude of $V_{hem} \simeq 1000$ cm^{-1} even by the zero-point vibration of the system, coupled with which electric-dipole transitions are allowed to have non-vanishing intensities.

§ 4. Comparison with Experiments

We now compare our results with the experimental data in accordance with the considerations in the previous section, and determine Dq and, redetermine, if necessary, B and C which may be somewhat different from those determined previously from data for free ions, so as to give good agreement with the observation.

Measurements on these absorption spectra have been performed by many experimentalists, but the numbers of observed lines and bands available in the individual complex are often not so rich that the arbitrariness of the values of these parameters will exist in some cases.

$3d$: In the case of a single d-electron or hole, only a single absorption band is allowed to appear in our treatment. According to Hartmann an Schläfer[13], $[Ti(H_2O)_6]^{3+}$ has a single band with intensity $\log \varepsilon \simeq 0.6$ and its absorption maximum is at $20300\,cm^{-1}$, which can be assigned to $(d\varepsilon)^2F_2 \to (d\gamma)^2E$ transition; the value of Dq thus determined is $Dq = 2030\,cm^{-1}$.

$3d^9$: Dreisch[14] found a single band on $Cu^{2+}{}_{aq.}$; its absorption maximum is at $12200\,cm^{-1}$, which can be assigned to $(d\varepsilon^6 d\gamma^3)^2E \to (d\varepsilon^5 d\gamma^4)^2F_2$ transition. Then Dq is given as $1220\,cm^{-1}$.

$3d^2$: Hartmann and Schläfer[15] found two bands in the $[V(H_2O)_6]^{3+}$ complex ion at room temperature:

	abs. max.	log ε
$V(ClO_4)_3$ in $HClO_4$	$17.3 \times 10^3\,cm^{-1}$	0.74
	25.0	0.92
$NH_4V(SO_4)_2 \cdot 12H_2O_{aq.}$	17.8	0.54
	25.6	0.81
$CsV(SO_4)_2 \cdot 12H_2O_{aq.}$	17.8	0.65
	25.6	0.81

We assign these two bands to $(d\varepsilon^2)^3F_1 \to (d\varepsilon d\gamma)^3F_2$, $(d\varepsilon^2)^3F_1 \to (d\varepsilon d\gamma)^3F_1$ transitions respectively. Calculated values are $17300\,cm^{-1}$ for ${}^3F_1 \to {}^3F_2$ and $25400\,cm^{-1}$ for ${}^3F_1 \to {}^3F_1$, when we take the value of B as $640\,cm^{-1}$ which is determined using the fact that the energy interval between ${}^3F_2$ and ${}^3F_1$ should be aporoximately equal to $12B$. (In our treatment γ is always assumed not to be changed from those of the free ion.) The corresponding Dq is about $1860\,cm^{-1}$. It must be remarked here that our assignment of the second band is different from that adopted by Hartmann et al.,[15] namely the second band in their treatment corresponds to $(d\varepsilon^2)^3F_1 \to (d\gamma^2)^3A_2$ transition in our case, which is considered to be strongly forbidden even if the coupling with vibrations were taken into account, for it corresponds to a two electron jump. Further, besides this broad band just mentioned, our treatment leads to the expectation that lines must be found corresponding to the transitions $(d\varepsilon^2)^3F_1 \to (d\varepsilon^2)^1E, {}^1F_2$ and $(d\varepsilon^2)^3F_1 \to (d\varepsilon^2)^1A_1$ at $9600\,cm^{-1}$ and $21000\,cm^{-1}$ respectiyely, which will be of course very weak.

$3d^8$: A. v. Kiss et al.[16] measured the absorption spectra of the $[Ni(H_2O)_6]^{2+}$ complex ion, and obtained the following results:

	abs. max.	log. $\mathcal{E}$
$[Ni(H_2O)_6]SO_{4\,aq.}$	$\begin{cases}13.9\times10^3\,cm^{-1}\\15.2\end{cases}$	$\begin{cases}0.32\\0.27\end{cases}$
	25.3	0.76
$[Ni(H_2O)_6]Cl_{2aq.}$	8.5	0.35
	$\begin{cases}13.9\\15.2\end{cases}$	$\begin{cases}0.36\\0.27\end{cases}$

But Dreisch and Trommer[17] reported the existence of three bands at 8400, 14100 and 25000 cm^{-1} respectively for this complex ion $NiSO_{4\,aq.}$, and Houston, Bose and Mookherji[18] also observed these three bands. So that we will take these values as the absorption maxima of the three bands.

Our assignment and calculated values are as follows:

transition	calc. val.	obs. max.
$(d\varepsilon^6 d\gamma^2)^3A_2 \rightarrow (d\varepsilon^5 d\gamma^3)^3F_2$	$8.2\times10^3\,cm^{-1}$	8.4
$\rightarrow (d\varepsilon^5 d\gamma^3)^3F_1$	13.9	14.1
$\rightarrow (d\varepsilon^4 d\gamma^4)^3F_1$	26.3	25.0

$$Dq = 824\ cm^{-1},$$

where we have used the same values of B and C as those of the free ion. Here again the third band corresponds to a two electron jump. However in this case, because of the smallness of Dq, the mixing of the two configurations $d\varepsilon^5 d\gamma^3$, $d\varepsilon^4 d\gamma^4$ occurs for the $^3F_1$ level and the intensity of the third band is thus considered to be stolen from that of the second band, while in the case of vanadium complex there was no counter-part of the state $d\gamma^2\,^3A_2$. For the first two bands of Ni tutton salts, the same assignment has been done by Griffth and Owen[19], using the value of the crystalline field parameter determined from their experimental data on the paramagnetic resonance absorption of Ni tutton salts. We have no resonable explanation as to the origin of the double maxima at 13900 and 15200 cm^{-1} found by A. v. Kiss at the position corresponding to the second band (14100 cm^{-1}).

For the optical absorption of $[Ni(NH_3)_6]^{2+}$, measurements are performed by Sone[20], Dreisch and Trommer[17], Bose and Mookherji[19] and for $[Ni(en)_3]Cl_{2\,aq.}$ by Sone[20]. Our results are the following, where the same assignments, the same B, C values were adopted as for $[Ni(H_2O)_6]^{2+}$:

	calc. val.	obs. max.	obs. log $\mathcal{E}$	
$[Ni(NH_3)_6]^{2+}$	$10.3\times10^3\,cm^{-1}$	10.6	0.92	Dreisch and Trommer
	17.0	17.5	0.75	Sone
	29.3	27.6	0.92	
		$Dq = 1030\ cm^{-1}$		
$[Ni(en)_3]^{2+}$	10.8	—	—	
	17.5	18.4	0.84	
	29.9	28.9	0.93	
		$Dq = 1080\ cm^{-1}$		

The first band must be found in the near infrared region for $[Ni(en)_3]^{2+}$ too. Further, a line due to the transition $(d\varepsilon^6 d\gamma^2)^3A_2 \rightarrow {}^1E$ must be found at 17500 cm^{-1} for the $[Ni(H_2O)_6]^{2+}$ complex ion (for the other two complexes, it will be masked by the broad bands). Actually Gielessen[21] reported the existence of the line in the region 17300–20500 cm^{-1} in the solid $[Ni(H_2O)_6]Cl_2$ at -189°C.

$3d^3$: For $[Cr(H_2O)_6]Cl_{3\,aq.}$, Tsuchida and Kobayashi[22] found two bands at 17500 cm^{-1} ($\log\mathcal{E}=1.02$) and 24500 cm^{-1} ($\log\mathcal{E}=1.20$) respectively, and later found one more absorption maximum at 36600 cm^{-1} ($\log\mathcal{E}=0.5$)[23]. This maximum is also found in the solution of $KCr(SO_4)_2\cdot12H_2O$, but its presence is obscured by the existence of the intense edge absorption. In solid states of chromic salts, however, it is clearly observed.

It is well known that the spectra of the chromic salts show sharp lines in many cases; Sauer[9], Joos and Schnetzler[24] have reported the existence of the lines in the crystals of $[Cr(H_2O)_6]Cl_3$ (14800 cm^{-1}, 14900 cm^{-1}), $[Cr(NH_3)_6]Cl_3\cdot H_2O$ (15200 cm^{-1}, 15400 cm^{-1}) and $[Cr(en)_3]Br_3$ (14900 cm^{-1}, 15400 cm^{-1}) at -180°C, and Spedding[25] further found a sharp absorption (21000 cm^{-1}) between the two bands. The same lines are also observed in the solution containing $[Cr(NH_3)_6]^{3+}$ or $[Cr(en)_3]^{3+}$ ion.

To compare our results with the observed spectra of $[Cr(H_2O)_6]^{3+}$, we will take the observed values of $KCr(SO_4)_2 \cdot 12H_2O$ crystal. The results are as follows:*

$[Cr(H_2O)_6]^{3+}_{crys.}$

	transition	calc. val.	obs. max.	obs. log α	
$(d\varepsilon^3)^4A_2$ (bands)	$\rightarrow(d\varepsilon^2d\gamma)^4F_2$	17.2×10^3 cm^{-1}	18.0	0.5	Tauchida at room temp.
	$\rightarrow(d\varepsilon^2d\gamma)^4F_1$	24.6	24.6	0.6	
	$\rightarrow(d\varepsilon d\gamma^2)^4F_1$	38.2	36.6	0.2	
(lines)	$\rightarrow(d\varepsilon^3)^2E$	14.9	14.9		Sauer at -190°C
	$\rightarrow(d\varepsilon^3)^2F_1$	15.3	15.1		
	$\rightarrow(d\varepsilon^3)^2F_2$	21.8	21.0		Spedding at liq. hydrg. temp.

$Dq=1720$ cm^{-1}

Here we changed the values of B and C from those of the free ion in the following way: The values of B and C of the free ion are determined previously to give the term values of ^4F, ^4P and ^2H correctly, so that if we use these B, C the agreement of the calculated term value of ^2G is not so good. This is unfortunate because as we see the lines originate from ^2G of the free ion in our treatment. So we replace γ so as to give the correct term value ^2G of the free ion (resulting γ being 4.0). But on comparing the result thus calculated with the observation we found that we further had to adjust the value of B, C in order to give the right positions of the lines. The value of B to be used here was 765 cm^{-1} (γ being the same as above 4.0) which is much smaller than those of the free ion 918 cm^{-1}. We cannot get better agreement with the observed spectra by adjusting only the parameter Dq as Finkelstein and Van Vleck[26] claimed in their paper on a similar calculation of the position of the doublet line of the chrome alum, since the energy interval between the states of the same configuration is almost independent of the crystalline field.

The third absorption band corresponds to the two electron excitation and the same consideration applies to this case as was mentioned for the two electron jump in the case of Ni complex. In this case, however Dq is rather large so that the degree of contamination may be small and consequently there remain doubts about our assignment of the observed third band to this transition. If this assignment were correct, this band would be nearly twice as broad as the first and second bands according to the considerations given at the end of § 3. The third band really seems rather broad, but it may be safe not to draw any definite conclusisn as to the origin of this third band.

It is doubtful whether the observed two lines do correspond to the transitions to ^2E and $^2F_1$ respectively or not. To make this clear, it would be necessary to take the deviation from the perfect cubic symmetry due to the Jahn-Teller effect[8] or to the trigonal field[27] exerted by the ions outside the complex ion in question, and the influence of the spin-orbit interaction into consideration. Moreover, if we consider the deformation of the $d\varepsilon$ orbital from the atomic d-character, the excited states ^2E and $^2F_1$ have no necessity to lie so close together. (Though the deformation of orbits may be taken into account in some degree in the adopted B, C values which is different from those of the free ion, we cannot include in this way the deformation that will remove the accidental degeneracy of ^2E and $^2F_1$ levels occurred in the strong crystalline field limit.)

For $[Cr(NH_3)_6]^{3+}$ complex ion, assuming the same assignment and the B, C values redetermined above to be correct, the comparison is as follows:

* These data were used here with Tsuchida's which was obtained at room temperature, because for these lines the temperature shift is small. According to Sauer's experiments the doublet line at 14900 cm^{-1} shifts only about 20Å towards the longer wave leugth side when the temperature is changed from -190° to 18°C, while the bands shift as much as 150 Å.

$[Cr(NH_3)_6]^{3+}$	calc. val.	obs. max.	obs. log ε	
(bands)	21.4×10^3 cm^{-1}	21.5	1.62	
	29.4	28.5	1.57	Linhard[28]
	46.6	—	—	
(lines)	14.9	15.2		
	15.3	15.4		Joos and Schnetzler[24]
	21.8	—		

$$Dq=2140 \text{ cm}^{-1}$$

For $[Cr(en)_3]^{3+}$, we shall omit the discussion, since its spectra are almost identical (though a slight decrease of the intensity is found) with those of $[Cr(NH_3)_6]^{3+}$.

$3d^7$: On the $[Co(H_2O)_6]Cl_2$ complex ion, many observations have been made (Houston[18], Dreisch[14], Dreisch and Trommer[17], Datta and Deb[29], Gielessen[21]) and, though there exist slight differences among them, the existence of the following the bands is generally accepted. The comparison is as follows.

trans.	calc. val.	obs. max.	obs. log ε	
$(d\varepsilon^5 d\gamma^2)^4F_1 \rightarrow (d\varepsilon^4 d\gamma^3)^4F_2$	8.1×10^3 cm^{-1}	8.13	0.56	
$\rightarrow (d\varepsilon^4 d\gamma^3)^4F_1$	20.6	19.6	0.62	Dreisch and Trommer[17]

$$Dq=840 \text{ cm}^{-1}$$

If we assume the value of Dq as above, the absorption lines to which the transitions $(d\varepsilon^5 d\gamma^2)^4F_1 \rightarrow (d\varepsilon^5 d\gamma^2)a^2F_1$, a^2F_2 and $^4F_1 \rightarrow b^2F_1$ have to be assigned must appear near 18000 cm^{-1} and $\sim$24000 cm^{-1} respectively. These positions of the lines, however, do not seem to be compatible with the line observed by Gielessen in the crystals at -189°C.

Transition $(d\varepsilon^5 d\gamma^2)^4F_1 \rightarrow (d\varepsilon^6 d\gamma)^2E$ may give rise to a broad band with much weaker intensity, and, if it could actually be found, it would be very interesting because the dependence of its excitation energy on the strength of the crystalline field is different from those hitherto discussed. That is, this band is considered to have a different temperature shift (Fig. 6), if the magnitude of the temperature shift of the absorption maximum is predominantly due to the increase of the atomic distance caused by the thermal expansion. The absorption maximum of this band will then shift towards shorter wave length side when temperature is increased (Dq becomes smaller), while those of hitherto discussed will shift towards longer side as is often experimentally observed.

$3d^4$: For $[Mn(H_2O)_6]^{3+}$, Hartmann and Schläfer measured the absorption spectra of $CsMn(SO_4)_2\cdot 12H_2O$[30], and found only one band at 21000 cm^{-1}. This band will correspond to the transition $(d\varepsilon^3 d\gamma)^5E \rightarrow (d\varepsilon^2 d\gamma^2)^5F_2$. If we determine Dq to be 2100 cm^{-1}, our calculation predicts the existence of the lines near 24000 cm^{-1} corresponding to the transitions to ^3E, $^3F_2$ etc.

Recently the absorption spectra of the aqueous solution of $CrSO_4\cdot 7H_2O$ was measured by Tsuchida et al.*. According to their measurement, $[Cr(H_2O)_6]^{2+}$ shows a single band at 13900 cm^{-1} (log $\varepsilon \simeq 0.8$). In this complex ion, we can also adopt the same assignment as above, and Dq value thus determined is 1390 cm^{-1}.

$3d^6$: The binding property of the $[Co(NH_3)_6]^{3+}$ complex ion has been considered to be covalent by the reasons that it has the diamagnetic normal state and that the atomic distance between the central ion and one of the ligands is relatively small. The same argument will apply for the $[Co(H_2O)_6]^{3+}$ complex, because a recent measurement on the susceptibility of this complex shows that this is also diamagnetic[31].

But an alternative explanation is possible in the ionic model that the small atomic distance means the strong crystalline field and at such strong field the change of the ground state could occur, i.e., the ground state could become singlet (Fig. 8)[32]. In our treatment we

* This was reported at the meeting of the Chemical Society of Japan held in April, 1954.

shall adopt the latter explanation, and assume that the ionic structure will predominate over the covalent structure, so that the latter can be neglected.

For $[Co(H_2O)_6]^{3+}$, our assignment is as follows:

trans.	calc. val.	obs. max.	obs. log ε	
$(d\varepsilon^6)^1A_1 \rightarrow (d\varepsilon^5 d\gamma)^1F_1$	16.0×10^3 cm^{-1}	16.3	1.62	Topp[33]
$(d\varepsilon^5 d\gamma)^1F_2$	28.1	25.0	1.70	

$$Dq = 1920 \text{ cm}^{-1},$$

where we use a little smaller value of γ, so that the ground state is singlet when the value of Dq is determined to give the position of the first band correctly. The B value was not changed from that for the free ion. Assuming the parameter values to be as above, the third band corresponding to the transition $(d\varepsilon^6)^1A_1 \rightarrow (d\varepsilon^4 d\gamma^2)^1F_2$ must appear (which might be masked by an intense edge absorption in solution) and, moreover, according to our energy diagram a weak band due to the transition $^1A_1 \rightarrow {}^3F_1$ and $^1A_1 \rightarrow {}^3F_2$ must be found at the long wave length side of the first band.

In a similar way, the comparison of our results with observations for the $[Co(NH_3)_6]^{3+}$ complex ion is

$[Co(NH_3)_6]^{3+}{}_{aq.}$	calc. val.	obs. max.	obs. log ε	
	20.4×10^3 cm^{-1}	21.1	1.78	
	33.6	29.5	1.74	Linhard[28]

$$Dq = 2260 \text{ cm}^{-1},$$

and for the $[Co(NH_3)_6]Cl_3$ crystal

$[Co(NH_3)_6]Cl_3$ crys.	calc. val.	obs. max.	obs. log ε	
	19.2	21.6	1.35	
	32.0	28.6	1.29	Tsuchida[34]
$(d\varepsilon^6)^1A_1 \rightarrow (d\varepsilon^4 d\gamma^2)^1F_2$	39.4	35.4	1.25	

$$Dq = 2260 \text{ cm}^{-1},$$

where the third band is observed. However, as was mentioned earlier, the third band is considered to have an intensity much smaller than the first two bands, because Dq is also here rather large. So that the assignment for the third band is doubtful, judging from the comparative intensity of this band with the first two bands.

Calculated values do not show so good agreement with those of the observed as is seen above. When we adjust the value of B to give the relative positions of the bands correctly, the calculate absolute positions shift far from those of the observed values towards the longer wave length side. These facts perhaps mean that for this complex ion the approximation that assumes this complex as completely ionic is not so good as to give a quantitative agreement. But it suggests the possibility that this complex could be treated successfully starting from the ionic model and then taking the covalent structure into account. From this point of view, it is very interesting that the absorption band of relatively low intensity has recently been found* at the longer wave length side of the first band which seems to be due to the transition $^1A_1 \rightarrow {}^3F_1$ or $^1A_1 \rightarrow {}^3F_2$ in our ionic model.

For $[Fe(H_2O)_6]^{2+}$, only one band is observed by Dreisch and Kallscheuer[35] at 10300 cm^{-1} in the near infrared region. (Though Dreisch reports two maxima for $FeCl_{2\,aq.}$ at 9300 cm^{-1} ($\varepsilon = 1.74$) and 10500 cm^{-1} ($\varepsilon = 1.86$), we suspect that these two maxima would correspond to the vibrational structure of the band peaked at 10300 cm^{-1}.) Our energy diagram gives the observed value if we assign the transition $(d\varepsilon^4 d\gamma^2)^5F_2 \rightarrow (d\varepsilon^3 d\gamma^3)^5E$ to the observed band and use the value of Dq as 1030 cm^{-1}.

$3d^5$ Absorption spectra of the complex ion $[Fe(H_2O)_6]^{3+}{}_{aq.}$ has been measured by Rabinowitch and Stockmayer[36] and A. v. Kiss[37], who found the same results, though the complex $[Fe(H_2O)_6]^{3+}$ is rather unstable and their results may possibly contain some error in details. Absorption spectra in $[Fe(H_2O)_6]^{3+}{}_{aq.}$

* Private communication from Dr. Y. Kondo.

found by them consist of two weak bands and a relatively sharp one. Though the latter is broader than the lines found in other complexes, it is sharper than the bands which ordinarily appear, and we will therefore treat it as a line taking the possible error of the experiments into consideration. The results are as follows:

$[Fe(H_2O)_6]^{3+}{}_{aq.}$	trans.	calc. val.	obs. max.	obs. $\log \varepsilon$
$(d\varepsilon^3 d\gamma^2)^6A_1$	$\rightarrow(d\varepsilon^4 d\gamma)\,{}^4F_1$	14.3	14.3	−1.21
(bands)	$\rightarrow(d\varepsilon^4 d\gamma)\,{}^4F_2$	18.8	18.2	−1.00
(lines)	$\rightarrow(d\varepsilon^4 d\gamma^2)^4A_1, {}^4E$	25.0	24.6	−0.48

$$Dq = 1350\ \mathrm{cm}^{-1},$$

where we recalculated the value of γ to give the term value of ${}^4G$ of the free ion correctly in the same way as was done for the chromium complexes, and adjusted the value of B to be 820 cm^{-1}—a value a little smaller than that of the free ion.

According to this assignment, just as was seen in the case of the transition $(d\varepsilon^5 d\gamma^2)^4F_1 \rightarrow (d\varepsilon^6 d\gamma)^2E$ in $3d^7$, the excitation energies of the two absorptions at 14300 cm^{-1} and 18200 cm^{-1}, show the opposite dependence on the crystalline field strength to the usual absorptions (Fig. 9), while a relatively sharp absorption found at 2400 cm^{-1} is insensitive to the crystalline field. Experimental study of this difference in temperature dependence will be interesting as was mentioned at the case of $3d^7$.

For $[Mn(H_2O)_6]^{2+}$, Kato[38] observed a band at 19600 cm^{-1} in $MnCl_{2\,aq.}$. Gielessen's measurement on $MnCl_2 \cdot 6H_2O$ crystal[21] shows that this has also line spectra near 25000 cm^{-1}. The comparison is as follows:

$[Mn(H_2O)_6]^{2+}$	trans.	calc. val.	obs. max.
(band)	${}^6A_1 \rightarrow {}^4F_2$	$19.6 \times 10^3\ \mathrm{cm}^{-1}$	19.6
(line)	$\rightarrow {}^4A_1, {}^4E$	26.2	25.0

$$Dq = 1230\ \mathrm{cm}^{-1},$$

where the value of γ was adjusted as before, but the B-value was assumed to be that of the free ion. This assignment leads to the existence of the weak band at 14600 cm^{-1} corresponding to the transition ${}^6A_1 \rightarrow {}^4F_1$ which is not reported yet.

§ 5. Conclusion

Though absorption spectra of many iron group complexes could be identified with our calculated energy diagram choosing the values of the parameters suitably and their observed behaviours were explained fairly well from the view-point of the crystalline field approximation, we do not feel the ground for justifying our treatment settled perfectly because of the shortage of available absorption data in the individual complex. In the following, however, we shall survey the results to show that the values of the parameters thus determined empirically are resonable ones and the arbitrariness of their values is not so large as one might suppose at first.

The calculated positions of lines and bands are rather in fair agreement with the experimental data in divalent ions, if we adjust only the crystalline field parameter Dq suitably. In trivalent ions, it was necessary, besides the adjustment of Dq, to use smaller values of B and C than those of the free ions to obtain better agreement with experiments. The values of B and Dq determined in this way are summarized below, where the corresponding values of B in free states are parenthesized. The values of C are not tabulated there because γ are not altered from those of the free ions.

	Ti^{3+}	V^{3+}	Cr^{3+}	Mn^{3+}	Fe^{3+}	Co^{3+}
Dq (cm^{-1})	2030	1860	1720	2100	1350	1920?
B (cm^{-1})		642 (862)	765 (918)	? (965)	820 (1015)	? (1065)

	Cr^{2+}	Mn^{2+}	Fe^{2+}	Co^{2+}	Ni^{2+}
Dq (cm^{-1})	1390	1230	1030	840	820
B (cm^{-1})	810 (810)	860 (860)	917 (917)	971 (971)	1030 (1030)

These values of Dq are of resonable magnitude, being close to those obtained in other ways, for instance, those obtained from the paramagnetic suceptibilities or the paramagnetic resonance absorption data. It is to be noticed that the values of Dq in trivalent complex ions are always larger than those of corresponding divalent complexes, and that in each isovalent complexes the Dq-value has a tendency to decrease as the atomic number Z increases. The Dq-value of Co^{3+} remarked by ? will have possibly a large error because of the particular binding property of Co^{3+} complex ion as was mentioned in § 4.

We suppose that the decrease of B and C compared with those of free ions will be connected with Stevens' recent suggestion[39] on the interpretation of the paramagnetic resonance absorption, and a further theoretical investigation on B and C and Dq would be necessary.

Experimentally more detailed observation on the behavior of the absorption in the individual complex ions is highly desirable; particularly the detection of weak bands and lines which are usually obscured by relatively intense and broad bands; the measurement of the temperature shift of the absorption maxima or the temperature dependence of their intensities and all that will give the keys to these puzzles. Especially for Co^{3+} complexes, the absorption spectra of $[CoF_6]^{3-}$ will be interesting, since it has the paramagnetic normal state and its spectra may be identified with our energy level diagram where the value of Dq we must adopt is smaller than those at which the alteration of the ground state occur.

Acknowledgement

We should like to express our gratitude to Prof. M. Kotani for his much instructive advice and continual encouragement during the course of this work. Particularly we are greatly indebted to Dr. Y. Kondo of Chemical Department, Tokyo Institute of Technology, for his helpful information about the experimental data and to Dr. K. Itoh of Nagoya University for his kind discussion on this problem. Appreciation is also due to the members of Prof. Kotani's Laboratory.

References

1) Atomic Energy Levels (*National Bureau of Standard*, 1952)
2) W. M. Cady: Phys. Rev. **43** (1933) 322.
3) S. Meshkov: Phys. Rev. **91** (1953) 871.
4) S. Meshkov: Phys. Rev. **93** (1954) 270.
5) M. A. Catalan: F. Rohrlich and A. G. Shenstone: Proc. Roy. Soc. **A 221** (1954) 421.
6) H. Sauer: Ann. d. Phys. **87** (1928) 197.
7) J. H. Van Vleck: J. Phys. Chem. **41** (1937) 67.
8) J. H. Van Vleck: J. Chem. Phys. **7** (1939) 72.
9) D. G. Hill and A. F. Rosenberg: J. Chem. Phys. **22** (1954) 148.
10) E. U. Condon and G. H. Shortley: The Theory of Atomic Spectra (*Cambridge University press*), 1935.
11) S. Freed: Phys. Rev. **38** (1931) 2122.
12) S. Freed and R. J. Mesirow: J. Chem. Phys. **5** (1937) 22.
13) H. Hartmann and H. L. Schläfer: Zs. f. Phys. Chem. **197** (1951) 115.
14) T. Dreisch: Zs. f. Phys. **40** (1927) 714.
15) H. Hartmann and H. L. Schläfer: Zs. f. Naturforschg. **6a** (1951) 754.
16) A. v. Kiss et al.: Z. anorg. allg. Chem. **245** (1941) 356.
17) T. Dreisch and W. Trommer: Z. Phys. Chem. **B 45** (1939) 37, *ibid.* **B 45** (1940) 19.
18) D. M. Bose and P. C. Mookherji: Phil. Mag. **26** (1938) 757.
A. G. Houston; Proc. Roy. Soc. Edinburgh **31** (1910) 530.
19) J. H. E. Griffth and J. Owen: Proc. Roy. Soc. **A 213** (1952) 459.
20) K. Sone: J. Chem. Soc. Japan **71** (1950) 271.
21) J. Gielessen: Ann. d. Phys. **22** (1935) 537.
22) R. Tsuchida and M. Kobayashi: Bull. Chem. Soc. Japan **13** (1938) 47, J. Chem. Soc. Japan **64** (1943) 1268.
23) R. Tsuchida: Chemistry (in Japanese) No. **6** (1951) 4.
24) G. Joos and K. Schnetzler: Z. Phys. Chem. **B 20** (1933) 1.
25) F. H. Spedding and G. C. Nutting: J. Chem. Phys. **2** (1934) 421.
26) R. Finkelstein and Van Vleck: J. Chem. Phys. **8** (1940) 790.
27) J. H. Van Vleck: J. Chem. Phys. **7** (1939) 61.
28) M. Linhard: Z. f. Elektrochem. **50** (1944) 224.
29) S. Datta and M. Deb: Phil. Mag. **20** (1935) 1121.
30) H. Hartmann and H. L. Schläfer: Zs. f. Naturforschg. **6a** (1951) 760.
31) H. L. Friedman aud J. P. Hunt: J. Amer. Chem. Soc. **73** (1951) 4028.
32) J. H. Van Vleck: J. Chem. Phys. **3** (1935) 807.
33) J. Topp: Dissert. Münst. (1928).
34) R. Tsuchida: The Structure and The Color of Complex Salts (in Japanese).
35) T. Dreisch and O. Kallscheuer: Zs. f. Elektrochem. **50** (1944) 224.
36) E. Rabinowitch and W. H. Stockmeyer: J. Amer. Chem. Soc. **64** (1942) 335.
37) A. v. Kiss et al.: Z. anorg. allg. Chem. **244** (1940) 99.
38) S. Kato: Sci. Pap. Inst. Phys. Chem. Reserch, Tokyo **13** (1930) 49.
39) K. W. H. Stevens: Proc. Roy. Soc. **A 219** (1953) 542.

Errata

On the Absorption Spectra in Complex Ions. II
by Y. Tanabe and S. Sugano

J. Phys. Soc. Japan **9** (1954) 766

1) On p. 770, on the 7 th and 8 th lines from the bottom, " $2.5\times\varepsilon_{max}\Delta\nu$ " and " $2.5\times k_{max}\Delta\nu$ " should be read as " $1.1\times\varepsilon_{max}\Delta\nu$ " and " $1.1\times k_{max}\Delta\nu$ ", respectively.
2) On p. 771, on the 13 th line from the top, " line (though..." should be read as " line of $[Fe(H_2O)_6]^{3+}$ (though..."
3) On p. 771, on the 15 th and 16 th lines from the bottom, "...we do not thuch it here because of its uncertainly " should be read as "...we do not touch it here because of its uncertainty ".
4) On p. 778, on the 8 th line from the top, " $(d\varepsilon^4 d\gamma^2)^4A_1$, ^4E " should be read as " $(d\varepsilon^3 d\gamma^2)^4A_1$, ^4E ".

Reprinted from *J. Chem. Phys.*, **23**, 1004–1014 (1955)

Spectra of Transition-Metal Complexes*

L. E. ORGEL
Gates and Crellin Laboratories of Chemistry, California Institute of Technology, Pasadena, California

(Received October 11, 1954)

19

The electronic transitions observed in complexes of the transition-metal ions are interpreted in terms of a slightly modified crystal-field theory. Parameters of chemical interest are derived.

THE absorption bands of transition-metal complexes are commonly of two kinds: those due to charge-transfer processes and those arising from transitions which, to a good approximation, can be considered as taking place within the *d*-shell of the ion. It is usually possible to decide to which class an observed band belongs[1,2] although in certain complexes, where the interaction between the metal ion and the ligands is particularly strong, the two types of transition are no longer even approximately distinct and the theoretical treatment is then more complicated e.g., in the MnO_4^- ion.[3] In this paper we shall discuss only those transitions which are localized on the metal ion.

The general theory of the energy-levels for *d*-electrons in transition-metal complexes was first developed by Van Vleck[4] and by Schlapp and Penney.[5] Since then many applications of their theory to the magnetic properties of complexes and a few to the optical properties have been made.[6,7] We shall extend the methods of a previous paper on this subject,[8] developing a slightly different interpretation of crystal-field theory, and then derive certain quantities of chemical interest from a detailed study of the spectra of transition-metal complex ions.

We shall calculate the energy-level diagrams of a number of complex ions as a function of a parameter Dq which depends on the geometry of the ion. The energy levels which must be compared with the calculated diagram are those which correspond to the stable geometric configuration of the ground-state of the ion being maintained in the excited states. We shall be concerned, therefore, with the maxima of absorption bands, not with 0—0 bands.

Since, in discussing the intensities of transitions our interests will be qualitative we shall quote values of ϵ_{max}, the value of the molecular extinction coefficient at the absorption maximum, rather than attempt to calculate oscillator strengths.

THE EFFECT OF ENVIRONMENT ON THE *d*-ORBITALS OF TRANSITION-METAL IONS

The degeneracy of the *d*-orbitals of a transition-metal ion is removed, more or less completely, when the ion becomes part of a crystal or is solvated in solution. Only those ions or molecules attached directly to the metal atom are normally important in determining the coarse structure of the energy-level scheme of the ion so that the broad features of the spectrum of a complex ion in solution are similar to those of the same ion in a crystal. For this reason we shall be concerned only with the central metal atom and its immediate neighbors.

It has been usual to regard the splitting of the degeneracy of the *d*-orbitals as a consequence of the electrostatic field set up by the ligands. However, in view of the complexity of chemical-bonding phenomena, it seems surprising that so crude a picture is capable of giving a good basis for detailed calculations of magnetic properties. The reason for this success is, however, readily understood: the magnetic and optical properties depend on the wave functions and stationary state energy-levels of the *d*-electrons and not on the detailed form of the interaction of the ion with its environment. In most of the calculations, group theory is used to determine the symmetry type of the orbitals, but the energies are derived from empirical data, sometimes supported by qualitative arguments. Thus although the approach adopted is formally based on electrostatics, in practice the results are quite general and apply to any mechanism which removes the degeneracy of the *d*-orbitals.

It has been shown that the description of transition-metal complexes in terms of covalent and ionic bonds is almost equivalent to the crystal-field theory in which polarization of the ligands is taken into account. It was also suggested that double-bonding may be important in determining the energy-levels of the complex, an approach recently developed by Stevens.[9] It follows that the empirical pattern of energy levels must give information which will be useful in the interpretation of bond properties.

* Contribution No. 1945 from Gates and Crellin Laboratories of Chemistry, California Institute of Technology, Pasadena, California.

1 E. Rabinowitch, Revs. Modern Phys. **14**, 112 (1942).
2 L. E. Orgel, Quarterly Revs. (London) **8**, 422 (1954).
3 M. Wolfsberg and L. Helmholz, J. Chem. Phys. **20**, 837 (1952).
4 J. H. Van Vleck, *Theory of Electric and Magnetic Susceptibilities* (Oxford University Press, London, 1932).
5 R. Schlapp and W. G. Penney, Phys. Rev. **42**, 666 (1932).
6 B. Bleaney and W. K. Stevens, Repts. Progr. Phys. **16**, 108 (1953).
7 H. Hartmann and H. L. Schläfer, Z. Naturforsch. **6A**, 760 (1951).
8 L. E. Orgel, J. Chem. Soc. 4756 (1952).
9 W. K. Stevens, Proc. Roy. Soc. (London) **219A**, 542 (1953).

THE STATE OF IONIZATION OF THE METAL IN TRANSITION-METAL COMPLEXES

The conventional description of a hydrated or complex ion used in crystal-field theory is that of a single metal ion carrying a positive charge equal to the formal valency of the metal, surrounded by a group of negative ions or neutral dipoles. The success of quite detailed calculations[10] in which the energy-level diagrams for the ion in the crystal is taken as identical with that of the free ion seems at first to argue against covalent bonding, even in such stable, highly charged complexes as $[Cr(H_2O)_6]^{+++}$. However, a detailed comparison of the intervals between the states arising from the configurations d^n, d^ns, and d^ns^2 shows that only slight changes occur as the charge decreases, at least for the elements with several *d*-electrons.[11] This shows that the 4*s* electrons do not shield the 3*d*-electrons to any important extent and, *a fortiori*, we should expect the same to be true of the 4*p*-electrons. It follows that covalent bonding in which the 4*s*- and 4*p*-orbitals are partially filled is not precluded by the success of calculations in which it is neglected. Conversely, we are justified in using the levels of the free ion as a basis for calculations, even if there is reason to believe that "covalent bonding" is involved. In more accurate calculations, particularly on ions of high charge which have few *d*-electrons, corrections for the slight changes in the intervals which occur between d^n-, d^ns-, and d^ns^2-configurations might be made e.g., in V^{+++}.

While we have presented arguments suggesting that, apart from the effect of the environment on the degeneracy of the *d*-orbitals, the energy-level diagram for the states of one configuration of the free ion is adequate in discussing the states of the complex, it must be emphasized that the relative positions of different configurations will be profoundly changed by complex formation. Just as we have to introduce a splitting of the degenerate *d*-orbitals, so we need to introduce a change in the basic separation between the *d*- and *s*-levels. This has many important consequences, e.g., simple cuprous salts are colorless although, in the free ion, the transition from the $d^{10}\,{}^1S$ state to the $d^9s\,{}^3D$ and ${}^1D$ states occur at 22 000 cm^{-1} and 26 000 cm^{-1} respectively.[12] If it were not for an increase in the interval between the d^{10} and d^9s configurations certain components of these transitions would occur in the visible. In general the interval between configurations d^n and $d^{n-1}s$ or $d^{n-1}p$ in tetrahedral or octahedral complexes is expected to be greater than in the free ions.

THE CONFIGURATIONS d^1 AND d^9

In the most general environment the degeneracy of the *d*-orbitals is removed completely. There are five separate levels and transitions are possible between the ground state and any of the other four states. Fortunately many of the most important complexes have some elements of symmetry or near symmetry which greatly simplify their spectra.

The simplest and commonest type of complex is that in which the central ion is surrounded by an octahedron of ligands. If the octahedron is regular and the six ligands are identical it has been shown by Schlapp and Penney[5] that the *d*-orbitals are split into a stable triply degenerate t_{2g} orbital and a less stable doubly degenerate e_g orbital. The only transition possible is from the T_{2g} to the E_{2g} state and it, like all transitions within the *d*-shell, is a $g-g$ transition and is therefore forbidden.

The only example of a regular octahedrally coordinated ion with one 3*d*-electron whose spectrum has been studied is $[Ti(H_2O)_6]^{+++}$[13] and the theory has been discussed in some detail.[14] The ion has a single, weak absorption band whose maximum is at 20 400 cm^{-1}. The source of the absorption intensity is not understood, but, arguing by analogy with the spectra of other tightly bound molecules, it is probably due to the coupling of electronic and vibrational motion rather then to a quadrupole or magnetic dipole transition. If, in order to follow the convention introduced by Schlapp and Penney,[5] we take the separation between the t_{2g} and e_g orbitals as 10 Dq, then $Dq=2040$ cm^{-1}. A number of ions such as $TiCl_3(C_2H_5OH)_3$ has also been studied. They have a single absorption maximum, at longer wavelengths than that of $[Ti(H_2O)_6]^{+++}$.

In tetrahedral complexes the simple crystal-field theory predicts a splitting similar to that in octahedral complexes, but of the opposite sign, i.e., the *e* level is below a t_1 level.[15] However, the assumption that we may restrict our discussion to the *d*-orbitals is no longer valid. In the absence of a center of symmetry the t_1 component of the *d*-orbitals can mix with the 4*p*-orbitals. Thus for a single *d*-electron the orbital for the ground state is nearly a pure 3*d*-orbital, but in the excited state there is a considerable 4*p*-component. We, therefore, expect a more intense transition than is found in octahedral complexes. The only substance of the kind which has been studied is VCl_4. It has an absorption maximum at about 9000 cm^{-1} with $\epsilon_{max}=110$.[16] It should be noted that the split between the *e* and t_1 orbital is much smaller than is usual for octahedrally coordinated ions of high valency. It is not at all clear that the quantity Dq has any significance for complexes in which *p* and *d*

[10] R. Finkelstein and J. H. Van Vleck, J. Chem. Phys. **8**, 790 (1940).

[11] L. E. Orgel, J. Chem. Phys. (to be published).

[12] Data on atomic spectra are taken from Atomic Energy Levels, National Bureau of Standards, Washington 1949, 1952.

[13] H. Hartmann and H. L. Schläfer, Z. physik. Chem. **197**, 116 (1951).

[14] F. E. Ilse and H. Hartmann, Z. physik. Chem. **197**, 239 (1951).

[15] It must be noted that the representations of the octahedral (O_h) and tetrahedral (T_d) groups are related in the following way:

O_h	A_{1g}	A_{1u}	A_{2g}	A_{2u}	E_g	E_u	T_{1g}	T_{1u}	T_{2g}	T_{2u}
T_d	A_1	A_1	A_2	A_2	E	E	T_2	T_1	T_1	T_2.

[16] A. G. Whittaker and Don M. Yost, J. Chem. Phys. **17**, 188 (1949).

orbitals are mixed, but a simple application of the formula gives $Dq = 900$ cm^{-1}. The intensity is quite large, particularly for an infrared band, but still very much weaker than that in certain other tetrahedral complexes, e.g., $[CoCl_4]^{--}$.

From a chemical viewpoint the mixing of d and p orbitals is the d-p hybridization postulated by Pauling.[17] Orbitals whose maximum concentrations are near the ligands are used in bonding, while the remaining hybrids, which are automatically removed as far as possible from the ligands, contain any nonbonding electrons. The very high intensities and anomalous positions of the absorption bands of tetrahedral complexes, particularly of Co^{++}, is evidence for d-p hybridization.

The d^9 configuration, i.e., the d-shell with one vacancy, behaves like a single d-electron when the degeneracy of the d-orbitals is removed except that all the splittings are reversed. A similar relation holds between other d^n and d^{10-n} configurations. In octahedral d^9 complexes we expect a single $E-T_{2g}$ transition. The cupric complexes, which are the only extensively studied ions with this configuration, do not have a simple octahedral environment, but the central atom is attached to four nearest neighbors in a plane and often to two more distant groups which complete a distorted octahedron.[18] The energy levels are split as shown in Fig. 1 by the deviation from a regular octahedron. Clearly as many as three transitions may occur in the visible and near infrared if the splitting is large enough.

The cupric ion in solution, which is presumably $[Cu(H_2O)_6]^{++}$ has a very broad absorption band with

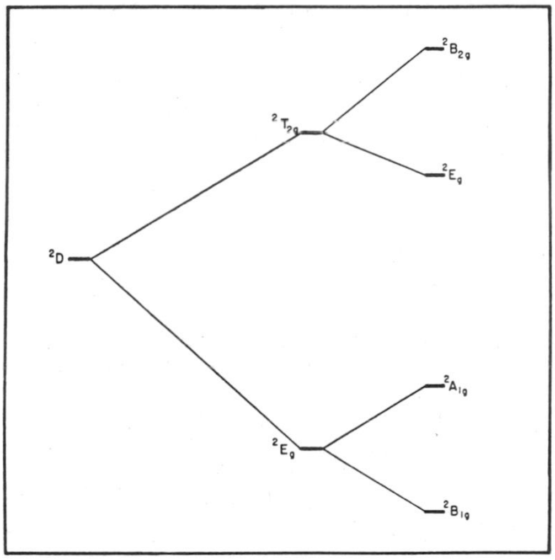

FIG. 1. Energy level diagram for the d^9 configuration.

a maximum close to 8000 A and $\epsilon_{max} = 11$.[19] Hydrated cupric sulfate crystals have a very similar absorption band, but recent work suggests that as many as three transitions of different polarizations may be involved.[20]

Replacement of coordinated water molecules by ammonia leads to a steady shift of the absorption maximum to shorter wavelengths until four ammonia molecules are present, when $\lambda_{max} = 5900$A and $\epsilon_{max} = 50$. The addition of a fifth ammonia molecule caused the maximum to move to 6600 A[21] i.e., to a longer wavelength; no evidence for the addition of a sixth ammonia molecule could be obtained even in liquid ammonia. Ethylene diamine complexes behave similarly, while pyridine does not seem to form complexes with an amine-copper ratio of greater than four to one. In octahedral complexes we usually find that successive replacement of water by ammonia causes the crystal field splitting to increase steadily, so that a special explanation of the effect of the fifth ammonia molecule on the cupric ion spectrum must be found. It seems probable that the first four ammonia molecules are added in a plane, say along the x- and y-axes of a coordinate system centered on the Cu^{++} ion. This results in a progressive destabilization of the $d_{x^2-y^2}$ orbital with respect to all other orbitals and also to a weakening of the bonding to the water molecules in the z-direction which stabilizes the d_{z^2}, d_{xz}, and d_{yz} orbitals. Thus the transitions from the d_{z^2}, d_{xz}, and d_{yz} to the half-empty $d_{x^2-y^2}$ move to shorter wavelengths. The fifth ammonia molecule must be attached in the z-direction and so will destabilize the d_{z^2}, d_{xz}, and d_{yz} orbitals while leaving the $d_{x^2-y^2}$ orbital little changed. Any second-order effect will be to weaken the bonds in the xy plane and so to stabilize the $d_{x^2-y^2}$ orbitals. In each case the result is to move the absorption band to longer wavelengths. Any distortion of the structure towards a tetragonal pyramid would have the same effect.

In presumably nonoctahedral complexes such as $[Cu(H_2O)_6]^{++}$ one is clearly not justified in associating the position of maximum absorption with the center of gravity of the states arising from the splitting of the T_{2g} level relative to the E_g level, for the latter level is itself split. Consequently the value of $Dq = 1250$ cm^{-1} obtained by the direct application of the simple theory is not comparable with values obtained for octahedral complexes. For the same reason we shall not discuss the spectra of other cupric complexes in detail. In general we find that the magnitude of the splitting produced by different ligands increases in the order

$$\text{chloride} < \text{water} < \text{pyridine} < \text{ammonia} < \text{ethylene diamine}.$$

In complexes whose stereochemistry deviates very markedly from the regular octahedral arrangement,

[17] L. Pauling, J. Am. Chem. Soc. **53**, 1367 (1931).
[18] A. F. Wells, *Structural Inorganic Chemistry* (Oxford University Press, New York, 1950).
[19] R. Mecke and H. Leg, Z. physik. Chem. **111**, 392 (1924).
[20] D. McClure, private communication.
[21] Z. Bjerrum and E. G. Nielsen, Acta Chem. Scand. **2**, 297 (1948).

particularly in planar complexes, it should be possible to resolve the visible absorption into two or more bands. This has in fact been done for a number of planar organic derivatives of cupric copper, e.g., cupric acetylacetonate by Calvin and Belfort.[22]

THE CONFIGURATIONS d^2 AND d^8

These configurations are the simplest in which we have to consider the simultaneous effects of the removal of the d-electron degeneracy and of the electrostatic interaction between the d-electrons. The way in which the crystal field splits the ^3F ground states of these configurations is well known.[5] We shall develop the slightly more general theory which includes the mixing together of the different atomic states by the crystal-field for all the singlet and triplet levels. The d^2 ions have not been studied at all extensively so we shall restrict our numerical calculations on them to the triplet states of V^{+++}; the Ni^{++} ion will be treated in more detail since a good deal of experimental evidence is available and calculations are made very easily.

We have followed closely the procedure of Finkelstein and Van Vleck.[10] The matrix elements of V expressed as multiples of Dq are for Ni^{++}

$$
\begin{array}{ccc}
{}^1A_{1g} & {}^1E_g & {}^1T_{2g} \\
\begin{array}{cc} {}^1S & {}^1G \end{array} & \begin{array}{cc} {}^1D & {}^1G \end{array} & \begin{array}{cc} {}^1D & {}^1G \end{array} \\
\begin{vmatrix} 0 & 4\sqrt{6} \\ 4\sqrt{6} & 4 \end{vmatrix} &
\begin{vmatrix} 24/7 & \dfrac{40\sqrt{3}}{7} \\ \dfrac{40\sqrt{3}}{7} & 4/7 \end{vmatrix} &
\begin{vmatrix} -16/7 & \dfrac{20\sqrt{3}}{7} \\ \dfrac{20\sqrt{3}}{7} & -26/7 \end{vmatrix}
\end{array}
$$

$$
\begin{array}{cccc}
{}^1T_{1g} & {}^3T_{1g} & {}^3T_{2g} & {}^3A_{2g} \\
{}^1G & \begin{array}{cc} {}^3F & {}^3P \end{array} & {}^3F & {}^3F \\
2 & \begin{vmatrix} -6 & 4 \\ 4 & 0 \end{vmatrix} & 2 & 12
\end{array}
$$

In the d^2 configurations the signs of the diagonal elements must be reversed.

We shall consider first the spectrum of the $[V(H_2O)_6]^{+++}$ ion.[23] In the V^{+++} ion the separation between the ^3F and ^3P states is 13 000 cm^{-1}, and using this quantity we are able to calculate the energy level diagram of Fig. 2 in which the position of each level is plotted against Dq. The transitions $^3T_{1g}-{}^3T_{2g}$ and $^3T_{1g}-{}^3A_{2g}$ depend on the $^3F-{}^3P$ separation of the free ion only to second order, while the $^3T_{1g}-{}^3T_{1g}$ transition depends on it to first order. We therefore choose Dq to fit the experimental separation between the $^3T_{1g}$ and $^3T_{2g}$ levels, a value of $Dq=1900$ cm^{-1} being appropriate. The transitions to the T_{1g} and A_{2g} states are then predicted at 28 500 cm^{-1} and 36 000 cm^{-1}. The transition observed is at 25 000 cm^{-1} while the 36 000 cm^{-1} neighborhood is completely covered by a strong charge-transfer transition. The discrepancy between the observed and predicted positions of the second transition is rather large. It is probable that part of it is due to the $^3F-{}^3P$ interval of the hydrated ion being less than that for the free ion. We have already remarked that shielding effects will be particularly important in highly charged ions with few d-electrons such as the V^{+++}.

Finkelstein and Van Vleck[10] have commented on the fact that even with comparatively large octahedral fields the Russell-Saunders coupling is not much affected in V^{+++} complexes. In tetrahedral fields the mixing of the two T_1 states is important even for fairly small values of Dq, as is shown by Fig. 2.

The singlet states of the Ni^{++} ion, whose positions are required to compute the energy-level diagram for nickelous complexes, have not been observed experimentally. However, we may extrapolate from values for Ni^+ d^8s and Ni d^8s^2 to obtain values of 12 000 cm^{-1} and 21 000 cm^{-1} for the energies of the ^1D and ^1G states above the ^3F ground state. The ^1S state has not been observed for any related ion, since it is far above the rest of the configuration. Using values of $F_2=1560$ cm^{-1} and $F_4=110$ cm^{-1} for the Slater-Condon parameters[11] we estimate that it lies 49 000 cm^{-1} above the ground state. Although this figure is subject to a considerable error it will affect our calculation only in second order, since we will not be interested in the upper A_{1g} level.

The final energy level diagram, excluding the upper A_{1g} level which lies far above the remaining levels, is shown in Fig. 3. We find that three triplet-triplet transitions, $^3A_{2g}-{}^3T_{2g}$, $^3A_{2g}-{}^3T_{1g}$, and $^3A_{2g}-{}^3T_{1g}$ should occur in that order for octahedral complexes. In addition there should be a number of much weaker triplet-singlet transitions.

The hydrated nickel ion, $Ni(H_2O)_6^{++}$ in aqueous solution has three main absorption bands at 8500 cm^{-1},[24] and 14 000 cm^{-1} and 26 000 cm^{-1} [25] with ϵ_{max} values of 2, 1.8, and 6 respectively. If we choose $Dq=850$ to fit the first transition we expect further transitions at 14 000 cm^{-1} and 27 000 cm^{-1} in good agreement with experiment.

A detailed examination of the published spectra of Ni^{++} in solution suggests that there is a further weak peak at 19 000 cm^{-1} and that the 26 000 cm^{-1} band lies over and partially obscures weak bands at 22 500 cm^{-1} and 30 000 cm^{-1}. Crystal absorption spectra[26] confirm the existence of weak discrete transitions at 22 500 cm^{-1} and in the region from 17 500–20 500 cm^{-1}. The peak at 30 000 cm^{-1} may correspond to a transition to the $^1E_{1g}$ state at 29 500 cm^{-1} while the 22 500 cm^{-1} band

[22] M. Calvin, private communication.
[23] S. C. Furman and S. Garner, J. Am. Chem. Soc. **72**, 1785 (1950).
[24] T. Dreisch and W. Trommer, Z. physik. Chem. **B37**, 37 (1937).
[25] A. Kiss and R. Szabo, Z. anorg. Chem. **252**, 172 (1943).
[26] G. Gielessen, Ann. Physik 5, **22**, 537 (1935).

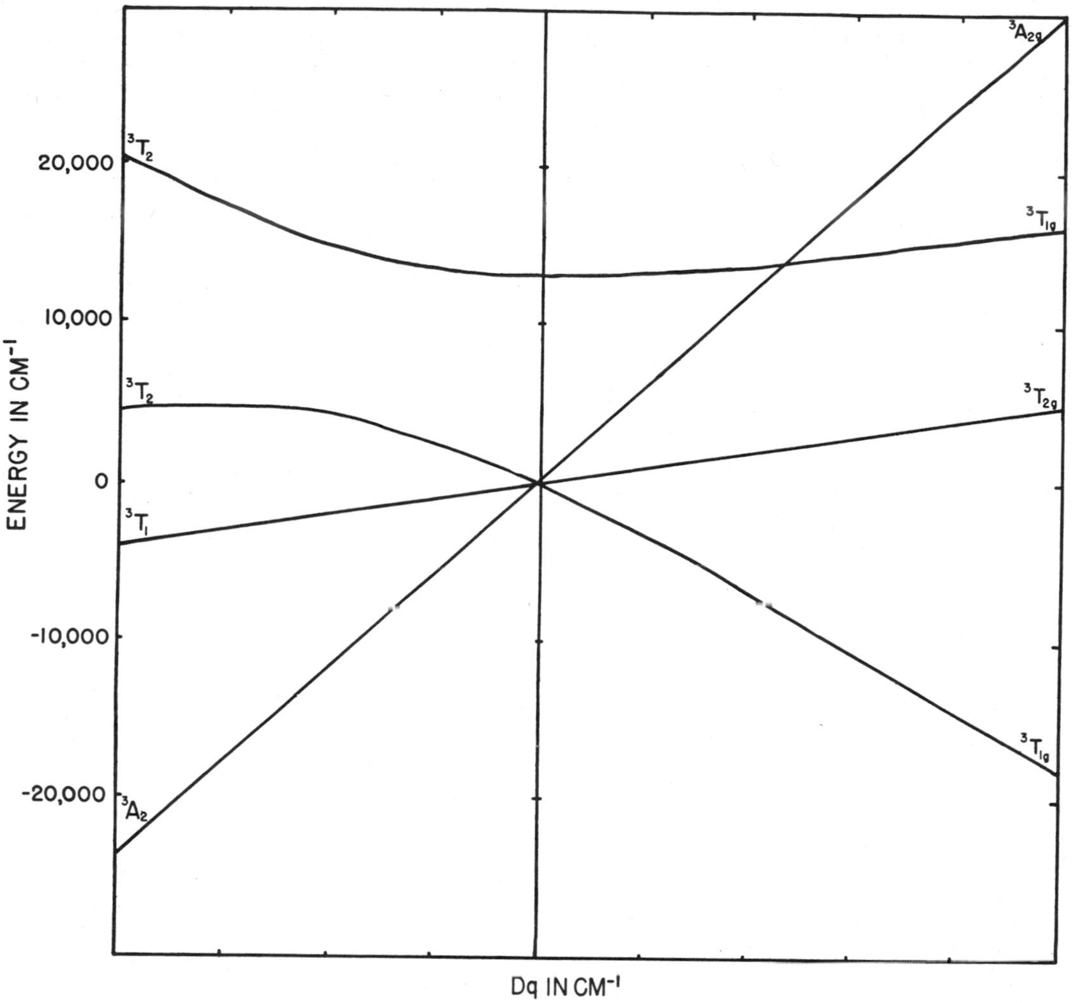

FIG. 2. Energy level diagram for the V^{+++} ion. The intervals indicated along the *Dq*-axis each represent 500 cm^{-1}.

would correspond exactly to the transition to a $^1T_{1g}$ state. The band close to 19 000 cm^{-1} does not correspond very well with any level on our diagram.

The identification of two of the bands as intercombinations suggested above is very tentative, since they are very much weaker than their triplet-triplet neighbor, and might represent vibrational structure in the latter. Further studies of a number of different complexes would be required to settle the assignments.

The infrared band of the hexamine nickel ion has its maximum at 10 600 cm^{-1}.[24] Taking $Dq=1060$ cm^{-1} to fit it, we predict further bands at 17 500 cm^{-1} and 29 200 cm^{-1}, in reasonable agreement with the observed bands at 16 700 cm^{-1} and 27 400 cm^{-1}.[27]

The corresponding peaks for tris ethylene-diamine nickel are observed at 11 200 cm^{-1}, 18 300 cm^{-1}, and 29 000 cm^{-1}, respectively.[27] Taking $Dq=1120$ cm^{-1} the calculated values are 11 200 cm^{-1}, 19 100 cm^{-1}, and 31 000 cm^{-1} respectively. In the tris-*o*-phenanthroline complex only two bands have been observed at 12 700 cm^{-1} and 19 000 cm^{-1}.[27] Bands are predicted at 12 700 cm^{-1} and 20 200 cm^{-1} if we use $Dq=1270$ cm^{-1}.

From the position of the first visible absorption peaks the following order of increasing *Dq* has been deduced for Ni^{++} complexes.[28]

Methyl alcohol < water < oxalate < pyridine
< ammonia < ethylenediamine < *o*-phenathroline.

[27] G. L. Roberts and F. H. Field, J. Am. Chem. Soc. **72**, 4232 (1950).

[28] Russell, Cooper, and Vosbergh, J. Am. Chem. Soc. **65**, 1301 (1943).

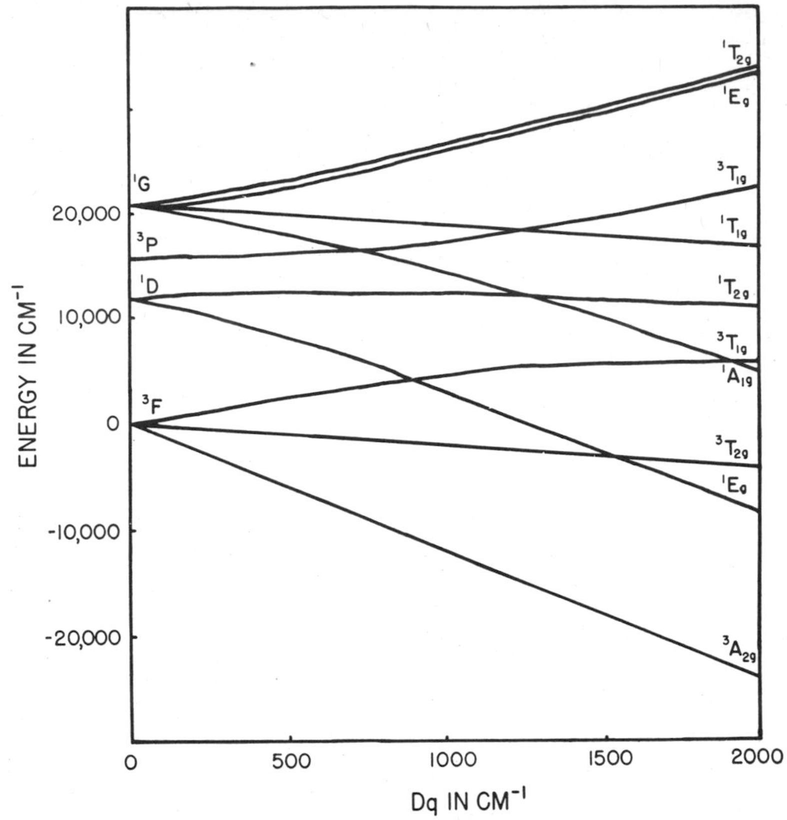

FIG. 3. Energy level diagram for the Ni^{++} ion.

A considerable number of octahedral compounds in which the ligands are not identical have been studied, in particular the ions $[Ni(H_2O)_n(NH_3)_{6-n}]^{++}$ with $n=0, 1, 2, 3, 4, 5, 6$.[29] Experimentally it is found that the visible-band shifts steadily to shorter wavelengths as the number of molecules of ammonia in the complex increases, but that no splitting of the band occurs. Theoretically we would expect that this band, like the analogous band in the Cr^{+++} ion, would be split for certain of the complexes. The absence of such a splitting is due to two factors. Firstly, the expected splitting is small compared with the total width of the band and secondly, due to the mobility of the system, a mixture of stereoisomers is always present for the complexes with $n=2$, 3, and 4. The observed shift to the blue with increasing n is the shift of the center of gravity of the bands of all the complexes with a given n.

The electronic spectra of tetrahedral nickel complexes have not been investigated. The planar, diamagnetic complexes of divalent nickel are characterized by strong absorption bands with their maxima close to 4000 A and with ϵ_{max} values ranging from 2000 to 10 000.[30] It seems unlikely that this transition is a simple transition within the d-shell in view of its very high intensity. Two other explanations seem plausible, namely that it is a transition which is not approximately localized on the nickel ion or that it is a transition from a $3d$-orbital to the unoccupied $4p$-orbital of the Ni^{++} ion. We are inclined to think that the latter explanation is the more probable one. It would be interesting to know whether there are any weak bands at longer wavelengths than the strong 4000 A band in these nickel complexes, corresponding to the expected transitions within the $3d$-shell.

THE CONFIGURATIONS d^3 AND d^7

The d^3 configuration has been discussed in detail by Finkelstein and Van Vleck, who have obtained the matrix elements of the cubic field for the quartet and doublet states.[10] They have applied their theory to the hydrated chromic ion and have shown that the lowest doublet state is brought to the observed position if a value of 1820 cm^{-1} is taken for Dq. We have identified the two strongest long wavelength absorption

[29] J. Bjerrum, *Metal Ammine Formation in Aqueous Solution* (Copenhagen, 1941).

[30] McKenzie, Mellon, Mills, and Short, Proc. Roy. Soc., New South Wales **58**, 70 (1944).

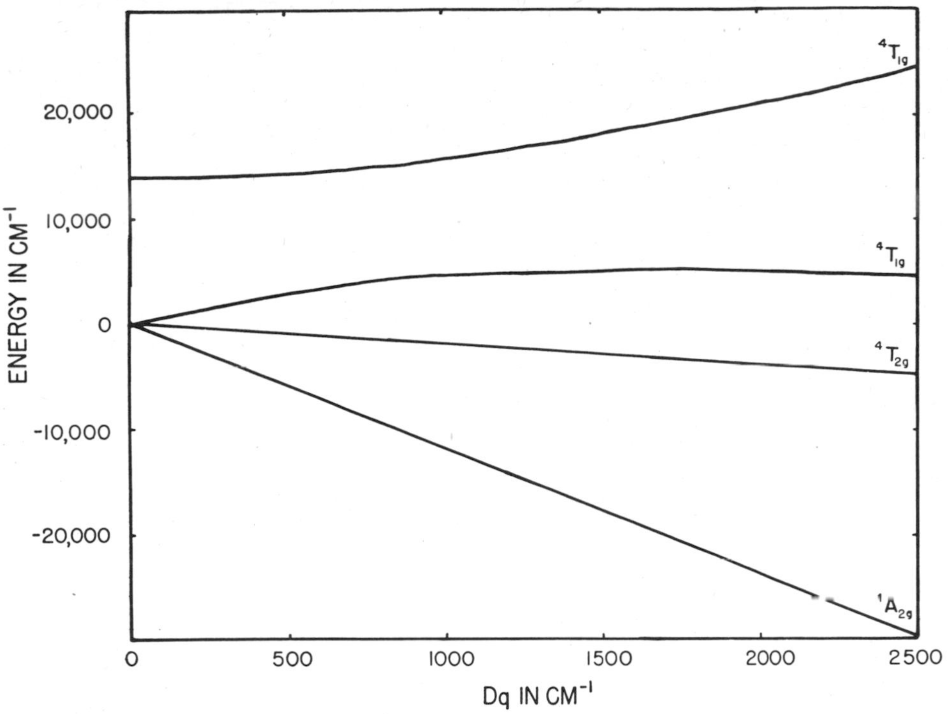

FIG. 4. Energy level diagram for the Cr^{+++} ion.

bands of trivalent chromium complexes as quartet-quartet transitions within the d^3 configuration.[8] The energy level diagram for the quartet states of the chromic ion in octahedral coordination is given in Fig. 4. Using a value of 1720 cm^{-1} for Dq we predict absorption maxima at 17 200 cm^{-1} and 25 700 cm^{-1} compared with the experimental values of 17 200 cm^{-1} and 25 600 cm^{-1}.[31] Thus a value of 1770 cm^{-1} for Dq would give good agreement with both the singlet and triplet states.

The energy level diagram of $[Cr(NH_3)_6]^{+++}$ does not agree quite so well with experiment. If we take $Dq = 2150$ cm^{-1}, to give the lower transition correctly at 21 500 cm^{-1}, then we predict an upper transition at 30 500 cm^{-1} compared with the observed value of 28 500 cm^{-1}.[32]

The successive substitution of ammonia molecules for water in $[Cr(H_2O)_6]^{+++}$ results in a progressive shift of the absorption maxima to longer wavelengths,[31] but there is no resolution of any of the absorption bands into two or more components.

A general survey of mono and polysubstituted chromium complexes[33,34] shows that the value of Dq increases in the order

Iodide < bromide < chloride < thiocyanate < oxalate < water < ammonia < ethylenediamine.

We have shown that in mono-substituted and particularly in *trans*-disubstituted chromic complexes the lowest absorption band should be split into two components while the upper band should, to a first approximation, remain unsplit.[8] With the knowledge of the relative size of Dq contributions which we have derived we can see that if the substituent has a smaller Dq value than the other ligands the short wavelength component of the split band, since it is degenerate, should be twice as strong as the long wavelength component. If the substituent has a larger Dq value than the other ligand the order of intensities should be reversed. We shall discuss this point in more detail in connection with cobaltic complexes, and here will only note that the theory seems to be qualitatively correct for chromium complexes.

The spectrum of the hydrated vanadous ion has been studied by Dreisch and Kallscheuer,[35] who found a band with its maximum at about 11 600 cm^{-1} and by Kato,[36] who found bands with maxima at 12 200 cm^{-1}

[31] R. I. Colmar and F. W. Schwartz, J. Am. Chem. Soc. **54**, 3204 (1932).
[32] M. Linhard, Z. Electrochem. **50**, 224 (1944).
[33] M. Linhard and M. Weigel, Z. anorg. Chem. **266**, 49 (1951).
[34] W. Kuhn, Z. anorg. Chem. **216**, 321 (1934).
[35] T. Dreisch and O. Kallscheuer, Z. physik. Chem. **B45**, 19 (1939).
[36] S. Kato, Sci. Papers Inst. Phys. Chem. Research (Tokyo) **13**, 49 (1930).

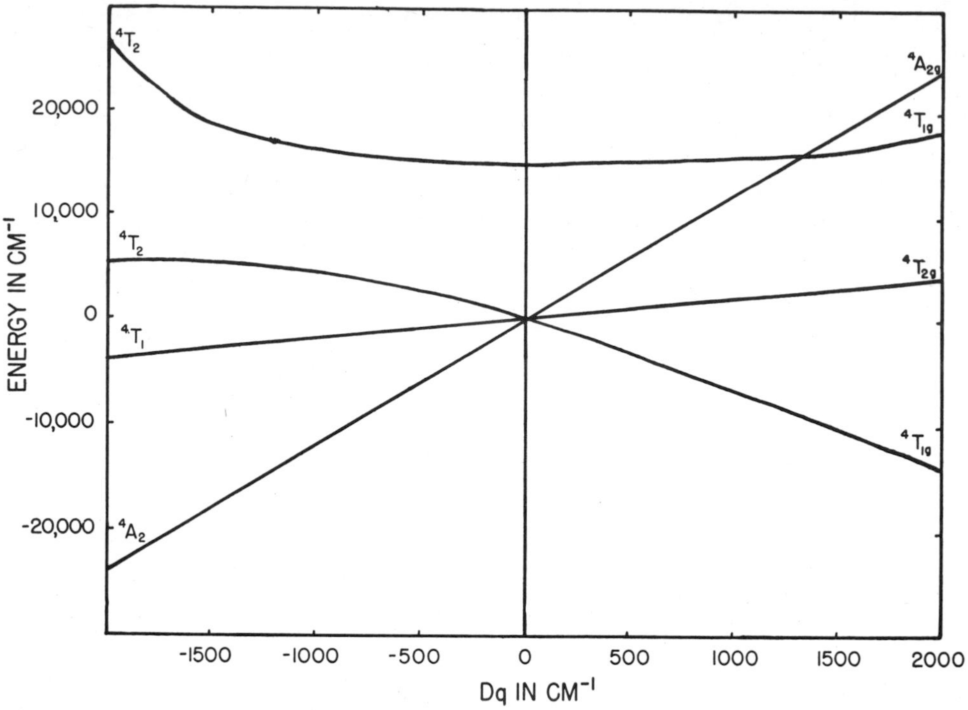

FIG. 5. Energy level diagram for the Co^{++} ion.

and 17 900 cm^{-1}. Taking a value of $Dq = 1220\ cm^{-1}$ we predict levels at 12 200 cm^{-1} and 18 700 cm^{-1}, in satisfactory agreement with experiment. No other vanadous complexes seem to have been studied.

The spectra of a number of cobaltous salts have been recorded. Those of the regular octahedral and tetrahedral complexes, in which the cobalt ion has a quartet ground state, are the most important from our point of view. The energy level diagram for the d^7 configuration may be obtained from the matrix-elements given by Finkelstein and Van Vleck. The energy-level diagram for the quartet states in both octahedral and tetrahedral complexes is shown in Fig. 5.

The hydrated cobaltous ion, $[Co(H_2O)_6]^{++}$, in aqueous solution has an absorption band in the infrared with its maximum at 8100 cm^{-1} [24] and a broad, complex band whose maximum is at 19 600 cm^{-1} and which stretches from about 15 000 cm^{-1} to about 28 000 cm^{-1}.[37] If we take $Dq = 970\ cm^{-1}$ we predict transitions at 8100 cm^{-1}, 17 500 cm^{-1}, and 22 000 cm^{-1}. We think it probable that the 19 600 cm^{-1} band of the cobaltous ion includes both of the shorter wavelength quartet-quartet transitions and, in addition, some quartet-doublet transitions.

The $[Co(NH_3)_6]^{++}$ ion has an absorption maximum at about 20 200 cm^{-1} while the corresponding ethylenediamine complex has two maxima at 20 800 cm^{-1} and 28 100 cm^{-1}.[27] In view of the great similarity of other spectra of ammonia and ethylenediamine complexes we assume that the two bands near 20 000 cm^{-1} have the same origin, namely in the $T_{1g}-A_{2g}$ transition. This would correspond to a value of 1130 cm^{-1} for Dq in the ethylenediamine complex and slightly less in that of ammonia.

The analysis of the spectra of the tetrahedral cobaltous complex ions $[COCl_4]^{--}$, $[CoBr_4]^{--}$, $[CoI_4]^{--}$ and $[Co(CNS)_4]^{--}$ presents a number of difficulties. Each of the halides has an absorption band in the infrared, the maxima being at 6300 cm^{-1}, 5300 cm^{-1}, and 5000 cm^{-1} with ϵ_{max}'s of 15 100 and 160 for the chloride, bromide, and iodide, respectively. In the visible there are a number of absorption bands, but one of these is always stronger than any other. This strong band is at 15 000 cm^{-1} in the chloride, at 13 700 cm^{-1} in the bromide, at 12 500 cm^{-1} in the iodide[37] and at 17 000 cm^{-1} in the thiocyanate.[38] The extinction coefficient is as high as 2500 in $[Co(CNS)_4]^{--}$ while in $[CoCl_4]^{--}$ ϵ_{max} is about 600.

The energy-level diagram for tetrahedral Co^{++} (Fig. 5) shows that the two strong bands do not correspond in position to the two lowest predicted transitions, for if we fit the lowest transition correctly in

[37] A. Kiss and P. Csokan, Z. physik. Chem. **A186**, 239 (1940).

[38] W. R. Brode and R. A. Morton, Proc. Roy. Soc. (London) **120**, 21 (1928).

$[CoCl_4]^{--}$ we predict the next transition at 11 000 cm^{-1}, which is much *below* the observed transition.

The two observed bands are much stronger than those of divalent transition-metal ions in octahedral coordination. This we believe to be due to the mixing together of the $3d_{xy}$, $3d_{xz}$, and $3d_{yz}$ orbitals and the $4p$ orbitals, which is possible since the tetrahedron of ligands has no center of symmetry and can therefore cause the mixing of orbitals which are of g and u symmetry in the free atom. The transition from the A_2 ground-state to the two T_2 upper states are formally allowed on symmetry grounds, while the transition to the T_1 state is forbidden. It seems plausible to assume that the two strong observed transitions are in fact the allowed transitions. If we do this we find $Dq=375$ cm^{-1} in $[CoCl_4]^{--}$ in order to account for the infrared band, and then predict a second band at 19 500 cm^{-1}. This value again disagrees rather badly with the experimental value. It is possible that this assignment is nevertheless correct and that the discrepancy is due to the depression of the upper T_2 state by a nearby state of the d^2p configuration. If this is so there should be a forbidden transition in the infrared to a state only about 3500 cm^{-1} above the ground state. The spectra of the bromide and iodide complexes interpreted in this way give values of 310 cm^{-1} and 290 cm^{-1} for Dq. Any other interpretation of these spectra would also require a small Dq value. A possible explanation of the low values of Dq will be given in the next paper.

THE CONFIGURATIONS d^4 AND d^6

Very little information is available concerning the spectra of the complexes which have the d^4 configuration, i.e., chromous and manganic complexes. Aqueous solutions of chromous chloride have an absorption band with its maximum at about 12 600 cm^{-1} which almost certainly belongs to the $[Cr(H_2O)_6]^{++}$ ion.[35] The energy level for the quintet states of d^4 is identical with that of d^9 and so we find a value of $Dq=1260$ cm^{-1}. It should be noted that although the chromous solutions studied were not very stable, the position of the band maximum would not be affected since the chromic ion does not absorb in the same region as the chromous.

The Mn^{+++} hydrated ion has an absorption maximum at 4750 A,[7] corresponding to a Dq value of 2100 cm^{-1}. No other manganic complexes seem to have been investigated.

The configuration d^6 occurs in the ferrous and cobaltic complexes. The energy diagram for the quintet states is identical with that for a single d-electron, so that the absorption maximum of the hydrated ferrous ion, which occurs at 10 500 cm^{-1} [35] corresponds to $Dq=1050$ cm^{-1}. No other ferrous complexes with quintet ground states have been examined.

The spectra of the few diamagnetic ferrous complexes which have been studied e.g., $[Fe(CN)_6]^{4-}$ and $[Fe(Phenanthroline)_3]^{++}$ do not fit into the present theoretical scheme at all well. This may be connected with the strong double bond formation which must occur in them.

The complexes of trivalent cobalt, with the exception of the $[CoF_6]^{3-}$ ion, are diamagnetic i.e., they have singlet ground states. Unfortunately the calculation of the energy-level diagram for the singlet states would be extremely lengthy, there being as many as seven ${}^1T_{2g}$ states. Instead we shall discuss the spectra in a qualitative way.

We have shown that the close similarity between the spectra of chromic and cobaltic complexes is due to the fact that when the splitting of the $3d$-orbitals becomes very large compared with the electrostatic interactions between $3d$-electrons the energy diagrams for the three lowest singlet states of Co^{+++} and for the three lowest quartet states of Cr^{+++} are essentially identical.[8] The two strong absorption bands correspond to transitions from the state $(t_{2g})^3\,{}^4A_{2g}$ to the states $(t_{2g})^2(e_g)^1\,{}^4T_{1g}$ and ${}^4T_{2g}$ in Cr^{+++}[39] and from the state $(t_{2g})^6\,{}^1A_{1g}$ to the states $(t_{2g})^5(e_g)^1\,{}^1T_{1g}$ and ${}^1T_{2g}$ in Cr^{+++}. The energy level diagram for Co^{+++} will differ somewhat from that of Cr^{+++} in fact, because "configuration interaction" is possible in the former e.g., between $(t_{2g})^6\,{}^1A_{1g}$ and $(t_{2g})^4(e_g)^2\,{}^1A_{1g}$, which is impossible in the latter.

The similarity between the strongest transitions in cobaltic and chromic complexes does not extend to the intercombinations. The reason for this is clearly that the lowest quartet-doublet transitions in Cr^{+++} are essentially those within the $(t_{2g})^3$ configuration. In Co^{+++} the lowest triplets belong to the $(t_{2g})^5(e_g)^1$ configuration.

A very large variety of cobaltic complexes has been studied, so that although we cannot get accurate values of Dq we can find the order of increasing Dq for a variety of ligands.

From the work of Kiss and Czegledy[40,41] we may deduce the following order of Dq contributions:

$$CO_3^{--} < OH^- < NO_3^- = SO_3^{--} < CNS^- < H_2O < \text{Oxalate} < NH_3 < \text{ethylene diamine} < NO_2^- < CN^-.$$

The measurements of Linhard and Weigel[42] give the following order of increasing Dq

$$I^- < Br^- < Cl^- < N_3^- < R\cdot CO_2^- < F^- < NH_3 < NO_2^-.$$

The two orders cannot be brought together with certainty since there are small differences between the positions of the maxima found by the different workers.

Perhaps the most interesting conclusion to be drawn from these spectra is that the cyanide ion produces a much bigger splitting than any other ion. The first

[39] We use small letters to designate orbitals in accordance with Professor Mulliken's recommendations on spectroscopic nomenclature. This involves a change from the notation of our earlier paper.

[40] A. Kiss and D. Czegledy, Z. anorg. Chem. **235**, 407 (1937).

[41] A. Kiss, Z. anorg. Chem. **246**, 28 (1941).

[42] M. Linhard and M. Weigel, Z. anorg. Chem. **264**, 321 (1951); **266**, 49 (1951); **267**, 113 (1952); **267**, 121 (1952).

maximum of $[Co(NH_3)_6]^{+++}$ is at 21 000 cm^{-1} while that of $[Co(CN)_6]^{---}$ is at 32 000 cm^{-1}. This is consistent with the very marked double-bond character of the transition-metal to cyanide bonds.

We have shown that the longest wavelength absorption band in monosubstituted cobaltic complexes is split into two components[8] and that if Dq for the substituent is less than it is for the other ligand then the long wavelength component should be twice as strong as the other. If Dq is greater for the substituent then the order of intensities should be reversed. Furthermore, it is readily seen that the magnitude of the splitting should be proportional to the difference in Dq values for the substituent and for the other ligand. These relations are in general found to be correct.[42]

In disubstituted complexes we would predict that the splitting of the longest wavelength transition would be much more marked in the *trans*-complexes than in the *cis*. This is completely confirmed by the measured spectra.[42]

THE CONFIGURATION d^5

The matrix-elements of the cubic field for the triplet states of the d^5 configuration have been calculated in the usual way. The elements which do not vanish through symmetry are given below.

E_1	^4P	^4F	^4G
^4P	0	0	$4\sqrt{5}$
^4F	0	0	$2\sqrt{5}$
^4G	$4\sqrt{5}$	$2\sqrt{5}$	0

E_2	^4D	^4F	^4G
^4D	0	$20/\sqrt{7}$	0
^4F	$20/\sqrt{7}$	0	$10\sqrt{3/7}$
^4G	0	$10\sqrt{3/7}$	0

E	^4D	^4G
^4D	0	0
^4G	0	0

It should be noted that all diagonal elements are zero. The energy level diagram for Mn^{++} in octahedral or tetrahedral coordination is shown in Fig. 6. An analogous diagram applies to Fe^{+++}. We will not reproduce it here as it may easily be obtained by solution of the appropriate secular equations using the extrapolated values of 32 000 cm^{-1}, 35 000 cm^{-1}, 38 500 cm^{-1}, and 52 000 cm^{-1} for the energies of the ^4G, ^4P, ^4D, and ^4F states of Fe IV(d^5).[43]

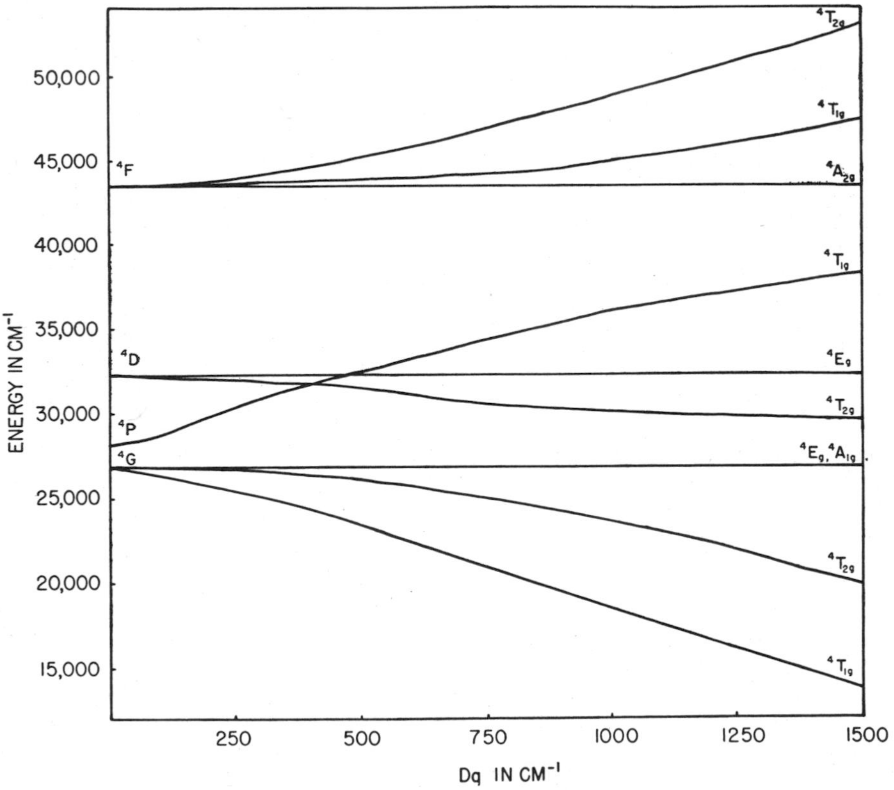

FIG. 6. Energy level diagram for the Mn^{++} ion.

[43] Edlen, *Actualities Scientifiques et Industrielles* (Hermann & Cie, Paris, 1941), No. 895, Part I.

The configuration d^5 is exceptional in that no spin-allowed transitions are to be expected, since the ^6S ground state is not split significantly by the field, and all excited states are of lower multiplicity. It follows that only very weak absorption corresponding to sextet-quartet transitions is to be expected. The sextet-doublet transitions are almost certainly too weak to appear in absorption.

It is found experimentally that $[Mn(H_2O)_6]^{++}$ exhibits very weak discrete absorption between 24 500 cm^{-1} and 25 500 cm^{-1}.[26] The ultraviolet region does not seem to have been investigated. If we assume a Dq value of about 1000 cm^{-1} we predict a number of transitions between 18 000 cm^{-1} and 35 000 cm^{-1} followed by a further series of transitions which begins at 43 000 cm^{-1}. The data available are not sufficient to make any definite assignment. We are inclined to think that the observed transition is not the lowest $^6S-{}^4G(T_{1g})$ transition but that the upper state is $^4G(A_{1g})$, $^4G(E_g)$ or possibly $^4G(T_{2g})$. A number of compounds in which the octahedron of ligands surrounding the manganous ion includes two or more species have been studied. They show a number of bands between 23 000 and 40 000 cm^{-1}.[26] This is consistent with the theoretical scheme, but the energy-level diagram is so complicated that no assignments can be made. An experimental study of the $[Mn(H_2O)_6]^{++}$ ion would be particularly valuable.

The spectrum of $[Fe(H_2O)_6]^{+++}$ has been studied in some detail by Rabinowitch and Stockmayer.[44] They find very weak bands at 24 500 and 18 400 cm^{-1} and a further band with a maximum at less than 14 000 cm^{-1}. Infrared studies show that maximum of the latter band is near 13 000 cm^{-1}. These observations are not easily explained since theoretically there should be only two transitions much below 32 000 cm^{-1}, the position of the ^4G state. Using a reasonable value of 2100 for Dq we predict $^4G(T_{1g})$ and $^4G(T_{2g})$ levels at 13 300 and 20 700 cm^{-1} respectively, in rough agreement with experiment, but the band at 24 500 remains completely unexplained. It might be a strongly forbidden charge-transfer band, but this seems unlikely in view of its sharpness. Until further evidence is available both the values of Dq and the assignments of the two long wavelength bands are tentative. It is interesting that complex-formation with ammonia or aliphatic amines should shift the two long wavelength bands of Fe^{+++} to *lower* energies.

[44] E. Rabinowitch and W. H. Stockmayer, J. Am. Chem. Soc. **64**, 335 (1942).

The tetrahedral ion $[FeCl_4]^-$ has been studied under a variety of conditions.[45] It has a broad very weak absorption band from 6000–8000 A with maxima at about 7500, 6900, and 6100 A, and sharper peaks at 5250 A and 6500 A. Like the spectra of other tetrahedral complexes these cannot be explained quantitatively by our theory. It is not clear whether the observed transitions are internal transitions of the iron atom or spin-forbidden charge-transfer bands.

CONCLUSIONS

We have shown that many features of the absorption spectra of transition-metal complexes may be understood in terms of crystal-field theory. A number of interesting regularities are evident when the Dq values are compared. In particular:

(1) The values of Dq for hydrated divalent ions are close to 1000 cm^{-1}, falling from 1220 cm^{-1} for V^{++} and 1260 cm^{-1} for Cr^{++} to 850 cm^{-1} for Ni^{++}.

(2) The values of Dq for hydrated trivalent ions are close to 2000 cm^{-1}.

(3) The common ligands may be arranged in a squence in such a way that the Dq values for their complexes with any metal ion increase as we go from left to right. Such a sequence is:

I^- Br^- Cl^- F^- H_2O oxalate pyridine NH_3 ethylene diamine NO_2^- CN^-.

(4) The little evidence available suggests that Dq is much smaller for tetrahedral complexes than for corresponding octahedral complexes.

We shall defer the discussion of the importance of these regularities in understanding the chemistry of transition-metal complexes to a subsequent paper.

The identification of excited electronic states of many complexes and the estimation of Dq values should prove helpful in interpreting paramagnetic susceptibility data and paramagnetic resonance spectra. In many cases the values of Dq assumed at present are very different from those derived in this work.

I am indebted to Dr. D. McClure for communicating results on the crystal spectra of hydrated transition-metal ions, prior to publication. His interpretation of these spectra is in substantial agreement with that presented in this paper.

[45] H. L. Friedman, J. Am. Chem. Soc. **74**, 5 (1952).

Reprinted from *Proc. Royal Soc.. London,* **A327**, 183–200 (1955)

20

The colours and magnetic properties of hydrated iron group salts, and evidence for covalent bonding

By J. Owen

Clarendon Laboratory, University of Oxford

(*Communicated by M. H. L. Pryce, F.R.S.—Received* 30 *June* 1954)

The typical colours of paramagnetic salts containing $[M(H_2O)_6]$, $M = 3d^n$, complexes, are generally assumed to arise from optical transitions between the orbital energy levels of the $3d$ ion which are split by the crystalline electric field due to the surrounding water dipoles. The available optical absorption data are analyzed using this ionic model and the associated crystal field theory, and it is shown that while experiment and theory agree fairly well, there are systematic discrepancies. In addition, the orbital level separations given by the optical spectrum are found to be smaller than those predicted from magnetic data. It is shown how these discrepancies can be accounted for by introducing weak covalent bonds into the $[M(H_2O)_6]$ complex, so that there is charge transfer between the paramagnetic ion and the attached water molecules.

1. Introduction

It is well known that the crystal field theory of Van Vleck (1932), Schlapp & Penney (1932) and others, accounts for the magnetic properties of hydrated iron group salts to a very good approximation. In this theory it is assumed that the $M(H_2O)_6$ complex is held together by electrostatic forces, and that the behaviour of the paramagnetic ion M depends on the electric field arising from the surrounding octahedron of water dipoles. This crystal field splits the orbital levels of the paramagnetic ion into groups with energy separations of about 10000 cm^{-1}, and since only the lowest levels are populated, there is a reduced orbital contribution to the magnetism. In some cases this orbital moment is related in a simple way to the separation Δ between the lowest and the next highest orbital levels, and, as will be shown below, it is then possible to test the theory rather critically by comparing values of Δ found directly from the optical spectrum of the salt, with values of the magnetic moment which are known accurately from paramagnetic resonance measurements. Even in cases where a simple test of this sort cannot be made, the relative separations between the orbital levels can be calculated from the theory, and these can be compared with the optical data. It is the purpose of the present paper to collect together some of the available information on hydrated iron group salts, and by making comparisons of these kinds, to show how far the magnetic data, optical data and crystal field theory are consistent.

Abragam & Pryce (1951*b*, *c*) first pointed out that such an analysis of the data for Co^{2+} and Cu^{2+} leads to discrepancies between experiment and theory, and Griffiths & Owen (1952) showed that the same is true for Ni^{2+}. As will be seen below, such discrepancies occur systematically for most hydrated iron group salts, and can be accounted for by introducing some charge transfer between the paramagnetic ion and the water molecules (see also Kleiner 1952; Stevens 1953), so that the complex is no longer purely ionic.

2. Crystal field and energy levels

It will first be shown to what extent a purely ionic model is successful in explaining the experimental results. The theory is extensively covered in the literature and will therefore be outlined only briefly. The type of complex that will be considered has the paramagnetic ion at the centre of an elongated or compressed octahedron of water molecules, i.e. four dipoles μ at $(\pm a, 0, 0)$, $(0, \pm a, 0)$, and two dipoles μ' at $(0, 0, \pm b)$. The negative ends of these dipoles (the oxygen atoms) are pointing inwards, and in the first approximation they are treated as point dipoles. The electric potential field at (x, y, z) near the ion at $(0, 0, 0)$ is then of the form up to fourth-power terms

$$\left.\begin{aligned} V &= K'(x^4+y^4+z^4-\tfrac{3}{5}r^4)+T_2'(2z^2-x^2-y^2)+T_4'(z^4+6x^2y^2-\tfrac{3}{5}r^4),\\ \text{where}\quad K' &= -\frac{25}{4}\left(\frac{4\mu}{a^6}+\frac{3\mu'}{b^6}\right),\quad T_2' = 3\left(\frac{\mu}{a^4}-\frac{\mu'}{b^4}\right),\quad T_4' = \frac{25}{4}\left(\frac{\mu}{a^6}-\frac{\mu'}{b^6}\right), \end{aligned}\right\} \tag{1}$$

and the term in K' is the part of the field with cubic symmetry, and those in T_2', T_4' have tetragonal symmetry. These coefficients are given by Polder (1942). It will be noticed that the tetragonal part of the field vanishes if the octahedron is regular, i.e. $T_2' = T_4' = 0$ if $a = b$ and $\mu = \mu'$.

To find the orbital level splittings produced by the crystal field V, it is necessary to find the matrix elements of the energy operator eV (e = electronic charge) for the orbital wave-functions belonging to the L-S term concerned, and thus set up the secular equation for the energies. The simplest method of doing this is to replace V by the equivalent orbital momentum operators as described by Stevens (see Bleaney & Stevens 1953). This leads to positions of the orbital levels which are shown below:

$3d^1$

Free ion: ground term is ^2D; no other terms of the same multiplicity lie very close.

Ion in crystal field:

$$
\begin{array}{cll}
 & \text{orbital level} & \text{energy} \\
 & d\gamma_1 & \frac{6}{21}K + \frac{4}{7}T_2 + \frac{12}{105}T_4 \\
 & d\gamma_2 & \frac{6}{21}K - \frac{4}{7}T_2 - \frac{28}{105}T_4 \\
{}^2D & d\epsilon_2, d\epsilon_3 \\
 & \text{(doublet)} & -\frac{4}{21}K + \frac{2}{7}T_2 - \frac{8}{105}T_4 \\
 & d\epsilon_1 & -\frac{4}{21}K - \frac{4}{7}T_2 + \frac{32}{105}T_4
\end{array}
\tag{2}
$$

where
$$K = \tfrac{2}{5}e\overline{r^4}K', \quad T_2 = e\overline{r^2}T_2', \quad T_4 = e\overline{r^4}T_4' \tag{3}$$

and $\overline{r^2}$ and $\overline{r^4}$ are average values for the radius of the $3d$ electrons. The energies for $3d^6$ are the same, and for $3d^4$ and $3d^9$ are the same but with reversed sign.

$3d^2$

Free ion: ^3F at 0,

^3P at E;

no other term of the same multiplicity lies very close.

Ion in crystal field:

$$
\begin{array}{cll}
 & \text{orbital level} & \text{approximate energy} \\
{}^3P & p_2\ \text{(doublet)} & E' + X + \frac{2}{5}T_2 \\
 & p_1 & E' + X - \frac{4}{5}T_2 \\
 & a & \frac{4}{7}K - \frac{16}{105}T_4 \\
 & b_1 & \frac{2}{21}K + \frac{44}{105}T_4 \\
{}^3F & b_2\ \text{(doublet)} & \frac{2}{21}K - \frac{26}{105}T_4 \\
 & c_1 & -\frac{2}{7}K - X + \frac{8}{35}T_2 - \frac{4}{35}T_4 \\
 & c_2\ \text{(doublet)} & -\frac{2}{7}K - X - \frac{4}{35}T_2 + \frac{6}{35}T_4
\end{array}
\tag{4}
$$

In these expressions K, T_2 and T_4 are again given by (3), and E' is the term separation for the ion in the crystal and for the present purpose is to be treated as an unknown variable not necessarily equal to the free ion value E. X is equal to

$$-\left(\frac{E' + \frac{2}{7}K}{2}\right) + \left[\left(\frac{E' + \frac{2}{7}K}{2}\right)^2 + \left(\tfrac{4}{21}K\right)^2\right]^{\frac{1}{2}},$$

and represents the effect of off-diagonal matrix elements of the cubic field, which couple the group of states p to the group c. Off-diagonal elements containing the smaller tetragonal terms T_2 and T_4 have been neglected. The energies for $3d^7$ (^4F and ^4P) are also given by (4), and those for $3d^3$, $3d^8$ are the same but with reversed sign of K, T_2 and T_4. These energies have already been derived for particular cases by previous authors (e.g. for d^3 by Finkelstein & Van Vleck (1940), for d^7 by Abragam & Pryce (1951*b*), for d^9 by Polder (1942)), but are given in a different form from equations (2) and (4).

The positions of the orbital levels given by these equations are shown in figure 1 for three examples. The values of K and E' have been chosen to fit the optical absorption data, on the assumption that the broad optical absorption lines which are found at 300° K arise from transitions between the lowest orbital level and the higher levels in equations (2) and (4). Transitions to other levels of different multiplicity from the ground state will be neglected, although it should be pointed out that such transitions have been found in the chromic alums at low temperatures, and give intense and very narrow lines in the visible part of the spectrum (Spedding & Nutting 1935; Van Vleck 1940).

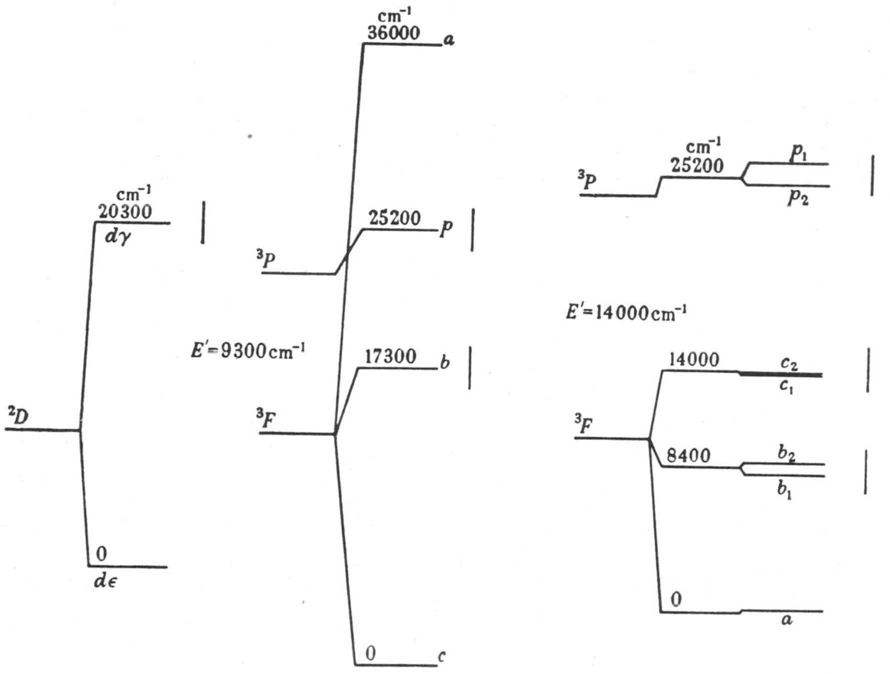

(*a*) $3d^1$, $[Ti(H_2O)_6]^{3+}$ (*b*) $3d^2$, $[V(H_2O)_6]^{3+}$ (*c*) $3d^8$, $[Ni(H_2O)_6]^{2+}$

FIGURE 1. The positions of the orbital energy levels in the cubic crystal field are drawn to scale for the examples of Ti^{3+}, $3d^1$; V^{3+}, $3d^2$; and Ni^{2+}, $3d^8$ (see equations (2) and (4) and tables 1 and 2). The positions of the broad optical absorption lines are indicated by the vertical lines on the right of each figure. The effect of a tetragonal field ($T_2 = T_4 = 1000\,cm^{-1}$) is shown for Ni^{2+} in figure (*c*); if the effect of spin-orbit coupling or any rhombic part of the field were included, there would be further small subsplittings of the levels.

The effect of the tetragonal part of the field is only shown in figure 1*c* for $Ni^{2+}(H_2O)_6$, using $T_2 = T_4 = 1000\,cm^{-1}$. This value of T_4 is fairly typical for the nickel Tutton salts, and is estimated from magnetic resonance data (see Griffiths & Owen 1952), and it is found that $T_2 \sim T_4$ by putting reasonable values of a, $\overline{r^2}$ and $\overline{r^4}$ in equations (1) and (3). The important point for the present purpose is that the non-cubic part of the crystal field has a very small effect on the positions of the broad optical absorption lines (see figure 1 *c*), and even though some of the later discussion is based on rather small discrepancies in these positions, it seems justifiable to neglect T_2 and T_4 when fitting the optical spectrum. This is also the case for other salts, e.g. in the chromic alums where the field is trigonal, and in most of the complexes that will now be discussed only the cubic part of the field will be considered.

3. Comparison of crystal field theory with experimental data

3·1. *Optical data*

By fitting the optical data in the way described above, values can be found for the cubic field constant K and the $^nF-{}^nP$ term separation E'. These values are listed in table 3 and the experimental evidence for them is given below.

Table 1

	salt	optical absorption (cm^{-1})	K (cm^{-1})
$3d^1$, ^2D	$CsTi(SO_4)_2 12H_2O$	20300 (1)	42600
$3d^4$, ^5D	$CsMn(SO_4)_2 12H_2O$	21000 (1)	44200
$3d^6$, ^5D	$[Fe(H_2O)_6]^{2+}$	10300 (3)	21600
$3d^4$, ^2D	$[Cu(H_2O)_6]^{2+}$	12300 (2)	25800 ($\sim$24000)

References: (1) Hartmann & Schlafer (1951); (2) Dreisch & Trommer (1937); (3) Dreisch & Kallscheuer (1939).

When the ground term is ^nD there is only a single absorption line in the visible with wave number Δ (see figure 1 *a*) corresponding to the separation between the $d\epsilon$ and $d\gamma$ levels of equation (2), so that $K = 2{\cdot}1\Delta$ if non-cubic terms are neglected. Examples are given in table 1. These optical data and those in table 2 are for $T \sim 300°$ K, and refer to hydrated solutions when only the complex is written and otherwise to crystals of the salt. There is considerable optical and magnetic evidence that the cubic field strength for a given complex is independent of the surroundings. The second value of $K \sim 24000\,cm^{-1}$ for Cu^{2+} in table 1 is calculated after making allowance for a tetragonal field of fairly typical strength for the Tutton salts, $T_2 \sim T_4 \sim \pm 1000\,cm^{-1}$. For $3d^9$ such a field causes rather a large depression of the lowest level (see equation (2)), so that $K < 2{\cdot}1\Delta$. Since $[Cu(H_2O)_6]^{2+}$ complexes are usually found to have approximately tetragonal symmetry, the smaller value of K is probably the better.

For ions where an ^nF term lies lowest there are two unknowns K and E' in (4) (T_2 and T_4 being neglected), and both can be determined if at least two of the three possible optical transitions have been measured. Examples are given in table 2 which includes the optical data, the values of K and E' which give the best fit to this data, and also the calculated positions of the levels to show how the optical lines have been identified. The levels for V^{3+} and Ni^{2+} from table 2 are shown in figures 1 *b* and 1 *c*, where the optical absorptions are represented by vertical lines to indicate that they are very broad.

The optical data for some of these complexes have been interpreted by previous authors in a rather similar way, but only the work of Abragam & Pryce (1951 *b*) on Co^{2+} will be quoted here, because only they have allowed for a difference between the term separation E' in the crystal and the value E for the free ion. Such differences are found for all the ions in table 2 and are listed in table 3. It should be pointed out that for V^{2+} there is some doubt about the interpretation of the spectrum, because the measurements of Dreisch & Kallscheuer (1939) down to 8300 cm^{-1} show no sign of the beginning of the expected line (see table 2) with centre at 7600 cm^{-1}. The absorption lines for the other ions in table 2 are fitted fairly closely, which helps to confirm that the interpretation of the spectra is correct. It should also be mentioned

TABLE 2. THE POSITIONS OF THE ENERGY LEVELS AND THE CRYSTAL FIELD PARAMETERS K AND E' (SEE EQUATION (4)) FOR COMPLEXES WHERE THE PARAMAGNETIC ION HAS AN ^nF TERM LOWEST. THE 'EXPERIMENTAL' VALUES ARE THE WAVE-NUMBERS OF THE OPTICAL ABSORPTION LINES.

$3d^2$, ^3F $[V(H_2O)_6]^{3+}$	levels	c	b	p	a	(cm^{-1})
$K = 39000$ cm^{-1}	expt. (1)	—	17240	25000	—	
$E' = 9300$ cm^{-1}	calc.	0	17300	25200	36000	
$3d^7$, ^4F $[Co(H_2O)_6]^{2+}$	levels	c	b	a	p	
$K = 21000$ cm^{-1}	expt. (2, 3)	—	8400	~20000		
$E' = 12500$ cm^{-1}	calc.	0	8800	18800	20100	
$3d^3$, ^4F $[Cr(H_2O)_6]^{3+}$	levels	a	b	c	p	
$K = 36700$ cm^{-1}	expt. (1)	—	17480	24500	—	
$E' = 10200$ cm^{-1}	calc.	0	17500	24650	38000	
$3d^3$, ^4F $[V(H_2O)_6]^{2+}$						
$K = 16000$ cm^{-1}	expt. (4)	—	—	12200	18200	
$E' = 9000$ cm^{-1}	calc.	0	7600	12100	19600	
$3d^8$, ^3F $[Ni(H_2O)_6]^{2+}$						
$K = 17700$ cm^{-1}	expt. (2, 5)	—	8400	14100	25000	
$E' = 14000$ cm^{-1}	calc.	0	8400	14000	25200	

References: (1) Hartmann & Schlafer (1951); (2) Dreisch & Trommer (1937); (3) Abragam & Pryce (1951*b*); (4) Dreisch & Kallscheuer (1939); (5) Houston (1911).

that many of the broad absorption lines for the ions in tables 1 and 2 show a superimposed many-line structure which is attributed to vibrational levels (see Schultz 1942). A theoretical investigation of these levels by Professor M. H. L. Pryce (unpublished) has shown that it is justifiable to assume that the centre of gravity of the complete absorption band gives the position of the orbital level. Such centres of gravity are the values given in tables 1 and 2.

3·2. *Magnetic data*

Having determined the orbital level splitting from the optical data, it is now possible for some complexes to compare the magnetic and optical data using crystal field theory.

The simplest cases are d^3 and d^8 where the orbital singlet 'a' lies lowest (e.g. figure 1*c*). To a good approximation the spectroscopic splitting factor of this level is isotropic with value (cf. Schlapp & Penney 1932)

$$g = 2{\cdot}0023 - \frac{8\lambda''}{\Delta}, \qquad (5)$$

where 2·0023 is the free spin contribution and $-8\lambda''/\Delta$ is the orbital contribution, Δ being the separation between levels *a* and *b*. On crystal field theory λ'' should equal the free-ion spin-orbit coupling constant λ, but by putting experimentally known values of *g* and Δ in equation (5) the calculated values of λ'' are in fact found to be different. These values of λ'' are given in table 3 and are calculated using the Δ's in table 2 and *g*-values which have been measured accurately by paramagnetic resonance methods and are as follows:

(i) For potassium chromic alum, $3d^3$, $g = 1{\cdot}976 \pm 0{\cdot}002$ (Bleaney & Bowers 1951); $\Delta = 17\,500\ \mathrm{cm}^{-1}$, hence $\lambda'' = 57{\cdot}5\ \mathrm{cm}^{-1}$.

(ii) For vanadous ammonium sulphate, $3d^3$, $g = 1{\cdot}956 \pm 0{\cdot}002$ (Bleaney, Ingram & Scovil 1951); these authors only give the value of g_z, and a small anisotropy correction has been applied to find the mean *g* value above; $\Delta \sim 7600\ \mathrm{cm}^{-1}$, hence $\lambda'' \sim 44\ \mathrm{cm}^{-1}$.

(iii) For hydrated Ni^{2+} salts, $3d^8$, it is difficult to measure *g* very accurately because of the relatively large Stark splitting of the spin levels. The average of values found in several different salts, including five nickel Tutton salts (Griffiths & Owen 1952), is $g = 2{\cdot}25_5$; $\Delta = 8400\ \mathrm{cm}^{-1}$, hence $\lambda'' = -270\ \mathrm{cm}^{-1}$.

(iv) For $[Ni(NH_3)_6]Br_2$, $g \doteqdot 2{\cdot}14$, and Δ is found to be $10\,800\ \mathrm{cm}^{-1}$ (giving $K = 22\,600\ \mathrm{cm}^{-1}$) from the optical measurements of Dreisch & Trommer (1937); hence $\lambda'' \sim -200\ \mathrm{cm}^{-1}$.

Equation (5) also applies for the *g*-value measured along the axis of distortion of an elongated $[Cu(H_2O)_6]^{2+}$, $3d^9$ complex. Using $g_z = 2{\cdot}45$, this being an average for several hydrated copper salts (at 90° K) with approximate tetragonal symmetry, and $\Delta = 12\,300\ \mathrm{cm}^{-1}$ (at 290° K), one finds $\lambda'' = -695\ \mathrm{cm}^{-1}$. For this particular complex it would be desirable to compare magnetic and optical data for the same salt at the same temperature, because, as discussed above, Δ is fairly sensitive to the non-cubic part of the field which is known to vary from one salt to another and is also temperature-dependent. An approximate value of λ'' for Cu^{2+} was first estimated by Abragam & Pryce (1951*c*), and a more complete analysis than the above has recently been made by Bleaney, Bowers & Pryce (1954), who give details of the most recent magnetic data.

All of these values of λ'' considered so far have depended on optical data, and are reliable only if these data have been correctly interpreted. For two cases, however, λ'' can be estimated from magnetic data alone. First, for ammonium vanadic alum, $3d^2$, Handel & Siegert (1937) deduced from their susceptibility measurements that $\lambda'' = 64\ \mathrm{cm}^{-1}$. These authors were the first to suggest that the spin-orbit coupling constant is reduced for an ion in a crystal. Secondly, for $[Ni(H_2O)_6]^{2+}$, $3d^8$, a comparison of magnetic resonance data and susceptibility data on five Tutton salts gives $\lambda'' = -250\ \mathrm{cm}^{-1}$ (Griffiths & Owen 1952). This is reasonably close to the value $\lambda'' = -270\ \mathrm{cm}^{-1}$ found from the optical spectrum, as it should be.

3·3. *Discussion*

It has now been shown how the optical spectra can be closely reconstructed by choosing suitable values for the cubic field strength *K* and the $^nF - {}^nP$ term separation E', and how the magnetic data then lead to values of the apparent spin-orbit coupling constant λ''. The values found for these three parameters are listed in table 3. The free ion values of E and λ in the table are from the spectroscopic data in *Atomic energy levels* 1949, 1952, E being the separation between the centre of gravity of the ground-term (^nF) multiplet and the centre of gravity of the ^nP multiplet. λ is estimated by equating the overall separation of the ^nF multiplet to $\lambda S(2L+1)$.

TABLE 3. DATA FOR IONS OF THE 3*d* GROUP

K = cubic crystal field constant, which is a measure of the splitting of the orbital energy levels (see equations (2) and (4)). r_{ion} = ionic radius of 3*d* ion when surrounded by octahedron of oxygen ions. $E = {}^nF - {}^nP$ term separation, λ = spin orbit coupling constant, for free ion (from spectroscopic data). E' and λ'' are calculated for the ion in the octahedral complex (see text).

complex	K (cm^{-1})	r_{ion} (Å)	E (cm^{-1})	E' (cm^{-1})	λ (cm^{-1})	λ'' (cm^{-1})	λ''/λ	E'/E
$3d^1$, ^2D $[Ti(H_2O)_6]^{3+}$	42600	0·69	—	—	154	—	—	—
$3d^2$, ^3F $[V(H_2O)_6]^{3+}$	39000	0·65	12920	9300	104	64 (*m*)	0·62	0·72
$3d^3$, ^4F $[Cr(H_2O)_6]^{3+}$	36700	0·64	13770	10200	91	57	0·63	0·74
$3d^4$, ^5D $[Mn(H_2O)_6]^{3+}$	44200	0·70	—	—	88	—	—	—
$3d^3$, ^4F $[V(H_2O)_6]^{2+}$	~16000	0·72	11320	~9000	55·5	~44	~0·8	~0·8
$3d^6$, ^5D $[Fe(H_2O)_6]^{2+}$	21600	0·84	—	—	−103	—	—	—
$3d^7$, ^4F $[Co(H_2O)_6]^{2+}$	21000	0·80	14500	12500	−178	—	—	0·86
$3d^8$, ^3F $[Ni(H_2O)_6]^{2+}$	17700	0·76	15840	14000	−324	−270 (−250, *m*)	0·83	0·88
$3d^9$, ^2D $[Cu(H_2O)_6]^{2+}$	~24000	~0·85	—	—	−829	−695	0·84	—

The approximate ionic radii of the ions in the complexes (Santen & Wieringen 1952) are included in the table because they give further qualitative evidence for the correctness of the values of K. Thus, for complexes with the same degree of ionization, differences in K follow differences in r_{ion}, which is just what might be expected theoretically since for these complexes $\overline{r^4}$ would be the most important variable in equations (1) to (4). However, although the results in table 3 show a number of features of this kind which are consistent with the ionic model and crystal field theory, and which give confidence that the optical spectra have been interpreted correctly, the discrepancies discussed below are direct evidence that this theory requires some modifications.

These discrepancies are simply that the term separations E, and the spin-orbit coupling constants λ, both appear to be smaller for ions in crystals than for free ions, as can be seen by comparing E with E' and λ with λ'' for the complexes in table 3. For V^{3+} and Ni^{2+} such apparent reductions in λ can be found from purely magnetic evidence (labelled '*m*' in the table) and are independent of optical data. The evidence also suggests that the reductions are larger for M^{3+} ions than for M^{2+}, and similarly, as K is increased from 17700 cm^{-1} in $[Ni(H_2O)_6]^{2+}$ to 22600 cm^{-1} in $[Ni(NH_3)_6]^{2+}$, λ'' drops from −270 cm^{-1} to about −200 cm^{-1} (see § 3·2).

These effects cannot be easily explained in terms of the crystal field theory as used so far, and suggest that it may not be correct to treat the complex on a purely ionic model. However, by allowing for a certain amount of covalent bonding between the paramagnetic ion and the surrounding water molecules, many of these discrepancies are explained quite naturally as is shown in the molecular orbital treatment outlined below.

4. The application of molecular orbitals

The method of molecular orbitals has been used by Van Vleck (1935) to investigate the magnetic properties of $[Fe(CN)_6]^{3-}$, $3d^5$, and a general discussion of the theoretical principles involved in the method is given in his paper. In the present application to $[M(H_2O)_6]$ complexes, where $M = 3d^n$, molecular orbital wave-functions identical with those of Van Vleck will be used, but the discussion will be confined to the case of fairly weak bonding, whereas the particular properties of the ferricyanides are the result of strong bonding.

The orbits on the central atom M which can participate in bond formation are $3d$, $4p$ and $4s$, of which the $3d$ orbits are of most interest for the present purpose because they determine the magnetic properties. These d orbits are of the form

$$\begin{array}{lll} d\gamma_1 & d_{x^2-y^2} & \frac{1}{\sqrt{2}}[|2\rangle + |-2\rangle] \\ d\gamma_2 & d_{3z^2-r^2} & |0\rangle \\ d\epsilon_1 & d_{xy} & \frac{1}{\sqrt{2}}[|2\rangle - |-2\rangle] \\ d\epsilon_2 & d_{yz} & \frac{1}{\sqrt{2}}[|1\rangle - |-1\rangle] \\ d\epsilon_3 & d_{zx} & \frac{1}{\sqrt{2}}[|1\rangle + |-1\rangle] \end{array} \tag{6}$$

where the first column gives the designation of the orbital levels used in equations (2) and figure 1*a*, the second column shows the angular dependence of the wave-functions, and the third column gives the proper wave-functions in terms of $|l_z\rangle$, which is a convenient form when considering magnetic properties (for an explanation of the notation see Bleaney & Stevens 1953).

Of these orbits, $d\gamma$ can combine with $\psi(\sigma)$ orbits on the attached atoms, σ denoting zero angular momentum about the bond axis. For the present case $\psi(\sigma)$ would include $2p_\sigma$ and $2s$ orbits on oxygen atoms, and the signs are chosen such that the $\psi(\sigma)$ wave-function is positive in the region of overlap with $d\gamma$ wave-function. Neglecting terms containing overlap, the allowed linear combinations of these atomic orbitals can be written (Van Vleck)

$$\left.\begin{array}{l} \phi_{3z^2-r^2} \equiv \alpha d_{3z^2-r^2} - (1-\alpha^2)^{\frac{1}{2}} (\tfrac{1}{12})^{\frac{1}{2}} [2\psi_3 + 2\psi_6 - \psi_1 - \psi_4 - \psi_2 - \psi_5]_\sigma, \\ \phi_{x^2-y^2} \equiv \alpha d_{x^2-y^2} - (1-\alpha^2)^{\frac{1}{2}} \tfrac{1}{2}[\psi_1 + \psi_4 - \psi_2 - \psi_5]_\sigma, \end{array}\right\} \text{(antibonding)} \tag{7}$$

$$\left.\begin{array}{l} \phi'_{3z^2-r^2} \equiv (1-\alpha^2)^{\frac{1}{2}} d_{3z^2-r^2} + \alpha (\tfrac{1}{12})^{\frac{1}{2}} [2\psi_3 + 2\psi_6 - \psi_1 - \psi_4 - \psi_2 - \psi_5]_\sigma, \\ \phi'_{x^2-y^2} \equiv (1-\alpha^2)^{\frac{1}{2}} d_{x^2-y^2} + \alpha \tfrac{1}{2}[\psi_1 + \psi_4 - \psi_2 - \psi_5]_\sigma, \end{array}\right\} \text{(bonding)} \tag{8}$$

where the suffixes 1, 2, 3, 4, 5 and 6 denote the oxygen atoms on the positive x, y, z and negative x, y, z axes of the octahedron respectively, and $(1-\alpha^2)$ is a measure of the degree of mixing of the orbits, or of the strength of the σ bonds which are formed. $\alpha^2 = 1$ for a purely ionic complex, $\alpha^2 = 0{\cdot}5$ for what might be called a purely covalent complex (the ferricyanides are in this region), and, as will be shown below, $\alpha^2 \sim 0{\cdot}8$ for an hydrated iron group complex such as $[Ni(H_2O)_6]^{2+}$. The oxygen $\psi(\sigma)$ orbits are presumed to be lower in energy than the central $d\gamma$ orbits, and when there is some admixing, i.e. when $\alpha < 1$, the $\psi(\sigma)$ and $d\gamma$ energy levels can be thought to 'repel' each other, so that the energy of $d\gamma$ is raised and that of $\psi(\sigma)$ is lowered (see figure 2). Equations (7) are thus taken to represent antibonding combinations of the orbits, and equations (8) to be bonding combinations (cf. Van Vleck).

Four more bonding molecular orbits similar to (8) and four more antibonding orbits similar to (7), can be built from linear combinations of the $\psi(\sigma)$ orbits with central $4s$, $4p_x$, $4p_y$ and $4p_z$ orbits. These are of no particular interest for the present purpose because they have no appreciable effect on the magnetic or optical properties. They are of importance in helping to lower the energy of the complex, because, as will be seen below (see figure 2), these four bonding orbits are filled with electrons whereas the four antibonding orbits are empty.

In an analogous way the central atom $d\epsilon$ orbits can combine with attached atom $\psi(\pi)$ orbits, π denoting unit angular momentum about the bond axis. These would be $2p_\pi$ orbits on oxygen atoms in the present case. There are three bonding and three antibonding combinations, one example of the latter being

$$\beta d_{xy} - (1-\beta^2)^{\frac{1}{2}} \tfrac{1}{2}[\psi_1 + \psi_2 - \psi_4 - \psi_5]_\pi, \tag{9}$$

where, as before, terms containing overlap are neglected. Although such π-bonding admixtures* with $\beta^2 \sim 0{\cdot}75$ are known to be of importance in octahedral complexes where the σ-bonding is very strong (Owen & Stevens 1953; Stevens 1953), it will be assumed that for the present case they are negligible as a first approximation, i.e. $\beta = 1$, so that the central atom d_{xy}, d_{yz}, d_{zy} orbits remain pure d orbits. This assumption is based on the usual criterion that the admixture depends on the overlap between central and attached orbits, and since these $d\epsilon$ wave-functions have lobes of high density pointing between the attached atoms, the overlap is assumed to be much smaller than that for the $d\gamma$ orbits which have lobes pointing towards the attached atoms.

It is now of interest to see how the electrons in $[M(H_2O)_6]$, $M = 3d^n$, fill the various orbits discussed above, which are shown diagrammatically in figure 2. If the complex is first treated as ionic, the central M atom has closed shells plus $3d^n$ and the attached oxygens (considered crudely as O^{2-} ions) have closed shells up to $2p^6$. The latter include two antiparallel electrons in $2p_\sigma$ (or $\psi(\sigma)$) orbits, which are represented in the molecular orbital scheme by having paired electrons in each of the six molecular orbits (d^2sp^3) of type (8), but with $\alpha = 1$ so that these are linear combinations of pure oxygen atomic orbits. When there is bonding, $\alpha < 1$, the $\psi(\sigma)$ orbits have admixtures of central atom orbits (which corresponds physically to the partial transfer of these paired electrons to the central atom) and their energy is lowered because (8) is taken to be a bonding combination. At the same time the $d\gamma$ orbits on the central atom are given admixtures of $\psi(\sigma)$ orbits (see (7)), and their

* The absolute sign of the admixture coefficient $(1-\beta^2)^{\frac{1}{2}}$ in (9) (cf. Stevens 1953) cannot be found from the measured magnetic properties which depend only on the square of this quantity. Similarly, in equations (7) and (8), only the value of α^2 can be found from the experimental results (see below).

energy is raised because (7) is taken to be an antibonding combination. These antibonding $d\gamma$ orbits, together with the approximately non-bonding $d\epsilon$ orbits, accommodate the n d-electrons which are responsible for the magnetic properties of the complex.

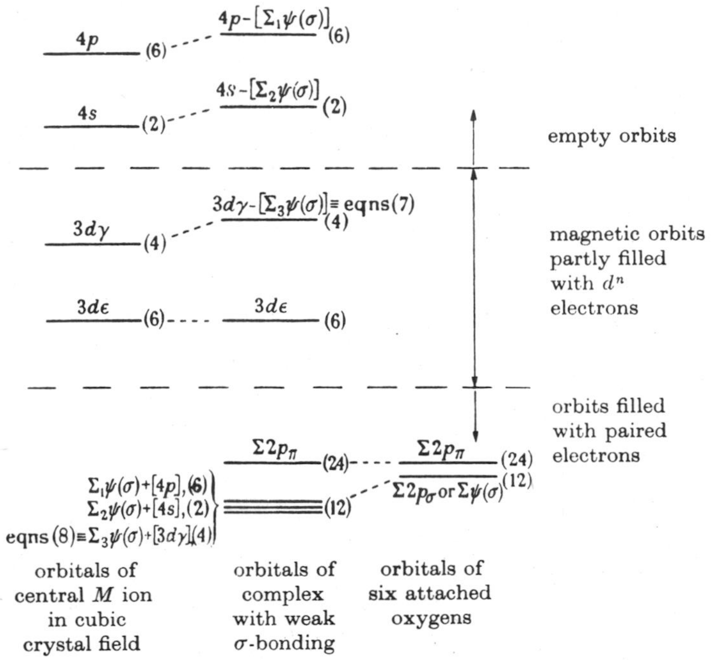

FIGURE 2. Schematic drawing showing the probable energy order of some of the orbits in an $[M(H_2O)_6]$ complex with weak σ-bonding. The diagram is not drawn to scale. The number of electrons that can be accommodated in each level is shown in brackets on the right, and the available electrons fill these orbits in the appropriate energy order (see text, especially §5).

Thus, from the magnetic point of view, the presence of bonding has two main effects:

(*a*) It increases the splitting between the $d\gamma$ doublet and the $d\epsilon$ triplet, and in this respect is similar in effect to an increase in the crystal field strength.

(*b*) It modifies the $d\gamma$ orbits so that electrons in these orbits are partially transferred to the attached atoms of the complex. The analogous modification to the $d\epsilon$ orbits is assumed to be much smaller and is neglected.

The first effect, if large, may result in the $d\gamma$ orbits being always unoccupied because of their high energy, which implies that the ground state for $n > 3$ may not be the state of maximum spin expected from Hund's rule. For example, for $n = 5$, the configuration $(d\epsilon)^5$ with $S = \frac{1}{2}$ may be lower in energy than $(d\epsilon)^3(d\gamma)^2$ with $S = \frac{5}{2}$ which is normally the ground state, because the high energy of the two $d\gamma$ electrons more than compensates for the lowering of energy which is achieved by keeping the electron spins parallel. This happens in the ferricyanides treated by Van Vleck and in many complexes of the 4*d* and 5*d* groups (see Griffiths, Owen &

Ward 1953), but in only one common hydrated iron group complex, viz. $[Co(H_2O)_6]^{3+}$, $(3d\epsilon)^6$, $S = 0$. For the hydrated complexes in table 3, it will be noticed that the energy separation between the $d\epsilon$ and $d\gamma$ single-electron states (which equals $K/2{\cdot}1$ for all configurations), is much bigger for M^{3+} ions than for M^{2+}. Although some increase in splitting from M^{2+} to M^{3+} would be expected on a purely ionic model, it can be shown by putting reasonable values in equations (1) and (3) that the increase given by this theory is smaller than that found experimentally. This suggests that the σ-bonding is stronger for M^{3+} ions—a result which is confirmed by the magnetic data (see § 5*b* below). This stronger bonding presumably arises because there is a greater tendency for electrons on the oxygen atoms to move into the central M ion in bonding molecular orbits of type (8), in order to even out the charge distribution.

The second effect (*b*) might reasonably be expected to reduce the coulomb interactions $\Sigma(e^2/r_{ij})$ between the d-electrons, because the cloud charge is more spread out over the complex. Since the term separations in the free ion are determined by the coulomb interactions, this suggests a qualitative explanation of the reductions of E to E' given in table 3. The modified orbits also lead to a reduced orbital magnetic moment of the complex (cf. Stevens 1953), and thus account for the apparent reductions in the spin-orbit coupling constant shown in table 3. This is discussed in detail below.

5. The effect of bonding on g-values

The simplest complexes to treat are those where an orbital singlet lies lowest, and where there is a measurable orbital contribution to the magnetic moment, i.e. $3d^3$, $3d^8$, and usually $3d^9$. The theory is very similar for all these cases, and is explained below with particular reference to the example of nickel.

(*a*) Ni^{2+}, $3d^8$, $^3F_4$, $L = 3$, $S = 1$

It will be helpful first to outline briefly the usual derivation of the g-value formula (5) for an ionic Ni^{2+} complex. For a fuller explanation of the method and for references, see, for example, Bleaney & Stevens (1953). The orbital levels belonging to $L = 3$ are split by a cubic crystal field as in equations (4) (with $T_2 = T_4 = 0$), and figure 1*c*. For the present purpose it is only necessary to know the orbital wave-functions of the lowest singlet $|a\rangle$ and the excited level $|b_1\rangle$, which, in terms of $|L_z\rangle$, are

$$|a\rangle \equiv \frac{1}{\sqrt{2}}(|2\rangle - |-2\rangle) \quad \text{and} \quad |b_1\rangle \equiv \frac{1}{\sqrt{2}}(|2\rangle + |-2\rangle).$$

When there is an external magnetic field H_z parallel to the z axis, the energies of the three spin levels belonging to $|a\rangle$ are found by applying to second order the perturbation

$$\mathscr{H}_z = \lambda L_z S_z + \beta H_z(L_z + 2S_z), \tag{10}$$

where β is the Bohr magneton and λ is the $L.S$ coupling constant. For the $S_z = +1$ spin level, first order gives $\langle a|\mathscr{H}_z|a\rangle = 2\beta H_z$, so that to this order the spectroscopic splitting factor has the free spin value $g = 2$, or more accurately 2·0023. Second order adds $-|\langle a|\mathscr{H}_z|b_1\rangle|^2/\Delta_{ab_1}$, which gives one term linear in βH_z, namely $(-8\lambda/\Delta)\beta H_z$. There are no matrix elements of $\mathscr{H}_z$ which couple $|a\rangle$ with any other orbital level except $|b_1\rangle$. Since the symmetry is cubic, x, y and z are equivalent, so that $g_z = g$ is isotropic, and $g = 2-(8\lambda/\Delta)$ as in equation (5).

To find how this formula is affected by the introduction of bonding, the problem must be treated in terms of single d-electron states. The various configurations can be estimated by putting the eight electrons of d^8 into the single-electron levels, i.e. the $d\epsilon$ orbital triplet and the higher $d\gamma$ doublet, in the appropriate energy order. Thus the lowest state $|a\rangle \equiv (d\epsilon)^6(d\gamma)^2$, and the excited state $|b_1\rangle$ has one electron promoted giving $|b_1\rangle \equiv (d\epsilon)^5(d\gamma)^3$. Alternatively, in terms of the two electron holes in the d-shell, $|a\rangle \equiv (d\gamma)^2$ and $|b_1\rangle \equiv (d\gamma)^1(d\epsilon)^1$. The exact form of these configurations can be found (see Santen & Wieringen 1952) by using the relations between states of total L and S, and the single-electron hole product states. For example

$$|L_z = 3, S_z = 1\rangle \equiv \{\overset{+}{l_z = 2}, \overset{+}{l_z = 1}\} \equiv |\overset{+(1)}{2}\rangle|\overset{+(2)}{1}\rangle - |\overset{+(2)}{2}\rangle|\overset{+(1)}{1}\rangle,$$

$$|L_z = 2, S_z = 1\rangle \equiv \{\overset{+}{2}, \overset{+}{0}\},$$

$$\cdots\cdots\cdots\cdots\cdots\cdots\cdots\cdots\cdots$$

$$|L_z = -2, S_z = 1\rangle \equiv \{\overset{+}{0}, \overset{+}{-2}\},$$

where in the second column + means $s_z = +\frac{1}{2}$ for each single-electron hole and the curly brackets denote that the product orbital state is antisymmetrical; in the third column this product state is written in full for one example, and the superscripts number the particular electron hole. Hence

$$|a, S_z = 1\rangle \equiv \frac{1}{\sqrt{2}}[|2\rangle - |-2\rangle]_{L_z}^{S_z=1} \equiv \frac{1}{\sqrt{2}}[\{\overset{+}{2}, \overset{+}{0}\} - \{\overset{+}{0}, \overset{+}{-2}\}]$$

$$\equiv \left\{\frac{1}{\sqrt{2}}(|\overset{+}{2}\rangle + |\overset{+}{-2}\rangle), |\overset{+}{0}\rangle\right\} \equiv \{\overset{+}{d}_{x^2-y^2}, \overset{+}{d}_{3z^2-r^2}\}$$

and

$$|b_1, S_z = 1\rangle \equiv \frac{1}{\sqrt{2}}[|2\rangle + |-2\rangle]_{L_z}^{S_z=1} \equiv \left\{\frac{1}{\sqrt{2}}(|\overset{+}{2}\rangle - |\overset{+}{-2}\rangle), |\overset{+}{0}\rangle\right\} \equiv \{\overset{+}{d}_{xy}, \overset{+}{d}_{3z^2-r^2}\}.$$

When weak bonding occurs the $d\gamma$ orbits are modified as in equations (7) and the $d\epsilon$ orbits (equation (9)) are assumed to be not much changed, so that these product states become approximately

$$|a, S_z = 1\rangle \equiv \{\overset{+}{\phi}_{x^2-y^2}, \overset{+}{\phi}_{3z^2-r^2}\}, \quad |b_1, S_z = 1\rangle \equiv \{\overset{+}{d}_{xy}, \overset{+}{\phi}_{3z^2-r^2}\}.$$

The g-value is again found from the perturbation (10) rewritten as sums over the single-electron holes

$$\mathscr{H}_z = \zeta \sum_i l_z^{(i)} s_z^{(i)} + \beta H_z(\sum_i l_z^{(i)} + 2\sum_i s_z^{(i)}), \tag{11}$$

where ζ is the $l.s$ coupling constant for one hole, and $\zeta = 2S\lambda'$ where λ' is the $L.S$ coupling constant for the nickel ion in the complex. Proceeding as before, first order gives terms

$$\langle a \mid 2\beta H_z \Sigma s_z \mid a\rangle = 2\beta H_z[\langle \overset{+}{\phi}_{x^2-y^2} \mid s_z \mid \overset{+}{\phi}_{x^2-y^2}\rangle + \langle \overset{+}{\phi}_{3z^2-r^2} \mid s_z \mid \overset{+}{\phi}_{3z^2-r^2}\rangle]$$
$$= 2\beta H_z(\tfrac{1}{2}+\tfrac{1}{2}),$$
$$\langle a \mid \Sigma l_z(\zeta s_z + \beta H_z) \mid a\rangle = 0,$$

where the method of Stevens (1953) is followed when operating with l_z on the attached $\psi(\sigma)$ orbitals belonging to ϕ. Hence in first order $g_z = 2$, or rather 2·0023, as before. The excited level $\mid b_1\rangle$ corresponds to the promotion of one electron hole to d_{xy} from $\phi_{x^2-y^2} \equiv \alpha d_{x^2-y^2} - \frac{(1-\alpha^2)^{\frac{1}{2}}}{2}\Sigma\psi(\sigma)$ (see (7)), so that the only second-order terms are

$$-\mid\langle \overset{+}{\phi}_{x^2-y^2} \mid \zeta l_z s_z + \beta H_z l_z + 2\beta H_z s_z \mid \overset{+}{d}_{xy}\rangle\mid^2/\Delta,$$
$$= \frac{\alpha^2}{\Delta}(\zeta + 2\beta H_z)^2,$$

where terms containing the overlap between $d_{x^2-y^2}$ and the attached $\psi(\sigma)$ orbits have been neglected. The contribution to g_z is thus $-\alpha^2\frac{4\zeta}{\Delta} = -\alpha^2\frac{8\lambda'}{\Delta}$. Since with cubic symmetry there is no preferred z axis, $g_z = g$, so that the approximate g-value for the bound Ni^{2+} ion is finally

$$g = 2{\cdot}0023 - \alpha^2\frac{8\lambda'}{\Delta}. \qquad (5')$$

In this expression λ' is the $L.S$ coupling constant for the bound ion, and since the term in λ' depends only on parts of the total wave-function which correspond to pure d orbits, i.e. on d_{xy} and $\alpha d_{x^2-y^2}$, it is likely that λ' is not greatly different from the free ion value λ (see §5c below). If this is so, the main difference between (5′) and the g-value for a purely ionic complex is that the orbital contribution is reduced by α^2, where α^2 is simply the probability of finding the magnetic electrons in orbits on the Ni^{2+} ion as can be seen from equations (7).

The 'apparent' spin-orbit coupling constant, $\lambda'' = 0{\cdot}83\lambda$ (see table 3), can now be explained by letting $\alpha^2 \doteqdot 0{\cdot}83$. Thus one can say that in $[Ni(H_2O)_6]^{2+}$ each of the two unpaired electrons spends about 83 % of the time in central $d\gamma$ orbits and about 3 % in a $\psi(\sigma)$ orbit on each attached oxygen atom. In $[Ni(NH_3)_6]^{2+}$ (see §3), the orbital level splittings are greater, and α^2 smaller than in $[Ni(H_2O)_6]^{2+}$, which is consistent with the bonding being stronger in the hexammine complex as might have been expected on chemical grounds.

The hyperfine structure arising from interaction between the magnetic electrons and the nuclear moment of the ^{61}Ni isotope would also be expected to be reduced by a factor α^2, because on a physical picture the magnetic electrons move in orbits near the central nucleus for only α^2 of the time. This hyperfine structure has not yet been measured in a paramagnetic resonance experiment.

(*b*) Cr^{3+}, V^{2+}, $3d^3$, $^4F_{\frac{3}{2}}$

The theory for this case is very similar to that for d^8 discussed above. The product states of the three electrons corresponding to the ground level $|\,a\rangle$ and the excited level $|\,b_1\rangle$ are

$$|\,a, S_z = \tfrac{3}{2}\rangle \equiv \frac{1}{\sqrt{2}}[|\,2\rangle - |\,-2\rangle]_{L_z}^{S_z=\frac{3}{2}} \rightarrow \{\overset{+}{d}_{xy}, \overset{+}{d}_{zx}, \overset{+}{d}_{yz}\},$$

$$|\,b_1, S_z = \tfrac{3}{2}\rangle \equiv \frac{1}{\sqrt{2}}[|\,2\rangle + |\,-2\rangle]_{L_z}^{S_z=\frac{3}{2}} \rightarrow \{\overset{+}{\phi}_{x^2-y^2}, \overset{+}{d}_{zx}, \overset{+}{d}_{yz}\}.$$

The orbital contribution to g_z then arises as before from the $l.s$ coupling between d_{xy} and $\phi_{x^2-y^2}$, so that the g-value is just as for d^8, i.e. equation (5′). Thus, even though the magnetic electrons are now in non-bonding orbits (since $|\,a\rangle = (d\epsilon)^3$), there is still a reduction $\lambda'/\lambda\alpha^2$ in the orbital part of the g-value.

The experimental results in table 3 show that for Cr^{3+} $\alpha^2 \sim 0{\cdot}6$, which is considerably smaller than for V^{2+} ($\alpha^2 \sim 0{\cdot}8$) and Ni^+ ($\alpha^2 \sim 0{\cdot}8$). This stronger bonding in Cr^{3+} might be expected because of the larger charge as discussed in §4. When the σ-bonding is as strong as this, some of the approximations made above, e.g. the neglect of π-bonding and of terms containing overlap, are probably no longer very good. If there were weak π-bonds (represented by $\beta < 1$ in equations of type (9)), then to a first approximation the orbital part of the g-value in (5′) would be reduced by the additional factor β^2, i.e. $g \doteqdot 2{\cdot}0023 - \alpha^2\beta^2 8\lambda'/\Delta$.

The hyperfine structure from the central nucleus would not be expected to be appreciably affected by the σ-bonds because the three magnetic electrons are in non-bonding, or strictly speaking π-bonding, $d\epsilon$ orbits. In this connexion it is of interest that the hyperfine structures in $[Cr(CN)_6]^{3-}$, $3d^3$, and $[V(CN)_6]^{4-}$, $3d^3$, are about 30 % smaller than in the analogous hydrated complexes (see Bowers 1952; Baker & Bleaney 1952). At first sight this suggests that in the cyanides, where the σ-bonding is stronger than in the hexahydrates, there may also be an appreciable amount of π-bonding which can be represented by putting $\beta^2 \sim 0{\cdot}7$ in equation (9). However, one cannot be sure that this decrease in hyperfine structure can be entirely attributed to π-bonding, because the main contribution to this structure is thought to arise from very small admixtures of unpaired s electrons in the central ion configuration (see Abragam & Pryce 1951*a*), and it is not clear how σ-bonding, or changes in σ-bond strength, would affect such admixtures. For the same reason the decrease in hyperfine structure in $3d^8$, discussed above, may not be exactly α^2.

(*c*) Cu^{2+}, $3d^9$, $^2D_{\frac{5}{2}}$

There is one electron hole in the d-shell, and, when the water octahedron is elongated, the lowest level (see equations (1) and (2)) is $d\gamma_1 \rightarrow \phi_{x^2-y^2}$. $\mathscr{H}_z$ couples this level to only one of the higher levels, namely, $d\epsilon_1 \equiv d_{xy}$, so that the value of g_z is given once again by (5′). Hence $\alpha^2 \doteqdot \lambda''/\lambda = 84$ % (see table 3), is the approximate time spent by the unpaired spin in the central $d_{x^2-y^2}$ orbit, and about 4 % is spent in a $\psi(\sigma)$ orbit on each of the four oxygen atoms in the xy plane. The hyperfine structure from the Cu nucleus is also reduced by a factor α^2 as for d^8.

A similar treatment of the bonding in $[Cu(H_2O)_6]^{2+}$ has been made by Bleaney *et al.* (1955). These authors have shown from a careful analysis of magnetic resonance data that the measured reduction α^2 in the hyperfine structure is nearly the same as the reduction $\lambda'/\lambda\alpha^2 = 0{\cdot}84$ in the orbital magnetic moment. This is the main reason for the assumption made in § 5 (*a*) that the spin-orbit coupling constant in the bound ion λ' is not greatly different from that in the free ion, λ.

A final point of interest in d^9 complexes is that the orbital contribution to the g value is larger for a $3d^9$, Cu^{2+} salt than for the isomorphous $4d^9$ Ag^{2+} salt, even though λ is much bigger for Ag^{2+} (Bowers 1953). This is consistent with the bonding being stronger in the heavier element because the d wave-functions extend farther from the central nucleus, and is analogous to the difference found between the bond strength in $4d^5$ and $5d^5$ complexes (see Griffiths *et al.* 1953).

(*d*) *Other configurations*

A detailed treatment for configurations other than the simple cases of d^3, d^8 and d^9 discussed above will not be given in this paper, althougth the theory would proceed along similar lines. The effect of σ-bonding on the orbital magnetic moment, and on the magnetic hyperfine structure from the central nucleus, would depend mostly on the amount of $d\gamma$ orbit in the ground state of the ion in the cubic crystal field. These ground state configurations are (Santen & Wieringen 1952):

for	d^1	d^2	d^3	d^4	d^5
ground state electrons	$(d\epsilon)^1$	$(d\epsilon)^{9/5}\,(d\gamma)^{1/5}$	$(d\epsilon)^3$	$(d\epsilon)^3\,(d\gamma)^1$	$(d\epsilon)^3\,(d\gamma)^2$
and for		d^6	d^7	d^8	d^9
ground state electron holes		$(d\epsilon)^2\,(d\gamma)^2$	$(d\epsilon)^{6/5}\,(d\gamma)^{9/5}$	$(d\gamma)^2$	$(d\gamma)^1$

For d^2 and d^7 the admixture of $|\,p\rangle$ to the ground state $|\,c\rangle$ (see equations (2), § 2) is neglected.

The only other case to have been analyzed is $[Ti(H_2O)_6]^{3+}$, $3d^1$, by Bleaney & O'Brien (to be published). Here there is a single $d\epsilon$ electron and the σ-bonding has no appreciable effect on the magnetic moment, but these authors find that there is evidence for weak π-bonding. As discussed in § 5 (*b*) above, this suggests that it may not be a good approximation to neglect π-bonding for any of the M^{3+} ions.

6. Conclusion

In the first part of this paper it is shown that there are systematic discrepancies in crystal field theory as applied to hydrated salts of the iron transition group, and, in particular, that the orbital magnetic moment in many cases is smaller than would be expected on a purely ionic model. In the second part of the paper it is shown how these discrepancies can be explained by introducing weak σ-bonds into the $[M(H_2O)_6]$ complex, so that there is some charge transfer between the paramagnetic ion and the surrounding oxygen atoms.

This interpretation of the evidence is a direct result of the recent investigations made in this laboratory of covalent complexes of the $3d$, $4d$ and $5d$ groups, where similar reductions in orbital magnetic moment have been found. In the particularly interesting case of $[IrCl_6]^{2-}$, $5d^5$, the microwave spectrum showed a complex hyperfine structure from both Ir and Cl nuclei, which was taken to be evidence that the magnetic electrons moved in molecular orbits over the whole complex (see Owen & Stevens 1953; Griffiths & Owen 1954). The detailed theoretical investigation by Stevens (1953) showed that such orbits accounted not only for the hyperfine structure, but also for the observed reductions in orbital magnetic moment. In complexes of this type, in which there are strong σ-bonds plus weaker π-bonds, all the magnetic electrons occupy π-antibonding molecular orbits, and the σ-antibonding orbits are unoccupied because they are of much higher energy. The transfer of the magnetic electrons thus arises only because of the π-bonds. In the present case of $[M(H_2O)_6]$, the σ-bonds are assumed to be fairly weak and the π-bonds to be practically negligible. The transfer of magnetic electrons then depends on how much σ-antibonding orbital is present in the wave function describing these electrons, and varies from one configuration to another. Direct evidence for such transfer in the form of a hyperfine structure from the attached oxygen nuclei is not found of course, because ^{16}O has no nuclear magnetic moment.

Finally, there are a number of interesting effects arising from the bonding, apart from those mentioned above. For example, there are the reduced term separations for ions in crystals discussed in §3. Also, there is considerable evidence that the bonding reduces the magnetic hyperfine structure from the central nucleus; the only example of this that has been dealt with quantitatively is $[Cu(H_2O)_6]^{2+}$ by Bleaney *et al.* (1955). The bonding picture is also likely to lead to a better understanding of the interactions between neighbouring paramagnetic ions in crystals. As a very simple example of this, one might expect the exchange interaction between V^{2+}, $3d^3$, ions in a vanadium Tutton salt, to be smaller than that between Ni^{2+}, $3d^8$ ions in the analogous nickel salt, because in the first case the magnetic electrons are localized on the V^{2+} ion in practically non-bonding orbits, while in the second case they are more spread out over the $[Ni(H_2O)_6]^{2+}$ complex, thus allowing greater overlap with neighbouring complexes. The available experimental evidence for making comparisons of this kind is at present rather limited.

The author would particularly like to thank Professor M. H. L. Pryce, F.R.S., for very valuable discussions on the interpretation of the results and for his continued interest in this work. He is also grateful to many of his colleagues, especially Miss M. C. M. O'Brien, Dr L. E. Orgel and Dr K. W. H. Stevens, for helpful discussions.

References

Abragam, A. & Pryce, M. H. L. 1951*a* *Proc. Roy. Soc.* A, **205**, 135.
Abragam, A. & Pryce, M. H. L. 1951*b* *Proc. Roy. Soc.* A, **206**, 173.
Abragam, A. & Pryce, M. H. L. 1951*c* *Proc. Roy. Soc.* A, **206**, 164.
Atomic energy levels 1949, vol. **1**; 1952, vol **2**. U.S. National Bureau of Standards Circular.
Baker, J. M. & Bleaney, B. 1952 *Proc. Phys. Soc.* A, **65**, 952.
Bleaney, B. & Bowers, K. D. 1951 *Proc. Phys. Soc.* A, **64**, 1135.
Bleaney, B., Bowers, K. D. & Pryce, M. H. L. 1955 (Submitted for publication in *Proc. Roy. Soc.*)
Bleaney, B., Ingram, D. J. E. & Scovil, H. E. D. 1951 *Proc. Phys. Soc.* A, **64**, 601.
Bleaney, B. & Stevens, K. W. H. 1953 *Rep. Progr. Phys.* **16**, 108.

Bowers, K. D. 1952 *Proc. Phys. Soc.* A, **65**, 860.
Bowers, K. D. 1953 *Proc. Phys. Soc.* A, **66**, 666.
Dreisch, Th. & Trommer, W. 1937 *Z. Phys. Chem.* B, **37**, 40.
Dreisch, Th. & Kallscheuer, O. 1939 *Z. Phys. Chem.* B, **45**, 19.
Finkelstein, R. & Van Vleck, J. H. 1940 *J. Chem. Phys.* **8**, 790.
Griffiths, J. H. E. & Owen, J. 1952 *Proc. Roy. Soc.* A, **213**, 459.
Griffiths, J. H. E. & Owen, J. 1954 *Proc. Roy. Soc.* A, **226**, 96.
Griffiths, J. H. E., Owen, J. & Ward, I. M. 1953 *Proc. Roy. Soc.* A, **219**, 526.
Handel, J. van den & Siegert, A. 1937 *Physica*, **4**, 871.
Hartmann, H. von & Schlafer, H. L. 1951 *Z. Naturf.* **6***a*, 760.
Houston, R. A. 1911 *Proc. Roy. Soc. Edinb.* **31**, 538.
Kleiner, W. H. 1952 *J. Chem. Phys.* **20**, 1784.
Owen, J. & Stevens, K. W. H. 1953 *Nature, Lond.*, **171**, 836.
Polder, D. 1942 *Physica*, **9**, 709.
Santen, J. H. & Wieringen, J. S. 1952 *Rec. Trav. chim. Pays-Bas*, **71**, 420.
Schlapp, R. & Penney, W. G. 1932 *Phys. Rev.* **42**, 666.
Schultz, M. L. 1942 *J. Chem. Phys.* **10**, 194.
Spedding, F. H. & Nutting, G. C. 1935 *J. Chem. Phys.* **3**, 369.
Stevens, K. W. H. 1953 *Proc. Roy. Soc.* A, **219**, 542.
Van Vleck, J. H. 1932 *Electric and magnetic susceptibilities.* Oxford University Press.
Van Vleck, J. H. 1935 *J. Chem. Phys.* **3**, 807.
Van Vleck, J. H. 1940 *J. Chem. Phys.* **8**, 787.

Reprinted from *J. Inorg. Nucl. Chem.*, **2**, 1–10 (1956)

21

ON THE STABILITIES OF TRANSITION METAL COMPLEXES—I

THEORY OF THE ENERGIES

J. STANLEY GRIFFITH
Department of Colloid Science, University of Cambridge*

(*Received* 14 *July* 1955)

Abstract—A molecular orbital theory of the energies of formation of transition metal complexes is presented. The energies of pairing of *d*-electrons in covalent regular octahedral d^n-complexes are calculated in terms of atomic spectral parameters. It is then shown, by means of a hole formalism, that the pairing energies for regular tetrahedral d^{10-n}-complexes are formally the same. A reason is suggested as to why regular complexes with certain spins, which are apparently possible in principle, are unknown experimentally. Certain aspects of the electrostatic crystal-field theory are briefly compared with the present treatment.

1. INTRODUCTION

AMONGST the complexes of transition series elements two kinds have been distinguished on the basis of their magnetic properties. In the first the total spin of the complex is the same as that of the constituent metal ion in its ground state in the gas phase. These have been called by PAULING[1] "ionic" or "essentially ionic" complexes. In the second the *d*-electrons are crowded as far as possible into those *d*-orbitals which cannot be used to form σ-bonds with the ligands. For elements in the middle of the transition series this means that the total spin is one or two units less than the spin of the corresponding essentially ionic complex. PAULING called these "covalent" or "essentially covalent" complexes. TAUBE[2] has suggested calling the two kinds outer and inner orbital complexes respectively. As in these papers we find no compelling reason for preferring the new notation, we use that of PAULING.

There are two main methods of theoretical treatment of these complexes—BETHE'S of the electrostatic crystal field and MULLIKEN'S of molecular orbitals. These, and also PAULING'S method of the directed bond, were compared by VAN VLECK[3] and shown to lead to broadly the same conclusions. The complexes concerned had a high degree of symmetry, and the conclusions depended more upon the nature of the symmetry group than upon the detailed assumptions of the theories. The electrostatic field model has been used by VAN VLECK and his co-workers[4, 5, 6] to explain the magnetic susceptibilities, and also later by other workers to interpret paramagnetic resonance absorption measurements (for review, see BLEANEY & STEVENS[7]).

*Now at the Department of Theoretical Chemistry, University of Cambridge.

(1) PAULING, L. (1940) *The Nature of the Chemical Bond*, p. 115. Oxford University Press.
(2) TAUBE, H. (1952) *Chem. Rev.*, **50**, 85.
(3) VAN VLECK, J. H. (1935) *J. Chem. Phys.*, **3**, 807.
(4) SCHLAPP, R., and PENNEY, W. G. (1932) *Phys. Rev.*, **42**, 666.
(5) JORDAHL, O. M. (1934) *Phys. Rev.*, **45**, 87.
(6) HOWARD, J. B. (1935) *J. Chem. Phys.*, **3**, 813.
(7) BLEANEY, B., and STEVENS, K. W. H. (1953) *Rep. Progr. Phys.*, **16**, 108.

The optical spectrum of a complex was first interpreted in detail by FINKELSTEIN and VAN VLECK.[8] For other spectra, see for example, OWEN.[9]

The general reasons for the quenching of the spin in covalent complexes are clear enough,[3] but detailed calculations do not appear to have been published. In the first paper of this series, a theoretical discussion of the quenching is presented. It is impracticable to solve the complete wave equation for the complex, and so any treatment must be approximate. We shall be interested mainly in the energies of the complexes and regard them as built up from two parts—the central metal ion and its ligands. Then we shall neglect π-bonding, and discuss the energy of the d-shell of the ion in terms of the normal theory of the spectra of many-electron atoms without configuration interaction. In fact we assume Russell-Saunders coupling and leave the spin-orbit coupling out of the Hamiltonian. This omission seems a reasonable first approximation for the first transition series; however, for the second and third series the spin-orbit coupling is larger, whilst the intra-atomic electronic repulsions are smaller, and so it may not even be a useful first approximation there. The interaction of the metal ion with its ligands is treated by means of an averaged field molecular orbital theory, and, at present, we consider only regular octahedral and regular tetrahedral complexes.

TABLE 1.—CHARACTER TABLE OF THE SYMMETRY GROUPS O AND T_d

	E	$8C_3$	$3C_2$	$6(C_2;\ \sigma_d)$	$6(C_4;\ S_4)$
A_1	1	1	1	1	1
A_2	1	1	1	-1	-1
E	2	-1	2	0	0
T_1	3	0	-1	-1	1
T_2	3	0	-1	1	-1

These assumptions are of course very restrictive; nevertheless it is felt that none of the essential physical features of the quenching of the spin have been omitted. There is, however, a consequent gain of generality which makes it possible to discuss all the complexes at once and not as a set of independent cases. This is partly due to the fact that the rotation group of a regular octahedron is simply isomorphic with the full symmetry group of a regular tetrahedron. Further, the five d functions on the metal form, for both groups, a representation which is the direct sum of two irreducible representations which have the dimensions 2 and 3 (see Tables 1 and 2).

2. THE BONDING IN THE COMPLEXES

We suppose the energy of a complex to consist of two distinct parts—that of bonding of the ligands and that, if any, needed to pair the d-electrons in quenching the spin. We treat the bonding first, by means of an averaged field molecular orbital theory with inclusion of nearest-neighbour overlap. For this we use the simplest possible model, taking only one valence orbital on each ligand, and considering only the σ-overlap and resonance integrals of these orbitals with the metal valence orbitals. These ligand

(8) FINKELSTEIN, R., and VAN VLECK, J. H. (1940) *J. Chem. Phys.*, **8**, 790.
(9) OWEN, J. (1955) *Proc. Roy. Soc.*, **A227**, 183.

orbitals may be regarded as lone-pair hybrids, typical examples being the lone pairs of H_2O or NH_3. We then consider the behaviour of these orbitals under the operations of the octahedral group O (without inversion) and the tetrahedral group T_d (with reflections). These groups are both simply isomorphic with the symmetric group on four symbols, and have the character table shown in Table 1. The behaviour of the orbitals on the central metal ion is shown in Table 2. There are four lone-pair orbitals in the tetrahedral and six in the octahedral case. They form reducible representations whose irreducible components are also shown in Table 2.

TABLE 2.—THE REPRESENTATIONS SPANNED BY THE ATOMIC ORBITALS. SMALL LETTERS ARE USED FOR REPRESENTATIONS HAVING ONE-ELECTRON FUNCTIONS AS A BASIS

	Octahedral	Tetrahedral
Metal s	a_1	a_1
Metal p	t_1	t_2
Metal d	$e + t_2$	$e + t_2$
Ligand lone pairs	$a_1 + e + t_1$	$a_1 + t_2$

We consider specifically the first transition-series ions, where the valence shell consists of $3d$, $4s$, and $4p$, with $4d$ and higher atomic orbitals of less importance. If we include the $4d$, many of the molecular orbitals (symmetry orbitals) of the system have to be determined as a combination of an orbital on the ligand molecules with several orbitals on the central atom. Thus, in the tetrahedral case, a lone-pair symmetry orbital belonging to t_2 can combine with $3d$, $4p$, and $4d$ on the metal. For this reason we consider first the general problem of the solution of the secular equations for a particular irreducible representation, for the interaction of one ligand orbital with several metal orbitals.

We take ϕ_0 on the ligands belonging to a particular irreducible representation X, say. Then each set of metal orbitals which forms a basis for X can be arranged to have just one orbital with a non-zero overlap with ϕ_0. We neglect the possibility of accidental degeneracy, and arrange these metal orbitals in order of increasing energy, writing them $\phi_1, \phi_2, \cdots, \phi_n$.

The secular equations are

$$f(E) \equiv |H_{ij} - ES_{ij}| = 0,$$

where $$H_{ii} < H_{jj} \text{ for } i < j \text{ and } i, j = 1, 2, \cdots, n. \quad (1)$$

In discussing (1), we take the metal orbitals orthonormal and also with their mutual resonance integrals zero. This is automatically true for those belonging to different representations of the three-dimensional rotation group, and is a natural requirement in general. Then $|H_{ij} - ES_{ij}|$ has zero elements except for its first row and column and its main diagonal, and can easily be expanded, giving

$$f(E) = \prod_{i=0}^{n} (H_{ii} - E) - \sum_{i=1}^{n} (H_{oi} - ES_{oi})^2 \prod_{j \neq o,i} (H_{jj} - E). \quad (2)$$

From (2) we have

$$f(H_{oo}) = -\sum_{i=1}^{n} (H_{oi} - H_{oo}S_{oi})^2 \prod_{j\neq o,i} (H_{jj} - H_{oo}),$$

$$f(H_{ii}) = -(H_{oi} - H_{ii}S_{oi})^2 \prod_{j\neq o,i} (H_{jj} - H_{ii})$$

$$(i = 1, \cdots, n). \tag{3}$$

We assume that none of the $(H_{oi} - H_{ii}S_{oi})$ are zero and then find, from (1) and (3), that $f(H_{ii})$ has the same sign as $(-1)^i$. For $|E|$ large:

$$f(E) = (-E)^{n+1}(1 - \sum_{i=1}^{n} S_{oi}^2) + O(|E|^n). \tag{4}$$

In (4) $\sum S_{oi}^2$ is never as large as 1, so $f(-\infty) = +\infty$. Also, for E large and positive, $f(E)$ has the same sign as $(-1)^{n+1}$. This means that the roots of $f(E) = 0$ are separated by $-\infty$, H_{ii} and $+\infty$. These findings are summed up in equation (5):

$$E_o < H_{11} < E_1 < \cdots H_{ii} < E_i < \cdots H_{nn} < E_n. \tag{5}$$

It is natural to suppose that $H_{oo} < H_{ii}$ for all $i \neq 0$, otherwise there would be too great a transfer of charge to the metal. We assume this, and can then deduce from (3) that

$$E_o < H_{oo} < H_{11} < E_1 . \tag{6}$$

Writing $\psi_i(X)$ for a molecular orbital of symmetry X corresponding to E_i in (5), the orbitals of lowest energy in the complex are the $\psi_o(X)$. There are the same number of these as there are ligands, and so the electrons initially in the lone-pair orbitals will go into these. This leaves in the valence shell only the 3*d*-electrons from the metal to be housed. We shall say that $\psi_i(X)$ corresponds to an atomic orbital $\phi_i(X)$ if their energies are related in the sense of E_i and H_{ii} in (5). If ϕ_i does not interact with ligand orbitals, then $\psi_i(X) \equiv \phi_i(X)$. Since, in the free ion, the 3*d*-orbitals lie lowest, we suppose that the 3*d*-electrons are housed in $\psi_1(e)$ and $\psi_1(t_2)$. We write Δ for the difference of energy between these representations, and λ for the number of electrons in the upper one. If Δ is small, HUND's rule will apply to the whole *d*-shell and the complex will be ionic. If Δ is large, the electrons will crowd into the lower level with a consequent gain of orbital energy. However, there will be additional coulomb repulsions and a loss of exchange energy. Then the energy of the complex may be written

$$E = p + P + \lambda\Delta, \tag{7}$$

where p represents the part of the energy which is not very sensitive to the value of λ—that is: the solvation energy of the whole complex; mutual interactions between the ligands; and the orbital energy of the ψ_o. P represents any energy required to pair *d*-electrons in the lower representation ψ_1. This lower representation consists of pure 3*d*-orbitals, and in the next section we evaluate P from the theory of atomic spectra.

Before doing this we discuss briefly in terms of (7) the meanings of the ground states of the various kinds of complex. For the ionic complex, P is zero whilst λ is as small as possible consistent with the complex having the same spin as the corresponding free ion. These values of λ are shown in Table 8 for octahedral complexes. In section 1 we also defined covalent complexes, and again λ has the minimum value consistent with the total spin. PAULING[1] pointed out that there is in principle a third class of complex possible. For octahedral complexes the spin of d^5- and d^6-ions could be quenched by only one unit, leaving spins of 3/2 and 1 respectively. A similar thing is possible for d^4 and d^5 tetrahedral complexes. We shall call these half-quenched complexes. The spins and electron arrangements for octahedral symmetry are shown in Table 3. In comparing stabilities we suppose p to be the same for the three types of the same complex.

TABLE 3.—THE THREE TYPES OF REGULAR OCTAHEDRAL COMPLEXES

(*a*) *Total spin*

	Ionic	Half-quenched	Covalent
d^4	2	—	1
d^5	$2\frac{1}{2}$	$1\frac{1}{2}$	$\frac{1}{2}$
d^6	2	1	0
d^7	$1\frac{1}{2}$	—	$\frac{1}{2}$

(*b*) *Arrangements of electrons for a d^6 complex.*

	t_2			e		Example
Ionic	(↑↓)	(↑)	(↑)	(↑)	(↑)	$Fe^{++}(H_2O)_6$
Half-quenched	(↑↓)	(↑↓)	(↑)	(↑)	○	unknown
Covalent	(↑↓)	(↑↓)	(↑↓)	○	○	$Fe(CN)_6^{4-}$

3. CALCULATION OF THE PAIRING ENERGY

We regard the calculation of the pairing energy P as a purely atomic problem closely analogous to the determination of the energy of a valence state relative to an atomic ground state. Thus, although the electrons are paired in directions away from those of the bonds, we do not consider any other effects of the ligands, for these have already been taken into account in the terms p and $\lambda\Delta$ in (7).

We take axes $OXYZ$ with O at the metal nucleus. The octahedral bonds are taken along these axes, and the tetrahedral bonds along the vectors (1, 1, 1), (1, −1, −1), (−1, 1, −1), (−1, −1, 1). Then, introducing atomic orbitals $|nlm\rangle$ quantized with respect to the Z-axis, we are interested only in the $3d$-orbitals and write

$$|m\rangle = |32m\rangle, \tag{8}$$

where the orbitals $|m\rangle$ are connected by the Dirac phase relation and with $|0\rangle$ real. Then $|1\rangle$, $|-1\rangle$ and $(1/\sqrt{2})(|2\rangle + |-2\rangle) \in t_2$ and $|0\rangle$ and $(1/\sqrt{2})(|2\rangle - |-2\rangle) \in e$, for both types of complex. It is convenient to use real orbitals, and so we write

$$\begin{aligned} |\xi\rangle &= -(1/\sqrt{2})(|1\rangle + |-1\rangle), \\ |\eta\rangle &= -(1/\sqrt{2})(|1\rangle - |-1\rangle), \\ |\zeta\rangle &= (1/\sqrt{2})(|2\rangle - |-2\rangle), \\ |\epsilon\rangle &= (1/\sqrt{2})(|2\rangle + |-2\rangle), \end{aligned} \qquad (9)$$

and have $(|\xi\rangle, |\eta\rangle, |\zeta\rangle) \in t_2$ and $(|O\rangle, |\epsilon\rangle) \in e$. In equation (9), $|\xi\rangle, |\eta\rangle, |\zeta\rangle$, and $|\epsilon\rangle$ are identical save for spatial orientation. We shall express P in terms of RACAH's parameters A, B, and C, these being related to the Slater-Condon F_i by the equations[10]

$$\begin{aligned} A &= F_0 - 49F_4, \\ B &= F_2 - 5F_4, \\ C &= 35F_4. \end{aligned} \qquad (10)$$

Coulomb and exchange integrals are normally defined in terms of the $|m\rangle$, but for our complexes it is much more convenient to work in terms of $|\xi\rangle, |\eta\rangle, |\zeta\rangle, |\epsilon\rangle$, and $|O\rangle$. The exchange integrals J and K, where

$$\begin{aligned} J(ab) &= \langle ab|V|ab\rangle, \\ K(ab) &= \langle ab|V|ba\rangle, \end{aligned} \qquad (11)$$

may be evaluated in terms of A, B, and C, the results being shown in Table 4. With the use of these we now calculate the energies of the ground states. These are well known

TABLE 4.—COULOMB AND EXCHANGE INTEGRALS WITHIN A d-SHELL

Orbital pair	J	K
$\xi\xi, \eta\eta, \zeta\zeta, \epsilon\epsilon, OO$	$A + 4B + 3C$	$A + 4B + 3C$
$\xi\eta, \eta\zeta, \zeta\xi, \xi\epsilon, \eta\epsilon$	$A - 2B + C$	$3B + C$
$\xi O, \eta O$	$A + 2B + C$	$B + C$
$\zeta O, \epsilon O$	$A - 4B + C$	$4B + C$
$\epsilon\zeta$	$A + 4B + C$	C

for the ions, the first column of Table 6 being taken from CATALAN, ROHRLICH, and SHENSTONE[11]. The other cases are calculated assuming a strong field, which means that for covalent complexes the electrons in a d^n-ion are in a configuration $t_2{}^n(n \leqslant 6)$ and $t_2{}^6e^{n-6}(n > 6)$ in the octahedral, and $e^n(n \leqslant 4)$ and $e^4t_2{}^{n-4}(n > 4)$ in the tetrahedral case. We shall see that a strong field is needed to quench the spin, so this seems a reasonable approximation. In Table 5, the symmetries of the lowest states are shown. It is seen that the covalent states all form irreducible representations and their energies can therefore be calculated immediately, but that the half-quenched

(10) RACAH. G., (1942) *Phys. Rev*. **62**, 452.
(11) CATALAN, M. A., ROHRLICH, F., and SHENSTONE, A. G. (1954) *Proc. Roy. Soc*., **A221**, 421.

TABLE 5.—SYMMETRIES OF GROUND STATES OF COMPLEXES

Tetrahedral	Octahedral	Ionic	Half-quenched	Covalent
d^6	d^4	E	—	T_1
d^5	d^5	A_1	$T_1 + T_2$	T_2
d^4	d^6	T_2	$T_1 + T_2$	A_1
d^3	d^7	T_1	—	E

states form reducible representations. The expression for the energy of a covalent ion is thus the familiar one

$$E = \sum_{k<1} J(k, 1) - \sum_{k<1} K(k, 1), \tag{12}$$

where the exchange integral $K(k, 1)$ is zero for pairs $(k, 1)$ with differing spin. The results of the calculations for the covalent states are also shown in Table 6.

TABLE 6.—ENERGIES OF GROUND STATES OF IONS

	Ionic	Octahedral Half-quenched	Octahedral Covalent	Tetrahedral Half-quenched	Tetrahedral Covalent
d^3	$3A - 15B$	—	—	—	$3A - 8B + 4C$
d^4	$6A - 21B$	—	$6A - 15B + 5C$	$6A - 16B + 5C$	$6A - 16B + 8C$
d^5	$10A - 35B$	$10A - 25B + 6C$	$10A - 20B + 10C$	$10A - 25B + 6C$	$10A - 20B + 10C$
d^6	$15A - 35B + 7C$	$15A - 30B + 12C$	$15A - 30B + 15C$	—	$15A - 29B + 12C$
d^7	$21A - 43B + 14C$	—	$21A - 36B + 18C$	—	—

This leaves the half-quenched states, and we illustrate the evaluation of the energies of the ground configurations of these in detail only for the octahedral d^5-ion. The total spin of the ground configuration is 3/2, and we take $S_z = 3/2$, giving as basic states

$$\begin{aligned} \psi_1 &= |\xi\alpha\xi\beta\eta\alpha\zeta\alpha O\alpha\rangle, & \psi_4 &= |\xi\alpha\xi\beta\eta\alpha\zeta\alpha\epsilon\alpha\rangle, \\ \psi_2 &= |\xi\alpha\eta\alpha\eta\beta\zeta\alpha O\alpha\rangle, & \psi_5 &= |\xi\alpha\eta\alpha\eta\beta\zeta\alpha\epsilon\alpha\rangle, \\ \psi_3 &= |\xi\alpha\eta\alpha\zeta\alpha\zeta\beta O\alpha\rangle, & \psi_6 &= |\xi\alpha\eta\alpha\zeta\alpha\zeta\beta\epsilon\alpha\rangle. \end{aligned} \tag{13}$$

Writing $V_{ij} = \langle\bar{\psi}_i|V|\psi_j\rangle$ for the matrix of the electrostatic energy, we find

$$\begin{aligned} V_{11} = V_{22} &= 10A - 19B + 6C, \\ V_{33} &= 10A - 25B + 6C, \\ V_{44} = V_{55} &= 10A - 23B + 6C, \\ V_{66} &= 10A - 17B + 6C. \end{aligned} \tag{14}$$

Knowing that the ψ_i between them span $T_1 + T_2$, it is unnecessary to evaluate any off-diagonal V_{ij}. Both ψ_3 and ψ_6 have a special relationship to the Z-axis. Calling ψ'_3, ψ'_6 and ψ''_3, ψ''_6 the corresponding functions having exactly the same relationship to the X- and Y-axes respectively, it is easy to verify that all the operations of O

transform $(\psi_3, \psi'_3, \psi''_3)$ amongst themselves, and also $(\psi_6, \psi'_6, \psi''_6)$ amongst themselves. In fact

$$(\psi_3, \psi'_3, \psi''_3) \in T_1, \qquad (\psi_6, \psi'_6, \psi''_6) \in T_2, \tag{15}$$

and so the energies are given by

$$E(T_1) = V_{33} = 10A - 25B + 6C, \qquad E(T_2) = V_{66} = 10A - 17B + 6C. \tag{16}$$

In a similar way the energies of the half-quenched d^6 octahedral ion are

$$E(T_1) = 15A - 30B + 12C, \qquad E(T_2) = 15A - 22B + 12C. \tag{17}$$

The energies of the various ground configurations of tetrahedral complexes could be calculated in the same way. This, however, is unnecessary, for we may deduce them from the energies for the octahedral case by a method due, in essence, to SHORTLEY[(12)], who showed that the matrix of electrostatic energy for a shell which was complete except for ϵ "missing" electrons was simply related to the matrix for the same shell containing just ϵ-electrons. We need only diagonal elements of the electrostatic energy, and correlate a state $\psi(O) = |d_1, \ldots, d_\epsilon\rangle$ for the octahedral with $\psi(T_d) = |d_{\epsilon+1}, \ldots, d_{10}\rangle$ for the tetrahedral case. If $\psi(O)$ belongs to the ground configuration for an octahedral ion of a given kind, then it is clear that $\psi(T_d)$ belongs to the ground configuration of a tetrahedral ion of the same kind (i.e., also ionic, half-quenched, or covalent). By considering the behaviour of the missing electrons from a wave function of the tetrahedral kind, it is clear that symmetry is conserved under the correlation. Thus, for example, the states of the ground configuration of a covalent d^4 octahedral complex are correlated with the states of the ground configuration of a covalent d^6 tetrahedral complex, and both have the symmetry T_1. Writing $V(O)$ and $V(T_d)$ for the electrostatic energies, we have:[(12, 13)]

$$V(T_d) = V(d^{10}) - \sum_{i=1}^{10} \sum_{j=1}^{\epsilon} (J(d_i d_j) - K(d_i d_j)) + V(O)$$

$$= 45A - 70B + 35C - \epsilon(9A - 14B + 7C) + V(O). \tag{18}$$

Here ϵ is the number of electrons missing from the d-shell of the tetrahedral ion. In (18) the first two terms depend only upon ϵ, and not upon whether the complex is ionic, half-quenched, or covalent. This means that the pairing energies for a d^ϵ tetrahedral ion are formally the same as those for a $d^{10-\epsilon}$-octahedral ion. These results are illustrated in Tables 5, 6, and 7. Table 5 shows the symmetries and Table 6 the energies of the ground states, whilst Table 7 gives the mean pairing energies defined in the next section.

(12) CONDON, E. U., and SHORTLEY, G. H. (1953) *The Theory of Atomic Spectra*, pp. 284, 295. Cambridge University Press.
(13) CONDON and SHORTLEY, 1953, p. 296.

4. THE NON-EXISTENCE OF REGULAR COMPLEXES WITH HALF-QUENCHED SPINS

It is seen from Table 3 that, as the spin of the ionic state is quenched by μ units, the value of λ also decreases by μ units, so from equation (7) the increase of energy of the complex is

$$\delta E = P - \mu\Delta\,. \tag{19}$$

Hence the reaction involving the quenching of the spin is exothermic if

$$\Delta > \mu^{-1}P\,. \tag{20}$$

It is therefore convenient to give $\mu^{-1}P$ a symbol and a name. We write it Π and call it the mean pairing energy. It is the same as P except for the octahedral covalent d^5 and d^6 and the tetrahedral covalent d^4 and d^5 complexes. The calculated values of Π are shown in Table 7.

TABLE 7.—MEAN PAIRING ENERGIES $\Pi(n)$

Tetrahedral	Octahedral	Half-quenched	Covalent
d^6	d^4	—	$6B + 5C$
d^5	d^5	$10B + 6C$	$7\frac{1}{2}B + 5C$
d^4	d^6	$5B + 5C$	$2\frac{1}{2}B + 4C$
d^3	d^7	—	$7B + 4C$

Now, in order to decide which of the three possible forms of a complex should be found experimentally, we must consider the free-energy changes involved in the reactions

$$(MA_n)_i \rightleftharpoons (MA_n)_h \rightleftharpoons (MA_n)_c. \tag{21}$$

In (21) we let ΔF_i, ΔF_h, and ΔF_c be the standard free energies of formation of the three species, with a corresponding notation for the heat contents H and the entropies S. Then

$$\begin{aligned}\Delta F_h - \tfrac{1}{2}(\Delta F_i + \Delta F_c) &= \Delta H_h - \tfrac{1}{2}(\Delta H_i + \Delta H_c) - T(\Delta S_h - \tfrac{1}{2}(\Delta S_i + \Delta S_c)) \\ &= 2\tfrac{1}{2}B + C - T(\Delta S_h - \tfrac{1}{2}(\Delta S_i + \Delta S_c)),\end{aligned} \tag{22}$$

from Table 7 in all four of the relevant cases. From atomic spectral data[11] $2\frac{1}{2}B + C$ is of the order of 20 kilocalories, whilst it is improbable that $T(\Delta S_h - \frac{1}{2}(\Delta S_i + \Delta S_c))$ is more than 2–3 kilocalories. This means that, whatever the value of Δ, either the ionic or the covalent complex has a considerably lower free energy than the corresponding half-quenched one. Accordingly, it appears that half-quenched spins are impossible for complexes with regular octahedral or tetrahedral symmetry. They are unknown experimentally for these symmetries. This conclusion is unlikely to be altered by a consideration of π-bonding or by small distortions such as would be expected for orbitally degenerate complexes. However, half-quenched spins are certainly possible in principle for complexes with a lower symmetry, such as certain haemoglobin complexes,[14] although it is improbable that they occur in fact.[15]

(14) GRIFFITH, J. S., (1955) Faraday Soc. Discussion on Microwave and Radio-frequency Spectroscopy, to be published.
(15) GRIFFITH, J. S., to be published.

5. RELATIONSHIP TO THE ELECTROSTATIC THEORY.

By taking $P = 0$ in equation (7), we have a formula for the energy of an ionic complex which appears to be rather different from that which would be obtained from an electrostatic field theory. The origin of the difference is that, in an electrostatic theory, the mean orbital energy of the d-shell is unchanged by the field, whilst in a molecular orbital theory it is raised. By shifting it, we can bring the two theories into coincidence. This is shown in equation (20),

TABLE 8.—λ AND THE EXTRA STABILIZATION ($\lambda - 2n/5$) FOR IONIC COMPLEXES

Complex	λ	$\lambda - 2n/5$
d^0	0	0
d^1	0	−2/5
d^2	0	−4/5
d^3	0	−6/5
d^4	1	−3/5
d^5	2	0
d^6	2	−2/5
d^7	2	−4/5
d^8	2	−6/5
d^9	3	−3/5
d^{10}	4	0

$$E = (p + 2n\Delta/5) + (\lambda - 2n/5)\Delta, \tag{20}$$

where $(p + 2n\Delta/5)$ still changes uniformly with n, but $(\lambda - 2n/5)\Delta$ gives just the extra stabilization given by an electrostatic theory. The values of λ and $\lambda - 2n/5$ are shown in Table 8. This comparison has been given for the strong-field case for which the additional stabilization for d^2 and d^7 is $-4\Delta/5$, but can easily be extended to cover the weak-field case in which this stabilization is $-3\Delta/5$.

The reason that the two theories give the same results is, fundamentally, that they both predict that the d-shell is split into the two parts, e and t_2, determined by the symmetry groups. So the discussions in Sections 3 and 4 would be virtually unaltered with an electrostatic theory. However, the physical origin of the splitting appears to be quite different in the two cases, and the theories should be regarded more as supplementary than alternative. We may expect the electrostatic effect to be large in K_3FeF_6, but the molecular orbital splitting to be predominant in $K_3Fe(CN)_6$. Yet there will be both types of splitting in both compounds. There is, however, a difference between the theories, because with the electrostatic theory there is, to a first order in perturbation, no charge transfer to the ligands. This transfer is a natural consequence of any molecular orbital theory, and has been used by STEVENS[16] in interpreting paramagnetic resonance measurements.

The author is indebted to Dr. P. GEORGE for his comments on the manuscript.

[16] STEVENS, K. W. H. (1953) *Proc. Roy. Soc.*. **A219**, 542.

Reprinted from *J. Chem. Phys.*, **26**, 1686–1694 (1957)

22

Optical Spectra of Hydrated Ions of the Transition Metals*†

OWEN G. HOLMES‡ AND DONALD S. MCCLURE§
Department of Chemistry and Chemical Engineering, University of California, Berkeley, California

(Received May 7, 1956)

The absorption spectra of crystalline hydrates of the first transition group ions have been measured with polarized light at room temperature, and with unpolarized light at a series of lower temperatures. Some new data on solution spectra have also been obtained. The values of Dq for these compounds are obtained, and are used to explain irregularities in the heats of hydration of the ions. The temperature dependence of the spectra shows that the transitions are vibration-induced electric-dipole transitions for the most part. The dichroism has been used to resolve some of the bands and to determine the effect of noncubic components of the crystal field.

I. INTRODUCTION

THE use of crystal field theory to explain the spectra of transition metal compounds is now well established.[1] We felt that three aspects of this field needed more study, and they are presented in this paper. These are: (1) a systematic review of aqueous solution and crystalline hydrate spectra to correct errors in the older literature and to determine the Dq values (cubic field splitting parameter) for the entire first transition group; (2) a study of the temperature dependence of the absorption spectra of crystalline hydrates to establish the mechanism of the transitions; (3) a study of the dichroism of the absorption bands to see what effect the noncubic components of the crystal field may have on the spectra.

II. ABSORPTION SPECTRA OF THE HYDRATES

The spectra of solutions and of single crystals were obtained with the Cary Model 11, the Beckman Model *DU* and the Perkin-Elmer Model 12C spectrophotometers. The molar extinction coefficient, ϵ, is used for solutions and crystals. The crystals were from 0.2 to 3 mm in thickness. Their ϵ values are not as accurate as for the solutions because of the structural imperfections present in them. We estimate $\pm 1\%$ error for solutions and $\pm 10\%$ error for the crystals. The preparation of some of the solutions and crystals is given in Sec. V. The common substances were always the highest grade chemicals commercially available.

Figure 1 shows the spectra obtained on a uniform scale so that the intensity differences are apparent. Both crystal and solution spectra are shown, and it is worth noting that they are virtually identical for each ion. This indicates that the environment of an ion in solution is the same as in the crystal, and that the fluctuations in the ionic environment are too slow to affect observably the breadth of the absorption bands. Representative data on the spectra, including f number and band width are given in Table I.

The heavy vertical lines in Fig. 1 are the band positions calculated from crystal field theory using only the parameter Dq, obtained from the band assignments. The excited state assignments are indicated by symmetry symbols (Eyring, Walter, and Kimball[2]) near each band; the symmetry symbol for the ground state is placed near the formula for the compound. Owen[3] has shown that the free-ion $F-P$ separation should be reduced in the hydrated ions. The amount of the reduction needed can be seen on the figure as the disagreement between the calculated and observed $T_1(P)$ state.

There are a few new points of interest in addition to the spectra of the crystalline hydrates. The spectra of Cr^{++} and V^{++} are appreciably different from those previously reported (see Sec. V for details). Extreme care must be taken to keep solutions of these ions free of higher oxidation states or to correct for their presence. Our assignment for V^{++} is different from Owen's,[3] and leads to a Dq value of 1180 cm^{-1}, instead of 656 cm^{-1}. Our assignment leads to good agreement of the spin-orbit coupling parameter with that found from magnetic susceptibility data, while the other does not.

The crystalline hydrates of the S-ground state ions Mn^{++} and Fe^{+++} were very nearly the same as the aqueous solutions. The very sharp 25 000 cm^{-1} band in the Mn^{++} spectrum is no sharper in the crystal at room temperature than in solution. There does not seem to exist in any of the literature on this subject a clear statement as to the band assignments for Fe^{+++}. In acidic solutions of $Fe(ClO_4)_3$, two very broad weak bands at 12 500 and 18 200 cm^{-1} and a double peaked band 24 250 and 24 600 cm^3 appear. These bands seem to be exactly analogous to the Mn^{++} bands at 18 900, 23 000, and the double band at 25 000. This assignment leads to a Dq value of 1400 cm^{-1} and requires the $^6S-^4G$ separation in the hydrated ion to be reduced by about 7500 cm^{-1} (using the value given by Moore[4] for the free ion separation).

* Supported by the Office of Naval Research under Contract N6ori 211-III.
† Presented at the Symposium on Molecular Structure and Molecular Spectra, The Ohio State University, June, 1955.
‡ Now at Regina College, Regina, Saskatchewan, Canada.
§ Now at RCA Laboratories, Princeton, New Jersey.

[1] W. Moffitt and C. J. Ballhausen, Ann. Rev. Phys. Chem. 7, 197 (1956).
[2] Eyring, Walter, and Kimball, *Quantum Chemistry* (John Wiley and Sons, Inc., New York, 1944), p. 388.
[3] J. Owen, Proc. Roy. Soc. (London) A227, 183 (1955).
[4] C. Moore, "Atomic Energy Levels," National Bureau of Standards Circular 467, Vol. II (1952).

One further point of interest is that we were unable to find any of the fine structure in the infrared bands as reported by Dreisch and his co-workers.[5,6] We were naturally quite interested to find this structure if it existed. We have no explanation for Dreisch's results. His resolving power, however, was apparently not very high, and much less than in our case. (See Sec. V for details.)

The Dq values obtained from the spectra are plotted against atomic number in Fig. 2. Regularities in this

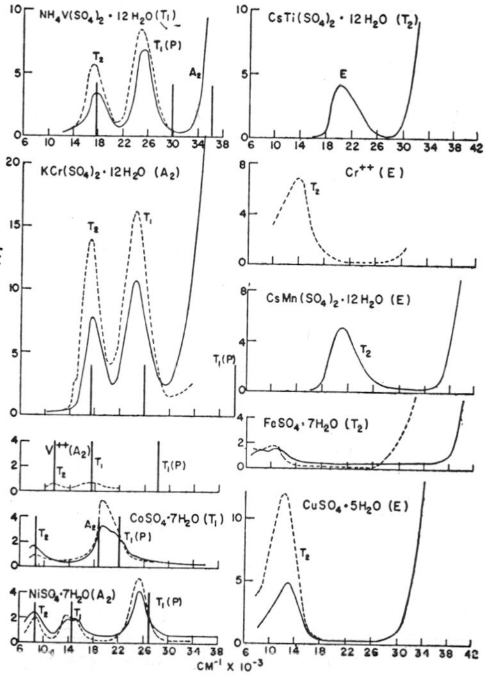

FIG. 1. The absorption spectra of hydrated transition metal ions and band assignments based upon the cubic field splitting of free ion levels. Dotted lines are aqueous solution spectra; solid lines are crystal spectra. The crystal spectra for Ti^{+++}, V^{+++}, and Mn^{+++} are from Hartmann's papers.[2,3] The vertical lines show calculated band positions according to cubic field theory. The ground state symmetry symbol is given next to the formula for each compound and the excited state symmetry is given near each band. The symbol P indicates that the state concerned arises mainly from the free ion P state.

figure are not very evident, although it does appear that the d^4 and d^9 configurations have higher than average values. Jorgensen[7] reports a regular downward trend with increasing atomic number with an anomaly only at Cu^{++} (but Mn^{++} is not considered). This downward trend may be real, but there are as many exceptions according to our data as there are points on the downward sloping line. These are "apparent Dq

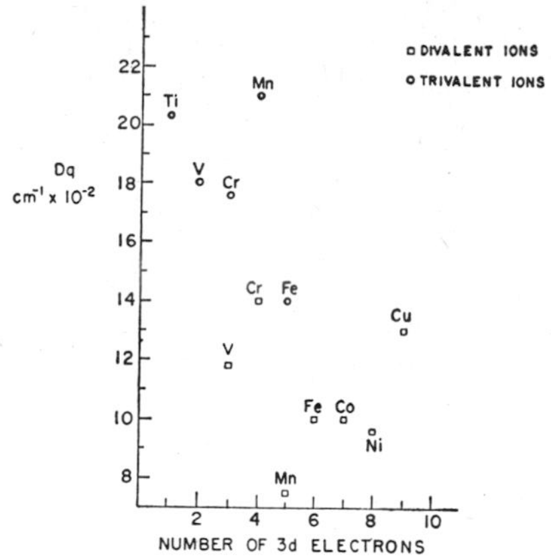

FIG. 2. Values of the parameter Dq. These values are taken directly from the spectra and are not corrected for effects of low symmetry fields.

values"; they have not been corrected for the effects of noncubic distortions (see Sec. IV).

The heats of hydration of these ions should be a smooth function of atomic number for ions of the same charge except for irregularities caused by crystal field splitting. Orgel[8] showed that this was at least approximately true. We have collected the thermodynamic data on heats of hydration and have used these data and our own Dq values to prepare Fig. 3, which shows the effect of subtracting the crystal field stabilization from $\Delta H_{hydr.}$. A remarkably smooth curve is obtained for the divalent ion hydrates, and only a small residual

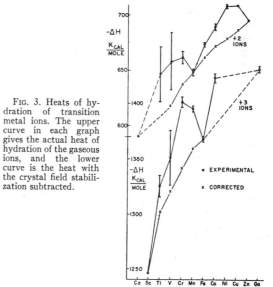

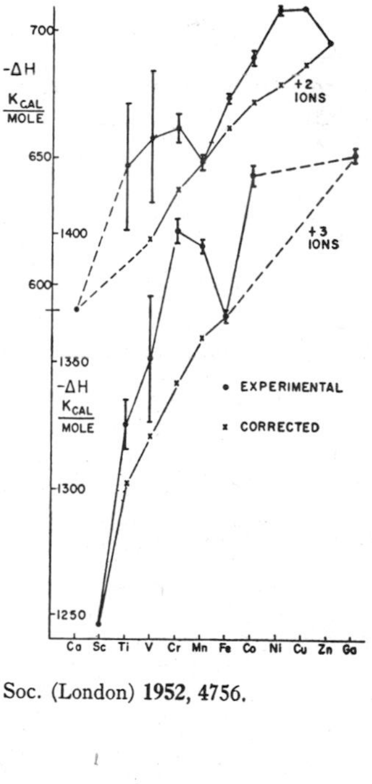

FIG. 3. Heats of hydration of transition metal ions. The upper curve in each graph gives the actual heat of hydration of the gaseous ions, and the lower curve is the heat with the crystal field stabilization subtracted.

[5] T. Dreisch and O. Kallscheuer, Z. physik. Chem. **B45**, 19 (1939).
[6] T. Dreisch and W. Trommer, Z. physik. Chem. **B37**, 37 (1937).
[7] C. K. Jorgensen, Acta Chem. Scand. **8**, 1502 (1954).
[8] L. E. Orgel, J. Chem. Soc. (London) **1952**, 4756.

TABLE I. Absorption spectra of hydrated transition metal ions.

Ion	System	Maximum ϵ	Energy at maximum cm^{-1}	$f\times10^4$	Band width at half-maximum extinction, cm^{-1}
Ti^{+++}					
	$CsTi(SO_4)_2\cdot12H_2O$	4	20 300	0.8	4600
$3d^1\ ^2D$					
V^{+++}					
	$NH_4V(SO_4)_2\cdot12H_2O$	3.5	17 800	0.6	3200
$3d^2\ ^3F$		6.6	25 700	1.1	3300
V^{++}					
	solution	0.45	11 800	0.1	4000
$3d^3\ ^4F$		0.65	17 500	0.2	5000
Cr^{+++}					
	$KCr(SO_4)_2\cdot12H_2O$	7.8	17 600	1.6	3600
$3d^3\ ^4F$		10.5	24 700	2.2	4400
Cr^{++}					
	solution	6.8	14 000	1.8	6000
$3d^4\ ^5D$					
Mn^{+++}					
	$CsMn(SO_4)_2\cdot12H_2O$	5	21 000	1.1	5000
$3d^4\ ^5D$					
Mn^{++}		0.008	18 900	9×10^{-4}	2500
		0.006	23 000	7×10^{-4}	2500
		0.017	25 000	5×10^{-4}	500 (double)
$3d^5\ ^6S$	solution	0.010	28 000	8×10^{-4}	1400
		0.009	29 750	6×10^{-4}	500
		0.006	32 400	7×10^{-4}	4000
Fe^{+++}		0.1	12 600	approx 20×10^{-4}	2200
	solution	0.1	18 200	approx 20×10^{-4}	3800
$3d^5\ ^6S$		0.4	24 500	approx 20×10^{-4}	1500 (double)
Fe^{++}					
	$FeSO_4\cdot7H_2O$	1.6	10 000	0.4	6000 (double)
$3d^6\ ^5D$					
Co^{++}					
	$CoSO_4\cdot7H_2O$	1.7	8350	0.3	3600
$3d^7\ ^4F$		3.4	19 800	0.9	5000
Ni^{++}		2.5	8600	0.45	3000
	$NiSO_4\cdot7H_2O$	1.8	14 700	0.35	4300
$3d^8\ ^3F$		4	25 500	0.6	3200
Cu^{++}					
	$CuSO_4\cdot5H_2O$	5	13 000	1.4	5300
$3d^9\ ^2D$					

correction is needed for the trivalent ion hydrates. The results show indeed that the irregularities in the heats of hydration are caused by the different amounts of crystal field stabilization of each ion. It has recently come to our attention that Hugus[9] has obtained similar results for the halides of this group of metals. It can be expected that other transition metal salts will show similar effects.

III. MECHANISM OF THE TRANSITION

The transitions we have been discussing are highly forbidden. Figure 1 shows that the molar extinction coefficient ranges from about 0.5 to 10 for transitions between states of the same multiplicity, and Table I gives the corresponding f numbers. Since the ion and its immediate environment have a center of symmetry, and the electronic states principally involve d electrons of the central ion, all the states of interest are even. The center of symmetry must be removed by a perturbation before a transition can occur (assuming it is to be an electric dipole transition). This perturbation may arise from a permanent non-centro-symmetric distortion or from vibrations. The T_{1u} and T_{2u} vibrations of a regular octahedral XY_6 molecule have the correct symmetry, but only the T_{1u} type involves motion of the central atom toward the ligands, so it is the type expected to be most effective.

In order to learn something about the transition

[9] Z. Z. Hugus (private communication).

mechanism, the temperature dependence of the absorption of several crystals was studied. Some of the data obtained are shown for $NiSO_4 \cdot 7H_2O$ in Fig. 4, for $KCr(SO_4)_2 \cdot 12H_2O$ in Fig. 5, and for $CoSO_4 \cdot 7H_2O$ in Fig. 6.

In every case the band intensity decreases by a large amount in going from 298°K to 77°K. The peak position shifts toward the violet by 200 to 500 cm^{-1}. The integrated intensity was calculated from the curves and plotted against temperature. Figures 7 and 8 are two such plots. Peak positions were also plotted against temperature. Such a plot for the two bands of chrome-alum is shown in Fig. 9. The temperature dependence of the intensity suggests a vibrational perturbation as the mechanism of the transition. The band shift, however, appeared to be greater than could be explained by freezing out of ground state vibrations. In order to give the data a more quantitative expression and to gain more insight into the effect of temperature, we performed the following analysis.

We assumed that part of the transition strength would be caused by ground state vibrations (having T_{1u} symmetry) and part by upper state vibrations, with only the former giving rise to temperature dependent absorption strength. The vibration spectrum is not known accurately enough, and some simple assumptions about it are therefore used instead of the actual spectrum. We took a set of levels with equidistant spacing, calling this θ and letting it be an adjustable parameter. (θ is the energy of the spacing in temperature units.) The same parameter serves for both states of the transition and represents both the symmetrical and the unsymmetrical vibrations. Further progress is made by noting that the bands which do not appear to be composite, such as the chrome-alum bands or the blue band of $NiSO_4 \cdot 7H_2O$, have nearly a Gaussian shape. The procedure adopted for dealing

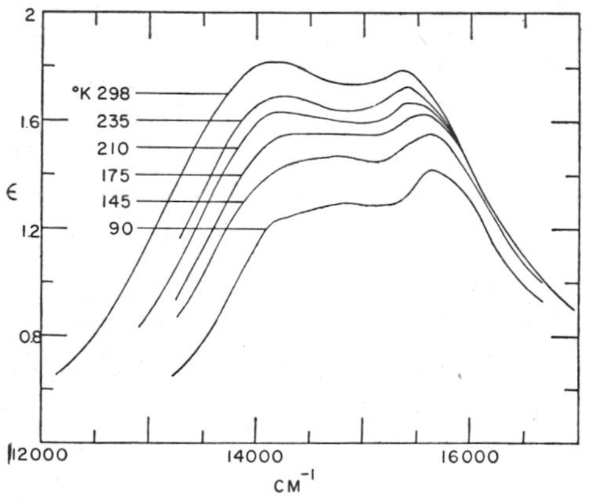

FIG. 4. The "red band" of $NiSO_4 \cdot 7H_2O$ at various temperatures. *ab* plane.

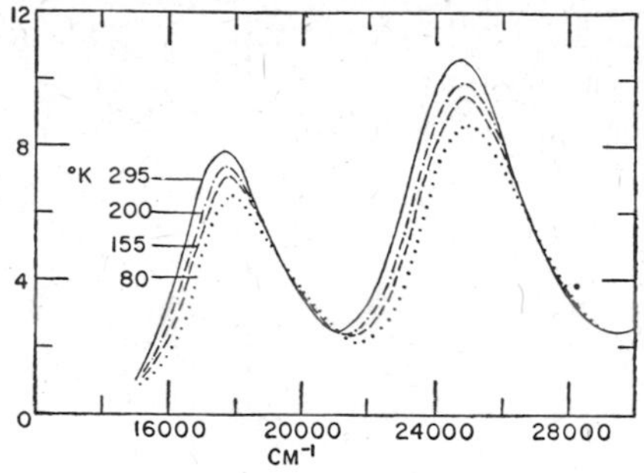

FIG. 5. The absorption spectrum of $KCr(SO_4)_2 \cdot 12H_2O$ at various temperatures.

with such bands used by Sulzer and Wieland[10] was modified for application to forbidden transitions. For the part of the transition induced by upper state vibrations the molar extinction coefficient as a function of frequency is given by

$$\epsilon^1_T(\nu) = \epsilon^m{}_0(\tanh\theta/2T)^{\frac{1}{2}} \exp\{-\tanh(\theta/2T) \cdot [(\nu-\nu_0)/\Delta\nu]^2\}, \quad (1)$$

where $\epsilon^m{}_0$ is the value of the molar extinction coefficient at the band maximum at 0°K, $\Delta\nu$ is the band half-width at $1/e$ of the band height at 0°K, and ν_0 is the frequency of the band maximum at 0°K. For the temperature-dependent part of the absorption strength, a formula similar to the above applies except for a multiplier of $\exp(-\theta/T)$ to supply the temperature dependence caused by the nontotally symmetric vibration of the ground state. The maximum of the temperature dependent transitions will be on the average $\omega''+\omega'$ to lower frequencies than the temperature independent part, where ω is the frequency of the non-

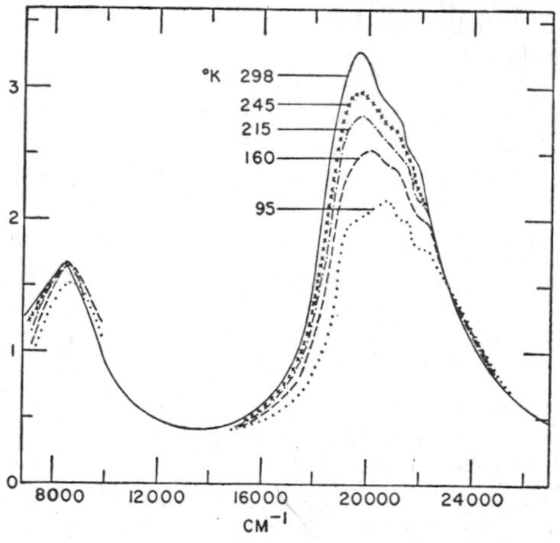

FIG. 6. The absorption spectrum of $CoSO_4 \cdot 7H_2O$ at various temperatures.

[10] P. Sulzer and K. Wieland, Helv. Phys. Acta 25, 653 (1952).

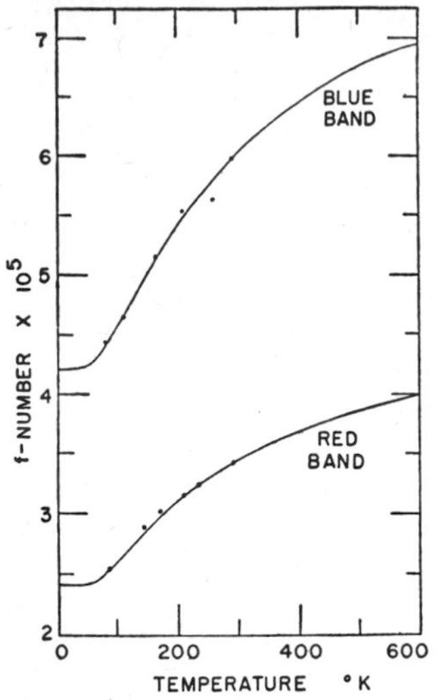

FIG. 7. Variation of f number of $NiSO_4 \cdot 7H_2O$ with temperature. The points are experimental and the line is drawn using Eq. (4) with $\theta = 250°K$.

totally symmetric vibration. It is taken equal to $\theta/1.44$ cm^{-1} for both states, and it is also assumed that the upper state vibration has the same effectiveness as the lower state vibration. We can, therefore, take $\epsilon^m{}_0$ the same for both the temperature dependent and temperature independent bands. Thus,

$$\epsilon^2{}_T(\nu) = \epsilon^m{}_0(\tanh\theta/2T)^{\frac{1}{2}} \cdot \exp(\theta/T) \cdot \exp\{-\tanh(\theta/2T) \cdot [(\nu-\nu_0)+2\omega/\Delta\nu]^2\}. \quad (2)$$

The total extinction coefficient is given by

$$\epsilon_T(\nu) = \epsilon^1{}_T(\nu) + \epsilon^2{}_T(\nu). \quad (3)$$

The f number is proportional to $\int \epsilon d\nu$, and we have

$$f = f_0(1+\exp(-\theta/T)) \quad (4)$$

with f_0 the value of f at 0°K. A curve of this form fits the data on the integrated extinction coefficient very well in several cases. The same parameter $\theta = 250°K$ fits both the blue and the red bands of $NiSO_4 \cdot 7H_2O$ equally well, as shown in Fig. 7.‖ Both chrome-alum bands are correctly fitted with $\theta = 400°K$ as shown in Fig. 8. The somewhat oversimplified theory is substantiated by these cases where two bands of the same substance can be fitted with the same parameter. The $CuSO_4 \cdot 5H_2O$ band intensity cannot be fitted to an equation of the form of 4. Two values of θ fit the data somewhat better, however, and a suggestion as to why this is so is that the large tetragonal field component splits the T_{1u} vibration into two components each of which may make the transition allowed. Even with two parameters, however, an equation of the form 4 does not give a good fit. The visible $CoSO_4 \cdot 7H_2O$ band intensity fits Eq. (4) with $\theta = 200°K$. The infrared band could not be reliably integrated for this salt nor for $NiSO_4 \cdot 7H_2O$ because of base line uncertainty.

The same approximate theory may be used to give the temperature dependence of the band maximum. Differentiating 3 with respect to ν gives approximately

$$\nu_{max} = \nu_0 + \theta/1.44 \cdot [\tanh(\theta/2T) - 1]. \quad (5)$$

The maximum band shift is $\theta/1.44$ cm^{-1}. Using the parameter θ obtained from the f-number data, several plots of $\nu_0 - \nu_m$ *vs* T were made. The one for chrome-alum is shown in Fig. 9. It is seen that both bands of this salt have a greater red shift with rising temperature than can be accounted for by exciting ground state vibrations. No choice of θ will give a temperature shift of the observed magnitude. Similar results were obtained for the other salts studied. These results indicate to us that as the temperature is reduced, the average effective field parameters increase. No data on the coefficient of thermal expansion is available for comparison. One might also expect that the shifts would correlate with $d(h\nu)/d(Dq)$. These quantities can be calculated easily from crystal field theory. The relative shifts of two bands of the same substance are not correctly given by the ratio of these quantities, however. This may be the result of rapidly changing anisotropies and changing effectiveness of different components of partially split degenerate vibrations, which could change the transition probability along the crystal axes as the temperature is lowered. Low-temperature anisotropy measurements could show whether or not this is true.

The temperature dependence study has shown that the distortion of the ionic environment by vibrations is

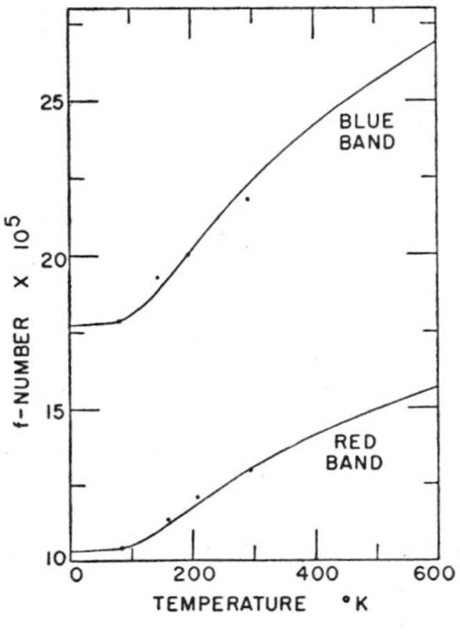

FIG. 8. Variation of f number of $KCr(SO_4)_2 \cdot 12H_2O$ with temperature. The points are experimental and the line is drawn using Eq. (4) with $\theta = 400°K$.

‖ The entire double peak of the "red band" was integrated at each temperature in order to find the f number. The two components change in relative intensity as the temperature is lowered and one might argue that a different vibration frequency must be used for each one. If Jorgensen's explanation of the double peak is correct, as we think it is [see Sec. IV(b)] the change in relative intensity only signifies a change in the singlet-triplet perturbation. This change may come about because the ^1D and ^3F components which interact are oppositely displaced by any change in the crystal field. The change in this case is the thermal contraction of the lattice.

an important factor in determining the intensity of the transitions. Approximately 40 percent of the intensity is lost in going from 300°K to 100°K in each case studied. The decrease in intensity is caused by the loss of ground state vibrations, and the remaining intensity, in our theory, is all ascribed to upper state vibrations. This is analogous to electronically forbidden spectra in molecules where approximately the same intensity is produced by both upper and lower state vibrations. There is still a possibility that a permanent noncentrosymmetric distortion causes part of the transition intensity. This portion of the intensity would be temperature independent except for the possible unequal thermal expansion of the lattice which could change the noncentrosymmetric part of the potential. Anisotropic thermal expansion could not explain the large intensity changes in cubic chrome alum, but it may be important in the highly anisotropic crystal of $CuSO_4 \cdot 5H_2O$.

If vibration rather than permanent distortion be the cause of the transition strength, then the theory of vibrational-electronic interaction should be applied to give the absolute f numbers. This of course is rather complicated, and cannot be carried out quantitatively. The transition strength must come from an admixture of p character in the predominantly d orbitals. If other factors were equal, then the intensity of the bands could be taken as a measure of the p character induced in the electronic states by vibration.

IV. EFFECTS OF THE NONCUBIC PARTS OF THE CRYSTAL FIELD

In several of the spectra illustrated in the figures, the bands are evidently composite. This is true for some bands of Fe^{++}, Co^{++}, and Ni^{++}. When the polarized spectra of the crystals are examined other bands are seen to be composite. We have attempted to interpret the composite structure in terms of the noncubic components of the crystal field. These low-symmetry components will cause a splitting of bands which arise from transitions to a degenerate upper level.

a. $CuSO_4 \cdot 5H_2O$

The detailed atomic arrangement in this crystal has been reported by Beevers and Lipson.[11] There are two nonequivalent ionic groups per unit cell of the triclinic crystal. The environment of each ion consists of four water molecules arranged in an approximate square with two polar sulfate oxygens at slightly greater distances from the Cu^{++}. The two types of ions are sufficiently similar so that we felt justified in neglecting the differences. The field experienced by the ion has a strongly tetragonal component along the axis joining the sulfate oxygens, and smaller rhombic components in the equatorial plane. Thus the doubly degenerate ground state (E_g), in the cubic field approximation and the triply degenerate upper state (T_{2g}) should split into two and three components, respectively. Since the tetragonal component is larger than the rhombic, two of the upper state components should lie close together with the third either lower or higher, depending on the sign of the tetragonal field parameters, A and Q.¶

Two cuts of the crystal were examined at room temperature under polarized light: the a and the c axis projections. The polarization directions are illustrated in Fig. 10. The results are shown as the absorption curves of Fig. 11.

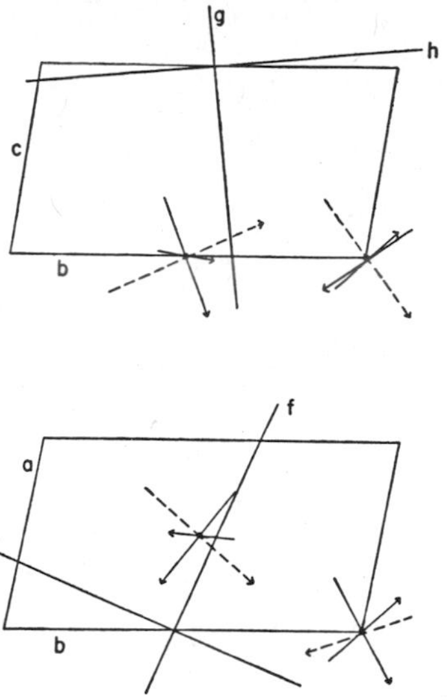

FIG. 10. The crystal planes used for the polarized absorption measurements on $CuSO_4 \cdot 5H_2O$. Upper drawing shows the a-axis projection; lower drawing c-axis projection. The orientations of the approximately tetragonal axes in the $Cu^{++}(H_2O)_6$ units, I and II are shown for each plane by the dotted arrows. The dielectric axes are labeled g and h in the a-axis projection, and e and f in the c-axis projection.

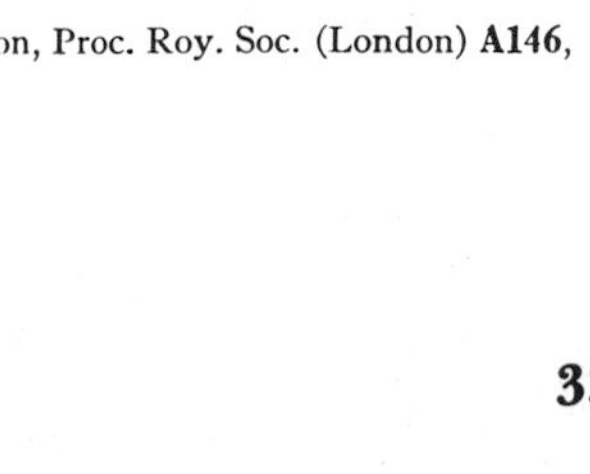

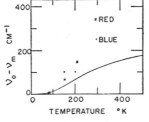

FIG. 9. Variation of the two peaks of $KCr(SO_4)_2 \cdot 12H_2O$ with temperature. The points are experimental; the line is drawn using Eq. (5) with the parameter $\theta = 400°K$.

[11] C. A. Beevers and L. Lipson, Proc. Roy. Soc. (London) **A146**, 570 (1934).

¶ The tetragonal field parameters A and Q are defined as the coefficients of the second and fourth power tetragonal potential terms:

$$V(\text{tetr.}) = A[(2z^2 - x^2 - y^2)] + Q[(z^4 + 6x^2y^2 - (3/5)r^4)]$$

where x, y, z are the coordinates of a d electron relative to the metal atom as origin, and the metal ligand bond lines as coordinate axes. The experimentally measured parameters are Ap and Qq, where $p = e\bar{r}^2/105$, $q = 2e\bar{r}^4/105$.

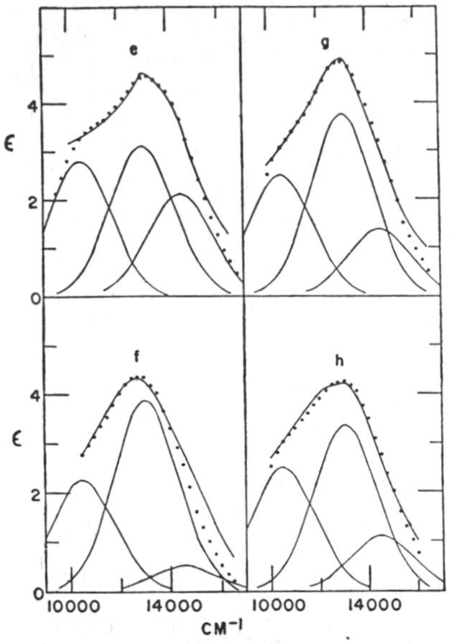

FIG. 11. Analysis of the $CuSO_4 \cdot 5H_2O$ polarized absorption into Gaussian curves. The solid line is the experimental curve and the dots show the points where the Gaussian curves below were added. The crystal directions are explained in Fig. 10.

If there is a large change in equilibrium position after the transition, the form of the absorption band should be given by Eq. (3). This is nearly Gaussian for values of the vibration frequency small compared to the band width. In order to see if the bands were composite, we analyzed them into Gaussian curves. A moderately good fit of the data is obtained with three Gaussians centered at 10 500, 13 000, and 14 500 cm^{-1}, each having a half-width of 3000 cm^{-1}. The data are not perfectly consistent; one of the four absorption curves of Fig. 11 should be derivable from the other three, but the agreement between the calculated and the observed curve is not very good. This disagreement comes mainly in the 14 500 region, and is probably caused by uncertainties in the base line of zero absorption.

The resolution into two components seems quite definite. Bjerrum *et al.*[12] have reported such a resolution for aqueous copper ion. The third component remains somewhat doubtful.

If the two upper Gaussian curves are assigned to the close-lying pair of upper levels (the degenerate *Eg* state in a purely tetragonal field) and the 10 500 cm^{-1} curve to the third component, and a *Dq* value about the same as for Ni^{++} is chosen, then one predicts a large splitting of the ground (E_g) level. We therefore searched for electronic absorption in $CuSO_4$ in D_2O solution. No such absorption was found between 4000 cm^{-1} (the cutoff for D_2O) and the near infrared band. Thus the transition between the components of the E_g level is either below 4000 cm^{-1} or is the cause of one of the peaks in the near infrared absorption spectrum. The latter alternative means nearly a square field. This alternative leads to a *Dq* value similar to that in Ni^{++}, and also explains the seemingly high "apparent" *Dq* value (see Fig. 2).

[12] Bjerrum, Ballhausen, and Jorgensen, Acta Chem. Scand. 8, 1275 (1954).

b. Nickel Salts

Spectroscopic data on the crystals $NiSO_4 \cdot 7H_2O$, $NiSiF_6 \cdot 6H_2O$, and $K_2(Zn,Ni)(SO_4)_2 \cdot 6H_2O$ (1% Ni) were obtained.

The anisotropy in the absorption of $NiSO_4 \cdot 7H_2O$ is shown in Fig. 12. The polarized absorption was measured along each of the principal axes of the orthorhombic crystal. It is interesting to note that the bands in the red and blue region of the spectrum have as their order of increasing absorption $b<a<c$, while the order in the infrared band is $a<c<b$. The first two bands were assigned to the T_{1g} class in O_h, while the infrared band was assigned to T_{2g}, so that the bands having the same symmetry have the same polarization properties.

The degree of polarization for a given direction of propagation in the crystal is never very high. There are two reasons for this. The distortions of the ionic environments are not ordinarily oriented in the same way for each ion of the unit cell, and the transitions are vibration induced.

In a perfectly cubic environment, the triply degenerate T_{1u} vibrations induce an isotropic transition. In the presence of a distortion into a lower symmetry, both electronic and vibrational splitting may occur, but the vibration components, as a detailed analysis shows, still induce a nearly isotropic transition. In a well-oriented single crystal of $NiSiF_6 \cdot 6H_2O$, the trigonal axes of each ion are parallel, so that the first depolarizing effect does not come in. The degree of polarization, however, is still very small.

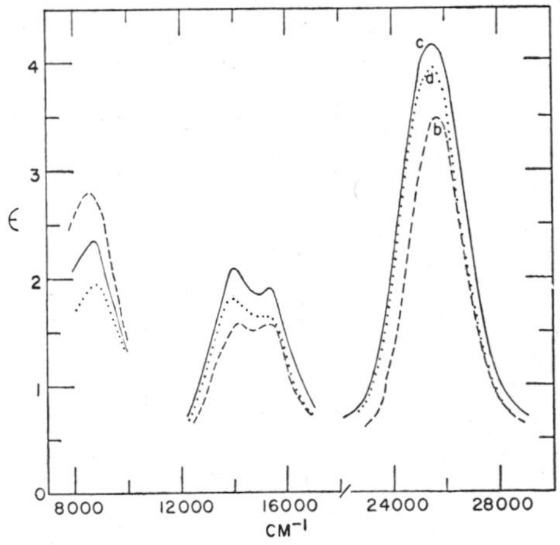

FIG. 12. The polarized absorption of $NiSO_4 \cdot 7H_2O$ at room temperature.

The spectra of Fig. 12 also show that the peaks of the b-polarized absorption are slightly displaced from the peaks of the a and c polarized absorption. For the "red" and "blue" bands $b>a$, c and for the infrared band $b<a$, c. The differences are small, but are believed to be real. The b axis is the direction in which the approximate axis of tetragonality for each of the four ions in the unit cell has its maximum projection. The difference in b and a, c peaks could then be interpreted as electronic splitting caused by the tetragonal distortion. This splitting is on the order of 200 cm^{-1} for each band.

Figure 13 shows that the spectra of the three crystals studied and the spectrum of a dilute aqueous solution are very similar to each other. This shows, in the first place, that the arrangement of nearest neighbors about the Ni^{++} ion is the same octahedral arrangement in solution as is known to exist in the crystal. One also notices that the splitting of the "red band" is nearly the same in the three crystals as it is in solution, and that the relative intensities of the two peaks are somewhat different in the four cases.

As possible reasons for the double peak of the "red band" one could consider the low symmetry field components, spin-orbit coupling,** and Jorgensen's explanation[13] that one peak arises from a singlet state intensified by singlet-triplet mixing. If the first explanation were correct, one would not expect the same splitting in all four spectra. In fact, after an extensive investigation, we have not been able to find a consistent set of field parameters based on the assumption that the red band is split because of distortions of the $Ni(H_2O)_6^{++}$ octahedra. Our data are consistent with the last two hypotheses. Jorgensen's paper gives evidence that spin-orbit coupling alone cannot explain the bands in other complexes such as $Ni(NH_3)_6^{++}$. Thus, we feel that Jorgensen's hypothesis is the correct one.

The energy difference between differently polarized absorption peaks shown in Fig. 12, therefore, probably arises from the distortion of the $Ni(H_2O)_6^{++}$ octahedra.

c. $CoSO_4\cdot 7H_2O$

The visible band of this salt shows a good deal of structure. Three components seem to be present at room temperature and four at 90°K (Fig. 6). The band is assigned to superposed $T_1(^4F)\rightarrow T_1(^4P)$ and $T_1(^4F)\rightarrow A_2(^4F)$ transitions. Four components would therefore be expected in the low-temperature spectrum since the field at the Co^{++} ion has appreciable rhombic character. The change in shape of this band with changing temperature may be interpreted by considering the effect of lattice contraction on the crystal

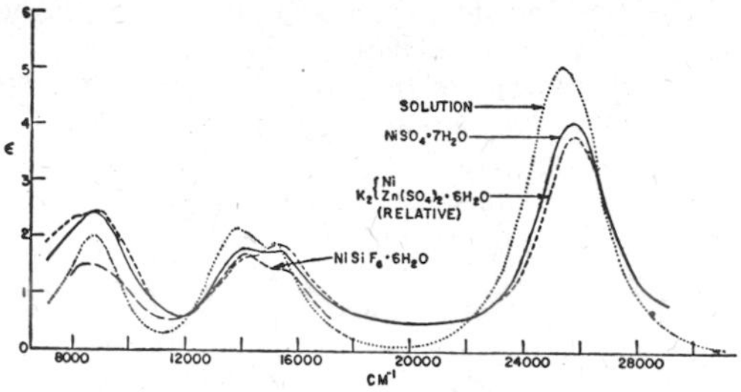

FIG. 13. The absorption spectra of $NiSO_4\cdot 7H_2O$, $NiSiF_6\cdot 6H_2O$, K_2(Zn, 1% Ni) $(SO_4)_2\cdot 6H_2O$, and aqueous Ni^{++}. The blue band of $NiSiF_6\cdot 6H_2O$ practically coincides with that for $NiSO_4\cdot 7H_2O$, and so was not put in the figure. The concentration of Ni^{++} in the mixed salt was not accurately determined, so the absolute height of the curve relative to the others is somewhat in error.

field splitting. For this purpose we consider only the change of band position with the change of Dq. For the $T_1(F)\rightarrow T_1(P)$ transition $d(h\nu)/d(Dq)=6$ while it is 18 for the $T_1(F)\rightarrow T_1(A_2)$ transition. One of the components of the visible absorption band should therefore move to the blue with respect to the others with decreasing temperature. This seems to be the correct interpretation of the band shapes seen in Fig. 6, where the second component moves more rapidly than the others.

No structure is observed in the infrared band. Since it is very broad, some of the breadth is probably contributed by the rhombic field, and each component transition must be broad.

The spin-orbit coupling is probably not strong enough in Co^{++} to account for over 3000 cm^{-1} of splitting in the visible band. It is more likely that the environment is distorted away from regular octahedral shape (Jahn-Teller effect in the T_{1g} ground state), and that this gives the observed splitting in the upper T_{1g} state. The second power field parameters may be derived from the structure in the visible band. The components are best resolved in the 95°K spectrum and lie at 19 000, 21 500, and 22 200 cm^{-1}, with the infrared band at 8600 cm^{-1}. The component at 20 700 cm^{-1} seems to be the one which moves most rapidly as the temperature rises and it is therefore tentatively assigned to the A_{2g} state. The energy levels in a rhombic field in terms of the field parameters†† are

$$E_1=+42(B+C)p,$$
$$E_2=-42Bp,$$
$$E_3=-42Cp.$$

If the splitting $22\,200-21\,500=E_3-E_2$ then the parameter values are $Bp=-14.3$: $Cp=-39.3$. Unfortunately the fourth power terms cannot be derived from the spectra at present. A detailed study of the

** Suggested by Professor M. H. L. Pryce (private communication, July, 1956). He shows that the $T_{1g}(F)$ state should split more than the others, and by approximately the correct amount.

[13] C. K. Jorgensen, Acta Chem. Scand. 9, 1362 (1955).

†† The orthorhombic field parameters are defined similarly to the tetragonal field parameters:

$$[Bx^2+Cy^2-(B+C)z^2]+[E(x^4+6y^2z^2-3/5r^4)+G(y^4+6x^2z^2-3/5r^4)].$$

infrared band would be necessary; if structure is found in it, the fourth power terms could be calculated.

Preliminary work on the visible band of $CoSiF_6 \cdot 6H_2O$ shows that the resolution of the components in the low-temperature spectrum is nearly complete. Our assumption about the temperature dependence of the $^3A_2$ state receives some support from this work.

d. $FeSO_4 \cdot 7H_2O$

The one absorption band of this compound has two peaks with a separation of 2000 cm^{-1}. This represents the complete removal of the degeneracy of the upper E_g state. This state is analogous to the E_g state of Cu^{++}. There is no way to tell from the spectrum if this splitting is tetragonal or rhombic. The tetragonal splitting is $-120\ Ap+20\ Qq$, but we do not have enough data to evaluate both parameters and so cannot calculate the ground state splitting. Since Dq seems normal, however, it is probable that the ground state splitting is less than 1000 cm^{-1}.

V. EXPERIMENTAL DETAILS

We shall mention the preparation of some of the substances used and compare with earlier work in cases where there is need for confirmation or contradiction of the earlier work.

a. Vanadous Ion

Vanadous ion V^{++} is a strong reducing agent so there is difficulty in preparing and preserving its solution and crystals. We were unsuccessful in obtaining crystals of the hydrated sulfate large and clear enough for absorption spectrum work. The spectrum of aqueous V^{++} was measured. Fisher's cp V_2O_5 was dissolved in H_2SO_4 solution by reduction with SO_2 to VO^{++}. This solution was further reduced to V^{++} by metallic zinc. The solution for optical study was about 0.4 M in V^{++}, 1 M in H_2SO_4, and contained an appreciable quantity of Zn^{++}.

Two weak bands were observed at 11 800 and 17 500 cm^{-1}. The extinction coefficients obtained are uncertain by as much as 50% because the concentration in the absorbing solution was not known accurately and contained unreduced V^{+++} absorption which had to be subtracted out. Strong absorption in the ultraviolet beyond 25 000 cm^{-1} completely obliterated any weak bands in that region. Dreisch and Kallscheuer[5] reported a double peak at about 11 350 cm^{-1} and Bose and Mukherji[14] reported maxima at 12 200 and 13 100 cm^{-1} for aqueous V^{++}. The latter's data are not reliable, however, because the solutions contained material which may have complexed with V^{++}, as well as higher oxidation states of vanadium.

[14] D. M. Bose and P. C. Mukherji, Phil. Mag. 26, 757 (1938).

b. Chromous Ion

Chromous ion Cr^{++} is an even stronger reducing agent than V^{++}. A solution for optical study was prepared by reduction of an acidic solution of Baker and Adamson's reagent $Cr_2(SO_4)_3$ with metallic zinc. Complete reduction to Cr^{++} was not attained but the absorption due to the Cr^{+++} fraction of the analyzed solution was subtracted out, leaving the spectrum shown in Fig. 1. The solution was approximately 0.031 M in Cr^{++}, 0.017 M in Cr^{+++}, 1 M in H_2SO_4, and contained an appreciable quantity of Zn^{++}. Bose and Mukherji[14] reported a broad band in the region 13 700–16 100 cm^{-1} and in addition bands at 22 200 and 24 400. The latter band was undoubtedly due to unreduced Cr^{+++} in their solution. The 22 200 band was not observed in the present work.

c. Ferrous Ion

Crystals of $FeSO_4 \cdot 7H_2O$ were grown from an acidic solution of Mallinckrodt's reagent $FeSO_4$, which was evaporated under an atmosphere of nitrogen, since Fe^{++} is oxidized by air. The spectra of $FeSO_4 \cdot 7H_2O$ and its acidic solution are shown in Fig. 1. Dreisch and Kallscheuer[5] examined solutions of ferrous sulfate and ferrous halides for which they reported broad bands stretching from 7000 to 15 000 cm^{-1}. Their absorption curves showed detailed fine structure with at least twelve distinct components to each band. The results of the present experiments are in disagreement with theirs. No such structure was observed in either heptahydrate or solution, although it should be easily observed with our resolution.

d. Cobaltous Ion

Crystals of $CoSO_4 \cdot 7H_2O$ were grown by evaporation of a saturated solution of Baker's reagent $CoSO_4$. The spectra of the crystal and its solution are illustrated in the figures. The hydrated ion absorbs in two regions centered at 8350 and 19 800 cm^{-1}. The visible band shows some fine structure in both solution and crystal, particularly at low temperature. Dreisch and Trommer[6] found six components in the infrared band with over-all separation of 1100 cm^{-1}, but we observed no such structure with the highly resolving Perkin Elmer spectrophotometer.

e. Nickel Ion

Crystals of $NiSO_4 \cdot 7H_2O$ were grown from a saturated solution of Baker's reagent $NiSO_4$. Again we failed to observe the fourfold structure of the infrared band which Dreisch and Trommer[6] have claimed.

ACKNOWLEDGMENTS

We would like to thank Dr. Leslie Orgel and Dr. John Owen for their helpful comments on the manuscript and for discussing their own results in this field with us.

Reprinted from *J. Inorg. Nucl. Chem.*, **8**, 143–148 (1958)

CLAUS E. SCHÄFFER AND C. KLIXBÜLL JØRGENSEN
Chemistry Department A, Technical University of Denmark, Copenhagen – Denmark

23

The nephelauxetic series of ligands corresponding to increasing tendency of partly covalent bonding

Summary: Ligand field theory has interpreted the Laporte forbidden absorption spectra of complexes the central ions of which have partly filled *d* shells. If octahedral symmetry *Oh*, the ligand field perturbations can be expressed by one parameter Δ (denoted also ($E_1 - E_2$) or 10 *Dq*). However, by adaptation of experimental energy levels to the predicted function of Δ it has been found nessesary to use smaller values of the term distances than those known from the gaseous ions.

This reduction of the term distances of the free ions will be discussed in the present communication. Numerous experiments have shown that the decrease grows more and more important in the series of metals.

Mn(II) *Ni*(II) *Cr*(III) *Co*(III) *Rh*(III) *Ir*(III)

and in the series of ligands

F^- H_2O NH_3 en SCN^- Cl^- CN^- Br^-.

All our experiments seem to indicate that the coefficients of intermixing x lie between those of the electrostatic model ($x = O$) and those of the Pauling d^2sp^3 case ($x = 0.5$). The variation of the radial function with a constant hydrogen like angular function and the relative importance of σ and π bonding is discussed.

In his lecture this morning, Professor Hartmann described the two common ways of approaching the crystal field problem. He mentioned the two main perturbations of equal order of magnitude which the electrostatic treatment had to take into account. One of them, the crystal field perturbation, is described in the case of O_h symmetry by the parameter Δ, 10 Dq or $E_1 - E_2$, these three symbols all representing the orbital energy difference between the e and the t_2 electron. The other perturbation is the electrostatic interaction between the electrons in the d shell. This perturbation is calculated by the theory of Slater, Condon and Shortley or of Racah giving the energy differences within the d shell in terms of F^2 and F^4 or B and C, respectively. Here the following interrelation exists:

$$B = \frac{1}{49} F^2 - \frac{5}{441} F^4$$

$$C = \frac{5}{63} F^4$$

The spectra of complexes with regularly octahedral symmetry has has been interpreted as transitions between energy levels characterized by their total spin quantum numbers and a group theoretical symbol Γ indicating that the wavefunction of the level has the same symmetry properties as the irreducible representation Γ of the O_h group. If q levels of characteristics $^{2S+1}\Gamma$ occur within a d^n configuration, their energy can be described as the eigenvalues of a determinant of the degree q. If there is only one level of $^{2S+1}\Gamma$, its energy will be given as a linear combination of a multiplet term energy (expressed in B and C) and Δ. If $q > 1$, two determinants of equal significance but of different form are obtained dependent on the order in which the two perturbations are taken into account.

From the comparison with experimental results of atomic spectroscopy the values of B and C can be evaluated for the gaseous ions.

In literature, these values usually have been considered fixed, and Δ has been chosen so as to adapt all the known energy levels to the theory in the best possible way. Tanabe and Sugano [1], who gave the « strong field » determinants for all levels of d^n systems, were the first authors to mention the possibility of making B and C free parameters. In this way they found that much better agreement could be obtained between the experimental values of the energy levels and the values calculated from their matrices. Owen related the relative decrease in these parameters in a series of different hexaaquo metal ions to the effect of increasing covalency. Later Klixbüll Jorgensen [3,4] developed these conceptions, and the present contribution will give a series of ligands for which the ratio $\beta = \frac{B_{\text{complex}}}{B_{\text{gaseous ion}}}$ decreases. In order to eliminate possible discrepancies caused by deviations from O_h symmetry due to the Jahn-Teller effect we have only studied complexes with orbitally non degenerate ground levels.

If the ligand field is accounted for first, you get « strong field » determinants. Here the diagonal elements corresponding to pure sub configurations $e^n t_2^m$ contain a contribution from the ligand field parameter Δ together with a linear combination of the Racah parameters distinguishing the levels within the sub configuration. The off-diagonal elements only contain linear combinations of Racah parameters mixing levels with the same Γ originating in different sub configurations.

When the electrostatic interaction is taken into account first, the « weak field » determinant is obtained. Here the diagonal elements correspond to multiplet terms of the gaseous ion characterized by their L and S. These elements are expressed as a sum of Racah parameters equal to the energy of the multiplet terms and a multiple of Δ corresponding to the slope $\frac{\delta E}{\delta \Delta}$ ($\Delta = 0$) of the energy function E, when the ligand field is zero. The off-diagonal elements being multiples of Δ describe how the levels of the gaseous ion are mixed as a function of the ligand field.

In *table I* are given the Δ values and in *table II* the β values for a number of complexes of this type.

We shall give a short account of how these tables have been deduced from the experimental material.

TABLE I.

	Br—	*Cl—*	*F—*	*ur*	*ox——*	H_2O	*SCN—*	NH_3	*en*	*CN—*
Mn(II)	—	—	—	—	—	8300	—	—	9900	—
Ni(II)	7000	7200	—	—	—	8500	—	10800	11500	—
Fe(III)	—	—	13900	13000	13600	14200	—	—	—	—
Cr(III)	—	13800	15200	16000	17500	17400	17800	21600	21900	26700
Co(III)	—	—	—	--	18100	18500	—	23000	23200	33500
Rh(III)	19000	20400	—	—	26400	27000	—	34100	34600	—
Ir(III)	23100	25000	—	—	—	—	—	—	41400	—

Ligands arranged in the order of the spectrochemical series.
The values of Δ (cm^{-1}) are given, *ur* = urea, *en* = ethylenediamine and *ox*—— = oxalate.

TABLE II.

	F—	H_2O	*ur*	NH_3	*en*	*ox——*	*SCN—*	*Cl—*	*CN—*	*Br—*
Mn(II)	—	.93	—	—	.88	—	—	—	—	—
Ni(II)	—	.89	—	.84	.81	—	—	.74	—	.72
Fe(III)	.79	.76	.72	—	—	.69	—	—	—	—
Cr(III)	.89	.79	.72	.71	.67	.68	.62	.56	.58	—
Co(*III*)	—	.7	—	.62	.59	.57	—	—	.40	—
Rh(III)	—	.73	—	.60	.59	—	—	.49	—	.40
Ir(III)	—	—	—	—	—	—	—	.41	—	.33

Ligands arranged in the order of the nephelauxetic series.
The values of $B_{\text{complex}} / B_{\text{gaseous ion}}$ are given.

We have applied the determinants of Tanabe and Sugano using for the ratio between the Racah parameters *C*/*B* always the average value of 4. Thus we have evaluated the figures of *B* and Δ by adapting wavenumbers of the absorption bands to the eigenvalues of the determinants. From B_{complex} and $B_{\text{gaseous ion}}$ the β values have been calculated.

With Cr(III) d^3 and Ni(II) d^8 the values of Δ are well defined as the wavenumber of the first spin allowed absorption band, and with d^3 we have made the second band fit the matrix in order to determine $B_{complex}$, while in the d^8 system we have used the sum of the second and the third band. The B values of the gaseous ions are here known from atomic spectroscopy.

With the d^6 systems of Co(III), Rh(III), and Ir(III) this is not the case. Here we have obtained the following figures by extrapolation from a rather sparse material of other metal ions:

$Co(III)$ . . . . B = 1100 cm^{-1}

Rh(III) . . . . B = 720 cm^{-1}

Ir(III) . . . . B = 650 cm^{-1}

The B values of the complexes are here largely determined by the distance between the two spin allowed absorption bands, but as the Δ values are greater than the wavenumber of the first absorption band by approximately the Racah parameter C, our ratio C/B will induce some uncertainty.

In the d^5 systems of Mn(II) and $Fe(III)$ we have not evaluated B because here we have an atomic term distance reflected in the complex ions. The distance ^4G — ^6S is found as the wavenumber of the transition from the ground level $^6A_1$ to $^4A_1$, ^4E. β is determined directly by the ratio between this wavenumber and the wavenumber of the $^6S \longrightarrow {}^4G$ transition of the gaseous ion. By this method we have also eliminated the error due to the fact that the Racah theory does not give very good results for the d^5 systems. In Mn(II) the ^4G — ^6S values are known to be 26800 cm^{-1}, and in the Fe(III) the extrapolated values of Moore-Sitterly and Edlén [2] of 32000 cm^{-1} have been used. Δ is difficult to estimate here, and in order to get comparable results we have still made the first absorption band $^6A_1 \longrightarrow {}^4T_1$ fit the matrices.

Table I gives a number of such complexes in the order of the spectrochemical series corresponding to increasing values of Δ. For each metal you get approximately the same series of ligands and for each ligand the same series of metals.

It has been emphasized before that these features could not be explained reasonably by a mere electrostatic ligand field. Some mixing of ligand and metal orbitals would have to be accounted for. If the e orbital of the metal becomes antibonding by being mixed with σ-combinations of ligand orbitals, Δ will increase, while a similar mixing of the metal t_2 orbitals with π-combinations will cause Δ to decrease. The latter phenomenon would tend to produce smaller Δ – values for halide and oxoanion complexes than for amine complexes. The metal d orbital might also mix with empty ligand orbitals of higher energy. In this way Δ would change in the opposite direction to that already mentioned. In general, however, the latter processes seem to have little importance. The considerations of covalency can be supported by studying the three orbital energy differences occurring in complexes of tetragonal symmetry, but the most essential evidence for partly covalent bonding, as derived from absorption spectra, is the decrease of β ($B_{\text{complex}} / B_{\text{gaseous ion}}$) below 1.

Table II lists the same complexes in the order of decreasing β values. There exists the same series of ligands for all the metal ions in question, but the order is quite different from that of the spectrochemical series. We shall propose to call this series the nephelauxetic (cloud expanding) series, as it corresponds to a development of the *d* shell in the region of the ligands. The neo-greek word was constructed by Professor Kaj Barr from the University of Copenhagen.

The variation of β with the central ion for a given ligand seems to go symbatically with that of Δ though *Fe(III)* and *Cr(III)* may be interchanged. The comparison of β values of different metals is somewhat more uncertain than that of different ligands. This is mainly because the Racah parameter *B* is not always known for the gaseous ions.

Except for the case of d^5 systems the values of β have been found from the spin allowed bands, the distances between which are mainly functions of B_{complex}. If we had studied the spin forbidden transitions (e.g. in *Cr*(III) and *Ni(II)*) we would have found that their wavenumbers were almost independent of Δ and mainly a linear combination of both Racah parameters *B* and *C*. The question would arise as to whether the ratio between *C* and *B* is invariant.

We shall show in a later publication in Acta Chem. Scand. that Δ and $1 - \beta$ separately can be described with a fair accuracy as a product of a ligand characteristic and a central ion characteristic.

The development of the ligand field theory in the past six years has shown that different assumptions about the wavefunctions have predicted essentially the same absorption spectra. It is therefore not possible from the successfulness of any model in predicting energy levels to draw conclusions as to the correctness of the wavefunctions of the model. Nevertheless the nephelauxetic series seem to have shown that the partly filled *d* shell of the central ions is expanded.

REFERENCES

[1] Tanabe, Y. and Sugano, S., « J. Phys. Soc. Japan. », **9**, 753 (1954).
[2] Moore, C. E., *Atomic Energy Levels.*, « Nat. Bur. Stand. », Circular 467, Vol. II.
[3] Jorgensen, C. Klixbüll, *Energy Levels of Complexes and Gaseous Ions.* Gjellerups Forlag, Copenhagen, 1957.
[4] Jorgensen, C. Klixbüll, « Acta Chem. Scand., » **11**, 53 (1957).

DISCUSSIONS

Adamson (*Los Angeles*) – In our photochemical work we find that the quantum yields for the aquation of complexes of type MA_5X varies with the nature of X in a manner similar to your nephelauxetic series. In our case the series can be explained in terms of the oxidation potentials for the ligands. Have you considered any similar interpretation of your series?

Schäffer – Such an intepretation is probably most reasonable in the cases where the covalency is due to a donation of the electrons from the ligand to the metal ion. But in the case of CN^- there is evidence for donation of t_2 electrons from the metal ions to the ligands (to-gether with the σ-bonding of opposite polarity). The experimental Δ values found for metal complexes generally exibit a marked dependence on the charge of the central metal ion. For the hexaaqvo ions of divalent and trivalent metals of the first transition series we find Δ values of 10000 and 17000 cm^{-1}, respectively. A similar charge dependence of Δ is not found in CN^- complexes and especially in d^6 systems the evidence for π-bonding is pronounced. The Co(III) complex has the large value of $\Delta = 33500\ cm^{-1}$, the Fe(II) has $\Delta = 32500$ and Mn(I) has $\Delta =$ about 30000 cm^{-1} (reported to be colourless). These extremely high values of Δ to-gether with the much lesser dependence on the electrical charge of the metal ion suggest the donation of t_2 electrons from the metal ions to the ligands.

Bjerrum (*Copenhagen*) – In response to Professor Adamson, Professor Bjerrum pointed out the resemblance of the « nephelauxetic » series with that of Abegg-Bodländer, which for a given (one may add sufficiently electronegative) metal was based on the oxidation-reduction potentials of the ligands.

Reprinted from *J. Mol. Spectr.*, **2**, 342–360 (1958)

24

Intensities in Inorganic Complexes*

Part II. Tetrahedral Complexes

C. J. Ballhausen† and Andrew D. Liehr

Bell Telephone Laboratories, Murray Hill, New Jersey

An account is given of the modern theories concerning the nature of the electron orbits in tetrahedral inorganic complexes of the first transition series. It is shown that, although the characteristic energy spectra of these complexes may be qualitatively understood on the basis of the Bethe-Van Vleck ionic (crystal field) model, the probabilities of electronic transitions among the various stationary energy states requires the introduction of covalency, in the manner prescribed by Van Vleck (molecular orbital or "ligand field" model). It is found that the transition probabilities predicted by use of the Van Vleck molecular orbital wave functions of labile complex ions, such as the cupric and cobaltous tetrachloride ions, are in excellent agreement with experiment; while those predicted for stable complex ions, such as the chromate and permanganate ions, require that these wave functions be augmented by ligand π-molecular orbitals. Estimates of the covalent character of the tetrahedral $CuCl_4^=$, $CoCl_4^=$, MnO_4^-, and $CrO_4^=$ ions are given and compared, where possible, with those obtained from other sources, such as paramagnetic resonance experiments. An interpretation of the chromate and permanganate spectra, alternative to that of Wolfsberg and Helmholtz, is presented.

INTRODUCTION

More than twenty years have passed since Van Vleck (*1–3*) demonstrated the superiority of the ligand field[1] approach in the discussion of the magnetic properties of inorganic complexes. In the intervening years great strides have been made in applying the Van Vleck method[2] to such diverse physical phenomena as

* Presented at the Symposium on Molecular Structure and Spectroscopy, Columbus, Ohio, June 16–20, 1958.

† Visiting scholar 1957–8. On leave from Chemical Laboratory A, Technical University of Denmark, Copenhagen, Denmark.

[1] Ligand field theory is the accepted nomenclature for the hybridization of the crystal field theory of Bethe (*4*) and Van Vleck (*5*) and the molecular orbital theory of Mulliken (*3, 6*). A recent review of these theories may be found in Ref. *7*.

[2] The application of the Bethe (*4*) crystal field theory to the interpretation of magnetic (*5, 8–13*) and spectral (*14–20*) phenomena has also come to be called the method of Van Vleck. To distinguish these latter applications from those enumerated in the above text, we shall refer to applications of the pure crystal field theory as the Bethe-Van Vleck method and those of the ligand field theory as the Van Vleck method.

1

the prediction of the visible absorption spectra of octahedral (*21–25*) and tetrahedral (*26, 27*) complexes and the explanation of the paramagnetic resonance spectra of hexa-coordinated inorganic complexes (*28, 29*). However, to date it has not been firmly established whether the sundry semiempirical parameters required for the detailed characterization of these complexes, such as the degree of covalency, are unambiguously determined by the divers experimental measurements which fix them. This circumstance could, of course, be resolved by the performance of detailed molecular orbital calculations to determine the energy levels and wave functions of such compounds (*21–27*). Indeed, for complexes possessing an inversion center this is the only method feasible, as symmetry considerations show that the parameters determined by absorption spectroscopy are vibronic in character (*16, 30–32*), while those determined by paramagnetic resonance, for example, are not (*28, 29*). For noncentrosymmetric complexes, however, there exist much simpler means of determining some of these parameters. In what follows we shall illustrate one such method: the determination of the degree of covalency from optical intensity data. Wherever possible, the results of this determination are compared with those obtained from other measurements.

A. THE BETHE-VAN VLECK IONIC MODEL

1. Theory

Let us suppose, for the sake of simplicity, that the noncentrosymmetric complex under consideration is both tetrahedral and "ionic". Let us further suppose that the metal atom involved is a first row transition metal in an [A] $(3d)^n$ electronic configuration.[3] If this be the case we may, according to the Bethe-Van Vleck theory, write the electronic Hamiltonian for the system as

$$\mathcal{H} = \mathcal{H}_0 + \mathcal{H}^{(1)}, \tag{1}$$

where

$$\mathcal{H}_0 = \frac{-\hbar^2}{2m} \sum_{k=1}^{n} \nabla_{\mathbf{r}k}^2 + U(\mathbf{r}_1, \cdots, \mathbf{r}_n), \tag{2}$$

$$\mathcal{H}^{(1)} = e^2 \sum_{k=1}^{n} \sum_{i=1}^{4} \frac{q_i}{|\mathbf{r}_k - \mathbf{a}_i|}. \tag{3}$$

In Eq. (3) we have designated the change on ligand i, located at the distance $\mathbf{a}_i$ from the metal atom, M, as q_i (see Fig. 1).

The solution to our problem is now straightforward. We need merely form those linear combinations of the degenerate solutions of our zero-order Hamiltonian, $\mathcal{H}_0$, which transform properly under the tetrahedral point group, T_d,

[3] We shall henceforth use the symbol [A] to designate the argon (core) electronic configuration.

2

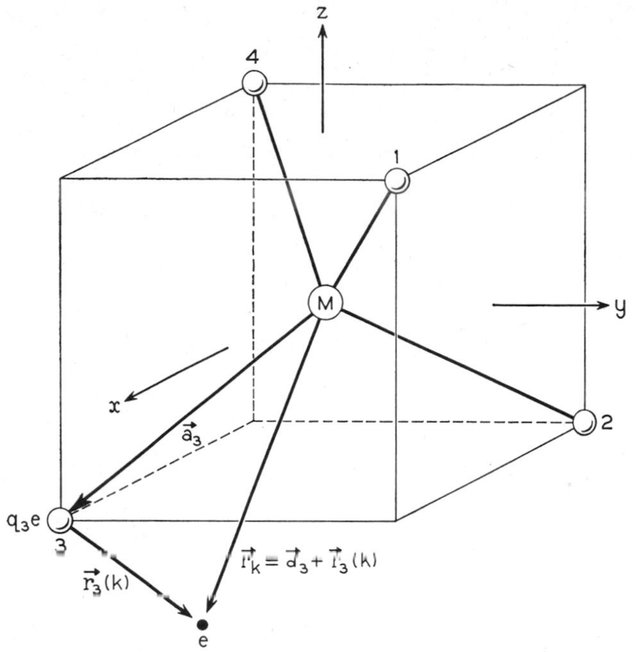

FIG. 1. Geometry assumed for the tetra-coordinated complexes.

and utilize these functions to construct the appropriate secular equations:[4]

$$|(\mathcal{H}^{(1)})_{\Gamma_{js},\Gamma_{jt}} - E(\Gamma_j)\cdot\delta_{s,t}| = 0. \tag{4}$$

The quantity Γ_{js} which appears in Eq. (4) represents the sth wave function belonging to the symmetry species Γ_j. The solutions of the zero-order Hamiltonian, $\mathcal{H}_0$, are readily obtained by use of the Condon-Shortley (*33*) technique, if one first expands the perturbation, $\mathcal{H}^{(1)}$, about the metal atom, in a series of normalized spherical harmonics:

$$\mathcal{H}^{(1)} = e^2 \sum_{k=1}^{n}\sum_{i=1}^{4}\sum_{l=0}^{\infty}\sum_{m=-l}^{+l} \frac{4\pi q_i}{2l+1}\frac{r_<^{\,l}}{r_>^{\,l+i}} Y_l^{m*}(\theta_k, \varphi_k) Y_l^{m}(\theta_i, \varphi_i), \tag{5}$$

where $r_<$ is the smaller and $r_>$ is the larger of the two distances $|\mathbf{r}_k|$ and $|\mathbf{a}_i|$.

Let us for the moment consider the case $n = 1$, that is, one $3d$ electron moving outside the argon core [A]. In this case the five degenerate $3d$ orbitals split into groups characterized by the symmetry "quantum numbers" $\Gamma_j = e$ (two-fold

[4] The construction of the symmetry functions, Γ_{js}, for tetrahedral symmetry yields a factoring of the secular equation which is exactly analogous to that obtained by the construction of eigenfunctions of $\mathbf{L}^2$ (or $\mathbf{J}^2$), $\mathbf{S}^2$, M_L (or M_J), and M_S for the case of spherical symmetry (i.e., free atoms and ions).

3

degenerate) and $\Gamma_j = t_2$ (three-fold degenerate) (*34*), with the t_2 level lying at the higher energy value (*35, 36*). Indeed, the energy separation of the e and t_2 levels is found to be simply $-4/9$ times the corresponding separation in the octahedrally coordinated [A] $(3d)^1$ complex (*36*). If we now rather naively divide the n $3d$ electrons between the two crystal field levels e and t_2, we obtain the so-called strong crystal field configurations (*4*); if we on the other hand, just as naively, decompose the known atomic states into states characterized by the symmetry "quantum numbers" Γ_j, we obtain the weak crystal field configurations (*4*). In general the true electronic configurations for the complex (in the Bethe-Van Vleck approximation) lie somewhere between these two extremes (*4, 8, 19*). It is only for the [A] $(3d)^1$ configuration and its conjugates (*37*), [A] $(3d)^m$, (m = 4, 6, and 9), that the weak and strong crystal field approximations yield identical level schemes. In Figs. 2 and 3 are depicted the weak-strong crystal field correlations for the configurations [A] $(3d)^n$, (n = 7, 9).

If the wave functions describing the e and t_2 energy levels were composed solely of [A] $(3d)^n$ electronic configurations, there would exist no allowed electric dipole transitions in the visible absorption spectrum of complexes having such electronic configurations (*30*). However, as there exist absorption bands in this region of the spectrum with oscillator strengths, f, of about 10^{-3}, we know that this can not be the case. Since vibronic perturbations yield absorption bands with intensities (*30–32*), $f \sim 10^{-4} - 10^{-5}$, such relatively intense absorptions can only

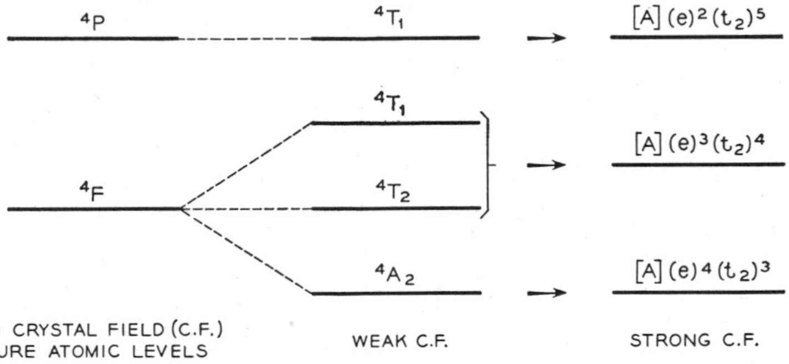

FIG. 2. Energy level correlations for the [A] $(3d)^7$ electronic configuration.

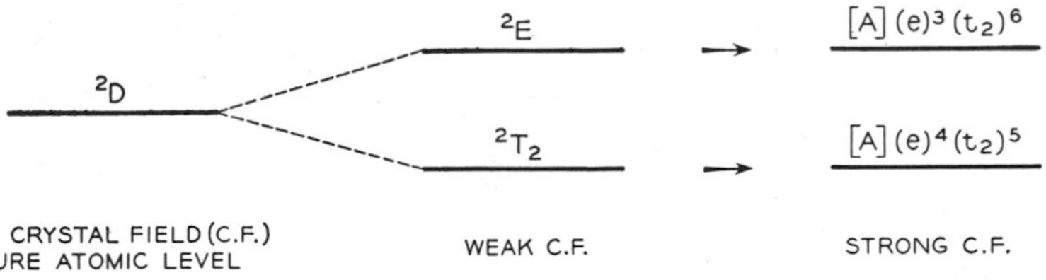

FIG. 3. Energy level correlations for the [A] $(3d)^9$ electronic configuration.

be due to two mechanisms: (1) mixing of the $3d$ and $4p$ wave functions due to the tetrahedral *crystal* field; and (2) mixing of the $3d$ and ligand wave functions due to the tetrahedral *ligand* field. In part A we shall be only concerned with mechanism 1. *As we shall see, it by itself is insufficient to account for the observed intensities.*

Let us assume that only mechanism 1 is operative. Then the correct "$3d$" wave functions to use in the construction of the Slater determinants (*33*) describing the electronic configuration of our complexes are

$$\psi(3d) = \psi^0(3d) + \sum_{4p} \frac{(\psi^0(3d) \mid \mathcal{H}^{(1)} \mid \psi^0(4p))}{E_p^0 - E_d^0} \cdot \psi^0(4p). \tag{6}$$

It may be readily verified that the use of Eq. (5) in connection with the Condon-Shortley (*33*) integral tables for products of three spherical harmonics yields the following results:

$$\left.\begin{aligned} \psi_{xy} &= (3d_{xy}) + \omega(4p_z) \\ \psi_{xz} &= (3d_{xz}) + \omega(4p_y) \\ \psi_{yz} &= (3d_{yz}) + \omega(4p_x) \end{aligned}\right\} \quad t_2,$$

$$\left.\begin{aligned} \psi_{x^2-y^2} &= (3d_{x^2-y^2}) \\ \psi_{z^2} &= (3d_{z^2}) \end{aligned}\right\} \quad e, \tag{7}$$

where

$$\omega = \frac{\frac{20}{21}\sqrt{\frac{3}{5}}\, qG^3_{3d,4p}}{E_p^0 - E_d^0} \cdot \frac{e^2}{a_0}, \tag{8}$$

and

$$G^3_{3d,4p} = \int R(3d) \frac{r_<^3}{r_>^4} R(4p) r^2\, dr. \tag{9}$$

In Eq. (9), $R(3d)$, and $R(4p)$ are the $3d$ and $4p$ hydrogenic radial functions and the distances r are measured in units of the Bohr radius a_0. The integrals $G^3_{3d,4p}$ have been tabulated (*38*) and are to be published shortly.

As we are concerned only with one electron transitions and as the transition probability operator,

$$\sum_{k=1}^{n} e\mathbf{r}_k,$$

is a one electron operator, all of our intensity expressions may be reduced to linear combinations of the six basic expressions,

$$f[a(e) \rightarrow b(t_2)], \qquad [a(e) = x^2 - y^2, z^2; b(t_2) = xy, xz, yz],$$

where (*39*)

$$f[a(e) \rightarrow b(t_2)] = 1.085 \cdot 10^{11} \cdot \nu(e \rightarrow t_2; \quad cm^{-1}) \cdot |[a(e) | \mathbf{r} | b(t_2)]|^2. \quad (10)$$

A simple calculation gives the following values for the one electron intensities $f[a(e) \rightarrow b(t_2)]$:

$$\begin{aligned}
f(z^2 \rightarrow xy) &= \frac{4}{3}\omega^2 \frac{\Delta E_d}{\Delta E_{dp}} f(xz \rightarrow p_z), \\
f(z^2 \rightarrow xy) &= \frac{1}{3}\omega^2 \frac{\Delta E_d}{\Delta E_{dp}} f(xz \rightarrow p_z), \\
f(z^2 \rightarrow yz) &= \frac{1}{3}\omega^2 \frac{\Delta E_d}{\Delta E_{dp}} f(xz \rightarrow p_z), \\
f(x^2 - y^2 \rightarrow xy) &= 0, \\
f(x^2 - y^2 \rightarrow xz) &= \omega^2 \frac{\Delta E_d}{\Delta E_{dp}} f(xz \rightarrow p_z), \\
f(x^2 - y^2 \rightarrow yz) &= \omega^2 \frac{\Delta E_d}{\Delta E_{dp}} f(xz \rightarrow p_z),
\end{aligned} \quad (11)$$

where ΔE_d is the energy difference $E(t_2) - E(e)$, ΔE_{dp} is the energy difference $E_p - E_d$, and $f(xz \rightarrow p_z)$ is the oscillator strength of a $3d_{xz} \rightarrow 4p_z$ electronic transition.

2. Intensities in the $CuCl_4^=$ Visible Spectrum

The absorption spectrum of the $CuCl_4^=$ complex (*36*, *40*) consists of one "low band" placed at about 10,000 cm^{-1}, and some "high bands" placed towards the blue. The low band has been interpreted as *the* crystal field band, while thehigh bands have been assigned to charge-transfer[5] transitions (*36*). Although the structure of the $CuCl_4^=$ ion in Cs_2CuCl_4 crystals is a "flattened" tetrahedron (*40*), we shall assume that, due to the removal of steric hindrances, the ion is rigorously tetrahedral in solution.[6]

As the crystal field theory predicts the ground state of this ion to be the $^2T_2$ state, $[A](e)^4(t_2)^5$, and its upper state to be the ^2E state, $[A](e)^3(t_2)^6$ (see Fig. 3), we may write the intensity of the crystal field band as

$$f(^2T_2 \rightarrow {}^2E) = \frac{4}{3}\omega^2 \frac{\Delta E_d}{\Delta E_{dp}} f(xz \rightarrow p_z). \quad (12)$$

If we take, as in our vibronic studies (*31*, *32*), $\Delta E_{dp} \approx \frac{1}{2}R$, $\Delta E_d = \frac{4}{9}(\frac{2}{11}R)$,

[5] This is the nomenclature recommended by Mulliken (*39*) for electronic transitions of the type $AB \rightarrow A^+B^-$.

[6] The crystal and solution spectra are nearly identical (*36*).

$f(xz \rightarrow p_z) \approx 0.2$, $q^2 = 15$, $Z_d = 7.85$, $Z_p = 4.25$ and $|\mathbf{a}_i| = 2.22$ A (*40*), where R is the Rydberg constant and Z_t, $(t = d, p)$ are the $3d$ and $4p$ level effective nuclear charges, we obtain the following values for the coupling constant $G^3_{3d,4p}$ and the intensity $f(^2T_2 \rightarrow {}^2E)$:

$$G^3_{3d,4p} = -4.41376 \cdot 10^{-3}, \tag{13}$$

$$f(^2T_2 \rightarrow {}^2E) = 1 \cdot 10^{-4}. \tag{14}$$

As the experimental value of f is (*40*) $2 \cdot 10^{-3}$, we see that the Bethe-Van Vleck ionic model does *not* yield a correct value of the intensity for the crystal field band of $CuCl_4^{=}$.

3. Intensities in the $CoCl_4^{=}$ Visible Spectrum

The crystal field theory predicts for the tetrahedral complex $CoCl_4^{=}$ the level scheme pictured in Fig. 2. If we assume that this complex is well characterized by the weak crystal field theory, we have for the three possible crystal field transitions (*41*):

$$\begin{aligned} {}^4A_2 \rightarrow {}^4T_2: &\quad \nu_1 = 40/9\, Dq(\text{cm}^{-1}), \\ {}^4A_2 \rightarrow {}^4T_1(F): &\quad \nu_2 = 24/3\, Dq(\text{cm}^{-1}), \\ {}^4A_2 \rightarrow {}^4T_1(P): &\quad \nu_3 = 14{,}000 + 10/3\, Dq(\text{cm}^{-1}), \end{aligned} \tag{15}$$

where Dq is the usual octahedral crystal field parameter (*7*). The corresponding weak field wave functions are

$$\begin{aligned} \Psi(^4A_2) &= [\text{A}]\ (e)^4(t_2)^3, \\ \Psi(^4T_2) &= [\text{A}]\ (e)^3(t_2)^4, \\ \Psi[^4T_1(F)] &= \sqrt{4/5}\ [\text{A}]\ (e)^2(t_2)^5 - \sqrt{1/5}\ [\text{A}]\ (e)^3(t_2)^4, \\ \Psi[^4T_1(P)] &= \sqrt{1/5}\ [\text{A}]\ (e)^2(t_2)^5 + \sqrt{4/5}\ [\text{A}]\ (e)^3(t_2)^4. \end{aligned} \tag{16}$$

Now since the electric dipole operator,

$$\sum_{k=1}^{7} e\mathbf{r}_k ,$$

transforms as T_2 in the tetrahedral point group, T_d, we see immediately that it is impossible to have an electric dipole allowed transition ${}^4A_2 \rightarrow {}^4T_2$. Furthermore, we see that in the approximation of single electron excitation, we have

$$\frac{f(\nu_3)}{f(\nu_2)} = \frac{4 \cdot \nu_3}{1 \cdot \nu_1} \approx 10, \tag{17}$$

if we identify the 15,000-cm^{-1} band of $CoCl_4^=$ with ν_3 and the 6300-cm^{-1} band with ν_2 (*41*).[7] Hence, we need only compute the oscillator strength of *either* the ν_2 or ν_3 band to get that of the other band. A simple computation yields for the intensity of the $^4A_2 \rightarrow {}^4T_1(P)$ transition the result

$$f[^4A_2 \rightarrow {}^4T_1(P)] = \frac{16}{5}\,\omega^2\,\frac{\Delta E_d}{\Delta E_{dp}}\,f(xz \rightarrow p_z). \tag{18}$$

Taking (*41*) $q^2 = \frac{1}{4}$, $Z_d = 6.55$, and $\Delta E_d = 15{,}000$ cm^{-1}; and setting $|\mathbf{a}_i|$ equal to its experimental value 2.34A (*42*) $Z_p = 3.25$, $\Delta E_{dp} = \frac{1}{2} R$, and $f(xz \rightarrow p_z) = 0.2$, we obtain

$$f[^4A_2 \rightarrow {}^4T_1(P)] = 2 \cdot 10^{-5}. \tag{19}$$

This is in very poor agreement with the experimental value of $f = 6 \cdot 10^{-3}$.

4. Comments on the Ionic Model

In the only two tetrahedral complexes of the type $[A](3d)^n$, $n \neq 0$, which have been well characterized, we have found that the Bethe-Van Vleck ionic model yields intensities which are from ten to one hundred times too small. We are thus forced to conclude that, at least for $CoCl_4^=$ and $CuCl_4^=$, the predominant contribution to the observed intensities is neither vibronic couplings nor crystal field couplings of the 3*d* and 4*p* levels. This circumstance implies that these complexes have some covalent character and that the Van Vleck model must be used in computing intensities. As we shall see in part *B*, this latter model gives an admirable account of the spectrum of both the cupric and the cobaltous tetrachloride ions.

B. VAN VLECK MOLECULAR ORBITAL MODEL

1. Theory

If, instead of regarding the metal and ligand electronic systems to be separable, we take explicit cognizance of the fact that the charge distributions describing these systems overlap, and solve the appropriate metal plus ligand secular equa-

[7] Unfortunately, the experimental studies thus far reported have not been of sufficient accuracy to verify the predicted intensity ratio given in Eq. (17).

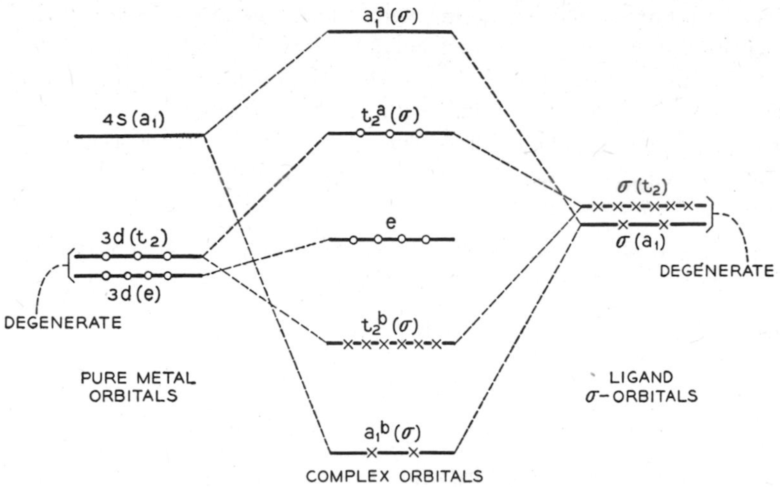

FIG. 4. Energy level diagram for the complex $CoCl_4^=$. The corresponding diagram for $CuCl_4^=$ is readily obtained from that pictured above by the addition of two more electrons to the metal and complex orbitals. A superscript *a* and *b* is used to distinguish the antibonding and bonding orbitals, respectively.

tion, we obtain tetrahedral wave functions of the form (3):[8]

$$\psi(e) = \begin{Bmatrix} (3d_{x^2-y^2}) \\ (3d_z^{\,2}) \end{Bmatrix} \text{(nonbonding)},$$

$$\psi_b(t_2) = N_b \begin{Bmatrix} \sqrt{1-\alpha^2}(3d_{yz}) + \alpha\cdot\tfrac{1}{2}(\sigma_1+\sigma_3-\sigma_2-\sigma_4) \\ \sqrt{1-\alpha^2}(3d_{xz}) + \alpha\cdot\tfrac{1}{2}(\sigma_1+\sigma_2-\sigma_3-\sigma_4) \\ \sqrt{1-\alpha^2}(3d_{xy}) + \alpha\cdot\tfrac{1}{2}(\sigma_1+\sigma_4-\sigma_2-\sigma_3) \end{Bmatrix} \text{(bonding)},$$

$$\psi_a(t_2) = N_a \begin{Bmatrix} \alpha(3d_{yz}) - \sqrt{1-\alpha^2}\cdot\tfrac{1}{2}(\sigma_1+\sigma_3-\sigma_2-\sigma_4) \\ \alpha(3d_{xz}) - \sqrt{1-\alpha^2}\cdot\tfrac{1}{2}(\sigma_1+\sigma_2-\sigma_3-\sigma_4) \\ \alpha(3d_{xy}) - \sqrt{1-\alpha^2}\cdot\tfrac{1}{2}(\sigma_1+\sigma_4-\sigma_2-\sigma_3) \end{Bmatrix} \text{(antibonding)},$$

$$\begin{aligned} \psi_b(a_1) &= N_b'\{\sqrt{1-\beta^2}(4s) + \beta\cdot\tfrac{1}{2}(\sigma_1+\sigma_2+\sigma_3+\sigma_4)\} \text{ (bonding)}, \\ \psi_a(a_1) &= N_a'\{\beta(4s) - \sqrt{1-\beta^2}\cdot\tfrac{1}{2}(\sigma_1+\sigma_2+\sigma_3+\sigma_4)\} \text{ (antibonding)}. \end{aligned} \tag{20}$$

In Eq. (20) the parameters α and β measure the degree of covalency present in the complex, and the functions σ_i, ($i = 1, 2, 3, 4$), represent the ligand orbitals

[8] We have here neglected π-bonding between the metal and the ligand systems as it is most probably of minor importance for complexes of the type $CuCl_4^=$ and $CoCl_4^=$, that is, [A] $(3d)^n$ complexes, ($n \neq 0$). See, however, Section 4 for complexes where this is definitely not the case.

which are appropriate for σ-bonding with the metal atom. When α and β take values very close to unity, the molecular orbital treatment reduces to the previously discussed (ionic) *crystal field* model. In general α and β take on values between unity and five-tenths, which implies that we are dealing with the (molecular orbital) *ligand field* model.[9] The energy level scheme corresponding to such a ligand field model is shown in Fig. 4. The quantities N_t and N_t', $(t = a, b)$, are the required normalizing factors. If we define the overlap integral, S, as

$$S = \int \psi_{\text{metal}} \cdot \tfrac{1}{2} \sum_i \sigma_i \, d\tau, \tag{21}$$

these normalizing factors are given by

$$N_b \text{ (or } N_b') = [1 + 2S\alpha(1 - \alpha^2)^{1/2}]^{-1/2}, \tag{22}$$

$$N_a \text{ (or } N_a') = [1 - 2S\alpha(1 - \alpha^2)^{1/2}]^{-1/2}. \tag{23}$$

The one-electron transitions in which we are interested for the ligand field model are the six $e \longrightarrow t_2^a$ transitions. According to Eq. (10), the intensities of these transitions are completely determined by the transition probability matrix elements $(e \mid \mathbf{r} \mid t_2^a)$:

$$\begin{aligned}
[3d_{x^2-y^2} \mid \mathbf{r} \mid \psi_a(yz)] &= -\gamma \cdot \frac{4}{\sqrt{3}} A\mathbf{i}, \\
[3d_{x^2-y^2} \mid \mathbf{r} \mid \psi_a(xz)] &= +\gamma \cdot \frac{4}{\sqrt{3}} A\mathbf{j}, \\
[3d_{x^2-y^2} \mid \mathbf{r} \mid \psi_a(xy)] &= 0, \\
[3d_{z^2} \mid \mathbf{r} \mid \psi_a(yz)] &= +\gamma \cdot \tfrac{4}{3} A\mathbf{i}, \\
[3d_{z^2} \mid \mathbf{r} \mid \psi_a(xz)] &= +\gamma \cdot \tfrac{4}{3} A\mathbf{j}, \\
[3d_{z^2} \mid \mathbf{r} \mid \psi_a(xy)] &= -\gamma \cdot \tfrac{8}{3} A\mathbf{k},
\end{aligned} \tag{24}$$

where

$$\gamma = \left\{ \frac{1 - \alpha^2}{1 - 2S\alpha(1 - \alpha^2)^{1/2}} \right\}^{1/2}. \tag{25}$$

The ligand coupling integral A and the overlap integral S, appearing in Eqs. (24) and (25) are algebraically evaluated in the Appendix. It may be readily verified that the matrix elements given in Eq. (24) yield the required isotropy for the transition $e \longrightarrow t_2^a$. The one electron intensities $f[a(e) \longrightarrow b(t_2^b)]$ may be easily derived from Eqs. (10) and (24).

[9] Values of α and β near zero do not seem to occur frequently for σ-bonded transition metal complexes. Such systems should exhibit visible absorption bonds of enormous intensity, as their spectrum would be a pure charge-transfer spectrum.

2. Intensities in the $CuCl_4^=$ Visible Spectrum

The intensity of the ligand field band $^2T_2 \rightarrow {}^2E$, that is the intensity of the transition[10]

$$[A]\,[L]\,(a_1^{\,b})^2\,(t_2^{\,b})^6\,(e)^4\,(t_2^{\,a})^5 \rightarrow [A]\,[L]\,(a_1^{\,b})^2\,(t_2^{\,b})^6\,(e)^3\,(t_2^{\,a})^6,$$

is, according to the previously developed theory and formulas,

$$f(^2T_2 \rightarrow {}^2E) = \frac{1 - \alpha^2}{1 - 2S\alpha(1 - \alpha^2)^{1/2}} \cdot \frac{64}{9} \cdot 1.085 \cdot 10^{11} \cdot A^2 \nu(^2T_2 \rightarrow {}^2E, \text{cm}^{-1}). \quad (26)$$

Using the experimental value of the ligand-metal distance $|\mathbf{a}_i|$, of 2.22 A (*40*), and setting Z_d and Z_σ equal to 7.865 and 5.72, respectively, in close agreement with Slater's (*43*) rules, we find that the square of the ligand coupling integral, A^2, is equal to $2.1429 \cdot 10^{-18}$ and that the overlap integral, S, takes on the value $-6.71 \cdot 10^{-2}$. Since $\nu(^2T_2 \rightarrow {}^2E)$ is equal to 13,000 cm^{-1} and $f(^2T_2 \rightarrow {}^2E)$ equals $2 \cdot 10^{-3}$ (*40*), we obtain as the degree of ionicity of the complex, α, a value of $(9/10)^{1/2}$.[11] This is a rather interesting result as it shows that, as far as energy calculations are concerned, the crystal field approximation is quite valid.

3. Intensities in the $CoCl_4^=$ Visible Spectrum

The computation of the intensity of the $CoCl_4^=$ visible absorption bands, in the ligand field approximation, is quite analogous to that of the $CuCl_4^=$ bands. One finds that the intensity of the $^4A_2 \rightarrow {}^4T_1(P)$ transition is given by

$$f[^4A_2 \rightarrow {}^4T_1(P)] = \frac{256}{15} \frac{1 - \alpha^2}{1 - 2S\alpha(1 - \alpha^2)^{1/2}} \cdot A^2 \cdot 1.085 \cdot 10^{11} \cdot \nu(^4A_2 \rightarrow {}^4T_1(P), \text{cm}^{-1}). \quad (27)$$

Taking for Z_σ and Z_d the values 5.7143 and 6.4286, respectively, in close agreement with Slater's (*43*) rules, and using the experimental values of $|\mathbf{a}_i|$, 2.34 A (*42*), we obtain for A and S the values $6.154 \cdot 10^{-10}$ and $-9.47 \cdot 10^{-2}$, respectively. We then find that if α is equal to $(4/10)^{1/2}$, we obtain the experimental value of $f[^4A_2 \rightarrow {}^4T_1(P)]$, $6 \cdot 10^{-3}$ (*41*) [α equal to $\sqrt{1/2}$ yields for $f[^4A_2 \rightarrow {}^4T_1(P)]$, $4.8 \cdot 10^{-3}$].

It is interesting to compare the above value of α with that obtained by fitting the observed paramagnetic resonance g-factor, $g_\parallel = 2.32$ (*41*), to the theory of Owen (*29*). Using first-order perturbation theory in the usual manner (*11–13, 28, 29*), we find that the wave functions given in Eq. (20) imply

$$g_\parallel = 2 - N_a^{\,2}\alpha^2 \cdot \frac{8\lambda}{\nu_1\,(\text{cm}^{-1})}. \quad (28)$$

[10] We have here designated the filled ligand (core) orbitals as [L].

[11] If we had assumed "full covalency" (that is, $\alpha = \sqrt{1/2}$) we would have obtained an oscillator strength for the $^2T_2 \rightarrow {}^2E$ transition which was ten times too large ($f = 10^{-2}$).

If we take for ν_1 (cm^{-1}) the value, 3000 cm^{-1}, obtained from Eq. (15) and the experimental data (*41*), and for the spin-orbit coupling parameter, λ, -178 cm^{-1} (*29*), we have that α is approximately $(2/3)^{1/2}$. The agreement with our value for α, determined from intensity measurements, is quite satisfactory, when one considers the experimental uncertainties present in any intensity measurement and the crudeness of Eq. (28).

4. Intensities in the $CrO_4^{=}$ and MnO_4^{-} Visible Spectra

A semiempirical calculation of the energy spectrum of the chromate and permanganate ions has been carried out by Wolfsberg and Helmholtz (*26*). Their results exhibit some rather surprising features. They find a remarkable stabilization of the non-σ-bonding metal orbital of symmetry e due to π-bonding. Indeed, the amount of this stabilization is *greater* than that expected from the usual valence bond picture of "resonating" double bonded structures (*44*). Furthermore, they find a puzzling inversion of the bonding energy levels on going from the permanganate to the chromate electronic configurations, even though these configurations are isoelectronic. We believe that these results are due to the (necessary) over-simplification of their secular equations, and hence, we shall adopt an alternative energy level scheme for our intensity calculations.

The visible splectra of both the permanganate and chromate ions exhibit two very intense bands with oscillator strengths, f, of about $10^{-2} - 10^{-1}$. As we have seen previously, bands of such high intensity are incompatible with an ionic model. We shall now demonstrate that they are also inconsistent with a Van Vleck σ-bonded model.

The electronic configuration of the $^1A_1$ ground state of the permanganate and chromate ions, according to the Van Vleck σ-bonded model, is [A] [L] $(a_1^b)^2$ $(t_2^b)^6$. The first allowed absorption band[12] is, on the basis of this model, due to the excitation of a σ-bonding t_2^b electron to a non-σ-bonding e orbital:

$$[A]\,[L]\,(a_1^b)^2\,(t_2^b)^6 \rightarrow [A]\,[L]\,(a_1^b)^2\,(t_2^b)^5\,(e)^1.$$

If we now proceed in the usual fashion and compute the oscillator strength expected for such a transition, we obtain

$$({}^1A_1\,|\,\mathbf{r}_1\,|\,{}^1T_2) = \frac{8\sqrt{2}}{3}\,\frac{\alpha}{\sqrt{1 + 2S\alpha\sqrt{1-\alpha^2}}}\cdot B\cdot(\mathbf{i} + \mathbf{j} + \mathbf{k}) \qquad (29)$$

and

$$f({}^1A_1 \rightarrow {}^1T_2) = \frac{128}{3}\cdot\frac{\alpha^2}{1 + 2S\alpha\sqrt{1-\alpha^2}}\cdot B^2\cdot 1.085\cdot 10^{11}\cdot \nu_1({}^1A_1 \rightarrow {}^1T_2, \text{cm}^{-1}). \qquad (30)$$

[12] The only electric dipole allowed transitions for the permanganate and chromate ions are of the type $^1A_1 \rightarrow {}^1T_2$. Thus, for example, transitions of the type $a_1^b \rightarrow e$ are symmetry forbidden.

The ligand coupling integral B is as defined in the Appendix. Before Eq. (30) may be numerically evaluated, a suitable choice must be made for the metal and ligand effective nuclear charges, Z_d and Z_σ. Now as α must undoubtedly be $\sim\sqrt{1/2}$ for these complexes, the eight σ-bonding electrons, $(a_1^b)^2\,(t_2^b)^6$, are distributed equally between the metal and ligand systems. Hence, Z_d must have a value near 5, and Z_σ a value near 3 [recall that the σ molecular orbitals are composed of $2p_\sigma$ oxygen functions]. If we set Z_σ and Z_d equal to 3.30 and 4.96, respectively, and give to the ligand-metal distance, $|\mathbf{a}_i|$, of $CrO_4^=$, its experimental value of 1.60 A (*45*), we obtain an oscillator strength, f, for the 27,000 cm^{-1} (*46*) band, of $7.3 \cdot 10^{-3}$. This result is relatively insensitive to rather wide variations of the parameters Z_d and Z_σ. Since the experimentally observed intensity is $8.9 \cdot 10^{-2}$ (*26*), we see that the Van Vleck σ-bonded model yields an intensity of the first chromate band which is too small by a factor of ten.

Explicit calculations for the second allowed chromate transition reveal it to have an oscillator strength of zero. Since similar computations for the first and second allowed bands of the isoelectronic permanganate ion give results exactly analogous to those for the chromate ion, we see that *a consideration of the σ-bonded electrons alone can not account for the observed spectra of* $CrO_4^=$ *and* MnO_4^-. As will be shown below, a consideration of the ligand π-molecular orbitals contained in the ligand (core) configuration [L], gives an admirable account of the observed spectra.

The construction of the appropriate linear combinations of ligand π-orbitals

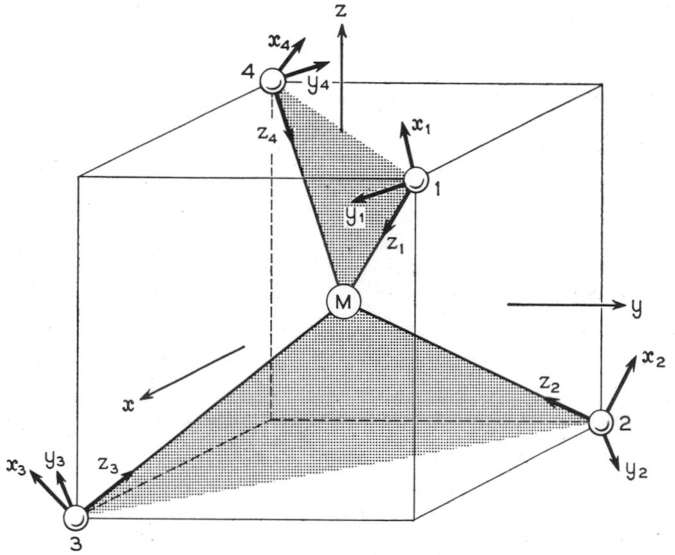

FIG. 5. Ligand geometry for π-bonding.

13

suitable for use in the description of inorganic complexes, follows from well-known techniques (*26–28*). If we choose our local coordinate system on each ligand as shown in Fig. 5, we obtain the following π-orbitals:

$$
\begin{aligned}
&\pi(e):\left\{\begin{array}{l}\tfrac{1}{2}(P_{x_1}-P_{x_2}-P_{x_3}+P_{x_4})\\ \tfrac{1}{2}(P_{y_1}-P_{y_2}-P_{y_3}+P_{y_4})\end{array}\right\}\text{(bonding)},\\
&\pi(t_2):\left\{\begin{array}{l}\tfrac{1}{2}(P_{x_1}+P_{x_2}+P_{x_3}+P_{x_4})\\ -\tfrac{1}{4}(P_{x_1}-P_{x_2}+P_{x_3}-P_{x_4})\\ \qquad -\sqrt{\tfrac{3}{4}}(P_{y_1}-P_{y_2}+P_{y_3}-P_{y_4})\\ -\tfrac{1}{4}(P_{x_1}+P_{x_2}-P_{x_3}-P_{x_4})\\ \qquad +\sqrt{\tfrac{3}{4}}(P_{y_1}+P_{y_2}-P_{y_3}-P_{y_4})\end{array}\right\}\text{(bonding)},\\
&\pi(t_1):\left\{\begin{array}{l}\pi_1=\tfrac{1}{2}(P_{y_1}+P_{y_2}+P_{y_3}+P_{y_4})\\ \pi_2=-\tfrac{1}{4}(P_{y_1}-P_{y_2}+P_{y_3}-P_{y_4})\\ \qquad +\sqrt{\tfrac{3}{4}}(P_{x_1}-P_{x_2}+P_{x_3}-P_{x_4})\\ \pi_3=-\tfrac{1}{4}(P_{y_1}+P_{y_2}-P_{y_3}-P_{y_4})\\ \qquad -\sqrt{\tfrac{3}{4}}(P_{x_1}+P_{x_2}-P_{x_3}-P_{x_4})\end{array}\right\}\text{(nonbonding)}.
\end{aligned}
\tag{31}
$$

We shall assume that these π-states fit into the energy level scheme of the complex ions $CrO_4^{=}$ and MnO_4^{-}, as indicated in Fig. 6. Note that for the per-

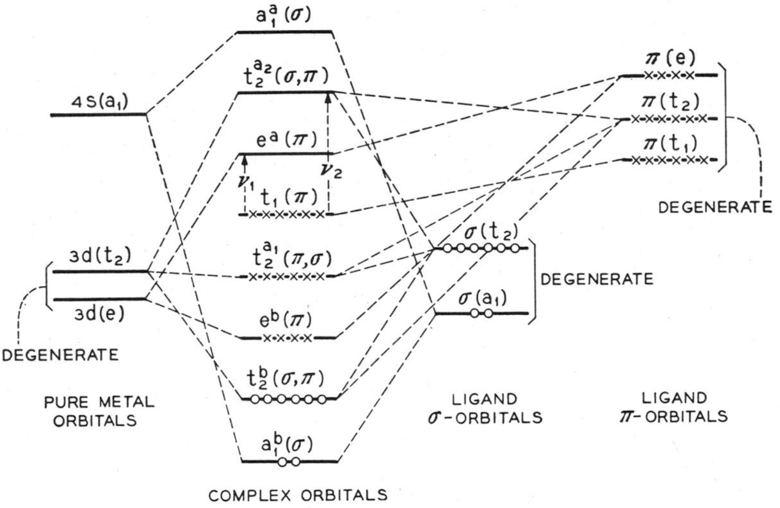

FIG. 6. Schematic energy level scheme for the complex ions $CrO_4^{=}$ and MnO_4^{-}, illustrating both σ- and π-bonding.

manganate ion the level scheme differs from that proposed by Wolfsberg and Helmholtz (*26*) only in the location of the π-antibonding levels of symmetry e. The ordering of energy levels given in Fig. 6 is more in line with the usual concepts of π-bonding than that of Wolfsberg and Helmholtz. Hence, the first strong band of the permanganate and chromate ions is assigned as $t_1(\pi) \rightarrow e^a(\pi)$.[13] The observed long wavelength satellites of this transition (*26*) are then due, at least in part, to the symmetry forbidden $^1A_1 \rightarrow {}^1T_1$ component of the transition $t_1(\pi) \rightarrow e^a(\pi)$.

Proceeding in customary manner, we obtain for the transition probability matrix elements of the $t_1(\pi) \rightarrow e^a(\pi)$ transition:

$$
\begin{aligned}
(\pi_1 \mid \mathbf{r} \mid z^2) &= 0, \\
(\pi_2 \mid \mathbf{r} \mid z^2) &= +\sqrt{3/2}\, C\mathbf{j}, \\
(\pi_3 \mid \mathbf{r} \mid z^2) &= -\sqrt{3/2}\, C\mathbf{i}, \\
(\pi_1 \mid \mathbf{r} \mid x^2 - y^2) &= +C\mathbf{k}, \\
(\pi_2 \mid \mathbf{r} \mid x^2 - y^2) &= -\tfrac{1}{2}\, C\mathbf{j}, \\
(\pi_3 \mid \mathbf{r} \mid x^2 - y^2) &= -\tfrac{1}{2}\, C\mathbf{i},
\end{aligned}
\tag{32}
$$

where C is the ligand coupling constant defined in the Appendix. We thus have that

$$
f(^1A_1 \rightarrow {}^1T_2) = 1.085 \cdot 10^{11} \cdot C^2 \cdot \nu_1(\mathrm{cm}^{-1}). \tag{33}
$$

Because of the π-bonding it seems reasonable to assume values of $Z_d \sim 3$ and $Z_\sigma \sim 2$. Thus interpreting the "first band", ν_1, at 27,000 cm^{-1} of $CrO_4^{=}$ (*46*) to be due to this transition, and setting Z_d and Z_σ equal to 3.22 and 2.15, respectively, we obtain for $f(^1A_1 \rightarrow {}^1T_2)$ a value of $6.7 \cdot 10^{-2}$, in good agreement with experiment [$f = 8.9 \cdot 10^{-2}$ (*26*)]. In an analogous manner, we find for the MnO_4^- band at 18,500 cm^{-1} (*26*) an f-value of $3.9 \cdot 10^{-2}$, in excellent agreement with the experimental value, $3.2 \cdot 10^{-2}$ (*26*). These results are not too sensitive to variations of the effective charges Z_d and Z_σ.

Now if we interpret the second band of permanganate and chromate as due

[13] A reversal of the $t_1(\pi)$ and $t_2^{a_1}(\pi, \sigma)$ levels cannot be distinguished on the grounds of intensity considerations alone. We have here followed Wolfsberg and Helmholtz and placed the $t_1(\pi)$ level above the $t_2^{a_1}(\pi, \sigma)$ level. Physical and chemical intuition supports this level arrangement.

to the transition $t_1(\pi) \rightarrow t_2^{a2}(\sigma, \pi)$, we find[14]

$$f_2(^1A_1 \rightarrow {}^1T_2) = 2\alpha^2C^2 \cdot 1.085 \cdot 10^{11} \cdot \nu_2(^1A_1 \rightarrow {}^1T_2, \text{cm}^{-1}). \tag{34}$$

Assuming "full σ-bond covalency" for the $t_2^{a2}(\sigma,\pi)$ levels that, is, setting α equal to $\sqrt{1/2}$, we obtain the following results:

$CrO_4^{=}$: $\nu_2 = 37{,}000$ cm^{-1}, $f_{exp.} = 0.11$ (Ref. *46*),

$f_{calc.} = 0.092$;

MnO_4^{-}: $\nu_2 = 32{,}000$ cm^{-1}, $f_{exp.} = 0.07$ (Ref. *46*),

$f_{calc.} = 0.073$.

The agreement with experiment is seen to be most gratifying.

CONCLUSION

From the comparison of the Bethe-Van Vleck ionic model (the crystal field model) with the Van Vleck molecular orbital model (ligand field model) here carried out, we have seen that the latter mode of description of tetrahedral inorganic complexes is by far the superior one, at least in so far as intensity calculations are concerned. Indeed, the calculation of the intensity of the permanganate and chromate absorption bands appears to indicate that inorganic complexes exhibiting primarily charge-transfer spectra might best be considered not only as having σ-molecular orbitals, but also as possessing π-molecular orbitals. It would be quite interesting to apply the methods outlined in this paper to paramagnetic π-complexes such as $MnO_4^{=}$ (one unpaired electron) and $FeO_4^{=}$ (two unpaired electrons). However, the dearth of appropriate experimental data pertaining to the manganate and ferrate ions makes such an application, at present, rather fruitless.[15]

APPENDIX

The various two-center integrals occurring in this paper were evaluated by the usual techniques (*47*). Using the auxilliary integrals

$$A_n(\alpha) = \int_1^{\infty} t^n e^{-\alpha t}\, dt, \tag{A-1}$$

$$B_n(\beta) = \int_{-1}^{1} t^n e^{-\beta t}\, dt, \tag{A-2}$$

[14] We have here neglected transition dipole integrals over two *different ligands*.

[15] The crystal spectra of $MnO_4^{=}$ and of the isoelectronic series $VO_4^{\equiv}$, $CrO_4^{=}$, MnO_4^{-} may be found in (*46*).

we find the ligand coupling integral A to be

$$A = \frac{Z_\sigma^{5/2}\bar{R}^7 a_0}{3\sqrt{2}\cdot 2^6\cdot Z_d^{7/2}} \left\{ \begin{aligned} &6\left[\begin{aligned} &A_0(B_2 - B_4) - A_2(B_0 + 4B_4 - B_6) - 4A_3B_5 \\ &+ A_4(B_0 + 4B_2 - B_6) + 4A_5B_3 - A_6(B_2 - B_4) \end{aligned}\right] \\ &-\tfrac{3}{2}\epsilon\bar{R}\left[\begin{aligned} &A_1(B_2 - B_4) - A_3(B_0 + 4B_4 - B_6) - 4A_4B_5 \\ &+ A_5(B_0 + 4B_2 - B_6) + 4A_6B_3 - A_7(B_2 - B_4) \\ &- A_0(B_3 - B_5) + A_2(B_1 + 4B_5 - B_7) + 4A_3B_6 \\ &- A_4(B_1 + 4B_3 - B_7) - 4A_5B_4 + A_6(B_3 - B_5) \end{aligned}\right] \end{aligned} \right\} \tag{A-3}$$

In Eq. (3) Z_σ is the effective charge for a $3p\sigma$ orbital, Z_d the effective charge for a $3d$ orbital, ϵ is the ratio Z_σ/Z_d, and $\bar{R} = (Z_d/3a_0)\cdot a$, where a is the ligand metal separation and a_0 is the Bohr radius. The argument, α, of the $A_n(\alpha)$ integrals appearing in Eq. (3) is $\alpha = \frac{1}{2}\,\bar{R}(1 + \epsilon)$; the argument β, of the $B_n(\beta)$ integrals is $\beta = \frac{1}{2}\,\bar{R}(1 - \epsilon)$. The analogous expressions for the overlap integral S, and the ligand coupling integrals B and C are

$$S = \left(\frac{Z_\sigma}{Z_d}\right)^{5/2}\cdot\frac{\bar{R}^6}{648\cdot\sqrt{2}} \left\{ \begin{aligned} &6\left[\begin{aligned} &A_0(B_4 - 3B_2) - A_1(B_3 + B_5) + A_2(3B_0 + B_4) \\ &+ A_3(B_1 + 3B_5) - A_4(B_0 + B_2) + A_5(B_1 - 3B_3) \end{aligned}\right] \\ &-\tfrac{3}{2}\epsilon\bar{R}\left[\begin{aligned} &A_0(3B_3 - B_5) + A_1(2B_4 - 3B_2 + B_6) \\ &- A_2(3B_1 + B_3 + 2B_5) \\ &+ A_3(3B_0 - B_2 + B_4 - 3B_6) \\ &+ A_4(2B_1 + B_3 + 3B_5) \\ &- A_5(B_0 + 2B_2 - 3B_4) + A_6(B_1 - 3B_3) \end{aligned}\right] \end{aligned} \right\}, \tag{A-4}$$

$$B = \frac{27}{2048}\cdot\frac{Z_\sigma^{5/2}\bar{R}^7 a_0}{Z_d^{7/2}}\left[(A_4 - A_2)(B_0 - B_6) - (A_6 - A_0)(B_2 - B_4)\right], \tag{A-5}$$

$$C = -\frac{5\sqrt{2}}{2}\,C_1 + \frac{4\sqrt{2}}{3}\,C_2, \tag{A-6}$$

where,

$$C_1 = \frac{27}{1024} \cdot \frac{Z_\pi^{5/2} \bar{R}^7 a_0}{Z_d^{7/2}} \left[\begin{aligned} &A_0(-B_2 + 2B_4 - B_6) + A_2(B_0 - 3B_4 + 2B_6) \\ &+ A_4(-2B_0 + 3B_2 - B_6) + A_6(B_0 - 2B_2 + B_4) \end{aligned} \right], \tag{A-7}$$

$$C_2 = \frac{27}{1024} \cdot \frac{Z_\pi^{5/2} \bar{R}^7 a_0}{Z_d^{7/2}}$$

$$\left[\begin{aligned} &A_0(B_4 - B_6) + A_1(2B_3 - 2B_5) + A_2(-B_4 + B_6) + A_3(-2B_1 + 2B_5) \\ &+ A_4(-B_0 + B_2) + A_5(2B_1 - 2B_3) + A_6(B_0 - B_2) \end{aligned} \right], \tag{A-8}$$

The definitions of the quantities appearing in Eqs. (5)–(8) are, with the exception of ϵ, the same as for Eqs. (3) and (4). The parameter ϵ which occurs in the arguments of the integrals $A_n(\alpha)$ and $B_n(\beta)$ is equal to $3Z_\sigma/2Z_d$ in Eq. (5) and equal to $3Z_\pi/2Z_d$ for Eqs. (6)–(8). Here Z_σ is the effective charge for a $2p_\sigma$ orbital and Z_π the effective charge for a $2p_\pi$ orbital.

The integrals A, B, C, and S have been computed by the use of hydrogen-like wave functions.

RECEIVED: March 28, 1958

REFERENCES

1. J. H. VAN VLECK, *J. Chem. Phys.* **3**, 803 (1935).
2. J. H. VAN VLECK, *J. Chem. Phys.* **3**, 807 (1935).
3. J. H. VAN VLECK AND A. SHERMAN, *Revs. Modern Phys.* **7**, 167 (1935).
4. H. BETHE, *Ann Physik* [5], **3**, 133 (1929).
5. J. H. VAN VLECK, "The Theory of Electric and Magnetic Susceptibilities." Oxford Univ. Press, London and New York, 1932.
6. R. S. MULLIKEN, *Phys. Rev.* **40**, 55 (1932).
7. W. E. MOFFITT AND C. J. BALLHAUSEN, *Ann. Rev. Phys. Chem.* **7**, 107 (1956).
8. R. SCHLAPP AND W. G. PENNEY, *Phys. Rev.* **42**, 666 (1932).
9. D. POLDER, *Physica* **9**, 709 (1942).
10. M. KOTANI, *J. Phys. Soc. (Japan)* **4**, 293 (1949).
11. A. ABRAGAM AND M. H. L. PRYCE, *Proc. Roy. Soc.* **A205**, 135 (1951).
12. B. BLEANEY AND K. W. H. STEVENS, *Repts. Progr. Phys.* **16**, 108 (1953).
13. K. D. BOWERS AND J. OWEN, *Repts. Progr. Phys.* **18**, 304 (1955).
14. J. H. VAN VLECK, *J. Chem. Phys.* **8**, 787 (1940).
15. R. FINKELSTEIN AND J. H. VAN VLECK, *J. Chem. Phys.* **8**, 790 (1940).
16. F. E. ILSE AND H. HARTMANN, *Z. Phys. Chem.* **197**, 239 (1951).
17. F. E. ILSE AND H. HARTMANN, *Z. Naturforsch.* **6a**, 751 (1951).
18. L. E. ORGEL, *J. Chem. Soc.* p. 4756 (1952).
19. L. E. ORGEL, *J. Chem. Phys.* **23**, 1004 (1955).
20. J. BJERRUM, C. J. BALLHAUSEN, AND C. K. JØRGENSEN, *Acta Chem. Scand.* **8**, 1275 (1954).
21. W. H. KLEINER, *J. Chem. Phys.* **20**, 1784 (1952).
22. Y. KURODA AND K. ITO, *J. Chem. Soc. (Japan)* **76**, 766 (1955).

23. Y. TANABE AND S. SUGANO, *J. Phys. Soc. (Japan)* **11**, 864 (1956).
24. H. YAMATERA, *J. Inst. Polytech. Osaka City Univ.* **5** (Series C), 163 (1956).
25. K. NAKAMOTO, J. FUJITA, M. KOBAYASHI, AND R. TSUCHIDA, *J. Chem. Phys.* **27**, 439 (1957).
26. M. WOLFSBERG AND L. HELMHOLTZ, *J. Chem. Phys.* **20**, 837 (1952).
27. L. HELMHOLTZ, H. BRENNAN, AND M. WOLFSBERG, *J. Chem. Phys.* **23**, 853 (1955).
28. K. W. H. STEVNES, *Proc. Roy. Soc.* **A219**, 542 (1953).
29. J. OWEN, *Proc. Roy. Soc.* **A227**, 183 (1955).
30. J. H. VAN VLECK, *J. Phys. Chem.* **41**, 67 (1937).
31. A. D. LIEHR AND C. J. BALLHAUSEN, *Phys. Rev.* **106**, 1161 (1957).
32. C. J. BALLHAUSEN AND A. D. LIEHR (to be published).
33. E. U. CONDON AND G. SHORTLEY, "Theory of Atomic Spectra." Cambridge Univ. Press, London and New York, 1935.
34. R. S. MULLIKEN, *Phys. Rev.* **43**, 279 (1933).
35. C. J. GORTER, *Phys. Rev.* **42**, 437 (1932).
36. C. J. BALLHAUSEN, *Kgl. Danske Videnskab, Selskab. Mat.-fys. Medd.* **29**, No. 4 (1954).
37. J. H. VAN VLECK, *Phys. Rev.* **41**, 208 (1932).
38. C. J. BALLHAUSEN AND E. M. ANCMON, "Tables of Ligand Field Integrals" (to be published).
39. R. S. MULLIKEN AND C. A. RIEKE, *Repts. Progr. Phys.* **8**, 231 (1941).
40. L. HELMHOLTZ AND R. F. KRUH, *J. Am. Chem. Soc.* **74**, 1176 (1952)
41. C. J. BALLHAUSEN AND C. K. JØRGENSEN, *Acta Chem. Scand.* **9**, 397 (1955).
42. H. M. POWELL AND A. F. WELLS, *J. Chem. Soc.* p. 359 (1935).
43. J. C. SLATER, *Phys. Rev.* **36**, 57 (1930).
44. L. PAULING, "The Nature of the Chemical Bond," Cornell Univ. Press, Ithaca, 1940.
45. H. W. SMITH, JR. AND M. Y. COLBY, *Z. Krist.* **103**, 90 (1940).
46. LANDOLT-BÖRNSTEIN, "Atom und Molekularphysik," Vol. 1, Parts 3 and 4. Springer, Berlin, 1951 and 1955.
47. M. KOTANI, A. AMEMIYA, E. ISHIGURO, AND T. KIMURA, "Table of Molecular Integrals." Maruzen, Tokyo, 1955.

19

Errata

C. J. Ballhausen and Andrew D. Liehr, Intensities in Inorganic Complexes. Part II. Tetrahedral Complexes. *J. Mol. Spectroscopy* **2,** 342 (1958).

(*1*) Page 344, Eq. (5): denominator should be $r_{>}^{l+1}$
(*2*) Page 348, Eq. (15): ${}^4A_2 \rightarrow {}^4T_1(P)$ should read $\nu_3 = 14000 + 16/3\ Dq$
(*3*) Page 352, paragraph three, the second line: the $g_{\parallel} = 2.32$ reference should read (*13*) and not (*41*).
(*4*) Page 358: the integral for A (Eq. A-3) should read:

$$\frac{Z_\sigma^{5/2}\bar{R}^7 a_0}{6\cdot 2^6\cdot Z_d^{7/2}}[6\{(A_0 - A_6)(B_2 - B_4) + (A_4 - A_2)(B_0 - B_6)\}$$

$$- 3/2\epsilon\bar{R}\{(A_1 - A_7)(B_2 - B_4) + (A_6 - A_0)(B_3 - B_5)$$

$$+ (A_5 - A_3)(B_0 - B_6) + (A_4 - A_2)(B_7 - B_1)\}$$

The error was due to a sign mistake, and to the wrong normalization constant given for $3p_z$, $3p_y$, and $3p_x$ in H. Eyring, J. Walter, and G. E. Kimball, "Quantum Chemistry," p. 90 (Wiley, New York, 1944).

(*5*) For the same reason the value of S given in (A-4) on p. 358 should be multiplied with $3/\sqrt{2}$.

(*6*) The values in §2 and §3 on p. 352 are then altered to:

$$A(\mathrm{CuCl_4}^{=}) = -4.01\cdot 10^{-10}$$

$$S(\mathrm{CuCl_4}^{=}) = -0.14$$

and

$$A(\mathrm{CoCl_4}^{=}) = -5.19\cdot 10^{-10}$$

$$S(\mathrm{CoCl_4}^{=}) = -0.20$$

Calling

$$\frac{1 - \alpha^2}{1 - 2S\alpha(1 - \alpha^2)^{1/2}} = k$$

we have:

$$f({}^2T_2 \rightarrow {}^2E)_{\mathrm{CuCl_4}^{=}} = k\cdot 1.61\cdot 10^{-3}$$

A comparison with the measured value of f shows that $k = 1.24$. As we can not for any value of α get a k equal to 1.24 we must conclude that either the

σ-bonding alone is insufficient to explain the intensity found or that it is not permissible to treat the $CuCl_4^{=}$ as a regular tetrahedron. (If α is taken to be $\sqrt{1/2}$, the computed intensity is 3 times too small — which is not too bad considering the approximations made.) [*Note:* Contrary to popular belief, Jahn-Teller distortions cannot occur in tetrahedral $CuCl_4^{=}$ (Liehr, *J. Phys. Chem.*, Jan. 1959), but ordinary electrostatic instabilities are still operative.]

For $CoCl_4^{=}$ we obtain

$$f[{}^4A_2 \rightarrow {}^4T_1(P)] = k \cdot 7.48 \cdot 10^{-3}$$

A comparison with the measured value gives $k = 0.8$ indicating $\alpha \sim \sqrt{4/10}$ as before.

(*7*) Page 358: a misprint occurs in (A-5) where the factor should read

$$\frac{27}{1024} \text{ instead of } \frac{27}{2048}.$$

(*8*) Page 358: a misprint occurs in (A-6) which should read

$$C = -\frac{5\sqrt{2}}{3}C_1 + \frac{4\sqrt{2}}{3}C_2$$

(*9*) Pages 356 and 357: a factor of 6 is missing both in Eqs. (33) and (34). This means that the calculated intensities are too small by a factor of 6. Consequently our results are altered in a way such that a larger amount of π-bonding is necessary to explain the intensity observed for the complexes $CrO_4^{=}$ and MnO_4^{-}.

(*10*) Page 360: in reference *28* read Stevens.

(*11*) A calculation of the $CuCl_4^{=}$ energy spectrum has been given by G. Felsenfeld, *Proc. Roy. Soc.* **A236,** 506 (1956).

(*12*) Page 356, paragraph two: read Z_π instead of Z_σ.

(*13*) Page 356, Eq. (32): the r.h.s. of the second and third lines should read

$$\frac{\sqrt{3}}{2}, \textit{ not } \sqrt{3/2}.$$

Reprinted from *Ann. Phys.*, **6**, 134–155 (1959)

25

Complete Theory of Ni(II) and V(III) in Cubic Crystalline Fields

ANDREW D. LIEHR AND C. J. BALLHAUSEN*

Bell Telephone Laboratories, Incorporated, Murray Hill, New Jersey

The generalized crystal field theory of Finkelstein and Van Vleck is augmented by the introduction of spin-orbit coupling and an attempt is made to thus account for both the spin-allowed and the spin-forbidden d^n, (n = 2,8), spectral transitions of Ni(II) and V(III) complexes. The secular equations are derived by the use of functions which span, in both the weak and strong crystalline field pictures, the proper spin-coordinate cubic symmetry classification of Bethe. These secular equations were solved numerically on the IBM 704 data processing machine at the Murray Hill Bell Telephone Laboratories and the results are illustrated by suitable graphs. The existing spectral data on tetrahedral and octahedral complexes of Ni(II) and V(III) have been interpreted theoretically and it is found that the calculations do not support Jørgensen's assignment of the double-peaked red band of hexaquo nickel systems. Spin-allowed infrared transitions are predicted at ~350, 1000, 1100, and 5300 cm^{-1} for tetrahedral Ni(II) complexes and at ~1.6, 168, and 250 cm^{-1} for octahedral V(III) complexes. Low's spectral data on divalent nickel dissolved in magnesium oxide and trivalent vanadium dissolved in aluminum oxide are given a rigorous theoretical interpretation. Some comments on the applicability of the derived transformation and energy matrices to the related magnetic problems (paramagnetic resonance and magnetic susceptibilities) and spectral problems (energy level splittings due to fields of lower symmetry) are given in the appendices. In particular, it is shown that the ground electronic state splitting of 0.49 cm^{-1} found by Holden *et al.* in nickel fluosilicate hexahydrate implies a 50-cm^{-1} trigonal field splitting of certain of the excited electronic states.

1. INTRODUCTION

In recent years a renewed interest has been taken in the Bethe-Kramers-Van Vleck (*1–3*) theory of crystalline fields. However, despite the vigor with which this theory has been applied to explain the magnetic and optical properties of transition metal, rare-earth, and actinide ion complexes,[1] relatively little effort has been expended to push the general theory to its logical conclusion—an "all inclusive"[2] calculation of the energy spectrum of such complex ions. The initial applications of the theory were concerned solely with the breakdown of *L-S* coupling in the lowest electronic states of the ions concerned and its consequent

* Visiting scholar, 1957–1958. Now at the Technical University of Denmark, Chemical Laboratory A, Copenhagen, Denmark.

[1] Recent reviews of these applications may be found in Refs. *4–7*.

[2] By "all inclusive" we mean a complete energy value computation within the framework set up by the fundamental assumptions of the crystalline theory.

magnetic implications (*8–12*). For systems containing either one electron or one hole in the *d* or *f* shell, this approximation entailed no limitation in accuracy, other than that imposed by the use of perturbation theory (*12*). The first attempt, by Finkelstein and Van Vleck (*13*), to extend the theory to include the effects of higher atomic configurations, and hence to predict the optical spectra of such ions, was of modest intent: spin-orbit coupling was neglected, cubic symmetry assumed, and the coulomb interaction parameters were kept fixed. Subsequent calculations dealing with the colors of inorganic complexes have, for the most part, retained these approximations, but have embellished the techniques somewhat by the inclusion of fields of lower symmetry (*14–35*). Recently some consideration has been given to the combined effects of spin-orbit coupling and the crystalline field upon optical spectra (*36*, *37*).

Although spin-orbit coupling plays a dominant role in the theory of f^n configurations (*9*, *11*), its role in the analogous d^n systems has heretofore been viewed only in a perturbative manner (*10*, *12*, *16*, *17*, *37–40*). It has been pointed out by Jørgensen (*41*) that for hexaquo nickel complexes this latter viewpoint is inadequate for the rationalization of the observed optical spectra. To clarify this situation,[3] we have undertaken a complete calculation of the configuration interaction effects for a d^8 electronic system immersed in an octahedral ligand field. The mathematical "equivalence" of octahedral and tetrahedral fields (*42*, *20*) coupled with that of d^2 and d^8 electronic systems (*8*, *11*) was utilized to solve "exactly" the related problems of the energy levels of tetrahedral Ni(II) complexes, and octahedral and tetrahedral V(III) complexes. The excellent agreement of our theoretical energy level curves with those observed for Ni(II) and V(III) in cubic fields of various strengths demonstrates the power of the crystalline field approach to problems in the physics and chemistry of inorganic complex ions. For example, these calculations definitely eliminate Jørgensen's (*41*) assignment of the 15,400-cm^{-1} band of the hexaquo nickel ion.

2. THEORY

The problem which we have set for ourselves is the solution of the differential equation

$$\mathcal{H}\Psi = E\Psi, \tag{1}$$

where $\mathcal{H}$ is given by

$$\mathcal{H} = \sum_i \left\{ -\frac{\hbar^2}{2m}\nabla_i^2 - \frac{Ze^2}{|\mathbf{r}_i|} + \xi(\mathbf{r}_i)\mathbf{l}_i\cdot\mathbf{s}_i \right\} + \sum_{j>i} \frac{e^2}{|\mathbf{r}_i - \mathbf{r}_j|} + V_{\text{C.F.}}. \tag{2}$$

The first four terms in the Hamiltonian, $\mathcal{H}$, yield the usual atomic Hamiltonian operator; the last term, $V_{\text{C.F.}}$, expresses the effects of the superimposed nonspherical (cubic) potential field of the ligands surrounding the atom in question. It has been recognized for quite sometime now that an exact solution is, for all

[3] The requirement of strong spin-orbit resonance effects to explain the spectra of hexaquo nickel complexes has placed the water molecule in an anomalous spectrochemical position.

practical purposes, unobtainable and that some scheme of approximation must be adopted within whose framework we may speak of an "exact" solution. The conventional approximation procedure is that of Slater *et al.* (*43*). It is this scheme which we shall utilize in the solution of Eq. (2). We shall carry out this solution by use of two approaches: (a) the use of the weak crystalline field functions as a basis, and (b) the use of the strong crystalline field functions as a basis. As these two sets of basis functions are connected by a unitary transformation, they yield identical results if each problem is solved exactly. However, it is informative to carry through the computation of the energy matrix elements in both schemes as such a procedure not only gives an additional algebraic check, but also illustrates the limitations of the usual weak and strong crystalline field approximations.

(a) Use of Weak Field Basis Functions

There are many advantages to using as basis functions for the solution of Eq. (2) atomic functions in the $\{S, L, J, M\}$ representation. In this representation one may make full use of Condon and Shortley's (*43*) tables and thus circumvent the repetition of the solution of the atomic problem. Also as Bethe (*1*) and Finkelstein and Van Vleck (*13*) have tabulated the unitary transformations which connect the spherical $\{J, M\}$ representation and the cubic $\{\Gamma_j\}$ representation, the matrix elements which need to be explicitly computed in the weak field basis are those involving the crystalline potential, $V_{\text{C.F.}}$. Here again this computation is simplified by use of Penney and Schlapp's (*9, 10*) tabulation of crystalline field matrix elements.

There does, however, arise a very tricky point in the utilization of these sundry compilations of matrix elements: the correct and consistent choice of phase. It is quite important to keep a sharp eye peeled for the appearance of the omnipresent and elusive factor of minus one. This factor will appear if one: (i) does not use Condon and Shortley's (*43*) Section 14^3 to obtain the eigenfunctions, $|S, L; M_S, M_L\rangle$, of the total orbital angular momentum, L; (ii) does not use the Condon and Shortley (*43*) choice of phase for the hydrogenic one electron functions; (iii) does not use the standard Condon and Shortley (*43*) ordering of one electron functions in the Slater determinants; and (iv) does not add **S** to **L** in the standard order, $\mathbf{S} + \mathbf{L}$, during the formation of the J eigenfunctions via Section 14^3 of Condon and Shortley (*43*). If one carefully avoids these pitfalls, one may use Condon and Shortley's (*43*) Table I^{11} for the spin-orbit coupling matrix elements.

Without loss of generality, we may take our cubic crystalline field potential to be (*13, 44*)

$$V_{\text{C.F.}} = \sum_i V_i, \quad \text{with} \quad V_i = D[x_i^4 + y_i^4 + z_i^4 - \tfrac{3}{5}r_i^4] + f(r_i), \tag{3}$$

where the sum is taken only over the $3d$ electrons. With this form for the crystalline field the required solutions of Eq. (2) are given by the roots of the secular determinants as shown on pages 138 and 139. In Eq. (4) λ is equal to one half the

Condon and Shortley (*43*) spin-orbit coupling constant, ζ_{3d}, i symbolizes $\sqrt{-1}$, Dq is the standard cubic crystalline field constant, and the F_k, ($k = 2, 4$), are the famous Slater-Condon-Shortley parameters (*43*). The solutions of Eq. (4) yield the eigenvalues and eigenfunctions for a d^2 electronic system immersed in an octahedral field. The corresponding solutions for a d^8 system are readily obtained by a simultaneous change in the sign of λ (*43*) and Dq (*8, 11*) in Eq. (4). To obtain the correct secular equations for a d^2 electronic system immersed in a tetrahedral field, one need only replace Dq by $-(4/9)Dq$ (*20, 42*) in Eq. (4);[4] and to obtain the analogous solutions for a d^8 system one must simultaneously change the sign of λ and replace Dq by $(4/9)Dq$.

(b) USE OF STRONG FIELD FUNCTIONS

The use of the strong field representation does not offer as many computational advantages as the weak field picture, but it does lend itself more readily to various approximation schemes. The correct one electron strong field functions are obtained by solving Eq. (2) minus the spin-orbit and electron correlation terms. This leads to orbitals characterized by the cubic "quantum numbers",[5] t_2 and e, rather than the spherical quantum numbers, m_l. If, as shall be done in this paper, the functions are taken to be hydrogenic, the triply degenerate set of orbitals, t_2, are given by the functions d_{xy}, d_{xz}, and d_{yz}; and the doubly degenerate set, e, are given by $d_{x^2-y^2}$ and d_{z^2}. We shall henceforth confine our explicit discussion to d^2 electronic systems in octahedral fields, as the appropriate secular equations for fields of tetrahedral symmetry are readily obtained from these by considerations similar to those enumerated in Section 2(a). The secular equations required for the conjugate d^8 problem also follow as in Section 2(a).

To simplify the computation of the strong field energy matrix, it is convenient to construct linear combinations of Slater functions which exhibit cubic symmetry in *both* spin and coordinate space. In this representation both the spin-orbit interaction and the electron correlation terms are diagonal in the cubic $\{\Gamma_j\}$ quantum number.[6] Although the construction of such linear combinations is quite straightforward, caution must be exercised in the use of spin transformation matrices as tabulated in various texts. For example, if one employs the spinor matrices of Goldstein (*45*), one must be sure to also use transformation matrices which keep points in space fixed, but redefine the coordinate system being utilized. This procedure yields the solutions of Eq. (2) as the roots of the secular determinants as shown on pages 140 and 141.

[4] The proof of this relation is quite simple. All one need note is that the values of the sum of the spherical harmonics $Y_{l,m}(\theta_j, \varphi_j)$, which occur in the potential $V_{C.F.}$, over the j ligands arranged tetrahedrally yields a numerical value which is exactly $-4/9$ that of the same sum for an octahedral ligand distribution.

[5] For fields of octahedral symmetry a subscript g is usually added to indicate evenness under inversion in the center of symmetry.

[6] This Γ_j is the joint spin-coordinate symmetry classification of Bethe (*1*).

$$\begin{vmatrix} {}^1S_0 & {}^1G_4 & {}^3P_0 & {}^3F_4 \\ 14F_2 + 126F_4 - E & 4\sqrt{6}Dq & -2\sqrt{6}\lambda & 0 \\ & 4F_2 + F_4 + 4Dq - E & 0 & 2\lambda \\ & & 7F_2 - 84F_4 - 2\lambda - E & -4Dq \\ (\Gamma_1) & & & -8F_2 - 9F_4 + 3\lambda - 6Dq - E \end{vmatrix} = 0,$$

$$\begin{vmatrix} {}^3F_3 \\ -\lambda - 8F_2 - 9F_4 + 2Dq - E \\ (\Gamma_2) \end{vmatrix} = 0,$$

$$\begin{vmatrix} {}^1G_4 & {}^1D_2 & {}^3P_2 & {}^3F_2 & {}^3F_4 \\ 4F_2 + F_4 + \frac{4}{7}Dq - E & -\frac{40\sqrt{3}}{7}Dq & 0 & 0 & 2\lambda \\ & -3F_2 + 36F_4 + 24/7\,Dq - E & \sqrt{42/5}\,\lambda & -4\sqrt{3/5}\,\lambda & 0 \\ & & 7F_2 - 84F_4 + \lambda - E & -6\sqrt{2/7}\,Dq & +2\sqrt{10/7}\,Dq \\ & & & -8F_2 - 9F_4 - 4\lambda - \frac{22}{7}Dq - E & \frac{12\sqrt{5}}{7}Dq \\ (\Gamma_3) & & & & \\ & & & & -8F_2 - 9F_4 + 3\lambda - 6/7\,Dq - E \end{vmatrix} = 0,$$

$$
\begin{vmatrix}
{}^1G_4 & {}^3P_1 & {}^3F_3 & {}^3F_4 \\
4F_2 + F_4 + 2Dq - E & 0 & 0 & 2\lambda \\
 & 7F_2 - 84F_4 - \lambda - E & -\sqrt{6}\,Dq & +i\sqrt{10}\,Dq \\
 & & -8F_2 - 9F_4 - \lambda - Dq - E & +i\sqrt{15}\,Dq \\
(\Gamma_4) & & & -8F_2 - 9F_4 + 3\lambda - 3Dq - E
\end{vmatrix} = 0,
$$

$$
\begin{vmatrix}
{}^1G_4 & {}^1D_2 & {}^3P_2 & {}^3F_3 & {}^3F_4 & {}^3F_2 \\
4F_2 + F_4 - \frac{26}{7}Dq - E & \frac{20\sqrt{3}}{7}Dq & 0 & 0 & 2\lambda & 0 \\
 & -3F_2 + 36F_4 - \frac{16}{7}Dq - E & \sqrt{42/5}\,\lambda & 0 & 0 & -4\sqrt{3/5}\,\lambda \\
 & & 7F_2 - 84F_4 + \lambda - E & -i\sqrt{10}\,Dq & -\sqrt{70/7}\,Dq & +4/7\sqrt{14}\,Dq \\
 & & & -8F_2 - 9F_4 - \lambda + \frac{1}{3}Dq - E & \frac{i15\sqrt{7}}{7}Dq & -\frac{i22\sqrt{35}}{21}Dq \\
 & & & & -8F_2 - 9F_4 + 3\lambda + \frac{39}{7}Dq - E & -\frac{6\sqrt{5}}{7}Dq \\
(\Gamma_5) & & & & & -8F_2 - 9F_4 - 4\lambda + \frac{44}{21}Dq - E
\end{vmatrix} = 0. \quad (4)
$$

$$
\begin{vmatrix}
{}^1A_{1g}(e_g^2) & {}^3T_{1g}(t_{2g}e_g) & {}^3T_{1g}(t_{2g}^2) & {}^1A_{1g}(t_{2g}^2) \\
8F_2 + 51F_4 + 12Dq - E & -2\sqrt{3}\lambda & 0 & \sqrt{6}(2F_2 + 25F_4) \\
 & 4F_2 - 69F_4 + 2Dq - \lambda - E & -6F_2 + 30F_4 + 2\lambda & -2\sqrt{2}\lambda \\
 & & -5F_2 - 24F_4 - 8Dq + 2\lambda - E & +2\sqrt{2}\lambda \\
(\Gamma_1) & & & 10F_2 + 76F_4 - 8Dq - E
\end{vmatrix} = 0.
$$

$$
\begin{vmatrix}
{}^3T_{2g}(t_{2g}e_g) \\
-8F_2 - 9F_4 + 2Dq - \lambda - E \\
(\Gamma_2)
\end{vmatrix} = 0,
$$

$$
\begin{vmatrix}
{}^1E_g(e_g^2) & {}^3T_{2g}(t_{2g}e_g) & {}^3T_{1g}(t_{2g}e_g) & {}^3T_{1g}(t_{2g}^2) & {}^1E_g(t_{2g}^2) \\
21F_4 + 12Dq - E & \sqrt{6}\lambda & \sqrt{6}\lambda & 0 & 2\sqrt{3}(F_2 - 5F_4) \\
 & -8F_2 - 9F_4 + 2Dq + \tfrac{1}{2}\lambda - E & -\tfrac{3}{2}\lambda & -3\lambda & 0 \\
 & & 4F_2 - 69F_4 + 2Dq + \tfrac{1}{2}\lambda - E & -6F_2 + 30F_4 - \lambda & -2\sqrt{2}\lambda \\
 & & & -5F_2 - 24F_4 - \lambda - 8Dq - E & -\sqrt{2}\lambda \\
 & (\Gamma_3) & & & F_2 + 16F_4 - 8Dq - E
\end{vmatrix} = 0,
$$

$$
(\Gamma_4)\qquad
\begin{array}{cccc}
{}^3T_{2g}(t_{2g}e_g) & {}^3T_{1g}(t_{2g}e_g) & {}^3T_{1g}(t_{2g}^2) & {}^1T_{1g}(t_{2g}e_g)
\end{array}
$$

$$
\begin{vmatrix}
-8F_2-9F_4+2Dq+\frac{1}{2}\lambda-E & \frac{\sqrt{3}}{2}\lambda & \sqrt{3}\lambda & \frac{\sqrt{6}}{2}\lambda \\
 & 4F_2-69F_4+2Dq-\frac{1}{2}\lambda-E & -6F_2+30F_4+\lambda & \frac{\sqrt{2}}{2}\lambda \\
 & & -5F_2-24F_4-8Dq+\lambda-E & \sqrt{2}\lambda \\
 & & & 4F_2+F_4+2Dq-E
\end{vmatrix} = 0,
$$

$$
(\Gamma_5)\qquad
\begin{array}{cccccc}
{}^3A_{2g}(e_g^2) & {}^3T_{2g}(t_{2g}e_g) & {}^3T_{1g}(t_{2g}e_g) & {}^1T_{2g}(t_{2g}e_g) & {}^3T_{1g}(t_{2g}^2) & {}^1T_{2g}(t_{2g}^2)
\end{array}
$$

$$
\begin{vmatrix}
-8F_2-9F_4+12Dq-E & 2\sqrt{2}\lambda & 0 & 2\lambda & 0 & 0 \\
 & -8F_2-9F_4+2Dq-\frac{1}{2}\lambda-E & \frac{\sqrt{3}}{2}\lambda & -\frac{\sqrt{2}}{2}\lambda & -\sqrt{3}\lambda & \sqrt{6}\lambda \\
 & & 4F_2-69F_4+2Dq+\frac{1}{2}\lambda-E & -\frac{\sqrt{6}}{2}\lambda & 6F_2-30F_4+\lambda & \sqrt{2}\lambda \\
 & & & 21F_4+2Dq-E & -\sqrt{6}\lambda & 2\sqrt{3}(F_2-5F_4) \\
 & & & & -5F_2-24F_4-8Dq-\lambda-E & \sqrt{2}\lambda \\
 & & & & & F_2+16F_4-8Dq-E
\end{vmatrix} = 0. \tag{5}
$$

The parameters appearing in Eq. (5) have been previously defined in Section 2(a). The unitary transformation which connects Eqs. (4) and (5) is given in Appendix 1.

3. SOLUTION OF THE SECULAR EQUATIONS

From Eqs. (4) and (5) we see that the requisite characteristic functions and values of Eq. (2) are parametrically dependent upon the four quantities F_2, F_4, λ, and Dq. It has been recognized for quite some time that the spin-orbit coupling constant λ and the electron correlation terms F_2 and F_4 are, in general, reduced in magnitude on passing from the free ion to the complex salt (*46*), and that the cubic crystalline field strength, Dq, though amenable to calculation from first principles, must, in practice, be determined from experiment (*47*, *48*). We have therefore regarded these four quantities as phenomenological parameters in the detailed numerical computations. The parametric set of secular equations, Eq. (5), were programed for the IBM 704 data processing machine at the Murray Hill Bell Telephone Laboratories by Victor A. Vyssotsky. To save machine time we assumed the relation

$$F_2 = 14F_4, \tag{6}$$

which is fairly well fulfilled within the elementary theory of atomic term values (*43*). Tables I and II summarize the values of F_4, λ, and Dq utilized in the computations. In Figs. 1 and 2 we have plotted the energy level system which best describes the fine structure of those aquo and oxo complexes of Ni(II) and V(III), respectively, which have cubic symmetry.

4. COMPARISON WITH EXPERIMENT

We shall confine our discussion to a consideration of the complexes of V(III) and Ni(II) as these are the only d^2 and d^8 systems which have been well characterized. To the authors' knowledge there are no known tetrahedral complexes

TABLE I

PARAMETRIC VALUES ASSUMED FOR A d^8 ELECTRONIC SYSTEM IN EQ. (5). POSITIVE VALUES OF Dq CORRESPOND TO TETRAHEDRAL COMPLEXES AND NEGATIVE VALUES TO OCTAHEDRAL COMPLEXES

F_4	λ	Dq
cm^{-1}	cm^{-1}	cm^{-1}
110	−325	0, ±1000, ±2000
100	−300	±250, ±1100, ±3000
90	−275	±500, ±1200, +400
	−250	±900, ±1500
	−200	

TABLE II

PARAMETRIC VALUES ASSUMED FOR A d^2 ELECTRONIC SYSTEM IN EQ. (5). POSITIVE VALUES OF Dq CORRESPOND TO OCTAHEDRAL COMPLEXES AND NEGATIVE VALUES TO TETRAHEDRAL COMPLEXES

F_4	λ	Dq
cm^{-1}	cm^{-1}	cm^{-1}
90	105	0, ±1500, ±3500
70	85	±250, ±2000, ±4000
50	65	±500, ±2500, −800
	45	±1000, ±3000, +1800
	25	

of V(III), except possibly for a few inverted spinels (*49*). Thus the left-hand side of Fig. 2 is mostly of academic interest. The right-hand side of Fig. 2, though not capable of detailed confirmation due to the dearth of known octahedral complexes of V(III),[7] will be shown to adequately account for the general spectral features of known hexacoordinated V(III) complexes. As there exist well documented complexes of Ni(II) of both tetrahedral and octahedral symmetry, a more critical test of the theory is possible. Indeed, it will be shown that even the spin-orbit fine structure predicted by Figure 1 finds confirmation in the band shapes of hydrated Ni(II) complexes.

(a) SPECTRA OF Ni(II) SYSTEMS

The most extensively studied of all transition metal ions has been Ni(II) (*32*, *49–59*). We shall not here compare our calculated results with all the recorded experimental observations, but shall rather discuss in detail three typical complexes: the hexaquo (and hexoxo), the hexammine, and the tris (o-phenanthroline).

An octahedral hexaquo Ni(II) ion exhibits four absorption bands (*32*, *51*, *54*, *55*, *59*) located at 8500, 13,500, 15,400, and 25,300 cm^{-1}. As all these absorption bands are vibronic in origin, and hence rather broad, too much emphasis should not be placed on the exact agreement of the calculated *electronic* energy with the observed *vibronic* band maxima. From Fig. 1, we see that a consistent assignment of the experimental spectrum requires Dq to be about -850 cm^{-1}. We then have that the observed absorptions are correlated with the following computed levels: (i) $\Gamma_5 \rightarrow \Gamma_j$, ($j$ = 2, 3, 4, 5), 8500 cm^{-1}; (ii) $\Gamma_5 \rightarrow \Gamma_j$, ($j$ = 1, 4), 13,500 cm^{-1}; (iii) $\Gamma_5 \rightarrow \Gamma_j$, ($j$ = 3, 5), 14,500 cm^{-1}; (iv) $\Gamma_5 \rightarrow \Gamma_j$, ($j$ = 1, 3, 4, 5), 24,000 cm^{-1}. *An inspection of the corresponding wave functions shows that all of these transitions are essentially pure triplet-triplet transitions amongst the perturbed* ^3F *and*

[7] The hexacoordinated complexes of V(III) which have been characterized to date are known to be subject to fields of low symmetry (e.g., trigonal).

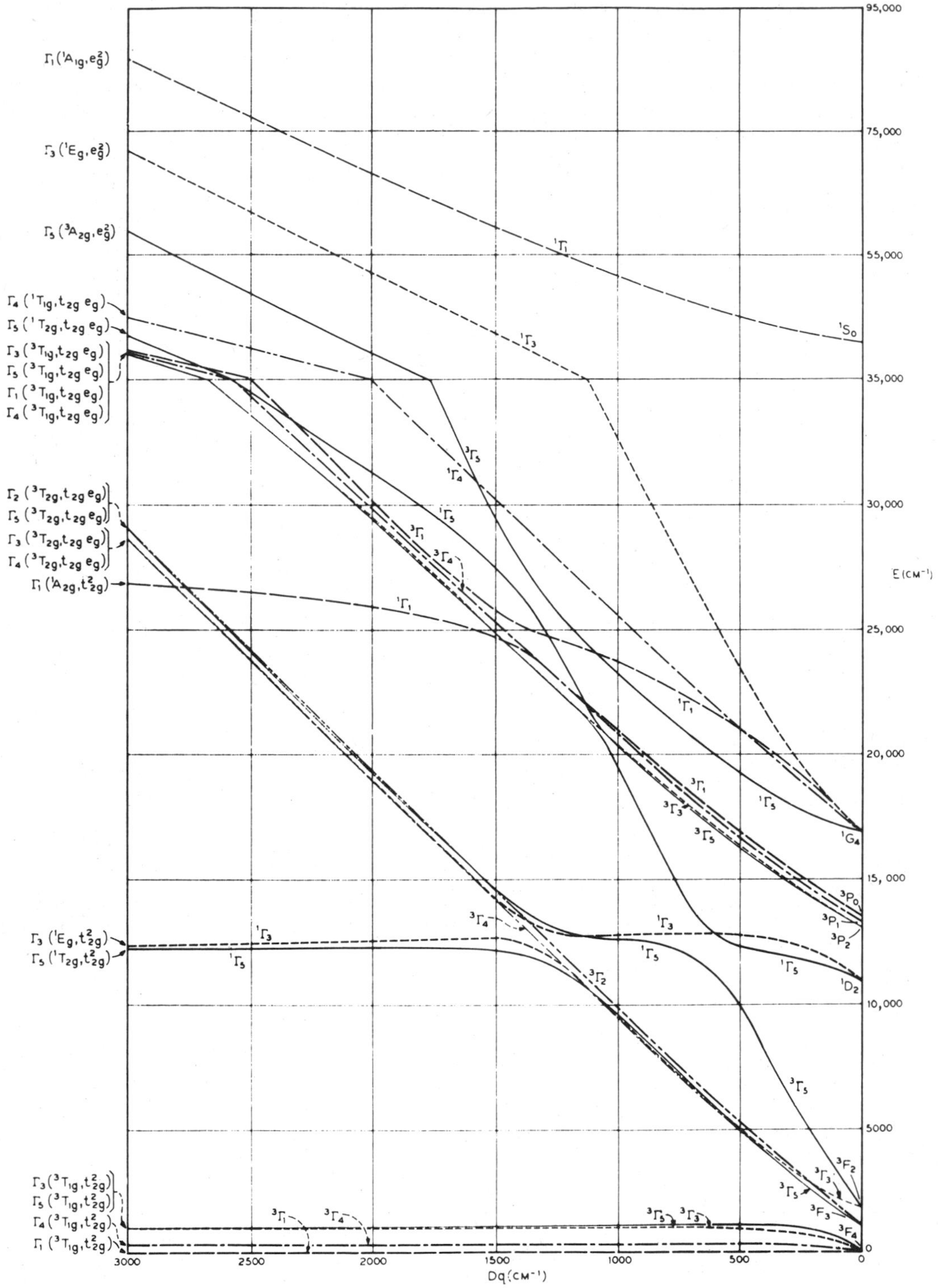
95,000
75,000
55,000
35,000
30,000
25,000
20,000
15,000
10,000
5000
0
E(CM⁻¹)
3000
2500
2000
1500
1000
500
0
Dq(CM⁻¹)
Γ1(¹A1g,eg²)
Γ3(¹Eg,eg²)
Γ5(³A2g,eg²)
Γ4(¹T1g,t2g eg)
Γ5(¹T2g,t2g eg)
Γ3(³T1g,t2g eg)
Γ5(³T1g,t2g eg)
Γ1(³T1g,t2g eg)
Γ4(³T1g,t2g eg)
Γ2(³T2g,t2g eg)
Γ5(³T2g,t2g eg)
Γ3(³T2g,t2g eg)
Γ4(³T2g,t2g eg)
Γ1(¹A2g,t2g²)
Γ3(¹Eg,t2g²)
Γ5(¹T2g,t2g²)
Γ3(³T1g,t2g²)
Γ5(³T1g,t2g²)
Γ4(³T1g,t2g²)
Γ1(³T1g,t2g²)
¹S0
¹G4
³P0
³P1
³P2
¹D2
³F2
³F3
³F4

FIG. 1.

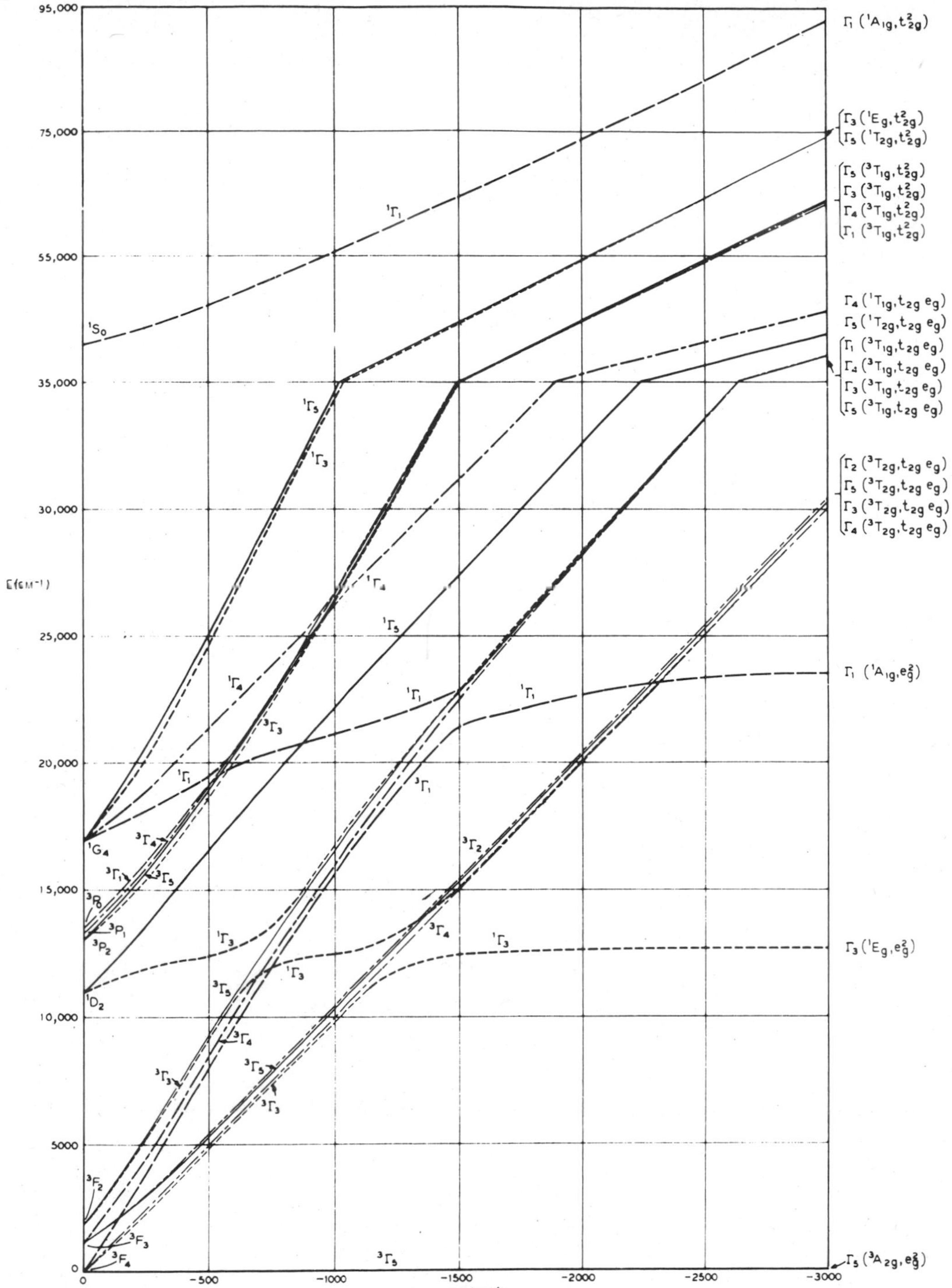

FIG. 1. Energy level curves for a d^8 electronic system immersed in fields of tetrahedral (left-hand side) and octahedral (right-hand side) symmetry for λ equal to -275 cm^{-1} and F_4 equal to 90 cm^{-1} ($F_2 = 14F_4$).

145

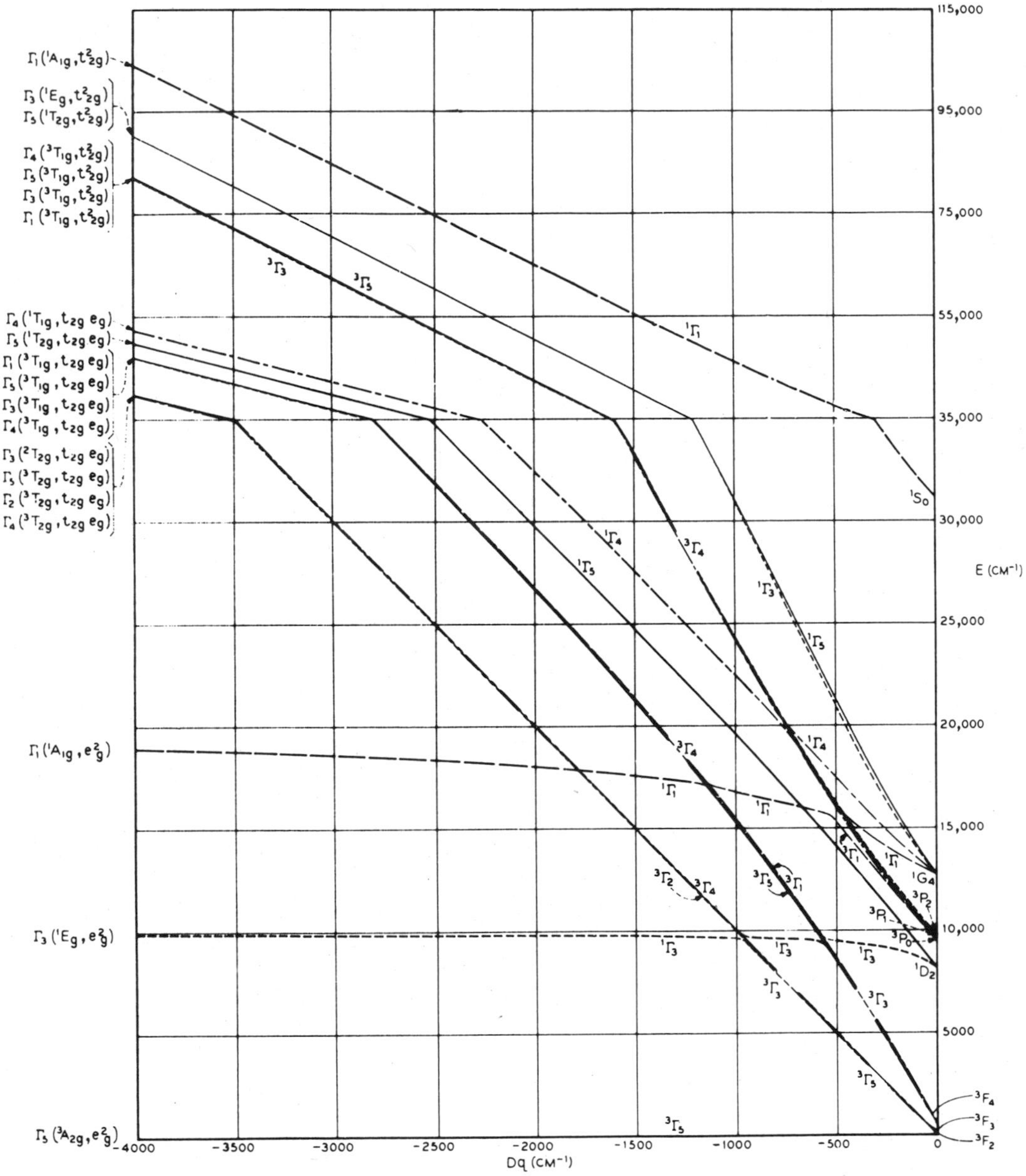
$\Gamma_1(^1A_{1g}, t_{2g}^2)$
$\Gamma_3(^1E_g, t_{2g}^2)$
$\Gamma_5(^1T_{2g}, t_{2g}^2)$
$\Gamma_4(^3T_{1g}, t_{2g}^2)$
$\Gamma_5(^3T_{1g}, t_{2g}^2)$
$\Gamma_3(^3T_{1g}, t_{2g}^2)$
$\Gamma_1(^3T_{1g}, t_{2g}^2)$
$\Gamma_4(^1T_{1g}, t_{2g}\, e_g)$
$\Gamma_5(^1T_{2g}, t_{2g}\, e_g)$
$\Gamma_1(^3T_{1g}, t_{2g}\, e_g)$
$\Gamma_5(^3T_{1g}, t_{2g}\, e_g)$
$\Gamma_3(^3T_{1g}, t_{2g}\, e_g)$
$\Gamma_4(^3T_{1g}, t_{2g}\, e_g)$
$\Gamma_3(^2T_{2g}, t_{2g}\, e_g)$
$\Gamma_5(^3T_{2g}, t_{2g}\, e_g)$
$\Gamma_2(^3T_{2g}, t_{2g}\, e_g)$
$\Gamma_4(^3T_{2g}, t_{2g}\, e_g)$
$\Gamma_1(^1A_{1g}, e_g^2)$
$\Gamma_3(^1E_g, e_g^2)$
$\Gamma_5(^3A_{2g}, e_g^2)$
115,000
95,000
75,000
55,000
35,000
30,000
25,000
20,000
15,000
10,000
5000
E (CM⁻¹)
$^1S_0$
$^1G_4$
$^3P_2$
$^3P_1$
$^3P_0$
$^1D_2$
$^3F_4$
$^3F_3$
$^3F_2$
-4000
-3500
-3000
-2500
-2000
-1500
-1000
-500
0
Dq (CM⁻¹)

Fig. 2.

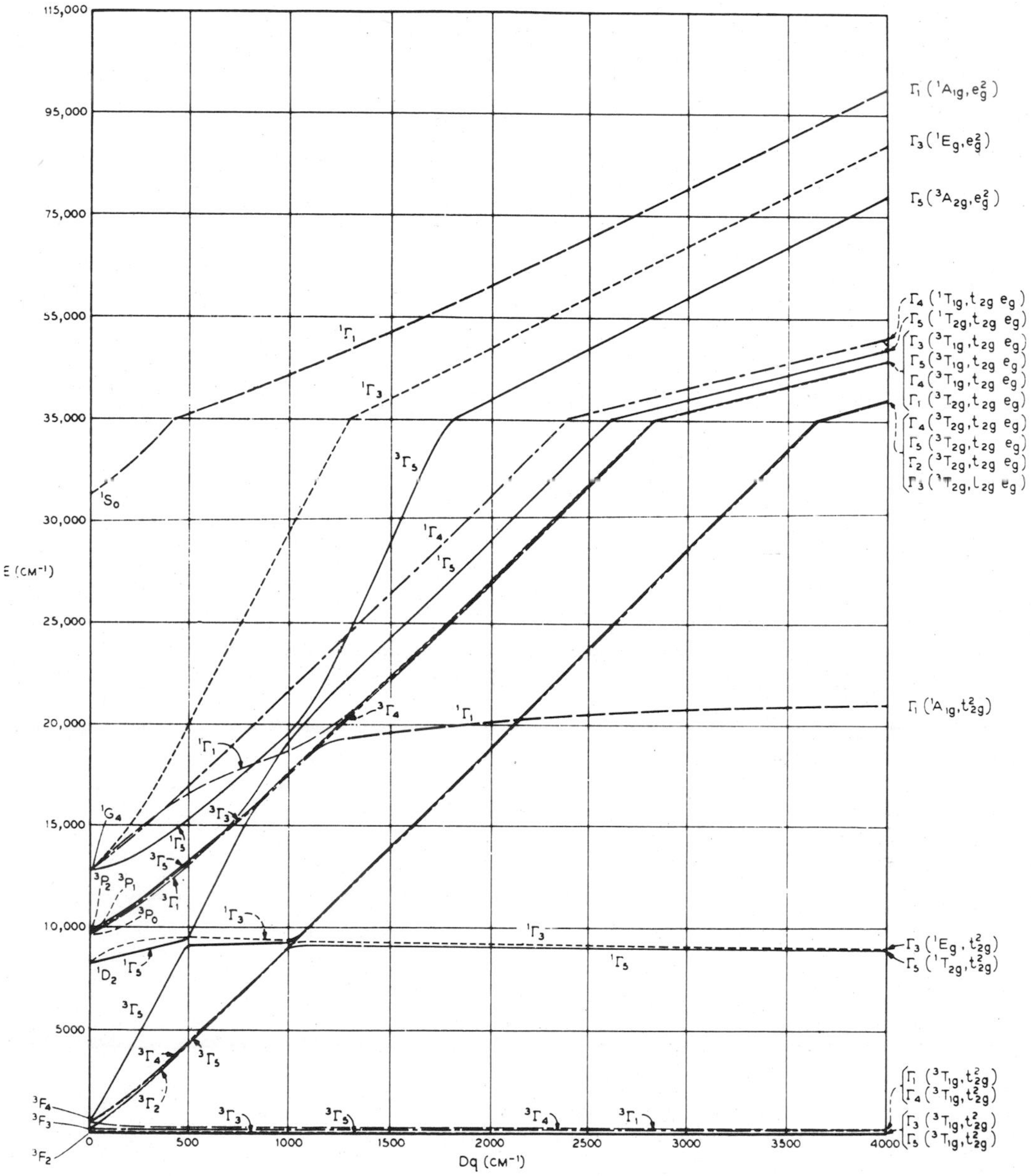

FIG. 2. Energy level curves for a d^2 electronic system immersed in fields of tetrahedral (left-hand side) and octahedral (right-hand side) symmetry for λ equal to 65 cm^{-1} and F_4 equal to 70 cm^{-1} ($F_2 = 14F_4$).

TABLE III

PREDICTED SPECTRUM FOR AN OCTAHEDRAL Ni(II) COMPLEX WITH $Dq = -900$, $\lambda = -275$, AND $F_4 = 90$ cm^{-1}. A LEVEL DESIGNATED WITH AN ASTERISK IS ESSENTIALLY A SPIN SINGLET

$\Gamma_5 \rightarrow \Gamma_3$: 8723 cm^{-1}	$\Gamma_5 \rightarrow \Gamma_5$*: 20,931
Γ_4 : 8886	Γ_3 : 24,452
Γ_5 : 9212	Γ_5 : 24,627
Γ_2 : 9354	Γ_4 : 24,703
Γ_3*: 12,338	Γ_1 : 24,895
Γ_1 : 14,027	Γ_4*: 25,181
Γ_4 : 14,441	Γ_3*: 32,115
Γ_5 : 15,052	Γ_5*: 32,446
Γ_3 : 15,287	Γ_1*: 53,396
Γ_1*: 20,763	

^3P *states of the free ion.*[8] *Hence Jørgensen's (51) assignment of the 15,400 cm^{-1} hexaquo bands cannot be correct.* Indeed, we would predict the "spin-forbidden" $\Gamma_5 \rightarrow \Gamma_3$ transition to lie at about 12,200 cm^{-1}.[9]

If the theoretical results at Dq equal to -850 cm^{-1} are compared with Low's (*56*, *57*) Ni^{++}:MgO data,[10] we have that the following additional predicted spectral transitions are seen experimentally: (v) $\Gamma_5 \rightarrow \Gamma_j$, ($j = 1, 5$), 20,500 cm^{-1}; (vi) $\Gamma_5 \rightarrow \Gamma_4$, 24,700 cm^{-1}; (vii) $\Gamma_5 \rightarrow \Gamma_j$, ($j = 3, 5$), 31,300 cm^{-1}. These transitions are all "spin-forbidden". The agreement with experiment for these bands is poorer,[10] but could be improved if the $^1G_4$ level of the free ion were displaced to higher energies by $\sim$1000 cm^{-1}. This could be accomplished by the introduction of a new parameter in Eqs. (4) and (5). The band reported at 28,300 cm^{-1} by Low (*56*, *57*) does not fit into the theoretical energy level scheme. It is possible that this absorption is spurious. In Table III we give, for illustrative purposes, the energy levels computed for $Dq = -900$, $\lambda = -275$, and $F_4 = 90$ cm^{-1}.

The Ni(II) hexammine and tris (o-phenanthroline) spectra as reported by Jørgensen (*51*) are presented in Table IV. *The detailed calculations agree with Jørgensen's assignment (51) of the "spin-forbidden" $\Gamma_5 \rightarrow \Gamma_3$ transition for the hexammine, but not for the tris (o-phenanthroline).* The rather large intensity observed for the $\Gamma_5 \rightarrow \Gamma_3$ transition for the tris (o-phenanthroline) complex (*51*), as contrasted with that for the hexammine complex, is nicely accounted for by the circumstance that the Γ_3 wave function is 23.6% triplet in character for Ni(II) tris (o-phenanthroline), but only 7.7% triplet in character for Ni(II) hexammine.

[8] The controversial Γ_3 band at 14500 cm^{-1} is calculated to have only 6.5% singlet character.

[9] Preliminary experimental results of McClure (*59*) on $NiSiF_6 \cdot 6H_2O$ indicate that such a band exists.

[10] Low (*56*) observes bands at 8600, 13,400, 14,800, 21,550, 24,500, 25,950, 28,300, and 34,500 cm^{-1}.

TABLE IV

PREDICTED AND OBSERVED SPECTRA FOR THE Ni(II) HEXAMMINE AND TRIS (o-PHENANTHROLINE) COMPLEXES. A LEVEL DESIGNATED BY AN ASTERISK IS ESSENTIALLY A SPIN SINGLET

Hexammine		Tris (o-phenanthroline)	
Observed (*51*)	Calculated ($Dq = -1100$, $\lambda = -200$, $F_4 = 90$)	Observed (*51*)	Calculated ($Dq = -1200$, $\lambda = -200$, $F_4 = 90$)
10,750 cm^{-1}	$\Gamma_5 \rightarrow \Gamma_j$, ($j$ = 2, 3, 4, 5), ~11,000 cm^{-1}	11,550 cm^{-1}	$\Gamma_5 \rightarrow \Gamma_j$, ($j$ = 2, 3, 4, 5), ~12,000 cm^{-1}
13,150	$\rightarrow \Gamma_3^*$, ~12,540	12,700	$\rightarrow \Gamma_3^*$, ~12,700
17,500	$\rightarrow \Gamma_j$, (j = 1, 3, 4, 5), ~17,200	19,300	$\rightarrow \Gamma_j$, (j = 1, 3, 4, 5), ~18,700
28,200	$\rightarrow \Gamma_j$, (j = 1, 3, 4, 5), ~27,900	—	$\rightarrow \Gamma_j$, (j = 1, 3, 4, 5), ~29,600

It is only for values of Dq larger than 1300 cm^{-1} that the Γ_3 "spin singlet" state may lie lower than the aggregate of "spin triplet" states Γ_j, (j = 2, 3, 4, 5).

The only complexes of Ni(II) which are definitely known to be tetrahedral are of the type $Ni(Ph_3P)_2X_2$, (X = I, Br, Cl) (*60*). Spectra of these compounds are not yet available. The correctness of the left-hand side of Fig. 1 can, however, be verified by a comparison with McClure's (*49*) spectrum of Ni^{++}:ZnO. The computed transitions (for Dq equal to 400 cm^{-1}) at 8300, 12,100, 12,700, 15,600, 16,000, 16,200, and 18,600 cm^{-1} agree well with experiment. The theory also predicts infrared bands at approximately 350, 900, 1100, and 4400 cm^{-1}. These have not as yet been observed, presumably due to experimental difficulties.

(b) SPECTRA OF V(III) SYSTEMS

Unfortunately the known hexacoordinated V(III) complexes are trigonally distorted,[11] and so the observed spectra (*29, 50, 61–64*) cannot be used to directly confirm some of the finer details of the theory. With the exception of a few fluoro complexes (*64*), all the thus far investigated V(III) systems are of the oxo type (*29, 50, 61–63*). As Low's spectra of V(III) dissolved in aluminum oxide are much more detailed than the other oxo spectra, we shall, for the most part, limit the discussion to the elucidation of the V^{+++}:Al_2O_3 system. Table V presents the detailed comparison of the theoretical expectations and the experimental findings. The agreement is quite good. The computed levels which disagree badly (by ~1000 cm^{-1}), as in the case of Ni(II), arise from the free ion $^1G_4$ level. The agreement could be presumably improved by the introduction of another parameter in Eqs. (4) and (5). Low's (*63*) very weak and narrow band at 25,400 cm^{-1} is difficult to rationalize on the basis of the theory. It is perhaps spurious.

[11] The similarity of the free V(III) hexaquo spectra to that in trigonal crystals indicates similar distortions for this ion also.

TABLE V

PREDICTED AND OBSERVED SPECTRUM FOR $V^{+++}:Al_2O_3$ UNDER THE ASSUMPTION OF OCTAHEDRAL SITE SYMMETRY. A LEVEL DESIGNATED BY AN ASTERISK IS ESSENTIALLY A SPIN SINGLET

Observed	Calculated (Dq = 1860, λ = 65, F_4 = 70)
17,400 cm^{-1}	$\Gamma_3 \rightarrow \Gamma_5$, ~1.5 cm^{-1}
21,000 (v.w.)	$\rightarrow \Gamma_4$, ~167
25,200	$\rightarrow \Gamma_1$, ~250
25,400 (v.w.)	$\rightarrow \Gamma_j^*$, (j = 3, 5), ~9,100
29,300 (v.w.)	$\rightarrow \Gamma_j$, (j = 2, 3, 4, 5), ~17,400
30,150 (v.w.)	$\rightarrow \Gamma_1^*$, ~20,000
34,500	$\rightarrow \Gamma_j$, (j = 1, 3, 4, 5), ~25,400
	$\rightarrow \Gamma_5^*$, ~27,600
	$\rightarrow \Gamma_4^*$, ~29,800
	$\rightarrow \Gamma_5$, ~35,800

Except for the reflection spectra of Ballhausen and Winther (*64*), no infrared data on V(III) complexes are available. Due to the experimental uncertainties inherent in the reflection method, the exact location of the band observed by these authors in the vicinity of ~10,000 cm^{-1} is not known. However, the occurrence of a weak absorption in this region of the spectrum may be taken as substantiation of the theoretical expectations.

6. CONCLUSION

The fruitfulness of the complete configuration interaction approach to crystal field theory has been underscored by the satisfactory account rendered by the theoretical energy levels in the interpretation of the existing spectra of cubic Ni(II) and V(III) complexes. In a later communication (*65*), we shall also demonstrate that the configuration interaction wave functions are equally useful in the comprehension of the observed spectral intensity distributions. Indeed, on the basis of the computed energy level scheme and wave functions, it has proved possible to resolve the puzzle of the exact nature of double-peaked red band of Ni(II) hexaquo complexes, and to thus show that Jørgensen's (*51*) interpretation of the structure of this band is incorrect. Moreover, a less empirical assignment of Low's Ni(II) (*56*, *57*) and V(III) (*63*) spectra has been made feasible by the use of the theoretical energy level curves. Other useful applications of the "exact" Ni(II) and V(III) eigenvalues and eigenfunctions to related spectral and magnetic problems are outlined in the appendices. In termination, it may be said that the results here presented illustrate quite clearly that caution must be exercised in the interpretation of experimental data by means of the simple crystal field theory, and that satisfactory conclusions, in general, may only be obtained from the "exact" approach.

APPENDIX 1

The matrix transformations which connect the weak field functions $| S, L, J, \Gamma_j >$ and the strong field functions $| X, S, t_{2g}{}^n e_g{}^m, \Gamma_j >$, where X is the Mulliken (66) orbital symmetry classification and Γ_j the Bethe (1) spin-orbital symmetry classification, are as follows:

$$\Gamma_1: \left\| \begin{array}{c|cccc} & {}^1A_{1g}(t_{2g}{}^2) & {}^1A_{1g}(e_g{}^2) & {}^3T_{1g}(t_{2g}{}^2) & {}^3T_{1g}(t_{2g}e_g) \\ \hline {}^1S_0 & -i\sqrt{3/5} & -i\sqrt{2/5} & 0 & 0 \\ {}^1G_4 & +i\sqrt{2/5} & -i\sqrt{3/5} & 0 & 0 \\ {}^3P_0 & 0 & 0 & +i\sqrt{1/5} & -i\sqrt{4/5} \\ {}^3F_4 & 0 & 0 & +i\sqrt{4/5} & +i\sqrt{1/5} \end{array} \right\|,$$

$$\Gamma_2: \left\| \begin{array}{c|c} & {}^3T_{2g}(t_{2g}e_g) \\ \hline {}^3F_3 & +i \end{array} \right\|,$$

$$\Gamma_3: \left\| \begin{array}{c|ccccc} & {}^1E_g(t_{2g}{}^2) & {}^1E_g(e_g{}^2) & {}^3T_{1g}(t_{2g}{}^2) & {}^3T_{1g}(t_{2g}e_g) & {}^3T_{2g}(t_{2g}e_g) \\ \hline {}^1G_4 & -\sqrt{4/7} & -\sqrt{3/7} & 0 & 0 & 0 \\ {}^1D_2 & -\sqrt{3/7} & +\sqrt{4/7} & 0 & 0 & 0 \\ {}^3P_2 & 0 & 0 & -\sqrt{1/5} & +\sqrt{4/5} & 0 \\ {}^3F_2 & 0 & 0 & -6\sqrt{1/70} & -3\sqrt{1/70} & -5\sqrt{1/70} \\ {}^3F_4 & 0 & 0 & 2\sqrt{1/14} & \sqrt{1/14} & -3\sqrt{1/14} \end{array} \right\|,$$

$$\Gamma_4: \left\| \begin{array}{c|cccc} & {}^1T_{1g}(t_{2g}e_g) & {}^3T_{1g}(t_{2g}{}^2) & {}^3T_{1g}(t_{2g}e_g) & {}^3T_{2g}(t_{2g}e_g) \\ \hline {}^1G_4 & -i & 0 & 0 & 0 \\ {}^3P_1 & 0 & \sqrt{1/5} & -\sqrt{4/5} & 0 \\ {}^3F_3 & 0 & \sqrt{\frac{3}{10}} & \frac{1}{2}\sqrt{\frac{3}{10}} & -\frac{5}{2}\sqrt{\frac{1}{10}} \\ {}^3F_4 & 0 & \frac{1}{i\sqrt{2}} & \frac{1}{2i\sqrt{2}} & \frac{1}{2i}\sqrt{\frac{3}{2}} \end{array} \right\|,$$

$$
\Gamma_5: \quad
\begin{array}{c|cccccc|}
 & {}^1T_{2g}(t_{2g}^2) & {}^1T_{2g}(t_{2g}e_g) & {}^3T_{1g}(t_{2g}^2) & {}^3A_{2g}(e_g^2) & {}^3T_{1g}(t_{2g}e_g) & {}^3T_{2g}(t_{2g}e_g) \\
{}^1G_4 & -i\sqrt{4/7} & -i\sqrt{3/7} & 0 & 0 & 0 & 0 \\
{}^1D_2 & i\sqrt{3/7} & -i\sqrt{4/7} & 0 & 0 & 0 & 0 \\
{}^3P_2 & 0 & 0 & i\sqrt{1/5} & 0 & i\sqrt{4/5} & 0 \\
{}^3F_3 & 0 & 0 & -\frac{1}{\sqrt{2}} & -\frac{1}{\sqrt{3}} & +\frac{1}{2\sqrt{2}} & +\frac{1}{2\sqrt{6}} \\
{}^3F_4 & 0 & 0 & i\sqrt{\frac{1}{14}} & -i\sqrt{\frac{6}{14}} & -\frac{i}{2}\sqrt{\frac{1}{14}} & -\frac{3i}{2}\sqrt{\frac{3}{14}} \\
{}^3F_2 & 0 & 0 & -\frac{2i}{5}\sqrt{\frac{30}{21}} & +i\sqrt{\frac{5}{21}} & +\frac{i}{5}\sqrt{\frac{30}{21}} & -i\sqrt{\frac{10}{21}}
\end{array}
\qquad \text{(A1-1)}
$$

If we denote the complex conjugate of the above transformation matrices as $T(\Gamma_j)^*$, the connection between the corresponding energy matrices is then given by

$$\mathcal{H}(X,\ S,\ t_{2g}{}^n e_g{}^m,\ \Gamma_j) = T(\Gamma_j)^{*-1}\mathcal{H}(S,\ L,\ J,\ \Gamma_j)T(\Gamma_j)^*. \qquad \text{(A1-2)}$$

It is readily apparent that the transformation matrix, $T(\Gamma_j)$, is also very useful in the determination of the requisite g-factors for the strong crystalline field picture in terms of the atomic g-factors (*43*),

$$g(S, L, J) = 1 + \frac{J(J+1) - L(L+1) + S(S+1)}{2J(J+1)}. \qquad \text{(A1-3)}$$

The true g-factors can then be obtained from those of the strong crystalline field by use of the unitary transformation which relates the strong field basis functions to the true eigenfunctions.[12] Once acquired, the sundry g-factors may be used to compute the thermal behavior of the magnetic susceptibilities of V(III) and Ni(II) complexes in cubic crystalline fields (*3, 67, 68*).

APPENDIX 2

The strong crystalline field energy matrix in the spin-coordinate symmetry representation Γ_j readily lends itself to a computation of the ground state splitting of Ni(II) in fields of lower symmetry. For example, in a trigonal field, as occurs in nickel hexahydrate fluosilicate, the splitting of the cubic ground electronic

[12] It is this latter transformation which was tabulated for us by the IBM 704 data possessing machine.

state Γ_{5s}, $(s = a, b, c)$, may be determined in terms of that of the excited electronic states of the same symmetry by means of second order perturbation theory in the following manner. Let k_j, $(j = 1, 2)$, represent the $\Gamma_{5a} - \Gamma_{5b,c}$ energy difference and Δ_j, $(j = 1, 2)$, the height of the Γ_{5a} levels above the ground state for the $\Gamma_5[^3T_{2g}(t_{2g}e_g)]$ and $\Gamma_5[^1T_{2g}(t_{2g}e_g)]$ states, respectively. Then by Eq. (5) we have the resultant ground state splitting, δ, as

$$\delta = \sum_j \frac{|H_{oa,ja}'|^2}{E_{oa} - E_{ja}} - \sum_j \frac{|H_{ob,jb}'|^2}{E_{ob} - E_{jb}}, \qquad \text{(A2-1)}$$

or

$$\delta = 4\lambda^2 \left[\frac{2}{\Delta_1 + k_1} + \frac{1}{\Delta_2 + k_2} - \frac{2}{\Delta_1} - \frac{1}{\Delta_2}\right]. \qquad \text{(A2-2)}$$

For Ni(II) complexes Δ_2 is approximately twice Δ_1. If we assume, as is eminently reasonable, that for the $\Gamma_5[^3T_{1g}(t_{2g}e_g)]$ and $\Gamma_5[^1T_{1g}(t_{2g}e_g)]$ states the trigonal splittings k_1 and k_2 are nearly equal and much less than the cubic splittings Δ_j, $(j = 1, 2)$, Eq. (A2-2) may be simplified to read

$$\delta \approx \frac{-9k_1\lambda^2}{\Delta_1^2}. \qquad \text{(A2-3)}$$

Now Holden *et al.* (*69*) find δ is equal to 0.49 cm^{-1} for $NiSiF_6 \cdot 6H_2O$. Thus as λ is -275 cm^{-1} for this complex (*46*) and Δ_1 is $\sim$9000 cm^{-1} (*59*), we see that the excited state splitting k_1 is predicted to be in the neighborhood of 50 cm^{-1}. It is not unreasonable to expect the trigonal field splittings of the other Ni(II) levels to also be of this order of magnitude. This circumstance may account for some of the mysterious fine structure observed by McClure (*59*) in the absorption spectra of nickel fluorosilicate hexahydrate.

ACKNOWLEDGMENTS

We are deeply indebted to Victor A. Vyssotsky for the machine solution of the secular equations. We should also like to thank Donald S. McClure for communicating to us his recent experimental results on $NiSiF_6 \cdot 6H_2O$ prior to publication.

RECEIVED: November 13, 1958

REFERENCES

1. H. A. BETHE, *Ann. Physik* [5], **3**, 133 (1929).
2. H. A. KRAMERS, *Proc. Amsterdam Acad.* **32**, 1176 (1929).
3. J. H. VAN VLECK, "The Theory of Electric and Magnetic Susceptibilities." Oxford Univ. Press, London and New York, 1932.
4. B. BLEANEY AND K. W. H. STEVENS, *Repts. Progr. Phys.* **16**, 108 (1953).
5. K. D. BOWERS AND J. OWEN, *Repts. Progr. Phys.* **18**, 304 (1955).
6. W. E. MOFFITT AND C. J. BALLHAUSEN, *Ann. Rev. Phys. Chem.* **7**, 107 (1956).

7. M. H. L. Pryce, Nuovo cimento, Suppl. 3 [10], **6**, 817 (1957).
8. J. H. Van Vleck, *Phys. Rev.* **41**, 208 (1932).
9. W. G. Penney and R. Schlapp, *Phys. Rev.* **41**, 194 (1932).
10. R. Schlapp and W. G. Penney, *Phys. Rev.* **42**, 666 (1932).
11. G. J. Kynch, *Trans. Faraday Soc.* **33**, 1402 (1937).
12. O. M. Jordahl, *Phys. Rev.* **45**, 87 (1934).
13. R. Finkelstein and J. H. Van Vleck, *J. Chem. Phys.* **8**, 790 (1940).
14. F. E. Ilse and H. Hartmann, *Z. Phys. Chem.* **197**, 239 (1951).
15. F. E. Ilse and H. Hartmann, *Z. Naturforsch.* **6a**, 751 (1951).
16. A. Abragam and M. H. L. Pryce, *Proc. Roy. Soc.* **A206**, 164 (1951).
17. A. Abragam and M. H. L. Pryce, *Proc. Roy. Soc.* **A206**, 173 (1951).
18. L. E. Orgel, *J. Chem. Soc.* p. 4756 (1952).
19. J. Bjerrum, C. J. Ballhausen, and C. K. Jørgensen, *Acta Chem. Scand.* **8**, 1275 (1954).
20. C. J. Ballhausen, *Kgl. Danske Videnskab, Selskab. Mat.-fys. Medd.* **29**, No. 4 (1954).
21. Y. Tanabe and S. Sugano, *J. Phys. Soc. (Japan)* **9**, 753 (1954).
22. Y. Tanabe and S. Sugano, *J. Phys. Soc. (Japan)* **9**, 766 (1954).
23. C. J. Ballhausen, *Kgl. Danske Videnskab, Selskab. Mat.-fys. Medd.* **29**, No. 8 (1955).
24. L. E. Orgel, *J. Chem. Phys.* **23**, 1004 (1955).
25. H. Hartmann and H. Fischer-Wasels, *Z. Phys. Chem. (Frankfurt)* [*N.F.*] **4**, 297 (1955).
26. H. Hartmann and H. H. Kruse, *Z. Phys. Chem. (Frankfurt)* [*N.F.*] **5**, 9 (1955).
27. C. J. Ballhausen and C. K. Jørgensen, *Kgl. Danske Videnskab, Selskab. Mat.-fys. Medd.* **29**, No. 14 (1955).
28. H. Hartmann, C. Furlani, and A. Bürger, *Z. Phys. Chem. (Frankfurt)* [*N.F.*] **9**, 62 (1956).
29. H. Hartmann and C. Furlani, *Z. Phys. Chem. (Frankfurt)* [*N.F.*] **9**, 162 (1956).
30. C. J. Ballhausen and W. E. Moffitt, *J. Inorg. Nucl. Chem.* **3**, 178 (1956).
31. J. S. Griffith and L. E. Orgel, *J. Chem. Soc.* p. 4981 (1956).
32. C. Furlani, *Z. Phys. Chem. (Frankfurt)* [*N.F.*] **10**, 291 (1957).
33. R. L. Belford, M. Calvin, and G. Belford, *J. Chem. Phys.* **26**, 1165 (1957).
34. C. Furlani and G. Sartori, *Gazz. Chim. ital.* **87**, 380 (1957).
35. G. Maki, *J. Chem. Phys.* **28**, 651 (1958).
36. Y. Tanabe and H. Kamimura, *J. Phys. Soc. (Japan)* **13**, 394 (1958).
37. S. Sugano and Y. Tanabe, *J. Phys. Soc. (Japan)* **13**, 880 (1958).
38. J. H. Van Vleck and W. G. Penney, *Phil. Mag.* **17**, 961 (1934).
39. A. Abragam and M. H. L. Pryce, *Proc. Roy. Soc.* **A205**, 135 (1951).
40. H. Watanabe, *Progr. Theoret. Phys. (Japan)* **18**, 405 (1957).
41. C. K. Jørgensen, *Acta Chem. Scand.* **9**, 1362 (1955).
42. C. J. Gorter, *Phys. Rev.* **42**, 437 (1932).
43. E. U. Condon and G. H. Shortley, "The Theory of Atomic Spectra." Cambridge Univ. Press, London and New York, 1953.
44. J. H. Van Vleck, *J. Chem. Phys.* **7**, 61 (1939).
45. H. Goldstein, "Classical Mechanics." Addison-Wesley, Cambridge, 1951.
46. J. Owen, *Proc. Roy. Soc.* **A227**, 183 (1955).
47. W. H. Kleiner, *J. Chem. Phys.* **20**, 1784 (1952).
48. Y. Tanabe and S. Sugano, *J. Phys. Soc. (Japan)* **11**, 864 (1956).
49. D. S. McClure, *J. Phys. Chem. Solids* **3**, 311 (1957).
50. H. Hartmann and H. L. Schläfer, *Angew. Chem.* **66**, 768 (1954).
51. C. K. Jørgensen, *Acta Chem. Scand.* **9**, 1362 (1955).
52. C. K. Jørgensen, *Acta Chem. Scand.* **10**, 887 (1956).

53. C. K. JØRGENSEN, *Acta Chem. Scand.* **11,** 1223 (1957).
54. O. G. HOLMES AND D. S. MCCLURE, *J. Chem. Phys.* **26,** 1686 (1957).
55. R. PAPPALARDO, *Nuovo cimento* [10], **6,** 392 (1957).
56. W. LOW, *Phys. Rev.* **109,** 247 (1958).
57. W. LOW, *Ann. N. Y. Acad. Sci.* **72,** 69 (1958).
58. G. MAKI, *J. Chem. Phys.* **29,** 162 (1958).
59. D. S. MCCLURE (private communication).
60. L. M. VENANZI, *J. Chem. Soc.* p. 719 (1958).
61. H. HARTMANN AND H. L. SCHLÄFER, *Z. Naturforsch.* **6a,** 754 (1951).
62. H. HARTMANN AND H. L. SCHLÄFER, *Z. Naturforsch.* **6a,** 760 (1951).
63. W. LOW, *Z. Phys. Chem. (Frankfurt)* [*N.F.*] **13,** 107 (1957).
64. C. J. BALLHAUSEN AND F. WINTHER (to be published).
65. C. J. BALLHAUSEN AND A. D. LIEHR, submitted to *Mol. Physics.*
66. R. S. MULLIKEN, *Phys. Rev.* **43,** 279 (1932).
67. M. KOTANI, *J. Phys. Soc. (Japan)* **4,** 293 (1949).
68. H. KAMIMURA, *J. Phys. Soc. (Japan)* **11,** 1171 (1956).
69. A. N. HOLDEN, C. KITTEL, AND W. YAGER, *Phys. Rev.* **75,** 1443 (1949).

Reprinted from *Nature,* **182**, 168–170 (1958)

26

STEREOCHEMISTRY OF COMPLEX HALIDES OF THE TRANSITION METALS

By Dr. NAIDA S. GILL, Prof. RONALD S. NYHOLM, F.R.S., and PETER PAULING

THE ligand (crystal) field theory, originally developed by Bethe[1] and Van Vleck[2], has been used extensively since 1951 for the interpretation of the spectra of transition metal complexes[3]. It had been employed earlier to explain the magnetic properties of compounds of the transition[4] and rare earth metals[5], and, more recently, those of the actinide metals[6]. It has also proved very valuable in providing an increased understanding of the stereochemistry of transition metal complexes[7,8].

The ligand field theory considers the effect of the electrostatic field due to the ligands on the energy of the non-bonding d-shell of the central metal atom and assumes that the arrangement of the ligands about the metal atom is determined primarily by this non-bonding d-shell. The successes of the ligand field theory suggest that in many molecules and complex ions the dominating energy terms are indeed those related to electrostatic effects on the inner non-bonding d-electrons. We wish to direct attention in this article to cases in which this is not true, and in doing so to delineate more precisely the value of ligand field and other theories in the prediction and interpretation of the stereochemistry of transition metal complexes. While recognizing the validity of the energy effects calculated by ligand field theory, we believe that other factors such as the stabilizing energy of the directed σ-bond, the principle of electroneutrality of the metal ion[9], the polarizability of the ligand, the possibility of π-bonding between metal and ligand and the steric effects which result from interaction between ligands, are potentially important as structure determinants. We suggest that sometimes these may overcome ligand field effects and give rise to co-ordination shapes different from those expected from simple ligand field theory. First, we regard the principle of electroneutrality as of major significance in deciding the co-ordination number of the metal ion, for a knowledge of this must precede discussion of stereochemistry. In general, the preferred co-ordination number of a metal ion increases with its valency, and Pauling[9] has suggested that, as a rough rule, the co-ordination number is twice the valency. It is well known, however, that two or more co-ordination numbers may occur for the same metal ion; for example, $[FeF_6]^{3-}$ and $[FeCl_4]^-$. We consider that this arises not so much from steric effects as from differences in the polarizability of the various ligands and consequent transfer of charge to the metal ion through the σ-bond. For example, Fe^{3+} takes up ligands until its effective charge is reduced to nearly zero; it will require a greater number of F^--ligands to achieve this than Cl^--ligands, because there is greater electron transfer from each Cl^- to Fe^{3+} than from each F^-. Another example of this is the comparison of $[Co(H_2O)_6]^{2+}$ with $[CoCl_4]^{2-}$; the co-ordination number decreases as we pass from the less easily polarized H_2O to the more readily polarized Cl^- ligand.

Before discussing the shapes of complexes of metals the atoms of which have a non-spherical electronic shell, it is helpful to summarize the shapes expected for ions which have perfect spherical symmetry. As discussed elsewhere[8], the stereochemistry of complexes of non-transition metals can be generalized in terms of repulsions between *bonds* (whether single, double or triple) and lone pairs of electrons in the valency shell. This follows the approach originally suggested by Sidgwick and Powell[10]. On this model, the shapes of CH_4, NH_3 and H_2O are all based upon the tetrahedral arrangement, there being a total of four (bonds + lone pairs) in each case. The steady fall in the $H - X - H$ angle in the above sequence is explained by assuming that the repulsion lone pair : lone pair > lone pair : bond pair > bond pair : bond pair[11]. One can extend this picture to those transition metals the non-bonding d-shells of which are spherically symmetrical, elements with configurations d^0, d^5 (spin-free) and d^{10}. For ions with these configurations the shapes of complex molecules having no lone pairs are expected to be linear, triangular planar, tetrahedral, trigonal bipyramidal, octahedral and pentagonal bipyramidal, according as the co-ordination number is 2, 3, 4, 5, 6 or 7.

To appreciate the stereochemistry of complex ions with other than 0, 5 or 10 non-bonding d-electrons, the shapes of d-orbitals and their effect on the arrangement of the ligands must be considered. The ligand field theory assumes a purely electrostatic model for the calculation of energy effects; for example, the ion $[CoCl_4]^{2-}$ is regarded as a Co^{2+} surrounded by four Cl^- at the corners of a regular tetrahedron. Referred to orthogonal axes, the five d-orbitals may be divided into two classes; one class, the d_γ-orbitals, have lobes of electron density along

Table 1. CRYSTAL FIELD STABILIZATION* AND STEREOCHEMISTRY EXPECTED ON LIGAND FIELD THEORY (ALL SPIN-FREE)

No. of d-electrons	0, 5 or 10	1 or 6	2 or 7	3 or 8	4 or 9
Element and valency	(0) Ca^{2+}, Sc^{3+} (5) Mn^{2+}, Fe^{3+} (10) Zn^{2+}	(1) Ti^{3+} (6) Fe^{2+}	(2) Ti^{2+}, V^{3+} (7) Co^{2+}	(3) V^{2+}, Cr^{3+} (8) Ni^{2+}	(4) Cr^{2+}, Mn^{3+} (9) Cu^{2+}
Crystal field stabilization for perfect tetrahedron	0	$0·6\ \Delta^1 = 0·27\ \Delta$	$1·2\ \Delta^1 = 0·53\ \Delta$	$0·8\ \Delta^1 = 0·36\ \Delta$	$0·4\ \Delta^1 = 0·18\ \Delta$
Crystal field stabilization for perfect octahedron	0	$0·4\ \Delta$	$0·8\ \Delta$	$1·2\ \Delta$	$0·6\ \Delta$
Preferred stereochemistry	For shapes see text	Octahedron (very slightly distorted)	Regular tetrahedron, *or* octahedron (very slightly distorted)	Regular octahedron	Square planar *or* tetragonal

* $\Delta^1 = 4/9\ \Delta$

the z-axis (d_{z^2}) and along the x- and y-axes ($d_{x^2-y^2}$). The second class, d_ε orbitals, have their lobes of electron density in between the x-, y- and z-axes. There are three d_ε orbitals (d_{xy}, d_{xz} and d_{yz}) in the xy-, xz- and yz-planes, respectively. In an octahedral complex, these two classes of orbitals have different energies, the d_γ-doublet being of higher energy than the degenerate d_ε triplet. The lobes of electron density in the d_γ-orbitals point *towards* the ligands and in the d_ε-orbitals they point *between* the ligands. The separation Δ, or 10 Dq, between these degenerate levels is a measure of the strength of the electrical field produced by the ligands[8]. This is shown in Fig. 1a. For a tetrahedral complex the d_ε- and d_γ-levels are inverted since it is the d_ε orbitals which now point towards the ligands. The energy-levels are shown in Fig. 1b. In all cases the ligands are regarded as negative point charges.

When comparing complex ions of the same shape but having different numbers of d-electrons, this splitting leads to certain stabilization. Consider an octahedral complex of a d^1-ion, for example, Ti^{3+}. If the single d-electron were to spend 1/5 of its time in each of the five d-orbitals, its energy would be given by the point A. However, in fact, it is located in the d_ε-orbitals and the complex is more stable than the above hypothetical mean by an amount 4 Dq. This is the crystal field stabilization energy. For a tetrahedral complex, the value of Δ^1 is only 4/9 of Δ, and crystal field stabilization energies are consequently smaller for these complexes. Table 1 gives theoretical values for these crystal field stabilization energies.

We must take into account this effect in addition to the interactions between bonding pairs of electrons (σ-pairs) discussed earlier for spherically symmetrical metal ions. One expects to obtain a regular tetrahedron with the configurations d^0, d^2, d^5 (spin-free), d^7 (spin-free) and d^{10}, that is, when d_ε- and d_γ-shells

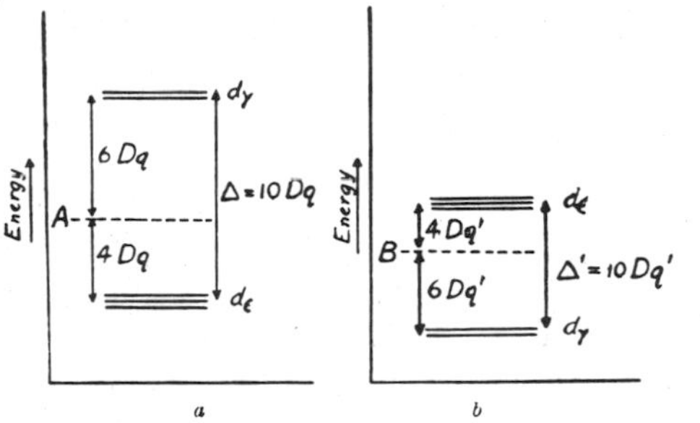

a *b*

Fig. 1. Splitting of orbital levels in (a) octahedral complex; (b) tetrahedral complex, $\Delta^1 = 4/9\ \Delta$

in Fig. 1b are empty, half full or full. Tetragonal or square planar arrangements are expected for d^4 (spin-free) and d^9 configurations. Finally, all spin-free configurations except d^4 and d^9 should also give rise to a perfect or very slightly distorted octahedron. From ligand field theory the tetrahedral arrangement is expected to occur less frequently than the octahedral, since the crystal field stabilization energy is always larger in the latter case. These values are sometimes used as a means of assessing the relative ease of obtaining tetrahedral *vis à vis* octahedral complexes. However, we consider that there are so many other factors influencing the question that the value of crystal field stabilization energies for this purpose is limited.

We note that Orgel[3] has pointed out that, although crystal field splittings appear to be the important factors determining the relative stabilities of the tetrahedral and octahedral configurations, nevertheless these qualitative arguments may well prove deceptive; and he stressed the need for further testing of the theory.

First, even when using the purely electrostatic model, it needs to be emphasized that the energy required to assemble the field-producing arrangement is not the same. In other words, the positions A and B in Fig. 1 do not represent the same energy, and, indeed, the difference between these could be of even greater significance than the crystal field stabilization energy which is measured relative to these points. Secondly, these calculations ignore repulsions between ligands, and the fact that metal-ligand bond distances in octahedral and tetrahedral complexes are not usually the same.

The theory suggests that for electronic configurations other than d^0, d^2, d^5, d^7, d^{10} and also spin-paired d^4, a regular tetrahedral arrangement is much less probable than an octahedral, tetragonal or square planar one. Thus four-covalent spin-free Ni^{II} (d^8) and Fe^{II} (d^6) complexes are not expected to occur in a regular tetrahedral arrangement; but rather, if they occur at all, in a grossly distorted tetrahedron. There has long been speculation concerning the existence of tetrahedral Ni^{II} complexes. The usually accepted example, Ni^{II} *bis*-acetyl acetone[12,13], is now known to be a trimer almost certainly involving octahedrally co-ordinated nickel. Recently, however, Venanzi[14] has reported that complexes of the type $NiCl_2.2Ph_3P$ are monomeric in benzene, and hence the nickel atom is four-covalent; X-ray studies by Venanzi and Powell indicate that it is at the centre of a distorted tetrahedron[14]. This result is supported by the magnetic moment (3·1 B.M.), which is only slightly in excess of the spin-only value, 2·83 B.M., the orbital contribution being small. The latter should be large if the Ni^{II} is surrounded by a regular tetrahedron of negative charges. This arises because,

Table 2. ARSONIUM TETRACHLORO COMPLEXES OF TRANSITION METALS
General formula $[Ph_3MeAs]_2[MCl_4]$

Property	Metal Mn^{II}	Fe^{II}	Co^{II}	Ni^{II}*	Cu^{II}	Zn^{II}
Colour	Pale green	Cream	Blue	Green blue	Yellow	White
Conductivity in $PhNO_2$ in r.o. (conc. ~ M/2,000) (20° C.)	57·0	Oxidizes	55·0	55·8	60·4	53·2
Magnetic moment at 20° C. (B.M.)	5·88	5·33	4·69	3·89	1·91	Diamag.

* The corresponding iodides of Mn^{II}, Fe^{II}, Co^{II} and Ni^{II}, for example, $[Ph_3MeAs]_2$ $[NiI_4]$, have been prepared and are isomorphous. The nickel complex is red and again has a high magnetic moment (3·49 B.M.). With the tetraethylammonium cation similar behaviour is observed, the magnetic moments of $[Et_4N]_2[NiCl_4]$ and $[Et_4N]_2[NiBr_4]$ being in the range 3·8–3·9 B.M.

Table 3. PREDICTED MAGNETIC MOMENTS FOR SPIN-FREE BIVALENT NICKEL COMPOUNDS

Environment of Ni^{II} atom	$\mu_{eff.}$ (calculated)
(1) Free Ni^{++} ion ($^3F_4$). No electrical field. $S = 1$, $L = 3$ (assuming $h\nu = 0$)	4·47 B.M.
(2) As in (1) but using accepted value of $\nu = 2{,}347$ cm.$^{-1}$	5·56 B.M.
(3) As in (1) but assuming $h\nu = \infty$	5·59 B.M.
(4) Ni^{2+} ion in octahedral ligand field assuming no spin-orbital coupling	2·83 B.M.
(5) Octahedral Ni^{II}, for example, $[Ni(H_2O)_6]^{2+}$ or $[Ni(NH_3)_6]^{2+}$, using accepted value of spin-orbit coupling constant λ (varies from 200 to 324)	~ 3·1 – 3·2 B.M.
(6) Ni^{2+} ion in tetrahedral ligand field, assuming no spin–orbital coupling (taking $L = 1$)	3·16 B.M.
(7) As in (6), using the same value of λ as in the octahedral case but assuming that Δ is very large, that is, that the only configuration to be considered is $d_\gamma^4 d_\varepsilon^4$	3·6 B.M.*

* We are indebted for this figure to Dr. B. N. Figgis, who also reports that if Δ is not large enough to preclude configuration interaction, then the moment may be as large as 4·2 B.M. (unpublished work).

in order to obtain a large orbital contribution to the magnetic moment, one must have the possibility of re-arrangement of electrons in d_ε-orbitals when a magnetic field is applied. Three degenerate d_ε-orbitals containing a total of four electrons, as occurs for the Ni^{II} atom in a regular tetrahedral complex, provides an ideal example of this. A probable example of tetrahedral Fe^{II} occurs in the compound[15] $FeI_2.2Ph_3P$.

We have recently been studying complexes of the type M(hal)$_2$.2 pyridine of those bivalent metals which should give rise to both regular tetrahedral and octahedral arrangements; for example, $Mn^{II}(d^5)$, $Co^{II}(d^7)$ and $Zn^{II}(d^{10})$. In this connexion we have prepared compounds of the type $[Ph_3CH_3As]_2$–M (hal)$_4^{2-}$, where hal = Cl, Br and I. These are obtained readily by adding the arsonium halide to the metal halide in alcoholic solution. In addition to the tetrahalide complexes of the metals listed above, we have obtained those of Fe^{II}, Ni^{II} and Cu^{II}. These are listed in Table 2. The Mn^{II}, Co^{II} and Zn^{II} complexes have the properties expected for four-covalent tetrahedral anions. The compounds are electrolytes in nitrobenzene, with the conductance expected for bi-univalent electrolytes. Only in the case of the Co^{II} compound is the orbital contribution to the magnetic moment helpful in deciding stereochemistry. The tetrahedral structure of the $[CoCl_4]^{2-}$ ion has been established in the compounds Cs_3CoCl_5 (ref. 16) and Cs_2CoCl_4 (ref. 17), the magnetic moments of which are close to that observed for $[Ph_3MeAs]_2$ $CoCl_4$. However, the magnetic moment of the Ni^{II} compound, 3·89 B.M., is unusually high; this points strongly to a tetrahedral arrangement of the ligands about the Ni^{II}. Moreover, for the orbital contribution to be so large, distortion from a regular tetrahedral arrangement is small. The expected magnetic moment for a tetrahedral Ni^{II} complex can only be calculated roughly owing to uncertainty as to the splitting, Δ, and the spin-orbit coupling constant. In Table 3 we list the calculated moments for Ni^{II} in various circumstances. The octahedral case has been carefully examined by Griffiths and Owen[18], who also emphasize that charge transfer between metal and ligand in metal hydrates occurs owing to weak σ-bonding.

The conclusion that these complex ions are tetrahedral is supported by X-ray diffraction investigations of single crystals of the compounds. With the exception of the Cu^{II} compound, which is expected to contain a tetragonal or square complex ion, the compounds of Mn^{II}, Fe^{II}, Co^{II}, Ni^{II}, and Zn^{II} are isomorphous, forming cubic crystals with cell edge 15·5 A. (Ni), space group $P2_13$, with four molecules in the unit cell. Though there are slight differences in cell constants, the intensities of corresponding diffraction maxima of the several compounds appear substantially the same, strongly indicating a regular tetrahedral arrangement for the Ni^{II} and Fe^{II} complexes, having accepted such arrangement for the Mn^{II}, Co^{II} and Zn^{II} complexes. Furthermore, the crystal symmetry is such that the metal atom lies on a three-fold axis, and taken in conjunction with the results of our other physical-chemical experiments, we conclude that it is surrounded by three equivalent chloride ions and one non-equivalent chloride ion also on the three-fold axis. The magnetic behaviour of $[Ph_4As]_2$ $[NiCl_4]$ in a suitable non-co-ordinating solvent should prove of interest, because under these conditions the constraining effects, if any, of the lattice of neighbouring ions will be removed.

These investigations are being extended to include other bivalent and tervalent metals, and to complete the crystal structure determination of $[Ph_3MeAs]_2$ $[NiCl_4]$. The important feature which emerges from these studies is that when considering cases where crystal field forces are opposed by others, particularly those arising from bonds, the crystal field forces may be swamped and the structure be that determined by bond and lattice interactions.

1 Bethe, H., *Ann. Phys.*, **3**, 1933 (1929); *Z. Phys.*, **60**, 218 (1930).
2 Van Vleck, J. H., *Phys. Rev.*, **41**, 208 (1932); *J. Chem. Phys.*, **3**, 807 (1935).
3 For references see Jorgensen, K., Tenth Solvay Council Proceedings (Brussels), 355 (1956); Orgel, L. E., *ibid.*, 289.
4 Van Vleck, J. H., and Howard, P., *J. Chem. Phys.*, **3**, 813 (1935). Penney, W., and Schlapp, R., *Phys. Rev.*, **41**, 194 (1932); **42**, 666 (1932); **43**, 486 (1933).
5 Schlapp, R., and Penney, W., "Reports on Progress of Physics", 2, 60 (1935), and references therein.
6 Eisenstein, J., and Pryce, M. H. L., *Proc. Roy. Soc.*, A, **229**, 20 (1955); *ibid.*, A, **238**, 31 (1956).
7 Orgel, L. E., ref. 3. Griffith, J. S., and Orgel, L. E., *Quart. Rev. Chem. Soc.*, **11**, 381 (1958).
8 Gillespie, R. J., and Nyholm, R. S., *Quart. Rev. Chem. Soc.*, **11**, 339 (1957), and references therein.
9 Pauling, L., *J. Chem. Soc.*, 1461 (1958).
10 Sidgwick, N. V., and Powell, H. M., *Proc. Roy. Soc.*, A, **176**, 153 (1940).
11 Dickens, P. G., and Linnett, J., *Quart. Rev. Chem. Soc.*, **11**, 291 (1957), and references therein.
12 Bullen, G. J., *Nature*, **177**, 537 (1956).
13 Shibata, S., Kishita, M., and Kubo, M., *Nature*, **179**, 320 (1957).
14 Venanzi, L., *J. Chem. Soc.*, 719 (1958); *Proc. Chem. Soc.*, 6 (1958). Powell (personal communication).
15 Hieber, W., and Floss, J. G., *Z. anorg. Chem.*, **291**, 314 (1957).
16 Powell, H. M., and Wells, A. F., *J. Chem. Soc.*, 359 (1935).
17 Porai-Koshits, M. A., *Kristallografiya* (*U.S.S.R*), **1**, 291 (1956).
18 Griffiths, J. H. E., and Owen, J., *Proc. Roy. Soc.*, A, **213**, 459 (1952). Owen, J., *ibid.*, A, **227**, 183 (1955).

Reprinted from *J. Chem. Phys.*, **29**, 930–937 (1958)

27

Effect of Pressure on the Spectra of Certain Transition Metal Complexes*

R. W. PARSONS AND H. G. DRICKAMER

Department of Chemistry and Chemical Engineering, University of Illinois, Urbana, Illinois

(Received February 28, 1958)

The effect of pressure (to 130 000 atmospheres) has been measured on the splitting of the 3d levels of the nickel and chromium ions complexed with water and ammonia. Data have also been obtained on $CoSO_4 \cdot 7H_2O$ and $K_3Fe(CN)_6$. In general, the effect of pressure is to increase the effect of the ligand field and to increase the splitting of the levels. This is interpreted in terms of the change in ligand-metal distance. A phase transition was noted in $[Ni(H_2O)_6]SO_4$ at 65 000 atmospheres.

A charge transfer band was also observed in $[NiNH_3)_6]Cl_2$. The effect of pressure is to increase the intensity with no apparent change in the location of the maximum.

THE electronic spectra of the complexes of the first transition group metals can be thought of as arising from two separate processes. First, there are a low number of absorption bands due to essentially internal transitions of the electrons in the incomplete 3d shell of the metal ion. The number of these bands is the same for all combinations of a given geometry and metal with various ligands, and they appear in the region from the near infrared, through the visible, and into the near ultraviolet. The position of these bands is highly dependent upon the particular ligand with which the metal is coordinated. By considering the anion or dipole ligands to exhibit an electric field of symmetry corresponding to the stereochemistry of the complex, it has been shown that the energy levels of the ground and excited states of the free ion are, in general, each split into various levels. The spectra can be nicely interpreted as the electronic transition from the lowest level to the various upper levels. Although some work has been done on the extremely weak transitions between states of differing multiplicities, only those transitions between states of the same multiplicity will be considered in this work.

Although qualitative calculations of the strength of the ligand field (commonly called the crystal field) do not give the proper order of magnitude of the spectra, the use of an empirically determined value does give good results, especially when other refinements are added. The discrepancy between theory and experiment is explained in part by the presence of chemical bonding. It is found that the empirically determined crystal field strength (E_1-E_2), historically termed 10 Dq, varies with changing ligands in approximately the same manner for each transition metal.

Another way of changing the crystal field strength, but in a continuous manner, is through the use of pressure. Using high-pressure bombs with sodium chloride as both the pressure transmitting medium and the light path, absorption spectra were taken to approximately 130 000 atmospheres in the visible ultraviolet range on transition metal [Ni(II) and Cr(III)] complexes using H_2O and NH_3 as the six ligands placed around the central atom in an essentially octahedral arrangement. In all cases the absorption maximum shifted blue indicating an increase in the crystal field strength. In two cases intensity measurements were also made, with the absorption of the peak increasing with pressure.

Less extensive measurements, mainly on band shapes, were made on two other systems. The observable band of $CoSO_4 \cdot 7H_2O$, which is a double band, was measured as a function of pressure. Also one "covalent" complex, $K_3Fe(CN)_6$, was studied to determine if its behavior was of a different character than the "normal" complexes.

The second type of electronic process giving rise to spectra has been termed change transfer and involves the transfer of an electron from the metal to the ligand or *vice versa*. These are broad, intense bands of higher energy than those previously mentioned, and usually

* This work was supported in part by U. S. Atomic Energy Commission contract AT(11-1)-67 Project 5.

only the edge of the band is observed. The effect of pressure upon one of these bands was observed with the main effect being an increase in absorption.

The high-pressure equipment and experimental methods were described elsewhere.[1]

The optical system consisted of a tungsten or hydrogen lamp, a Beckman DUR monochromator with homemade wavelength drive, an optical system for focusing the light in the bomb, and a 1P21 or 1P28 photomultiplier detector. The light was chopped at 13 cycles per second, and the photomultiplier signal passed to a Perkin-Elmer Model 81A amplifier and then to a Brown Electronik recorder.

The following list contains the chemicals used, the source of supply, any treatment given them, and the probable crystal form.

1. $NiSO_4 \cdot 6H_2O$: Mallinckrodt—reagent grade; dissolved in water with subsequent slow evaporation to give single crystals of a workable size; tetragonal; used as single crystal.

2. $[Ni(NH_3)_6]Cl_2$:$NiCl_2 \cdot 6H_2O$: J. T. Baker Chemical Company—reagent grade; dissolved in water then treated with DuPont reagent NH_3OH to give the crystals which were filtered and vacuum dried; cubic; used as pellet of pure material and as dilute pellet in NaCl matrix.

3. $KCr(SO_4)_2 \cdot 12H_2O$: Merk—reagent grade; treated similarly to 1; cubic; used as single crystal and pellet made from pure material.

4. $[Cr(NH_3)_6]Cl_3$: obtained from students of Dr. J. C. Bailar, Jr., and used without further treatment; cubic; used as pellet made from pure material.

5. $CoSO_4 \cdot 7H_2O$: Baker and Adamson—reagent grade; treated similar to 1; monoclinic; used as single crystal.

6. $K_3Fe(CN)_6$: J. T. Baker Chemical Company—reagent grade; no further treatment; monoclinic; used as dilute pellet in NaCl matrix.

Orgel[2] has calculated a relationship between the splitting of the α levels and the crystal field $E_1 - E_2$ for various transition metals. From his diagrams and the measured spectra $E_1 - E_2$ for various transition metals. From his diagrams and the measured spectra $E_1 - E_2$ can be established for any pressure.

Van Vleck,[3] using water dipoles placed in an octahedron around the metal atom and the whole cluster placed in a trigonal field due to non-nearest neighbors, derived an equation relating the crystal field strength to the metal-ligand distance,

$$Dq = \frac{(E_1 - E_2)}{10} = \frac{5e\mu\langle r^4 \rangle}{6R^6}, \qquad (1)$$

where e= electronic charge, μ= ligand dipole moment, r= radius of the $3d$ shell, and R= distance from center of metal ion to ligand. With changing metal-ligand distance the $\langle r^4 \rangle$ term and probably the μ term should not change as rapidly as $1/R^6$. Therefore to a first approximation $(E_1 - E_2)$ varies as $1/R^6$.

Jorgensen[4] has given a similar expression for the dependence of $(E_1 - E_2)$ on R. Using atomic units

$$(E_1 - E_2) = \frac{5q\langle r^4 \rangle}{3R^5}, \qquad (2)$$

where q is the charge of the ligand and would be found from the ligand electronic distribution.

From considerations of the purely vibrational dissymmetries of an octahedral complex regarded in the light of the crystal field theory, the following equation can be derived for the intensity of the absorption band of an "internal" $3d$ transition[5]:

$$P = \nu[(E_1 - E_2)^2/\nu^+]^2 e \times 10^{-7}, \qquad (3)$$

where P= oscillator strength, ν= frequency of maximum absorption, ν^+= distance to nearest odd level and, e= degeneracy of upper level. This equation is derived for the weak field approximation and therefore should apply to the common Ni^{++} and Co^{++}, but not the Cr^{+++}, complexes. Although it contains several assumptions and estimates as to certain numerical values, it gives the correct order of magnitude for several "weak field" complexes.

SHIFTS OF ABSORPTION PEAKS

Data were obtained for four systems using Ni(II) and Cr(III) for the central metal atoms and H_2O and NH_3 as the six octahedrally arranged ligands. Two absorption peaks were observed and the change in the position of the maximum was obtained as a

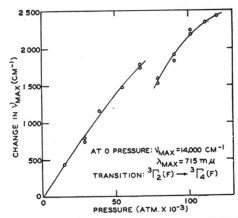

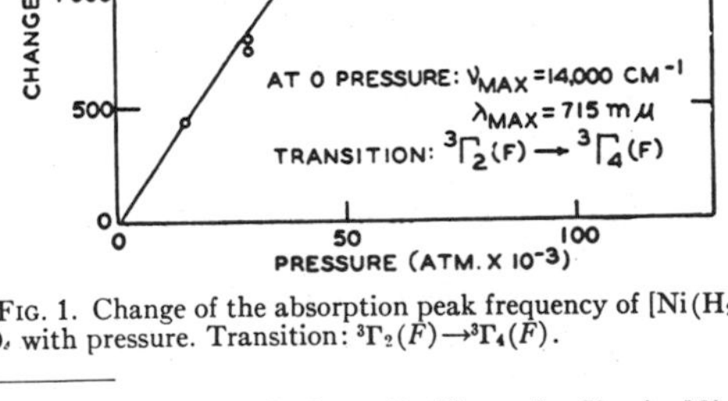

FIG. 1. Change of the absorption peak frequency of $[Ni(H_2O)_6]SO_4$ with pressure. Transition: $^3\Gamma_2(F) \rightarrow {}^3\Gamma_4(F)$.

[1] Fitch, Slykhouse, and Drickamer, J. Opt. Soc. Am. **74**, 1015 (1957).

[2] L. E. Orgel, J. Chem. Phys. **23**, 1004 (1955).

[3] J. H. Van Vleck, J. Chem. Phys. **7**, 72 (1939).

[4] C. K. Jorgensen, *Quelques Problèmes de Chemie Minerale* (R. Stoops, Brussels, 1956), p. 355.

[5] C. J. Ballhausen, Acta Chem. Scand. **9**, 821 (1955).

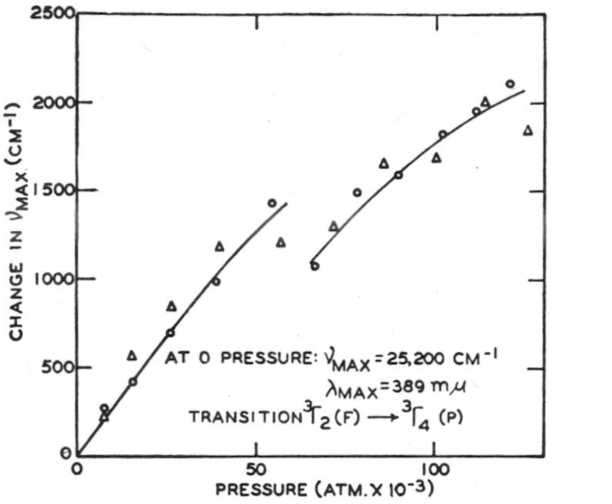

FIG. 2. Change of the absorption peak frequency of $[Ni(H_2O)_6]SO_4$ with pressure. Transition: $^3\Gamma_2(F)\rightarrow{}^3\Gamma_4(P)$.

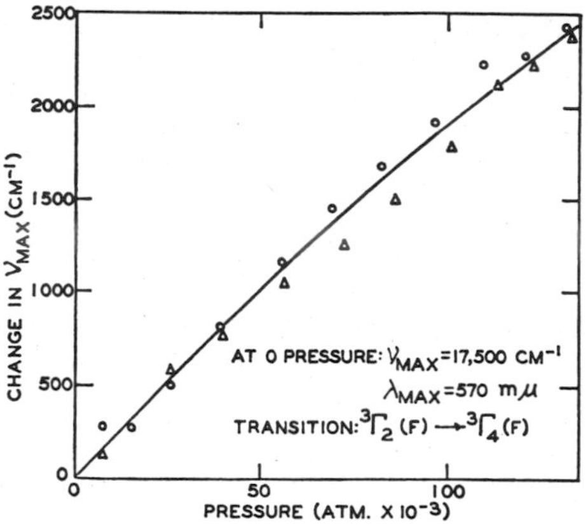

FIG. 4. Change of the absorption peak frequency of $[Ni(NH_3)_6]Cl_2$ with pressure. Transition: $^3\Gamma_2(F)\rightarrow{}^3\Gamma_4(F)$.

function of pressure. These data and the change in the parameter (E_1-E_2) are shown in Figs. 1–12. This latter was obtained from the diagrams of Orgel.[2] According to the simple picture the change in the energy for both absorption peaks should lead to the same dependence of (E_1-E_2) upon pressure. Therefore, on the third plot data from both bands are used.

All of the data can be represented as smooth continuous curves with the exception of those of

$$[Ni(H_2O)_6]SO_4,$$

whose discontinuity will be discussed later. The change in the position of maximum absorption is toward the blue, higher energy, in every case. When these shifts are converted to crystal field strengths, the resulting two curves for each compound do not coincide. The qualitative agreement between the curves representing the two bands is good, but there is a discrepancy of up to 15% of the shift.

This discrepancy might be due to a systematic experimental error, but it also could be the result of the noncubic nature of the crystalline field. Tanabe and Sugamo[6] and Owen[7] have presented more refined analyses of the energy differences between the various states. In the nomenclature of Tanabe and Sugamo, Δ is equivalent to our E_1-E_2, and $15B$ is an effective measure of covalency.

For Ni^{++}:

$$E(^3\Gamma_2(F)\rightarrow{}^3\Gamma_4(F)) = 3/2\Delta-15/2B-1/2[(\Delta-9B)^2+144B^2]^{\frac{1}{2}}, \quad (4)$$

$$E(^3\Gamma_2(F)\rightarrow{}^3\Gamma_4(P)) = 3/2\Delta-15/2B+1/2[(\Delta-9B)^2+144B^2]^{\frac{1}{2}} \quad (5)$$

For Cr^{+++}:

$$E(^4\Gamma_2(F)\rightarrow{}^4\Gamma_5(F)) = \Delta, \quad (6)$$

$$E(^4\Gamma_2(F)\rightarrow{}^4\Gamma_4(F)) = 3/2\Delta-15/2B-1/2[(\Delta-9B)^2+144B^2]^{\frac{1}{2}} \quad (7)$$

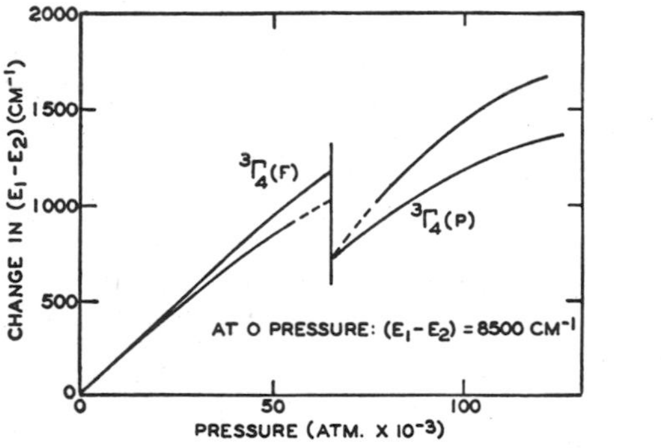

FIG. 3. Change in crystal field strength (E_1-E_2) *vs* pressure for $[Ni(H_2O)_6]SO_4$.

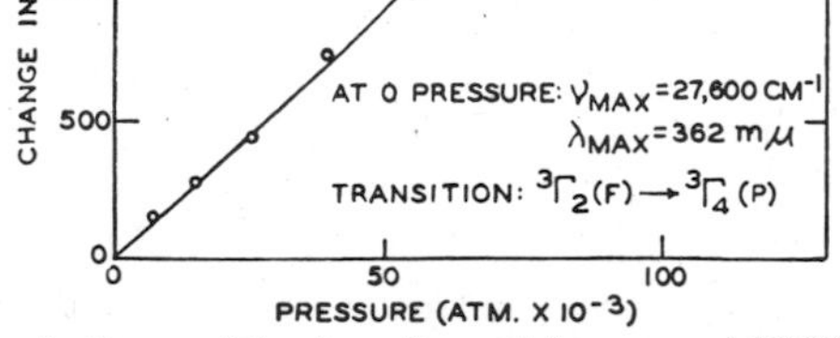

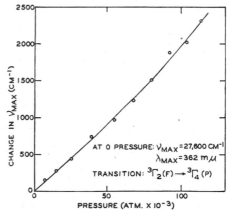

FIG. 5. Change of the absorption peak frequency of $[Ni(NH_3)_6]Cl_2$ with pressure. Transition: $^3\Gamma_2(F)\rightarrow{}^3\Gamma_4(P)$.

[6] Y. Tanabe and S. Sugano, J. Phys. Soc. Japan **9**, 753, 766 (1954).
[7] J. Owen, Proc. Roy. Soc. (London) **A227**, 183 (1955).

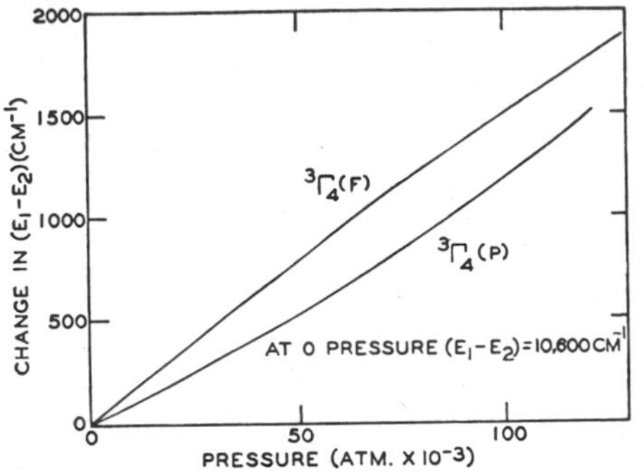

FIG. 6. Change of crystal field strength (E_1-E_2) *vs* pressure for $[Ni(NH_3)_6]Cl_2$.

From Eqs. (4)–(7) and our data, (E_1-E_2) and $15B$ can be calculated as a function of pressure. The results for the two systems involving Ni^{++} and the two involving Cr^{+++} are included in Table I.

(1) System $NI(6H_2O)SO_4$

The values of E_1-E_2 check those from Orgel's diagrams. No reasonable assumed error in the data would change the trend for $15B$ to decrease with increasing pressure, i.e., for the covalency to increase with pressure.

(2) System $Ni(NH_3)_6Cl_2$

The values of E_1-E_2 are somewhat higher than those obtained from Orgel's diagrams. The value of $15B$ increases with pressure. If ν max for zero pressure for the transition ${}^3\Gamma_2(F)\rightarrow{}^3\Gamma_4(F)$ is lowered to 16 600, the lowest value found in the literature, $(E_1-E_2)_{p=0}$ is lowered to 11 200, but the trend with pressure for both E_1-E_2 and $15B$ remains the same. Apparently the covalency decreases with increasing pressure for this system.

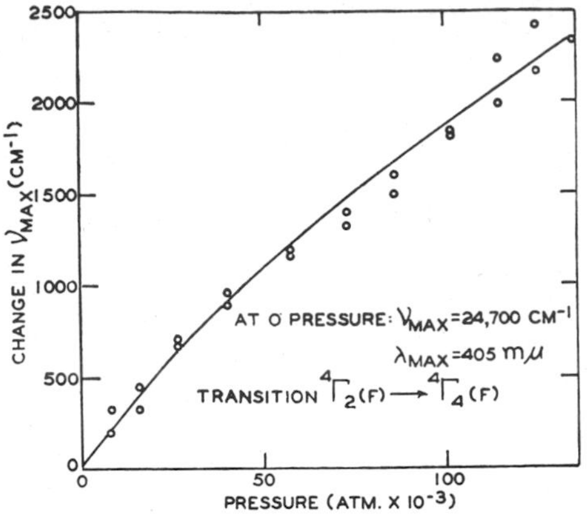

FIG. 8. Change of the absorption peak frequency of $KCr(SO_4)_2\cdot12H_2O$ with pressure. Transition: ${}^4\Gamma_2(F)\rightarrow{}^4\Gamma_4(F)$.

(3) System $KCr(SO_4)_2\cdot12H_2O$

The change $15B$ with pressure is most difficult to establish unequivocally for this system because small differences are involved in both numerator and denominator. No reasonable error in the data would give a decrease of $15B$ with increasing pressure, but the increase shown may not be significant. The results indicate a possible small decrease in covalency with pressure.

(4) System $Cr(NH_3)_6Cl_3$

The calculations indicate a decrease in $15B$ with increasing pressure. The direction of the change, although not the magnitude, seem outside of experimental error.

In general there are changes in $15B$ in the order of $\pm5-10\%$ in 120 000 atmospheres. There appears to be no consistency in the direction of the change with

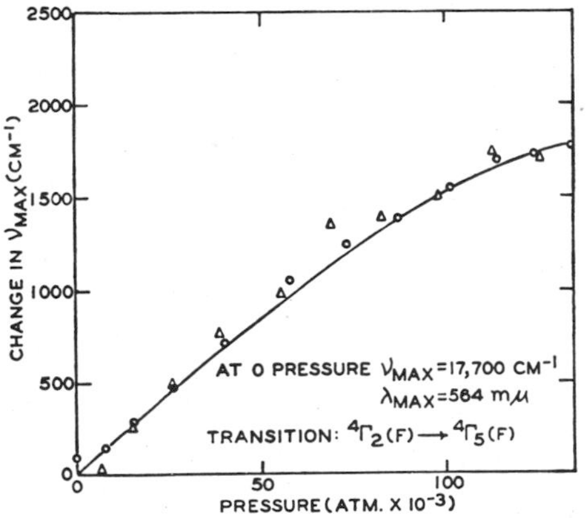

FIG. 7. Change of the absorption peak frequency of $KCr(SO_4)_2\cdot12H_2O$ with pressure. Transition: ${}^4\Gamma_2(F)\rightarrow{}^4\Gamma_5(F)$.

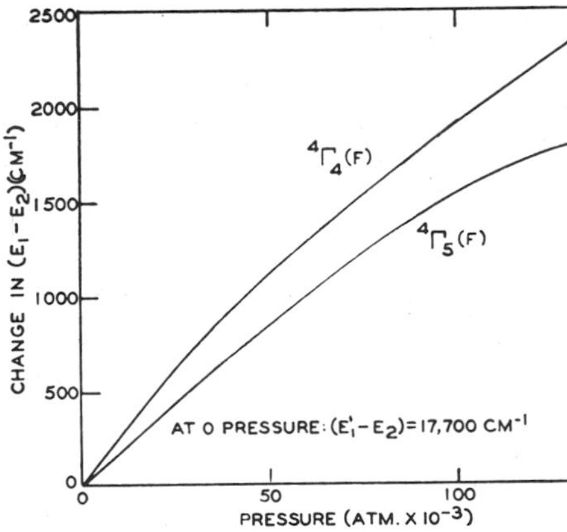

FIG. 9. Change in crystal field strength (E_1-E_2) *vs* pressure for $KCr(SO_4)_2\cdot12H_2O$.

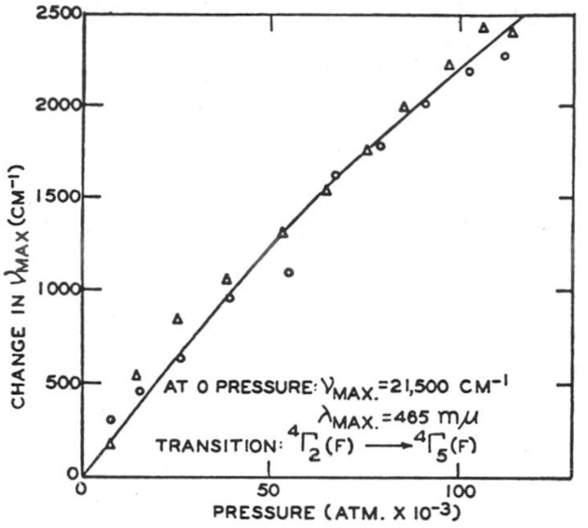

FIG. 10. Change of the absorption peak frequency of $Cr(NH_3)_6Cl_3$ with pressure. Transition: $^4\Gamma_2(F)\rightarrow{}^4\Gamma_5(F)$.

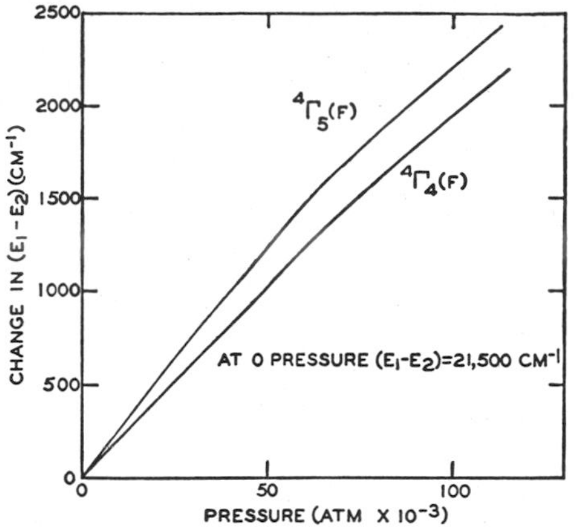

FIG. 12. Change in crystal field strength (E_1-E_2) *vs* pressure for $[Cr(NH_3)_6]Cl_3$.

pressure, and thus in the change of the degree of covalency with pressure.

From Eqs. (1) and (2) one can approximate the changes in the metal-ligand bond distance with pressure, using the average data for the (E_1-E_2) change. From these equations a first approximation gives (E_1-E_2) as being proportional to $1/R^6$ or $1/R^5$. The fractional change in bond length *versus* pressure is shown in Fig. 13 with both dependencies being shown. The range of the data is from 1.5 to 3.5% change in bond length in 120 000 atmospheres. Whether the dependence of R on (E_1-E_2) is to the -5 or -6 power does not make too much difference, especially in view of the discrepancy of the data arising from the two bands of the same compound.

For the most part these data verify what would be expected from qualitative considerations. For a given complex pressure would be expected to cause a compression of the metal-ligand distance, which in turn would cause the electrical field in the vicinity of the metal ion to be increased, resulting in further splitting of the levels. It might also be noted that the divalent Ni complexes showed a greater metal-ligand compressibility than the trivalent Cr complexes. In general the trivalent complex ions have larger $(E_1-E_2)_0$ values than comparable divalent ones.

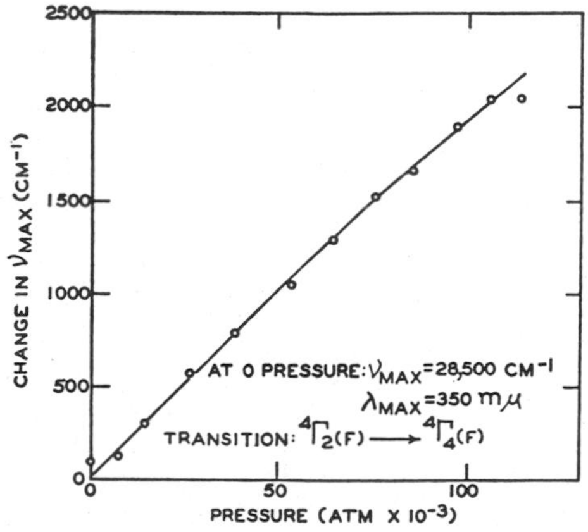

FIG. 11. Change of the absorption frequency of $Cr(NH_3)_6Cl_3$ with pressure. Transition: $^4\Gamma_2(F)\rightarrow{}^4\Gamma_4(F)$.

$Ni(H_2O)_6SO_4$ PHASE TRANSITION

The data for both of the absorption bands observed for $Ni(H_2O)_6SO_4$ show a definite discontinuity in the

TABLE I. Calculation of E_1-E_2 and $15B$ [from Eqs. (4) to (7)].

P in atmos$\times 10^3$
$\Delta=(E_1-E_2)$ in cm^{-1}
$15B$ in cm^{-1}

System $Ni(6H_2O)SO_4$			System $Ni(NH_3)_6Cl_2$		
P	$\Delta=E_1-E_2$	$15B$	P	$\Delta=E_1-E_2$	$15B$
0	8360	14 120	0	11 705	9400
20	8800	13 940	20	11 910	9500
40	9270	13 660	40	12 120	9650
60	9540	13 670	60	12 310	9860
80	9600	13 510	80	12 550	9930
100	9900	13 410	100	12 750	10 100
120	9970	13 400	120	13 000	10 300

System $KCr(SO_4)_2 12H_2O$			System $Cr(NH_3)_6Cl_3$		
P	$\Delta=E_1-E_2$	$15B$	P	$\Delta=E_1-E_2$	$15B$
0	17 700	10 500	0	21 500	9830
20	18 050	10 700	20	22 000	9780
40	18 380	10 800	40	22 500	9550
60	18 700	10 800	60	22 950	9480
80	18 980	10 810	80	23 350	9330
100	19 220	10 900	100	23 720	9330
120	19 400	11 000	120	24 050	9250

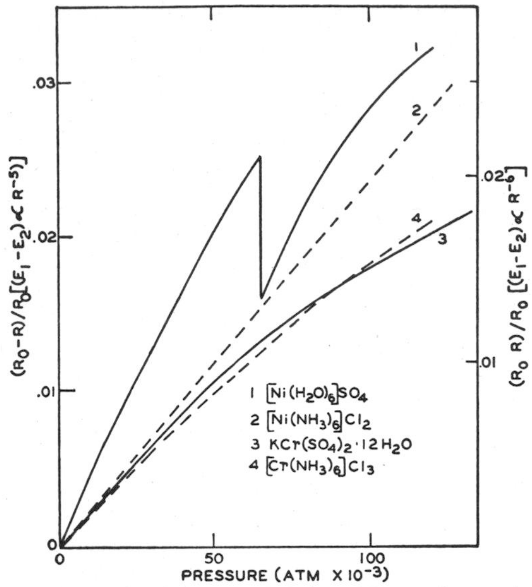

FIG. 13. Fractional decrease of the metal-ligand distance (R) *vs* pressure.

shift of maximum wave number with pressure (see Figs. 1 and 2). Since the highest experimental point of the lower portion of Fig. 1 is at 67 000 atmospheres and the lowest point of the upper portion of Fig. 2 is at 66 000 atmospheres it is clear that the discontinuity is fairly sharp. Because of the experimental error in the determination of pressure the point of discontinuity will be taken to be 65 000 *atmospheres*.

The interpretation of this discontinuity is a phase change of the crystal. Since the crystal field, and consequently the spectrum, depends upon the surrounding ligands to the first order and upon the rest of the crystal structure to the second order, one would not expect too drastic a change in the spectrum with a change in crystal structure, as long as the same ligands remain coordinated with the central metal ion in the same general geometrical configuration. If this is true, the phase change could affect the crystal field strength in two ways, through the direct contribution of the non-nearest neighbors to the field and through the indirect action in which the ligands are influenced by the surrounding crystal structure, thus altering their contribution to the field.

For the case at hand the discontinuity represents an energy decrease of the absorption peaks. If it is assumed that the complex ion is still octahedral Ni $(H_2O)_6{+}{+}$, which is a good assumption since both bands shift about the same amount in (E_1-E_2), the effect of the phase transition upon the complex ion is to lengthen the metal-ligand distance.

$CoSO_4 \cdot 7H_2O$

The only transition examined for this compound was ${}^4\Gamma_4(F) \to {}^4\Gamma_4(P)$, which is reported as a double band with unresolved maxima at 19 500 and 21 500 cm^{-1}. The spectra were taken in a 1/8 inch bomb up to 42 000 atmospheres (see Fig. 14). If the atmospheric spectrum were resolved into the component bands, the maxima would correspond closely with the reported values. As the pressure is increased the doublet shape of the band disappears. Whether this is due to a change in the intensity or position, apparently mainly of the lower wavelength band, of the components is not certain. The position of the maximum of the higher wavelength component is estimated to shift about 900 cm^{-1} to the blue upon going from 0 to 41 700 atmospheres. This corresponds to a change of 1050 cm^{-1}, or 11.5% in (E_1-E_2). This is of the proper order of magnitude as compared with the other complexes studied.

INTENSITY MEASUREMENTS

The intensity of the peaks of two absorption bands were measured as a function of pressure and the results are shown in Fig. 15. For comparison purposes the primary data were normalized so that the atmospheric value of I_0/I was 1 for both bands.

It is assumed that the intensity, as measured, is only due to the particular band under consideration. There is a possibility of the edges of other bands contributing somewhat to the absorption but it is too difficult to determine to what extent they may be present.

A semitheoretical estimation of the change in intensity for the two bands is also shown. These lines are

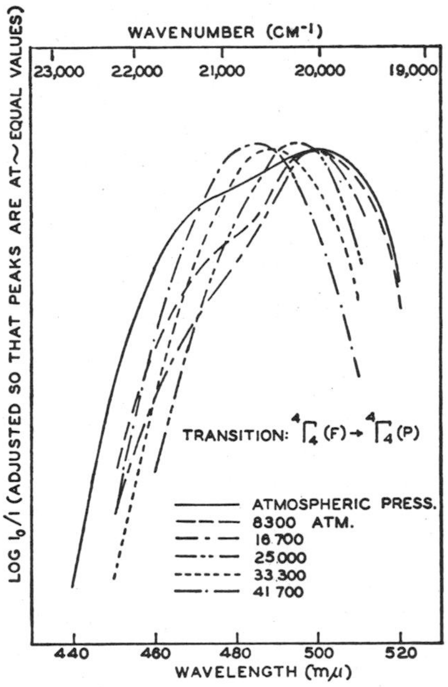

FIG. 14. Absorption spectra of $CoSO_4 \cdot 7H_2O$ as a function of pressure.

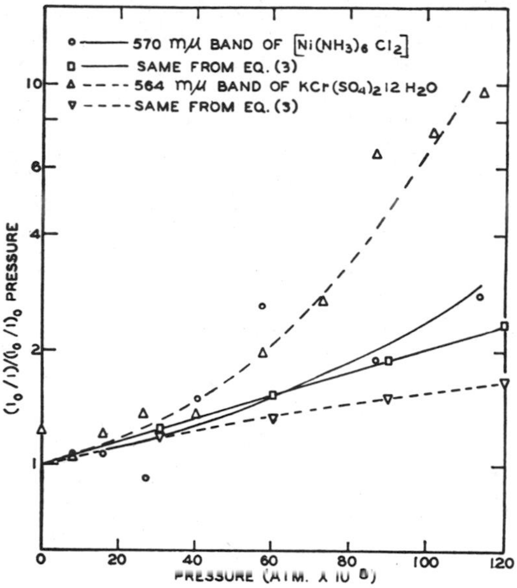

FIG. 15. Relative absorption at band peak *vs* pressure.

obtained from Eq. (3), which relates the oscillator strength to the crystal field strength and energy of the absorption peak. The change of oscillator strength, which is proportional to $\ln I_0/I$, can thus be estimated by the change in $(E_1 - E_2)$ and ν max.

It is seen that the two sets of points for the 570-mμ

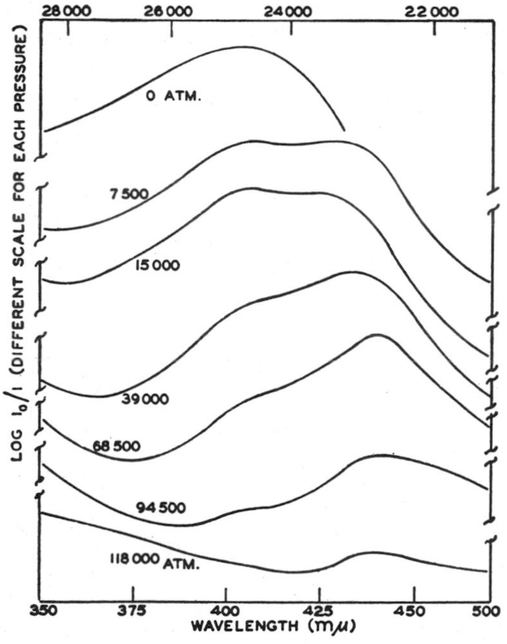

FIG. 16. Absorption spectra of $K_3Fe(CN)_6$ at various pressures.

TABLE II. Intensity of $K_3Fe(CN)_6$ absorption *vs* pressure.

Pressure	I_0/I (at band max)
0	7.7
7 500	3.22
15 000	3.55
39 000	2.81
65 500	2.95
94 500	2.53
118 000	1.93

band of $Ni(NH_3)_6Cl_2$ are in close correspondence, while those for the 546-mμ band of $KCr(SO_4)_2 \cdot 12H_2O$ are of widely divergent character. This would indicate that the CrIII aquocomplex contains considerably more covalent binding than does the Ni(II) complex with ammonia.

$K_3Fe(CN)_6$

Spectra for $K_3Fe(CN)_6$ were run at various pressures up to 118 000 atmospheres and the results are shown in Fig. 16 for the band whose maximum is about 400 mμ. It is estimated that the absorption coefficient is in the order of 100 times larger than those of the previously discussed bands, as is evidenced by the necessity of using a very dilute sample [1% $K_3Fe(CN)_6$ in NaCl pellet]. This salt differs from others investigated in that it is a "covalent" complex, according to Pauling's magnetic criterion. One would expect the empirical crystal field strength $(E_1 - E_2)$ to be quite large mainly because of the double bond character of the cyanide-metal linkage. No theoretical interpretation will be made of the results.

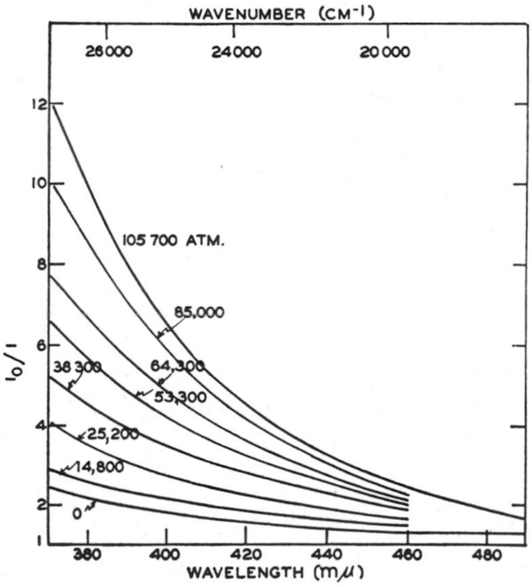

FIG. 17. Change in absorption, I_0/I, with pressure for a dilute pellet of $[Ni(NH_3)_6]Cl_2$.

The maximum absorption can be seen to exhibit a definite shift to lower energies (red shift), being some 2100 cm^{-1} in 118 000 atmospheres. The shape of the band also appears to change considerably, exhibiting a doublet appearance, with a relative change in the contribution from each component with changing pressure. It is possible, and even probable, that some of the irregularity in band shape is due to the use as a source of hydrogen discharge lamp, which exhibits some discrete emission spectra in this region, coupled with the unreliability of the condition of the salt windows from spectrum to spectrum (in this case from blank to sample.) However, this uncertainty is not great enough to invalidate the previously mentioned red shift. Another factor which may affect the peak position is the overlapping of adjacent bands and the subsequent behavior of these bands with pressure.

Aside from the reverse direction of the shift of absorption peak, it is also interesting to note the relative intensities of the absorption bands. Table II shows this information, and generally the intensity is seen to decrease with increasing pressure. This is in the reverse direction from that shown for the "normal" complexes.

CHARGE TRANSFER SPECTRA

It was noticed when obtaining data on $Ni(NH_3)_6Cl_2$ that the higher energy band was practically obscured by an intense absorption, presumably the beginning of a broad charge transfer band. In order to study this further a dilute pellet was made containing about 8% of the complex in sodium chloride. At these conditions the broad absorption could be measured, with the contribution to I_0/I of the normal bands appearing only a slight indentations on the absorption curve. The results are plotted in Fig. 17 showing the change in I_0/I with pressure over the wavelength range 370 to 490 mμ.

The obvious change with pressure is an increased absorption at a given wavelength. For example, at 105 700 atmospheres the fractional change in $\ln I_0/I$ above the atmospheric pressure value is 1.8, 2.2, and 2.3 for the wavelengths 370 400 and 450 mμ, respectively. This is proportional to the change in extinction or absorption coefficients. Since these values are fairly constant and since the curves seem to be converging to a common point, it is concluded that the edge of the charge transfer absorption band does not change positions but only increases in intensity.

From the small values and the small changes of I_0/I above 500 mμ, plus the possible experimental error, it is deduced the intensities reported for the 570-mμ band of $Ni(NH_3)_6Cl_2$ are due entirely to the internal 3d transition.

ACKNOWLEDGMENTS

The authors would like to express appreciation to Dr. M. H. L. Pryce for fruitful suggestions. R. W. Parsons would like to acknowledge financial assistance from a Minnesota Mining and Manufacturing Company Fellowship.

Reprinted from *J. Amer. Chem. Soc.*, **82**, 5771–5774 (1960)

28

[CONTRIBUTION FROM THE DEPARTMENT OF CHEMISTRY, MASSACHUSETTS INSTITUTE OF TECHNOLOGY, CAMBRIDGE, MASSACHUSETTS]

New Tetrahedral Complexes of Nickel(II)

BY F. ALBERT COTTON AND DAVID M. L. GOODGAME

RECEIVED MARCH 9, 1960

The Ni(II) complexes $[(C_6H_5)_3PO]_2NiX_2$ where X = Cl, Br and I have been prepared. Their physical properties, especially their electronic spectra and high magnetic moments, leave little doubt that the nickel ion is in each case surrounded by two oxygen atoms and two halogen ions in an essentially tetrahedral array.

Introduction

The occurrence of tetrahedrally coördinated nickel(II) ions evidently is rather rare, considerably rarer than was believed only a few years ago. It is now well established that, contrary to the presumption long made on the basis of valence bond theory, but in accord with the suggestion early made by Ballhausen,[1] paramagnetic nickel(II) complexes may be planar.[2–4] Thus it cannot now be certain, and is perhaps even unlikely, that many of the complexes formerly assumed to be tetrahedral simply because they are paramagnetic are, actually, tetrahedral.

Moreover, calculations by the methods of ligand field theory have provided criteria by which tetrahedral complexes may be identified with fair certainty. Liehr and Ballhausen[5] have provided definitive calculations of the energy levels of a Ni(II) ion in fields of T_d symmetry which provide a basis for ascertaining by spectral studies whether the ion is tetrahedrally coördinated. Figgis[6] has established the range in which the magnetic moments of nickel ions in fields of T_d symmetry may be expected to lie at room temperature and also has shown what the intrinsic temperature dependence of the moment should be.

Within the past few years a few authentic examples of tetrahedrally coördinated Ni(II) ions have been reported. Tetrahalo nickel(II) ions have been shown to exist in certain molten salt mixtures[7,8] and in salts with very large cations,[9,10] and Ni(II) ions have been trapped in tetrahedral interstices in oxides[11,12] and glasses.[13] In all these cases, the coördination is "truly" tetrahedral, in the sense that all four ligand atoms are identical. It has been suggested[14,15] that even in such cases the geometry could not remain exactly tetrahedral because of a Jahn–Teller distortion expected for a T_2 electronic ground state. However, Liehr[16] has pointed out that when the effects of spin-orbit coupling are considered, the ground state is not degenerate (it has A_1 symmetry) and Jahn–Teller forces are inoperative. Moreover, since the closest electronic state, which under nuclear displacements might perturb the ground state conformational regularity, is ~300 cm.$^{-1}$ distant,[5] there should not be any large pseudo Jahn–Teller distortions either, and the regular tetrahedral configuration should be stable. The only certain

(1) C. J. Ballhausen, *Kgl. Danske Videnskab. Selskab, Mat.-fys. Medd.*, **29**, No. 9 (1955).

(2) C. J. Ballhausen and A. D. Liehr, THIS JOURNAL, **81**, 538 (1959).

(3) G. Maki, *J. Chem. Phys.*, **28**, 651 (1958); **29**, 162 (1958).

(4) C. Furlani, *Gazz. chim. Ital.*, **88**, 279 (1958).

(5) A. D. Liehr and C. J. Ballhausen, *Ann. phys.*, **6**, 134 (1959).

(6) B. N. Figgis, *Nature*, **182**, 1568 (1958).

(7) D. M. Gruen and R. L. McBeth, *J. Phys. Chem.*, **63**, 393 (1959).

(8) B. R. Sundheim and G. Harrington, *J. Chem. Phys.*, **31**, 700 (1959).

(9) N. S. Gill and R. S. Nyholm, *J. Chem. Soc.*, 3997 (1959).

(10) F. A. Cotton and R. Francis, THIS JOURNAL, **82**, 2986 (1960); *ibid.*, in press.

(11) D. S. McClure, *Phys. Chem. Solids*, **3**, 311 (1957).

(12) O. Schmitz-DuMont, H. Gössling and H. Brokopf, *J. anorg. u. allgem. Chem.*, **300**, 159 (1959).

(13) W. A. Weyl, Proceedings of the Xth Solvay Conference, Brussels, 1956.

(14) L. Venanzi, *J. Chem. Soc.*, 719 (1958); R. S. Nyholm, *J. Inorg. Nuclear Chem.*, **8**, 401 (1958).

(15) F. A. Cotton, E. Bannister, R. Barnes and R. H. Holm, *Proc. Chem. Soc.*, 158 (1959).

(16) A. D. Liehr, Symposium on the Synthesis and Properties of Coördination Compounds, 137th Meeting of the American Chemical Society, April 5–14, 1960, Cleveland, Ohio.

"pseudotetrahedral" nickel(II) complexes to be reported heretofore are compounds $(C_6H_5P)_2NiX_2$ (X = Cl, Br, I). We use the prefix pseudo in those cases where the nickel(II) ion is surrounded by several different kinds of ligand atoms, either two of one kind and two of another or three of one kind and one of another kind, which stand exactly or approximately at the apices of a tetrahedron with the nickel(II) ion at its center. Such complexes cannot have T_d symmetry of the ligand field but only C_{2v} (in the 2, 2, case) or C_{3v} (in the 3, 1 case). Venanzi and Powell[14] have demonstrated the pseudotetrahedral character of these molecules, which is in marked contrast to the *cis*-planar or dimeric nature of all other $(R_3P)_2NiX_2$ compounds so far investigated[17–22] with the possible exception of $[(C_2H_5)_3P]_2Ni(NO_3)_2$.[14,18,19,22]

In this paper we report the preparation and thorough characterization of the new pseudotetrahedral nickel(II) complexes $[(C_6H_5)_3PO]_2NiX_2$, where X = Cl, Br and I. Attempts to obtain corresponding compounds with acetate and thiocyanate as anions gave yellow materials which have not as yet been characterized further. They are presumably planar as is $Ni(OPC_6H_5)_2(NO_3)_2$ which has already been reported.[23]

Experimental

Preparation. Bis-(triphenylphosphine oxide)-dichloronickel.—This compound has not been prepared in a completely pure state. A solution of triphenylphosphine oxide (4.40 g., 0.0158 mole) and hexa-aquo nickel chloride (1.78 g., 0.0075 mole) in absolute ethanol (10 ml.), was placed in an evacuated desiccator over sulfuric acid for several days. The dark green residue was heated on a steam-bath to remove the last traces of ethanol, when a blue solid contaminated with some yellow product was obtained. Excess triphenylphosphine oxide was removed by treating the solid with two portions of 25 ml. of hot cyclohexane. The pale blue product was filtered, washed with 25 ml. of cold cyclohexane and dried *in vacuo*. The yield was practically quantitative. The compound melted at 190°.

Anal. Calcd. for $C_{36}H_{30}Cl_2NiO_2P_2$: C, 63.01; H, 4.41; Cl, 10.33; Ni, 8.55. Found: C, 61.16; H, 4.57; Cl, 9.68; Ni, 8.20.

The compound is soluble in alcohols to give green solutions. It is very soluble in dimethylformamide to give solutions which are blue when hot but green when cold. A blue solution also is formed on heating with acetonitrile, but a yellow solid is formed in cooling this solution. The compound is insoluble in but not decomposed by ligroin, *n*-hexane or cyclohexane. All other solvents (fourteen were tried) decomposed the compound with the formation of mainly yellow solids. The compound could not therefore be recrystallized and no stable blue solution could be obtained.

Bis-(triphenylphosphine oxide)-dibromonickel.—A solution of triphenylphosphine oxide (4.70 g., 0.0169 mole) and nickel bromide (2.05 g., 0.0075 mole) in absolute ethanol (45 ml.) was evaporated on a steam-bath until blue crystals began to separate out from the deep blue-green solution. The mixture then was kept overnight in an evacuated desiccator over sulfuric acid. The resulting mixture of blue crystals and viscous blue-green solution was treated with about 50 ml. of hot diethyl ether to remove excess triphenylphosphine oxide. The deep blue crystalline solid was filtered off, washed with cold ether and dried *in vacuo*. The compound melted at 208°.

Anal. Calcd. for $C_{36}H_{30}Br_2NiO_2P_2$: C, 55.78; H, 3.90; Br, 20.62; Ni, 7.57; P, 7.99. Found: C, 55.60; H, 3.87; Br, 20.57; Ni, 7.72; P, 7.79.

The compound is slightly soluble in nitrobenzene but readily soluble in cold acetone to give a blue solution. It is decomposed by nitromethane and yields orange-yellow solids on treatment with benzene or chloroform.

Bis-(triphenylphosphine oxide)-di-iodonickel.—A solution of triphenylphosphine oxide (4.69 g., 0.0169 mole) and nickel iodide (2.35 g., 0.0075 mole) in absolute ethanol (37.5 ml.) was heated to boiling and then cooled and placed in an evacuated desiccator over sulfuric acid. After a week a dark red viscous product remained. On treating this with hot ethyl acetate (40 ml.) and filtering, a dark olive-green crystalline solid was obtained. The solid product was washed at the pump with cold ethyl acetate (25 ml.) and dried *in vacuo*. The yield was 2.66 g. (41%). The compound melted at 209.5°.

Anal. Calcd. for $C_{36}H_{30}I_2NiO_2P_2$: C, 49.75; H, 3.48; I, 29.21; Ni, 6.75; P, 7.13. Found: C, 49.49; H, 3.40; I, 28.85; Ni, 6.77; P, 6.92.

The compound is insoluble in carbon tetrachloride, cyclohexane and ligroin. It is decomposed by nitrobenzene, dioxane and chloroform. On treatment with cold acetonitrile the compound becomes dark red and then readily dissolves to give a very pale green solution. Dark green solutions are formed in benzene, chlorobenzene, toluene and hot ethyl acetate. Red-brown solutions are given by acetone, methyl ethyl ketone, nitromethane and 2-nitropropane.

Measurement of Electrolytic Conductances.—Electrolytic conductance measurements were carried out using a Serfass bridge and a conventional cell previously calibrated with an aqueous solution of potassium chloride.

Bis-(triphenylphosphine oxide)-dibromonickel was found to be a non-electrolyte in nitrobenzene solution. Bis-(triphenylphosphine oxide)-di-iodonickel had a molar conductance of 33.9 mhos at 23.8° for a 10^{-3} molar solution in nitromethane.

Magnetic Measurements.—Bulk susceptibility measurements were made at room temperature using a Gouy method as previously described.[24] Mohr's salt and copper sulfate pentahydrate were used to calibrate the Gouy tubes. Duplicate determinations were carried out.

	T, °K.	$\chi^M_{corr.} \times 10^6$	Diamagnetic corr. $\times 10^6$	μ (B.M.)
$[Ni((C_6H_5)_3PO)_2Cl_2]$	298.5	5694	−406	3.7 ± 0.1
$[Ni((C_6H_5)_3PO)_2Br_2]$	297.3	6615	−427	3.98 ± .05
$[Ni((C_6H_5)_3PO)_2I_2]$	296.5	6171	−455	3.84 ± .05

The diamagnetic corrections were calculated using the measured value for the susceptibility of triphenylphosphine oxide given by Foex.[25]

Infrared Absorption Spectra.—Infrared absorption spectra were taken on a Perkin–Elmer 21 spectrophotometer, fitted with a rock-salt prism. Nujol mulls were used. The positions of the peaks due to the P=O stretching vibration of the coördinated phosphine oxide ligands are shown below together with the P=O frequency of triphenylphosphine oxide itself.

	P=O stretching freq., cm.$^{-1}$	Shift, cm.$^{-1}$
$(C_6H_5)_3PO$	1195	..
$[Ni((C_6H_5)_3PO)_2Cl_2]$	1160	−35
$[Ni((C_6H_5)_3PO)_2Br_2]$	1154	−41
$[Ni((C_6H_5)_3PO)_2I_2]$	1151	−44

These results are in good agreement with those previously reported for other complexes of triphenylphosphine oxide.[26]

Electronic Spectra.—The reflectance spectra of the solid compounds were measured using a Beckman DU spectro-

(17) L. Venanzi, *et al.*, International Conference on Coördination Chemistry, London, April 6–9, 1959 (Abstract No. 82).

(18) K. A. Jensen, *Z. anorg. Chem.* **229**, 265 (1936).

(19) R. W. Asmussen, *et al.*, *Acta Chem. Scand.* **9**, 1391 (1955).

(20) A. Turco, V. Scatturin and G. Giacometti, *Gazz. chim. Ital.*, **89**, 2005 (1959).

(21) A. Turco and G. Giacometti, *Ricerc. Sci.*, **29**, 1057 (1959).

(22) G. Giacometti, V. Scatturin and A. Turco. *Gazz. chim. Ital.*, **88**, 434 (1958).

(23) F. A. Cotton and E. Bannister, *J. Chem. Soc.*, 2276 (1960)

(24) R. H. Holm and F. A. Cotton, *J. Chem. Phys.*, **31**, 788 (1959).

(25) G. Foex, "Constantes Selectionées Diamagnetisme et Paramagnetisme." Masson et Cie, Paris. 1957.

(26) F. A. Cotton, R. D. Barnes and E. Bannister, *J. Chem. Soc.*, 2199 (1960).

TABLE I

ELECTRONIC ABSORPTION SPECTRA OF THE COMPLEXES

Compound	Medium	Color	Position of absorption bands, mμ — Molar extinction coefficients, for solns.; sh = shoulder				
$((C_6H_5)_3PO)_2NiCl_2$	Solid	Blue	440		615	700	>1200
$((C_6H_5)_3PO)_2NiBr_2$	Solid	Blue	487		635	700	>1200
	0.01 *M* in acetone	Blue	487(21.5)		645(101)	750(sh)	1370(19)
$((C_6H_5)_3PO)_2NiI_2$	Solid	Green	~400(broad)	545	670	750(sh)	>1200
	0.005 *M* in PhCl	Green	440(470)	~540(sh)	662(137)	730(sh)	1415(18.2)
	0.01 *M* in acetone	Red	515(535)	~590(sh)	715(136)	775(sh)	1365(19.2)

photometer with the standard Beckman reflectance accessory and magnesium carbonate as the reference sample. The solution spectra were measured with a Beckman DK2 recording spectrophotometer. The results are shown in Table I and Fig. 1.

Discussion

The data quoted in Table I and in the Experimental section and the spectra given in Fig. 1 suffice to show that each of the compounds reported here is a true pseudotetrahedral nickel(II) complex, and, moreover, that the C_{2v} component necessarily superimposed on the main cubic hemihedral symmetry of the ligand field is, in these cases, of no major importance insofar as the spectra and magnetic moments at room temperature are concerned.

The evidence for these conclusions is essentially perfect in the case of $((C_6H_5)_3PO)_2NiBr_2$ and we shall discuss this case first. The compound is a non-electrolyte in nitrobenzene, which rules out any possibility of its being ionic in nature, *e.g.*, $[((C_6H_5)_3PO)_4Ni][NiBr_4]$, although this kind of ionic structure is common among dimethyl sulfoxide complexes[10] and the $[((C_6H_5)_3PO)_4Ni]^{+2}$ cation is known.[27] Thus evidence of the presence of tetrahedrally coördinated nickel(II) must also show that it is the molecular species $[Ni(OP(C_6H_5)_3)_2Br_2]$ which is tetrahedral. There are three principal lines of evidence for the presence of this tetrahedral molecule. (1) The energies of the bands in the electronic spectrum (Fig. 1 and Table I) agree very well with the energy level scheme calculated by Liehr and Ballhausen,[5] using $B = 810$ cm.$^{-1}$, $\lambda = -275$ cm.$^{-1}$ and $Dq \approx 250$ cm.$^{-1}$. We assign the doublet at 635–700 mμ to the $\Gamma_1(^3T_{1g}) \rightarrow {}^3P$ transition and the symmetrical band at 1370 mμ to the $\Gamma_1(^3T_{1g}) \rightarrow {}^3\Gamma_5(^3F)$ transition. From the Liehr–Ballhausen nomograph, it seems reasonable to assign the weak band at 487 mμ to the $\Gamma_1(^3T_{1g}) \rightarrow {}^1\Gamma_3(^1G)$ transition. (2) The intensity of the $\Gamma_1(^3T_{1g}) \rightarrow {}^3P$ absorption is so high that from this datum alone it is fairly certain that we must be dealing with a tetrahedral complex. Our spectra also may be compared with spectra previously reported for nickel(II) known or believed to be tetrahedrally coördinated.[7–12] (3) The very high magnetic moment, 3.98 B.M. at 297°K., of the nickel (II) ion in this compound is also by itself strong evidence for the presence of tetrahedrally coördinated nickel(II), when compared with the theoretical predictions of Figgis.[6] Again, on a purely empirical basis, our moment may be compared with those reported for other known or presumed tetrahedral nickel(II) complexes.[9,10]

It also appears that the C_{2v} component of the ligand field is sufficiently small that it does not appreciably alter the level pattern expected to subsist in a truly tetrahedral ligand field which is, in the sense of Figgis,[6] weak. This is credible since oxygen and bromide are not much separated in the spectrochemical series. Were the C_{2v} component of the ligand field very large, we should expect to see splitting in the electronic absorption bands and we should expect the magnetic moment to be substantially lower than that found. Such effects of a really strong C_{2v} perturbation have been observed in $((C_6H_5)_3P)_3CoX_2$ and $((C_6H_5)_3P)_2NiX_2$ compounds.[28]

The next best characterized of the complexes reported here is the iodide, $[Ni(OP(C_6H_5)_3)_2I_2]$. The equivalent conductance in nitromethane, ~34 mho at 24° in 10^{-3} molar solution leads us to believe that the substance is a non-electrolyte in the solid state but that there is some electrolytic dissociation, probably in the nature of

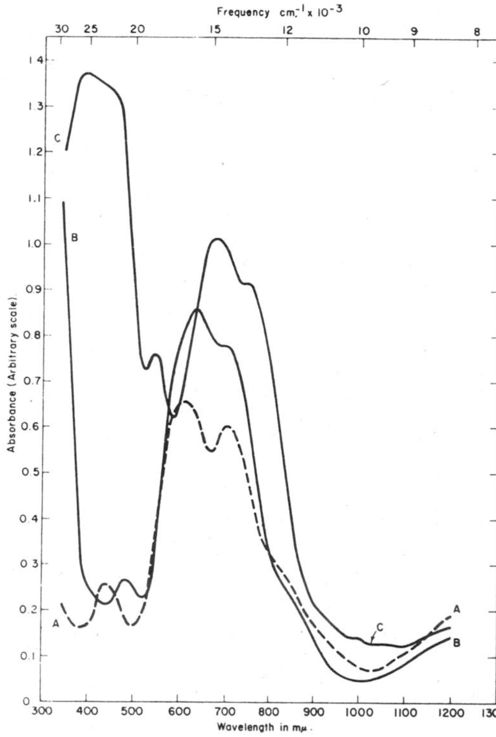

Fig. 1.—The electronic reflectance spectra of the solid compounds: A, $[(C_6H_5)_3PO]_2NiCl_2$; B, $[(C_6H_5)_3PO]_2NiBr_2$; C, $[(C_6H_5)_3PO]_2NiI_2$.

(27) F. A. Cotton and E. Bannister *J. Chem. Soc.*, 1873 (1960).

(28) F. A. Cotton, D. M. L. Goodgame, R. H. Holm and O. D. Faut, to be published.

$$[Ni(OP(C_6H_5)_3)_2I_2] = [Ni(OP(C_6H_5)_3)_2I]^+ + I^-$$

in nitromethane solution. Were the compound actually ionic, we should expect a much higher conductance, probably ~180 mho for a di-divalent electrolyte. Assuming, then, that the complex is a non-electrolyte, its tetrahedral conformation is attested by the electronic absorption spectra (Table I and Fig. 1) and its high magnetic moment (3.84 B.M. at 297°) following the same lines of reasoning as those given above for the homologous bromide. Again, the effects of the C_{2v} component of the ligand field are not marked.

The compound $[Ni(OP(C_6H_5)_3)_2Cl_2]$ has not been obtained in a state of high purity, but the spectral and magnetic data provide, here again, quite strong evidence for its pseudotetrahedral constitution. The reflectance spectrum (Fig. 1) agrees well with the Liehr–Ballhausen nomograph taking $D_q \approx 300$ cm.$^{-1}$ and the magnetic moment is high.

It has already been shown that the cobaltous compound, $Co(OP(C_6H_5)_3)_2Br_2$, is tetrahedral.[29] However, this and the corresponding Ni(II) complex are not isomorphous according to X-ray powder diagrams.[30] The diagrams are rather similar in the low angle region, suggesting similar short-range order in both cases, but in the higher angle region the two patterns are diverse indicating lack of identity of the lattices.

Acknowledgment.—The financial support of the United States Atomic Energy Commission, under Contract No. AT(30-1)-1965, is gratefully acknowledged.

(29) R. H. Holm and F. A. Cotton, *J. Chem. Phys.*, **32**, 1168 (1960).

(30) Kindly obtained for us by Dr. I. Simon of Arthur D. Little Co.

Reprinted from *J. Amer. Chem. Soc.*, **82**, 5005 (1960)

29

ELECTRONIC STRUCTURE AND MOLECULAR ASSOCIATION OF SOME BIS-(β-DIKETONE)-NICKEL(II) COMPLEXES

Sir:

Recently[1,2] it was shown from ligand field arguments that one may expect planar, tetracoördinate complexes of Ni(II) to be either paramagnetic or diamagnetic depending on the strength of the surrounding field. Bis-acetylacetonato-Ni(II) has been cited as a prominent example of a planar paramagnetic material. Studies in this laboratory make it increasingly apparent that the spin-free ground state may be caused by intermolecular associations, instead of being characteristic of the free monomeric molecule.

While bis-(acetylacetonato)-Ni(II), I, is planar[3] and monomeric[4] in the vapor phase, an incomplete X-ray analysis of the sublimed solid[5] indicates intermolecular associations. The material is known to form a dihydrate readily, suggesting a strong tendency for the Ni(II) to become octahedrally coördinated. While the visible spectrum and magnetic properties of the complex in hydrocarbon solvents exclude a tetrahedral arrangement, octahedral coördination by means of polymerization is not excluded. Molecular weight determinations in dichloromethane indicate that I is substantially polymerized.[6] The preparation and characterization of the complexes reported offers additional evidence for this view.

Bis-(2,2,6,6-tetramethyl-3,5-heptanediono)-Ni-(II), II, ($R_1 = R_2 = -C(CH_3)_3$, $R_3 = H$) was found

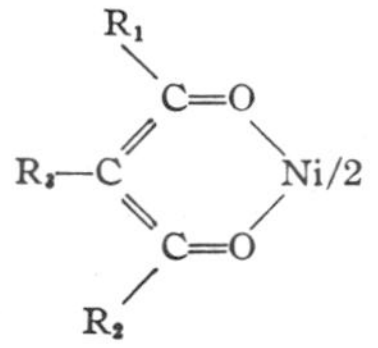

to be diamagnetic. Hydrocarbon solutions of the material exhibit a single absorption peak in the visible at 535 mμ, with a molar extinction coefficient of 60, giving a red coloration. The material readily forms a blue-green paramagnetic dihydrate.

Anal. Calcd. for $NiC_{22}H_{38}O_4$: C, 62.14; H, 8.92. Found: C, 61.98; H, 8.92.

Bis-(2,2-dimethyl-3,5-heptanediono)-Ni(II), III ($R_1 = R_2 = -CH(CH_3)_2$, $R_3 = H$), was found to be paramagnetic, μ_{eff} (297.4° K.) = 3.41 B.M., in the solid phase. Toluene solutions of the anhydrous material indicate that the magnetic moments of the solutions are concentration and temperature dependent. The color of the solutions changes from green near 0° to red at around 50°, corresponding to the growth of a band at 535 mμ. All attempts to explain the magnetic properties of these solutions and the growth of the 535 mμ band by means of a Boltzmann distribution of the molecules in the singlet and triplet states[2] failed. The concentration dependence of both the moment and the intensity of the 535 mμ band excluded this assumption and led to an attempt to correlate the data assuming a temperature-dependent distribution among monomers and polymers in solution. Assuming the monomer to be diamagnetic and to be the only species absorbing at 535 mμ, it was possible to fit the data at several concentrations and temperatures to a set of equilibrium expressions.

Anal. Calcd. for $NiC_{18}H_{30}O_4$: C, 58.57; H, 8.11. Found: C, 58.48; H, 8.29.

The spin-paired ground state in II can be attributed to the fact that the bulky *t*-butyl groups prevent intermolecular association. Scale molecular models indicate that only slight strain is incurred when the Ni(II) atom in III is placed near the oxygens or chelate ring of a neighboring molecule. The tendency for Ni(II) to become octahedrally coördinate apparently overcomes this strain in the solid.

Preliminary studies on a third compound, bis-(3-phenyl-2,4-pentanediono)-Ni(II), IV, ($R_1 = R_2 = CH_3$, $R_3 = C_6H_5$) gives additional support to the idea that the paramagnetism in anhydrous β-diketone complexes of Ni(II) can be attributed to intermolecular interactions. Models indicate that the phenyl group in IV cannot rotate into coplanarity with the chelate ring due to the proximity of the methyl groups. Close intermolecular association of the Ni(II) atom with a neighboring molecule would be hindered by the phenyl groups perpendicular to the chelate ring. The anhydrous material, from drying the blue-green hydrate, is red and diamagnetic, a single band being observed in the visible spectrum of the solid at 535 mμ. In toluene the material appears to behave similarly to III, the color of the solution changing from green to red as the temperature is increased from 0 to 50°. It also appears that a green crystalline modification of the material can be formed by melting the red powder.

Studies are also being carried out on the visible spectrum of I in hydrocarbon solutions at fairly high temperatures. It has been observed visually that the color of solutions of I in dibenzyl become

(1) G. Maki, *J. Chem. Phys.*, **28**, 651 (1958); **29**, 162 (1959); **29**, 1129 (1959).

(2) A. H. Liehr and C. Ballhausen, THIS JOURNAL, **81**, 538 (1959).

(3) S. Shibata, *Bull. Chem. Soc. Japan*, **30**, 753 (1957).

(4) F. Gach, *Monatsh. Chemie*, **21**, 98 (1900).

(5) G. J. Bullen, *Nature*, **177**, 537 (1956). Drs. R. Mason and P. Pauling at University College London recently have re-examined Bullen's three-dimensional data and find that the molecules in the trimer are so arranged that each nickel ion is surrounded by six oxygen atoms in a slightly distorted octahedron (private communication).

(6) F. A. Cotton and R. H. Soderberg, to be published.

reddish-brown near 200° and reversibly return to the green color upon cooling.

Financial support was provided by the U. S. Atomic Energy Commission under Contract No. AT(30–1)-1965 and by the Research Corporation to whom grateful acknowledgment is made. J.P.F. wishes to acknowledge a Fellowship received from the Allied Chemical and Dye Corporation, Semet-Solvay Division.

DEPARTMENT OF CHEMISTRY J. P. FACKLER, JR.
MASSACHUSETTS INSTITUTE OF TECHNOLOGY
CAMBRIDGE 39, MASSACHUSETTS F. A. COTTON

RECEIVED JULY 12, 1960

Reprinted from *Inorg. Chem.*, **6**, 924–929 (1967)

30

CONTRIBUTION FROM THE DEPARTMENT OF CHEMISTRY,
MASSACHUSETTS INSTITUTE OF TECHNOLOGY, CAMBRIDGE MASSACHUSETTS 02139

Molecular Orbital Calculations for Complexes of Heavier Transition Elements. III. The Metal–Metal Bonding and Electronic Structure of $Re_2Cl_8^{2-}$ [1]

BY F. A. COTTON AND C. B. HARRIS[2]

Received November 3, 1966

The metal–metal and metal–chlorine bonding in $(Re_2Cl_8)^{2-}$ are treated by "extended" Hückel molecular orbital theory. The calculation suggests that the π-bonding contribution to the Re–Re bond is five times that of the δ bonding and almost three times that of the σ bonding. The Re–Re bond stabilization and the rotational barrier in $(Re_2Cl_8)^{2-}$ are calculated as 366 and 51 kcal, respectively, by comparing a hypothetical $(ReCl_4)^-$ anion (C_{4v}) with the $(Re_2Cl_8)^{2-}$ anion (D_{4h}). The ordering of the molecular orbitals is discussed with respect to the magnetic properties and the observed and calculated spectral properties.

Introduction

The preparation[3] and structure[4] of the $[Re_2Cl_8]^{2-}$ anion have been discussed. Its chemistry[5] appears to be consistent with the proposed[6] quadruple metal–metal bond. It therefore provides an excellent opportunity to study the effects of the various factors contributing to metal–metal bonding. This approach, however, requires a detailed knowledge of the individual orbital contributions to the metal–metal bond; therefore, an extensive molecular orbital calculation for $[Re_2Cl_8]^{2-}$ has been undertaken. All Re–Re, Re–Cl, and Cl–Cl interactions have been considered in a semiempirical approach of the type generally called an extended Hückel calculation. It is recognized that for various reasons, to be discussed in detail below, such a calculation cannot provide results which can be taken literally. It is our belief, however, that the results obtained, when interpreted properly, do provide a semiquantitative picture of the main features of the metal–metal bonding and their relative importance. For this reason, we believe that the study reported here provides a useful advance beyond the level of the simple overlap treatment which has already been given[6] for $Re_2Cl_8^{2-}$.

Method of Calculation

Choice of the Basis Set.—A basis set of fifty atomic orbitals, χ_i (i = 1, 2, ..., 50), was used to construct the molecular orbitals ψ^j (j = 1, 2, ..., 50), in the LCAO–MO approximation[7] (eq 1). This basis set included the 5d, 6s, and 6p orbitals of each Re atom and the 3s and 3p orbitals of each Cl atom. It was assumed that the nonvalence atomic orbitals on both Re atoms and the eight Cl atoms did not participate in bonding, but formed a core potential that was unaltered by interactions of the valence electrons. The atomic orbitals, χ_i's, were expressed as single term Slater-type orbitals[8] (STO's) as given in eq 2, where α_i is the shielding parameter, N_i is the normalization co-

$$\psi^j = \sum_i C_i^j \chi_i \qquad (1)$$

(1) Work supported by the U. S. Atomic Energy Commission.
(2) Predoctoral Fellow of the National Institutes of Health, 1964–1966; AEC Postdoctoral Fellow, 1966–1967.
(3) F. A. Cotton, N. F. Curtis, B. F. G. Johnson, and W. R. Robinson, *Inorg. Chem.*, **4**, 326 (1965).
(4) F. A. Cotton and C. B. Harris, *ibid.*, **4**, 330 (1965).
(5) (a) F. A. Cotton, N. F. Curtis, and W. R. Robinson, *ibid.*, **4**, 1696 (1965); (b) F. A. Cotton, C. Oldham, and R. A. Walton, *ibid.*, **5**, 1798 (1966).
(6) F. A. Cotton, *ibid.*, **4**, 334 (1965).
(7) J. H. Van Vleck, *J. Chem. Phys.*, **2**, 22 (1934).
(8) J. C. Slater, *Phys. Rev.*, **36**, 57 (1930).

efficient, n_i is the principle quantum number, and $Y_l^m(\theta,\varphi)$ is the usual spherical harmonic. All atomic

$$\chi_i = N_i r^{n_i - 1} \exp(-\alpha_i r) Y_l^m(\theta,\varphi) \qquad (2)$$

orbitals were expressed in a right-handed coordinate system as shown in Figure 1.

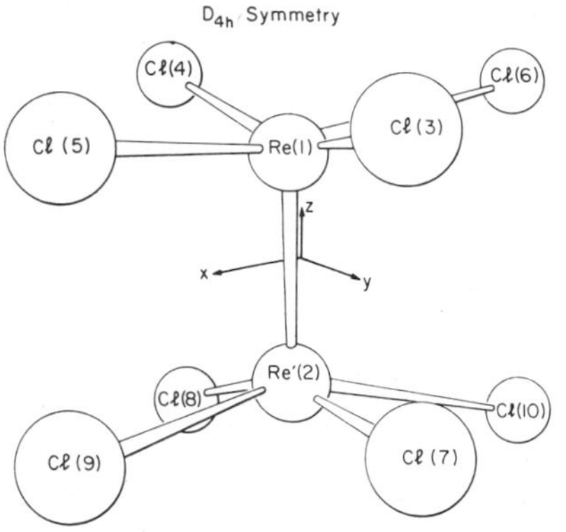

Figure 1.—$Re_2Cl_8^{2-}$ structure and coordinate system used to calculate overlap integrals. All atoms in a right-hand coordinate system. Numbers in parentheses correspond to atom numbers in Table I.

Evaluation of the Overlap Integrals.—The shielding parameters, α_i's, for the Re orbitals were determined by adjusting α_i to fit the overlap integrals between STO orbitals centered on Re and Re′ (Figure 1) to a numerical overlap integral between self-consistent field (SCF)[9] Re wave functions. This method has been discussed more fully earlier.[10] The α_i's for the Re 5d and 6s orbitals were obtained in this fashion from the overlap integrals of the type Re–Re′ 5d–5d and 6s–6s, respectively. Since the SCF Re 6p wave function was not available, the Re 6p radial wave function distribution was assumed to be slightly more diffuse than that of an Re 6s orbital.

The Cl α_i's were determined by numerically integrating the SCF Re orbital with an SCF[11,12] Cl orbital and fitting the resultant overlap with an overlap integral between STO Re and Cl wave functions. The Re α_i's had already been fixed by the Re–Re′ overlaps; thus, only the Cl α_i's needed to be varied. The Cl 3s and 3p α_i's listed with the Re α_i's in Table I were obtained by averaging the α_i's from Re–Cl 5d–3s, 6s–3s, 5d–3p, and 6s–3p overlaps. In this fashion all Re–Re and Re–Cl overlap integrals are essentially the same as overlap integrals between SCF wave functions. Only the very small Cl–Cl overlaps are in error. We assumed that the shielding parameters did not change with charge configuration changes. The fact that only a very small error is introduced by such an assumption has been discussed earlier.[10]

Evaluation of the Diagonal Matrix Elements, H_{ii}.—

(9) F. Herman and S. Skillman, "Atomic Structure Calculations," Prentice-Hall, Englewood Cliffs, N. J., 1963.
(10) F. A. Cotton and C. B. Harris, *Inorg. Chem.*, **6**, 369 (1967).
(11) R. E. Watson and A. J. Freeman, *Phys. Rev.*, **123**, 521 (1961).
(12) R. E. Watson and A. J. Freeman, *ibid.*, **120**, 1125 (1960).

TABLE I

INPUT PARAMETERS

Orbital	Atom	Orbital type	Coordinates, A X	Y	Z	Orbital exp	H_{ii}
1	1	6s	0	0	1.12	1.56	−9.65
2	1	$6p_z$	0	0	1.12	1.50	−5.20
3	1	$6p_x$	0	0	1.12	1.50	−5.20
4	1	$6p_y$	0	0	1.12	1.50	−5.20
5	1	$5d_{z^2}$	0	0	1.12	2.11	−10.46
6	1	$5d_{xz}$	0	0	1.12	2.11	−10.46
7	1	$5d_{x^2-y^2}$	0	0	1.12	2.11	−10.46
8	1	$5d_{yz}$	0	0	1.12	2.11	−10.46
9	1	$5d_{xy}$	0	0	1.12	2.11	−10.46
10	2	6s	0	0	−1.12	1.56	−9.65
11	2	$6p_z$	0	0	−1.12	1.50	−5.20
12	2	$6p_x$	0	0	−1.12	1.50	−5.20
13	2	$6p_y$	0	0	−1.12	1.50	−5.20
14	2	$5d_{z^2}$	0	0	−1.12	2.11	−10.46
15	2	$5d_{xz}$	0	0	−1.12	2.11	−10.46
16	2	$5d_{x^2-y^2}$	0	0	−1.12	2.11	−10.46
17	2	$5d_{yz}$	0	0	−1.12	2.11	−10.46
18	2	$5d_{xy}$	0	0	−1.12	2.11	−10.46
19	3	3s	0	2.22	1.66	3.37	−23.01
20	3	$3p_z$	0	2.22	1.66	2.46	−13.85
21	3	$3p_x$	0	2.22	1.66	2.46	−13.85
22	3	$3p_y$	0	2.22	1.66	2.46	−13.85
23	4	3s	0	−2.22	1.66	3.37	−23.01
24	4	$3p_z$	0	−2.22	1.66	2.46	−13.85
25	4	$3p_x$	0	−2.22	1.66	2.45	−13.85
26	4	$3p_y$	0	−2.22	1.66	2.46	−13.85
27	5	3s	2.22	0	1.66	3.37	−23.01
28	5	$3p_z$	2.22	0	1.66	2.46	−13.85
29	5	$3p_x$	2.22	0	1.66	2.46	−13.85
30	5	$3p_y$	2.22	0	1.66	2.46	−13.85
31	6	3s	−2.22	0	1.66	3.37	−23.01
32	6	$3p_z$	−2.22	0	1.66	2.46	−13.85
33	6	$3p_x$	−2.22	0	1.66	2.46	−13.85
34	6	$3p_y$	−2.22	0	1.66	2.46	−13.85
35	7	3s	0	2.22	−1.66	3.37	−23.01
36	7	$3p_z$	0	2.22	−1.66	2.46	−13.85
37	7	$3p_x$	0	2.22	−1.66	2.46	−13.85
38	7	$3p_y$	0	2.22	−1.66	2.46	−13.85
39	8	3s	0	−2.22	−1.66	3.37	−23.01
40	8	$3p_z$	0	−2.22	−1.66	2.46	−13.85
41	8	$3p_x$	0	−2.22	−1.66	2.46	−13.85
42	8	$3p_y$	0	−2.22	−1.66	2.46	−13.85
43	9	3s	2.22	0	−1.66	3.37	−23.01
44	9	$3p_z$	2.22	0	−1.66	2.46	−13.85
45	9	$3p_x$	2.22	0	−1.66	2.46	−13.85
46	9	$3p_y$	2.22	0	−1.66	2.46	−13.85
47	10	3s	−2.22	0	−1.66	3.37	−23.01
48	10	$3p_z$	−2.22	0	−1.66	2.46	−13.85
49	10	$3p_x$	−2.22	0	−1.66	2.46	−13.85
50	10	$3p_y$	−2.22	0	−1.66	2.46	−13.85

As noted earlier,[10] H_{ii}, the energy of an electron of the ith atomic orbital moving in the field of the nuclei and other electrons of the molecule, can be expressed in terms of the one-center atomic energy integrals, A_{ii}, and the multicentered molecular energy integrals, M_{ii}. This is expressed in eq 3. The A_{ii}'s can be estimated

$$H_{ii} = A_{ii} + M_{ii} \qquad (3)$$

by the valence state ionization potentials (VSIP) of the ith atomic orbital.

However, since most of the spectral states and processes[13] for Re(0) and Re(I) have not been assigned,

(13) C. E. Moore, "Atomic Energy Levels," U. S. National Bureau of Standards Circular 467, U. S. Government Printing Office, Washington D. C., 1949 and 1952.

the VSIP's for Re(0) cannot be determined with any accuracy by this method. Therefore, the VSIP's for the Re(0) 6s, 6p, and 5d orbitals utilized in this calculation were set equal to those used in the $ReCl_6^{2-}$ calculation.[14] This can be justified, to a large extent, by the *a posteriori* success in obtaining reasonable Re–Cl interactions in $ReCl_6^{2-}$ as shown by the agreement of the results with various experimental data, particularly by the satisfactory correlation of quadrupole coupling constant data[15] with the calculated charge distribution[14] using a recently published theoretical relationship.[16]

While one might expect, therefore, that the calculation of the Re–Cl interactions in $Re_2Cl_8^{2-}$ could be carried out employing these VSIP's, there remains, however, the effect of the Re–Re interactions or the metal–metal bonding on these parameters. Little can be said quantitatively about this, but qualitatively, the Re orbitals can only become more stable (more negative values for the VSIP's) because of the Re–Re interactions. This is a consequence of the increased electron-nuclear Coulombic interaction resulting from both a short Re–Re bond[2] and an effective positive charge on the Re atoms in $Re_2Cl_8^{2-}$.

Semiquantitatively, this stabilization should be proportional to the extent of penetration by the wave function on one Re atom into the core of the other Re atom; consequently the σ orbitals would be stabilized more than the π orbitals which would be stabilized more than the δ orbitals. It is not feasible to try to calculate the amount of stabilization energy for each orbital type (σ, π, δ). Therefore, rather than make arbitrary guesses at these energies the calculations were carried out making no changes in the Re VSIP's from the values[14] used in $ReCl_6^{2-}$ with the intention of making allowances for penetration effects in interpreting the results.

Basically, there should be two effects on the final MO diagram (Figure 2). First, the separations between the σ–σ^*, π–π^*, and δ–δ^* molecular orbitals ought to be greater than those calculated since the bonding states should be more stable than indicated. Second, the nonbonding $\sigma_n(1)$ molecular orbital should have lower energy than that calculated. Because of the large positive effective charge (1.25) calculated for Re, the sensitivity of penetration corrections,[17] and the diffuseness of the 6s and 6p wave function,[9] we believe that the $\sigma_n(1)$ could be stabilized by about 2 ev and the δ–δ^* separation could be increased by 1 ev.

The M_{ii}'s can be approximated by a point charge expression and an additional penetration correction. The same dependence on charge was assumed here for M_{ii} as was assumed in the $PtCl_4^{2-}$ molecular orbital calculation.[10]

The Cl 3s and 3p A_{ii}'s used were those calculated by Hinze and Jaffé[18] for $sp^2p^2p^2$ and $s^2p^2p^2p$ states,

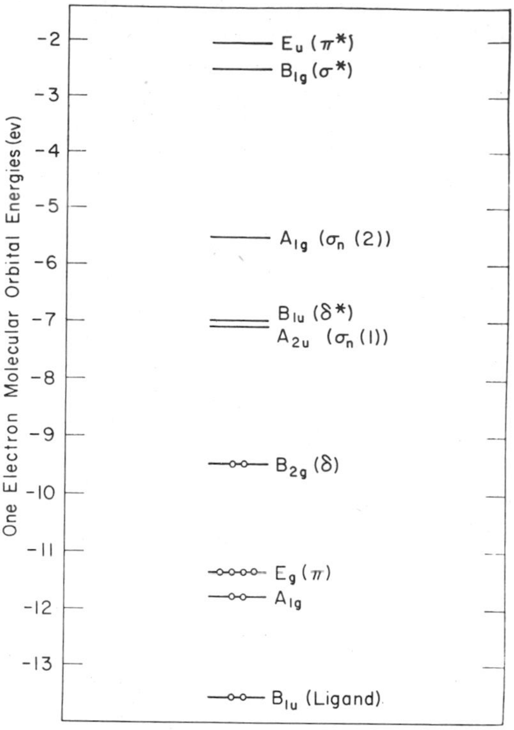

Figure 2.—A partial one-electron MO diagram for orbitals having predominant metal d character.

respectively. All input parameters are summarized in Table I.

Evaluation of the Off-Diagonal Matrix Elements, H_{ij}.—The H_{ij}'s, the Hamiltonian matrix elements between the ith and jth atomic orbitals, were evaluated by the Mulliken–Wolfsberg–Helmholz approximation[19,20]

$$H_{ij} \approx KS_{ij}(H_{ii} + H_{jj})/2 \quad (4)$$

K is the Wolfsberg–Helmholz factor and S_{ij} is the overlap integral between the ith and jth atomic orbitals.

A K equal to 1.80 was used for all H_{ij}'s. This value is the same as the optimum value determined from previous calculations[8] and is close to the "best fit" value for small molecules.[21,22]

Calculation Procedure.—The extended Hückel molecular orbital theory was employed, in which all overlap integrals were evaluated.[23] No assumption was made as to the extent of hybridization of either the ligand or metal wave functions. These conditions and the LCAO-MO assumption lead directly to the familiar constraint on the secular determinant

$$\det|H_{ij} - ES_{ij}| = 0 \quad (5)$$

(14) F. A. Cotton and C. B. Harris, *Inorg. Chem.*, **6**, 376 (1967).
(15) R. Ikeda, D. Nakamura, and M. Kubo, *J. Phys. Chem.*, **69**, 2101 (1965).
(16) F. A. Cotton and C. B. Harris, *Proc. Natl. Acad. Sci. U. S.*, **56**, 12 (1966).
(17) H. Pohl, R. Rein, and K. Appel, *J. Chem. Phys.*, **41**, 3385 (1964).
(18) J. Hinze and H. H. Jaffé, *J. Am. Chem. Soc.*, **84**, 540 (1962).
(19) R. Mulliken, *J. Chem. Phys.*, **46**, 497 (1949).
(20) M. Wolfsberg and L. Helmholz, *ibid.*, **20**, 837 (1952).
(21) M. D. Newton, F. P. Boer, W. E. Palke, and W. N. Lipscomb, *Proc. Natl. Acad. Sci. U. S.*, **53**, 1089 (1965).
(22) R. Hoffman, *J. Chem. Phys.*, **39**, 1397 (1963); *ibid.*, **40**, 2474 (1963)
(23) A modification of the program written by R. Hoffmann was used All α_i's were determined by a MAD program WAVEF written by C. B. Harris

The calculation was refined on charge to a self-consistency[10] of 0.01 electronic charge unit on Re.

Results

A table giving all overlap integrals, eigenvectors, and eigenvalues for the molecular orbitals has been filed with the American Documentation Institute.[24] A diagram giving the positions of the one-electron MO's most likely to be involved in discussing the spectrum and the bonding is given in Figure 2.

Interpretation of the Spectrum.—The calculated energies for both electric-dipole-allowed and vibronically-induced transitions together with the observed transition energies and estimates of the oscillator strengths of the observed transitions are collected in Table II.

TABLE II

CALCULATED AND OBSERVED ELECTRONIC TRANSITIONS

Electric dipole allowed	Polarization	Calcd transition energy, cm^{-1}	Experimental spectrum Energy, cm^{-1}	Oscillator strength
${}^1b_{2g} \rightarrow {}^1b_{1u}$ ($\delta \rightarrow \delta^*$)	z	19,700	32,800	0.31
${}^1e_g \rightarrow {}^1a_{2u}$ ($\pi \rightarrow \sigma_n(1)$)	x,y	34,800	39,200	0.65
${}^1e_g \rightarrow {}^1b_{1u}$ ($\pi \rightarrow \delta^*$)	x,y	35,000		
${}^1a_{1g} \rightarrow {}^1a_{2u}$ ($\sigma \rightarrow \sigma_n(1)$)	z	43,100		
${}^1b_{2g} \rightarrow {}^1e_u$ ($\delta \rightarrow \pi^*$)	x,y	59,700		
Electric dipole forbidden				
${}^1b_{2g} \rightarrow {}^1a_{2u}$ ($\delta \rightarrow \sigma_n(1)$)		18,700	14,500	0.023
${}^1b_{2g} \rightarrow {}^1a_{1g}$ ($\delta \rightarrow \sigma_n(2)$)		30,900		
${}^1e_g \rightarrow {}^1a_{1g}$ ($\pi \rightarrow \sigma_n(2)$)		46,100		
${}^1a_{1g} \rightarrow {}^1a_{1g}$ ($\sigma \rightarrow \sigma_n(2)$)		49,400		

The low-energy, electric-dipole-forbidden transition at 14,500 cm^{-1} has been previously[6] assigned as the ${}^1b_{2g} \rightarrow {}^1a_{2u}$ ($\delta \rightarrow \sigma_n(1)$) transition. Such an assignment was based on very qualitative theoretical considerations[6] and also on the comparison of the $Re_2Cl_8^{2-}$ spectrum with the spectra of $Re_2(O_2CR)_4X_2$ systems.[5b,6] In the latter systems there are halide ions, X, coordinated at each end of the molecule, on the fourfold axis.[25] In order to bond these halide ions, the rhenium σ orbitals used to form the $\sigma_n(1)$ and $\sigma_n(2)$ MO's would have to be used, thus eliminating (or drastically shifting) the ${}^1b_{2g} \rightarrow {}^1a_{2u}$ transition in these molecules. Experimental results[5b] are in accord with this.

The difference between the observed and calculated transition energies, ~4000 cm^{-1}, is probably due in part to neglect of penetration effects on the Re atomic orbitals, as noted earlier, and in part to neglect of interelectronic repulsion energies (that is, to the fact that the actual transitions are between states of the system whereas the calculations pertain only to one-electron orbitals). Both of these factors would tend to make the calculated energy greater than that observed, as is the case.

The two transitions at 32,800 and 39,200 cm^{-1} have oscillator strengths approaching unity. As such, they should be electric-dipole-allowed transitions. The 32,800 cm^{-1} band was previously[6] assigned to a ${}^1b_{2g} \rightarrow {}^1b_{1u}$ ($\delta \rightarrow \delta^*$) transition. Although the agreement between the calculated and observed transition energy for this assignment is poor, primarily because of the neglect of the penetration correction, no other reasonable assignment is possible. Both the ${}^1b_{2g} \rightarrow {}^1e_u$ and the ${}^1e_g \rightarrow {}^1b_{1u}$ assignments can be ruled out as possibilities. Such assignments would require an inversion of the b_{2g} and e_g (δ and π) molecular orbitals. In other words, the interaction of the Re δ orbitals would be greater than that of the Re π orbitals. Clearly this is unreasonable. If, as suggested earlier, the penetration effect might lower the δ orbital by about 1 ev relative to the δ^* orbital, the calculated separation would become ~28,000 cm^{-1}, which is in reasonable agreement with the observed transition energy of ~33,000 cm^{-1}.

Assigning this transition as ${}^1a_{1g} \rightarrow {}^1a_{2u}$ would mean that the σ to nonbonding σ separation would be less than the $\delta \rightarrow \delta^*$ separation. This is highly unlikely in view of the very large overlap difference between σ and δ orbitals.

The only other possible assignment for this transition would be ${}^1e_g \rightarrow {}^1a_{2u}$. If this were, in fact, the case one would expect this transition to be absent, which it is not, in the $Re_2(O_2CR)_4X_2$ molecules, since the σ_n's (Figure 2) would be greatly shifted by the interaction with the X's in such molecules. Furthermore such an assignment would require an inversion in the order of the e_g and b_{2g} orbitals. A measurement of the polarization of the band would certainly distinguish between the ${}^1b_{2g} \rightarrow {}^1b_{1u}$ and ${}^1e_g \rightarrow {}^1a_{2u}$ assignments since the former is z polarized while the latter is x,y polarized. However, such experiments have not yet proven feasible.

The transition appearing at 39,200 cm^{-1} cannot be assigned unambiguously, although in all likelihood it is a ${}^1e_g \rightarrow {}^1a_{2u}$ transition.

It should again be emphasized that the MO diagram is used only as a qualitative guide in assigning the spectrum. As a quantitative tool, extended Hückel theory is far from satisfactory. This point has been discussed in earlier papers.[10,14,26]

Nature of the Re–Re Bond.—The Re 5d molecular orbitals have an occupation of $(a_{1g})^2(e_g)^4(b_{2g})^2$. This ${}^1A_{1g}$ ground state is in agreement with the observed diamagnetism of $[Re_2Cl_8]^{2-}$.

Within the framework of the electron pair definition of a single bond, the Re–Re bond is a quadruple bond. It is even possible to represent the Re–Re bonding in terms of an equivalent (in the Lennard-Jones sense) set of orbitals as four separate but equivalent, bent single bonds.[27] However, it seems more appropriate

(24) Overlap integrals, eigenvectors, and eigenvalues have been deposited as Document No. 9304 with the American Documentation Institute, Auxiliary Publications Project, Photoduplication Service, Library of Congress, Washington 25, D. C. Copies may be secured by citing the document number and remitting in advance $1.25 for photoprints or $1.25 for 35-mm microfilm, payable to Chief, Photoduplication Service, Library of Congress.

(25) This statement is based upon structural results obtained by W. R. Robinson for $Re_2(O_2CC_6H_5)_4Cl_2$; the assumption that such a structure is general for $Re_2(O_2CR)_4X_2$ molecules seems reasonable.

(26) F. A. Cotton, C. B. Harris, and J. J. Wise, *Inorg. Chem.*, **6**, 909 (1967).

(27) F. A. Cotton, *Rev. Pure Appl. Chem.*, **16**, 175 (1966).

to describe the Re–Re bond directly in terms of the occupied molecular orbitals which contribute to it.

Bond order, as defined by Mulliken,[28] is given in eq 6. $N(k)$ is the number of electrons in the kth molecular

$$BO_{\alpha\beta} = \sum_i \sum_j N(k) C^{\alpha}{}_{ik} C^{\beta}{}_{jk} S^{\alpha\beta}{}_{ij} \qquad (6)$$

orbital, $C^{\alpha}{}_{ik}$ and $C^{\beta}{}_{jk}$ are coefficients of the ith and jth atomic orbitals in the kth molecular orbital, $S^{\alpha\beta}{}_{ij}$ is the overlap between the ith and jth atomic orbitals, and the sum over i and j is restricted to orbitals on the α and β atoms. Thus the Re–Re bond order can be expressed as

$$BO_{\mathrm{Re,Re}} = \sigma + \pi + \delta \qquad (7)$$

where σ, π, and δ are the designations for σ, π, and δ contributions. Table III lists the various contributions to the Re–Re bond. In an analogous way the Re–Cl bond is also described in Table IV.

Table III

Re–Re Bond Contributions[a–d]

σ Bond[a]

	6s	6pσ	5dσ
6s	0.2243	0.0286	−0.0117
6pσ	0.0286	0.0019	−0.0094
5dσ	−0.0117	−0.0094	0.1551

π Bond[b]

	6pπ(2)	5dπ(2)
6pπ(2)	0.0318	0.0894
5dπ(2)	0.0894	0.9424

δ Bond[c]

	5d$_{xy}$ (δ)	5d$_{x^2-y^2}$ (δ')
5d$_{xy}$ (δ)	0.2213	0.0
5d$_{x^2-y^2}$ (δ')	0.0	0.0041

[a] σ bond order = 0.3923. [b] π bond order = 1.1530. [c] δ bond order = 0.2254. [d] Total Re–Re bond order = 1.7707.

Table IV

Re–Cl Bond Contributions[a]

	3s	3p$_y$ (σ)	3p$_z$ (π(‖))	3p$_x$ (π(⊥))
6s	0.0422	0.0182	0.0001	. . .
6p$_y$ (σ)	0.0522	0.0127	0.0002	. . .
6p$_z$ (π(‖))	0.0028	0.0037	0.0174	. . .
6p$_x$ (π(⊥))	. . .	. . .	. . .	0.0164
5d$_{x^2-y^2}$ (σ)	0.0292	0.0940	0.0153	. . .
5d$_{z^2}$ (σ)	0.0145	0.0244	0.0203	. . .
5d$_{yz}$ (π(‖))	0.0049	0.0257	−0.0039	. . .
5d$_{xy}$ (π(⊥))	. . .	. . .	. . .	0.0061

[a] Total Re–Cl bond order = 0.3986.

It is apparent that the major components of the Re–Re bond are the π bonds. Furthermore, the total π bonding is about five times as strong as the δ bond and three times as strong as the σ bond. The $BO_{\mathrm{Re,Re}}/BO_{\mathrm{Re,Cl}}$ ratio of 4.4 is the Re–Re electron pair bond order assuming the Re–Cl bonds to be single bonds.

Because the Re–Re π and σ bonds are symmetric about the fourfold rotation axis in $[Re_2Cl_8]^{2-}$, they should not preferentially stabilize[29] the molecule in D_{4h} (eclipsed) as opposed to D_{4d} (staggered) symmetry. Only the δ bond has the proper symmetry characteristics to stabilize the molecule in the observed eclipsed D_{4h} symmetry. However, the Cl–Cl′ repulsions would be higher for a D_{4h} (as opposed to D_{4d}) configuration. Therefore, the ultimate stabilization of the molecule in D_{4h} requires that the δ bonding stabilization energy be greater than the Cl–Cl′ repulsion energy.

In order to calculate the δ bond stabilization energy, either the energy of the δ molecular orbital in a staggered D_{4d} $[Re_2Cl_8]^{2-}$ configuration must be known or the energy of the δ molecular orbital in a hypothetical $[ReCl_4]^-$ molecule must be known. The latter approach was preferred because it not only gives the δ bond stabilization energy but, in fact, gives an approximate *total* Re–Re′ bond energy.

The one-electron molecular orbitals were calculated[30] for a hypothetical $[ReCl_4]^-$ ($^1/_2$ of $[Re_2Cl_8]^{2-}$) molecule. The correlation between the $[ReCl_4]^-$ and $[Re_2Cl_4]^{2-}$ molecular orbitals is given in Table V for the δ, π, and σ orbitals. Assuming a high-spin d^4 system for $[ReCl_4]^-$, eq 8 is the Re–Re′ bond stabilization energy for individual orbital interactions (i = σ, π, and δ).

$$\text{stabilization energy} = [2N(i)\epsilon_i(\mathrm{ReCl_4})^- - \epsilon_i N(i)(\mathrm{Re_2Cl_8})^{2-}] \qquad (8)$$

$N(i)$ is the number of electrons in the ith molecular orbital and the ϵ_i's are the molecular orbital energies of the ith orbital for the species indicated. The sum of these stabilization energies is the total Re–Re′ bond energy.[31] These results are listed in Table V.

Table V

Contributions to the Re–Re Bond Energy[a]

MO symmetry	$ReCl_4^-$ (C_{4v}), ev	$Re_2Cl_8^{2-}$ (D_{4h}), ev	Re–Re bond energies, kcal/mole
δ	−8.35	−9.45	51
π	−8.87	−11.35	114
σ	−9.87	−11.75	87

[a] Total Re–Re bond energy = 366 kcal/mole (Re–Re bond = $\sigma + 2\pi + \delta$).

If we assume that the fourfold rotation barrier arising from the Cl–Cl′ repulsions in $[Re_2Cl_8]^{2-}$ is the same order of magnitude as the threefold rotation barrier in C_2Cl_6 ($\geq$7 kcal/mole), then the δ bond stabilization is *seven times* the energy of the fourfold rotational barrier. Thus, the δ bond provides a potential well far in excess of the Cl–Cl′ barrier to rotation and the molecule is stabilized in an eclipsed D_{4h} rather than a staggered D_{4d} configuration.

Nature of the Re–Cl Bond.—It is apparent from Table IV that the σ contribution to the Re–Cl bond order is about three times that of the π. Since the Re_1 (Figure 1) is not in the plane of the 4 Cl (Cl_1, Cl_2, Cl_3,

(28) R. W. Mulliken, *J. Chem. Phys.*, **23**, 1833 (1955).

(29) Since the Cl–Cl′ overlap integrals would change under a C_4 rotation, there could be a very slight stabilization in one or the other symmetry. However, these effects would be energetically small compared to the energy differences between the δ bond in the two symmetries.

(30) The same parameters were used in this calculation as were used in the $[Re_2Cl_8]^{2-}$ calculation.

(31) No claim is made for the absolute accuracy of these energies. They are, in fact, only upper limits. Their significance can be only in the ratios, σ/π, π/δ, etc.

Cl_4), the σ and π contributions to the Re–Cl bonds are not rigorously separable by molecular symmetry. Thus, the values of σ and π Re–Cl bonds in Table IV are accurate only insofar as they represent an upper limit of π bonding and lower limit of σ bonding within the framework of rigorous σ–π separability. The total Re–Cl bond order, 0.40, is, of course, not subject to the separability of σ and π orbitals.

Vibrational Spectroscopy

VI

Editor's Comments on Papers 31 Through 37

31 Wigner: *Concerning the Elastic Characteristic Vibrations of Symmetric Systems*

32 Wilson: *The Normal Modes and Frequencies of Vibrations of the Regular Plane Hexagon Model of the Benzene Molecule*

33 Wilson: *The Degeneracy, Selection Rules, and Other Properties of the Normal Vibrations of Certain Polyatomic Molecules*

34 Wilson: *Symmetry Considerations Concerning the Splitting of Vibration-Rotation Levels in Polyatomic Molecules*

35 Meister, Cleveland, and Murray: *Interpretation of the Spectra of Polyatomic Molecules by Use of Group Theory*

36 Meister and Cleveland: *Application of Group Theory to the Calculation of Vibrational Frequencies of Polyatomic Molecules*

37 Claasen, Chernick, and Malm: *Vibrational Specta and Structure of Xenon Tetrafluoride*

At approximately the same time that Van Vleck, Mulliken, and others were developing the theory dealing with the electronic structure of molecules, Eugene Wigner, a truly remarkable person [30], was applying group theoretical principles to vibrational behavior. He grasped the power of symmetry and group theory in dealing with quantum mechanical problems long before other physicists and chemists accepted the concepts.

As he states in the preface to the English translation [4] of his book *Group Theory:* "When the original German version was first published, in 1931, there was a great reluctance among physicists toward accepting group theoretical arguments and the group theoretical point of view. It pleases the author that this reluctance has virtually vanished in the meantime and that, in fact, the younger generation does not understand the causes and the basis for this reluctance."

His 1930 paper, presented here in translation from the German, rivals Bethe's 1929 paper in importance to the entire field.

Fortunately Wigner's work was read by some people, notably Bethe, van Vleck, Mulliken, and a young chemist named Wilson. E. B. Wilson showed how Wigner's method (Paper 31) may be applied to the vibrational spectroscopy of benzene (Paper 32) and other polyatomic molecules (Papers 33 and 34). As a result of this work, group theoretical techniques were accepted relatively early by chemists working in the field of vibrational spectroscopy. The review articles (Papers 35 and 36) by Meister et al., because of the clarity with which they are written, have become the papers to which the neophyte vibrational spectroscopist is first directed.

The three papers by Wilson apply the group theoretical techniques of Wigner to the normal modes of vibration of molecules. Wilson also showed how selection rules readily allow one to determine which bands are permitted in the infrared and Raman spectra. He deduced that the molecular symmetry could be ascertained from a knowledge of the number of allowed infrared and Raman bands found experimentally for a given molecule. Paper 33 emphasizes this point. In Paper 34, symmetry arguments are used to determine splittings which may occur in degenerate vibration–rotation levels.

The papers by Wigner, Wilson, and others established the value of group theoretical techniques in vibrational spectroscopy. However, they assumed some familiarity with group theory itself, as did the review by J. E. Rosenthal and G. M. Murphy, [*Rev. Mod. Phys.*, **8,** 317 (1936)]. The two review articles by Meister et al. from the *American Journal of Physics* reformulated the concepts in such a way as to make them readily understood by the practicing chemist unable (or unwilling) to grasp the details of group theory. As a result, these papers became the "How To Do It" papers referred to by chemists prior to incorporation of the work into texts on spectroscopy. They still represent a rather painless introduction to the field of vibrational spectroscopy and the group theory used.

The section on vibrational spectroscopy cannot be completed without presenting a good example of the application of the theory to the deduction of the symmetry of an important inorganic molecule. The number of vibrational analyses available from which to choose is obviously very large. In fact, G. Herzberg produced a book in 1945, *Molecular Spectra and Molecular Structure II. Infrared and Raman Spectra of Polyatomic Molecules* (Van Nostrand Reinhold, New York), a classic which deals with both the theory and many valuable examples. Two other books, K. Nakamoto, *Infrared Spectra of Inorganic and Coordination Compounds* (Wiley-Interscience, New York, 1963, 1970), and L. H. Jones, *Inorganic Vibrational Spectroscopy* (Marcel Dekker, New York, 1971)' are useful additions to the vibrational spectroscopy literature in inorganic chemistry.

Since the discovery of xenon reactivity by Neil Bartlett was a very significant mid-twentieth-century event in chemistry, I have chosen Paper 37, dealing with the vibrational spectra, of XeF_4, as an example of the application of vibrational theory to structure elucidation.

Translated from *Göttingen Nachrichten*, 133–146 (1930)

31

Concerning the Elastic Characteristic Vibrations of Symmetric Systems

E. WIGNER

Translation by R. Haberkorn, Department of Chemistry, Case Western Reserve University, of the paper presented by M. Born at the May 23, 1930, meeting

1. It is well known that one is able in quantum mechanics to make use of the symmetry characters of a system to determine its type of movement. From the data a final multiplicity "whose energy[1] is E" is characterized from a condition in a manner invariant to rotation. This is impossible in classical mechanics in general, because the initial conditions are important.

The elastic vibrations of a point system about its equilibrium configuration are an exception to this. The equations of motion are linear (i.e., the superposition of two possible vibrations is again a possible vibration); therefore, the circumstances are wholly analogous to those in quantum mechanics. Therefore, it is to be expected that the same means which one is able to use there can be used here also.

A consideration of the elastic vibrations of symmetric structures, which we want to investigate here, is also the subject of an older work by Brester.[2] He has solved the problem completely using entirely elementary means. If this task is repetitious here, then let this be excused, because the aforementioned group theoretical technique is wholly fit to be published. With its aid one also obtains a better overview of Brester's conclusions.

In the following we consider the vibrations of a system of n points which are bound together by means of elastic forces.[3] The center of mass of the system is found at the origin of the coordinate system. In the rest position the coordinates

[1]If E is a position of the discrete spectrum with only a finite number of eigenfunctions.
[2]C. J. Brester, Diss., Utrecht, 1923; reproduced in Göttingen Institute for Theoretical Physics S. 8-90. Also D. M. Dennison treats a special case [*Astrophys. J.* **62,** 82 (1925)], that of the vibrations of a CH_4 molecule, which we shall use as an example.
[3]One thinks of a CH_4 molecule with the atoms idealized as points.

of the atoms are the components of the vectors[4] $r_1, r_2, \ldots, r_n$, while the displacements from the equilibrium positions are given by the vectors $s_1, s_2, \ldots, s_n$. The coordinates of the particles are the components of $r_1 + s_1,\ r_2 + s_2,\ \ldots,\ r_n + s_n$. The s_k also depend on time; however, they are always small compared to r_k, for the characteristic vibrations the time dependence is given to s_k through a factor $\sin(t - t_0)$. One can also summarize the n vectors $s_1, s_2, \ldots, s_n$, of a displacement by a $3n$-dimensional vector s.

The whole system should have a symmetry [i.e., there should exist a group G of three-dimensional rotations $(R_{\alpha\beta})$[5]] which transforms the equilibrium configuration into itself. Equation (1) is valid for all transformations R of the group and for every k and α:

$$\sum_{\beta=1}^{3} R_{\alpha\beta} r_{k\beta} = r_{l\alpha} \qquad \text{or simply } Rr_k = r_l. \tag{1}$$

The kth and lth particles must be of the same kind (e.g., both hydrogen atoms).[6] The number l of the particle, on whose site the kth particle will be shifted by means of R, we call $R(k)$. Then Eq. (1) reads

$$\sum_{\beta=1}^{3} R_{\alpha\beta} r_{k\beta} = r_{R(k),\alpha}; \qquad Rr_k = r_{R(k)}. \tag{1a}$$

A characteristic vibration of the system can be described by means of the n vectors $s_1, s_2, \ldots, s_n$, which indicate the displacement of the atoms from their equilibrium positions when the amplitude of the vibration is just at the maximum. $s_1, s_2, \ldots, s_n$ are the amplitudes of a characteristic (normal) vibration, and if one subjects these through the given configuration of the system to a transformation R of the group G, then one again conserves the configuration of a characteristic vibration of the same frequency, because, of course, one has by no means altered the relative configuration of the particles. The coordinates of the kth point are now $Rr_k + Rs_k = r_{R(k)} + Rs_k$, a point in the proximity of any equilibrium position.

With this, therefore, one can also generate an identical configuration by means of small displacements of the points out of their equilibrium positions. One must for this reason displace the $l = R(k)$th particle around $Rs_k = Rs_{R^{-1}(l)}$. The displacements

[4]Through this the vectors r are also not uniquely determined because there is also a free rotation of the whole system. In the following we imagine it somehow to be fixed arbitrarily (perhaps r_1 lies on the Z axis, r_2 in the YZ plane).

[5]The indices α and β always refer to the coordinates $X_1 = X$, $X_2 = Y$, $X_3 = Z$ and become attached to the character of the components of a vector as lower indices.

[6]In crystallography one designates the points which will transform into each other by means of symmetry elements as "equivalent." Equivalent points are always of the same kind, but the inverse is not necessarily true.

$$R s_{R^{-1}(1)}, R s_{R^{-1}(2)}, \ldots, R s_{R^{-1}(n)}$$

(2)
$$\bar{R} s_l = R s_{R^{-1}(l)}; \qquad R s_{l\alpha} = \sum_{\beta} R_{\alpha\beta} s_{R^{-1}(l),\beta}$$

form an amplitude system $\bar{R}s$ of a characteristic vibration, which has the same frequency as the characteristic vibration s. The operations $\bar{R}$ replace the "rotations" and "interchange of electrons" of quantum mechanics.

One designates the linearly independent characteristic vibrations of a definite frequency with $s^{(1)}, s^{(2)}, \ldots, s^{(f)}$; therefore, $\bar{R}s^{(x)}$ must be linearly expressible through these,

$$\bar{R}s^{(\kappa)} = \sum_{\lambda=1}^{f} D(R)_{\lambda\kappa} s^{(\lambda)} \tag{3}$$

Similarly, one deduces, as in quantum mechanics, through application of another operation $\bar{S}$, an element S of the group G,

$$\bar{S}\bar{R}s^{(\kappa)} = \sum_{\lambda=1}^{f} D(R)_{\lambda\kappa} S s^{(\lambda)} = \sum_{\lambda=1}^{f}\sum_{\mu=1}^{f} D(R)_{\lambda\kappa} D(S)_{\mu\lambda} s^{(\mu)}$$
$$= \sum_{\mu=1}^{f} D(SR)_{\mu\kappa} s^{(\mu)}.$$

Therefore, the s-dimensional matrix $[D(R)_{\lambda\kappa}]$ forms a representation of the group G. In this way every characteristic vibration can be associated with a certain representation. One deduces further, as in quantum mechanics, that these representations are assumed to be irreducible. There are, therefore, as many types of characteristic vibrations[7] as G has irreducible representations (i.e., classes). The number of linearly independent characteristic vibrations of a frequency is given by the dimensions of the associated representations.

2. There are still interesting questions. How many frequencies belong to a given representation? How many characteristic frequencies exist of a certain type? The analogous questions cannot be formulated in quantum mechanics, because all the numbers there are infinite.

As is known, each movement of the system is allowed to consist of characteristic vibrations in which the equilibrium positions will not be changed. Furthermore, one

[7]Actually this applies only for real irreducible representations. Eigenfrequencies to which a complex representation belongs always fall together with an eigenfrequency to which the complex-conjugate representation belongs. In the case of the groups with real irreducible representations which are not also irreducible in the complex representation, the "accidental degeneracy" of position occurs. Nevertheless, this is easy to perceive and is insignificant for the following.

must include the three parallel displacements of the three coordinates and the rotations around the three axes (it will be assumed that all the points do not lie on a straight line). These displacements could also be assumed to be characteristic vibrations; only their frequencies are zero. The associated representation is in the first case $D^{(v)}(R)$, the representation of the polar vectors, in the second case $D^{(v')}(R)$, the rotation of the axial vectors.

If we designate the amplitude system of the displacements (by which only the kth particle has altered its site and this only by unity in the X_α-direction) with $e^{(k\alpha)}$ [i.e., let $e^{(k\alpha)}_{l\beta} = \delta^{kl}\delta^{\alpha\beta}$]. One can express all linear displacements by means of $e^{(k\alpha)}$:

$$s^{(\kappa)} = \sum_{k=1}^{n} \sum_{\alpha=1}^{3} s^{(\kappa)}_{(k\alpha)} e^{(k\alpha)}. \tag{4}$$

Conversely, one can also express the e's by means of the $3n$ coefficients[8] of s (if one also takes the displacement vectors of translation and rotation):

$$e^{(k\alpha)} = \sum_{\substack{\text{all} \\ \text{characteristic} \\ \text{vibrations}}}^{\kappa} S_{k\alpha;\kappa} s^{(\kappa)}. \tag{4a}$$

We now imagine for a moment that we know the amplitudes of all $3n$ characteristic vibrations. For those belonging to the same frequency, (3) is valid for the entire system. Therefore,

$$\bar{R}s^{(\kappa)} = \sum_{\lambda} \Delta(R)_{\lambda\kappa} s^{(\lambda)}, \tag{5}$$

where $\Delta(R)$ is a representation of the form

$$\Delta(R) = \begin{pmatrix} D^{(1)}(R) & 0 & \cdots \\ 0 & D^{(2)}(R) & \cdots \\ \cdots & \cdots & \cdots \end{pmatrix}. \tag{5a}$$

In (5a), $D^{(1)}(R)$, $D^{(2)}(R)$, . . . , are the representations which are assigned to the individual frequencies in the sense of (3). Therefore, when we want to know how many frequencies certain representations have in common, we must determine how often this representation occurs in $\Delta(R)$. For this purpose, as is known, it suffices to know the characters of $\Delta(R)$, and the sum $\Sigma\Delta(R)_{\kappa\kappa}$, for all R. These we can determine in the following way.

When we introduce e in (5) with the aid of (4), (4a), in place of s and express $Re^{(k\alpha)}$ by means of $e^{(j\beta)}$

$$\bar{R}e^{(k\alpha)} = \sum_{j=1}^{n} \sum_{\beta=1}^{3} \bar{\Delta}(R)_{j\beta:k\alpha} e^{(j\beta)}, \tag{6}$$

---

[8]Because of the orthogonality of the s' this is true when one normalizes the s' assuming $s^{(\kappa)} = s_{k\alpha;\kappa}$.

the coefficients form a representation $\bar{\Delta}(R)$ of the group G. This is equivalent to $\Delta(R)$ and follows from it by means of a similarity transformation with the $3n$-dimensional matrix $S_{k\alpha;\kappa}$. Its character $\chi(R) = \sum_{j\beta} \bar{\Delta}(R)_{j\beta;j\beta}$ is equal therefore to that of $\Delta(R)$ and it suffices to determine the first set.

Now $e^{(k\alpha)}$ is the unit displacement which the kth particle shifts in the X_α-direction, which leaves other particles in the equilibrium position. $\bar{R}e^{(k\alpha)}$ originates from this displacement by rotation of the whole system about R. Then every equilibrium configuration is occupied except $R(k)$. This, according to (2), has the displacement components $R_{1\alpha}, R_{2\alpha}, R_{3\alpha}$. Thus we have

$$\bar{R}e^{(k\alpha)} = \sum_{\beta} R_{\alpha\beta} e^{(R^{-1}(k),\beta)} = \sum_{j\beta} \bar{\Delta}(R)_{j\beta;k\alpha} e^{(j\beta)}. \tag{7}$$

Equation (7) determines the representation $\bar{\Delta}(R)$. In order to calculate its character, we must form the sum of the diagonal elements. Now a zero certainly is in the k_α column on the principal diagonal when $k \neq R^{-1}(k)$. In this case, of course, one does not need $e^{(k\alpha)}$ at all to express $\bar{R}e^{(k\alpha)}$. On the other hand, when $k = R^{-1}(k)$, $R_{\alpha\alpha}$ stands in $\Delta(R)$ in the kth column on the principal diagonal. It is, therefore,

$$\bar{\Delta}(R)_{k\alpha;k\alpha} = \begin{cases} 0 & \text{for } R(k) \neq k, \\ R_{\alpha\alpha} & \text{for } R(k) = k, \end{cases} \tag{8}$$

and, therefore,

$$\sum_{\alpha} \bar{\Delta}(R)_{k\alpha;k\alpha} = \begin{cases} 0 & \text{for } R(k) \neq k, \\ \sum_{\alpha} R_{\alpha\alpha} = \text{trace}(R) & \text{for } R(k) = k. \end{cases} \tag{8a}$$

We also designate the number of equilibrium positions which are unchanged by R with u_R; we have then

$$\chi(R) = \sum_{k} \sum_{\alpha} \bar{\Delta}(R)_{k\alpha;k\alpha} = u_R \text{ trace}(R) = \pm u_R(1 + 2 \cos \phi_R), \tag{9}$$

where $\text{trace}(R) = \pm(1 + 2 \cos \phi_R)$ has been put in. ϕ_R is the angle of rotation of R with a positive or negative sign according to whether R is a pure rotation or a rotation–reflection. From (9) one can calculate the character from $\chi(R)$; naturally it is enough to do this for just one element of each class.

One writes $\chi(R)$ as a linear combination of the characters $\chi^{(1)}(R), \chi^{(2)}(R), \ldots,$ of different irreducible representations of G,

$$\chi(R) = a_1'\chi^{(1)}(R) + a_2'\chi^{(2)}(R) + \cdots, \tag{10}$$

so the coefficients $a_1', a_2', \ldots$ (they are nonnegative whole numbers) give the number of characteristic frequencies which belong to the different representations $D^{(1)}(R)$, $D^{(2)}(R), \ldots$. One can determine the numbers $a_1', a_2', \ldots,$ easily with the aid of the explicit formula

$$(11) \qquad a_p' = \frac{1}{h} \sum_R \chi(R)\chi^{(P)}(R),$$

where the summation is to be extended over all h group elements of G. However, it is to be noticed that in this way one also forms "spurious" vibrations, which are associated with translation and rotation. If one does not want to have this, then one must subtract the characters of these from $\chi(R)$ as determined by (9). For the translation this is the character of $D^{(v)}(R)$ [i.e., $\pm(1 + 2\cos\phi_R)$]; for the rotation, that of $D^{(v')}(R)$ [i.e., $+(1 + 2\cos\phi_R)$]. On the whole, therefore, one must subtract from the character the pure rotations $2(1 + 2\cos\phi_R)$, and leave the characters of rotation–reflection unchanged. For the number, a, of nondegenerate, nonspurious characteristic frequencies are

$$(10a) \qquad a_1\chi^{(1)}(R) + a_2\chi^{(2)}(R) + \cdots = \Xi(R) = \begin{cases} (u_R - 2)(1 + 2\cos\phi_R) & R \text{ pure rotation} \\ (-u_R)(1 + 2\cos\phi_R) & R \text{ rotation–reflection} \end{cases}$$

or explicitly:

$$(11a) \qquad \sum_R \chi^{(P)}(R)\Xi(R) = \left\{\sum{}' (u_R - 2)(1 + 2\cos\phi_R) - \sum{}'' u_R(1 + 2\cos\phi_R)\right\} \frac{\chi^{(p)}(R)}{h} \quad a_p = \frac{1}{h},$$

where the first summation is to be extended over the pure rotations and the second over the rotation–reflections of G. With this, u_R is the number of particles which do not become displaced by R. $\chi^{(p)}(R)$ is the character of the irreducible representation of this type. The number of linearly independent vibrations belonging to a characteristic vibration of a certain type is the dimension of the representation to which it belongs.

H. Bethe[9] has determined the characters of the irreducible representations of most symmetry groups, since they are important for various problems in quantum mechanics. They are compiled at the end of this work. [*Editor's Note:* These character tables have been omitted.]

The symmetry group of methane is the tetrahedral group, T_d; it consists of the identity E, four threefold rotations C_3 (eight group elements), three twofold C_2 (three elements), six mirror planes σ_d (six elements), and two fourfold rotation reflections, S_4 (six elements). The characters of the five irreducible representations are tabulated in Table 1 (the group is homomorphic with the permutation group of order four). In order to determine the character from (R), we place together in a second

[9] *Ann. Physik,* (5), **3**, 133 (1929).

table [Table 2] the numbers u_R of the atoms remaining unchanged; the angle of rotation ϕ_R; the quantity $\pm(1 + 2\cos\phi_R)$; the characters $\chi(R)$, which include also the degenerate vibrations; and finally the character $\Xi(R)$, where these are already subtracted. With the aid of (10a) or even directly from (11a) we obtain

$$\Xi(R) = \chi^{(1)}(R) + \chi^{(2)}(R) + 2\chi^{(3)}(R).$$

The characteristic vibrations of methane have (except for the zero frequencies) four different frequencies. One is not degenerate, two belong to another frequency, and the last two frequencies have three characteristic vibrations each (triply degenerate).

The last part of the calculation can be considerably simplified, in that one does not establish Table 2 for the whole system all at once, but for all equivalent points separately (therefore, separately for the C atom and then separately for the four H atoms). One must consider $\chi(R)$ in all the tables but subtract the characters of translation and rotation in only one of them [i.e., calculate $\Xi(R)$]. In order for this selection to be possible, the table must contain points which do not all lie on a straight line; until now this has been required only for the totality of all points. The simplification lies in this, that one can split the several $\chi(R)$ [or simply $\Xi(R)$] into a sum of irreducible characters more easily than the sum which appears in Table 2. One obtains the complete a_p in this case by addition from the several tables.

3. It is easy to determine the "active" characteristic vibrations associated with excitation by dipole radiation. If we designate the total polarization vector by

$$G_\alpha^{(\kappa)} = e_1 s_1^{(\kappa)} + e_2 s_2^{(\kappa)} + \cdots + e_n s_{n\alpha}^{(\kappa)}, \tag{12}$$

then from (3)

$$RG_\alpha^{(\kappa)} = \sum_{\lambda=1}^{f} D(R)_{\lambda\kappa}(e_1 s_{1\alpha}^{(\lambda)} + e_2 s_{2\alpha}^{(\lambda)} + \cdots + e_n s_{n\alpha}^{(\lambda)} = \sum_{\lambda=1}^{f} D(R)_{\lambda\kappa} G_\alpha^{(\lambda)}. \tag{13}$$

On the other hand, G obviously transforms as a vector

$$\overline{R}G_\alpha^{(\kappa)} = \sum_{\beta=1}^{3} R_{\alpha\beta} G_\beta^{(\kappa)}. \tag{13a}$$

Table 1

	$E(1)$	$C_3(8)$	$C_2(3)$	$\sigma_d(6)$	$S_4(6)$
$\chi^{(1)}$	1	1	1	1	1
$\chi^{(2)}$	2	−1	2	0	0
$\chi^{(3)}$	3	0	−1	1	−1
$\chi^{(4)}$	3	0	−1	−1	1
$\chi^{(5)}$	1	1	1	−1	−1

Table 2

	$E(1)$	$C_3(8)$	$C_2(3)$	$\sigma_d(6)$	$S_4(6)$
u_R	5	2	1	3	1
ϕ_R	0	$2\pi/3$	π	π	$\pi/4$
$\pm(1 + 2\cos\phi_R)$	3	0	−1	1	−1
$\chi(R)$	15	0	−1	3	−1
$\Xi(R)$	9	0	1	3	−1

This yields

(14) $$\sum_{\beta=1}^{3} R_{\alpha\beta} G_{\beta}^{(\kappa)} = \sum_{\lambda=1}^{s} D(R)_{\lambda\kappa} G_{\alpha}^{(\lambda)}; \qquad G_{\beta}^{(\lambda)} = \sum_{\lambda}\sum_{\alpha} R_{\alpha\beta} D(R)_{\lambda\kappa} G_{\alpha}^{(\lambda)}.$$

If one sums the right-hand side over all group elements R of G, then this is zero if the matrices $(R_{\alpha\beta})$ do not contain the irreducible representation $D(R)$ of the corresponding characteristic frequency.[10] The condition for this is that

(15) $$\sum_{\substack{\text{pure}\\ \text{rotation}}}' (1 + 2\cos\phi_R)\chi^{(p)}(R) - \sum_{\substack{\text{rotation}\\ \text{reflection}}}'' (1 + 2\cos\phi_R)\chi^{(p)}(R)$$

disappear. In this case the vibration is inactive; otherwise, it is active.[11] In the preceding example only the two threefold characteristic vibrations are active.

In the same way one can account for the existence or disappearance of higher moments.

For the explicit determination of the amplitudes of separate characteristic vibrations one proceeds best from the comparison of formulas (2) and (3). This yields

(16) $$\overline{R} s_{l\alpha}^{(\kappa)} = \sum_{\beta} R_{l\alpha} s_{R^{-1}(l),\beta}^{(\kappa)} = \sum_{\lambda} D(R)_{\lambda\kappa} s_{l\alpha}^{(\lambda)},$$

[10]The largest number of active characteristic vibrational types is therefore three; these must then all be "nondegenerate" (i.e., each characteristic vibration can only belong to an eigenfrequency of these types). It can also be active for two types; two characteristic vibrations then belong to the eigenfrequencies of each of these types. Finally, it is possible that only a triply degenerate characteristic vibrational type is active (as in the case of tetrahedral symmetry). It follows that $(R_{\alpha\beta})$ can consist of either three one-dimensional, or a one- and a two-dimensional, or a three-dimensional irreducible representation from G.

[11]It is to be noticed that Tables 1 and 2 contain only one column for each class. Therefore, one must take each product as often as the corresponding class has elements, e.g., $\sum\chi^4(R)(1 + 2\cos\phi_R) = (1)(3)(3) + (8)(0)(0) + (3)(-1)(-1) + (6)(-1)(1) + (6)(1)(-1) + (6)(1)(-1) = 0.$

from which one obtains

(17)
$$s^{(\kappa)}_{R^{-1}(l),\beta} = \sum_{\alpha=1}^{3}\sum_{\lambda=1}^{s} R_{\alpha\beta} D(R)_{\lambda\kappa} s^{(\lambda)}_{l\alpha}.$$

This is an equation among the displacements of purely equivalent points, which in many cases even allows the characteristic vibrations to be determined entirely. Otherwise, one is forced also to use the equations of motion.

In order to introduce the characters in place of the representation coefficients $D(R)_{\lambda\kappa}$, we multiply (16) by the character $\chi^{(p')}(R)$ of an irreducible representation from G and sum over all group elements. One obtains, with the aid of the orthogonalization relation for the representation coefficients,

(18)
$$\sum_{R}\sum_{\beta} \chi^{(p')}(R) R_{\alpha\beta} s^{(\kappa)}_{R^{-1}(l),\beta} = \sum_{R}\sum_{\lambda} \chi^{(p')}(R) D^{(p)}(R)_{\lambda\kappa} s^{(\lambda)}_{l\alpha} = \frac{h}{s_p}\delta_{p'p} s^{(\kappa)}_{l\alpha}$$

In particular, zero is obtained if the character $\chi^{(p')}(R)$ of a representation is different from $D^{(p)}(R)$.

4. The characteristic frequencies are the square roots of the eigenvalues of a $3n$-dimensional symmetric matrix $H_{k\beta;l\alpha}$, which is associated with the eigenvectors and forms the amplitude system of the corresponding characteristic vibrations. We designate the characteristic frequencies belonging to the representation $D^{(p)}(R)$ by $\nu_{p1}, \nu_{p2}, \ldots, \nu_{pa}, \ldots$, the s_p characteristic vibrations belonging to ν_{pa} by $s^{(pa1)}, \ldots, s^{(pas)}$. So it gives

(19)
$$\sum_{l\alpha} H_{k\beta;l\alpha} s^{(pa\kappa)}_{l\alpha} = \nu^2_{pa} s^{(pa\kappa)}_{k\beta},$$

and if one assumes $s^{(pa\kappa)}$ is normalized, $\sum_{l\alpha}\left[s_{l\alpha}^{(pa\kappa)}\right]^2 = 1$,

(19a)
$$\sum_{p}\sum_{a}\sum_{\kappa} \nu^2_{pa} s^{(pa\kappa)}_{k\beta} s^{(pa\kappa)}_{la} = H_{k\beta;la}.$$

If one replaces $R^{-1}(l)$ by k, multiplies by $R_{\alpha\beta}$ and the character $\chi^{(p')}(R)$ of an irreducible representation, and sums over β and over all group elements, one obtains

(20)
$$\sum_{p}\sum_{a}\sum_{\kappa} \nu^2_{pa} s^{(pa\kappa)}_{l\alpha} \sum_{\beta}\sum_{R} R_{\alpha\beta}\chi^{(p')}(R) s^{(pa\kappa)}_{R^{-1}(l),\beta} = \sum_{\beta}\sum_{R} R_{\alpha\beta}\chi^{(p')}(R) H_{R^{-1}(l)\beta}.$$

According to (18), this is

(20a) $$\sum_p \sum_\alpha \sum_\kappa \nu^2_{p\alpha} s^{(p\alpha\kappa)}_{l\alpha} \frac{h}{s_p} \delta_{pp'} s^{(p\alpha\kappa)}_{l\alpha} = \sum_p \sum_R R_{\alpha\beta} \chi^{(p')}(R) H_{R^{-1}(l)\beta,l\alpha},$$

which is summed over l and α because of the normalization of the eigenvectors

(21) $$\frac{h}{s_{p'}} \sum_\alpha v^2_{p'\alpha} = \sum_{\alpha\beta} \sum_R \sum_l R_{\alpha\beta} \chi^{(p')}(R) H_{R^{-1}(l);l\alpha}$$

obtained for the sum of all the squares for the representation $D^{(p')}(R)$ of the associated characteristic frequencies. The sum is calculated in (21) if the equations of motion of the system and therefore the matrix $H_{k\beta;1\alpha}$ are known. Equation (21) corresponds exactly to Heitler's average value formula.[12] This is sufficient in the calculation of the frequencies of characteristic vibrations of which not more than one other characteristic vibration belongs to the representation.

If this is not the case, then one must, corresponding to Heisenberg's procedure, [13] form the square, the third power, etc., of $H_{k\beta;1\alpha}$, of which the eigenvalues are the fourth, sixth, etc., powers of the characteristic frequencies, while the assigned eigenvectors remain the $s^{(p\alpha\kappa)}$'s. For these the reader is referred to the procedure described previously for $H_{k\beta;1\alpha}$. One obtains the formula for the sum of the fourth, sixth, etc., powers, which assign the characteristic frequencies to the same representation:

(21a) $$\frac{h}{s_p} \sum \nu^{2t}_{p\alpha} = \sum_{\alpha\beta} \sum_\kappa \sum_l R_{\alpha\beta} \chi^{(p)}(R) H^{(t)}_{R(l^{-1}),\beta;l\alpha},$$

where the $H^t_{k\beta;1\alpha}$ are the matrix elements of the tth power matrix from $H_{k\beta;1\alpha}$.

From these equations (they are the equivalent of the "irreducible" secular equation of quantum mechanics) one can calculate the characteristic frequencies relatively easily.

In the following table [*Editors Note:* table not included] the irreducible representations of the 32 crystal classes are tabulated.[14] The characters of the crystal classes C, which are the direct product of another crystal class C_0 with the reflection group $C_i = (E, i)$ [or $C_s = (E, \sigma_h)$], $C = C_0 \times C_i$ (or $C = C_0 \times C_C$), are not given separately. In this way one obtains from every representation from C_0 two each from C. Then one assigns to the elements A from C, which also appear in C_0, the same matrix as in the representation from C_0. The elements which do not appear in C are always written Ai (or $A\sigma_h$), where A appears in C_0. One assigns the same matrix which one has assigned to A to them, with either a positive or a negative sign throughout.

---

[12] W. Heitler, *Z. Physik,* **46,** 47 (1927).
[13] W. Heisenberg, *Z. Physik,* **49,** 619 (1928).
[14] The notation is the same as that which is used in P. P. Ewald's article in Greiger-Scheel's *Handbuch der Physik,* Band XXIV.

Therefore, one obtains all the representations of C, and has exactly twice as many as C_0.

The crystal classes, which are written as direct products, as is known, are the following:

$$C_{2h} = C_2 \times C_s = C_2 \times C_i; \quad C_{3h} = C_3 \times C_s; \quad C_{4h} = C_4 \times C_s = C_4 \times C_i;$$
$$C_{2v} = C_s \times C_s; \quad C_{6h} = C_6 \times C_s = C_6 \times C_i$$
$$D_{3h} = D_3 \times C_s; \quad D_{4h} \times C_s = D_4 \times C_i; \quad D_{6h} = D_6 \times C_s = D_6 \times C_i$$
$$C_{2v} = C_s \times C_s; \quad C_{6h} = C_6 \times C_s = C_6 \times C_i$$
$$C_{3i} = C_3 \times C_i; \quad V_h = V \times C_s = V \times C_i; \quad D_{3d} = C_{sv} \times C_i$$
$$T_h = T \times C_i; \quad O_h = O \times C_i$$

[*Ed. Note:* $V \equiv D_2$.]

Reprinted from *Phys. Rev.*, **5**, 706–714 (1934)

32

The Normal Modes and Frequencies of Vibration of the Regular Plane Hexagon Model of the Benzene Molecule*

E. Bright Wilson, Jr., *Gates Chemical Laboratory, California Institute of Technology*

(Received February 6, 1934)

The thirty modes of vibration of the regular plane hexagon model for the benzene molecule, including both the hydrogen and carbon atoms, are derived by the group theory method described by Wigner. From these the twenty frequencies of vibration are calculated in terms of a simple potential function involving six force constants. Selection rules for the Raman and infrared spectra are listed. Seven fundamentals are permitted in the Raman spectrum and four fundamentals in the infrared. Both analytical and graphical descriptions of the modes of vibration are given. These depend largely on the symmetry of the molecule and are only in part influenced by the choice of potential function adopted.

Introduction

In interpreting the infrared and Raman spectra of polyatomic molecules, it is customary to assume that the potential function for the forces between the atoms can be represented with sufficient accuracy by retaining only the quadratic terms in its Taylor expansion in powers of the coordinates.[1] With this restriction, which is equivalent from the classical mechanical standpoint to the assumption that the vibrations are of small amplitude compared to the interatomic distances, it is possible to obtain the fundamental frequencies and normal modes of vibration of the molecule as functions of the force constants. This has been done by a straightforward procedure for a number of simple cases,[2] but complicated molecules, even with high symmetry, are very difficult to treat without more powerful methods.

Wigner[3] has shown that group theory greatly simplifies the process of obtaining the normal modes of vibration of symmetric molecules. In this paper I have applied his method to benzene, obtaining the normal vibrations and frequencies; in addition, the selection rules for the infrared and Raman spectra are given.

Mathematical Methods

The positions of the atoms in the molecule are described by giving the Cartesian coordinates of each atom referred to the equilibrium position of the atom as origin. It is convenient to change the scale of the coordinates by the relation

$$q_i = m_i^{\frac{1}{2}} q_i', \tag{1}$$

in which q_i' is the ith coordinate in ordinary units, m_i the mass of the atom whose coordinates are concerned, and q_i the new coordinate. For small oscillations, the kinetic and potential energies are

$$2T = \sum_i \dot{q}_i^2, \quad 2V = \sum_{ij} a_{ij} q_i q_j. \tag{2}$$

The problem of finding the normal modes of vibration and frequencies is solved when the transformation

$$q_i = \sum_k l_{ik} Q_k \tag{3}$$

has been obtained which transforms T and V into

$$2T = \sum_k \dot{Q}_k^2, \quad 2V = \sum_k \lambda_k Q_k^2. \tag{4}$$

The Q_k's are the normal coordinates and the λ_k's are related to the frequencies ν_k by the equation

$$\lambda_k = 4\pi^2 \nu_k^2. \tag{5}$$

To obtain the coefficients l_{ik} and the frequencies

---

* Contribution No. 396.

[1] D. M. Dennison, Rev. Mod. Phys. **3**, 280 (1931).

[2] Some of these are given by K. W. F. Kohlrausch, *Der Smekal-Raman Effekt*. Springer, Berlin, 1931.

[3] E. Wigner, *Göttinger Nachrichten*, p. 133, 1930.

it is necessary to solve the secular equation

$$|a_{ij}-\delta_{ij}\lambda|=0, \quad (6)$$

in which δ_{ij} is unity if i equals j and zero otherwise.

The solution of the secular equation is facilitated by a knowledge of the symmetry of the molecule, which can be specified by giving the *symmetry operations* which transform the molecule into itself. Thus there are twenty-four symmetry operations which transform the regular plane hexagon model of benzene into itself. These include rotations about symmetry axes, reflections in symmetry planes, inversion through a center of symmetry, and rotations about axes combined with reflections in planes perpendicular to the axes. The set of such operations for a molecule forms a *point group*, which has the characteristic group property that the effect of any two operations applied successively is equivalent to the operation of some one member of the group.

Since the application of a symmetry operation R of the undistorted molecule leaves the molecule in a condition indistinguishable from its original position, the potential and kinetic energies are invariant under all such operations. As may be seen by considering Eq. (4), this restricts the possible modes of vibration Q_k. If Q_k is a non-degenerate vibration; i.e., if λ_k has a value distinct from the other λ's, then Q_k must be either symmetric or antisymmetric with respect to the symmetry operations of the point group G of the molecule. If Q_k is degenerate; i.e., if there are several vibrations, say Q_1, Q_2, Q_3, with the same frequency, then the application of a symmetry operation will transform each of these degenerate vibrations into a linear combination of the three.

$$RQ_k \rightarrow Q_k \quad (Q_k \text{ is symmetric to } R), \quad (7)$$

$$RQ_k \rightarrow -Q_k \quad (Q_k \text{ is antisymmetric to } R), \quad (8)$$

$$RQ_k \rightarrow \sum_n R_{kn}Q_n \quad \text{(degenerate)}. \quad (9)$$

The sum is over all the Q_n's with the same λ.

The symmetry of the molecule restricts still further the normal coordinates. In the theory of groups it is proved that each group G has only a certain number of *irreducible representations*[4] $\Gamma_1, \Gamma_2, \cdots \Gamma_r$, each of which represents a type of symmetry which is allowed for the normal coordinates of a molecule with point group G. Every Q_k must belong to a symmetry type which is correlated with one of the Γ_i's belonging to G, but for a given molecule every symmetry type Γ_i may not be represented among the Q_k's and more than one Q_k may have the same symmetry.[3]

The sum of the diagonal coefficients R_{kk} in the transformation of Eq. (9) is called the *character* χ_R of the operation R when applied to the set Q_k which have the same λ.

$$\chi_R=\sum_k R_{kk}. \quad (10)$$

From this definition and Eqs. (7) and (8) it is clear that the character of an operation for a non-degenerate Q_k must be equal to $+1$ or -1. The manner in which the symmetry of a given normal coordinate is restricted by the knowledge of the Γ_i to which it belongs is illustrated by Table I. The numbers in this table are the characters $\chi_R^{(i)}$ for each symmetry operation R when applied to a Q_k or set of Q_k's with the same λ which belong to the symmetry type of Γ_i. Table I is for the point group D_{6h} to which benzene is assumed to belong. Wigner[3] has published such tables for each of the crystallographic point groups and their construction is described in standard works on group theory.[4]

[4] A. Speiser, *Theorie der Gruppen von endlicher Ordnung.* Springer, Berlin, 1927.

TABLE I. *Values of* $\chi_i^{(i)}$ *for the point group* D_{6h} *and of* χ_i' *for benzene.*

$R=$	E	C_2	C_3	C_6	C_2'	C_2''	i	σ_h	S_6	S_3	σ_v''	σ_v'
A_{1g} Γ_1	1	1	1	1	1	1	1	1	1	1	1	1
A_{2g} Γ_2	1	1	1	1	−1	−1	1	1	1	1	−1	−1
B_{1g} Γ_3	1	−1	1	−1	1	−1	1	−1	1	−1	1	−1
B_{2g} Γ_4	1	−1	1	−1	−1	1	1	−1	1	−1	−1	1
E_{2g} Γ_5	2	2	−1	−1	0	0	2	2	−1	−1	0	0
E_{1g} Γ_6	2	−2	−1	1	0	0	2	−2	−1	1	0	0
A_{1u} Γ_7	1	1	1	1	1	1	−1	−1	−1	−1	−1	−1
A_{2u} Γ_8	1	1	1	1	−1	−1	−1	−1	−1	−1	1	1
B_{1u} Γ_9	1	−1	1	−1	1	−1	−1	1	−1	1	−1	1
B_{2u} Γ_{10}	1	−1	1	−1	−1	1	−1	1	−1	1	1	−1
E_{2u} Γ_{11}	2	2	−1	−1	0	0	−2	−2	1	1	0	0
E_{1u} Γ_{12}	2	−2	−1	1	0	0	−2	2	1	−1	0	0
h_i	1	1	2	2	3	3	1	1	2	2	3	3
χ_i'	36	0	0	0	−4	0	0	12	0	0	0	4
$h_i\chi_i'$	36	0	0	0	−12	0	0	12	0	0	0	12

The number n_i of Q_k's which have the characters corresponding to Γ_i is given by the group theory formula

$$n_i=(1/N)\sum_R \chi_R^{(i)}\chi_R'. \tag{11}$$

Here N is the number of symmetry operations of G, $\chi_R^{(i)}$ is the character of R in Γ_i, and χ_R' is a new quantity, the character of R when applied to the original set of coordinates q_i. Thus if

$$Rq_i=\sum_j \tau_{ij}q_j \tag{12}$$

summed over all the q_j's, then

$$\chi_R'=\sum_i \tau_{ii}. \tag{13}$$

There is a simple theorem which shortens the labor of the calculation: the symmetry operations of a group G can be divided into *classes* such that χ_R for every member R of the same class is the same. This applies both to $\chi_R^{(i)}$ and to χ_R'. If h_j is the number of operations and χ_j is the character of the operations in the jth class, then Eq. (11) becomes

$$n_i=(1/N)\sum_j h_j\chi_j^{(i)}\chi_j' \tag{14}$$

and we need to calculate χ_j' for only one member of each class. The n_i's give the number of Q_k's which have the symmetry properties ($\chi_R^{(i)}$) of Γ_i. We may symbolize this by writing:

$$\Gamma_{\text{benzene}}=\sum_i n_i\Gamma_i. \tag{15}$$

Six Q_k's will correspond to translation and rotation of the molecule as a whole. The symmetry classes to which these belong may be found by inspection, and a new set of numbers n_i' obtained which gives the number of purely internal motions of each symmetry class.

By using the symmetry of the molecule it is possible to factor the secular Eq. (6) into a number of equations of lower degree. Group theory provides the theorem: the degree of the factor of the secular equation from which the Q_k's belonging to Γ_i are obtained is n_i' and this factor is repeated $\chi_E^{(i)}$ times, where $\chi_E^{(i)}$ is the character of the identity operation E for Γ_i. A corollary is, therefore, that the roots λ_k of this factor will have a multiplicity $\chi_E^{(i)}$.

Application to Benzene

The model assumed for benzene is the usual regular plane hexagon with the carbon and hydrogen atoms lying in the same plane as is shown in Fig. 1. The thirty-six q's are shown in Fig. 2 and form systems of rectangular coordinates with origin at the equilibrium position of each atom. R_i, Y_i, Z_i and r_i, y_i, z_i are the coordinates of the ith carbon and the ith hydrogen atom, respectively. The R_i, r_i point radially outward as shown in Fig. 2, while the Z_i, z_i are perpendicular to the plane of the ring.

The symmetry operations of this model are also shown in Fig. 2; they fall into the following twelve classes. $E(h_1=1)$ identity; $C_2(h_2=1)$ rotation by π about the six-fold axis; $C_3(h_3=2)$ rotation by $\pm 2\pi/3$ about the six-fold axis; $C_6(h_4=2)$ rotation by $\pm\pi/3$ about the six-fold axis; $C_2'(h_5=3)$ rotation by π about axes P, Q, R; $C_2''(h_6=3)$ rotation by π about axes T, U, V; $i(h_7=1)$ inversion through the center of symmetry; $\sigma_h(h_8=1)$ reflection through the plane of symmetry in the XY plane; $S_6(h_9=2)$ rotatory-reflection about the six-fold axis by $\pm\pi/3$; $S_3(h_{10}=2)$ rotatory-reflection about the six-fold axis by $\pm 2\pi/3$; $\sigma_v''(h_{11}=3)$ reflection through planes D, F, G; $\sigma_v'(h_{12}=3)$ reflection through planes A, B, C. These operations form the point group D_{6h}.

Table I gives the characters $\chi_j^{(i)}$ for the twelve irreducible representations of D_{6h}, which has twenty-four operations as listed above. In addi-

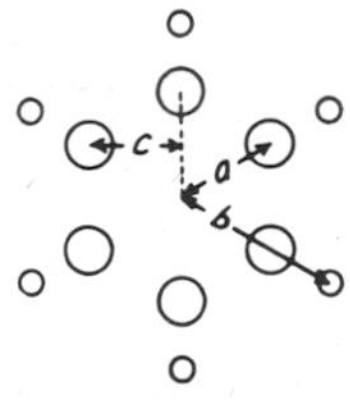

FIG. 1. Model of benzene.

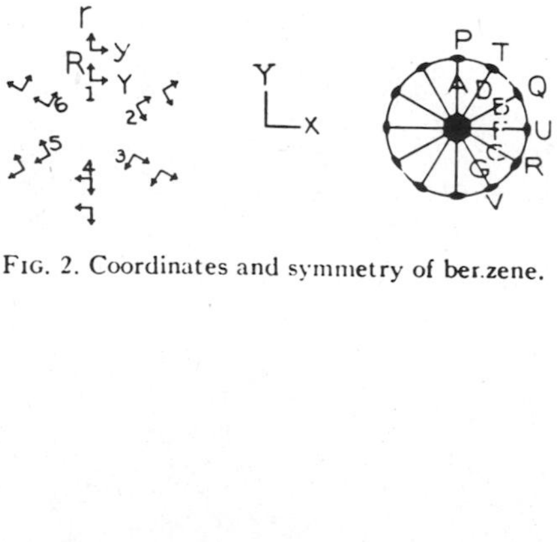

FIG. 2. Coordinates and symmetry of benzene.

tion to the symbols $\Gamma_1, \Gamma_2, \cdots \Gamma_{12}$ for the irreducible representations which were used in the general discussion, Tisza's[5] symbols, A_g, A_u, etc., are given. These are more useful for specific applications since they are more descriptive. Thus A and B refer to non-degenerate, E_1, E_2 to degenerate types of vibration. The subscript g means that the vibration is symmetric with respect to the inversion i, while u means that it is anti-symmetric to i. A refers to vibrations that are symmetric to rotation by $2\pi/n$ about the n-fold axis; B labels vibrations which are anti-symmetric to this operation. Table I also gives the characters χ_j' of the *reducible representation;* i.e., the q_i's or R_i, Y_i, Z_i, r_i, y_i, z_i. The operation of C_2, for example, on the molecule moves every atom to a new position. Therefore, in the transformation Eq. (12), each of the τ_{ii} must be zero so that χ_R' is zero for C_2. The same is true for the operations belonging to the classes C_3, C_6, C_2'', i, S_6, S_3, and σ_v''. This illustrates that it is only the unshifted atoms which contribute to χ_j'. Thus reflection of the molecule through the plane of symmetry A shifts all but the atoms C_1, C_4, H_1, H_4. For these:

$$AR_1 \rightarrow R_1, \quad Ar_1 \rightarrow r_1,$$
$$AY_1 \rightarrow -Y_1, \quad Ay_1 \rightarrow -y_1, \qquad (16)$$
$$AZ_1 \rightarrow Z_1, \quad Az_1 \rightarrow z_1.$$

with similar equations for C_4, H_4. The sum of the coefficients τ_{ii} is thus $+4$.

When the values of $h_j\chi_j'$ and $\chi_j^{(i)}$ from Table I are substituted in Eq. (14), the result is obtained that:

$$\Gamma_{\text{benzene}} = \underset{(1)}{2\Gamma_1} + \underset{(1)}{2\Gamma_2} + \underset{(1)}{2\Gamma_4} + \underset{(2)}{4\Gamma_5} + \underset{(2)}{2\Gamma_6}$$
$$+ \underset{(1)}{2\Gamma_8} + \underset{(1)}{2\Gamma_9} + \underset{(1)}{2\Gamma_{10}} + \underset{(2)}{2\Gamma_{11}} + \underset{(2)}{4\Gamma_{12}}. \qquad (17)$$

The figures in parentheses indicate the multiplicities of the frequencies derived from the corresponding Γ_i.

Before interpreting this in terms of frequencies and factors of the secular equation, it is con-

[5] L. Tisza, Zeits. f. Physik **82**, 48 (1933).

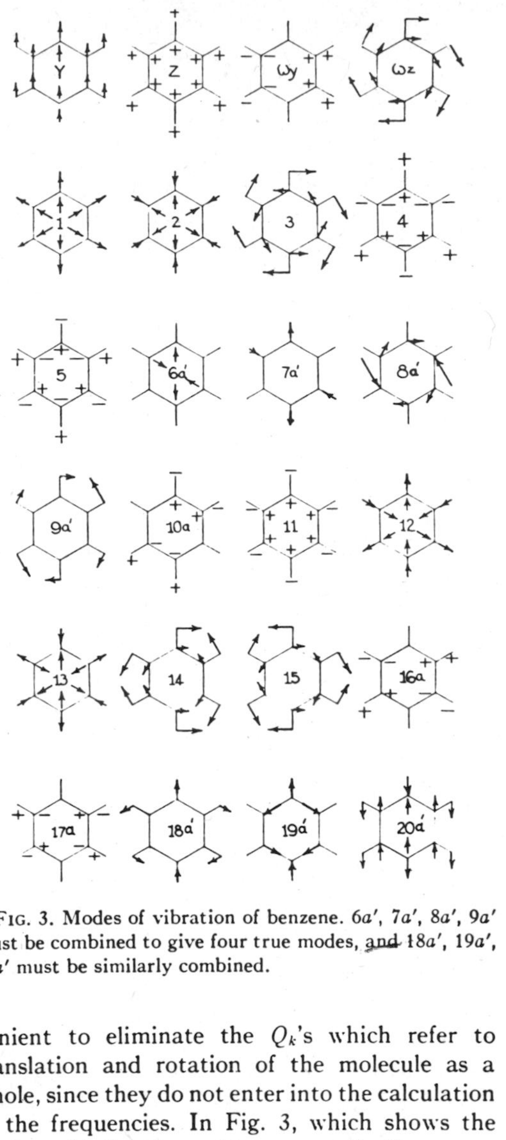

FIG. 3. Modes of vibration of benzene. $6a'$, $7a'$, $8a'$, $9a'$ must be combined to give four true modes, and $18a'$, $19a'$, $20a'$ must be similarly combined.

venient to eliminate the Q_k's which refer to translation and rotation of the molecule as a whole, since they do not enter into the calculation of the frequencies. In Fig. 3, which shows the modes of vibration of benzene, Y, Z, ω_Y, ω_Z represent translation and rotation with respect to the Y and Z axes, and the remaining translation and rotation are of course related to Y and ω_Y by a rotation about the six-fold axis. The symmetry properties of these motions, obtained by inspection, show that Q_z belongs to Γ_8 and Q_{ω_z} to Γ_2, while Q_x, Q_y and Q_{ω_x}, Q_{ω_y} are two sets of degenerate modes belonging to Γ_{12} and Γ_6, respectively. When these are taken out, we obtain:

$$\Gamma'_{\text{benzene}}=2\Gamma_1+\Gamma_2+2\Gamma_4+4\Gamma_5+\Gamma_6+\Gamma_8+2\Gamma_9+2\Gamma_{10}+2\Gamma_{11}+3\Gamma_{12}, \quad (18)$$

or, using the other type of symbols:

$$\Gamma'_{\text{benzene}}=2A_{1g}+A_{2g}+2B_{2g}+4E_{2g}+E_{1g}+A_{2u}+2B_{1u}+2B_{2u}+2E_{2u}+3E_{1u}. \quad (18a)$$

From this we immediately obtain the following information: There are 20 different frequencies, of which 10 are double. The secular equation may be factored by the use of symmetry and without regard to the type of forces involved, into four linear factors of which two are equal (Γ_2, Γ_8, $\Gamma_6(2)$), six quadratic factors of which two are equal (Γ_1, Γ_4, Γ_9, Γ_{10}, $\Gamma_{11}(2)$), two equal cubic factors ($\Gamma_{12}(2)$), and two equal quartic factors ($\Gamma_5(2)$).

TABLE II. *Selection rules for fundamentals in the infrared and Raman spectra.*

Frequencies permitted in Raman spectrum:
ν_1, ν_2, ν_6*, ν_7*, ν_8*, ν_9*, ν_{10}*.
Frequencies permitted in infrared spectrum:
ν_{11}, ν_{18}, ν_{19}, ν_{20}.

* These lines should have a depolarization of $\frac{6}{7}$.

Furthermore, this information is sufficient to determine the modes of vibration shown in Fig. 3 and described in the next section. Additional information which comes from the symmetry discussion is contained in Table II, which gives the selection rules for the Raman and infrared spectra. These were obtained directly from the general table given by Placzek.[5a]

NORMAL COORDINATES FOR BENZENE

The coefficients l_{ki} of the transformation of Eq. (3) have certain properties which assist in their determination, such as, for example,

$$\sum_k l_{ki}l_{kj}=\begin{cases}0 \text{ if } i\neq j.\\ 1 \text{ if } i=j.\end{cases} \quad (19)$$

The inverse transformation to Eq. (3) is

$$Q_i=\sum_k l_{ki}q_k. \quad (20)$$

By using the symmetry requirements of Eq. (17) and Table I, together with the restrictions of Eq. (19), we obtain the following combinations Q_i', which are either the normal coordinates Q_i themselves (if they belong to a linear factor of the secular equation) or functions which must be combined with one or more others in order to obtain the normal coordinates (if they belong to factors of higher degree). Q_1 indicates the normal coordinate with frequency ν_1. Q_{10a}, Q_{10b} indicate two normal coordinates with the same frequency ν_{10}. Any linear combination of these two will also have the same frequency, Q_4', Q_5' connected by a parenthesis indicate that two independent combinations of these, with coefficients dependent on the force constants and determined by solving a quadratic factor of the secular equation, are the normal coordinates for ν_4 and ν_5.

$$\left.\begin{aligned}Q_X&=m^{\frac12}[3^{\frac12}(r_2+r_3-r_5-r_6)+2y_1+y_2-y_3-2y_4-y_5+y_6]\\&\quad+M^{\frac12}[3^{\frac12}(R_2+R_3-R_5-R_6)+2Y_1+Y_2-Y_3-2Y_4-Y_5+Y_6].\\Q_Y&=m^{\frac12}[2r_1+r_2-r_3-2r_4-r_5+r_6-3^{\frac12}(y_2+y_3-y_5-y_6)]\\&\quad+M^{\frac12}[2R_1+R_2-R_3-2R_4-R_5+R_6-3^{\frac12}(Y_2+Y_3-Y_5-Y_6)].\end{aligned}\right\}(\Gamma_{12}(2))$$

$$Q_Z=m^{\frac12}(z_1+z_2+z_3+z_4+z_5+z_6)+M^{\frac12}(Z_1+Z_2+Z_3+Z_4+Z_5+Z_6). \quad (\Gamma_8)$$

$$\left.\begin{aligned}Q_{\omega_x}&=bm^{\frac12}(2z_1+z_2-z_3-2z_4-z_5+z_6)+aM^{\frac12}(2Z_1+Z_2-Z_3-2Z_4-Z_5+Z_6).\\Q_{\omega_y}&=bm^{\frac12}(z_2+z_3-z_5-z_6)+aM^{\frac12}(Z_2+Z_3-Z_5-Z_6).\end{aligned}\right]\,(\Gamma_6(2))$$

[5a] G. Placzek, *The Structure of Molecules* (edited by P. Debye), Blackie and Son, London, 1932.

$$Q_{\omega_z}=bm^{\frac{1}{2}}(y_1+y_2+y_3+y_4+y_5+y_6)+aM^{\frac{1}{2}}(Y_1+Y_2+Y_3+Y_4+Y_5+Y_6). \qquad (\Gamma_2)$$

$$\left.\begin{aligned} Q_1'&=R_1+R_2+R_3+R_4+R_5+R_6.\\ Q_2'&=r_1+r_2+r_3+r_4+r_5+r_6.\end{aligned}\right] \qquad (2\Gamma_1)$$

$$Q_3=aM^{\frac{1}{2}}(y_1+y_2+y_3+y_4+y_5+y_6)-bm^{\frac{1}{2}}(Y_1+Y_2+Y_3+Y_4+Y_5+Y_6). \qquad (\Gamma_2)$$

$$\left.\begin{aligned} Q_4'&=Z_1-Z_2+Z_3-Z_4+Z_5-Z_6.\\ Q_5'&=z_1-z_2+z_3-z_4+z_5-z_6.\end{aligned}\right] \qquad (2\Gamma_4)$$

$$\left.\begin{aligned} Q_{6a}'&=R_1-R_3+R_4-R_6. & Q_{6b}'&=R_1-2R_2+R_3+R_4-2R_5+R_6.\\ Q_{7a}'&=r_1-r_3+r_4-r_6. & Q_{7b}'&=r_1-2r_2+r_3+r_4-2r_5+r_6.\\ Q_{8a}'&=Y_1-2Y_2+Y_3+Y_4-2Y_5+Y_6. & Q_{8b}'&=Y_1-Y_3+Y_4-Y_6.\\ Q_{9a}'&=y_1-2y_2+y_3+y_4-2y_5+y_6. & Q_{9b}'&=y_1-y_3+y_4-y_6.\end{aligned}\right] \qquad (4\Gamma_5(2))$$

$$\left.\begin{aligned} Q_{10a}&=aM^{\frac{1}{2}}(z_1+z_2-z_4-z_5)-bm^{\frac{1}{2}}(Z_1+Z_2-Z_4-Z_5).\\ Q_{10b}&=aM^{\frac{1}{2}}(z_1-z_2-2z_3-z_4+z_5+2z_6)-bm^{\frac{1}{2}}(Z_1-Z_2-2Z_3-Z_4+Z_5+2Z_6).\end{aligned}\right] \qquad (\Gamma_6(2))$$

$$Q_{11}=M^{\frac{1}{2}}(z_1+z_2+z_3+z_4+z_5+z_6)-m^{\frac{1}{2}}(Z_1+Z_2+Z_3+Z_4+Z_5+Z_6). \qquad (\Gamma_8)$$

$$\left.\begin{aligned} Q_{12}'&=R_1-R_2+R_3-R_4+R_5-R_6.\\ Q_{13}'&=r_1-r_2+r_3-r_4+r_5-r_6.\end{aligned}\right](2\Gamma_9) \qquad \left.\begin{aligned} Q_{14}'&=Y_1-Y_2+Y_3-Y_4+Y_5-Y_6.\\ Q_{15}'&=y_1-y_2+y_3-y_4+y_5-y_6.\end{aligned}\right] \quad (2\Gamma_{10})$$

$$\left.\begin{aligned} Q'_{16a}&=Z_2-Z_3+Z_5-Z_6. & Q'_{16b}&=2Z_1-Z_2-Z_3+2Z_4-Z_5-Z_6.\\ Q'_{17a}&=z_2-z_3+z_5-z_6. & Q'_{17b}&=2z_1-z_2-z_3+2z_4-z_5-z_6.\end{aligned}\right] \qquad (2\Gamma_{11}(2))$$

$$\left.\begin{aligned} Q'_{18a}&=2r_1+r_2-r_3-2r_4-r_5+r_6+3^{\frac{1}{2}}(y_2+y_3-y_5-y_6).\\ Q'_{19a}&=2R_1+R_2-R_3-2R_4-R_5+R_6+3^{\frac{1}{2}}(Y_2+Y_3-Y_5-Y_6).\\ Q'_{20a}&=M^{\frac{1}{2}}[2r_1+r_2-r_3-2r_4-r_5+r_6-3^{\frac{1}{2}}(y_2+y_3-y_5-y_6)]\\ &\quad-m^{\frac{1}{2}}[2R_1+R_2-R_3-2R_4-R_5+R_6-3^{\frac{1}{2}}(Y_2+Y_3-Y_5-Y_6)].\\ Q'_{18b}&=3^{\frac{1}{2}}(r_2+r_3-r_5-r_6)-(2y_1+y_2-y_3-2y_4-y_5+y_6).\\ Q'_{19b}&=3^{\frac{1}{2}}(R_2+R_3-R_5-R_6)-(2Y_1+Y_2-Y_3-2Y_4-Y_5+Y_6).\\ Q'_{20b}&=M^{\frac{1}{2}}[3^{\frac{1}{2}}(r_2+r_3-r_5-r_6)+(2y_1+y_2-y_3-2y_4-y_5+y_6)]\\ &\quad-m^{\frac{1}{2}}[3^{\frac{1}{2}}(R_2+R_3-R_5-R_6)+(2Y_1+Y_2-Y_3-2Y_4-Y_5+Y_6)].\end{aligned}\right] \qquad (3\Gamma_{12}(2))$$

The above equations should be multiplied by appropriate factors in order that they may be normalized in the sense of Eq. (19). a and b are the radii of the carbon and hydrogen rings, respectively; while M and m are the masses of the carbon and hydrogen atoms.

Fig. 3 shows these modes diagrammatically. The component along any coordinate direction of the arrow attached to a given atom gives the coefficient of that coordinate in the above expressions. Since these numbers are also the coefficients in the reverse transformation of Eq. (3), they represent the amplitudes of the motions of the atoms in the corresponding modes of vibration. Plus and minus signs refer to motions perpendicular to the plane of the paper.

Potential Energy Function

When the modes of vibration are known and the potential energy as a quadratic function of the coordinates is given, it is possible to calculate the corresponding frequencies. In this treatment a much simplified potential function was used, involving only six force constants: K, q, κ, h, H, k. Fig. 4 shows some of the coordinates of the distorted molecule in terms of which the potential energy is expressed.

K is the constant for the stretching of the carbon-carbon bond and q is the corresponding constant for the carbon-hydrogen bond.

$$\therefore \quad 2V_1 = K\{(\Delta R_{12})^2+(\Delta R_{23})^2+(\Delta R_{34})^2 + (\Delta R_{45})^2+(\Delta R_{56})^2+(\Delta R_{61})^2\}. \quad (21)$$

$$2V_2 = q\{(\Delta S_1)^2+(\Delta S_2)^2+(\Delta S_3)^2 + (\Delta S_4)^2+(\Delta S_5)^2+(\Delta S_6)^2\}. \quad (22)$$

According to the quantum-mechanical view of the structure of benzene,[6] each of the carbon-carbon bonds is of partially double bond character, which would give the bond a resistance to twisting such as is found in ethylene. κ is the force constant for this twisting and enters into the equations whenever the carbon atoms move out of a plane. This is seen if any four adjacent carbon atoms are considered and the effect on the central bond of motion out of the plane is calculated.

$$\therefore \quad 2V_3 = \kappa c^2(\varphi_{12}^2+\varphi_{23}^2+\varphi_{34}^2+\varphi_{45}^2 + \varphi_{56}^2+\varphi_{61}^2). \quad (23)$$

φ_{12} is the angle of twist of the bond between the first and second carbon atoms, while c is the length indicated in Fig. 1. h is the force constant for the bending of the carbon-hydrogen bond out of the plane of the three adjacent carbon atoms. μ_i is the angle of this bending.

$$\therefore \quad 2V_4 = h(b-a)^2(\mu_1^2+\mu_2^2+\mu_3^2+\mu_4^2 + \mu_5^2+\mu_6^2). \quad (24)$$

b and a are given in Fig. 1.

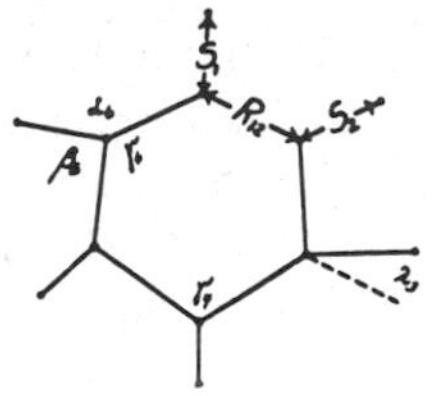

Fig. 4. Distorted molecule and coordinates used in potential energy function.

There are several methods of representing the other angular distortions of the bonds, one being given by

$$2V_5 = ka^2\{(\Delta\gamma_1)^2+(\Delta\gamma_2)^2+(\Delta\gamma_3)^2 + (\Delta\gamma_4)^2+(\Delta\gamma_5)^2+(\Delta\gamma_6)^2\}. \quad (25)$$

$$2V_6 = H(b-a)^2(\lambda_1^2+\lambda_2^2+\lambda_3^2+\lambda_4^2 + \lambda_5^2+\lambda_6^2), \quad (26)$$

in which $\Delta\gamma_i$ is the change in the carbon-carbon bond angle at the ith carbon and λ_i is the angle of deviation of the carbon-hydrogen bond from the bisector of the carbon-carbon bond angle (all lying in the same plane). This is equivalent to

$$2V_6 = H'(b-a)^2 \sum_i \{(\Delta\alpha_i)^2+(\Delta\beta_i)^2\} \quad (27)$$

with α and β given by Fig. 4, with the restriction that $\Delta\alpha_i+\Delta\beta_i+\Delta\gamma_i=0$.

The methods used to obtain the frequencies are not restricted to the above choice of potential function but could be applied to any homogeneous quadratic function of the coordinates.

Calculation of the Frequencies

The procedure used to calculate the frequencies from a knowledge of the modes of vibration and the potential function is best illustrated by an example. From Fig. 3 or the analytical equations for the normal coordinates, Q_3 is seen to come from a linear factor of the secular equation, which means that the positions of the atoms during this motion can be completely described by giving the value of one coordinate, say y_1. In an infinitesimal vibration of this type the only force constant which enters is H, since the carbon and hydrogen rings merely rotate with respect to each other and are not deformed. If we call the amplitude of the vibration at any instant

---

[6] Linus Pauling and G. W. Wheland, J. Chem. Phys. **1**, 362 (1933).

δ, then the hydrogen atoms have each undergone a displacement $y_i'=a(M/m)^{\frac{1}{2}}\delta$, the carbon atoms a displacement $Y_i'=-b(m/M)^{\frac{1}{2}}\delta$, in ordinary units. From this it can be calculated that the potential energy of the distortion, using Eq. (26) and the geometry of the distorted molecule, is

$$2V=6H\delta^2(a^2M+b^2m)^2/(a^2mM), \qquad (28)$$

while the kinetic energy of the motion is

$$2T=6(a^2M+b^2m)(d\delta/dt)^2. \qquad (29)$$

Substitution of these in the equation of motion in terms of δ and solution of this equation gives for the frequency:

$$\nu_3=(1/2\pi)[H(a^2M+b^2m)/(a^2mM)]^{\frac{1}{2}}. \qquad (30)$$

In case the frequencies come from a quadratic equation the method is similar, except that two amplitudes δ and ϵ must be used. Thus ν_1 and ν_2 come from a quadratic factor involving Q_1' and Q_2'. If the amplitude of Q_1' is δ, every carbon atom is displaced a distance δ outward; if ϵ refers to Q_2', every hydrogen atom is moved radially outward a distance ϵ. None of the angles is altered by these displacements, the only effects being to stretch the carbon-carbon bond an amount δ and the hydrogen-carbon bond an amount $(\epsilon-\delta)$. The potential energy for this motion is therefore, from Eqs. (21) and (22),

$$2V=6K\delta^2+6q(\epsilon-\delta)^2$$
$$=6(K+q)\delta^2-12q\epsilon\delta+6q\epsilon^2. \qquad (31)$$

The kinetic energy is:

$$2T=6M(d\delta/dt)^2+6m(d\epsilon/dt)^2. \qquad (32)$$

The equations of motion become

$$\left.\begin{aligned}6M(d^2\delta/dt^2)+6(K+q)\delta-6q\epsilon=0\\ 6m(d^2\epsilon/dt^2)+6q\epsilon-6q\delta=0\end{aligned}\right\}. \qquad (33)$$

These have a solution of the form

$$\delta=A\sin\lambda^{\frac{1}{2}}t;\quad \epsilon=B\sin\lambda^{\frac{1}{2}}t \qquad (34)$$

if the secular determinant vanishes.

$$\therefore\quad \begin{vmatrix}K+q-M\lambda & -q\\ -q & q-m\lambda\end{vmatrix}=0. \qquad (35)$$

This is satisfied for

$$\lambda_{1,\,2}=(1/2mM)\{mK+q(m+M)\pm[\{mK+q(m+M)\}^2-4mMqK]^{\frac{1}{2}}\}. \qquad (36)$$

$$\nu^2=\lambda/4\pi^2. \qquad (37)$$

The other frequencies were obtained in an exactly similar manner and are listed below, with the exception of those which come from the cubic and quartic factors of the secular equation. These higher degree equations are also given and may be solved numerically for any given values of the force constants.

ν_{18}, ν_{19} and ν_{20} are the roots of the equation.

$$\begin{vmatrix}3(K+k)+q+H\zeta^2-2M\lambda & -q-H\zeta & ql-H\zeta l\\ -q-H\zeta & q+H-2m\lambda & -ql+Hl\\ ql-H\zeta l & -ql+Hl & ql^2+Hl^2-2Ml\lambda\end{vmatrix}=0, \qquad (38)$$

in which $\zeta=(3b-a)/2a$, $l=(M+m)/m$.

ν_6, ν_7, ν_8 and ν_9 are the roots of the equation.

$$\begin{vmatrix}q-m\lambda & 0 & -q & 0\\ 0 & 4H-4m\lambda & 3^{\frac{1}{2}}H\eta & -H\xi\\ -16q & 12(3^{\frac{1}{2}})H\eta & 4K+108k+16q+9\eta^2H-16M\lambda & -3^{\frac{1}{2}}\{4K+12k+H\xi\eta\}\\ 0 & -4H\xi & -3^{\frac{1}{2}}\{4K+12k+H\xi\eta\} & 36K+12k+H\xi^2-16M\lambda\end{vmatrix}=0, \qquad (39)$$

in which $\eta=(b-a)/a$, $\xi=(b+3a)/a$.

$$\lambda_{4,\,5}=(1/2mM)\{36\kappa m+(3-4b/a)^2hm+Mh\pm[\{36\kappa m+(3-4b/a)^2hm+Mh\}^2-144mM\kappa h]^{\frac{1}{2}}\}. \quad (40)$$

$$\lambda_{10}=h(b^2m+a^2M)/a^2mM. \quad (41)$$

$$\lambda_{11}=h(m+M)/mM. \quad (42)$$

$$\lambda_{12,\,13}=(1/2mM)\{12km+q(m+M)\pm[\{12km+q(m+M)\}^2-48mMkq]^{\frac{1}{2}}\}. \quad (43)$$

$$\lambda_{14,\,15}=(1/2mM)\{3Km+H(m+M)\pm[\{3Km+H(m+M)\}^2-12mMKH]^{\frac{1}{2}}\}. \quad (44)$$

$$\lambda_{16,\,17}=(1/2mM)\{12\kappa m+mhf+Mh\pm[(12\kappa m+mhf+Mh)^2-48mM\kappa h]^{\frac{1}{2}}\}. \quad (45)$$

where $f=(3b-2a)^2/a^2$.

It is planned to apply the above results to the analysis of the experimental data for benzene.

I should like to express my appreciation to Professors Linus Pauling and R. M. Badger for their suggestions and criticisms.

Reprinted from *J. Chem. Phys.*, **2**, 432–439 (1934)

The Degeneracy, Selection Rules, and Other Properties of the Normal Vibrations of Certain Polyatomic Molecules

33

E. BRIGHT WILSON, JR., *Gates Chemical Laboratory, California Institute of Technology*

(Received April 2, 1934)

The number, degeneracies, and symmetries of the normal modes of vibration are given for molecules in which several atoms are bonded to a central atom. Ninety-nine symmetries, including all possible structures with three to seven atoms and the more important structures of eight and nine atoms are listed in tables. The selection rules for the Raman and infrared spectra are included as well as the polarization properties of the Raman lines and the rotational structure of the vibrational bands. In a number of more important cases figures are given showing the normal modes of vibration. Many examples of molecules which are believed to possess certain of the structures are included in the tables, together with references to experimental papers dealing with them.

IN drawing structural conclusions from observed Raman and infrared spectra, it is necessary to know the number of normal frequencies of vibration and the selection rules which characterize the various possible configurations of atoms. The theory of the number and degeneracy of the normal vibrations of symmetric molecules, first developed by Brester,[1] was very concisely expressed by Wigner,[2] using group theory, while the selection rules for the Raman and infrared have been discussed by Placzek,[3] Tisza[4] and others.

In order to provide a tabular presentation of the important results of these theories for a special class of molecules, so arranged that they may be used without any knowledge of group theory, the degeneracies, number of permitted Raman and infrared fundamentals, and the polarization properties of the Raman lines have been calculated for molecules in which two or more atoms (which may be identical or different) are attached to a single central atom. The treatment is intended to be complete up to six attached atoms; that is, every possible symmetry for such molecules is listed, even those which are not very reasonable physically. For molecules with seven and eight outer atoms, only a few of the more reasonable structures are given, since there are very few molecules of these classes known and a very large number of possible symmetries.

The group-theoretical method used to find the number and degeneracies of the normal vibrations was described in a previous paper,[5] where it was used to obtain the normal coordinates and frequency formulas for benzene. The selection rules have also been obtained by the application of group theory[4] and the results checked wherever possible with the table given by Placzek.[3] (The following divergences from his results were found: p. 81 of the English translation,[6] point group S_{pu}, line labelled B_1, fifth column, read "all zero but C_{xy}"; line labelled B_2, fifth column, "all zero but $C_{xx} = -C_{yy}$"; line labelled C, fifth column, "$C_{xy} = C_{ii} = 0$"; line labelled D, sixth column, "$C_{xx} = -C_{yy}$, $C_{ii} = 0$.")

In using these results in connection with experimentally observed spectra, it is necessary to take certain precautions. Thus the number of Raman lines found may not coincide with that theoretically required for two reasons: first, lines which are not forbidden by symmetry restrictions may still be so weak that they do not appear under ordinary conditions, and second, overtones and combinations may occur in special cases. The latter difficulty is one which is rather frequently observed in spite of the fact that overtones and combinations should be much less important in Raman spectra than in infrared spectra. Another phenomenon which sometimes

[1] C. J. Brester, Kristallsymmetrie und Reststrahlen, Utrecht, 1923; Zeits. f. Physik **24**, 324 (1924).

[2] E. Wigner, Gott. Nach. 133 (1930).

[3] G. Placzek, *Leipziger Vorträge*, p. 71, 1931; Zeits. f. Physik **70**, 84 (1931). G. Placzek and E. Teller, Zeits. f. Physik **81**, 209 (1933).

[4] L. Tisza, Zeits. f. Physik **82**, 48 (1933).

[5] E. Bright Wilson, Jr., Phys. Rev. **45**, 706 (1934).

[6] P. Debye, *The Structure of Molecules*, Blackie and Sons, Ltd., London, 1932.

occurs, for example in CO_2 and CCl_4, and which causes deviations from the simple selection rules, is quantum-mechanical resonance between certain of the energy states of the molecule due to accidental coincidence or approximate coincidence of the levels corresponding to different types of vibration. This effect usually allows otherwise forbidden lines to appear in the spectrum, but it is probably not very common.

Explanation of Tables

The results are given in separate tables, arranged according to the number of atoms in the molecule. Thus Table I lists triatomic molecules of the type AB_2 and ABB', Table II includes AB_3, AB_2B', and $ABB'B''$, etc. A is not equivalent by symmetry to any of the B's (although it may be an atom of the same element, as for example in N N O).

The numbering, given in the first column of the tables, is designed to indicate the number and character of the attached atoms. The figure in front of the dot gives the total number of attached atoms while the figures following the dot indicate the numbers of symmetrically related atoms. Thus 6.6 means that the molecule is AB_6 in which all six B atoms are equivalent from a symmetry standpoint. 6.42 refers to AB_4B_2', etc. When there are several molecules of the same type but of different symmetries, they are distinguished by using a, b, c, etc. The name, given in the second column, is meant to be descriptive enough to characterize the structure if possible, but in addition the coordinates of the atoms are given under the discussion of each type. The notation R and L following the name of a structure indicates that right and left-handed forms are possible.

The column labelled I gives the point-group symmetry of the structure, the symbols being those of Schönflies,[7] which represent the axes, planes and other elements of symmetry possessed by the molecule. The column headed II tells whether the molecule is a linear rotator L, a spherical top Sp (which has three equal moments of inertia), a symmetric top S (with one distinct and two equal moments of inertia), or an asymmetric top A (which has three unequal moments of inertia).

The columns labelled ν_1, ν_2, ν_3 contain the number of singly, doubly, and triply degenerate normal frequencies of vibration. Only molecules with cubic point-group symmetries can have triply degenerate frequencies. Under R is listed the number of fundamental frequencies of vibration which are allowed to appear in the Raman spectrum, although a line so permitted may be of very small intensity. Following this number, in parentheses, is the number of these Raman fundamentals which have the special depolarization $\rho=\frac{3}{4}$, when plane polarized incident light is used. ρ is the ratio of the intensities of the perpendicular and parallel components of the scattered light. For unpolarized incident light this becomes $\rho=6/7$.

Under I.R. is given the number of fundamental frequencies which are active in the infrared absorption spectra, the figures in parentheses being the number of these bands which are of the $\|$ type[8] if the molecule is a symmetric top. Finally, the last column tabulates the symmetries of the possible normal vibrations, the degrees of the factors into which the secular equation for the normal coordinate problem may be factored by the use of the symmetry of the molecule, the degeneracies of these vibrations, and the ones which are active in the Raman and infrared spectra. To illustrate: 2^{R}, 0, 2^{RI}, 1^{R}, 2_2^{RI} after 4.4d means that there are two frequencies with the symmetry of Γ_1 or A_1, the first irreducible representation of the point-group V_d, none with symmetry Γ_2 or A_2, two with Γ_3 or B_2, one with Γ_4 or B_1, and two doubly degenerate frequencies with symmetry Γ_5 or E_1, the degeneracy being indicated by the subscript. The meaning of the term "irreducible representation" is explained in the paper on benzene already referred to.[5] The order Γ_1, Γ_2, etc. is that used by Wigner[2] in his tabulation of the irreducible representations of the crystallographic point groups. Several point groups which do not occur in crystallography were used: namely $D_{\infty h}$, $C_{\infty v}$, D_{5h}, C_{5v}, C_{7h}, C_{7v}, C_{8h}, C_{8v} and D_{4d}, the characters for their irreducible representations being taken from the general tables given by Tisza.[4]

[7] H. Hilton, *Mathematical Crystallography*, Chapters V, VI. Oxford, 1903.

[8] D. M. Dennison, Rev. Mod. Phys. **3**, 314 (1931).

The numbers with superscript R indicate the frequencies which are active in the Raman spectrum while a superscript I on a number means that this frequency is infrared active. The numbers themselves, besides the significance given to them above, also represent the degrees of the factors of the factored secular equation. For the linear molecules, to which Wigner's tables do not apply, the notation of Tisza[4] is used for the irreducible representations.

Discussion of Results

In the discussion of individual cases which is given below, the number and name of the structure is followed by the Cartesian coordinates of the different atoms, with A always at the origin. u, v, w represent parameters which are not fixed by the symmetry, while the letter o has been printed instead of zero. These were mostly taken from a tabulation by Nowacki[9] of the equivalent positions for the point-groups. (Two slight misprints were found in this paper; namely, in Table III, p. 28, after $D_{2d} \equiv V$ (4) the second line should read $|xxz|\bar{x}\bar{x}\bar{z}|\bar{x}\bar{x}z|x\bar{x}\bar{z}|$ (C_s). After D_{3d} (6) (b) the second line should read $|\bar{x}0x\bar{z}|x\bar{x}0\bar{z}|0x\bar{x}\bar{z}|$ (C_s).) Wherever possible examples are given of molecules which have been found or which might be supposed to belong to the given type. In a number of examples which seem likely to be of importance and which, so far as I know, have not been previously treated, figures are given showing the normal vibrations, derived from the symmetry restrictions yielded by the group theory. The order used in the diagrams is the same as that in the last column of the tables. Only one representative of a set of degenerate vibrations is shown. An example is Fig. 2 which shows the normal vibrations for 5.32a, the trigonal bipyramid model of AB_3B_2'. The last column of Table IV reads: 2^R, 0, 3_2^{RI}; 0, 2^I, 1_2^R. Therefore the first two motions of Fig. 2, labelled ν_1 and ν_2, belong to Γ_1 or A_1', the first irreducible representation of Wigner's table for D_{3h}, and since 2^R has the superscript R, they are Raman active. The next three motions, ν_3', ν_4', ν_5' belong to Γ_3 or E_1'. They are active in both the Raman and infrared spectra and are doubly degenerate; i.e., for each of them there exists another similar motion with the same frequency, differing only in spatial orientation of the directions of motion. The primes on the ν's indicate that to obtain the three true normal vibrations these three motions enclosed in the square brackets must be compounded with proportions determined by the force constants.

Many of the simpler structures listed in Tables I–VII have been previously studied, and although it is impracticable to give all the references it is hoped that those listed will enable the others to be found.

I am very much indebted to Professor Linus Pauling for many valuable suggestions in connection with this paper.

Table I. *Molecules with three atoms.*

No.	Name	I	II	ν_1	ν_2	R	I.R.	Irreducible representations
2.2a	Symmetrical collinear	$D_{\infty h}$	L	2	1	1(0)	2	A_{1g}^R, A_{2u}^I, $E_{1u}^I(2)$.
2.2b	Symmetrical bent	C_{2v}	A	3		3(1)	3	2^{RI}, 0, 0, 1^{RI}.
2.11a	Unsymmetrical collinear	$C_{\infty v}$	L	2	1	3(1)	3	$2A_1^{RI}$, $E_1^{RI}(2)$.
2.11b	Unsymmetrical bent	C_s	A	3		3(0)	3	3^{RI}, 0.

Molecules with Three Atoms

2.2a. Symmetrical collinear. B : oow, oo$\bar{w}$. CO_2,[10, 11] CS_2.[10, 12] Normal coordinates.[10]

2.2b. Symmetrical bent. B : uow, $\bar{u}$ow. H_2O,[10] SO_2,[10] H_2S.[10] Normal coordinates.[10]

2.11a. Unsymmetrical collinear. B : oow. B′ : oow′. N_2O,[10, 13] COS.[14] Normal coordinates.[10]

2.11b. Unsymmetrical bent. B : uow. B′ : u′ow′. H^1H^2O,[15] H^1H^2S. Normal coordinates.[10, 16]

[9] W. Nowacki, Zeits. f. Kristallographie (A) **86**, 19 (1933).

[10] K. W. F. Kohlrausch, *Der Smekal-Raman-Effekt*, Springer, Berlin, 1931, pp. 169–186.

[11] A. Adel and D. M. Dennison, Phys. Rev. **44**, 99 (1933).

[12] G. Placzek, reference 6, p. 88.

[13] E. K. Plyler and E. F. Barker, Phys. Rev. **38**, 1827 (1931).

[14] C. R. Bailey and A. B. D. Cassie, Proc. Roy. Soc. **A135**, 375 (1932).

[15] R. W. Wood, Nature **132**, 970 (1933).

[16] P. C. Cross and J. H. Van Vleck, J. Chem. Phys. **1**, 350 (1933).

TABLE II. *Molecules with four atoms.*

No.	Name	I	II	ν_1	ν_2	R	I.R.	Irreducible representations
3.3a	Plane equilateral triangle	D_{3h}	S	2	2	3(2)	3(1)	1^R, 0, $2_2{}^{RI}$; 0, 1^I, 0.
3.3b	Regular pyramid	C_{3v}	S	2	2	4(2)	4(2)	2^{RI}, 0, $2_2{}^{RI}$.
3.21a	Plane isosceles triangle	C_{2v}	A	6		6(3)	6	3^{RI}, 1^{RI}, 0, 2^{RI}.
3.21b	Pyramid	C_s	A	6		6(2)	6	4^{RI}, 2^{RI}.
3.111a	Linear	$C_{\infty v}$	L	3	2	5(2)	5	$3A_1{}^{RI}$, $2E_1{}^{RI}(2)$.
3.111b	Plane scalene triangle	C_s	A	6		6(1)	6	5^{RI}, 1^{RI}.
3.111c	Pyramid (R+L)	C_1	A	6		6(0)	6	6^{RI}.

MOLECULES WITH FOUR ATOMS

3.3a. Plane equilateral triangle. B : (ovo)$(\frac{1}{2}\sqrt{3}v, -\frac{1}{2}v, o)(-\frac{1}{2}\sqrt{3}v, -\frac{1}{2}v, o)$. $CO_3{}^{--}$,[17] $NO_3{}^-$.[17] Normal coordinates.[17, 18]

3.3b. Regular pyramid. B : $(o, v, w)(\frac{1}{2}\sqrt{3}v, -\frac{1}{2}v, w)(-\frac{1}{2}\sqrt{3}v, -\frac{1}{2}v, w)$. NH_3,[19] $AsCl_3$,[17, 20] PCl_3,[17, 20] AsF_3,[20] etc.[21] Normal coordinates.[17]

3.21a. Plane isosceles triangle. B : uvo, $\bar{u}$vo. B′ : ov′o. H_2CO.[22] Approximate normal coordinates.[22, 23]

3.21b. Pyramid with isosceles base. B : uvw, $\bar{u}$vw. B′ : ov′w′. PCl_2Br,[24] $PClBr_2$,[24] NH_2Cl.

3.111a. Linear. B : oow. B′ : oow′. B″ : oow″.

3.111b. Plane scalene triangle. B : uvo. B′ : u′v′o. B″ : u″v″o. H^1H^2CO.

3.111c. Pyramid with scalene base. B : uvw. B′ : u′v′w′. B″ : u″v″w″.

TABLE III. *Molecules with five atoms.*

No.	Name	I	II	ν_1	ν_2	ν_3	R	I.R.	Irreducible representations
4.4a	Regular tetrahedron	T_d	Sp	1	1	2	4(3)	2	1^R, 0, $1_2{}^R$, $2_3{}^{RI}$, 0.
4.4b	Plane square	D_{4h}	S	5	2		3(2)	3(1)	1^R, 0, 1^R, 1^R, 0; 0, 1^I, 0, 1, $2_2{}^I$.
4.4c	Square pyramid	C_{4v}	S	5	2		7(5)	4(2)	2^{RI}, 0, 2^R, 1^R, $2_2{}^{RI}$.
4.4d	Tetragonal sphenoid	V_d	S	5	2		7(5)	4(2)	2^{R}[illegible]0, 2^{RI}, 1^R, $2_2{}^{RI}$.
4.4e	Plane rectangle	V_h	A	9			3(1)	5	2^R, 0, 1^R, 0; 1, 2^I, 1^I, 2^I.
4.4f	Orthorhombic sphenoid (R+L)	V	A	9			9(6)	6	3^R, 2^{RI}, 2^{RI}, 2^{RI}.
4.4g	Rectangular pyramid	C_{2v}	A	9			9(6)	7	3^{RI}, 2^{RI}, 2^R, 2^{RI}.
4.31a	Trigonal pyramid	C_{3v}	S	3	3		6(3)	6(3)	3^{RI}, 0, $3_2{}^{RI}$.
4.22a	Linear	$D_{\infty h}$	L	4	3		3(1)	4	$2A_{1g}{}^R$, $2A_{2u}{}^I$, $E_{1g}{}^R(2)$, $2E_{1u}{}^I(2)$.
4.22b	Plane rhombus	V_h	A	9			3(1)	6	2^R, 0, 1^R, 0; 0, 2^I, 2^I, 2^I.
4.22c	Plane trapezoid	C_{2v}	A	9			9(5)	8	4^{RI}, 1^{RI}, 1^R, 3^{RI}.
4.22d	Rhombic pyramid	C_{2v}	A	9			9(5)	8	4^{RI}, 2^{RI}, 1^R, 2^{RI}.
4.22e	Plane parallelogram	C_{2h}	A	9			3(0)	6	3^R, 0, 2^I, 4^I.
4.22f	Trapezoidal pyramid	C_s	A	9			9(4)	9	5^{RI}, 4^{RI}.
4.22g	Monoclinic sphenoid (R+L)	C_2	A	9			9(4)	9	5^{RI}, 4^{RI}.
4.211a	Plane	C_{2v}	A	9			9(5)	9	4^{RI}, 2^{RI}, 0, 3^{RI}.
4.211b	Monoclinic sphenoid	C_s	A	9			9(3)	9	6^{RI}, 3^{RI}.
4.1111a	Linear	$C_{\infty v}$	L	4	3		7(3)	7	$4A_1{}^{RI}$, $3E_1{}^{RI}(2)$.
4.1111b	Plane	C_s	A	9			9(2)	9	7^{RI}, 2^{RI}
4.1111c	General (R+L)	C_1	A	9			9(0)	9	9^{RI}.

MOLECULES WITH FIVE ATOMS

4.4a. Regular tetrahedron. B : www, $\bar{w}$w$\bar{w}$, $\bar{w}\bar{w}$w, w$\bar{w}\bar{w}$. CCl_4,[25] CBr_4,[25] $SiCl_4$,[25] etc. Normal coordinates.[25, 26]

[17] Kohlrausch, reference 10, pp. 195–205.

[18] H. H. Nielsen, Phys. Rev. **32**, 773 (1928).

[19] C. M. Lewis and W. V. Houston, Phys. Rev. **44**, 903 (1933).

[20] D. M. Yost and J. E. Sherborne, J. Chem. Phys. **2**, 125 (1934).

[21] J. H. Hibben, Chem. Rev. **13**, 345 (1933).

[22] Kohlrausch, reference 10, pp. 211–212.

[23] F. Matossi and H. Aderhold, Zeits. f. Physik **68**, 683 (1931).

[24] B. Trumpy, Zeits. f. Physik **68**, 675 (1931).

[25] Kohlrausch, reference 10, pp. 212–218.

[26] D. M. Yost, C. C. Steffens and S. T. Gross, J. Chem. Phys. **2**, 311 (1934).

4.4b. Plane square. B : woo, owo, $\bar{w}$oo, o$\bar{w}$o. $PtCl_4{}^{--}$, $PdCl_4{}^{--}$.

4.4c. Pyramid with square base. B : wov, owv, $\bar{w}$ov, o$\bar{w}$v. SCl_4, $TeCl_4$. Normal coordinates.[27]

4.4d. Tetragonal sphenoid. B : wwv, $\bar{w}$w$\bar{v}$, $\bar{w}\bar{w}$v, w$\bar{w}\bar{v}$.

4.4e. Plane rectangle. B : uvo, $\bar{u}\bar{v}$o, $\bar{u}$vo, u$\bar{v}$o.

4.4f. Orthorhombic sphenoid. B : uvw, u$\bar{v}\bar{w}$, $\bar{u}$v$\bar{w}$, $\bar{u}\bar{v}$w.

4.4g. Pyramid with rectangular base. B : uvw, $\bar{u}\bar{v}$w, $\bar{u}$vw, u$\bar{v}$w.

4.31. Trigonal pyramid. B : $(o, v, w)(\frac{1}{2}\sqrt{3}v, -\frac{1}{2}v, w)(-\frac{1}{2}\sqrt{3}v, -\frac{1}{2}v, w)$. B′ : oow′. CH_3Cl,[28] CH_3Br,[28] etc.

[27] V. Guillemin, Ann. d. Physik **81**, 173 (1926).

[28] Kohlrausch, reference 10, p. 208.

4.22a. *Linear.* B : oow, oow̄. B′ : oow′, oow̄′. C_3O_2.[29]

4.22b. *Plane rhombus.* B : uoo, ūoo. B′ : ov′o, ov̄′o.

4.22c. *Plane regular trapezoid.* B : uvo, ūvo. B′ : u′v′o, ū′v′o.

4.22d. *Rhombic pyramid.* B : uow, ūow. B′ : ov′w′, ov̄′w′. If the two B′ atoms are placed above A with 2B below, this structure is a distorted tetrahedron. CH_2Cl_2,[30] CF_2Cl_2.[31]

4.22e. *Plane parallelogram.* B : uvo, ūv̄o. B′ : u′v′o, ū′v̄′o.

4.22f. *Pyramid with trapezoidal base.* B : uvw, ūvw. B′ : u′v′w′, ū′v′w′.

4.22g. *Monoclinic sphenoid.* B : uvw, ūv̄w. B′ : u′v′w′, ū′v̄′w′. If B and B′ are on same side of A this is a pyramid with a parallelogram for a base.

4.211a. *Plane.* B : uow, ūow. B′ : oow′. B″ : oow″. CH_2CO.

4.211b. *Monoclinic sphenoid.* B : uvw, uvw̄. B′ : u′v′o. B″ : u″v″o. $CHFCl_2$.[31]

4.1111a. *Linear.* B : oow. B′ : oow′. B″ : oow″. B‴oow‴.

4.1111b. *Plane.* B : uvo. B′ : u′v′o. B″ : u″v″o. B‴ : u‴v‴o.

4.1111c. *General.* No symmetry.

TABLE IV. *Molecules with six atoms.*

No.	Name	I	II	ν_1	ν_2	R	I.R.	Irreducible representations
5.5a	Plane pentagon	D_{5h}	S	2	5	3(2)	3(1)	1^R, 0, $2_2{}^I$, $2_2{}^R$; 1^I, 0, 0, 1_2.
5.5b	Pentagonal pyramid	C_{5v}	S	2	5	7(5)	4(2)	2^{RI}, 0, $2_2{}^{RI}$, $3_2{}^R$.
5.41a	Square pyramid	C_{4v}	S	6	3	9(6)	6(3)	3^{RI}, 0, 2^R, 1^R, $3_2{}^{RI}$.
5.41b	Rectangular pyramid	C_{2v}	A	12		12(8)	10	4^{RI}, 3^{RI}, 2^R, 3^{RI}.
5.32a	Trigonal bipyramid	D_{3h}	S	4	4	6(4)	5(2)	2^R, 0, $3_2{}^{RI}$; 0, 2^I, $1_2{}^R$.
5.311a	Polar trigonal bipyramid	C_{3v}	S	4	4	8(4)	8(4)	4^{RI}, 0, $4_2{}^{RI}$.
5.221a	Plane	C_{2v}	A	12		12(7)	11	5^{RI}, 4^{RI}, 1^R, 2^{RI}.
5.221b	Triangular bipyramid	C_{2v}	A	12		12(7)	11	5^{RI}, 3^{RI}, 1^R, 3^{RI}.
5.221c	Trapezoidal pyramid	C_s	A	12		12(5)	12	7^{RI}, 5^{RI}.
5.221d	(R+L)	C_2	A	12		12(6)	12	6^{RI}, 6^{RI}.
5.2111a	Triangular bipyramid	C_s	A	12		12(4)	12	8^{RI}, 4^{RI}.
5.11111a	Linear	$C_{\infty v}$	L	5	4	9(4)	9	$5A_1{}^{RI}$, $4E_1{}^{RI}(2)$.
5.11111b	Plane	C_s	A	12		12(3)	12	9^{RI}, 3^{RI}.
5.11111c	General (R+L)	C_1	A	12		12(0)	12	12^{RI}.

MOLECULES WITH SIX ATOMS

5.5a. *Plane pentagon.* B : (o, v, o)(av, bv, o)(cv, −dv, o)(−cv, −dv, o)(−av, bv, o). Here a = sin 72°, b = cos 72°, c = sin 36°, d = cos 36°.

5.5b. *Pentagonal pyramid.* B : (o, v, w)(av, bv, w)(cv, −dv, w)(−cv, −dv, w)(−av, bv, w). a, b, c, d as in 5.5a.

5.41a. *Square pyramid.* B : uuw, uūw, ūūw, ūuw. B′ : oow′. PF_5.[32] Modes of vibration, Fig. 1.

5.41b. *Rectangular pyramid.* B : uvw, ūv̄w, ūvw, uv̄w. B′ : oow′.

5.32a. *Trigonal bipyramid.* B : (o, v, o)($\frac{1}{2}\sqrt{3}$v, $-\frac{1}{2}$v, o)($-\frac{1}{2}\sqrt{3}$v, $-\frac{1}{2}$v, o). B′ : (oow′)(oow̄′). PF_3Cl_2. Modes of vibration, Fig. 2.

5.311a. *Polar trigonal bipyramid.* B : (o, v, w)($\frac{1}{2}\sqrt{3}$v, $-\frac{1}{2}$v, w)($-\frac{1}{2}\sqrt{3}$v, $-\frac{1}{2}$v, w). B′ : (oow′). B″ : (oow″). PF_3ClI, CH_3CN.

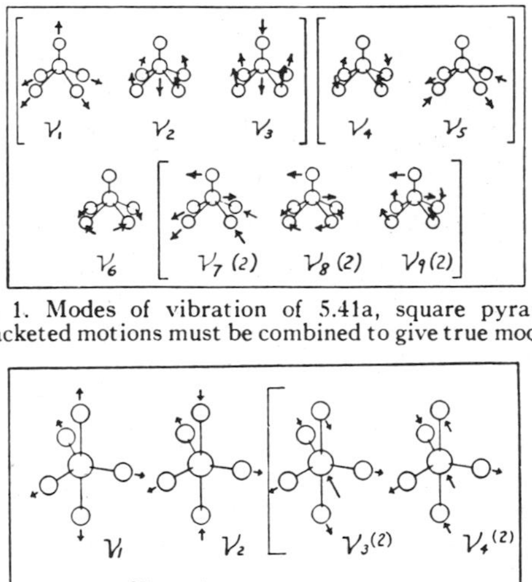

FIG. 1. Modes of vibration of 5.41a, square pyramid. Bracketed motions must be combined to give true modes.

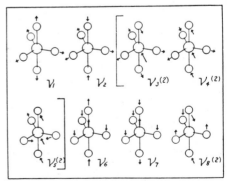

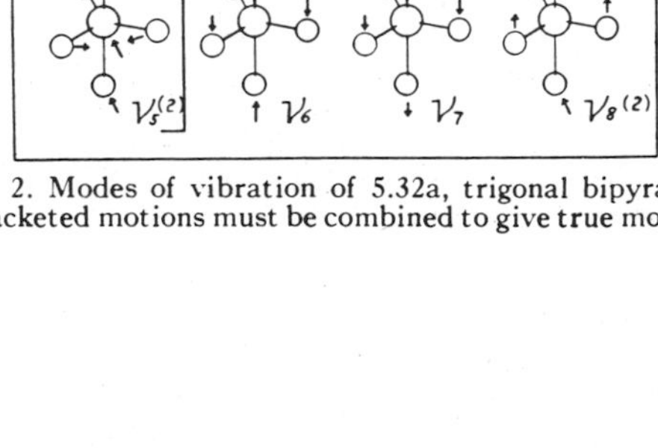

FIG. 2. Modes of vibration of 5.32a, trigonal bipyramid. Bracketed motions must be combined to give true modes.

[29] L. O. Brockway and Linus Pauling, Proc. Nat. Acad. Sci. **19**, 860 (1933).

[30] Kohlrausch, reference 10, p. 305.

[31] C. A. Bradley, Jr., Phys. Rev. **40**, 908 (1932).

[32] L. O. Brockway. Private communication. Electron diffraction studies give this structure for PF_5.

5.221a. Plane. B : uow, ūow. B′ : u′ow′, ū′ow′. B″ : oow″.

5.221b. Triangular bipyramid. B : uow, ūow. B′ : ov′w′, ov̄′w′. B″ : oow″.

5.221c. Trapezoidal pyramid. B : uvw, uvw̄. B′ : u′v′w′, u′v′w̄′. B″ : u″v″o.

5.221d. B : uvw, ūv̄w. B′ : u′v′w′, ū′v̄′w′. B″ : oow″.

5.2111a. Triangular bipyramid. B : uvw, uvw̄. B′ : u′v′o. B″ : u″v″o. B‴ : u‴v‴o.

5.11111a. Linear. $B^{(i)}$: oow_i.

5.11111b. Plane. $B^{(i)}$: u_iv_io.

5.11111c. General. No symmetry.

TABLE V. *Molecules with seven atoms.*

No.	Name		I	II	ν_1	ν_2	ν_3	R	I.R.	Irreducible representations
6.6a	Regular octahedron		O_h	Sp	1	1	4	3(2)	2	$1^R, 0, 1_2^R, 1_3^R, 0; 0, 0, 0, 1_3, 2_3^I$.
6.6b	Regular plane hexagon		D_{6h}	S	5	5		3(2)	3(1)	$1^R, 0, 0, 1, 2_2^R, 0; 0, 1^I, 1, 1, 1_2, 2_2^I$.
6.6c	Hexagonal pyramid		C_{6v}	S	5	5		7(5)	4(2)	$2^{RI}, 0, 2, 1, 3_2^R, 2_2^{RI}$.
6.6d	Trigonal prism		D_{3h}	S	5	5		7(5)	5(2)	$2^R, 0, 3_2^{RI}; 1, 2^I, 2_2^R$.
6.6e	Trigonal plane hexagon		D_{3h}	S	5	5		7(5)	5(1)	$2^R, 1, 4_2^{RI}; 1, 1^I, 1_2^R$.
6.6f	Twisted trigonal prism		D_{3d}	S	5	5		4(2)	5(2)	$2^R, 0, 2_2^R; 2^I, 1, 3_2^I$.
6.6g	Twisted trigonal prism	(R+L)	D_3	S	5	5		8(5)	7(2)	$3^R, 2^I, 5_2^{RI}$.
6.6h	Trigonal hexangular pyramid		C_{3v}	S	5	5		8(5)	8(3)	$3^{RI}, 2, 5_2^{RI}$.
6.51a	Pentagonal pyramid		C_{5v}	S	3	6		9(6)	6(3)	$3^{RI}, 0, 3_2^{RI}, 3_2^R$.
6.42a	Tetragonal octahedron		D_{4h}	S	7	4		5(3)	5(2)	$2^R, 0, 1^R, 1^R, 1_2^R; 0, 2^I, 0, 1, 3_2^I$.
6.42b			V_d	S	7	4		11(8)	7(3)	$3^R, 0, 3^{RI}, 1^R, 4_2^{RI}$.
6.42c	Rectangular bipyramid		V_h	A	15			6(3)	8	$3^R, 1^R, 1^R, 1^R; 1, 3^I, 2^I, 3^I$.
6.42d	Plane		V_h	A	15			6(3)	8	$3^R, 1^R, 2^R, 0, 1, 3^I, 2^I, 3^I$.
6.42e		(R+L)	V	A	15			15(11)	11	$4^R, 4^{RI}, 3^{RI}, 4^{RI}$.
6.42f			C_{2v}	A	15			15(10)	12	$5^{RI}, 3^{RI}, 3^R, 4^{RI}$.
6.42g			C_{2h}	A	15			6(2)	9	$4^R, 2^R, 4^I, 5^I$.
6.411a	Polar tetragonal octahedron		C_{4v}	S	7	4		11(7)	8(4)	$4^{RI}, 0, 2^R, 1^R, 4_2^{RI}$.
6.411b	Polar rectangular bipyramid		C_{2v}	A	15			15(10)	13	$5^{RI}, 4^{RI}, 2^R, 4^{RI}$.
6.33a	Plane		D_{3h}	S	5	5		7(5)	6(2)	$2^R, 1, 4_2^{RI}; 0, 2^I, 1_2^R$.
6.33b	Plane		C_{3h}	S	5	5		8(5)	6(2)	$3^R, 4^{RI}, 4^{RI}; 2^I, 1^R, 1^R$.
6.33c	Pyramid or prism		C_{3v}	S	5	5		9(5)	9(4)	$4^{RI}, 1, 5_2^{RI}$.
6.33d	Pyramid or prism	(R+L)	C_3	S	5	5		10(5)	10(5)	$5^{RI}, 5^{RI}, 5^{RI}$.
6.3111a	Trigonal bipyramid		C_{3v}	S	5	5		10(5)	10(5)	$5^{RI}, 0, 5_2^{RI}$.
6.222a	Linear		$D_{\infty h}$	L	6	5		5(2)	6	$3A_{1g}^R, 3A_{2u}^I, 3E_{1u}^I, 2E_{1g}^R$.
6.222b	Orthorhombic octahedron		V_h	A	15			6(3)	9	$3^R, 1^R, 1^R, 1^R; 0, 3^I, 3^I, 3^I$.
6.222c	Trapezoidal bipyramid		C_{2v}	A	15			15(9)	13	$6^{RI}, 4^{RI}, 2^R, 3^{RI}$.
6.222d	Plane		C_{2h}	A	15			6(1)	9	$5^R, 1^R; 3^I, 6^I$.
6.222e	Monoclinic octahedron		C_{2h}	A	15			6(2)	9	$4^R, 2^R, 3^I, 6^I$.
6.222f	Triangular prism		C_s	A	15			15(7)	15	$8^{RI}, 7^{RI}$.
6.222g	Monoclinic octahedron	(R+L)	C_2	A	15			15(7)	15	$8^{RI}, 7^{RI}$.
6.222h	Triclinic octahedron		C_i	A	15			6(0)	9	$6^R, 9^I$.
6.2211a	Orthorhombic octahedron		C_{2v}	A	15			15(9)	14	$6^{RI}, 4^{RI}, 1^R, 4^{RI}$.
6.2211b	Heptahedron		C_s	A	15			15(6)	15	$9^{RI}, 6^{RI}$.
6.2211c	Monoclinic octahedron	(R+L)	C_2	A	15			15(8)	15	$7^{RI}, 8^{RI}$.
6.21111a			C_s	A	15			15(5)	15	$10^{RI}, 5^{RI}$.
6.111111a	Linear		$C_{\infty v}$	L	6	5		11(5)	11	$6A_1^{RI}, 5E_1^{RI}$.
6.111111b	Plane		C_s	A	15			15(4)	15	$11^{RI}, 4^{RI}$.
6.111111c	General		C_1	A	15			15(0)	15	15^{RI}.

MOLECULES WITH SEVEN ATOMS

6.6a. Regular octahedron. B : uoo, ūoo, ouo, oūo, oou, ooū. SF_6[26], SeF_6[26], TeF_6[26]. Modes of vibration, Fig. 3. Normal frequencies.[26]

6.6b. Regular plane hexagon. B : (o, v, o)($\frac{1}{2}\sqrt{3}$v, $\frac{1}{2}$v, o)($\frac{1}{2}\sqrt{3}$v, $-\frac{1}{2}$v, o)(o, v̄, o)($-\frac{1}{2}\sqrt{3}$v, $-\frac{1}{2}$v, o)($-\frac{1}{2}\sqrt{3}$v, $\frac{1}{2}$v, o).

6.6c. Hexagonal pyramid. B : (ovw)($\frac{1}{2}\sqrt{3}$v, $\frac{1}{2}$v, w)($\frac{1}{2}\sqrt{3}$v, $-\frac{1}{2}$v, w)(o, v̄, w)($-\frac{1}{2}\sqrt{3}$v, $-\frac{1}{2}$v, w)($-\frac{1}{2}\sqrt{3}$v, $\frac{1}{2}$v, w).

6.6d. Trigonal prism. B : (o, v, w)($\frac{1}{2}\sqrt{3}$v, $-\frac{1}{2}$v, w)($-\frac{1}{2}\sqrt{3}$v, $-\frac{1}{2}$v, w)(o, v, w̄)($\frac{1}{2}\sqrt{3}$v, $-\frac{1}{2}$v, w̄)($-\frac{1}{2}\sqrt{3}$v, $-\frac{1}{2}$v,w̄).

6.6e. Trigonal plane hexagon. B : (ū, v, o)(u, v, o)($\frac{1}{2}\sqrt{3}$v$+\frac{1}{2}$u, $-\frac{1}{2}$v$+\frac{1}{2}\sqrt{3}$u, o)($\frac{1}{2}\sqrt{3}$v$-\frac{1}{2}$u, $-\frac{1}{2}$v $-\frac{1}{2}\sqrt{3}$u, o)($-\frac{1}{2}\sqrt{3}$v$+\frac{1}{2}$u, $-\frac{1}{2}$v$-\frac{1}{2}\sqrt{3}$u, o)($-\frac{1}{2}\sqrt{3}$v$-\frac{1}{2}$u, $-\frac{1}{2}$v$+\frac{1}{2}\sqrt{3}$u, o).

6.6f. Twisted trigonal prism. B : (o, v, w)($\frac{1}{2}\sqrt{3}$v, $-\frac{1}{2}$v, w)($-\frac{1}{2}\sqrt{3}$v, $-\frac{1}{2}$v, w)(o, v̄, w̄)($-\frac{1}{2}\sqrt{3}$v, $\frac{1}{2}$v, w̄)($\frac{1}{2}\sqrt{3}$v, $\frac{1}{2}$v, w̄).

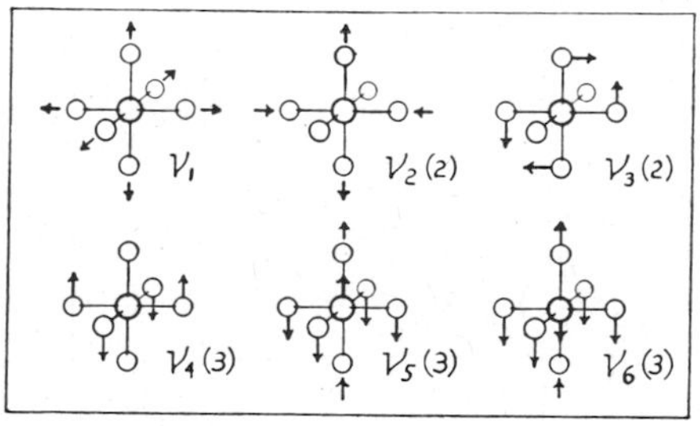

FIG. 3. Modes of vibration of 6.6a, regular octahedron.

6.6g. Twisted trigonal prism. $B:(\bar{u}, v, w)(u, v, \bar{w})(\frac{1}{2}\sqrt{3}v+\frac{1}{2}u, -\frac{1}{2}v+\frac{1}{2}\sqrt{3}u, w)(\frac{1}{2}\sqrt{3}v-\frac{1}{2}u, -\frac{1}{2}v-\frac{1}{2}\sqrt{3}u, \bar{w})(-\frac{1}{2}\sqrt{3}v+\frac{1}{2}u, -\frac{1}{2}v-\frac{1}{2}\sqrt{3}u, w)(-\frac{1}{2}\sqrt{3}v-\frac{1}{2}u, -\frac{1}{2}v+\frac{1}{2}\sqrt{3}u, \bar{w})$.

6.6h. Trigonal hexangular pyramid. $B:(\bar{u}, v, w)(u, v, w)(\frac{1}{2}\sqrt{3}v+\frac{1}{2}u, -\frac{1}{2}v+\frac{1}{2}\sqrt{3}u, w)(\frac{1}{2}\sqrt{3}v-\frac{1}{2}u, -\frac{1}{2}v-\frac{1}{2}\sqrt{3}u, w)(-\frac{1}{2}\sqrt{3}v+\frac{1}{2}u, -\frac{1}{2}v-\frac{1}{2}\sqrt{3}u, w)(-\frac{1}{2}\sqrt{3}v-\frac{1}{2}u, -\frac{1}{2}v+\frac{1}{2}\sqrt{3}u, w)$.

6.51a. Pentagonal pyramid. $B:(o, v, w)(av, bv, w)(cv, -dv, w)(-cv, -dv, w)(-av, bv, w)$. $B':(oow')$. a, b, c, d as in 5.5a.

6.42a. Tetragonal octahedron. $B:uoo, \bar{u}oo, ouo, o\bar{u}o$. $B':oou', oo\bar{u}'$.

6.42b. $B:uow, \bar{u}ow, ou\bar{w}, o\bar{u}\bar{w}$. $B':oow', oo\bar{w}'$.

6.42c. Rectangular bipyramid. $B:uvo, \bar{u}\bar{v}o, \bar{u}vo, u\bar{v}o$. $B':oow', oo\bar{w}'$.

6.42d. Plane. $B:uvo, \bar{u}\bar{v}o, \bar{u}vo, u\bar{v}o$. $B':u'oo, \bar{u}'oo$.

6.42e. $B:uvw, u\bar{v}\bar{w}, \bar{u}v\bar{w}, \bar{u}\bar{v}w$. $B':oow', oo\bar{w}'$.

6.42f. $B:uvw, u\bar{v}w, \bar{u}\bar{v}w, \bar{u}vw$. $B':u'ow', \bar{u}'ow'$. $(CH_2)_2O$.

6.42g. $B:uvw, u\bar{v}w, \bar{u}\bar{v}\bar{w}, \bar{u}v\bar{w}$. $B':u'ow', \bar{u}'o\bar{w}'$.

6.411a. Polar tetragonal octahedron. $B:uow, \bar{u}ow, ouw, o\bar{u}w$. $B':oow'$. $B'':oow''$.

6.411b. Polar rectangular bipyramid. $B:uvw, \bar{u}\bar{v}w, \bar{u}vw, u\bar{v}w$. $B':oow'$. $B'':oow''$.

6.33a. Plane. $B:(o, v, o)(\frac{1}{2}\sqrt{3}v, -\frac{1}{2}v, o)(-\frac{1}{2}\sqrt{3}v, -\frac{1}{2}v, o)$. $B':(o, \bar{v}, o)(-\frac{1}{2}\sqrt{3}v, \frac{1}{2}v, o)(\frac{1}{2}\sqrt{3}v, \frac{1}{2}v, o)$.

6.33b. Plane. $B:(o, v, o)(\frac{1}{2}\sqrt{3}v, -\frac{1}{2}v, o)(-\frac{1}{2}\sqrt{3}v, -\frac{1}{2}v, o)$. $B':(u', v', o)(\frac{1}{2}\sqrt{3}v'-\frac{1}{2}u', -\frac{1}{2}v'-\frac{1}{2}\sqrt{3}u', o)(-\frac{1}{2}\sqrt{3}v'-\frac{1}{2}u', -\frac{1}{2}v'+\frac{1}{2}\sqrt{3}u', o)$.

6.33c. Pyramid or prism. $B:(o, v, w)(\frac{1}{2}\sqrt{3}v, -\frac{1}{2}v, w)(-\frac{1}{2}\sqrt{3}v, -\frac{1}{2}v, w)$. $B':(o, v', w')(\frac{1}{2}\sqrt{3}v', -\frac{1}{2}v', w')(-\frac{1}{2}\sqrt{3}v', -\frac{1}{2}v', w')$.

6.33d. Pyramid or prism. $B:(o, v, w)(\frac{1}{2}\sqrt{3}v, -\frac{1}{2}v, w)(-\frac{1}{2}\sqrt{3}v, -\frac{1}{2}v, w)$. $B':(u', v', w')(\frac{1}{2}\sqrt{3}v'-\frac{1}{2}u', -\frac{1}{2}v'-\frac{1}{2}\sqrt{3}u', w')(-\frac{1}{2}\sqrt{3}v'-\frac{1}{2}u', -\frac{1}{2}v'+\frac{1}{2}\sqrt{3}u', w')$.

6.3111a. Trigonal bipyramid. $B:(o, v, w)(\frac{1}{2}\sqrt{3}v, -\frac{1}{2}v, w)(-\frac{1}{2}\sqrt{3}v, -\frac{1}{2}v, w)$. $B':oow'$. $B'':oow''$. $B''':oow'''$. CH_3NCO.

6.222a. Linear. $B:oow, oo\bar{w}$. $B':oow', oo\bar{w}'$. $B'':oow'', oo\bar{w}''$.

6.222b. Orthorhombic octahedron. $B:uoo, \bar{u}oo$. $B':ov'o, o\bar{v}'o$. $B'':oow'', oo\bar{w}''$.

6.222c. Trapezoidal bipyramid. $B:uow, \bar{u}ow$. $B':u'ow', \bar{u}'ow'$. $B'':ov''w'', o\bar{v}''w''$.

6.222d. Plane. $B:uvo, \bar{u}\bar{v}o$. $B':u'v'o, \bar{u}'\bar{v}'o$. $B'':u''v''o, \bar{u}''\bar{v}''o$.

6.222e. Monoclinic octahedron. $B:uvo, \bar{u}\bar{v}o$. $B':u'v'o, \bar{u}'\bar{v}'o$. $B'':oow'', oo\bar{w}''$.

6.222f. Triangular prism. $B:uvw, uv\bar{w}$. $B':u'v'w', u'v'\bar{w}'$. $B'':u''v''w'', u''v''\bar{w}''$.

6.222g. Monoclinic octahedron. $B:uvw, \bar{u}\bar{v}w$. $B':u'v'w', \bar{u}'\bar{v}'w'$. $B'':u''v''w'', \bar{u}''\bar{v}''w''$.

6.222h. Triclinic octahedron. $B:uvw, \bar{u}\bar{v}\bar{w}$. $B':u'v'w', \bar{u}'\bar{v}'\bar{w}'$. $B'':u''v''w'', \bar{u}''\bar{v}''\bar{w}''$.

6.2211a. Orthorhombic octahedron. $B:uow, \bar{u}ow$. $B':ov'w', o\bar{v}'w'$. $B'':oow''$. $B''':oow'''$.

6.2211b. Heptahedron. $B:uvw, uv\bar{w}$. $B':u'v'w', u'v'\bar{w}'$. $B'':u''v''o$. $B''':u'''v'''o$.

6.2211c. Monoclinic octahedron. $B:uvw, \bar{u}\bar{v}w$. $B':u'v'w', \bar{u}'\bar{v}'w'$. $B'':oow''$. $B'''oow'''$.

6.21111a. $B:uvw, uv\bar{w}$. $B':u'v'o$. $B'':u''v''o$. $B''':u'''v'''o$. $B'''':u''''v''''o$.

6.111111a. Linear. $B^{(i)}:oow_i$.

6.111111b. Plane. $B^{(i)}:u_iv_io$.

6.111111c. General. No symmetry. $B^{(i)}:u_iv_iw_i$.

TABLE VI. *Molecules with eight atoms.*

No.	Name	I	II	ν_1	ν_2	R	I.R.	Irreducible representations
7.7a	Plane heptagon	D_{7h}	S	2	8	3(2)	3(1)	$1^R, 0, 2_2^I, 2_2^R, 2_2; 0, 1^I, 0, 1_2, 1_2$.
7.7b	Heptagonal pyramid	C_{7v}	S	2	8	7(5)	4(2)	$2^{RI}, 0, 2_2^{RI}, 3_2^R, 3_2$.
7.52a	Pentagonal bipyramid	D_{5h}	S	4	7	5(3)	5(2)	$2^R, 0, 3_2^I, 2_2^R; 2^I, 0, 1_2^R, 1_2$.
7.331a		C_{3v}	S	6	6	11(6)	11(5)	$5^{RI}, 1, 6_2^{RI}$.
7.331b	Truncated trigonal bipyramid	C_{3v}	S	6	6	11(6)	11(5)	$5^{RI}, 1, 6_2^{RI}$.
7.421a		C_{2v}	A	18		18(12)	15	$6^{RI}, 5^{RI}, 3^R, 4^{RI}$.

Molecules with Eight Atoms

Only the type AB_7 is treated completely; the other structures listed were chosen because they seem the most probable physically. IF_7 presumably belongs to one of the classes below.

7.7a. Plane heptagon. $B:(o, v, o)(gv, hv, o)(jv, kv, o)(ev, fv, o)(-ev, fv, o)(-jv, kv, o)(-gv, hv, o)$. $g=\sin\varphi, h=\cos\varphi, j=\sin 2\varphi, k=\cos 2\varphi, e=\sin 3\varphi, f=\cos 3\varphi$. $\varphi=2\pi/7$.

7.7b. Heptagonal pyramid. $B:(o, v, w)(gv, hv,$

w)(jv, kv, w)(ev, fv, w)(−ev, fv, w)(−jv, kv, w)(−gv, hv, w). g, h, j, k, e, f as in 7.7a.

7.52a. Pentagonal bipyramid. B : (o, v, o)(av, bv, o)(cv, −dv, o)(−cv, −dv, o)(−av, bv, o). B′ : (oow′)(oow̄′).

7.331a, b. B : (o, v, w)($\frac{1}{2}\sqrt{3}$v, $-\frac{1}{2}$v, w)($-\frac{1}{2}\sqrt{3}$v, $-\frac{1}{2}$v, w). B′ : (o, v′, w′)($\frac{1}{2}\sqrt{3}$v′, $-\frac{1}{2}$v′, w′)($-\frac{1}{2}\sqrt{3}$v′, $-\frac{1}{2}$v′, w′). B″ : oow″.

7.421a. B : uvw, ūv̄w, ūvw, uv̄w. B′ : u′ow′, ū′ow′. B″ : oow″.

TABLE VII. *Molecules with nine atoms.*

No.	Name	I	II	ν_1	ν_2	ν_3	R	I.R.	Irreducible representations
8.8a	Cube	O_h	Sp	2	2	5	4(3)	2	1^R, 0, $1_2{}^R$, $2_3{}^R$, 0; 0, 1, 1_2, 1_3, $2_3{}^I$.
8.8b	Plane octagon	D_{8h}	S	5	8		4(3)	3(1)	1^R, 0, 1, 1, $1_2{}^R$, 0, $2_2{}^R$; 0, 1^I, 0, 1, $2_2{}^I$, 2_2, 1_2.
8.8c	Octagonal pyramid	C_{8v}	S	5	8		7(5)	4(2)	2^{RI}, 0, 1, 2, $2_2{}^{RI}$, 3_2, $3_2{}^R$.
8.8d	Archimedian anti-prism	D_{4d}	S	5	8		7(5)	5(2)	2^R, 0, 1, 2^I, $3_2{}^I$, $2_2{}^R$, $3_2{}^R$.
8.8e	Plane tetragonal octagon	D_{4h}	S	11	5		7(5)	5(1)	2^R, 1, 2^R, 2^R, $1_2{}^R$; 1, 1^I, 1, 1, $4_2{}^I$.
8.8f	Tetragonal parallelopiped	D_{4h}	S	11	5		7(5)	5(2)	2^R, 0, 1^R, 2^R, $2_2{}^R$; 1, 2^I, 2, 1, $3_2{}^I$.
8.8g	Twisted cube	D_4	S	11	5		14(11)	7(2)	3^R, 2^I, 3^R, 3^R, $5_2{}^{RI}$.
8.8h	Tetragonal pyramid	C_{4v}	S	11	5		14(11)	8(3)	3^{RI}, 2, 3^R, 3^R, $5_2{}^{RI}$.
8.8i	Twisted tetragonal parallelopiped	V_d	S	11	5		14(11)	8(3)	3^R, 2, 3^{RI}, 3^R, $5_2{}^{RI}$.
8.8j	Rectangular parallelopiped	V_h	A	21			9(6)	9	3^R, 2^R, 2^R, 2^R; 3, 3^I, 3^I, 3^I.

MOLECULES WITH NINE ATOMS

Only the type AB_8 is considered. OsF_8[33] is said to belong to 8.8a or 8.8d.

8.8a. Cube. B : uuu, ūūū, ūuū, uūu, ūūu, uuū, uūū, ūuu.

8.8b. Plane octagon. B : (o, v, o)(mv, mv, o)(v, o, o)(mv, −mv, o)(o, v̄, o)(−mv, −mv, o)(v̄, o, o)(−mv, mv, o). m = $\frac{1}{2}\sqrt{2}$.

8.8c. Octahedral pyramid. B : (o, v, w)(mv, mv, w)(v, o, w)(mv, −mv, w)(o, v̄, w)(−mv, −mv, w)(v̄, o, w)(−mv, mv, w). m = $\frac{1}{2}\sqrt{2}$.

8.8d. Archimedian anti-prism. B : (uuw)(ūūw)(ūuw)(uūw)(pu, o, w̄)(o, pū, w̄)(pū, o, w̄)(o, pu, w̄). p = $\sqrt{2}$.

[33] H. Braune and S. Knoke, Naturwiss. **21**, 349 (1933).

8.8e. Plane tetragonal octagon. B : uvo, ūv̄o, ūvo, uv̄o, vūo, v̄uo, vuo, v̄ūo.

8.8f. Tetragonal parallelopiped. B : uow, ūow̄, ūow, uow̄, oūw, ouw, ouw̄, oūw̄.

8.8g. Twisted cube. B : uvw, ūvw̄, ūv̄w, uv̄w̄, vūw, v̄uw, vuw̄, v̄ūw̄.

8.8h. Tetragonal octangular pyramid. B : uvw, ūv̄w, ūvw, uv̄w, vūw, v̄uw, vuw, v̄ūw.

8.8i. Twisted tetragonal parallelopiped. B : uvw, ūvw̄, ūv̄w, uv̄w̄, vūw̄, v̄uw̄, vuw, v̄ūw.

8.8j. Rectangular parallelopiped. B : uvw, ūv̄w̄, ūvw̄, uv̄w, ūv̄w, uvw̄, uv̄w̄, ūvw.

Reprinted from *J. Chem. Phys.*, **3**, 818–821 (1935)

Symmetry Considerations Concerning the Splitting of Vibration-Rotation Levels in Polyatomic Molecules

34

E. BRIGHT WILSON, JR.,* *Mallinckrodt Chemical Laboratory, Harvard University*
(Received September 30, 1935)

The interaction of rotation and vibration and perhaps other effects may split degenerate vibration-rotation energy levels of symmetrical polyatomic molecules into a number of components. The permutation symmetry of molecules containing several identical atoms provides certain restrictions on this splitting. This paper discusses the maximum number of fine-structure components, their quantum weights when nuclear spins are taken into account, and the selection rule for transitions. All arguments are based solely on symmetry considerations so that no estimate of the magnitude of the splitting is given.

DEFINITION OF COORDINATES

In discussing the motions of the atoms in a polyatomic molecule (as distinct from the motion of the electrons) it is useful to use the Eulerian angles θ, φ, χ to describe the orientation of the molecule in space and the normal coordinates $Q_1, Q_2, Q_3, \cdots, Q_N$ to describe the mutual positions of the atoms in the molecule. It is important to define these coordinates more definitely. Let the angles θ, φ, χ define a set of rotating Cartesian axes X, Y, Z, with origin at the center of gravity of the molecules. The position of any atom i can be specified by the Eulerian angles and the coordinates x, y, z relative to the rotating system. However, there are three too many coordinates so that three relations between the coordinates must be written down in order to define the rotating system uniquely. These relations are conveniently chosen to be[1a]

$$\begin{aligned}\Sigma m_i(a_iy_i-b_ix_i)&=0,\\ \Sigma m_i(b_iz_i-c_iy_i)&=0,\\ \Sigma m_i(c_ix_i-a_iz_i)&=0,\end{aligned}\qquad(1)$$

where a_i, b_i, c_i are the coordinates (in terms of the rotating system) of the equilibrium position of the ith atom and m_i is the mass of the atom. A transformation from the coordinates x_i, y_i, z_i to normal coordinates $Q_1, Q_2, \cdots, Q_N$ can then be carried out in the usual manner and it will be found that the relations (1) are equivalent to those usually made in normal coordinate treatments; namely, that to the first approximation there is no rotational angular momentum with respect to the rotating system of coordinates.

AN APPROXIMATE WAVE EQUATION

It is natural to set up the wave equation in terms of these coordinates. This involves first expressing the kinetic energy in terms of θ, φ, χ, $Q_1, Q_2, \cdots, Q_N$ and then using the result to construct the wave equation in the usual manner.[1] This is a possible, but complicated, procedure. However, if we assume that we can neglect in the classical Lagrangian expression for the kinetic energy those terms involving the normal coordinates to the first or higher powers (which means classically that we are assuming small vibrations), then the wave equation resulting from this approximate Lagrangian will be simply the sum of two parts; one the equation for the rigid rotator (coordinates θ, φ, χ), the other the equation for harmonic vibrations (coordinates $Q_1, Q_2, \cdots, Q_N$).*

The solutions of this equation are products of the type

$$\psi=R(\theta,\varphi,\chi)V(Q_1,\cdots,Q_N),\qquad(2)$$

where $R(\theta, \varphi, \chi)$ is the rotational wave function and $V(Q_1, \cdots, Q_N)$ is the vibrational wave function.

For molecules whose equilibrium configurations possess a certain minimum symmetry; namely, at least a threefold axis of symmetry,

* Junior Fellow of the Society of Fellows.
[1a] C. Eckart, Phys. Rev. **47**, 552 (1935).

[1] B. Podolsky, Phys. Rev. **32**, 812 (1928).
* We assume here that the potential function is harmonic but this assumption does not change the final conclusions. This method of justifying the usual separation into rotational and vibrational parts does, however, depend on the choice of the rotating coordinate system defined by (1).

both the rotational and vibrational energy levels may be degenerate when the approximation considered here is used. Thus for CH_3F the functions R are the symmetrical top solutions with the energy levels

$$W_{J,\,K}=(h^2/8\pi^2)\{J(J+1)/A+K^2(1/C-1/A)\}, \quad (3)$$

so that levels with $|K|>0$ are doubly degenerate. For this molecule certain of the fundamental vibrational levels are also doubly degenerate so that, to this approximation, CH_3F possesses some fourfold degenerate energy levels because of its threefold symmetry axis.

It will be shown that this high degeneracy is not really required by the symmetry so that these energy levels may be split by terms which have been omitted in obtaining this simple approximation. Some degeneracy, however, will remain on account of the symmetry.

Symmetry Considerations[2]

The true energy operator, H, is unchanged by any permutation of identical particles; it is therefore invariant under the group of permutations which are equivalent to rotations of the molecule. The wave functions RV are not in general invariant under these permutations since the coordinates θ, φ, χ, Q_1, $\cdots$, Q_N, are not. Thus the permutation (123) of the three H atoms in CH_3F changes χ into $\chi+2\pi/3$ and therefore $R(\chi)$ into $e^{2\pi iK/3}R(\chi)$.

The functions RV belonging to a given approximate energy level form a *representation*[3] of the group of permutations equivalent to rotations. This representation is in general *reducible* so that by forming the correct linear combinations of these functions RV the representation can be *reduced* and each combination will correspond to one of the *irreducible representations* of the group. The technique of finding the number of wave functions with each type of symmetry has been treated in detail in a previous paper[2] and will be assumed here.

If ψ_A, ψ_B, etc., are the linear combinations of the functions RV which reduce the representation then the perturbed energy levels are determined by integrals of the type $\int\psi_A H\psi_B d\tau$. It is a consequence of group theory that if ψ_A and ψ_B belong to different irreducible representations the integral is identically zero. Therefore the energy matrix, if set up in terms of ψ_A, ψ_B, etc., will factor into diagonal blocks, the elements of each block involving only functions of one symmetry type or irreducible representation. All other elements of the matrix will vanish. Furthermore, blocks corresponding to doubly degenerate representations of the group will occur twice, so that the energy levels which are the roots of such a block of the secular equation will be doubly degenerate. Similar remarks apply to triply degenerate representations.

CH_3F as an Example

These statements may be better understood if illustrated by an example, CH_3F. From reference 2, Table X, it is found that if the rotational quantum number K has a value not divisible by three the rotational functions R for that level have the symmetry E with respect to the group of permutations equivalent to rotations. If the vibrational energy level under consideration is also doubly degenerate with the symmetry E, there will be four wave functions of the type RV, having the symmetry $E\times E=2A+E$; that is, there can be formed four independent linear combinations of the four functions RV, such that two of them are symmetrical with respect to permutations equivalent to rotations whereas the other two form a degenerate pair. Applying the theorems of the previous paragraph we obtain the energy matrix (or secular equation) shown in Fig. 1, in which only the elements in the heavy squares are different from zero and the elements of the two squares marked E are equal. As far as the symmetry is concerned, therefore, the originally fourfold level may split, under the influence of coupling terms, etc., into three levels, two of which are the roots of the quadratic equation

---

[2] This paper is an extension of a previous publication (J. Chem. Phys. **3**, 276 (1935)), which considered the same symmetry questions but which obtained the total statistical weights of the unsplit energy levels. Here exactly the same methods are used, only the split levels will be considered and their separate statistical weights determined.

[3] For the meaning and use of the group theory terms employed here see E. Wigner, *Gruppentheorie* (Vieweg and Sohn); B. L. van der Waerden, *Die gruppentheoretische Methode in der Quantenmechanik* (Springer, Berlin).

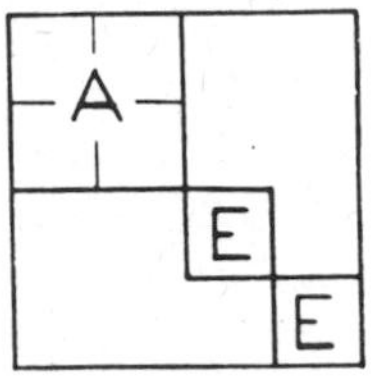

FIG. 1.

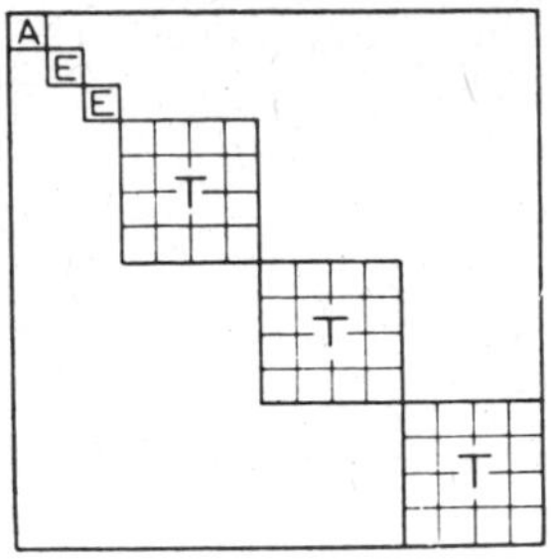

FIG. 2.

FIG. 1. Energy matrix for a sample approximate level of CH_3F. Heavy lines enclose non-zero elements.
FIG. 2. Energy matrix for a sample approximate level of CH_4.

obtained from the upper block of Fig. 1, while the remaining level is doubly degenerate and comes from either of the two equal lower blocks of the secular equation.

METHANE AS AN EXAMPLE

Methane may be similarly treated. As a sample let us consider the level with J equal to 2 and with one quantum of one of the triply degenerate vibrations excited. From Table III, reference 2 we find that the symmetry of this level is $(E+T)\times T=A+E+4T$. The secular equation for this level therefore has the form shown in Fig. 2, in which the lower three blocks are identical and the two blocks above these are identical. This energy level may therefore split into one single level (A), one double level (E), and four triple levels (T).

In connection with methane it may be pointed out that it is possible, but not proved, that even the nonvibrating rotational levels of methane may be split. For example, the level with J equal to 5 has the symmetry $E+3T$ so that in the nonvibrating state it is possible for a perturbation with tetrahedral symmetry (such as conceivably the stretching terms) to split this eleven-fold level into one double level and three triple levels. It requires further study of a more quantitative nature in order to decide whether or not such a splitting actually occurs to an appreciable degree or not. This treatment merely shows that the symmetry of the molecule does not require such a high degree of degeneracy.

SELECTION RULES

The electric moment of a molecule is a function of the coordinates of the nuclei and is unaffected by interchange of coordinates of identical nuclei. It is thus symmetrical. The integral $\int \psi_A \mu \psi_B d\tau$, determining the transition probability, is therefore zero if ψ_A and ψ_B are in different irreducible representations. The selection rule is that transitions occur only between levels of the same symmetry. This rule does not replace the ordinary selection rules but is an additional restriction.

STATISTICAL WEIGHTS OF COMPONENT LEVELS[2]

The calculation of the statistical weights of the components of a level (important for the computation of relative intensities) is carried out in exactly the same manner as in reference 2. Thus for methane the nuclear-spin functions have the symmetry $5A+E+3T$. The total wave function must have the symmetry A. Therefore the components of the sample vibration-rotation level previously discussed will have the weights shown in Fig. 3 (multiplied by $2J+1$).

MOLECULES WITH SEVERAL FRAMEWORKS

Both of the examples above are molecules possessing two "frameworks" in the sense used in reference 2. As discussed in that reference, however, this fact does not need to be taken into account in calculating the statistical weights of the levels since the total wave function must have one definite symmetry with respect to interchanges of identical particles and there will

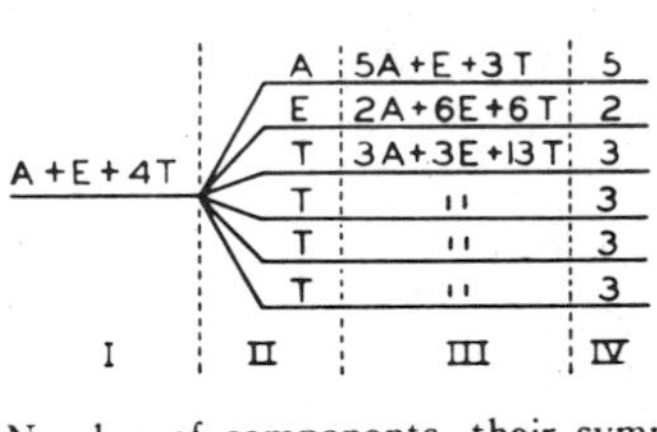

FIG. 3. Number of components, their symmetries and statistical weights (factor $2J+1$ omitted) for a sample level of methane. Regions I and II give the symmetries before and after splitting, spins not included. Region III gives the symmetries when the spin function $\psi_S(5A+E+3T)$ is included. Region IV gives the statistical weights of the components when the exclusion principle is applied (A only). The vertical order and relative spacing of the components has no significance.

be only one linear combination of the functions based on the various frameworks which will have the correct symmetry.

Conclusion

It is to be emphasized that only symmetry considerations have been used in this paper so that the results are necessarily incomplete since they do not yield any information regarding the magnitudes of the splittings to be expected. Actual energy levels may not be split to an observable extent or they may show the splitting indicated here only incompletely. It is possible also that certain of the perturbations, especially the rotation-vibration coupling, may split the approximate energy levels to such an extent that a component of one approximate level may nearly coincide with a component of another approximate level. In spite of these complications, however, the symmetry does restrict the maximum number of components which are possible and gives the selection rules and statistical weights involved. Experimental results[4] exist which may be capable of interpretation in terms of these effects but further work is necessary.

[4] W. B. Steward and H. H. Nielsen, Phys. Rev. **47**, 828 (1935).

Reprinted from *Amer. J. Phys.*, **11**, 239–247 (1943)

35

Interpretation of the Spectra of Polyatomic Molecules by Use of Group Theory

ARNOLD G. MEISTER, FORREST F. CLEVELAND, *Department of Physics*

AND

M. J. MURRAY, *Department of Chemistry, Illinois Institute of Technology, Chicago, Illinois*

GROUP theory methods have been used to obtain relations that are helpful in analyzing the Raman and infra-red spectra of symmetrical polyatomic molecules. These relations have been derived by Brester,[1] Wigner[2] and Tisza,[3] and are summarized by Rosenthal and Murphy,[4] but their papers are somewhat advanced for one who is not familiar with group theory. However, it is possible to use the results without much study of group theory itself. The purpose of this paper is to give an elementary explanation of the use of the formulas in obtaining the number of fundamental vibrations belonging to each vibration type (or state) and the activity of the fundamentals, combinations, and overtones in the Raman and infra-red spectra.

SELECTION RULES

Since this paper is intended for those having no previous knowledge of group theory, only those considerations necessary for obtaining the selection rules for the fundamentals, combinations and overtones of methylacetylene ($CH_3-C\equiv C-H$) will at first be introduced; further details will be given in a later section.

One assumes for the equilibrium configuration of the molecule that the nuclei are located at fixed positions in space. Because of the geometric arrangement of the nuclei, the molecule possesses symmetry and there is a definite number of geometric operations, called *covering operations*, which can be performed on the molecule so that equivalent nuclei occupy the same points in space as in the originally assumed equilibrium configuration. There are only two kinds of covering operations—proper rotations and improper rotations (or rotary reflections). A proper rotation is simply a rotation through an angle $\pm\varphi$ about some axis of symmetry. An improper rotation is a rotation followed by a reflection in a plane perpendicular to the axis of rotation; thus a reflection is an improper rotation through an angle of 0°.

For methylacetylene the principal symmetry axis Z (always considered to be vertical) passes through the three carbon atoms and the acetylenic hydrogen atom (Fig. 1). The covering operations for this geometric arrangement of nuclei are: (1) the identity operation E, which is

[1] C. J. Brester, Dissertation, Utrecht (1923).
[2] E. Wigner, Göttingen Nachrichten, 133 (1930); *Gruppentheorie* (Braunschweig, 1931).
[3] L. Tisza, Zeits. f. Physik **82**, 48 (1933).
[4] J. E. Rosenthal and G. M. Murphy, Rev. Mod. Phys. **8**, 317 (1936).

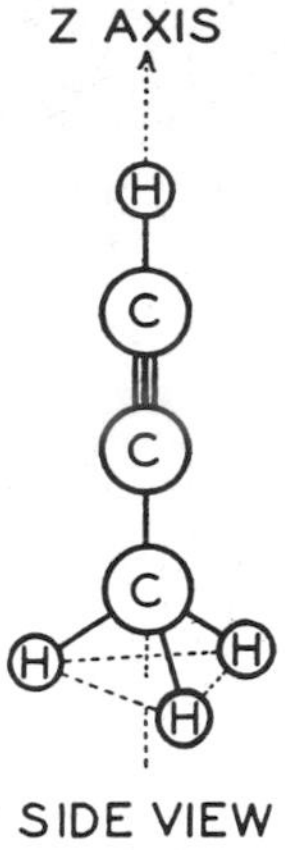

FIG. 1

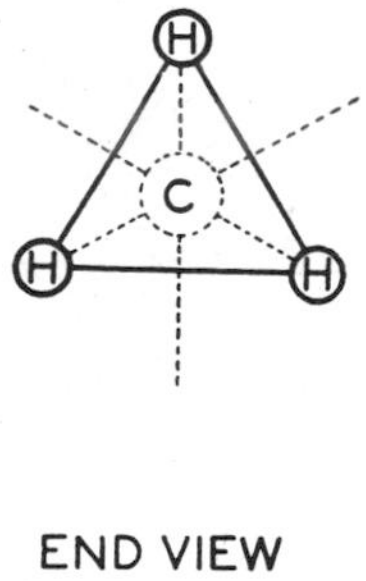

FIG. 2

FIGS. 1 and 2. Side and end views of the methylacetylene molecule.

a rotation through 0° about the Z axis, (2) the operation C_3, which means a rotation through an angle $\pm 2\pi/3$ rad about the Z axis and (3) the operation σ_v, which means a reflection in a plane passing through the Z axis and one of the H atoms. (The axis for this improper rotation is not the Z axis but an axis perpendicular to this plane.) Thus there are three *classes* of operations: The first class E contains one element; the second class C_3 contains two elements since there are two possible rotations, $+2\pi/3$ or $-2\pi/3$ rad; and the third class σ_v contains three elements since there are three equivalent vertical reflection planes (Fig. 2).

The afore-described set of operations constitutes a point *group* designated by the symbol $\mathbf{C}_{3v}$; the subscript $3v$ means that there are three vertical planes of symmetry and that the covering operations consist of three reflections and three rotations.[5]

It can be shown by group theory methods that only three types of vibration are possible for any molecule of symmetry $\mathbf{C}_{3v}$: (1) nondegenerate vibrations which are symmetrical with respect to C_3 and also with respect to σ_v; this type is designated by the symbol A_1. (2) Nondegenerate vibrations which are symmetrical with respect to C_3, but are antisymmetrical with respect to σ_v; these are designated by the symbol A_2. (3) Doubly degenerate vibrations—vibrations that are isotropic in two directions at right angles to each other—which are antisymmetrical with respect to C_3; these are designated[6] by the symbol E.

A quantity called the *character* (obtained by use of group theory) is necessary for the determination of the selection rules and the number of fundamentals of each vibration type. For a given vibration type there is a separate character for each class of symmetry operations; thus for a molecule of symmetry $\mathbf{C}_{3v}$, there are three characters for type A_1, three for type A_2, and three for type E since there are three classes of operations for this group. These characters are listed in the upper part of Table I; the vibration types are written at the left and the classes are written at the top; the number in the parenthesis before each class symbol is the number of elements (or operations) in that class. The characters under the class symbol E give the degeneracy of the vibration: 1 for singly degenerate, 2 for doubly degenerate. The characters under the other classes are $+1$ for symmetrical vibrations and -1 for antisymmetrical vibrations.

TABLE I.

C_{3v}	$(1)E$	$(2)C_3$	$(3)\sigma_v$	
A_1	1	1	1	
A_2	1	1	-1	
E	2	-1	0	
φ	0°	120°	0°	
$2\cos\varphi$	2	-1	2	
$\pm 1+2\cos\varphi$	3	0	1	$\chi_M(R)$
2φ	0°	240°	0°	
$2\cos 2\varphi$	2	-1	2	
$2\pm 2\cos\varphi$ $+2\cos 2\varphi$	6	0	2	$\chi_\alpha(R)$
$\chi_E(R^2)$	2	-1	2	
$\chi_E(R^3)$	2	2	0	
$\chi_E(R^4)$	2	-1	2	
$\chi_E(R^5)$	2	-1	0	
u_R	7	4	5	
$\Xi(R)$	15	0	5	

[5] For further details concerning this and other group symbols, see reference 4, pp. 325–329.

[6] Unfortunately, it is customary to use the symbol E to represent both the identity operation and also the doubly degenerate type of vibration.

Infra-Red Spectra

To determine which vibration types are active or inactive in the infra-red, one needs the character $\chi_M(R)$ of the dipole moment. This is always given by

$$\chi_M(R) = \pm 1 + 2\cos\varphi, \quad (1)$$

where φ is the angle associated with the proper or improper rotation R; the plus sign is used for the proper rotations (E, C_3), the minus sign for the improper rotations (σ_v). The quantities involved are listed in Table I. The character $\chi_M(R)$ for a given class is always a linear combination of the characters of the vibration types for that class. Thus, for a molecule of symmetry $\mathbf{C}_{3v}$, $\chi_M(R) = \chi_{A_1}(R) + \chi_E(R)$, the characters for A_1 and E here occurring once as shown below:

	E	C_3	σ_v	
$\chi_{A_1}(R)$	1	1	1	
$\chi_E(R)$	2	-1	0	(adding)
$\chi_M(R)$	3	0	1	

In general it is difficult to determine by inspection how many times the characters of the vibration types (here A_1, A_2, E) occur in the character $\chi_M(R)$ of the dipole moment. The task is simplified by means of the reduction formula,

$$N_i = \frac{1}{N_G} \sum n_e \chi_M(R) \chi_i(R), \quad (2)$$

where N_G is the number of elements in the group (the sum of the number of elements in each class), n_e is the number of elements in each class, $\chi_i(R)$ is the character of the vibration type, N_i is the number of times the character $\chi_i(R)$ of the vibration type appears in $\chi_M(R)$, and the summation extends over all the classes of the group. In the present example,

$$N_{A_1} = \tfrac{1}{6}\{1(3)(1) + 2(0)(1) + 3(1)(1)\} = 1,$$
$$N_{A_2} = \tfrac{1}{6}\{1(3)(1) + 2(0)(1) + 3(1)(-1)\} = 0,$$
$$N_E = \tfrac{1}{6}\{1(3)(2) + 2(0)(-1) + 3(1)(0)\} = 1,$$

where $N_G = 1+2+3 = 6$. Hence $\chi_{A_1}(R)$ and $\chi_E(R)$ appear once and $\chi_{A_2}(R)$ does not appear in $\chi_M(R)$; so $\chi_M(R) = \chi_{A_1}(R) + \chi_E(R)$. This is often written as $M \sim A_1 + E$ or as $M = A_1 + E$.

Since only vibration types whose characters are contained in $\chi_M(R)$ are active in the infra-red, only the fundamental vibrations of the types A_1 and E will be active for a molecule whose symmetry is $\mathbf{C}_{3v}$. The reduction of $\chi_M(R)$ by Eq. (2) is unique, that is, only one integral value of N_i is obtained for each $\chi_i(R)$.

Raman Spectra

To determine which vibrations are active or inactive in the Raman spectra the character $\chi_\alpha(R)$ of the polarizability α is necessary. This is always given by

$$\chi_\alpha(R) = 2 \pm 2\cos\varphi + 2\cos 2\varphi. \quad (3)$$

The necessary values are listed in Table I. Again, as in Eq. (1), the plus and minus signs apply to proper and improper rotations, respectively, and φ is the angle associated with the operation. The character $\chi_\alpha(R)$ must likewise be some linear combination of the $\chi_i(R)$. In this case one sees readily by inspection that

	E	C_3	σ_v	
$2\chi_{A_1}(R)$	2	2	2	
$2\chi_E(R)$	4	-2	0	(adding)
$\chi_\alpha(R)$	6	0	2	

or $\chi_\alpha(R) = 2\chi_{A_1}(R) + 2\chi_E(R)$. In general, however, one must use Eq. (2), replacing $\chi_M(R)$ by $\chi_\alpha(R)$. Thus,

$$N_i = \frac{1}{N_G} \sum n_e \chi_\alpha(R) \chi_i(R) \quad (4)$$

or, for the present example,

$$N_{A_1} = \tfrac{1}{6}\{1(6)(1) + 2(0)(1) + 3(2)(1)\} = 2,$$
$$N_{A_2} = \tfrac{1}{6}\{1(6)(1) + 2(0)(1) + 3(2)(-1)\} = 0,$$
$$N_E = \tfrac{1}{6}\{1(6)(2) + 2(0)(-1) + 3(2)(0)\} = 2.$$

Hence $\chi_\alpha(R) = 2\chi_{A_1}(R) + 2\chi_E(R)$.

Since only vibration types whose characters appear in $\chi_\alpha(R)$ are active in the Raman effect, only the fundamental vibrations of the types A_1 and E will be active for a molecule of symmetry $\mathbf{C}_{3v}$. Again, as for $\chi_M(R)$, the reduction of $\chi_\alpha(R)$ by Eq. (4) is unique. The selection rules for the fundamentals of a molecule with symmetry $\mathbf{C}_{3v}$ are summarized in Table II. Active vibrations are designated by a and inactive vibrations by ia.

TABLE II. Selection rules for the fundamental vibrations of a molecule with symmetry C_{3v}.

Type	Activity Raman	Activity Infra-red
A_1	*a*	*a*
A_2	*ia*	*ia*
E	*a*	*a*

Combinations

To obtain the selection rules for combination frequencies $\nu_i \pm \nu_j$, it is necessary to form the *direct product* of the characters of the vibration types to which ν_i and ν_j belong. This is done by multiplying the characters of the vibration types, class by class. Thus in the present example the direct product $A_1 \times E$ of A_1 and E is obtained as fo''ows:

	E	C_3	σ_v	
$\chi_{A_1}(R)$	1	1	1	
$\chi_E(R)$	2	−1	0	(multiply)
$\chi_{A_1\times E}(R)$	2	−1	0	

It is seen that the character of the combination frequency $\chi_{A_1\times E}(R)$ is the same as $\chi_E(R)$. This is often written briefly as $A_1\times E \sim E$ or as $A_1 \times E = E$, which means that if a frequency ν_i of type E combines with a frequency ν_j of type A_1, the resulting combination frequency $\nu_i \pm \nu_j$ will be active wherever an E-type frequency is active, namely, for the present case, in both the Raman and infra-red spectra.

Selection rules for other combination frequencies are obtained in the same way. In general it is necessary to use the reduction formula, Eq. (2), $\chi_M(R)$ being replaced by the character of the combination frequency. This may be illustrated by the combination of a frequency ν_i of type E with another ν_j also of type E. One has

$\chi_E(R)$	2	−1	0	
$\chi_E(R)$	2	−1	0	(multiply)
$\chi_{E\times E}(R)$	4	1	0	

This can be uniquely reduced by means of the formula,

$$N_i = \frac{1}{N_G}\sum n_e \chi_{E\times E}(R)\chi_i(R). \quad (5)$$

Hence

$$N_{A_1} = \tfrac{1}{6}\{1(4)(1)+2(1)(1)+3(0)(1)\} = 1,$$
$$N_{A_2} = \tfrac{1}{6}\{1(4)(1)+2(1)(1)+3(0)(-1)\} = 1,$$
$$N_E = \tfrac{1}{6}\{1(4)(2)+2(1)(-1)+3(0)(0)\} = 1.$$

Thus in the present case one has

$\chi_{A_1}(R)$	1	1	1	
$\chi_{A_2}(R)$	1	1	−1	
$\chi_E(R)$	2	−1	0	(add)
$\chi_{E\times E}(R)$	4	1	0	

This is usually expressed briefly as $E\times E \sim A_1 + A_2 + E$ or as $E \times E = A_1 + A_2 + E$, which means that if a frequency ν_i of type E combines with another frequency ν_j of type E to give the combination frequency $\nu_i \pm \nu_j$, the combination frequency will be active wherever any one of the three components A_1, A_2 or E of $E\times E$ is active. Thus in the present case this combination will be active in both the Raman and infra-red.

The selection rules, obtained in the foregoing manner, for all the possible binary combination frequencies of any molecule of symmetry C_{3v} are summarized in Table III. It is seen that all combinations except $A_1 \times A_2$ are active in both the Raman and infra-red.

Overtones

To obtain the selection rules for the overtones of nondegenerate vibrations, one proceeds in a manner similar to that used for combination frequencies. The character of the $(n-1)$th overtone is given by

$$\chi_i^n(R) = [\chi_i(R)]^n. \quad (6)$$

Thus for the first overtone of an A_2 frequency ν_i, the character is obtained as follows:

$\chi_{A_2}(R)$	1	1	−1	
$\chi_{A_2}(R)$	1	1	−1	(multiply)
$\overset{2}{\chi}_{A_2}(R)$	1	1	1	

Since the character $\overset{2}{\chi}_{A_2}(R)$ of the overtone is the same as the character of A_1, the character of the overtone contains only A_1 as a component and hence the overtone will be active where and only where A_1 is active; in the present example A_1 is active in both the Raman and infra-red, hence the first overtone of an A_2 frequency will likewise be active in both the Raman and infra-red.

The character of the second overtone $\overset{3}{\chi}_{A_2}(R)$ is obtained as follows:

$\overset{2}{\chi}_{A_2}(R)$	1	1	1	
$\chi_{A_2}(R)$	1	1	-1	(multiply)
$\overset{3}{\chi}_{A_2}(R)$	1	1	-1	

It is seen that the character of the second overtone is the same as the character of A_2 and hence contains only A_2 as a component. The second overtone is thus inactive in both the Raman and infra-red since A_2 is inactive in both cases. The selection rule for overtones of frequencies of type A_2 is thus seen to be

$$A_2^n = \begin{cases} A_1 \text{ for } n \text{ even} \\ A_2 \text{ for } n \text{ odd} \end{cases}$$

and in the same manner one gets for the A_1 overtones,

$$A_1^n = A_1.$$

Consequently, for any molecule of symmetry $\mathbf{C}_{3v}$, all overtones of A_1 frequencies are active in both Raman and infra-red, but only the A_2 overtones for which n is even are active in the Raman and infra-red.

Obtaining the selection rules for the overtones of the degenerate vibrations is somewhat more involved. For the doubly degenerate vibrations (type E), one must use the formula,

$$\chi_{E^n}(R) = \tfrac{1}{2}[\chi_{E^{n-1}}(R)\chi_E(R) + \chi_E(R^n)]. \quad (7)$$

For the first overtone this becomes

$$\chi_{E^2}(R) = \tfrac{1}{2}\{[\chi_E(R)]^2 + \chi_E(R^2)\}; \quad (8)$$

$\chi_E(R^2)$ is the character corresponding to the operation R performed twice in succession. For $\mathbf{C}_{3v}$ the classes of operations R are E, C_3 and σ_v. Consequently it is necessary to determine $\chi_E(E^2)$, $\chi_E(C_3^2)$ and $\chi_E(\sigma_v^2)$.

The identity operation E performed twice places each nucleus in its initial position, the same result as would be obtained by performing the identity operation once; hence, $\chi_E(E^2) = \chi_E(E) = 2$ (from Table I). The operation $C_3 = C(\pm 2\pi/3)$ performed twice places the nuclei in the same position as would have resulted by performing the operation $C(\mp 2\pi/3)$ once; hence $\chi_E(C_3^2) = \chi_E(C_3) = -1$. The operation σ_v performed twice places the nuclei in the same positions as would have resulted by performing the operation E once; hence $\chi_E(\sigma_v^2) = \chi_E(E) = 2$.

TABLE III. Selection rules for binary combination frequencies of a molecule with symmetry $\mathbf{C}_{3v}$; a = active, ia = inactive.

Combination	A_1	A_2	E	Activity Raman	Activity Infra-red
$A_1 \times A_1$	1	0	0	a	a
$A_1 \times A_2$	0	1	0	ia	ia
$A_1 \times E$	0	0	1	a	a
$A_2 \times A_2$	1	0	0	a	a
$A_2 \times E$	0	0	1	a	a
$E \times E$	1	1	1	a	a

In a similar manner it is easy to show that

$$\chi_E(E^3) = \chi_E(E) = 2,$$
$$\chi_E(C_3^3) = \chi_E(E) = 2,$$
$$\chi_E(\sigma_v^3) = \chi_E(\sigma_v) = 0.$$

The values of $\chi_E(R^n)$ up to $n=5$ are listed in the lower part of Table I.

The quantity $[\chi_E(R)]^2$ is obtained simply by squaring $\chi_E(R)$. The character of the first overtone $\chi_{E^2}(R)$ is thus obtained from Eq. (8) as follows:

$\chi_E(R)$	2	-1	0	
$\chi_E(R)$	2	-1	0	(multiply)
$[\chi_E(R)]^2$	4	1	0	
$\chi_E(R^2)$	2	-1	2	(add)
	6	0	2	(divide by 2)
$\chi_{E^2}(R)$	3	0	1	

Since $\chi_{E^2}(R)$ is not the same as the character of A_1, A_2 or E, it must contain more than one component. These components can be found by use of Eq. (2), substituting $\chi_{E^2}(R)$ for $\chi_M(R)$; thus one obtains,

$$N_i = \frac{1}{N_G} \sum n_e \chi_{E^2}(R) \chi_i(R) \quad (9)$$

or, for $\mathbf{C}_{3v}$,

$$N_{A_1} = \tfrac{1}{6}\{1(3)(1) + 2(0)(1) + 3(1)(1)\} = 1,$$

$$N_{A_2}=\tfrac{1}{6}\{1(3)(1)+2(0)(1)+3(1)(-1)\}=0,$$
$$N_E=\tfrac{1}{6}\{1(3)(2)+2(0)(-1)+3(1)(0)\}=1.$$

Hence $E^2=A_1+E$. This may be verified by adding the characters of A_1 and E (Table I). Since the A_1 and E frequencies are active in both Raman and infra-red, so is the first overtone of a frequency of the E type.

For the second overtone, Eq. (7) becomes

$$\chi_{E^3}(R)=\tfrac{1}{2}\{\chi_{E^2}(R)\chi_E(R)+\chi_E(R^3)\}, \quad (10)$$

and to obtain the character of the second overtone $\chi_{E^3}(R)$ one proceeds as follows:

$\chi_{E^2}(R)$	3	0	1	
$\chi_E(R)$	2	−1	0	(multiply)
$\chi_{E^2}(R)\chi_E(R)$	6	0	0	
$\chi_E(R^3)$	2	2	0	(add)
	8	2	0	(divide by 2)
$\chi_{E^3}(R)$	4	1	0	

The number of components in $\chi_{E^3}(R)$ may now be found from Eq. (9), replacing $\chi_{E^2}(R)$ by $\chi_{E^3}(R)$. The result is $E^3=A_1+A_2+E$. Hence the second overtone is likewise active in both the Raman and infra-red.

The selection rules obtained in the foregoing manner for the overtones of any molecule of symmetry $\mathbf{C}_{3v}$ are summarized in Table IV.

For triply degenerate frequencies[7] (type T) one must use the equation,

$$\chi_{T^n}(R)=\tfrac{1}{3}\{2\chi_T(R)\chi_{T^{n-1}}(R)-\tfrac{1}{2}\chi_{T^{n-2}}(R)[\chi_T(R)]^2 + \tfrac{1}{2}\chi_T(R^2)\chi_{T^{n-2}}(R)+\chi_T(R^n)\}. \quad (11)$$

TABLE IV. Selection rules for the overtones of a molecule with the symmetry $\mathbf{C}_{3v}$.

Overtone	A_1	A_2	E	Activity Raman	Activity Infra-red
A_1^n	1	0	0	*a*	*a*
A_2^n(n even)	1	0	0	*a*	*a*
A_2^n(n odd)	0	1	0	*ia*	*ia*
E^2	1	0	1	*a*	*a*
E^3	1	1	1	*a*	*a*
E^4	1	0	2	*a*	*a*
E^5	1	1	2	*a*	*a*
...	...	...	...	...	...

[7] For higher degeneracies, which seldom occur, see Tisza, reference 3.

Number of Fundamentals of Each Type

While the selection rules just discussed apply to any molecule of symmetry $\mathbf{C}_{3v}$, the number of vibrations of each type depends upon the number of atoms in the molecule. To find the number of fundamentals of each type, one needs a quantity $\Xi(R)$ which is given by

$$\Xi(R)=\begin{cases}(u_R-2)(1+2\cos\varphi) & \text{for proper rotations,}\\ (u_R)(-1+2\cos\varphi) & \text{for improper rotations,}\end{cases} \quad (12)$$

where φ is the angle associated with the proper or improper rotation and u_R is the number of nuclei unchanged by the operation R. The necessary quantities for methylacetylene are listed in Table I. The number of frequencies of each type is given by

$$N_i=\frac{1}{N_G}\sum n_e\Xi(R)\chi_i(R), \quad (13)$$

where the summation extends over all the classes of the group. Thus one has for the present example,

$$N_{A_1}=\tfrac{1}{6}\{1(15)(1)+2(0)(1)+3(5)(1)\}=5,$$
$$N_{A_2}=\tfrac{1}{6}\{1(15)(1)+2(0)(1)+3(5)(-1)\}=0,$$
$$N_E=\tfrac{1}{6}\{1(15)(2)+2(0)(-1)+3(5)(0)\}=5.$$

Hence the methylacetylene molecule has five nondegenerate A_1 fundamentals and five doubly degenerate E fundamentals, a total of fifteen. This is sometimes written in the form,

$$\Gamma=5A_1+5E.$$

There should be $3N-6$ fundamentals for a nonlinear molecule containing N atoms. For methylacetylene $N=7$, so $3N-6=15$; hence the above number of fundamentals is seen to be correct. Since all the fundamentals are of type A_1 or E, each of which is active in both Raman and infra-red, one has the result that for methylacetylene all fundamentals, binary combinations and overtones are active in both the Raman and infra-red.

Ammonia is another molecule of symmetry $\mathbf{C}_{3v}$ and hence the selection rules for NH_3 are the same as those already derived for methyl-

TABLE V. Characters for various groups.

$\mathbf{C}_1$	E
A	1

$\mathbf{C}_i$	$\mathbf{C}_2$	$\mathbf{C}_{1h}\equiv\mathbf{C}_s$	E	I
			E	C_2
			E	σ_h
A_g	A	A'	1	1
A_u	B	A''	1	-1

$\mathbf{C}_3$	E	C_3	C_3^2
A	1	1	1
E	$\{1$	ω	$\omega^2\}$
	$\{1$	ω^2	$\omega\}$

$\omega=e^{2\pi i/3}$
$\mathbf{C}_{3h}=\mathbf{C}_3\times\sigma_h$
$\mathbf{C}_{3i}\equiv\mathbf{S}_6=\mathbf{C}_3\times I$

$\mathbf{C}_4$ / $\mathbf{S}_4$	E / E	C_2 / C_2	C_4 / S_4	C_4^3 / S_4^3
A	1	1	1	1
B	1	1	-1	-1
E	$\{1$	-1	i	$-i\}$
	$\{1$	-1	$-i$	$i\}$

$\mathbf{C}_{4h}=\mathbf{C}_4\times I$

$\mathbf{C}_6$	E	C_6	C_3	C_2	C_3^2	C_6^5
A	1	1	1	1	1	1
B	1	-1	1	-1	1	-1
E_1	$\{1$	ω^2	ω^4	1	ω^2	$\omega^4\}$
	$\{1$	ω^4	ω^2	1	ω^4	$\omega^2\}$
E_2	$\{1$	ω	ω^2	-1	$-\omega$	$-\omega^2\}$
	$\{1$	$-\omega^2$	$-\omega$	-1	ω^2	$\omega\}$

$\omega=e^{2\pi i/6}=-\omega^4$
$\mathbf{C}_{6h}=\mathbf{C}_6\times I$

$\mathbf{C}_{2h}$	$\mathbf{C}_{2v}$	$\mathbf{V}\equiv\mathbf{D}_2$	E	C_2	σ_h	I
			E	C_2	σ_v	σ_v'
			E	C_2^z	C_2^y	C_2^x
A_g	A_1	A_1	1	1	1	1
B_g	B_2	B_3	1	-1	-1	1
A_u	A_2	B_1	1	1	-1	-1
B_u	B_1	B_2	1	-1	1	-1

$\mathbf{V}_h\equiv\mathbf{D}_{2h}=\mathbf{V}\times I$

$\mathbf{D}_3$ / $\mathbf{C}_{3v}$	E / E	$2C_3$ / $2C_3$	$3C_2'$ / $3\sigma_v$
A_1	1	1	1
A_2	1	1	-1
E	2	-1	0

$\mathbf{D}_{3d}=\mathbf{D}_3\times I$

$\mathbf{D}_4$ / $\mathbf{C}_{4v}$ / $\mathbf{V}_d\equiv\mathbf{D}_{2d}$	E / E / E	C_2 / C_2 / C_2	$2C_4$ / $2C_4$ / $2S_4$	$2C_2$ / $2\sigma_v$ / $2C_2$	$2C_2'$ / $2\sigma_d$ / $2\sigma_d$
A_1	1	1	1	1	1
A_2	1	1	1	-1	-1
B_1	1	1	-1	1	-1
B_2	1	1	-1	-1	1
E	2	-2	0	0	0

$\mathbf{D}_{4h}=\mathbf{D}_4\times I$

$\mathbf{D}_6$	$\mathbf{C}_{6v}$	$\mathbf{D}_{3h}$	E	C_2	$2C_3$	$2C_6$	$3C_2$	$3C_2'$
			E	C_2	$2C_3$	$2C_6$	$3\sigma_d$	$3\sigma_v$
			E	σ_h	$2C_3$	$2S_6$	$3C_2$	$3\sigma_v$
A_1	A_1	A_1'	1	1	1	1	1	1
A_2	A_2	A_2'	1	1	1	1	-1	-1
B_1	B_2	A_1''	1	-1	1	-1	1	-1
B_2	B_1	A_2''	1	-1	1	-1	-1	1
E_2	E_2	E'	2	2	-1	-1	0	0
E_1	E_1	E''	2	-2	-1	1	0	0

$\mathbf{D}_{6h}=\mathbf{D}_6\times I$

$\mathbf{T}$	E	$3C_2$	$4C_3$	$4C_3'$
A	1	1	1	1
E	$\{1$	1	ω	$\omega^2\}$
	$\{1$	1	ω^2	$\omega\}$
T	3	-1	0	0

$\omega=e^{2\pi i/3}$
$\mathbf{T}_h=\mathbf{T}\times I$

$\mathbf{O}$ / $\mathbf{T}_d$	E / E	$8C_3$ / $8C_3$	$3C_2$ / $3C_2$	$6C_2$ / $6\sigma_d$	$6C_4$ / $6S_4$
A_1	1	1	1	1	1
A_2	1	1	1	-1	-1
E	2	-1	2	0	0
T_2	3	0	-1	1	-1
T_1	3	0	-1	-1	1

$\mathbf{O}_h=\mathbf{O}\times I$

$\mathbf{D}_\infty$ / $\mathbf{C}_{\infty v}$	E / E	$2C(\varphi)$ / $2C(\varphi)$	C_2 / σ_v
A_1	1	1	1
A_2	1	1	-1
E_1	2	$2\cos\varphi$	0
E_2	2	$2\cos 2\varphi$	0
...	...	...	...
E_k	2	$2\cos k\varphi$	0
...	...	...	...

$\mathbf{D}_{\infty h}=\mathbf{D}_\infty\times I$

$\mathbf{D}_{3h}'$	E	$2C_3$	$3C_2$	σ_h	$2S_3$	$3\sigma_v$
A_1	1	1	1	1	1	1
A_2	1	1	-1	1	1	-1
$\bar{A}_1$	1	1	1	-1	-1	-1
$\bar{A}_2$	1	1	-1	-1	-1	1
E	2	-1	0	2	-1	0
$\bar{E}$	2	-1	0	-2	1	0

$\mathbf{S}_{8u}$	E	$2S_8$	$2C_4$	$2S_8^3$	C_2	$4\sigma_v$	$4C_2'$
A_1	1	1	1	1	1	1	1
A_2	1	1	1	1	1	-1	-1
B_1	1	-1	1	-1	1	1	-1
B_2	1	-1	1	-1	1	-1	1
E_1	2	$\sqrt{2}$	0	$-\sqrt{2}$	-2	0	0
E_2	2	0	-2	0	2	0	0
E_3	2	$-\sqrt{2}$	0	$\sqrt{2}$	-2	0	0

acetylene. However, since the number of atoms is different it will not have the same number of A_1 and E fundamentals. Using Eq. (13) as previously described, one obtains

$$\Gamma=2A_1+2E.$$

Thus the ammonia molecule has two nondegenerate vibrations of type A_1 and two doubly degenerate vibrations of type E.

Accidental Degeneracy

Sometimes an overtone or combination frequency has nearly the same value as the frequency of some fundamental. In such a case two lines rather than one may be observed.[8, 9] The

[8] E. Fermi, Zeits. f. Physik **71**, 250 (1931).
[9] D. M. Dennison, Phys. Rev. **41**, 304 (1932).

symmetry condition that this be possible is that the overtone or combination frequency have a component which has the same symmetry as the fundamental. An example of this is the case of dimethylacetylene,[10] for which there is a resonance splitting of the A_1 fundamental near 725 cm^{-1} due to interaction with the first overtone of the 371-cm^{-1} $\bar{E}$ fundamental. The symmetry of dimethylacetylene is $\mathbf{D}_{3h}'$; for this symmetry $\bar{E}^2 = A_1 + E$. Since the character of the first overtone contains the character of A_1 as a component, the resonance splitting is permitted in accordance with the symmetry condition. Another case of resonance splitting occurs in the same molecule; this is the splitting of the A_1 fundamental near 2270 cm^{-1} by interaction with the first overtone of the 1126-cm^{-1} $\bar{A}_2$ fundamental. This is permitted by the symmetry condition since $\bar{A}_2^2 = A_1$.

ADDITIONAL POINT GROUPS

In order that the foregoing methods may be applied to other molecules, tables of characters for additional point groups are given in Table V.

In the upper left-hand corner of each table is given the symbol of the group ($\mathbf{C}_1$, $\mathbf{C}_i$, $\mathbf{C}_3$, etc.). The classes of the covering operations are listed at the top; C_n^k denotes a proper rotation of $\pm 2\pi k/n$ rad about a symmetry axis, S_n^k denotes an improper rotation of $\pm 2\pi k/n$ rad about a symmetry axis, σ_h is a reflection plane perpendicular to the Z axis, σ_v is a reflection plane passing through the Z axis, σ_d is a diagonal reflection plane containing a symmetry axis and bisecting the angle between two other symmetry axes, and I is an inversion, that is, an improper rotation through 180°. The number which stands before the symbol of the class indicates the number of elements in that class. For $\mathbf{C}_{2v}$ there are two nonequivalent, vertical reflection planes, σ_v and σ_v', which must be considered as separate classes since the two reflections affect a different number of nuclei. For $\mathbf{D}_2$, C_2^x, C_2^y and C_2^z represent rotations of π rad about the X, Y and Z axes, respectively. C_n' represents a rotation of $2\pi/n$ rad about a symmetry axis other than the principal symmetry axis (Z axis).

The vibration types are listed at the left: types A and B are nondegenerate and types E and T are doubly and triply degenerate, respectively. Types A and B are symmetrical (character $= +1$) and antisymmetrical (character $= -1$), respectively, to rotation through $\pm 2\pi/n$ rad about the Z axis. Types that are symmetrical or antisymmetrical to σ_h are designated by ′ or ″, respectively, except in the case of $\mathbf{D}_{3h}'$, where

TABLE VI. Formation of the characters of $\mathbf{D}_{3d}$ from the direct product of the characters of $\mathbf{D}_3$ and $\mathbf{C}_i$.

$\mathbf{C}_i$	E	I
A_g	1	1
A_u	1	−1

$\mathbf{D}_3$	E	$2C_3$	$3C_2'$
A_1	1	1	1
A_2	1	1	−1
E	2	−1	0

$\mathbf{D}_{3d}$	E	$2C_3$	$3C_2'$	I	$2S_6$	$3\sigma_d$
A_{1g}	1	1	1	1	1	1
A_{1u}	1	1	1	−1	−1	−1
A_{2g}	1	1	−1	1	1	−1
A_{2u}	1	1	−1	−1	−1	1
E_g	2	−1	0	2	−1	0
E_u	2	−1	0	−2	1	0

[10] B. L. Crawford, Jr., J. Chem. Phys. **7**, 555 (1939).

TABLE VII. Molecules belonging to various groups.

Group	Molecules
$\mathbf{C}_1$	Any molecule having only the class E, for example, CH_3OH.
$\mathbf{C}_s$	HOD, HSD, PCl_2Br, PBr_2Cl, NH_2Cl, HDCO, $C_6H_3D_3$ (unsym.).
$\mathbf{C}_{3h}$	C_3H_6 (cyclopropane).
$\mathbf{C}_{2v}$	H_2O, D_2O, SO_2, H_2S, H_2CO, CH_2Cl_2, CF_2Cl_2, CH_2CO (ketene), $(CH_2)_2O$ (ethylene oxide), CCl_2O (phosgene), $(CH_2)_2S$ (ethylene sulfide), C_6H_5D, CD_5H, C_6H_5Cl, m-$C_6H_4D_2$, vic-$C_6H_3D_3$, m-$C_6H_2D_4$, o-$C_6H_4D_2$, o-$C_6D_4H_2$, m-$C_6H_4Cl_2$, ClO_2, Cl_2O.
$\mathbf{V}_h \equiv \mathbf{D}_{2h}$	C_2H_4, C_2D_4, C_2Cl_4, p-$C_6H_4D_2$, p-$C_6D_4H_2$, p-$C_6H_4Cl_2$, N_2O_4.
$\mathbf{C}_{3v}$	NH_3, $AsCl_3$, PCl_3, AsF_3, CH_3Cl, CH_3D, PF_3ClI, CH_3CN, CH_3NCO, $BrCCl_3$, $CHCl_3$, $CDCl_3$, $CH_3C\equiv CH$, $CH_3C\equiv CD$, $CH_3C\equiv CCl$, $CH_3C\equiv C-C\equiv C-H$, $CH_3C\equiv C-C\equiv CD$, $CH_3C\equiv C-C\equiv CCl$.
$\mathbf{V}_d \equiv \mathbf{D}_{2d}$	$H_2C=C=CH_2$.
$\mathbf{D}_{4h}$	$PtCl_4^=$, $PdCl_4^=$.
$\mathbf{D}_{3h}$	$CO_3^=$, NO_3^-, PF_3Cl_2, $B_3N_3H_6$ (triborane triamine), 1,3,5-$C_6H_3D_3$, 1,3,5-$C_6H_3Cl_3$.
$\mathbf{D}_{3h}'$	$CH_3C\equiv CCH_3$.
$\mathbf{D}_{3d}$	Cl_3C-CCl_3.
$\mathbf{D}_{6h}$	C_6H_6, C_6D_6, C_6Cl_6.
$\mathbf{S}_{8u}$	S_8.
$\mathbf{T}_d$	CH_4, CCl_4, $SiCl_4$, P_4, GeH_4, SiH_4, $TiCl_4$, $SnCl_4$, $GeBr_4$, $WO_4^=$, $SO_4^=$, $PO_4^{\equiv}$, ClO_4^-, IO_4^-.
$\mathbf{O}_h$	SF_6, SeF_6, TeF_6.
$\mathbf{C}_{\infty v}$	N_2O, COS, ClCN, HCN.
$\mathbf{D}_{\infty h}$	CO_2, CS_2, $O=C=C=C=O$, $HC\equiv CH$, $DC\equiv CD$, $H-C\equiv C-C\equiv C-H$, $D-C\equiv C-C\equiv C-D$.

unbarred and barred symbols are used. Vibrations that are symmetrical or antisymmetrical to a center of inversion are designated by the subscripts g and u. The subscripts 1 and 2 indicate that the vibration is symmetrical or antisymmetrical with respect to some of the other classes of the group not mentioned above.

The characters for all groups are not given explicitly in Table V; some must be obtained by forming the direct product of the characters of two groups whose characters are given. For example, the characters for $\mathbf{D}_{3d}$ are obtained by forming the direct product of $\mathbf{D}_3$ and $\mathbf{C}_i$. Thus $\mathbf{D}_{3d}=\mathbf{D}_3\times\mathbf{C}_i=\mathbf{D}_3\times I$. To obtain the direct product, one proceeds as follows: To obtain the classes of $\mathbf{D}_{3d}$, one multiplies each class of $\mathbf{D}_3$ by each class of $\mathbf{C}_i$, obtaining E, $2C_3$, $3C_2'$, I, $2C_3I$ (or $2S_6$), $3C_2'I$ (or $3\sigma_d$). Each vibration of $\mathbf{D}_3$ now yields two classes with subscripts g and u; the vibration types of $\mathbf{D}_{3d}$ are thus A_{1g}, A_{1u}, A_{2g}, A_{2u}, E_g and E_u, as shown in Table VI. The characters of the type A_{1g}, A_{2g} and E_g vibrations of $\mathbf{D}_{3d}$ are obtained by multiplying the characters of the A_1, A_2 and E types of $\mathbf{D}_3$ successively by the A_g characters of $\mathbf{C}_i$. Similarly, the characters of the A_{1u}, A_{2u} and E_u vibration types of $\mathbf{D}_{3d}$ are obtained by multiplying the A_1, A_2 and E types of $\mathbf{D}_3$ successively by the A_u characters of $\mathbf{C}_i$. The characters so obtained are listed in the lower part of Table VI. One could obtain the characters for $\mathbf{C}_{3h}=\mathbf{C}_3\times\mathbf{C}_s=\mathbf{C}_3\times\sigma_h$ in a similar manner.

To find the group associated with a molecule one determines all the classes of proper and improper rotations that constitute the covering operations of the molecule. These classes are then compared with the classes given for the various groups in Table V in order to find the group having these classes. As an aid to classification, a list of molecules of various symmetries is given in Table VII.[11]

[11] Cf. E. B. Wilson, Jr., J. Chem. Phys. **2**, 432 (1934).

Reprinted from *Amer. J. Phys.*, **14**, 13–27 (1946)

36

Application of Group Theory to the Calculation of Vibrational Frequencies of Polyatomic Molecules[1]

ARNOLD G. MEISTER AND FORREST F. CLEVELAND
Illinois Institute of Technology, Chicago, Illinois

WILSON[2] has devised a method for obtaining the vibrational frequencies of polyatomic molecules in which group theory is used to simplify the calculations. The method is especially good for molecules having considerable symmetry and several equivalent atoms, that is, atoms with identical nuclei that transform into one another for all operations of the point group of the molecule. A further advantage of the method is that it requires no coordinate system, but only bond distances, interbond angles and unit vectors directed along the bonds.

Since one who is beginning calculations of vibrational frequencies may find the symbolism of Wilson's papers difficult, and since other papers involving the method omit many of the details, it seems worth while to give an elementary treatment of a few typical molecules for those desiring to start work in this field. The H_2O molecule is considered first because it has only a small number of atoms, has no degenerate frequencies, and permits the reader to concentrate on the method without being confused by the complexity of the molecule.[3] Then the CH_3Cl, CH_4 and CD_4 molecules are treated to show how the method is applied when doubly or triply degenerate frequencies are present.[4]

[1] Communication No. 43 from the *Spectroscopy Laboratory.*

[2] E. B. Wilson, Jr., *J. Chem. Physics* **7**, 1047 (1939); **9**, 76 (1941).

[3] A more complicated molecule, CH_2Cl_2, involving only nondegenerate frequencies has been discussed by G. Glockler, *Rev. Mod. Physics* **15**, 125 (1943).

[4] A treatment in outline form of the CH_3Cl molecule is given at the end of Wilson's second paper, reference 2.

THE H_2O MOLECULE

Symmetry Coordinates

The methods given in a previous paper[5] are used to determine the point group of the molecule as well as the number of fundamental vibrations of each type. It is found that the H_2O molecule belongs to the point group $\mathbf{C}_{2v}$ and that there are two vibrations of type A_1 and one of type B_2. Since a nonlinear molecule containing N atoms has $3N-6$ vibrational degrees of freedom, $3N-6$ coordinates are necessary to describe the vibrations of the molecule. To attain the simplification made possible by the use of group theory, it is

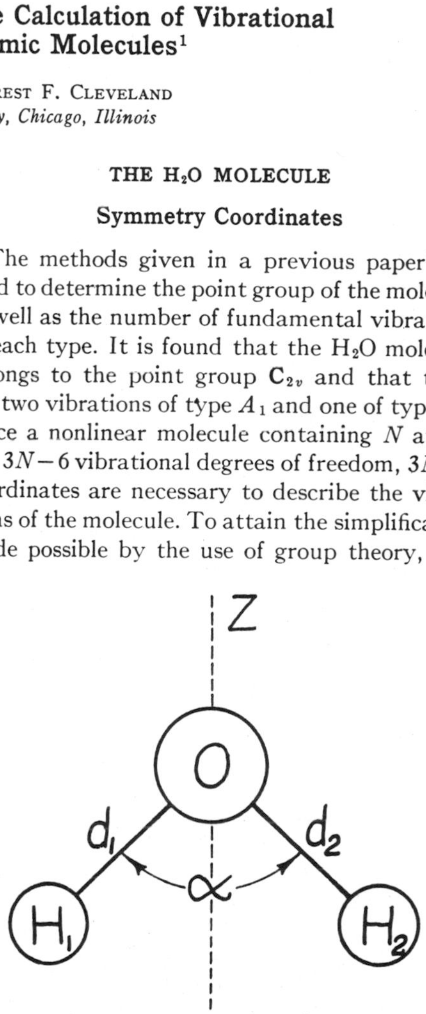

FIG. 1. Bond distances d_1 and d_2, interbond angle α, and principal symmetry axis Z for the H_2O molecule.

[5] A. G. Meister, F. F. Cleveland and M. J. Murray, *Am. J. Physics* **11**, 239 (1943).

TABLE I. Characters for the point group C_{2v}.

C_{2v}	E	C_2	σ_v	σ_v'
A_1	1	1	1	1
B_2	1	−1	−1	1
A_2	1	1	−1	−1
B_1	1	−1	1	−1

necessary that these $3N-6$ coordinates be *symmetry coordinates* which are linear combinations of the *internal coordinates*. The internal coordinates are the changes in bond distances and the changes in interbond angles. It is convenient to construct the symmetry coordinates from equivalent internal coordinates only. Moreover, the choice of linear combinations is not arbitrary but must be made in such a way that the symmetry coordinate transforms according to the characters for the vibration type concerned. Also, the symmetry coordinates must be normalized and orthogonal.

For example, for the H_2O molecule, $3(3)-6$, or three, symmetry coordinates are required. The internal coordinates (see Fig. 1) are Δd_1 and Δd_2, the changes in the bond distances d_1 and d_2 and $\Delta\alpha$, the change in the interbond angle α. In this case the equivalent internal coordinates are Δd_1 and Δd_2. Since there are two A_1 vibrations and one B_2 vibration, a suitable set of symmetry coordinates would be: for the A_1 vibrations,

$$R_1=(1/\sqrt{2})\Delta d_1+(1/\sqrt{2})\Delta d_2, \quad (1)$$

$$R_2=\Delta\alpha; \quad (2)$$

and for the B_2 vibration,

$$R_3=(1/\sqrt{2})\Delta d_1-(1/\sqrt{2})\Delta d_2. \quad (3)$$

These are of the form

$$R_j=\textstyle\sum_k U_{jk}r_k, \quad (4)$$

where R_j is the jth symmetry coordinate ($j=1, 2, \cdots, 3N-6$), U_{jk} is the coefficient of the kth internal coordinate r_k, and the summation is taken over all of the equivalent internal coordinates. [Eq. (4) can be written more concisely in the matrix notation as $\mathbf{R}=\mathbf{Ur}$.]

The condition for normalization of the jth symmetry coordinate is that

$$\textstyle\sum_k(U_{jk})^2=1. \quad (5)$$

Thus for R_1, R_2 and R_3, one has

$$(1/\sqrt{2})^2+(1/\sqrt{2})^2=1, \quad (1)^2=1,$$

and

$$(1/\sqrt{2})^2+(-1/\sqrt{2})^2=1;$$

so R_1, R_2 and R_3 are normalized.

For the symmetry coordinates to be orthogonal it is necessary that

$$\textstyle\sum_k U_{jk}U_{lk}=0, \quad (6)$$

where j and l refer to two different symmetry coordinates. Thus for R_1 and R_2, one has

$$(1/\sqrt{2})(0)+(1/\sqrt{2})(0)+(0)(1)=0.$$

Similarly for R_1 and R_3, and R_2 and R_3, one has

$$(1/\sqrt{2})(1/\sqrt{2})+(1/\sqrt{2})(-1/\sqrt{2})+(0)(0)=0$$

and

$$(0)(1/\sqrt{2})+(0)(-1/\sqrt{2})+(1)(0)=0.$$

Hence R_1, R_2 and R_3 are orthogonal.

It remains to be shown that the symmetry coordinates transform properly. For this purpose it is necessary to refer to the table of characters of the point group of the molecule. The characters of the point group $\mathbf{C}_{2v}$ are given in Table I. If the covering operations of the group are applied to the internal coordinates (Fig. 1), one finds that the internal coordinates are transformed as shown in Table II. Since only nondegenerate frequencies are involved, the result of applying a covering operation to a symmetry coordinate will be a transformation of the coordinate into itself or its negative, as indicated by the characters in Table I.

Consider now the symmetry coordinate R_3. When the identity operation E is performed, each internal coordinate is transformed into itself (see Table II). Hence one may write

$$(E)R_3=(1/\sqrt{2})\Delta d_1-(1/\sqrt{2})\Delta d_2=(+1)R_3.$$

However, when the operation C_2 is performed, Δd_1 is transformed into Δd_2 and Δd_2 is transformed into Δd_1. Hence one must write

$$(C_2)R_3=(1/\sqrt{2})\Delta d_2-(1/\sqrt{2})\Delta d_1=(-1)R_3,$$

TABLE II. Transformations of the internal coordinates by the covering operations of the group C_{2v}.

	E	C_2	σ_v	σ_v'
$\Delta d_1\rightarrow$	Δd_1	Δd_2	Δd_2	Δd_1
$\Delta d_2\rightarrow$	Δd_2	Δd_1	Δd_1	Δd_2
$\Delta\alpha\rightarrow$	$\Delta\alpha$	$\Delta\alpha$	$\Delta\alpha$	$\Delta\alpha$

which is obtained by replacing Δd_1 by Δd_2, and Δd_2 by Δd_1, in Eq. (3). In a similar manner, one may write

$$(\sigma_v)R_3=(1/\sqrt{2})\Delta d_2-(1/\sqrt{2})\Delta d_1=(-1)R_3$$

and

$$(\sigma_v')R_3=(1/\sqrt{2})\Delta d_1-(1/\sqrt{2})\Delta d_2=(+1)R_3.$$

Thus for the operations E, C_2, σ_v and σ_v', one gets $+1$, -1, -1, $+1$, respectively, as the characters for the transformations; and a comparison of these with the characters of the type B_2 vibration, shown in Table I, indicates that R_3 transforms according to the characters for the type B_2 vibration.

In a similar way one can show that R_1 and R_2 transform according to the characters for the type A_1 vibrations. Hence R_1, R_2 and R_3 satisfy all the requirements for symmetry coordinates given in the first paragraph.

The F Matrices

To get the equations from which the vibrational frequencies can be determined, one must calculate the elements of a matrix **F**, related to the potential energy, and a matrix **G**, related to the kinetic energy. If one assumes harmonic motion of small amplitude for the nuclei, the expression for the potential energy V of the molecule can be written in the form[6]

$$2V=\sum f_{ik}r_ir_k, \tag{7}$$

where $f_{ik}=f_{ki}$, and i and k extend over all the internal coordinates. For the H_2O molecule, Eq. (7) becomes

$$2V=f_d[(\Delta d_1)^2+(\Delta d_2)^2]+f_\alpha'(\Delta\alpha)^2 +2f_{d\alpha}'(\Delta d_1+\Delta d_2)(\Delta\alpha)+2f_{dd}(\Delta d_1)(\Delta d_2). \tag{8}$$

Since it is customary to express all of the force constants f_{ik} in terms of the dyne per centimeter, all of the internal coordinates involving a change in angle must be multiplied by a length expressed in centimeters, usually chosen to be the equilibrium bond length of one of the bonds forming the angle. If this is done, Eq. (8) becomes

$$2V=f_d[(\Delta d_1)^2+(\Delta d_2)^2]+f_\alpha(d\Delta\alpha)^2 +2f_{d\alpha}(\Delta d_1+\Delta d_2)(d\Delta\alpha)+2f_{dd}(\Delta d_1)(\Delta d_2), \tag{9}$$

[6] For a detailed discussion see E. T. Whittaker, *Analytical dynamics* (Cambridge, ed. 2, 1927), p. 178.

where $d[=d_1=d_2]$ is the equilibrium value of the O–H bond distance.

But the potential energy can also be expressed in terms of the symmetry coordinates by means of the equation

$$2V=\sum F_{jl}R_jR_l, \tag{10}$$

where $F_{jl}=F_{lj}$, and j and l extend over all the symmetry coordinates.

In matrix notation,[7] Eq. (7) and Eq. (10) become

$$2\mathbf{V}=\mathbf{r}'\mathbf{f}\mathbf{r} \tag{11}$$

and

$$2\mathbf{V}=\mathbf{R}'\mathbf{F}\mathbf{R}, \tag{12}$$

where $\mathbf{r}'$ and $\mathbf{R}'$ are the transposes of the $\mathbf{r}$ and $\mathbf{R}$ matrices, respectively. From Eqs. (11) and (12),

$$\mathbf{r}'\mathbf{f}\mathbf{r}=\mathbf{R}'\mathbf{F}\mathbf{R}. \tag{13}$$

But, as already mentioned in the preceding section,

$$\mathbf{R}=\mathbf{U}\mathbf{r}, \tag{14}$$

or

$$\mathbf{r}=\mathbf{U}^{-1}\mathbf{R}, \tag{15}$$

where $\mathbf{U}^{-1}$ is the inverse of $\mathbf{U}$. But since the R_j's are orthogonal and normalized, $\mathbf{U}^{-1}=\mathbf{U}'$. Hence Eq. (15) becomes

$$\mathbf{r}=\mathbf{U}'\mathbf{R}, \tag{16}$$

whence

$$\mathbf{r}'=(\mathbf{U}'\mathbf{R})'=\mathbf{R}'\mathbf{U}, \tag{17}$$

since the transpose of a product is equal to the product of the transposes taken in reverse order. So Eq. (13) may be written as

$$\mathbf{R}'(\mathbf{U}\mathbf{f}\mathbf{U}')\mathbf{R}=\mathbf{R}'\mathbf{F}\mathbf{R}, \tag{18}$$

whence

$$\mathbf{F}=\mathbf{U}\mathbf{f}\mathbf{U}'. \tag{19}$$

Equation (19) gives **F** in terms of **U** and **f**, which is the relation necessary for obtaining the **F** matrix.

For the H_2O molecule the **f** matrix is

	Δd_1	Δd_2	$\Delta\alpha$
Δd_1	f_d	f_{dd}	$df_{d\alpha}$
Δd_2	f_{dd}	f_d	$df_{d\alpha}$
$\Delta\alpha$	$df_{d\alpha}$	$df_{d\alpha}$	d^2f_α

[7] For an introduction to matrices, see H. Margenau and G. M. Murphy, *The mathematics of physics and chemistry* (Van Nostrand, 1943), p. 288.

and the **U** matrix for the type A_1 vibrations is

$$\begin{array}{c|ccc} A_1 & \Delta d_1 & \Delta d_2 & \Delta\alpha \\ \hline R_1 & \multicolumn{3}{c}{\multirow{2}{*}{}} \end{array}\quad \begin{matrix} R_1 \\ R_2 \end{matrix}\begin{pmatrix} 1/\sqrt{2} & 1/\sqrt{2} & 0 \\ 0 & 0 & 1 \end{pmatrix}.$$

Consequently, **U′** is

$$\begin{pmatrix} 1/\sqrt{2} & 0 \\ 1/\sqrt{2} & 0 \\ 0 & 1 \end{pmatrix}.$$

Therefore **fU′**, which is obtained by matrix multiplication, is

$$\begin{pmatrix} (1/\sqrt{2})(f_d+f_{dd}) & df_{d\alpha} \\ (1/\sqrt{2})(f_d+f_{dd}) & df_{d\alpha} \\ \sqrt{2}df_{d\alpha} & d^2f_\alpha \end{pmatrix}.$$

Hence, from Eq. (19), the **F** matrix for the type A_1 vibrations is

$$\begin{pmatrix} F_{11} & F_{12} \\ F_{21} & F_{22} \end{pmatrix} = \begin{pmatrix} f_d+f_{dd} & \sqrt{2}df_{d\alpha} \\ \sqrt{2}df_{d\alpha} & d^2f_\alpha \end{pmatrix}.$$

For the type B_2 vibration,

$$\mathbf{U}=(1/\sqrt{2}\quad -1/\sqrt{2}\quad 0)\quad\text{and}\quad \mathbf{U}'=\begin{pmatrix} 1/\sqrt{2} \\ -1/\sqrt{2} \\ 0 \end{pmatrix},$$

whence

$$\mathbf{fU}'=\begin{pmatrix} (1/\sqrt{2})(f_d-f_{dd}) \\ (1/\sqrt{2})(f_{dd}-f_d) \\ 0 \end{pmatrix}$$

and

$$\mathbf{F}=(F_{33})=(f_d-f_{dd}),$$

a somewhat trivial case since there is only one matrix element.

The G Matrices

Wilson[2] has shown that the elements of a modified form of the kinetic energy matrix can be obtained, when only nondegenerate vibrations are present, from the equation

$$G_{jl}=\textstyle\sum_p \mu_p g_p \mathbf{S}_j^{(t)}\cdot\mathbf{S}_l^{(t)}, \tag{20}$$

where j and l refer to the symmetry coordinates used in determining the **S** vectors; p refers to a set of equivalent atoms, a typical one of the set being t; μ_p is the reciprocal of the mass of the typical atom t; g_p is the number of equivalent atoms in the pth set; and the summation extends over all the sets of equivalent atoms in the molecule. (For example, the two H atoms in the H_2O molecule form an equivalent set.)

The **S** vectors are given by

$$\mathbf{S}_j^{(t)}=\textstyle\sum_k U_{jk}\mathbf{s}_{kt}, \tag{21}$$

where j, U_{jk} and $\sum_k$ have the same meaning as in Eq. (4). The $\mathbf{s}_{kt}$ vectors can be expressed in terms of unit vectors directed along the chemical bonds. Referring to Fig. 2, if $\mathbf{v}'$ is a unit vector directed along the bond from atom t to atom t′, and if r_k is the change in length of this bond from its equilibrium value, then, according to Wilson,[2]

$$\mathbf{s}_{kt}=-\mathbf{v}',\quad \mathbf{s}_{kt'}=\mathbf{v}'. \tag{22}$$

But if r_k is the change in the interbond angle α (Fig. 2), then the vectors for the end atoms are

$$\begin{aligned} \mathbf{s}_{kt'}&=(\cos\alpha\mathbf{v}'-\mathbf{v}'')/(d'\sin\alpha),\\ \mathbf{s}_{kt''}&=(\cos\alpha\mathbf{v}''-\mathbf{v}')/(d''\sin\alpha); \end{aligned} \tag{23}$$

and for the apex atom,

$$\mathbf{s}_{kt}=[(1/d''-\cos\alpha/d')\mathbf{v}' + (1/d'-\cos\alpha/d'')\mathbf{v}'']/\sin\alpha. \tag{23}$$

For the H_2O molecule, t′ = H_1, t = O, t″ = H_2, and $d'=d''=d_1=d_2=d$. Hence the $\mathbf{s}_{kt}$ vectors for H_1 and O are

$$\begin{aligned} \mathbf{s}_{d_1H_1}&=\mathbf{v}',\\ \mathbf{s}_{\alpha H_1}&=(\cos\alpha\mathbf{v}'-\mathbf{v}'')/(d\sin\alpha),\\ \mathbf{s}_{d_1O}&=-\mathbf{v}',\\ \mathbf{s}_{d_2O}&=-\mathbf{v}'',\\ \mathbf{s}_{\alpha O}&=[(1-\cos\alpha)(\mathbf{v}'+\mathbf{v}'')]/(d\sin\alpha). \end{aligned}$$

Since H_1 and H_2 are equivalent atoms, it is not necessary to calculate the $\mathbf{s}_{kt}$ vectors for H_2.

In order to get the $\mathbf{S}_j^{(t)}$ vectors from Eq. (21), one must first find the U_{jk} values. From Eqs. (1)–(3), it is seen that they are

$$\begin{aligned} &U_{1d_1}=1/\sqrt{2}, && U_{2d_1}=0, && U_{3d_1}=1/\sqrt{2},\\ &U_{1d_2}=1/\sqrt{2}, && U_{2d_2}=0, && U_{3d_2}=-1/\sqrt{2},\\ &U_{1\alpha}=0, && U_{2\alpha}=1, && U_{3\alpha}=0. \end{aligned}$$

Hence the $\mathbf{S}_1^{H_1}$ vector is

$$\begin{aligned} \mathbf{S}_1^{H_1}&=U_{1d_1}\mathbf{s}_{d_1H_1}+U_{1d_2}\mathbf{s}_{d_2H_1}+U_{1\alpha}\mathbf{s}_{\alpha H_1}\\ &=(1/\sqrt{2})(\mathbf{v}')+(1/\sqrt{2})(0)\\ &\quad+(0)(\cos\alpha\mathbf{v}'-\mathbf{v}'')/(d\sin\alpha)=(1/\sqrt{2})\mathbf{v}', \end{aligned}$$

$s_{d_2H_1}$ being zero since a change in d_2 does not involve H_1. Similarly,

$$\mathbf{S}_2^{H_1}=(\cos\alpha\mathbf{v}'-\mathbf{v}'')/(d\sin\alpha),$$
$$\mathbf{S}_3^{H_1}=\mathbf{v}'/\sqrt{2},$$
$$\mathbf{S}_1^{O}=-(\mathbf{v}'+\mathbf{v}'')/\sqrt{2},$$
$$\mathbf{S}_2^{O}=(1-\cos\alpha)(\mathbf{v}'+\mathbf{v}'')/(d\sin\alpha),$$
$$\mathbf{S}_3^{O}=(\mathbf{v}''-\mathbf{v}')/\sqrt{2}.$$

The **G** matrix for the type A_1 vibrations will contain only the four elements G_{11}, G_{12}, G_{21} and G_{22}, since only R_1 and R_2 belong to the A_1 vibrations. From Eq. (20), the value of the element G_{11} is

$$\begin{aligned}G_{11}&=\mu_H g_H \mathbf{S}_1^{H_1}\cdot\mathbf{S}_1^{H_1}+\mu_O g_O \mathbf{S}_1^{O}\cdot\mathbf{S}_1^{O}\\&=\mu_H(2)(\mathbf{v}'/\sqrt{2})\cdot(\mathbf{v}'/\sqrt{2})\\&\quad+\mu_O(1)[-(\mathbf{v}'+\mathbf{v}'')/\sqrt{2}]\cdot[-(\mathbf{v}'+\mathbf{v}'')/\sqrt{2}]\\&=\mu_H(\mathbf{v}'\cdot\mathbf{v}')+\tfrac{1}{2}\mu_O[(\mathbf{v}'\cdot\mathbf{v}')\\&\quad+(\mathbf{v}'\cdot\mathbf{v}'')+(\mathbf{v}''\cdot\mathbf{v}')+(\mathbf{v}''\cdot\mathbf{v}'')]\\&=\mu_H+\mu_O(1+\cos\alpha),\end{aligned}$$

since $g_H=2$, $g_O=1$, $\mathbf{v}'\cdot\mathbf{v}'=\mathbf{v}''\cdot\mathbf{v}''=1$, and $\mathbf{v}'\cdot\mathbf{v}''=\mathbf{v}''\cdot\mathbf{v}'=\cos\alpha$.

The other matrix elements are obtained from Eq. (20) in a similar manner. The **G** matrix for the type A_1 vibrations is thus

$$\begin{pmatrix}G_{11} & G_{12}\\ G_{21} & G_{22}\end{pmatrix}=\begin{pmatrix}\mu_H+\mu_O(1+\cos\alpha) & (-\mu_O\sqrt{2}\sin\alpha)/d\\ (-\mu_O\sqrt{2}\sin\alpha)/d & 2[\mu_H+\mu_O(1-\cos\alpha)]/d^2\end{pmatrix}.$$

Note that $G_{12}=G_{21}$; in general, $G_{jl}=G_{lj}$.

The **G** matrix for the type B_2 vibration will contain only the element G_{33}, since only R_3 belongs to this vibration type. Using Eq. (20), one gets

$$(G_{33})=(\mu_H+\mu_O[1-\cos\alpha]).$$

Calculation of the Frequencies

Having obtained the **F** and **G** matrices, one can get the frequencies in the following manner. For the type A_1 frequencies the secular equation in expanded form is[2]

$$(-\lambda)^2+(F_{11}G_{11}+2F_{12}G_{12}+F_{22}G_{22})(-\lambda)+\begin{vmatrix}F_{11} & F_{12}\\ F_{21} & F_{22}\end{vmatrix}\cdot\begin{vmatrix}G_{11} & G_{12}\\ G_{21} & G_{22}\end{vmatrix}=0, \quad (24)$$

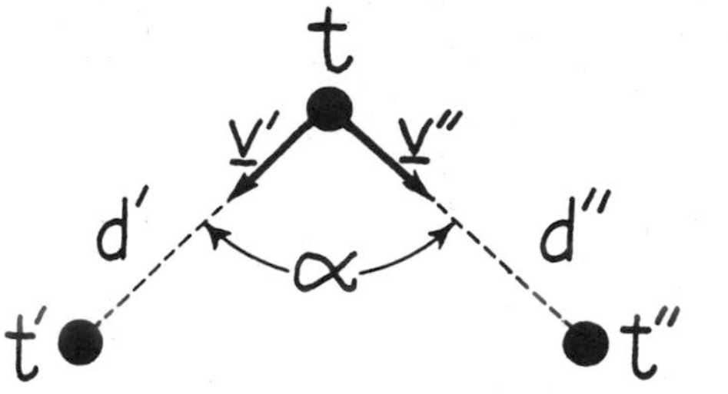

FIG. 2. Unit vectors **v**′ and **v**″ directed from atom t to atoms t′ and t″, respectively, along the bonds d' and d'', the interbond angle being α.

where the coefficient of the λ-term is the sum of the products of the corresponding one-rowed minors of the determinants of the **F** and **G** matrices, and the constant term is the product of the determinants of the **F** and **G** matrices.

At this point it is better to introduce the numerical values of the **F** and **G** matrix elements into Eq. (24). If the force constants, bond distances and interbond angles are known, the values of the type A_1 frequencies can be obtained from the λ's of Eq. (24) since

$$\lambda=4\pi^2c^2\nu^2, \quad (25)$$

where c (cm/sec.) is the velocity of light and ν (cm^{-1}) is the frequency. The values used here are those given by Glockler,[3] namely, $\alpha=104°31'$, $d=0.9580$A, $f_d=8.428\times10^5$ dyne/cm,

$$f_{dd}=-0.1051\times10^5 \text{ dyne/cm},$$
$$f_\alpha=0.7678\times10^5 \text{ dyne/cm},$$

and $f_{d\alpha}=0.2521\times10^5$ dyne/cm. The values used for μ_H and μ_O were

$$\mu_H=(6.022\times10^{23})/1.008=5.974\times10^{23}\text{ gm}^{-1}$$

and

$$\mu_O=(6.022\times10^{23})/16.00=3.764\times10^{22}\text{ gm}^{-1}.$$

When these quantities are used to obtain the numerical values of the **F** and **G** matrix elements, Eq. (24) becomes

$$\lambda^2-(6.159\times10^{29}\text{ sec}^{-2})\lambda+5.032\times10^{58}\text{ sec}^{-4}=0. \quad (26)$$

Solving this equation, one gets

$$\lambda_1=5.189\times10^{29}\text{ sec}^{-2}$$

and

$$\lambda_2=9.700\times10^{28}\text{ sec}^{-2}.$$

TABLE III. Comparison of the present calculated frequencies of the H_2O molecule with the values obtained by Dennison (see reference 8) using a different method.

Vibration Type	Symbol	Frequency (cm^{-1}) Present value	Dennison's value
A_1	ν_2	1653	1653.91
	ν_1	3825	3825.32
B_2	ν_3	3937	3935.59

Substituting these values of λ into Eq. (25) and using 2.998×10^{10} cm/sec. as the value of c, one obtains

$$\nu_1=3825\ \text{cm}^{-1},\quad \nu_2=1653\ \text{cm}^{-1}.$$

For the type B_2 frequency the secular equation in expanded form is very simple. It is

$$\lambda_3-F_{33}G_{33}=0. \tag{27}$$

From Eq. (25), one gets

$$\nu_3=(F_{33}G_{33})^{\frac{1}{2}}/(2\pi c). \tag{28}$$

Using the constants previously given, one finds $\nu_3=3937$ cm^{-1}.

In Table III, the present values are compared with those obtained by Dennison.[8]

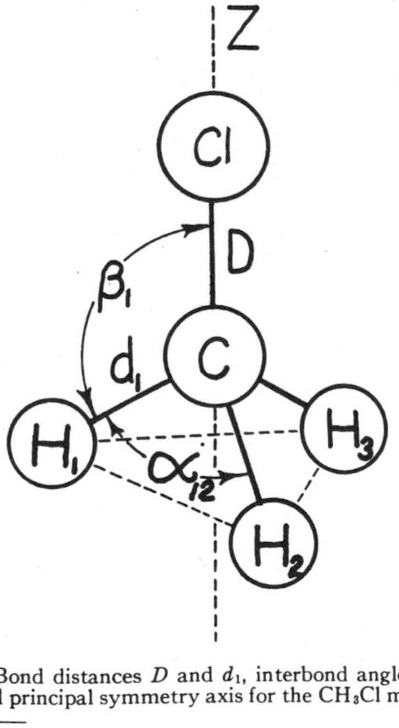

FIG. 3. Bond distances D and d_1, interbond angles β_1 and α_{12}, and principal symmetry axis for the CH_3Cl molecule.

[8] D. M. Dennison, *Rev. Mod. Physics* 12, 192 (1940).

Equations (24) and (28) are, of course, applicable to any nonlinear XY_2 molecule, such as H_2S or SO_2, if appropriate values of the constants[3] are used.

THE CH_3Cl MOLECULE

Symmetry Coordinates

The CH_3Cl molecule has the symmetry $\mathbf{C}_{3v}$ and has three nondegenerate type A_1 vibrations and three doubly degenerate type E vibrations.[5] Since there are $3(5)-6$, or nine, vibrational frequencies, only nine symmetry coordinates are necessary. However, there are (Fig. 3) ten internal coordinates; they are ΔD, Δd_1, Δd_2, Δd_3, $\Delta\beta_1$, $\Delta\beta_2$, $\Delta\beta_3$, $\Delta\alpha_{12}$, $\Delta\alpha_{23}$ and $\Delta\alpha_{31}$. Hence one of these is not independent of the others. Instead of ignoring one of the internal coordinates, which would destroy the symmetry, ten symmetry coordinates will be constructed, and then one of these will be considered as redundant. This is permissible, since the redundant coordinate can be chosen in such a way that it is identically zero and thus makes no contribution to the potential or kinetic energies.[2]

The symmetry coordinates chosen were the following: for the type A_1 vibrations,

$$R_1=\Delta D, \tag{29}$$

$$R_2=(\Delta d_1+\Delta d_2+\Delta d_3)/\sqrt{3}, \tag{30}$$

$$R_3=(\Delta\alpha_{12}+\Delta\alpha_{23}+\Delta\alpha_{31}-\Delta\beta_1-\Delta\beta_2-\Delta\beta_3)/\sqrt{6}, \tag{31}$$

$$R_4=(\Delta\alpha_{12}+\Delta\alpha_{23}+\Delta\alpha_{31}+\Delta\beta_1+\Delta\beta_2+\Delta\beta_3)/\sqrt{6}\equiv0\ \text{(redundant coordinate)}; \tag{32}$$

and for the type E vibrations,

$$R_{1a}=(2\Delta d_1-\Delta d_2-\Delta d_3)/\sqrt{6}, \tag{33}$$

$$R_{1b}=(\Delta d_2-\Delta d_3)/\sqrt{2}, \tag{34}$$

$$R_{2a}=(2\Delta\beta_1-\Delta\beta_2-\Delta\beta_3)/\sqrt{6}, \tag{35}$$

$$R_{2b}=(\Delta\beta_2-\Delta\beta_3)/\sqrt{2}, \tag{36}$$

$$R_{3a}=(2\Delta\alpha_{23}-\Delta\alpha_{12}-\Delta\alpha_{31})/\sqrt{6}, \tag{37}$$

$$R_{3b}=(\Delta\alpha_{31}-\Delta\alpha_{12})/\sqrt{2}. \tag{38}$$

The coordinate R_4 was chosen as the redundant coordinate, since it is clear that the sum of all the changes in the interbond angles must be zero. The remaining coordinates were made orthogonal to R_4 and all were normalized, using Eqs. (5) and

(6). Since each of the three type E vibrations is doubly degenerate, it is necessary to have two symmetry coordinates, designated by the subscripts $_a$ and $_b$, for each of these vibrations.

The transformations of the internal coordinates under the covering operations of the group $\mathbf{C}_{3v}$ are shown in Table IV. The characters for the group $\mathbf{C}_{3v}$ are given in Table V. Proceeding in the same manner as for the H_2O molecule, using Table IV, one can show that the A_1 symmetry coordinates R_1, R_2, R_3 and R_4 transform according to the characters for the type A_1 vibrations, given in Table V.

When a covering operation is applied to one of a pair of the E coordinates, one obtains not this coordinate or its negative, as was the case for the coordinates associated with the nondegenerate frequencies, but rather a linear combination of the two coordinates forming the pair. For example, when the operation C_3^+ is applied to the pair R_{1a}, R_{1b}, one has

$$(C_3^+)R_{1a}=(2\Delta d_2-\Delta d_3-\Delta d_1)/(6)^{\frac{1}{2}}=AR_{1a}+BR_{1b}$$

and

$$(C_3^+)R_{1b}=(\Delta d_3-\Delta d_1)/\sqrt{2}=A'R_{1a}+B'R_{1b},$$

where A, B, A' and B' are constants. Substituting for R_{1a} and R_{1b} from Eqs. (33) and (34), one gets

$$(2\Delta d_2-\Delta d_3-\Delta d_1)/(6)^{\frac{1}{2}}$$
$$=A(2\Delta d_1-\Delta d_2-\Delta d_3)/(6)^{\frac{1}{2}}+B(\Delta d_2-\Delta d_3)/\sqrt{2}$$

and

$$(\Delta d_3-\Delta d_1)/\sqrt{2}=A'(2\Delta d_1-\Delta d_2-\Delta d_3)/(6)^{\frac{1}{2}}$$
$$+B'(\Delta d_2-\Delta d_3)/\sqrt{2}.$$

Equating coefficients, one finds that

$$A=-\tfrac{1}{2},\qquad B=\sqrt{3}/2,$$
$$A'=-\sqrt{3}/2,\quad B'=-\tfrac{1}{2}.$$

If one now forms the matrix,

$$\begin{pmatrix} A & B \\ A' & B' \end{pmatrix}=\begin{pmatrix} -\frac{1}{2} & \sqrt{3}/2 \\ -\sqrt{3}/2 & -\frac{1}{2} \end{pmatrix},$$

the sum of the elements along the principal diagonal, $-\frac{1}{2}-\frac{1}{2}=-1$, gives the character under C_3 in Table V for the type E vibrations. The same character would have been obtained if C_3^- had been used.

In a similar manner, when the operation E is applied to R_{1a} and R_{1b} one gets the matrix

$$\begin{pmatrix} A & B \\ A' & B' \end{pmatrix}=\begin{pmatrix} 1 & 0 \\ 0 & 1 \end{pmatrix},$$

from which the character, 2, is obtained, in agreement with Table V. For the operation σ_{v1}, the corresponding matrix is

$$\begin{pmatrix} A & B \\ A' & B' \end{pmatrix}=\begin{pmatrix} 1 & 0 \\ 0 & -1 \end{pmatrix},$$

the sum of the elements along the principal diagonal being zero, in agreement with Table V. Had σ_{v2} or σ_{v3} been used, the same character would have been obtained.

One has proved, therefore, that the E coordinate pair, R_{1a}, R_{1b}, transforms according to the characters of the type E vibrations. In a similar manner, one can show that the coordinate pairs R_{2a}, R_{2b} and R_{3a}, R_{3b} also transform in this way.

Hence it has been demonstrated that the symmetry coordinates satisfy all the requirements given in the discussion of the symmetry coordinates for the H_2O molecule.

TABLE IV. Transformations of the internal coordinates of the CH_3Cl molecule under the covering operations of the group $\mathbf{C}_{3v}$.*

	E	C_3^+	C_3^-	σ_{v1}	σ_{v2}	σ_{v3}
$\Delta D\rightarrow$	ΔD	ΔD	ΔD	ΔD	ΔD	ΔD
$\Delta d_1\rightarrow$	Δd_1	Δd_2	Δd_3	Δd_1	Δd_3	Δd_2
$\Delta d_2\rightarrow$	Δd_2	Δd_3	Δd_1	Δd_3	Δd_2	Δd_1
$\Delta d_3\rightarrow$	Δd_3	Δd_1	Δd_2	Δd_2	Δd_1	Δd_3
$\Delta\alpha_{12}\rightarrow$	$\Delta\alpha_{12}$	$\Delta\alpha_{23}$	$\Delta\alpha_{31}$	$\Delta\alpha_{31}$	$\Delta\alpha_{23}$	$\Delta\alpha_{12}$
$\Delta\alpha_{23}\rightarrow$	$\Delta\alpha_{23}$	$\Delta\alpha_{31}$	$\Delta\alpha_{12}$	$\Delta\alpha_{23}$	$\Delta\alpha_{12}$	$\Delta\alpha_{31}$
$\Delta\alpha_{31}\rightarrow$	$\Delta\alpha_{31}$	$\Delta\alpha_{12}$	$\Delta\alpha_{23}$	$\Delta\alpha_{12}$	$\Delta\alpha_{31}$	$\Delta\alpha_{23}$
$\Delta\beta_1\rightarrow$	$\Delta\beta_1$	$\Delta\beta_2$	$\Delta\beta_3$	$\Delta\beta_1$	$\Delta\beta_3$	$\Delta\beta_2$
$\Delta\beta_2\rightarrow$	$\Delta\beta_2$	$\Delta\beta_3$	$\Delta\beta_1$	$\Delta\beta_3$	$\Delta\beta_2$	$\Delta\beta_1$
$\Delta\beta_3\rightarrow$	$\Delta\beta_3$	$\Delta\beta_1$	$\Delta\beta_2$	$\Delta\beta_2$	$\Delta\beta_1$	$\Delta\beta_3$

* C_3^+ = rotation by $+2\pi/3$, C_3^- = rotation by $-2\pi/3$ about the Z axis; σ_{v1}, σ_{v2} and σ_{v3} are reflection planes through the Z axis and through H_1, H_2 and H_3, respectively

TABLE V. Characters for the group $\mathbf{C}_{3v}$.

$\mathbf{C}_{3v}$	E	$2C_3$	$3\sigma_v$
A_1	1	1	1
A_2	1	1	−1
E	2	−1	0

The F Matrices

For the CH_3Cl molecule it is not possible to use the most general quadratic expression for the potential energy, Eq. (7), as was done for the H_2O molecule, since more force constants would be involved than could be determined from the observed frequencies. Consequently, one is forced to neglect some of the terms which, upon the basis of experience, one suspects would make only small contributions to the potential energy. Crawford and Brinkley[9] have given a rule that is helpful in deciding which terms to neglect.

The potential function used for the CH_3Cl molecule was

$$\begin{aligned}2V=&f_d[(\Delta d_1)^2+(\Delta d_2)^2+(\Delta d_3)^2]+f_\alpha[(d\Delta\alpha_{12})^2+(d\Delta\alpha_{23})^2+(d\Delta\alpha_{31})^2]+2f_{\alpha\beta}[(d\Delta\beta_1)(d\Delta\alpha_{12}+d\Delta\alpha_{31})\\&+(d\Delta\beta_2)(d\Delta\alpha_{12}+d\Delta\alpha_{23})+(d\Delta\beta_3)(d\Delta\alpha_{23}+d\Delta\alpha_{31})]+f_\beta[(d\Delta\beta_1)^2+(d\Delta\beta_2)^2+(d\Delta\beta_3)^2]\\&+f_D(\Delta D)^2+2f_{D\alpha}[(\Delta D)(d\Delta\alpha_{12}+d\Delta\alpha_{23}+d\Delta\alpha_{31})]+2f_{D\beta}[(\Delta D)(d\Delta\beta_1+d\Delta\beta_2+d\Delta\beta_3)].\end{aligned}\tag{39}$$

The corresponding **f** matrix is

$$\begin{array}{c|cccccccccc}
 & \Delta D & \Delta d_1 & \Delta d_2 & \Delta d_3 & \Delta\beta_1 & \Delta\beta_2 & \Delta\beta_3 & \Delta\alpha_{12} & \Delta\alpha_{23} & \Delta\alpha_{31}\\ \hline
\Delta D & f_D & 0 & 0 & 0 & df_{D\beta} & df_{D\beta} & df_{D\beta} & df_{D\alpha} & df_{D\alpha} & df_{D\alpha}\\
\Delta d_1 & 0 & f_d & 0 & 0 & 0 & 0 & 0 & 0 & 0 & 0\\
\Delta d_2 & 0 & 0 & f_d & 0 & 0 & 0 & 0 & 0 & 0 & 0\\
\Delta d_3 & 0 & 0 & 0 & f_d & 0 & 0 & 0 & 0 & 0 & 0\\
\Delta\beta_1 & df_{D\beta} & 0 & 0 & 0 & d^2f_\beta & 0 & 0 & d^2f_{\alpha\beta} & 0 & d^2f_{\alpha\beta}\\
\Delta\beta_2 & df_{D\beta} & 0 & 0 & 0 & 0 & d^2f_\beta & 0 & d^2f_{\alpha\beta} & d^2f_{\alpha\beta} & 0\\
\Delta\beta_3 & df_{D\beta} & 0 & 0 & 0 & 0 & 0 & d^2f_\beta & 0 & d^2f_{\alpha\beta} & d^2f_{\alpha\beta}\\
\Delta\alpha_{12} & df_{D\alpha} & 0 & 0 & 0 & d^2f_{\alpha\beta} & d^2f_{\alpha\beta} & 0 & d^2f_\alpha & 0 & 0\\
\Delta\alpha_{23} & df_{D\alpha} & 0 & 0 & 0 & 0 & d^2f_{\alpha\beta} & d^2f_{\alpha\beta} & 0 & d^2f_\alpha & 0\\
\Delta\alpha_{31} & df_{D\alpha} & 0 & 0 & 0 & d^2f_{\alpha\beta} & 0 & d^2f_{\alpha\beta} & 0 & 0 & d^2f_\alpha
\end{array}.$$

For the type A_1 vibrations the **U** and **U′** matrices are

$$\mathbf{U}=\begin{array}{c|cccccccccc}
 & \Delta D & \Delta d_1 & \Delta d_2 & \Delta d_3 & \Delta\beta_1 & \Delta\beta_2 & \Delta\beta_3 & \Delta\alpha_{12} & \Delta\alpha_{23} & \Delta\alpha_{31}\\ \hline
R_1 & 1 & 0 & 0 & 0 & 0 & 0 & 0 & 0 & 0 & 0\\
R_2 & 0 & 1/\sqrt{3} & 1/\sqrt{3} & 1/\sqrt{3} & 0 & 0 & 0 & 0 & 0 & 0\\
R_3 & 0 & 0 & 0 & 0 & -1/\sqrt{6} & -1/\sqrt{6} & -1\sqrt{6} & 1/\sqrt{6} & 1/\sqrt{6} & 1/\sqrt{6}
\end{array}$$

and

$$\mathbf{U}'=\begin{pmatrix}1 & 0 & 0\\ 0 & 1/\sqrt{3} & 0\\ 0 & 1/\sqrt{3} & 0\\ 0 & 1/\sqrt{3} & 0\\ 0 & 0 & -1/\sqrt{6}\\ 0 & 0 & -1/\sqrt{6}\\ 0 & 0 & -1/\sqrt{6}\\ 0 & 0 & 1/\sqrt{6}\\ 0 & 0 & 1/\sqrt{6}\\ 0 & 0 & 1/\sqrt{6}\end{pmatrix}.$$

[9] B. L. Crawford, Jr. and S. R. Brinkley, Jr., *J. Chem. Physics* 9, 71 (1941).

From Eq. (19), the **F** matrix for the type A_1 vibrations is found to be

$$\begin{bmatrix} F_{11} & F_{12} & F_{13} \\ F_{21} & F_{22} & F_{23} \\ F_{31} & F_{32} & F_{33} \end{bmatrix} = \begin{bmatrix} f_D & 0 & [d\sqrt{3}(f_{D\alpha}-f_{D\beta})]/2 \\ 0 & f_d & 0 \\ [d\sqrt{3}(f_{D\alpha}-f_{D\beta})]/2 & 0 & [d^2(f_\alpha+f_\beta-4f_{\alpha\beta})]/2 \end{bmatrix}.$$

Using R_{1b}, R_{2b} and R_{3b}, the **U** and **U′** matrices for the type E vibrations are found to be

$$\mathbf{U} = \begin{bmatrix} 0 & 0 & 1/\sqrt{2} & -1/\sqrt{2} & 0 & 0 & 0 & 0 & 0 & 0 \\ 0 & 0 & 0 & 0 & 0 & 1/\sqrt{2} & -1/\sqrt{2} & 0 & 0 & 0 \\ 0 & 0 & 0 & 0 & 0 & 0 & 0 & -1/\sqrt{2} & 0 & 1/\sqrt{2} \end{bmatrix}.$$

and

$$\mathbf{U}' = \begin{bmatrix} 0 & 0 & 0 \\ 0 & 0 & 0 \\ 1/\sqrt{2} & 0 & 0 \\ -1/\sqrt{2} & 0 & 0 \\ 0 & 0 & 0 \\ 0 & 1/\sqrt{2} & 0 \\ 0 & -1/\sqrt{2} & 0 \\ 0 & 0 & -1/\sqrt{2} \\ 0 & 0 & 0 \\ 0 & 0 & 1/\sqrt{2} \end{bmatrix}.$$

From Eq. (19), the **F** matrix for the type E vibrations is found to be

$$\begin{bmatrix} F_{11} & F_{12} & F_{13} \\ F_{21} & F_{22} & F_{23} \\ F_{31} & F_{32} & F_{33} \end{bmatrix} = \begin{bmatrix} f_d & 0 & 0 \\ 0 & d^2 f_\beta & -d^2 f_{\alpha\beta} \\ 0 & -d^2 f_{\alpha\beta} & d^2 f_\alpha \end{bmatrix}.$$

The same **F** matrix could have been obtained in the same way by use of R_{1a}, R_{2a} and R_{3a}.

The G Matrices

The elements of the **G** matrix for the non-degenerate type A_1 vibrations can be obtained from Eq. (20). However, for the doubly degenerate type E vibrations, the corresponding equation for the elements of the **G** matrix is[2]

$$G_{jl} = \tfrac{1}{2} \textstyle\sum_p \mu_p g_p (\mathbf{S}_{ja}^{(t)} \cdot \mathbf{S}_{la}^{(t)} + \mathbf{S}_{jb}^{(t)} \cdot \mathbf{S}_{lb}^{(t)}), \quad (40)$$

where $\mathbf{S}_{ja}^{(t)}$ and $\mathbf{S}_{jb}^{(t)}$ are the vectors, for the atom t, obtained from the U's appearing in the symmetry coordinate pair R_{ja}, R_{jb}, and the other symbols have the same meaning as in Eq. (20).

Using Eqs. (22) and (23) and the unit vectors, bond distances, and interbond angles indicated in Fig. 4, and assuming tetrahedral angles ($\alpha=\beta=109°28'$), the $\mathbf{s}_{kt}$ vectors for the typical atoms are found to be: for the H_1 atom, letting $\epsilon = 1/d$ and $\tau = 1/D$,

$$\mathbf{s}_{d_1 H_1} = \mathbf{v}_1,$$

$$\mathbf{s}_{\alpha_{12} H_1} = -\tfrac{3}{4}\sqrt{2}\epsilon(\mathbf{v}_1/3+\mathbf{v}_2),$$

$$\mathbf{s}_{\alpha_{31} H_1} = -\tfrac{3}{4}\sqrt{2}\epsilon(\mathbf{v}_1/3+\mathbf{v}_3),$$

$$\mathbf{s}_{\beta_1 H_1} = -\tfrac{3}{4}\sqrt{2}\epsilon(\mathbf{v}_1/3+\mathbf{v}),$$

$$\mathbf{s}_{\alpha_{23} H_1} = \mathbf{s}_{d_2 H_1} = \mathbf{s}_{d_3 H_1} = \mathbf{s}_{D H_1} = \mathbf{s}_{\beta_2 H_1} = \mathbf{s}_{\beta_3 H_1} = 0;$$

for the C atom

$$\mathbf{s}_{DC} = -\mathbf{v},$$

$$\mathbf{s}_{d_1 C} = -\mathbf{v}_1,$$

$$\mathbf{s}_{d_2 C} = -\mathbf{v}_2,$$

$$\mathbf{s}_{d_3 C} = -\mathbf{v}_3,$$

$$\mathbf{s}_{\alpha_{12} C} = \epsilon\sqrt{2}(\mathbf{v}_1+\mathbf{v}_2),$$

$$\mathbf{s}_{\alpha_{23} C} = \epsilon\sqrt{2}(\mathbf{v}_2+\mathbf{v}_3),$$

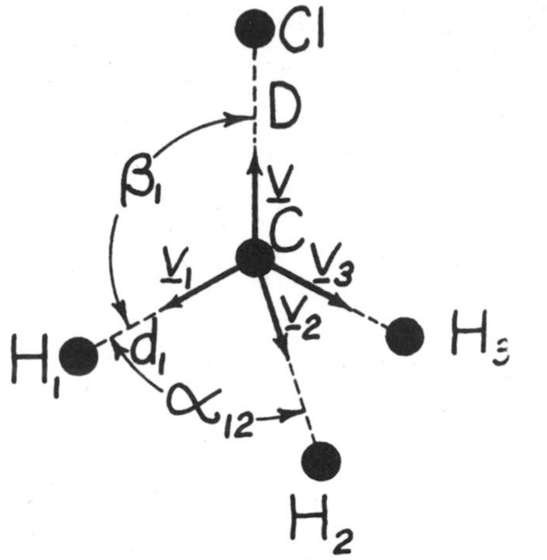

FIG. 4. Unit vectors for the CH_3Cl molecule.

$$s_{\alpha_{31}C}=\epsilon\sqrt{2}(\mathbf{v}_3+\mathbf{v}_1),$$
$$s_{\beta_1C}=\tfrac{3}{4}\sqrt{2}[(\epsilon+\tau/3)\mathbf{v}+(\tau+\epsilon/3)\mathbf{v}_1],$$
$$s_{\beta_2C}=\tfrac{3}{4}\sqrt{2}[(\epsilon+\tau/3)\mathbf{v}+(\tau+\epsilon/3)\mathbf{v}_2],$$
$$s_{\beta_3C}=\tfrac{3}{4}\sqrt{2}[(\epsilon+\tau/3)\mathbf{v}+(\tau+\epsilon/3)\mathbf{v}_3];$$

for the Cl atom

$$s_{DCl}=\mathbf{v},$$
$$s_{\beta_1Cl}=-\tfrac{3}{4}\sqrt{2}\tau(\mathbf{v}/3+\mathbf{v}_1),$$
$$s_{\beta_2Cl}=-\tfrac{3}{4}\sqrt{2}\tau(\mathbf{v}/3+\mathbf{v}_2),$$
$$s_{\beta_3Cl}=-\tfrac{3}{4}\sqrt{2}\tau(\mathbf{v}/3+\mathbf{v}_3),$$
$$s_{d_1Cl}=s_{d_2Cl}=s_{d_3Cl}=s_{\alpha_{12}Cl}=s_{\alpha_{23}Cl}=s_{\alpha_{31}Cl}=0.$$

From Eqs. (29) to (38), the U's are found to be: for the A_1 symmetry coordinates,

$U_{1D}=1$, all other $U_{1k}=0$,

$U_{2d_1}=U_{2d_2}=U_{2d_3}=1/\sqrt{3}$, all other $U_{2k}=0$,

$U_{3\beta_1}=U_{3\beta_2}=U_{3\beta_3}=-1/\sqrt{6}$,

$U_{3\alpha_{12}}=U_{3\alpha_{23}}=U_{3\alpha_{31}}=1/\sqrt{6}$, and all other $U_{3k}=0$;

for the E symmetry coordinates,

$U_{1ad_1}=2/\sqrt{6}$, $U_{1ad_2}=U_{1ad_3}=-1/\sqrt{6}$, all other $U_{1ak}=0$,

$U_{2a\beta_1}=2/\sqrt{6}$, $U_{2a\beta_2}=U_{2a\beta_3}=-1/\sqrt{6}$, all other $U_{2ak}=0$,

$U_{3a\alpha_{23}}=2/\sqrt{6}$, $U_{3a\alpha_{12}}=U_{3a\alpha_{31}}=-1/\sqrt{6}$, all other $U_{3ak}=0$,

$U_{1bd_2}=1/\sqrt{2}$, $U_{1bd_3}=-1/\sqrt{2}$, all other $U_{1bk}=0$,

$U_{2b\beta_2}=1/\sqrt{2}$, $U_{2b\beta_3}=-1/\sqrt{2}$, all other $U_{2bk}=0$,

$U_{3b\alpha_{31}}=1/\sqrt{2}$, $U_{3b\alpha_{12}}=-1/\sqrt{2}$, and all other $U_{3bk}=0$.

Now using Eq. (21), one finds the **S** vectors to be: for the A_1 vibrations,

$$\mathbf{S}_1^{H_1}=0,$$
$$\mathbf{S}_2^{H_1}=(1/\sqrt{3})\mathbf{v}_1,$$
$$\mathbf{S}_3^{H_1}=(-\epsilon\sqrt{3}/12)(\mathbf{v}_1+3\mathbf{v}_2+3\mathbf{v}_3-3\mathbf{v}),$$
$$\mathbf{S}_1^{C}=-\mathbf{v},$$
$$\mathbf{S}_2^{C}=(-1/\sqrt{3})(\mathbf{v}_1+\mathbf{v}_2+\mathbf{v}_3),$$
$$\mathbf{S}_3^{C}=(7\epsilon\sqrt{3}/12-\sqrt{3}\tau/4)(\mathbf{v}_1+\mathbf{v}_2+\mathbf{v}_3)-(\sqrt{3}/4)(3\epsilon+\tau)\mathbf{v},$$
$$\mathbf{S}_1^{Cl}=\mathbf{v},$$
$$\mathbf{S}_2^{Cl}=0,$$
$$\mathbf{S}_3^{Cl}=(\tau\sqrt{3}/4)(\mathbf{v}+\mathbf{v}_1+\mathbf{v}_2+\mathbf{v}_3);$$

for the E vibrations,

$$\mathbf{S}_{1a}^{H_1}=(2/\sqrt{6})\mathbf{v}_1,$$
$$\mathbf{S}_{2a}^{H_1}=(-\epsilon\sqrt{3}/2)(\mathbf{v}_1/3+\mathbf{v}),$$
$$\mathbf{S}_{3a}^{H_1}=(\epsilon\sqrt{3}/4)(2\mathbf{v}_1/3+\mathbf{v}_2+\mathbf{v}_3),$$
$$\mathbf{S}_{1a}^{C}=(1/\sqrt{6})(-2\mathbf{v}_1+\mathbf{v}_2+\mathbf{v}_3),$$
$$\mathbf{S}_{2a}^{C}=(\sqrt{3}/4)(\tau+\epsilon/3)(2\mathbf{v}_1-\mathbf{v}_2-\mathbf{v}_3),$$
$$\mathbf{S}_{3a}^{C}=(\epsilon\sqrt{3}/3)(-2\mathbf{v}_1+\mathbf{v}_2+\mathbf{v}_3),$$
$$\mathbf{S}_{1a}^{Cl}=0,$$
$$\mathbf{S}_{2a}^{Cl}=(\tau\sqrt{3}/4)(-2\mathbf{v}_1+\mathbf{v}_2+\mathbf{v}_3),$$
$$\mathbf{S}_{3a}^{Cl}=0,$$
$$\mathbf{S}_{1b}^{H_1}=0,$$
$$\mathbf{S}_{2b}^{H_1}=0,$$
$$\mathbf{S}_{3b}^{H_1}=\tfrac{3}{4}\epsilon(\mathbf{v}_2-\mathbf{v}_3),$$
$$\mathbf{S}_{1b}^{C}=(1/\sqrt{2})(\mathbf{v}_3-\mathbf{v}_2),$$
$$\mathbf{S}_{2b}^{C}=\tfrac{3}{4}(\tau+\epsilon/3)(\mathbf{v}_2-\mathbf{v}_3),$$
$$\mathbf{S}_{3b}^{C}=\epsilon(\mathbf{v}_3-\mathbf{v}_2),$$
$$\mathbf{S}_{1b}^{Cl}=0,$$
$$\mathbf{S}_{2b}^{Cl}=\tfrac{3}{4}\tau(\mathbf{v}_3-\mathbf{v}_2),$$
$$\mathbf{S}_{3b}^{Cl}=0.$$

The **G** matrix for the type A_1 vibrations is obtained by use of Eq. (20), just as for the H_2O molecule; it is

$$\begin{bmatrix} G_{11} & G_{12} & G_{13} \\ G_{21} & G_{22} & G_{23} \\ G_{31} & G_{32} & G_{33} \end{bmatrix} = \begin{bmatrix} \mu_{\text{Cl}}+\mu_{\text{C}} & (-\sqrt{3}/3)\mu_{\text{C}} & (4\epsilon\sqrt{3}/3)\mu_{\text{C}} \\ (-\sqrt{3}/3)\mu_{\text{C}} & \mu_{\text{H}}+\mu_{\text{C}}/3 & (-4\epsilon/3)\mu_{\text{C}} \\ (4\epsilon\sqrt{3}/3)\mu_{\text{C}} & (-4\epsilon/3)\mu_{\text{C}} & 2\epsilon^2\mu_{\text{H}}+(16/3)\epsilon^2\mu_{\text{C}} \end{bmatrix}.$$

The **G** matrix elements for the type E vibrations are obtained by use of Eq. (40). For example, the G_{23} element would be

$$G_{23}=\tfrac{1}{2}[\mu_{\text{H}}(3)(\mathbf{S}_{2a}^{\text{H}_1}\cdot\mathbf{S}_{3a}^{\text{H}_1}+\mathbf{S}_{2b}^{\text{H}_1}\cdot\mathbf{S}_{3b}^{\text{H}_1})+\mu_{\text{C}}(1)(\mathbf{S}_{2a}^{\text{C}}\cdot\mathbf{S}_{3a}^{\text{C}}+\mathbf{S}_{2b}^{\text{C}}\cdot\mathbf{S}_{3b}^{\text{C}})$$
$$+\mu_{\text{Cl}}(1)(\mathbf{S}_{2a}^{\text{Cl}}\cdot\mathbf{S}_{3a}^{\text{Cl}}+\mathbf{S}_{2b}^{\text{Cl}}\cdot\mathbf{S}_{3b}^{\text{Cl}})]. \quad (41)$$

The dot products are

$$\mathbf{S}_{2a}^{\text{H}_1}\cdot\mathbf{S}_{3a}^{\text{H}_1}=\epsilon^2/3,\quad \mathbf{S}_{2a}^{\text{C}}\cdot\mathbf{S}_{3a}^{\text{C}}=-2\epsilon(\tau+\epsilon/3),\quad \mathbf{S}_{2a}^{\text{Cl}}\cdot\mathbf{S}_{3a}^{\text{Cl}}=0,$$
$$\mathbf{S}_{2b}^{\text{H}_1}\cdot\mathbf{S}_{3b}^{\text{H}_1}=0,\quad \mathbf{S}_{2b}^{\text{C}}\cdot\mathbf{S}_{3b}^{\text{C}}=-2\epsilon(\tau+\epsilon/3),\quad \mathbf{S}_{2b}^{\text{Cl}}\cdot\mathbf{S}_{3b}^{\text{Cl}}=0.$$

Hence, $G_{23}=\frac{1}{2}\epsilon^2\mu_{\text{H}}-2\epsilon(\tau+\epsilon/3)\mu_{\text{C}}$. The complete matrix for the type E vibrations is

$$\begin{bmatrix} G_{11} & G_{12} & G_{13} \\ G_{21} & G_{22} & G_{23} \\ G_{31} & G_{32} & G_{33} \end{bmatrix} = \begin{bmatrix} \mu_{\text{H}}+(4/3)\mu_{\text{C}} & -\sqrt{2}(\tau+\epsilon/3)\mu_{\text{C}} & (4\epsilon\sqrt{2}/3)\mu_{\text{C}} \\ -\sqrt{2}\mu_{\text{C}}(\tau+\epsilon/3) & (3\tau^2/2)\mu_{\text{Cl}}+\epsilon^2\mu_{\text{H}}+(3/2)(\tau+\epsilon/3)^2\mu_{\text{C}} & (\epsilon^2/2)\mu_{\text{H}}-2\epsilon(\tau+\epsilon/3)\mu_{\text{C}} \\ (4\epsilon\sqrt{2}/3)\mu_{\text{C}} & (\epsilon^2/2)\mu_{\text{H}}-2\epsilon(\tau+\epsilon/3)\mu_{\text{C}} & (5\epsilon^2/2)\mu_{\text{H}}+(8\epsilon^2/3)\mu_{\text{C}} \end{bmatrix}.$$

Calculation of the Frequencies

The secular equation in expanded form[2] for the type A_1 frequencies obtained from the **F** and **G** matrices for the A_1 vibrations is

$$(-\lambda)^3+(F_{11}G_{11}+F_{22}G_{22}+F_{33}G_{33}+2F_{12}G_{12}+2F_{13}G_{13}+2F_{23}G_{23})(-\lambda)^2$$
$$+\left\{\begin{vmatrix}F_{11}&F_{12}\\F_{21}&F_{22}\end{vmatrix}\cdot\begin{vmatrix}G_{11}&G_{12}\\G_{21}&G_{22}\end{vmatrix}+\begin{vmatrix}F_{12}&F_{13}\\F_{22}&F_{23}\end{vmatrix}\cdot\begin{vmatrix}G_{12}&G_{13}\\G_{22}&G_{23}\end{vmatrix}+\begin{vmatrix}F_{11}&F_{13}\\F_{21}&F_{23}\end{vmatrix}\cdot\begin{vmatrix}G_{11}&G_{13}\\G_{21}&G_{23}\end{vmatrix}\right.$$
$$+\begin{vmatrix}F_{11}&F_{12}\\F_{31}&F_{32}\end{vmatrix}\cdot\begin{vmatrix}G_{11}&G_{12}\\G_{31}&G_{32}\end{vmatrix}+\begin{vmatrix}F_{12}&F_{13}\\F_{32}&F_{33}\end{vmatrix}\cdot\begin{vmatrix}G_{12}&G_{13}\\G_{32}&G_{33}\end{vmatrix}+\begin{vmatrix}F_{11}&F_{13}\\F_{31}&F_{33}\end{vmatrix}\cdot\begin{vmatrix}G_{11}&G_{13}\\G_{31}&G_{33}\end{vmatrix}$$
$$\left.+\begin{vmatrix}F_{21}&F_{22}\\F_{31}&F_{32}\end{vmatrix}\cdot\begin{vmatrix}G_{21}&G_{22}\\G_{31}&G_{32}\end{vmatrix}+\begin{vmatrix}F_{22}&F_{23}\\F_{32}&F_{33}\end{vmatrix}\cdot\begin{vmatrix}G_{22}&G_{23}\\G_{32}&G_{33}\end{vmatrix}+\begin{vmatrix}F_{21}&F_{23}\\F_{31}&F_{33}\end{vmatrix}\cdot\begin{vmatrix}G_{21}&G_{23}\\G_{31}&G_{33}\end{vmatrix}\right\}(-\lambda)$$
$$+\begin{vmatrix}F_{11}&F_{12}&F_{13}\\F_{21}&F_{22}&F_{23}\\F_{31}&F_{32}&F_{33}\end{vmatrix}\cdot\begin{vmatrix}G_{11}&G_{12}&G_{13}\\G_{21}&G_{22}&G_{23}\\G_{31}&G_{32}&G_{33}\end{vmatrix}=0. \quad (42)$$

It will be observed that the coefficient of $(-\lambda)^2$ is the sum of the products of all the corresponding one-rowed minors of the determinants of the **F** and **G** matrices; the coefficient of $(-\lambda)$ is the sum of the products of all the corresponding two-rowed minors; and the constant term is, as always, the product of the determinants of the **F** and **G** matrices.

The numerical values of the matrix elements can now be calculated, using the following values of the force constants,[9] bond distances[9] and reciprocals of masses:

$f_D=3.64\times10^5$ dyne/cm,

$f_d=4.790\times10^5$ dyne/cm,

$f_\alpha=0.46\times10^5$ dyne/cm,

$f_\beta=0.58\times10^5$ dyne/cm,

$f_{\alpha\beta}=0.01\times10^5$ dyne/cm,

$f_{D\alpha}-f_{D\beta}=-0.49\times10^5$ dyne/cm,

$d=1.093\text{A}=1/\epsilon$,

$D=1.66\text{A}=1/\tau$,

$\mu_{\text{Cl}}=1.698\times10^{22}$ gm^{-1},

TABLE VI. Comparison of the calculated and observed values of the vibrational frequencies of the CH_3Cl molecule.

Type	Calculated (cm^{-1}) Present values	Crawford and Brinkley (see reference 9)	Observed (see reference 9) (cm^{-1})
A_1	758	732*	732
	1371	1355*	1355
	2883	2902	2920
E	1012	1019	1020
	1468	1478	1460
	3001	3021	3047

* Used in obtaining the force constants.

$$\mu_C = 5.018\times 10^{22}\ \text{gm}^{-1},$$
$$\mu_H = 5.974\times 10^{23}\ \text{gm}^{-1}.$$

Upon substitution of the numerical values, Eq. (42) becomes

$$\lambda^3-(3.819\times 10^{29}\ \text{sec}^{-2})\lambda^2+(2.703\times 10^{58}\ \text{sec}^{-4})\lambda -4.006\times 10^{86}\ \text{sec}^{-6}=0. \quad (43)$$

Solution of this equation gives

$$\lambda_1 = 2.948\times 10^{29}\ \text{sec}^{-2},$$
$$\lambda^2 = 0.2037\times 10^{29}\ \text{sec}^{-2},$$
$$\lambda_3 = 0.6668\times 10^{29}\ \text{sec}^{-2},$$

and the corresponding values of ν from Eq. (25) are

$$\nu_1 = 2883\ \text{cm}^{-1},\quad \nu_2 = 758\ \text{cm}^{-1},\quad \nu_3 = 1371\ \text{cm}^{-1}.$$

The secular equation in expanded form for the type E frequencies is identical with Eq. (42), except that now the **F** and **G** matrix elements obtained for the E vibrations must be used. Evaluation and substitution of these, using the previously listed constants, gives

$$\lambda^3-(4.322\times 10^{29}\ \text{sec}^{-2})\lambda^2+(3.881\times 10^{58}\ \text{sec}^{-4})\lambda \dot{-}8.875\times 10^{86}\ \text{sec}^{-6}=0. \quad (44)$$

Solution of this equation gives

$$\lambda_4 = 3.195\times 10^{29}\ \text{sec}^{-2},$$
$$\lambda_5 = 0.3634\times 10^{29}\ \text{sec}^{-2},$$
$$\lambda_6 = 0.7643\times 10^{29}\ \text{sec}^{-2};$$

the corresponding values of ν are

$$\nu_4 = 3001\ \text{cm}^{-1},\quad \nu_5 = 1012\ \text{cm}^{-1},\quad \nu_6 = 1468\ \text{cm}^{-1}.$$

The present calculated values of the A_1 and E frequencies are compared with the values calculated by Crawford and Brinkley[9] and with the observed values in Table VI.

It should be noted that Eq. (42) and the **F** and **G** matrices are valid for any XY_3Z molecule whose symmetry is $\mathbf{C}_{3v}$, such as CD_3Cl, CH_3Br, CCl_3H, if the angles are tetrahedral and if appropriate values of the constants are used.

THE CH_4 AND CD_4 MOLECULES

Symmetry Coordinates

The CH_4 molecule has the symmetry $\mathbf{T}_d$ and therefore has one nondegenerate type A_1 vibration, one doubly degenerate type E vibration, and two triply degenerate type T_2 vibrations.[5] Although there are the same number of vibrational frequencies for this molecule as for CH_3Cl, the greater symmetry allows triply degenerate vibrations to occur. Again there are ten internal coordinates, but now all the inter-bond angles as well as all the bond distances are equivalent; the internal coordinates are (see Fig. 5) Δd_1, Δd_2, Δd_3, Δd_4, $\Delta\alpha_{12}$, $\Delta\alpha_{23}$, $\Delta\alpha_{31}$, $\Delta\alpha_{14}$, $\Delta\alpha_{24}$ and $\Delta\alpha_{34}$.

As before, ten symmetry coordinates, one of which is redundant, must be constructed from these internal coordinates. The symmetry coordinates used were: for the type A_1 vibrations,

$$R_1 = \tfrac{1}{2}(\Delta d_1+\Delta d_2+\Delta d_3+\Delta d_4), \quad (45)$$

$$R_2 = (1/\sqrt{6})(\Delta\alpha_{12}+\Delta\alpha_{23}+\Delta\alpha_{31}+\Delta\alpha_{14} +\Delta\alpha_{24}+\Delta\alpha_{34}) \equiv 0 \text{ (redundant coordinate)}; \quad (46)$$

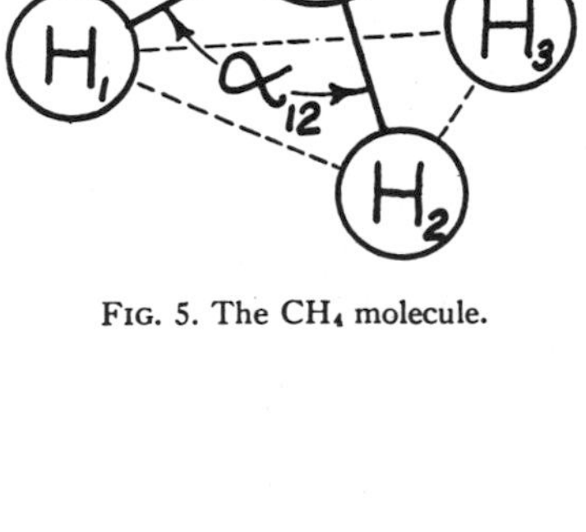

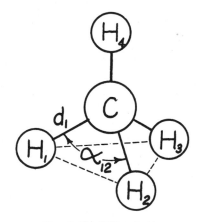

FIG. 5. The CH_4 molecule.

for the type E vibrations,

$$R_{1a}=(1/\sqrt{12})(2\Delta\alpha_{12}-\Delta\alpha_{23}-\Delta\alpha_{31}-\Delta\alpha_{14}-\Delta\alpha_{24}+2\Delta\alpha_{34}), \quad (47)$$

$$R_{1b}=\tfrac{1}{2}(\Delta\alpha_{14}-\Delta\alpha_{31}+\Delta\alpha_{23}-\Delta\alpha_{24}); \quad (48)$$

and for the type T_2 vibrations,

$$R_{1a}=(1/\sqrt{12})(2\Delta\alpha_{12}-\Delta\alpha_{23}-\Delta\alpha_{31}+\Delta\alpha_{14}+\Delta\alpha_{24}-2\Delta\alpha_{34}), \quad (49)$$

$$R_{1b}=(1/\sqrt{6})(\Delta\alpha_{12}+\Delta\alpha_{23}+\Delta\alpha_{31}-\Delta\alpha_{14}-\Delta\alpha_{24}-\Delta\alpha_{34}), \quad (50)$$

$$R_{1c}=\tfrac{1}{2}(\Delta\alpha_{23}-\Delta\alpha_{31}-\Delta\alpha_{14}+\Delta\alpha_{24}), \quad (51)$$

$$R_{2a}=(1/\sqrt{6})(\Delta d_1+\Delta d_2-2\Delta d_3), \quad (52)$$

$$R_{2b}=(1/\sqrt{12})(\Delta d_1+\Delta d_2+\Delta d_3-3\Delta d_4), \quad (53)$$

$$R_{2c}=(1/\sqrt{2})(\Delta d_2-\Delta d_1). \quad (54)$$

The symmetry coordinates for the A_1 and E vibrations can be shown to transform properly by the same method that was used for the CH_3Cl molecule. A similar method is used for the type T_2 vibrations, except that when a covering operation is applied to one of a set of three T_2 symmetry coordinates (say R_{1a}, R_{1b}, R_{1c}), the result is a linear combination of the three symmetry coordinates forming the set. For example, for the covering operation σ_d, one would have

$$(\sigma_d)R_{1a}=AR_{1a}+BR_{1b}+CR_{1c},$$
$$(\sigma_d)R_{1b}=A'R_{1a}+B'R_{1b}+C'R_{1c},$$
$$(\sigma_d)R_{1c}=A''R_{1a}+B''R_{1b}+C''R_{1c}.$$

The transformation matrix is therefore of the form

$$\begin{bmatrix} A & B & C \\ A' & B' & C' \\ A'' & B'' & C'' \end{bmatrix}.$$

Hence the character for the covering operation σ_d would be $A+B'+C''$, and this must be the same as the character for σ_d given in the table[5] for the point group $\mathbf{T}_d$. The determination of the transformation properties of the symmetry coordinates for the other covering operations of the group would be carried out in the same way.

Wilson[2] has given some rules that may be used in choosing the coefficients for the symmetry coordinates in complicated cases.

The F Matrices

For the present purpose it is desired to calculate force constants from the observed frequencies for the CH_4 molecule and subsequently to use these force constants to calculate the frequencies of the CD_4 molecule. Since there are only four distinct frequencies for CH_4, only four force constants can be used in expressing the potential energy. The expression used was

$$\begin{aligned}2V=&f_d[(\Delta d_1)^2+(\Delta d_2)^2+(\Delta d_3)^2+(\Delta d_4)^2]\\&+f_\alpha[(d\Delta\alpha_{12})^2+(d\Delta\alpha_{23})^2+(d\Delta\alpha_{31})^2+(d\Delta\alpha_{14})^2+(d\Delta\alpha_{24})^2+(d\Delta\alpha_{34})^2]\\&+2f_{\alpha\alpha}[(d\Delta\alpha_{12})(d\Delta\alpha_{31}+d\Delta\alpha_{14}+d\Delta\alpha_{23}+d\Delta\alpha_{24})+(d\Delta\alpha_{23})(d\Delta\alpha_{24}+d\Delta\alpha_{34}+d\Delta\alpha_{31})\\&+(d\Delta\alpha_{31})(d\Delta\alpha_{14}+d\Delta\alpha_{34})+(d\Delta\alpha_{14})(d\Delta_{24}+d\Delta\alpha_{34})+(d\Delta\alpha_{24})(d\Delta\alpha_{34})]\\&+2f_{d\alpha}[\Delta d_1(d\Delta\alpha_{12}+d\Delta\alpha_{31}+d\Delta\alpha_{14})+\Delta d_2(d\Delta\alpha_{12}+d\Delta\alpha_{23}+d\Delta\alpha_{24})\\&+\Delta d_3(d\Delta\alpha_{31}+d\Delta\alpha_{23}+d\Delta\alpha_{34})+\Delta d_4(d\Delta\alpha_{14}+d\Delta\alpha_{24}+d\Delta\alpha_{34})], \quad (55)\end{aligned}$$

where d is the equilibrium C−H bond distance.

By the same procedure used for CH_3Cl, the following **F** matrices were obtained: for type A_1,

$$(F_{11})=(f_d);$$

for type E,

$$(F_{11})=(d^2[f_\alpha-2f_{\alpha\alpha}]);$$

and for type T_2,

$$\begin{pmatrix} F_{11} & F_{12} \\ F_{21} & F_{22} \end{pmatrix}=\begin{pmatrix} d^2f_\alpha & \sqrt{2}df_{d\alpha} \\ \sqrt{2}df_{d\alpha} & f_d \end{pmatrix}.$$

In obtaining the **F** matrix for the E vibrations, one may use either R_{1a} or R_{1b}; and in obtaining the **F** matrix for the T_2 vibrations, one may use any one of the pairs: R_{1a}, R_{2a}; R_{1b}, R_{2b}; R_{1c}, R_{2c}.

The G Matrices

The elements for the **G** matrix for the nondegenerate type A_1 vibration can be obtained from Eq. (20), and the **G** matrix elements for the type E vibration can be obtained from Eq. (40). The corresponding equation for the ele-

TABLE VII. Fundamental frequencies for CH_4.

Type	Frequency (cm^{-1})	Designation
A_1	3029.8	ν_1
E	1390.2	ν_2
T_2	1357.6	ν_3
	3156.9	ν_4

ments of the **G** matrix for the type T_2 vibrations is[2]

$$G_{jl}=(1/3)\sum_p \mu_p g_p(\mathbf{S}_{ja}^{(t)}\cdot\mathbf{S}_{la}^{(t)}+\mathbf{S}_{jb}^{(t)}\cdot\mathbf{S}_{lb}^{(t)}+\mathbf{S}_{jc}^{(t)}\cdot\mathbf{S}_{lc}^{(t)}). \quad (56)$$

By proceeding in the same manner as for CH_3Cl, the following **G** matrices were obtained: for type A_1, $(G_{11})=(\mu_H)$; for type E, $(G_{11})=(3\mu_H/d^2)$; and for type T_2,

$$\begin{pmatrix} G_{11} & G_{12} \\ G_{21} & G_{22} \end{pmatrix}=\begin{pmatrix} [2\mu_H+(16/3)\mu_C]/d^2 & -8\mu_C/(3d) \\ -8\mu_C/(3d) & \mu_H+4\mu_C/3 \end{pmatrix}.$$

Force Constants for CH_4

The fundamental frequencies[8] for CH_4 are given in Table VII. For the type A_1 vibration,

$$\lambda_1=F_{11}G_{11}=(f_d)(\mu_H)=4\pi^2c^2\nu_1^2, \quad (57)$$

whence

$$f_d=(4\pi^2\nu_1^2)/\mu_H=5.4514\times10^5 \text{ dyne/cm}.$$

For the type E vibration,

$$\lambda_2=F_{11}G_{11}=[d^2(f_\alpha-2f_{\alpha\alpha})][3\mu_H/d^2]=4\pi^2c^2\nu_2^2, \quad (58)$$

whence

$$f_\alpha-2f_{\alpha\alpha}=0.38255\times10^5 \text{ dyne/cm}. \quad (59)$$

For the type T_2 vibrations, the secular equation in expanded form is

$$\lambda^2-(F_{11}G_{11}+2F_{12}G_{12}+F_{22}G_{22})\lambda+\begin{vmatrix} F_{11} & F_{12} \\ F_{21} & F_{22} \end{vmatrix}\cdot\begin{vmatrix} G_{11} & G_{12} \\ G_{21} & G_{22} \end{vmatrix}=0. \quad (60)$$

Using

$$\mu_H=5.9742\times10^{23} \text{ g}^{-1}, \quad \mu_C=5.0183\times10^{22} \text{ g}^{-1}$$

and $d=1.093\times10^{-8}$ cm, G_{11}, G_{12}, and G_{22} were found to be 1.2241×10^{40} gm^{-1} cm^{-2}, -1.2243×10^{31} gm^{-1} cm^{-1} and 6.6433×10^{23} gm^{-1}, respectively, and F_{22} has already been obtained from Eq. (57). Substituting these values into Eq. (60), one obtains

$$\lambda^2-[(1.2241\times10^{40})F_{11}-(2.4486\times10^{31})F_{12}+3.6215\times10^{29}]\lambda+(4.3514\times10^{69})F_{11}-(7.9821\times10^{63})F_{12}^2=0. \quad (61)$$

From Eq. (25) and the frequencies given in Table VII,

$$\lambda_3=4\pi^2c^2(1357.6)^2=6.5389\times10^{28} \text{ sec}^{-2} \quad (62)$$

and

$$\lambda_4=4\pi^2c^2(3156.9)^2=3.5357\times10^{29} \text{ sec}^{-2}. \quad (63)$$

But, since λ_3 and λ_4 are roots of Eq. (61), one can write

$$(\lambda-\lambda_3)(\lambda-\lambda_4)=0. \quad (64)$$

Expanding and substituting the values for λ_3 and λ_4 from Eqs. (62) and (63), one gets

$$\lambda^2-(4.1896\times10^{29})\lambda+2.3120\times10^{58}=0. \quad (65)$$

Equating coefficients in Eqs. (61) and (65), one obtains the two relations

$$(4.3514\times10^{69})F_{11}-(7.9821\times10^{63})F_{12}^2=2.3120\times10^{58}, \quad (66)$$

$$(1.2241\times10^{40})F_{11}-(2.4486\times10^{31})F_{12}=5.6810\times10^{28}. \quad (67)$$

Eliminating F_{11} from Eqs. (66) and (67), one gets

$$F_{12}^2-(1.0905\times10^{-3})F_{12}+3.6652\times10^{-7}=0. \quad (68)$$

Unfortunately, the values of F_{12} obtained from Eq. (68) are not real. This may be due to the fact that it was not possible to use the most general potential energy expression.

In order to obtain the values of the remaining force constants, it is therefore necessary to transfer f_α from the CH_3Cl molecule. Such a transfer of force constants between molecules of similar structure has been shown by Crawford and Brinkley[9] to be possible for a number of molecules. Substituting this value, $f_\alpha=0.46\times10^5$ dyne/cm, into Eq. (67), one gets

$$F_{12}=\sqrt{2}df_{d\alpha}=4.2702\times10^{-4} \text{ dyne}$$

or $f_{d\alpha}=0.2763\times10^5$ dynes/cm. These values of f_α and $f_{d\alpha}$ may be tested by substitution in Eq. (66).

Finally, the force constant $f_{\alpha\alpha}$ can be obtained from Eq. (59).

Calculation of the CD_4 Frequencies

Substituting μ_D for μ_H in the **G** matrices for the CH_4 molecule, and using the new matrix elements in Eqs. (57), (58) and (60), the frequencies for CD_4 were calculated. In Table VIII, the values obtained are compared with those calculated by Dennison[8] using a different method. Despite the approximate potential energy expression used, the agreement is reasonably satisfactory. A somewhat better agreement might be obtained by slight adjustments in the force constants.

SUMMARY OF THE METHOD

It may be helpful to summarize briefly the principal steps in applying the Wilson method to the calculation of the vibrational frequencies of a molecule. First, orthonormal—that is, both orthogonal and normalized—symmetry coordinates, which are linear combinations of equivalent internal coordinates and which transform according to the characters of the vibration type concerned are formed. From the **U** matrices obtained from these symmetry coordinates and from the **f** matrix obtained from the potential energy expressed in terms of the internal coordinates, one gets the **F** matrices for the different vibration types. In order to get the **G** matrices, one first calculates the $\mathbf{s}_{kt}$ vectors from the general expressions involving unit vectors directed along the chemical bonds. Next the $\mathbf{S}_j^{(t)}$ vectors are formed from the U_{jk} coefficients and the $\mathbf{s}_{kt}$ vectors. The **G** matrix elements are then obtained from the dot products of the $\mathbf{S}_j^{(t)}$ vectors. Finally, using these **F** and **G** matrices, the secular equations for the various vibration types are written in the expanded form, and the vibrational frequencies are then calculated from these equations.

TABLE VIII. Comparison of the present calculated frequencies of the CD_4 molecule with the values obtained by Dennison[8] using a different method.

Vibration		Frequency (cm^{-1})	
Type	Symbol	Present value	Dennison's value
A_1	ν_1	2151	2143.2
E	ν_2	987	983.4
T_2	ν_3	1019	1026.4
	ν_4	2335	2336.9

* * *

The authors wish to express their appreciation to Mr. Leonard Reiffel for preparing the drawings for Figs. 1 to 5.

Reprinted from *J. Amer. Chem. Soc.*, **85**, 1927–1928 (1963)

37

Vibrational Spectra and Structure of Xenon Tetrafluoride[1]

BY HOWARD H. CLAASSEN,[2] CEDRIC L. CHERNICK AND JOHN G. MALM

RECEIVED MARCH 18, 1963

The infrared spectrum of XeF_4 vapor has strong bands at 123, 291, and 586 cm.$^{-1}$. The Raman spectrum of the solid has very intense peaks at 502 and 543 cm.$^{-1}$ and weaker ones at 235 and 442 cm.$^{-1}$. These data show that the molecule is planar and of symmetry D_{4h}. The seven fundamental frequencies have been assigned as 543 (a_{1g}), 291 (a_{2u}), 235 (b_{1g}), 221 (b_{1u}), 502 (b_{2g}), 586 (e_u), and 123 (e_u). The (b_{1u}) frequency value is quite uncertain.

Introduction

The preparation of XeF_4 has been described previously[3] and the results of a preliminary study of its vibrational spectra reported briefly.[4] We report here the results of a more complete study of the Raman spectrum of the solid phase and the infrared spectrum of the vapor.

Experimental Procedures

Preparation.—The purity of the XeF_4, prepared as described elsewhere,[3] was checked by infrared analysis. The probable impurities are XeF_6 and XeF_2 which have absorption peaks at 612 and 566 cm.$^{-1}$, respectively. The sample was found to contain small amounts of the more volatile XeF_6 but this was easily removed since its vapor pressure is higher by a factor of 10. Pumping the equilibrium vapor rapidly out of the storage can several times removed the XeF_6 so that none of the 612 cm.$^{-1}$ absorption could be detected in the bulk of sample remaining.

Infrared Spectra.—The vapor pressure of XeF_4 (approximately 2 mm. at 20°) was sufficient to allow the observation of the fundamentals at or slightly above room temperature in a 10-cm. cell. For weaker bands a 60-cm. absorbing path was obtained by use of the mirror cell designed at this Laboratory and previously described.[5] The cells were made of nickel and were used with either AgCl or polyethylene windows. The spectra were obtained with a Beckman IR-7 with CsI prism and Perkin–Elmer 421 and 301 spectrophotometers. We are indebted to the Perkin–Elmer Corporation for the opportunity to use the 301 instrument at Norwalk, Conn., and to Charles Helms and Robert Anacreon for their help with the operation of that spectrophotometer.

The reproducibility of the spectrum and uniform composition of the sample were established by scanning several samples. A nickel can containing about 1 g. of XeF_4 was connected to the cell and to a similar can. The whole sample was transferred batchwise to the second can and vapor samples of each batch were taken into the cell. In most cases just the two most intense bands were examined to look for possible changes which would be indicative of impurities, but several complete spectra were also observed.

Raman Spectra.—The sample used was approximately 1 g. of XeF_4 that had grown to a single crystal in a sealed quartz tube. The spectrum was obtained using a Cary 81 photoelectric instrument with the lens system designed for solids.

Results and Interpretation

Figure 1 shows tracings of the regions of the infrared spectrum where bands were observed, and Fig. 2 is a tracing of the Raman spectrum. Judging from their positions and intensities the three infrared bands at 123, 291, and 586 cm.$^{-1}$ are probably fundamentals. Of the four bands observed in the Raman spectrum the one at 442 cm.$^{-1}$ is the least intense and may not represent a fundamental. In fact, the reality of the 442 cm.$^{-1}$ frequency is doubtful since the 543 cm.$^{-1}$ vibration excited by 4339 Å. and the 502 cm.$^{-1}$ one excited by 4337 Å. would occur at apparent shifts of 442 and 445 cm.$^{-1}$, respectively, from 4358 Å.

In considering the information the spectral data furnish on the molecular symmetry, it must be noted that the Raman measurements are for the solid compound and the infrared ones are for the vapor. Some solid–vapor shifts in frequencies are to be expected.

From the infrared spectrum alone one can conclude that there is high symmetry in the XeF_4 molecule. Only one band is observed in the region where bond stretching motions occur (500–700 cm.$^{-1}$). Of all the symmetries possible for a YZ_4-molecule, only for T_d (tetrahedral) and D_{4h} (square-planar) would there be just one infrared-active bond stretching fundamental. The infrared spectrum also allows the distinction to be made between these two symmetries since a T_d molecule would have one bending mode that would be infrared active while a D_{4h} molecule would have two. As two are observed for XeF_4 the D_{4h} model is the preferred one. Strong support for this is provided in the Raman spectrum, also.

The fundamental vibrations of a D_{4h}, YZ_4, molecule are described in Fig. 3 as to their symmetries, numbering, spectral activity and modes of atomic motions. The assignment of ν_2 is definite from the band contours expected according to Gerhard and Dennison.[6] Only for the out-of-plane motion, ν_2, should there be a very intense Q-branch and this is observed at 291 cm.$^{-1}$. The other two infrared fundamentals are then assigned without ambiguity.

The Raman spectrum of the solid fits very well and lends strong support for the planar model. The two very intense bands at 543 and 502 cm.$^{-1}$ must be due to the two stretching vibrations. Although polarization measurements could not be made it is quite certain that the symmetric vibration is the higher one because any significant repulsion between fluorines would almost require this. Further support for this interpretation of the Raman spectrum of the solid has recently been obtained in this Laboratory.[7] The 235 cm.$^{-1}$ Raman band is then assigned to ν_3 and this leaves 442 cm.$^{-1}$ to be assigned. If the band is real it cannot be a fundamental and it must be an overtone or combination band and the only plausible assignment is that it is $2\nu_4$. This gives the value of 221 cm.$^{-1}$ for ν_4 that is listed with a question mark since the assignment is not certain. The infrared absorption peaks at 1105 and 1136 cm.$^{-1}$ may be assigned as $\nu_5 + \nu_6 = 1088$ cm.$^{-1}$ and $\nu_1 + \nu_6 = 1129$ cm.$^{-1}$. The fit is satisfactory when account is taken of corrections needed due to vapor to solid shift of frequencies.

There is one feature of the infrared spectrum that we do not understand and that is the doublet appearance of ν_6. This has been traced many times and the peaks reproducibly found at 581 and 591 cm.$^{-1}$. Expected is a triplet band with all three peaks of about equal intensity and with a P–R separation of approximately 14 cm.$^{-1}$.[6] The observed splitting is much too large to ascribe to isotopes of xenon, and may be due to a Corio-

(1) Based on work performed under the auspices of the U. S. Atomic Energy Commission.

(2) Permanent address: Wheaton College, Wheaton, Ill.

(3) H. H. Claassen, H. Selig and J. G. Malm, *J. Am. Chem. Soc.*, **84**, 3593 (1962).

(4) C. L. Chernick, *et al.*, *Science*, **138**, 136 (1962).

(5) B. Weinstock, H. H. Claassen and C. L. Chernick, *J. Chem. Phys.*, **38**, 1470 (1963).

(6) S. L. Gerhard and D. M. Dennison, *Phys. Rev.*, **43**, 197 (1933).

(7) H. H. Hyman and L. A. Quarterman observed a Raman band at 553 cm.$^{-1}$ of XeF_4 in HF solution. This must correspond to the 543 cm.$^{-1}$ band for the solid. The ν_4 band was so broadened, however, that it was not definitely observed in the very dilute solution. That the higher frequency band remained sharp is good indication that it represents the totally symmetric vibration.

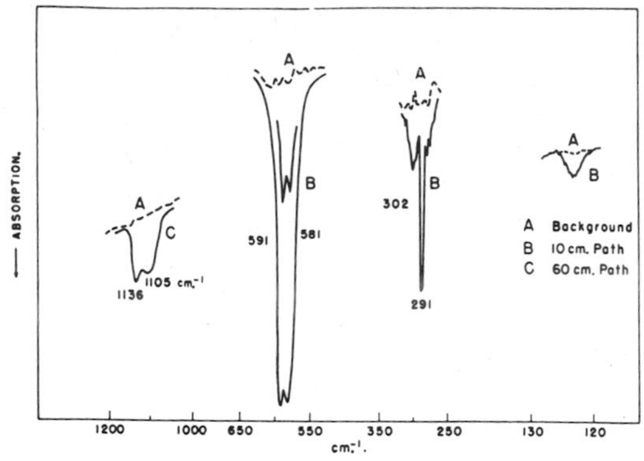

Fig. 1.—Infrared spectrum of XeF_4 vapor.

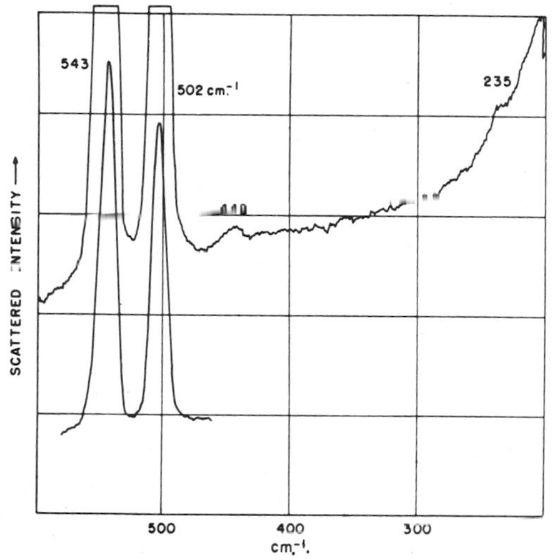

Fig. 2.—Raman spectrum of solid XeF_4.

lis coupling between the doubly degenerate vibration and rotation.

The structure of XeF_4 has been obtained by X-ray diffraction by Ibers and Hamilton.[8] They find that for the solid, also, the molecule is square planar within experimental error.

Several theoretical discussions[9-11] have stated that the square planar model best fits the theory and one of them[9] suggests that the molecule could possibly be distorted by coulomb repulsion. Therefore, it seems interesting to question whether the vibrational data require an exactly planar molecule or whether the "ring" of fluorines might be slightly puckered. If the latter were true the Raman-active ν_5 would be infrared active, but a slight distortion would, of course, result in a very weak infrared band. One can set a rough upper limit to the amount of possible puckering if one looks at the infrared spectrum in the region of 502 cm.$^{-1}$ and makes the plausible assumption that the rate of change of bond

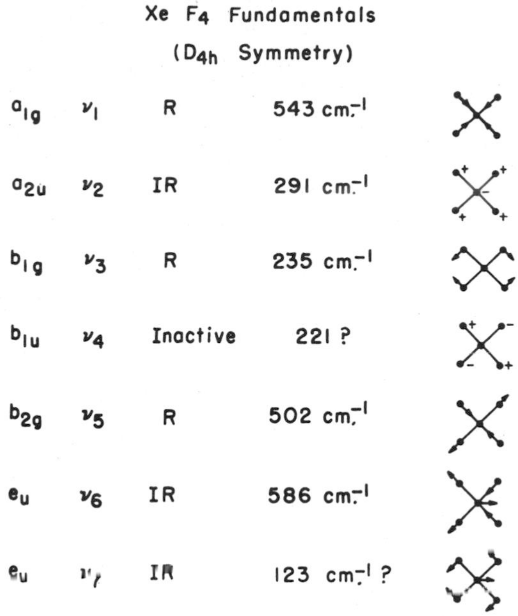

Fig. 3.—Numbering, spectral activity, and assignment of fundamentals.

moment with stretching is approximately the same for ν_5 and ν_6. The result is that an upper limit can be set for deviation of the Xe–F bond from the plane of about 0.5 degree, or fluorine distances of 0.02 Å. from the plane.

The Q–R separation of 11 ± 1 cm.$^{-1}$ in the 291 cm.$^{-1}$ band can be used to calculate a bond length. This gives 1.85 ± 0.2 Å. for the Xe–F bond, in good agreement with the value of 1.92 Å. for the solid obtained from X-ray diffraction.[8] Since a more precise value of the bond length for the vapor molecule will probably be available soon from electron diffraction studies, we have not calculated thermodynamic functions.

Preliminary force constant calculations using a valence plus interaction terms type of potential function similar to that used by Claassen for hexafluorides[12] gave a value of 3.00 mdynes/Å. for the bond stretching constant and 0.12 for the interaction constant between bonds at right angles. The interaction constant for opposite bonds cannot be determined accurately, but is approximately 0.06 mdyne/Å. These may be compared with values given by Smith[13] for XeF_2 and with those for PuF_6,[12] a molecule that also has fluorine bonds at right angles and a comparable bond length.

Molecule	XeF_4	XeF_2	PuF_6
Bond length, Å.	1.92[8]	2.00[14]	1.972
Stretching force constant, mdynes/Å.	3.00	2.85	3.59
Interaction constant for perpendicular bonds	0.12	..	0.22
Interaction constant for opposite bonds	~0.06	0.11	−0.08

(8) J. A. Ibers and W. C. Hamilton, *Science*, **139**, 106 (1963).
(9) R. E. Rundle, *J. Am. Chem. Soc.*, **85**, 112 (1963).
(10) L. L. Lohr and W. N. Lipscomb, *ibid.*, **85**, 240 (1963).
(11) L. C. Allen, *Science*, **138**, 892 (1962).
(12) H. H. Claassen, *J. Chem. Phys.*, **30**, 968 (1959).
(13) D. F. Smith, *ibid.*, **38**, 276 (1963).
(14) P. A. Agron, G. M. Begun, H. A. Levy, A. A. Mason, C. F. Jones and D. F. Smith, *Science*, **139**, 842 (1963).

Chemical Rearrangements and Reactivity

VII

Editor's Comments on Papers 38 Through 42

The employment of group theory to deal with time-dependent chemical behavior is of fairly recent origin. It is difficult to grasp an area such as this one from a historical perspective since the field is advancing so rapidly. Indeed, we are still learning the techniques in this area. Furthermore, the greatest impact so far has been on the theory of organic reaction mechanisms. However, as techniques such as nuclear magnetic resonance spectroscopy are applied more completely to time-dependent chemical reactions and rearrangements in inorganic systems, the concepts which place symmetry restrictions on paths that the rearrangements may take will be enumerated more clearly.

With the development of new, sophisticated instrumentation enabling the chemist to study rearrangements and reactions where a molecule of a given symmetry has a lifetime in the range 10^{-4} to 10^{-13} sec, some group theoretical techniques have been developed to deal with the continually changing symmetries. The H. C. Longuet-Higgins contributions (Papers 38 and 40) and those by R. B. Woodward and R. Hoffmann (Papers 39, 41, and 42) are certainly notable. Longuet-Higgins brilliantly establishes the symmetry requirements for chemical systems undergoing permutational rearrangements. In this article the symmetry properties of nonrigid molecules are fitted into the framework of group theory.

The final four short communications form the basis for applications of symmetry to chemical reactions. They clearly develop the idea that for concerted reactions to be feasible by thermal processes, orbital symmetries for products must be related to the orbital symmetries of reactants. If the ground-state symmetry of the combined reactants is associated with an excited state symmetry of the products, restrictions are placed upon the mechanism. Woodward and Hoffmann approached the problem using orbital representations of symmetry; Longuet-Higgins and Abrahamson emphasized the symmetry of the molecular wavefunctions themselves.

In spite of the dangers inherent in placing relatively recent papers such as those included here in a Benchmark series, I believe that the papers chosen are indeed classics. Other papers already published [31] may ultimately achieve the same status in this field, but they are yet not clearly identified.

Reprinted from *Mol. Phys.*, **6**, 445–460 (1963)

The symmetry groups of non-rigid molecules

by H. C. LONGUET-HIGGINS
Department of Theoretical Chemistry, University Chemical Laboratory, Cambridge

38

(*Received* 25 *November* 1962)

The concept of molecular symmetry is extended to molecules such as ethane and hydrazine which can pass from one conformation to another. The symmetry group of such a molecule is the set of (i) all feasible permutations of the positions and spins of identical nuclei and (ii) all feasible permutation-inversions, which simultaneously invert the coordinates of all particles in the centre of mass. According to the representations of this group one can classify not only the spin states and states of motions of the nuclei, but even the electronic states of the molecule. Examples are given to illustrate the use of this concept in determining the statistical weights of individual levels and selection rules for electric dipole transitions between them.

1

Ever since the early nineteen-thirties [1] physical chemists have been using group theory as a tool for classifying the states of polyatomic molecules. Particularly useful for this purpose are the point groups of rigid polyatomic molecules—rigid in the sense that they never depart very far from a unique symmetrical configuration. But apart from the pioneer work of Wilson and his colleagues [2–4] little attention has yet been paid to the symmetry properties of non-rigid molecules, which can change easily from one conformation to another. In the present paper I attempt to fill this gap.

Four examples will show the need to sharpen our ideas about molecular symmetry.

(i) The NH_3 molecule has symmetry C_{3v} in its most stable configuration. But if one is investigating its inversion spectrum one must use D_{3h} for classifying the quantum states.

(ii) C_2H_6 has symmetry D_{3d} in its staggered configuration, but the eclipsed configuration has symmetry D_{3h}. But by what symmetry group should one classify the levels of ethane when torsional tunnelling is important? The dilemma is even more acute in dimethylacetylene, where internal rotation is virtually unhindered.

(iii) Hydrazine, N_2H_4, has a rather unsymmetrical equilibrium configuration, with merely a twofold axis of symmetry. But the molecule may be converted into its mirror image by twisting about the N–N bond, and may also be inverted, like NH_3, at either nitrogen atom. What then is its symmetry group?

(iv) Boron trifluoride, BF_3 has symmetry D_{3h}. But what symmetry group should be used for classifying the stationary states of boron trimethyl, $B(CH_3)_3$?

The short answers to these questions are that dimethylacetylene, hydrazine and boron trimethyl have symmetry groups of order 36, 16 and 324 respectively.

Such a bald statement is, however, useless until one has defined precisely the symmetry group of a non-rigid molecule and shown how to use its character table in solving physical problems. This is the purpose of this essay. But one must begin by studying closely the familiar point groups of rigid molecules, and the following remarks owe much to a recent paper by J. T. Hougen [5], to whom I am indebted for some stimulating conversations.

A molecule is an aggregate of atomic nuclei and electrons. Its Hamiltonian is invariant under the following types of transformation:

(*a*) Any permutation of the positions and spins of the electrons.

(*b*) Any rotation of the positions and spins of all particles (electrons *and* nuclei) about any axis through the centre of mass.

(*c*) Any overall translation in space.

(*d*) The reversal of all particle momenta and spins.

(*e*) The simultaneous inversion of the positions of all particles in the centre of mass.

(*f*) Any permutation of the positions and spins of any set of identical nuclei.

The *complete group of the Hamiltonian* is thus the direct product of several groups. But for most molecules not all of the elements of the complete group need be taken into account. This is because the time scale of a given laboratory experiment may be too short to allow certain nuclear permutations ever to occur. One therefore restricts attention to 'feasible' transformations, those which can be achieved without passing over an insuperable energy barrier. The *molecular symmetry group* is then composed of feasible elements only, but it does not comprise *all* the feasible elements of the complete group of the Hamiltonian. Its clearest and most useful definition seems to be as follows:

Let P be any permutation of the positions and spins of identical *nuclei*, or any product of such permutations. Let E be the identity, E^* the inversion of all particle positions, and P^* the product $PE^* = E^*P$. Then the molecular symmetry group is the set of

(i) all feasible P, including E,

(ii) all feasible P^*, not necessarily including E^*.

We shall now show that this definition accords with our intuitive ideas of molecular symmetry, except for linear molecules, which would require special discussion.

First we must reconcile the present definition with the customary assertion that molecular symmetry groups comprise 'rotations', 'reflections', etc. A moment's thought will convince one that this assertion cannot be taken literally. The bodily rotation of a molecule can make no difference to its internal coordinates—the distances between the constituent particles—whereas vibrational symmetry coordinates and electronic wave functions are not *necessarily* invariant under the elements of the symmetry group. If they were, the symmetry group would be quite useless. No, it is not the *positions* of the particles which are 'rotated' but their *relative* coordinates, and this result is certainly achieved by permuting the positions of identical nuclei. Even the electronic wave function is susceptible to nuclear permutations, since it is a function of the coordinates of the electrons *relative to the nuclear framework*, not of their absolute positions in space. But an example is likely to be more convincing than any general remarks, so let us consider the particular case of H_2O.

Figure 1 shows how an arbitrary configuration of the water molecule is transformed by various kinds of symmetry element. The arrows at the H atoms, labelled 1 and 2, indicate an arbitrary distortion of the molecule, and the symbols ⊕ and ⊖ indicate the presence of an electron above and below the plane of the paper. The symmetry elements (12), E^* and $(12)^*$ convert the initial configuration E into the configurations shown at the left in each box. Subsequent bodily rotations of the whole molecule lead to configurations which can also be reached by rotating or reflecting the nuclear and electronic *displacements*. One could, of course, define the molecular symmetry group as the set of composite elements labelled E, C_2, σ_{xy} and σ_{yz}, but this convention has two disadvantages. First, one must exercise great care in fixing the axes, x y, z in the molecule, so that their orientation does not become indeterminate when the molecule is seriously distorted. This is only a mild nuisance, and Hougen has shown how it may be dealt with [5]. But much more seriously, such a definition of the molecular symmetry elements cannot be extended to a non-rigid molecule.

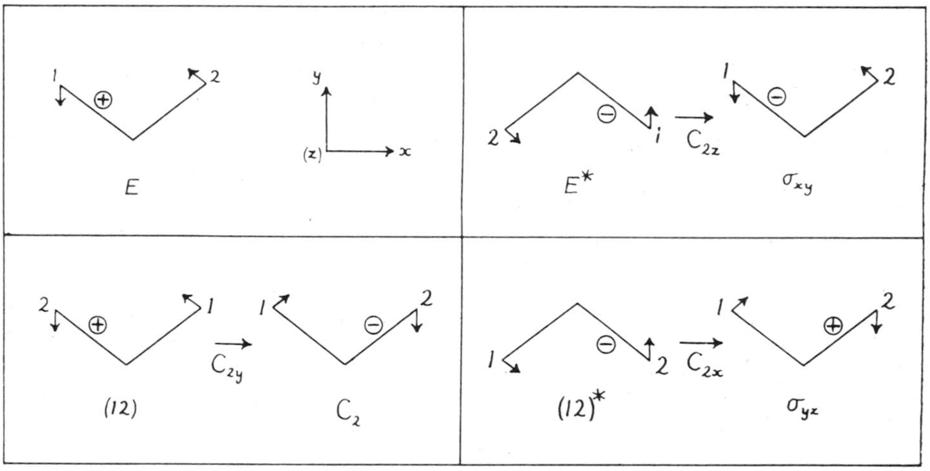

Figure 1.

This is because there are certain symmetry elements of non-rigid molecules—for example the twisting of ethane—which lead to configurations which *cannot* be brought into near-coincidence with the original one by a simple rotation of the whole molecule in space. It is for this important reason that I choose to define the symmetry group of H_2O as the set of elements E, (12), E^* and $(12)^*$ rather than the set E, $C_{2y}(12)$, $C_{2z}E^*$ and $C_{2x}(12)^*$. What matters, anyhow, is the effect of a symmetry element upon the *internal* coordinates, and figure 1 shows that in this respect there is nothing to choose between the two definitions as far as rigid molecules are concerned. But we shall see in the next section that the definition here proposed can be immediately extended to non-rigid molecules, whereas the more familiar interpretation breaks down completely when the molecular group has more elements of symmetry than any possible 'structure' for the molecule.

One final remark, before leaving rigid molecules. It may not always be obvious what permutation or permutation-inversion is to be associated with a given 'rotation' or 'reflection' belonging to the point group of a rigid molecule. The example of H_2O may again be helpful here. If one were to rotate the (undistorted) water molecule about its twofold symmetry axis one would interchange nuclei 1 and 2; ‘C_2’ is indeed the permutation (12). Again, reflecting all particle positions in the xy plane would not permute any nuclei, but merely reverse the sense of the molecular configuration; ‘σ_{xy}’ is understood to mean E^*. Finally, a reflection in the yz plane would interchange 1 and 2 and also reverse the sense; ‘σ_{yz}’ is to be interpreted as (12)*. This example may help the reader to avoid the pitfall of supposing that E^* has any direct connection with the ‘inversion’ belonging to the point group of a centrosymmetric molecule —H_2O is *not* centrosymmetric!

2

It is now time to consider in detail some non-rigid molecules, and a convenient first example is CH_3BF_2, which contains a nearly free rotating methyl group. Wilson *et al.* [4] discovered that one could classify the torsional–rotational levels of this molecule according to a group isomorphous with C_{6v}.

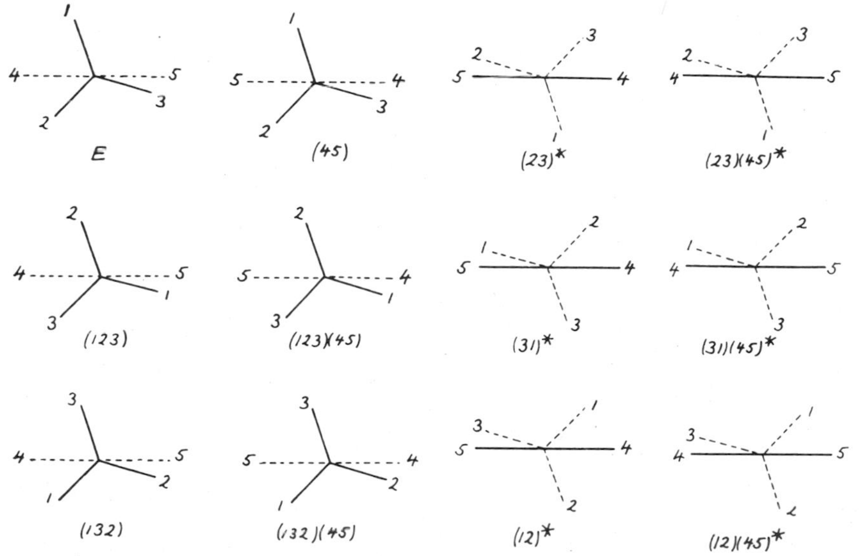

Figure 2.

We may therefore anticipate that according to the present definition of its symmetry group, this molecule will have many elements of symmetry; but it is obvious that no reasonable structure for the molecule can have more than two elements of *geometrical* symmetry (at best *either* a ‘vertical’ *or* a ‘horizontal’ plane of symmetry). This paradox is characteristic of non-rigid molecules, but if we stick faithfully to the definition I am proposing the paradox holds no terrors. The molecular symmetry group has in fact twelve elements, whose effects upon an unsymmetrical starting configuration are illustrated in figure 2†.

The protons have been labelled 1, 2 and 3 (anti-clockwise as viewed from the methyl end of the molecule); 4 and 5 are the fluorine nuclei. The six permutation-inversions bring the fluorine nuclei to the front of the picture;

† The group introduced by Wilson *et al.* is that generated by the elements ‘C_6’ $= C_{2x}$(132)(45), ‘σ_v’ $= C_{2z}$(23)*, where C_{2x} and C_{2z} are rotations through π about the x and z axes shown in figure 3.

the six permutations leave the methyl group in front. We may note, in passing, that E^* is absent from the list of permutation-inversions.

The multiplication table of the twelve elements is easy to work out, and one soon discovers that the group is isomorphous with the familiar point group D_{3h}. It is therefore convenient to call its irreducible representations by the same names as those of D_{3h}. The character table is as follows:

	E	(123) (132)	(23)* (31)* (12)*	(45)	(123)(45) (132)(45)	(23)(45)* (31)(45)* (12)(45)*
A_1'	1	1	1	1	1	1
A_2'	1	1	−1	1	1	−1
E'	2	−1	0	2	−1	0
A_1''	1	1	1	−1	−1	−1
A_2''	1	1	−1	−1	−1	1
E''	2	−1	0	−2	1	0

Table 1.

The primes and double primes have, of course, nothing to do with reflection in a plane of symmetry; they indicate the character with respect to the unique element (45) which one might be tempted to think of as a rotation (but this temptation should be resisted).

One of the simplest uses of this character table is for calculating the statistical weights of the lowest rotational and torsional–rotational levels of CH_3BF_2. The exclusion principle, applied to the protons and fluorine nuclei, imposes restrictions on the transformation properties of the complete wave function, and these in turn imply certain relations of compatibility between the species of the 'motional' wave function and the species of the nuclear spin functions. Let us look into these matters in turn.

According to the exclusion principle, if a symmetry element has the *sole* effect of permuting the positions and spins of identical particles, then it must multiply the wave function by the factor $\Pi(-1)^p$, where the product is over all sets of identical particles of half-odd spin and p denotes the parity of the permutation which the symmetry element induces in a typical set. Now the elements listed in table 1 are of three kinds; permutations for which the factor $\Pi(-1)^p$ is $+1$, those for which it is -1, and permutation-inversions, about which the exclusion principle is non-committal, since they do not *merely* permute identical particles. According to the exclusion principle, then, the overall wave function must belong to one or other of the irreducible representations A_1'' or A_2''.

Next, the nuclear spin states. ^1H and ^{19}F both have spin $\frac{1}{2}$, so the spin situation may be specified by assigning to each proton or fluorine nucleus a spin symbol α or β. Simple products of such symbols transform in an elementary way under the permutations of the group; they are, of course, invariant under E^*, and are therefore transformed by P^* in the same way as by P. Table 2 gives the characters of the reducible representations generated by the proton and fluorine spin functions.

	E	(123)	(23)*	(45)	(123)(45)	(23)(45)*	
$\alpha_1\alpha_2\alpha_3$	1	1	1	1	1	1	A_1'
$\alpha_1\alpha_2\beta_3$, $\alpha_1\beta_2\alpha_3$, $\beta_1\alpha_2\alpha_3$	3	0	1	3	0	1	$A_1'+E'$
$\alpha_1\beta_2\beta_3$, $\beta_1\alpha_2\beta_3$, $\beta_1\beta_2\alpha_3$	3	0	1	3	0	1	$A_1'+E'$
$\beta_1\beta_2\beta_3$	1	1	1	1	1	1	A_1'
$\alpha_4\alpha_5$	1	1	1	1	1	1	A_1'
$\alpha_4\beta_5$, $\beta_4\alpha_5$	2	2	2	0	0	0	$A_1'+A_1''$
$\beta_4\beta_5$	1	1	1	1	1	1	A_1'

Table 2.

For the protons, therefore, the possible spin states are $^4A_1'$ and $^2E'$; for the fluorine nuclei we have $^3A_1'$ and $^1A_1''$.

Turning to the rotational problem, we begin by noting that the mass distribution in CH_3BF_2 is such that the molecule is nearly an oblate symmetric top like BF_3. Adopting internal axes as indicated in figure 3, we may associate with each element of the molecular symmetry group an equivalent 'rotation' which

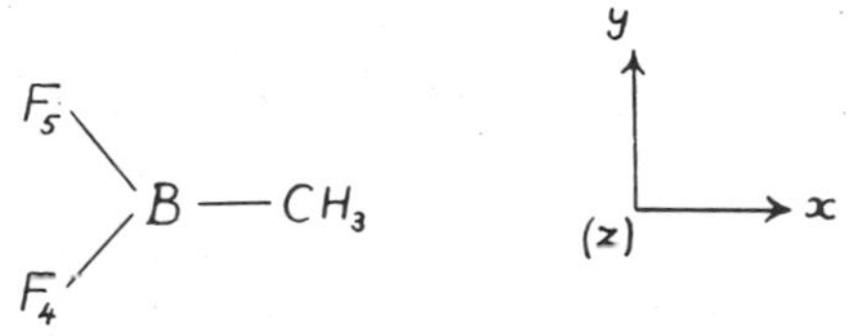

Figure 3.

takes the heavy nuclei into the same spatial orientation. These 'equivalent rotations' are as follows (table 3):

Symm. element	E	(123)	(23)*	(45)	(123)(45)	(23)(45)*
Equiv. rotn.	E	E	C_{2z}	C_{2x}	C_{2x}	C_{2y}

Table 3.

Now the wave functions of a symmetric top may be characterized by three integral quantum numbers J, K and M. J determines the magnitude of the total angular momentum (excluding spin, which we are assuming not to be coupled to the spatial motions); M measures the component of the total angular momentum about an arbitrary space-fixed axis, and K its component about the molecule-fixed z axis. Hougen has shown how one may determine the characters of the pair of functions $|J, \pm K, M\rangle$ under rotations of the kind listed in table 3; they are given in table 4.

It is now easy to find the species of the rotational wave functions, by amalgamating tables 3 and 4, and comparing the entries with those in the character table 1. For $K=0$ we obtain rotational functions of species A_1' or A_1'' according

	E	C_{2x}	C_{2y}	C_{2z}
$K=0$	1	$(-1)^J$	$(-1)^J$	1
$K>0$	2	0	0	$2(-1)^K$

Table 4.

as J is even or odd; for $K>0$ the species are $A_2'+A_2''$ or $A_1'+A_1''$ according as K is odd or even.

We are now in a position to calculate the statistical weights of the lowest rotational levels assuming that the internal (electronic–vibrational–torsional) wave function is totally symmetric—of species A_1'. The compatibility conditions may be summarized thus:

Proton spin species		Fluorine spin species		Rotational species		Overall species
$({}^4A_1' \;\; {}^2E')$	$\times$	$({}^3A_1' \;\; {}^1A_1'')$	$\times$	$(A_1' \;\; A_1'' \;\; A_2' \;\; A_2'')$	$=$	$(A_1''$ or $A_2'')$

The first obvious conclusion is that one cannot have the protons in a doublet spin state if the internal wave function is to be totally symmetric; we shall return to this point later. Secondly the fluorine triplet state is compatible only with the rotational species A_1'' or A_2'', and the singlet state with the rotational species A_1' and A_2'. Hence the statistical weights of the pure rotational states of CH_3BF_2 are

$$A_1'(4) \quad A_1''(12) \quad A_2'(4) \quad A_2''(12).$$

We have not yet investigated the transformation properties of the internal coordinates, and a coordinate of special interest is the torsional angle of the methyl group. Let τ be the angle through which one must twist the methyl group (in a right-handed sense) in order to bring the bond C–H_1 into a 'cis' position relative to the bond B–F_4. By consulting figure 2 we see that the twelve elements of symmetry transform τ as follows:

E τ	(45) $\tau+\pi$	(23)* $-\tau$	(23)(45)* $-\tau+\pi$
(123) $\tau+2\pi/3$	(123)(45) $\tau-\pi/3$	(31)* $-\tau-2\pi/3$	(31)(45)* $-\tau+\pi/3$
(132) $\tau-2\pi/3$	(132)(45) $\tau+\pi/3$	(12)* $-\tau+2\pi/3$	(12)(45)* $-\tau-\pi/3$

Table 5.

Consider now the pair of functions $\exp(\pm im\tau)$ where m is a positive integer. From table 5 we infer that each function transforms into a multiple of itself under a pure permutation and into a multiple of the other under a permutation-inversion. They therefore generate a twofold representation with the following characters (table 6).

Value of m	E	(123)	(23)*	(45)	(123)(45)	(23)(45)*	
0	1	1	1	1	1	1	A_1'
1	2	−1	0	−2	1	0	E''
2	2	−1	0	2	−1	0	E'
3	2	2	0	−2	−2	0	$A_1''+A_2''$
4	2	−1	0	2	−1	0	E'
5	2	−1	0	−2	1	0	E''
6	2	2	0	2	2	0	$A_1'+A_2'$

Table 6.

We may assume with confidence that the torsional wave functions, though not exactly of the form $\exp(\pm im\tau)$, will at least have the same transformation properties as these functions. In order of energy, then, successive torsional wave functions will be of species A_1', E'', E', etc., so that the $^2E'$ state of the protons is possible in a torsionally excited molecule, though not in an internally unexcited one.

The group character table 1 is also useful in the derivation of optical selection rules. First, there is a rigorous selection rule relating the overall species of levels which can combine in electric dipole transitions. The absolute dipole moment of the molecule—thought of as a vector in a laboratory coordinate system—is invariant under all elements of type P, but is reversed by all elements of type P^*. This statement applies to each of its three components, so the absolute dipole moment is of species $3A_2'$. The dipole selection rules therefore require $A_1'' \longleftrightarrow A_2''$ for the overall species. Additional restrictions arise from the fact that the nuclear spins are virtually uncoupled to the other degrees of freedom. This implies that in an optical transition the nuclear spin functions do not change. Hence the species of the 'motional' wave function (for all other degrees of freedom) can only change in the following ways:

$$A_1' \longleftrightarrow A_2' \quad A_1'' \longleftrightarrow A_2'' \quad E' \longleftrightarrow E' \quad E'' \longleftrightarrow E''.$$

More detailed information about internal transitions of the molecule may be obtained by calculating the species of the internal dipole moment, that is, of the projections of the dipole moment on the inertial axes. This is most easily done by noting that every element of the molecular symmetry group either rotates or reflects the dipole moment with respect to the molecule-fixed axes shown in figure 2. The associated elements of C_{2v}, and their effects on M_x, M_y and M_z, are given in table 7.

Element of symmetry	E	(123)	(23)*	(45)	(123)(45)	(23)(45)*	
Associated element of C_{2v}	E	E	σ_{xy}	C_{2x}	C_{2x}	σ_{xz}	
Effect on M_x	1	1	1	1	1	1	A_1'
M_y	1	1	1	−1	−1	−1	A_1''
M_z	1	1	−1	−1	−1	1	A_2''

Table 7.

In the last column of the table are given the species of the three components of the molecule-fixed dipole moment. We arrive at the following selection rules for internal transitions polarized along the x, y and z axes:

$$
\begin{array}{llll}
x & A_1' \longleftrightarrow A_1', & A_2' \longleftrightarrow A_2', & E' \longleftrightarrow E', \text{ etc.,} \\
y & A_1' \longleftrightarrow A_1'', & A_2' \longleftrightarrow A_2'', & E' \longleftrightarrow E'', \\
z & A_1' \longleftrightarrow A_2'', & A_2' \longleftrightarrow A_1'', & E' \longleftrightarrow E''.
\end{array}
$$

Finally, one may derive selection rules for the rotational transitions which may accompany internal transitions of these three types. To show how this is done let us assume that in the initial state the molecule is non-rotating (rotational species A_1'). For a z-polarized transition the product of the internal species of the initial and final states is A_2''. But the product of the overall species of the two states must be A_2', by our electric dipole selection rule. Therefore the product of the rotational species must be A_1''. But of the final rotational states with $J=1$, only that with $K=0$ is of species A_1''. Hence in a z-polarized transition, the rotational state cannot change from $J=0$, $K=0$ to $J=1$, $K=1$.

For x or y-polarized transitions the product of the internal species of the initial and final states is A_1' or A_2'', which are the species of the two rotational states with $J=1$, $K=1$. Hence in an x or y-polarized internal transition the rotational state cannot change from $J=0$, $K=0$ to $J=1$, $K=0$. These results are in keeping with the rotational selection rules for parallel and perpendicular transitions in a rigid symmetric top molecule.

3

Before passing on to consider other types of non-rigid molecule it is amusing to consider briefly one with a very much larger symmetry group, namely boron trimethyl, $B(CH_3)_3$. I mention it merely to show how very rapidly the effective symmetry of a molecule increases with its number of internal rotations.

The heavy-atom skeleton of boron trimethyl has symmetry D_{3h}, of order 12. But for each orientation of the skeleton there are $3 \times 3 \times 3 = 27$ possible orientations of the three methyl groups. The order of the molecular symmetry group is thus $12 \times 27 = 324$. An optimist would expect this group to be the direct product of much simpler subgroups; but it is not†. (Ethane, of which such a statement

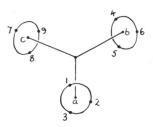

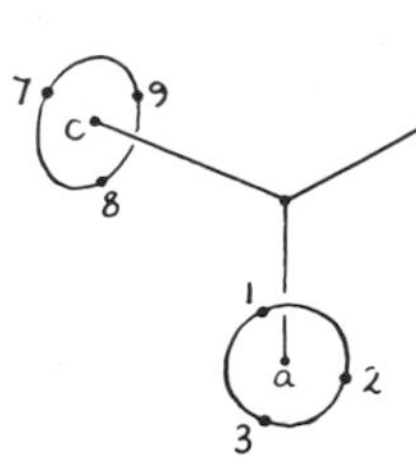

Figure 4.

can be made, is not typical of non-rigid molecules.) With a clockwise labelling convention for each of the three methyl groups (see figure 4) the character table is as shown in table 8. The reader should have no difficulty (!) in solving,

† If it were, the number of irreducible representations would be composite; table 8 shows that this number is prime.

	1 E	2 (123)(456)(789)	6 (123)(465)	6 (123)	6 (123)(456)	6 (132)(456)(789)	18 (147)(258)(369)(*abc*)	36 (147258369)(*abc*)	9 (47)(58)(69)(*bc*)	18 (132)(475869)(*bc*)	18 (475869)(*bc*)	18 (123)(47)(58)(69)(*bc*)	18 (123)(475869)(*bc*)	27 (23)(47)(59)(68)(*bc*)*	54 (23)(475968)(*bc*)*	27 (23)(56)(89)*	54 (147)(268359)(*abc*)*
A_1	1	1	1	1	1	1	1	1	1	1	1	1	1	1	1	1	1
A_2	1	1	1	1	1	1	1	1	−1	−1	−1	−1	−1	1	1	−1	−1
A_3	1	1	1	1	1	1	1	1	1	1	1	1	1	−1	−1	−1	−1
A_4	1	1	1	1	1	1	1	1	−1	−1	−1	−1	−1	−1	−1	1	1
E_1	2	2	2	2	2	2	−1	−1	0	0	0	0	0	0	0	2	−1
E_2	2	2	2	2	2	2	−1	−1	0	0	0	0	0	0	0	−2	1
E_3	2	2	2	−1	−1	−1	2	−1	2	2	−1	−1	−1	0	0	0	0
E_4	2	2	2	−1	−1	−1	2	−1	−2	−2	1	1	1	0	0	0	0
G	4	4	4	−2	−2	−2	−2	1	0	0	0	0	0	0	0	0	0
I_1	6	−3	0	0	3	3	0	0	2	−1	−1	2	−1	0	0	0	0
I_2	6	−3	0	0	−3	3	0	0	−2	1	1	−2	1	0	0	0	0
I_3	6	−3	0	3	0	−3	0	0	2	−1	2	−1	−1	0	0	0	0
I_4	6	−3	0	3	0	−3	0	0	−2	1	−2	1	1	0	0	0	0
I_5	6	−3	0	−3	3	0	0	0	2	−1	−1	−1	2	0	0	0	0
I_6	6	−3	0	−3	3	0	0	0	−2	1	1	1	−2	0	0	0	0
I_7	6	6	−3	0	0	0	0	0	0	0	0	0	0	2	−1	0	0
I_8	6	6	−3	0	0	0	0	0	0	0	0	0	0	−2	1	0	0

Table 8.

with the aid of this character table, the same problems for boron trimethyl as were solved for CH_3BF_2 in the preceding section.

Neopentane, $C(CH_3)_4$, has an even larger symmetry group, of order 1944. But the torsional barrier in this molecule is undoubtedly much higher than in boron trimethyl, so that for most practical purposes one may regard neopentane as a rigid molecule belonging to the point group T_d.

4

The ethane molecule has been subjected to careful study by several spectroscopists, particularly Howard [2] and Wilson [3] who found that its torsional levels could be classified by a symmetry group of higher order than D_{3d}. With apologies to these authors, let us briefly discuss the symmetry group of C_2H_6 from the present view-point. We adopt the clockwise labelling convention

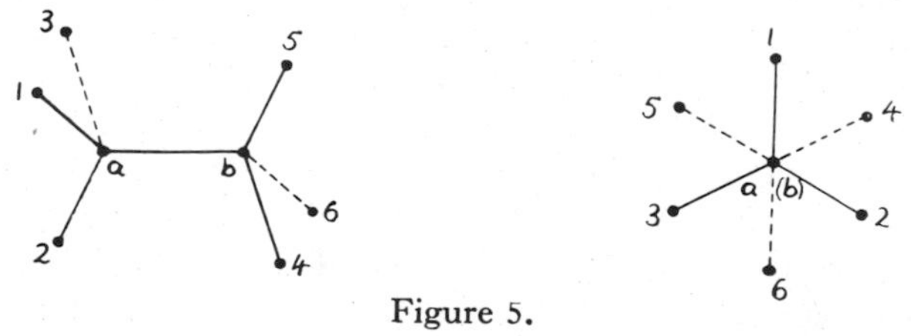

Figure 5.

shown in figure 5. The molecular symmetry group is found to be the direct product of two groups of order 6. Their elements are, respectively,

(i) E, (123)(465), (14)(25)(36)(ab),
(132)(456), (15)(26)(34)(ab),
(16)(24)(35)(ab),

and

(ii) E, (123)(456), (14)(26)(35)(ab)*,
(132)(465), (15)(24)(36)(ab)*,
(16)(25)(34)(ab)*, .

and each is isomorphous with the group of permutations of three symbols. (The reader may verify, if he wishes, that every element of group (i) commutes with every element of group (ii).) The character tables of the two factor groups are shown in tables 9 and 10. The irreducible representations of the molecular symmetry group are the direct products (see table 11) of those of the factor groups, and the same applies to its classes.

The character table of the molecular symmetry group is thus the direct product of tables 9 and 10; it is given in table 12.

	E	(123)(465)	(14)(25)(36)(ab)
A_1(i)	1	1	1
A_2(i)	1	1	−1
E (i)	2	−1	0

Table 9.

	E	(123)(456)	(14)(26)(35)(ab)*
A_1(ii)	1	1	1
A_2(ii)	1	1	−1
E (ii)	2	−1	0

Table 10.

	A_1(i)	A_2(i)	E(i)
A_1(ii)	A_1	A_2	E_1
A_2(ii)	A_3	A_4	E_2
E (ii)	E_3	E_4	G

Table 11.

Because of this factorization of the molecular symmetry group, one can define separately the rotational and the torsional species of a given level. To demonstrate this, let us consider dimethylacetylene, which has the same symmetry group as ethane and in which internal rotation is virtually unhindered.

	1 E	2 (123)(456)	3 (14)(26)(35)$(ab)^*$	2 (123)(465)	4 (123)	6 (142635)$(ab)^*$	3 (14)(25)(36)(ab)	6 (142536)(ab)	9 (12)(45)*
A_1	1	1	1	1	1	1	1	1	1
A_2	1	1	1	1	1	1	−1	−1	−1
A_3	1	1	−1	1	1	−1	1	1	−1
A_4	1	1	−1	1	1	−1	−1	−1	1
E_1	2	2	2	−1	−1	−1	0	0	0
E_2	2	2	−2	−1	−1	1	0	0	0
E_3	2	−1	0	2	−1	0	2	−1	0
E_4	2	−1	0	2	−1	0	−2	1	0
G	4	−2	0	−2	1	0	0	0	0

Table 12

A sensible trial wave function for one of the torsional–rotational levels of this molecule would be a combination of functions of the type

$$\Theta_{JKM}(\theta)\exp(iM\phi)\exp(iK_a\chi_a)\exp(iK_b\chi_b),$$

where (θ, ϕ, χ_a) and (θ, ϕ, χ_b) are the Eulerian angles of the two individual methyl groups, and $K=|K_a+K_b|$. To find the transformation properties of such wave functions, one must see how the angles θ, ϕ, χ_a and χ_b are affected by the elements of the molecular symmetry group. One finds, in fact, that the elements of group (i) leave $\chi_a-\chi_b$ invariant, while the elements of group (ii) leave θ, ϕ and $\chi_a+\chi_b$ unchanged. Writing the above wave function in the form

$$\Theta_{JKM}(\theta)\exp(iM\phi)\exp[\tfrac{1}{2}i(K_a+K_b)(\chi_a+\chi_b)] \times \exp[\tfrac{1}{2}i(K_a-K_b)(\chi_a-\chi_b)],$$

we deduce that the overall species is the product of a rotational species dependent only on $K=|K_a+K_b|$ and on J, and a torsional species dependent only on $|K_a-K_b|=L$, say. Detailed investigation yields the following rotational and torsional species for various values of J and K, and of L:

Value of K	0	1	2	3	4	5	6
Rotational species	A_1 (J even) A_2 (J odd)	E_1	E_1	A_1+A_2	E_1	E_1	A_1+A_2
Value of L	0	1	2	3	4	5	6
Torsional species	A_1	E_3	E_3	A_1+A_3	E_3	E_3	A_1+A_3

Table 13.

The overall torsional–rotational species is just the product of the species for the two kinds of motion.

It should be emphasized that ethane and dimethylacetylene are exceptional in possessing a symmetry group which can be factorized in this way. In general a molecule with a non-rigid inertial framework has a symmetry group which does not possess this property so that one cannot speak of the 'rotational species' or the 'torsional species' of a level. To illustrate this point we will now consider a more typical non-rigid molecule.

5

The hydrazine molecule N_2H_4 has already been mentioned†. Its equilibrium configuration has no more than a twofold axis of symmetry; the corresponding

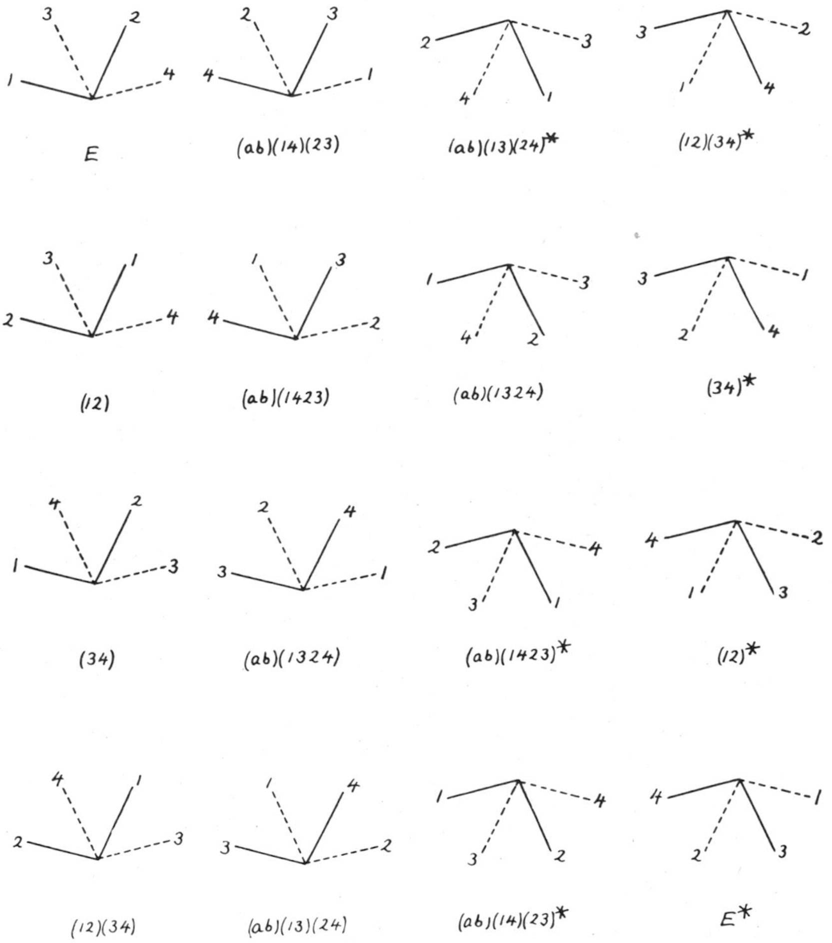

Figure 6.

point group consists of the two elements E and $(ab)(14)(23)$ of figure 6. The possibility of hindered internal rotation adds two more elements—$(ab)(13)(24)^*$

† The microwave spectrum of hydrazine has been very elegantly interpreted by Kasuya [6], who classified the levels according to a group of order 8, which is a subgroup of that described in this section.

and (12)(34)*—to the symmetry group. With the further possibility of inversion at either nitrogen atom we obtain a molecular symmetry group of 16 elements, enumerated in figure 6.

	E	(12) / (34)	$(ab)(13)(24)$ / $(ab)(14)(23)$	$(ab)(1324)$ / $(ab)(1423)$	(12)(34)	E^*	(12)* / (34)*	$(ab)(13)(24)^*$ / $(ab)(14)(23)^*$	$(ab)(1324)^*$ / $(ab)(1423)^*$	(12)(34)*
A_1^+	1	1	1	1	1	1	1	1	1	1
A_2^+	1	−1	−1	1	1	1	−1	−1	1	1
B_1^+	1	1	−1	−1	1	1	1	−1	−1	1
B_2^+	1	−1	1	−1	1	1	−1	1	−1	1
E^+	2	0	0	0	−2	2	0	0	0	−2
A_1^-	1	1	1	1	1	−1	−1	−1	−1	−1
A_2^-	1	−1	−1	1	1	−1	1	1	−1	−1
B_1^-	1	1	−1	−1	1	−1	−1	1	1	−1
B_2^-	1	−1	1	−1	1	−1	1	−1	1	−1
E^-	2	0	0	0	−2	−2	0	0	0	2

Table 14.

The group is isomorphous with D_{4h}. Its character table is table 14. The symbols + and − have been used to indicate the parity of each representation, that is, the character with respect to the inversion E^*.

Suppose now that we wish to determine the species of the lowest energy levels of hydrazine. For simplicity let us assume the molecule to be in its vibrational–electronic ground state so that the only degrees of freedom to be considered are those associated with rotation, twisting and inversion. If twisting and inversion were impossible, the molecule could exist in any one of eight distinct conformations, and each of these 'structures' would have its own rotational wave function, symmetric or antisymmetric about the twofold rotation axis. Taken together, these eight functions will generate a reducible representation of the complete molecular symmetry group; the characters of this representation will now be determined. Let R be one of the elements (not E) listed in figure 6. Then the character of R is just $\pm n$ where n is the number of structures for which R induces a rotation about the twofold axis; the sign is + or − according as the rotational wave functions for the individual structures are symmetric or anti-symmetric about their twofold axes. We thus arrive at the following characters for the two possible cases (table 15).

Symmetry under C_2	E, etc.					E^*, etc.				
symmetric	8	0	4	0	0	0	0	0	0	0
antisymmetric	8	0	−4	0	0	0	0	0	0	0

Table 15.

Tunnelling between the eight structures, by twisting and inversion, therefore splits a symmetric rotational level into sub-levels of species

$$A_1^+ + B_2^+ + E^+ + A_1^- + B_2^- + E^-,$$

and an antisymmetric rotational level into sub-levels of species

$$A_2^+ + B_1^+ + E^+ + A_2^- + B_1^- + E^-.$$

Thus a rotational level of given symmetry under C_2 cannot be assigned a unique species in the molecular symmetry group. This situation need not worry us unduly, but we must be aware of it.

With the aid of the character table 14 one can quite easily find the species of the nuclear spin states. The possible proton spin functions are found to be ${}^5A_1^+$, ${}^3B_1^+$, ${}^3E^+$, ${}^1A_1^+$ and ${}^1B_2^+$; those of the two ^{14}N nuclei are ${}^5A_1^+$, ${}^3B_1^+$ and ${}^1A_1^+$. Since ^{14}N nuclei are bosons, the overall species must be B_2^+ or B_2 , since these are the representations in which the character for an odd permutation of the protons is -1. The exclusion principle thus allows the following combinations of motional and nuclear spin species in $^{14}N_2H_4$ (table 16).

Motional species	^{14}N spin species	^{1}H spin species	Overall species	Statistical weight
A_1^+, A_1^-	$\{{}^5A_1^+, {}^1A_1^+\}$	${}^1B_2^+$	B_2^+, B_2^-	6
A_2^+, A_2^-	$\{{}^5A_1^+, {}^1A_1^+\}$	${}^3B_1^+$	B_2^+, B_2^-	18
,,	${}^3B_1^+$	$\{{}^5A_1^+, {}^1A_1^+\}$	B_2^+, B_2^-	18
B_1^+, B_1^-	${}^3B_1^+$	${}^1B_2^+$	B_2^+, B_2^-	3
B_2^+, B_2^-	$\{{}^5A_1^+, {}^1A_1^+\}$	$\{{}^5A_1^+, {}^1A_1^+\}$	B_2^+, B_2^-	36
,,	${}^3B_1^+$	${}^3B_1^+$	B_2^+, B_2^-	9
E^+, E^-	$\{{}^5A_1^+, {}^3B_1^+, {}^1A_1^+\}$	${}^3E^+$	B_2^+, B_2^-	27

Table 16.

Hence the various species of motional state have the statistical weights $A_1(6)$, $A_2(36)$, $B_1(3)$, $B_2(45)$ and $E(27)$, irrespective of parity.

The electric dipole selection rules on the motional species are easily found to be

$$A_1^+ \longleftrightarrow A_1^-, \quad A_2^+ \longleftrightarrow A_2^-, \quad B_1^+ \longleftrightarrow B_1^-, \quad B_2^+ \longleftrightarrow B_2^- \quad \text{and} \quad E^+ \longleftrightarrow E^-.$$

It is clear that these rules are quite independent of nuclear spins or statistics; they must be, since $^{14}N_2H_4$ and $^{15}N_2D_4$ cannot possibly have different optical selection rules.

6

Enough has been said, perhaps, to show the utility of the present definition of the symmetry group of a non-rigid molecule. Essentially its elements are *permutations* of the positions and spins of identical *nuclei* (with or without inversion of all particle positions in the centre of mass). According to the representations of this group one can classify not only the spin states and states of motion of the nuclei, but even the electronic states of the molecule, for the reasons already given.

In conclusion it should be added that the present definition can be extended to linear molecules, and to molecules where spin–orbit coupling is strong; but these topics are best dealt with separately.

References

[1] Mulliken, R. S., 1933, *Phys. Rev.*, **43**, 279.
[2] Howard, J. B., 1937, *J. chem. Phys.*, **5**, 442.
[3] Wilson, E. B., 1938, *J. chem. Phys.*, **6**, 740.
[4] Wilson, E. B., Lin, C. C., and Lide, D. R., 1955, *J. chem. Phys.*, **23**, 136.
[5] Hougen, J. T., 1962, *J. chem. Phys.*, **7**, 1433.
[6] Kasuya, T., *Sci. Papers, I.P.C.R.*, 1962, **56**, 1.

Reprinted from *J. Amer. Chem. Soc.*, **87**, 395–397 (1965)

39

Stereochemistry of Electrocyclic Reactions

Sir:

We define as *electrocyclic* transformations the formation of a single bond between the termini of a linear system containing k π-electrons (I → II), and the

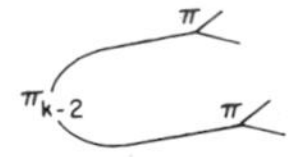

I

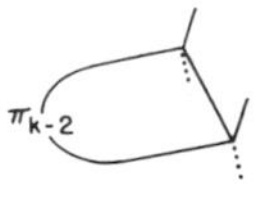

II

converse process. In such changes, fixed geometrical isomerism imposed upon the open-chain system is related to rigid tetrahedral isomerism in the cyclic array. *A priori*, this relationship might be *disrotatory* (III → IV or *vice versa*), or *conrotatory* (V → VI, or

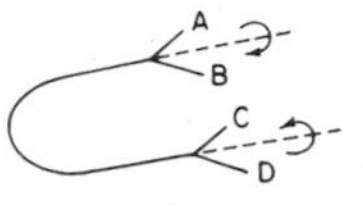

III

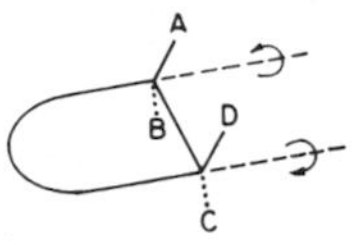

IV

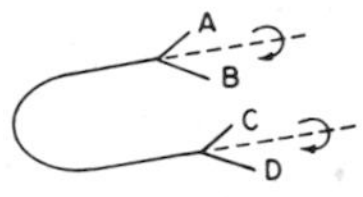

V

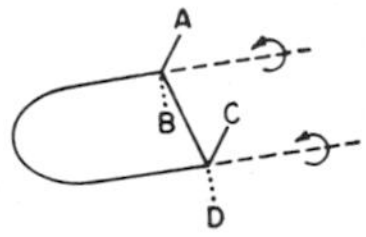

VI

vice versa). In practice, transformations of this type have been brought about thermally, or photochemically, and *all known cases proceed in a highly stereospecific manner.* For example, the thermal isomerization of cyclobutenes is cleanly *conrotatory* (VII ← VIII).[1]

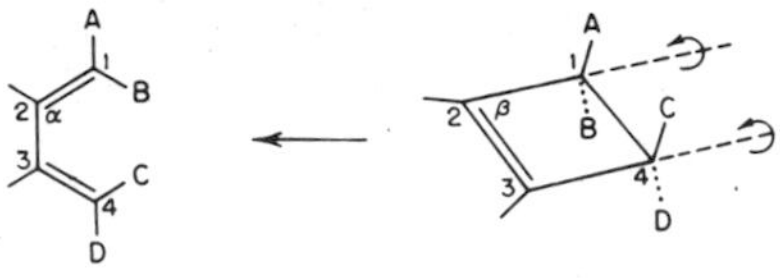

VII VIII

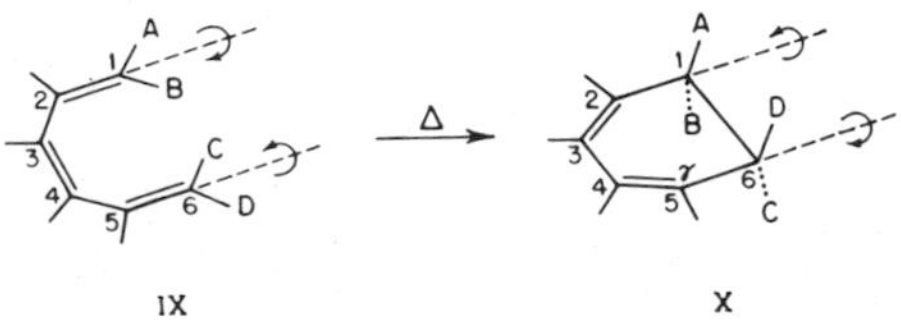

IX X

(1) *cis*-3,4-Dicarbomethoxycyclobut-1-ene: E. Vogel, *Ann.*, **615**, 14 (1958). *cis*- and *trans*-1,2,3,4-tetramethylcyclobut-1-enes: R. Criegee and K. Noll, *ibid.*, **627**, 1 (1959). W. Adam [*Chem Ber*, **97**, 1811 (1964)] describes several cases and cites others which proceed stereospecifically, but whose products are of as yet undetermined configuration.

By contrast, the thermal cyclization of hexatrienes is uniquely *disrotatory* (IX → X)[2]; this case is the more striking in view of the fact that factors of steric demand and angle strain clearly suggest that a conrotatory process should be followed.[3] Finally, the stereospecific *conrotatory* process (XI → XII) is observed when

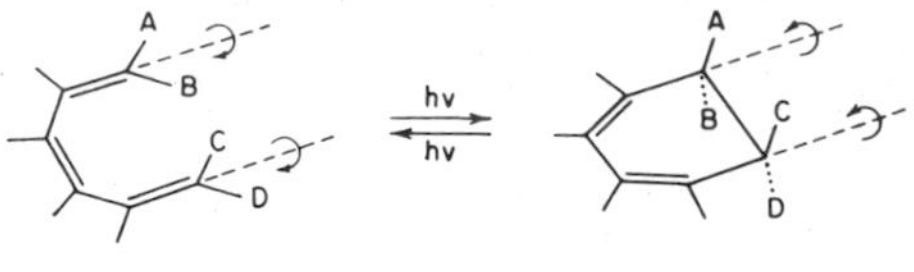

XI XII

hexatrienes are subjected to photochemical cyclization to cyclohexadienes, and *vice versa.*[4]

It is the purpose of this communication to suggest that the steric course of electrocyclic transformations is determined by the symmetry of the highest occupied molecular orbital of the open-chain partner in these changes.[5] Thus, in an open-chain system containing $4n$ π-electrons, the symmetry of the highest occupied ground-state orbital is such that a bonding interaction between the termini *must involve overlap between orbital envelopes on opposite faces of the system*, and this can only be achieved in a conrotatory process (*cf.* XIII). Conversely, in open systems containing $4n + 2$ π-electrons, terminal bonding interaction within ground-state molecules *requires overlap of orbital envelopes on the same face of the system*, attainable only by disrotatory displacements (*cf.* XIV). On

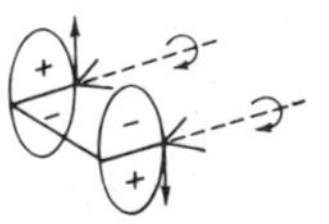

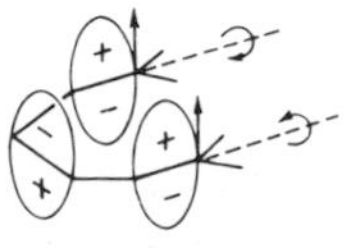

XIII XIV

the other hand, promotion of an electron to the first excited state leads to a reversal of terminal symmetry relationships in the orbitals mainly involved in bond redistribution, with the consequence that a system which undergoes a thermally induced disrotatory

(2) Precalciferol → pyro- and isopyrocalciferols: E. Havinga and J. L. M. A. Schlatmann, *Tetrahedron*, **16**, 146 (1961). *trans,cis,trans*-1,6-dimethylhexa-1,3,5-triene → *cis*-1,2-dimethylcyclohexa-3,5-diene; *trans,cis,cis*-1,6-dimethylhexa-1,3,5-triene → *trans*-1,2-dimethylcyclohexa-3,5-diene: E. Vogel, E. Marvell, private communications.

(3) *Cf.* K. E. Lewis and H. Steiner [*J. Chem. Soc.*, 3080 (1964)], who do not even consider the disrotatory course which is in fact very probably followed in the thermal cyclization of hexa-1-*cis*-3,5-triene itself.

(4) Precaliferol ⇄ ergosterol, tachysterol → lumisterol → precalciferol: E. Havinga, R. J. de Kock, and M. P. Rappoldt, *Tetrahedron*, **11**, 276 (1960); and E. Havinga and J. L. M. A. Schlatmann, ref. 2. *trans, cis,trans*-1,6-Dimethylhexa-1,3,5-triene ⇄ *trans*-1,2-dimethylcyclohexa-3,5-diene: G. J. Fonken, *Tetrahedron Letters*, 549 (1962).

(5) Professor L. J. Oosterhoff (Leiden) clearly deserves credit for having first put forward the suggestion that orbital symmetries might play a role in determining the course of the stereochemical phenomena attendant upon triene cyclizations (private communication to Professor Havinga, quoted in E. Havinga and J. L. M. A. Schlatmann, *Tetrahedron*, **16**, 151 (1961)). The suggestion was described so succinctly that it has received no currency, and it has not been generalized to include other cases.

electrocyclic transformation in the ground state should follow a conrotatory course when photochemically excited, and *vice versa*.[6]

It should be emphasized that our hypothesis specifies in any case which of two types of geometrical displacements will represent a favored process, but does not exclude the operation of the other under very energetic conditions. Thus, *cis*-1,2,3,4-tetramethylcyclobut-1-ene (XV) is smoothly transformed to

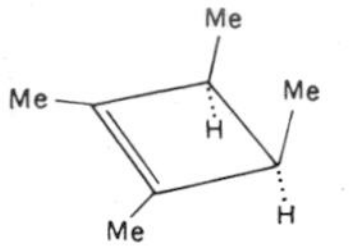

XV

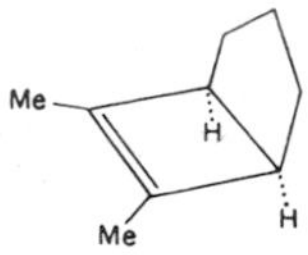

XVI

cis,trans-tetramethylbutadiene in a conrotatory process at 200°. In the dimethylbicyclo[0.2.3]heptene derivative (XVI), the presence of the five-membered ring makes a conrotatory process impossible, and the disrotatory opening is observed, but only slowly at 400°.[7]

Our hypothesis accommodates the known dramatic stereospecificities in electrocyclic reactions. It further permits clear predictions of the outcome in numerous interesting cases which have not yet been scrutinized. Some of these predictions are summarized in Table I; it may be noted specially that electrocyclic transformations within odd-electron systems should follow the same stereochemical course as the even-electron systems containing one further electron, and that charged systems should behave in the same manner as neutral systems containing the same number of electrons.

Table I

Predicted ground-state reactions	Type
Cyclopropyl cation → allyl cation	Disrotatory
Cyclopropyl radical → allyl radical	Conrotatory
Cyclopropyl anion → allyl anion	Conrotatory
Cyclopentenyl cation ← pentadienyl cation	Conrotatory

The simple symmetry argument presented above is supported by our results in a study of several cases by the extended Hückel theory.[8] Although the energetic preferences revealed by these calculations cannot be associated entirely with single energy levels, the major directive factor for displacements within ground states arises from energy variations within the highest, doubly occupied molecular orbital, and in excited states within the two highest, partially occupied orbitals, of which the higher level is dominant.

In the study of the butadiene cyclization, the initial conformation was the planar, *s-cis* form VII, with $d_{12} = d_{34} = 1.34$ Å., $d_{23} = 1.48$ Å., $d_{CH} = 1.10$ Å., and a range of values for the internal angles, α. The terminal methylene groups were twisted in disrotatory and conrotatory modes through a range of angles, while retaining their trigonal conformation. This calculation indicated that in the ground state, for $\alpha \gtrsim 117°$, the disrotatory displacement was slightly favored.[9] However, in the disrotatory motion the 1,4 bond order becomes more negative as twisting increases, while in the conrotatory mode a positive bond order develops. As α was decreased a sharply increasing preference for the conrotatory mode was found. In the first excited state these relationships are precisely reversed.

Approaching the transition state for a $k = 4$ case from the cyclic form, a model cyclobutene geometry (VIII) with $d_{23} = 1.34$ Å., $d_{12} = d_{34} = d_{14} = 1.54$ Å., $\beta = 93.7°$, was chosen. Hydrogen atoms at C-1 and C-4 were so placed that the C–H bonds formed tetrahedral angles with each other, and with C-1–C-2 or C-3–C-4. Disrotatory and conrotatory modes of twisting, retaining tetrahedral hydrogen dispositions, were studied as a function of β. There was a clear preference for conrotatory twisting in the ground state, disrotatory motion in the first excited state. In this case the preferred ground-state conrotatory process was associated with a much more rapidly decreasing 1,4 overlap population.

For the study of the hexatriene–cyclohexadiene transformations, it was decided to approach the transition states from the cyclic side, since the initial geometry of the hexatriene partner cannot be specified easily. A model geometry (X) was chosen, with $d_{34} = 1.48$ Å., $d_{45} = d_{23} = 1.34$ Å., $d_{12} = d_{16} = d_{56} = 1.54$ Å. A slightly unrealistic simplification was made in assuming coplanarity of the six carbon atoms. Disrotatory and conrotatory processes were examined in the above geometry, as well as for one with $\gamma = 150°$ ($d_{16} = 2.42$ Å.). In both cases it was found that the disrotatory mode was favored in the ground state, while the conrotatory process was preferred in the first excited state.

In the case of the cyclopropyl–allyl transformations, the contrasting twisting motions were considered in an intermediate geometry (XVII) with tetrahedral hydrogen

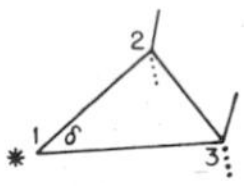

XVII

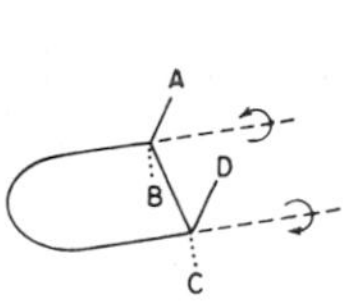

XVIII

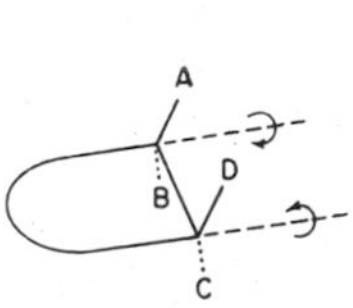

XIX

atoms, and $d_{12} = d_{13} = 1.50$ Å., $\delta = 90°$. The disrotatory process was found to be favored in the ground state of the cation, the conrotatory mode in the radical or anion.

(6) The situation in photochemical reactions is of course subject to many complications which are absent in the thermal cases. The generalization suggested here will apply only to electrocyclic transformations which actually take place within an excited state, before energy degradation occurs, with transformation to a new geometry, or to a different state.

(7) R. Criegee and H. Furrer, *Chem. Ber.*, **97**, 2949 (1964).

(8) R. Hoffmann, *J. Chem. Phys.*, **39**, 1397 (1963); **40**, 2480 (1964); identical parameters were used in this work.

(9) For large values of α, the 1,4-interaction is minimal, and the energetically decisive factor is the extent to which the π-system is uncoupled by a given rotation; the conrotatory motion uncouples the π-orbitals considerably more than the disrotatory mode.

Except in cases possessing special symmetry axes, two alternative conrotatory or disrotatory processes are possible and physically differentiable (*cf.* XVIII and XIX). Ordinarily, simple steric factors will be expected to direct the changes preferentially along one of the two paths, but in some cases, very interesting special stereoelectronic factors may be definitive. Thus, when a cyclopropyl cation is produced by ionization of a group X, and suffers concerted electrocyclic transformation to an allyl cation, our calculations indicate that the favored processes are XX and XXI.

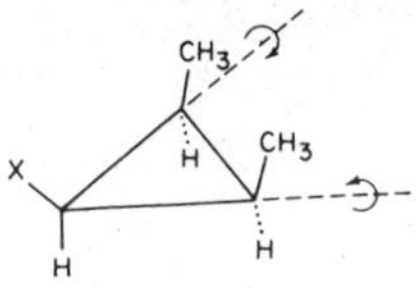

XX

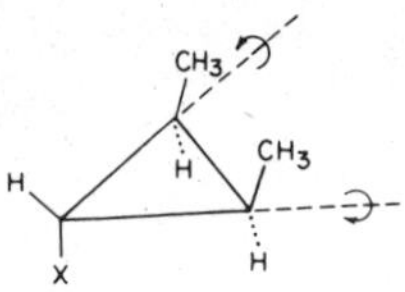

XXI

(10) Junior Fellow, Society of Fellows, Harvard University.

R. B. Woodward, Roald Hoffmann[10]
Department of Chemistry, Harvard University
Cambridge, Massachusetts 02138
Received November 30, 1964

Reprinted from *J. Amer. Chem. Soc.*, **87**, 2045–2046 (1965)

40

The Electronic Mechanism of Electrocyclic Reactions

Sir:

In a recent communication[1] Woodward and Hoffmann have offered a theoretical interpretation of the stereospecificity of "electrocyclic reactions," such as the conversion of cyclobutene to butadiene or the cyclization of hexatrienes to cyclohexadienes. Such reactions may be induced by either heat or light, and both the thermal and the photochemical reactions are highly stereospecific. In the ring opening of cyclobutenes the thermal reaction is "conrotatory" (see Figure 1a) while the photochemical reaction is "disrotatory" (Figure 1b); by contrast the photochemical cyclization of hexatrienes is a conrotatory process. Woodward and Hoffmann point out that the stereochemical course of an electrocyclic ring closure is determined by the symmetry of the highest occupied molecular orbital of the open-chain reactant; their calculations based on the extended Hückel theory support this generalization and show that it also applies to the ring-opening reactions.

In a sense the work of Woodward and Hoffmann disposes of the problem, but two questions might remain in the mind of the physical chemist. First, why should the ring-opening reactions be guided by the electronic structure of the product rather than the reactant, and second, why should the extended Hückel theory lead to the same predictions as a rule based on the symmetry of only one molecular orbital? Perhaps in this note we may be allowed to recommend a point of view from which these questions are more easily answered.

Let us begin by considering the electrocyclic conversion of cyclobutene to butadiene. In this process four molecular orbitals undergo a radical change: namely, σ, σ^*, π, and π^*, where the first two are the bonding and antibonding orbitals of the bond which is to be broken, and the last two refer to the carbon–carbon double bond in the cyclobutene ring. Eventually these four orbitals become the molecular π-orbitals of butadiene, namely, ψ_1, ψ_2, ψ_3, and ψ_4 in order of increasing energy. Each of the first set of orbitals passes adiabatically into one of the second set, and we now ask: how are the orbitals of the two sets correlated?

The answer depends upon whether the isomerization proceeds in a conrotatory or a disrotatory manner. In the conrotatory mode the system preserves a twofold axis of symmetry throughout the reaction, so that each orbital may be rigorously classified as A or B according as it is symmetric or antisymmetric about that axis. In the disrotatory mode a symmetry plane is maintained, and each orbital remains either symmetric (A′) or antisymmetric (A″) about this plane. The situation is illustrated in Figure 1, and Table I gives the symmetries of the orbitals in the two cases. Thus in the conrotatory mode σ correlates with ψ_2, π with ψ_1, etc., but in the disrotatory mode the correlations are different, σ correlating with ψ_1 and π^* with ψ_2, etc.

(1) R. B. Woodward and R. Hoffmann, *J. Am. Chem. Soc.*, 87, 395 (1965).

Table I

	Symmetry	Cyclobutene	*s-cis*-Butadiene
Conrotatory mode	A	σ, π^*	ψ_2, ψ_4
	B	π, σ^*	ψ_1, ψ_3
Disrotatory mode	A′	σ, π	ψ_1, ψ_3
	A″	π^*, σ^*	ψ_2, ψ_4

Having correlated the orbitals of cyclobutene with those of butadiene *via* the conrotatory and disrotatory modes, we can now correlate the electron configurations of the two molecules, with reference to the same alternative modes. The result is given in Figure 2. Here the straight lines indicate what would happen to a given electron configuration if there were no configurational interaction; the full lines represent a more realistic

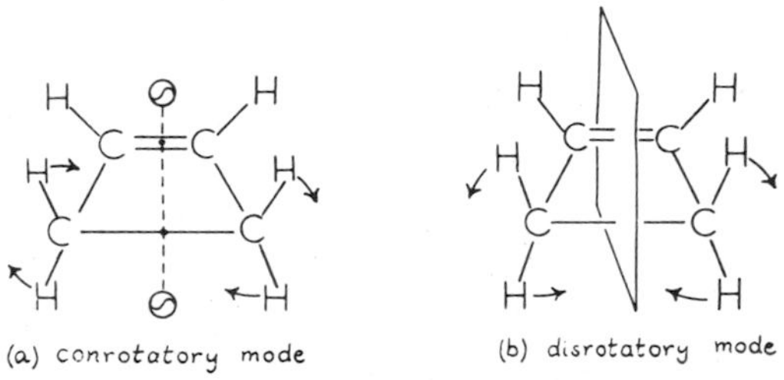

Figure 1. Alternative modes for the conversion of cyclobutene to *s-cis*-butadiene.

view of the correlations between states since electron repulsion prevents states of the same symmetry from crossing. It is immediately clear why both the forward and the reverse reactions should be conrotatory in the ground state and disrotatory in the first excited state, as indicated by the quantitative computations of Woodward and Hoffmann. An analogous treatment of the C_6 systems to an exactly converse result, again in agreement with their analysis.

Besides studying the even-membered rings and chains, Woodward and Hoffmann examined the C_3 and C_5 systems and made predictions about the stereochemical courses of their electrocyclic transformations. We think it may be of interest to give the orbital and state correlations for the interconversion of cyclopropyl and allyl, to bring out a point which was not mentioned in their paper. Table II gives the orbital correlations for the two species in the conrotatory and disrotatory modes; π stands for the 2p-orbital of the trivalent carbon atom.

Figures 3–5 represent the correlations between the ground states and first excited states of the cations, the neutral radicals, and the anions in the two modes. For the ions the stereochemical predictions are clear-cut and agree precisely with those given by Woodward and

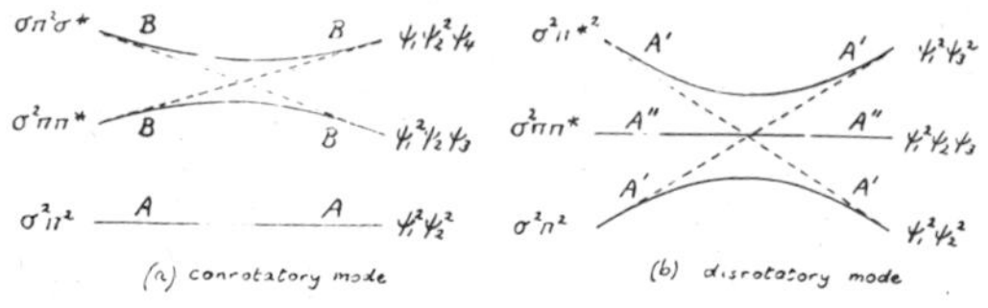

Figure 2. Correlation diagrams for the interconversion of cyclobutene and butadiene.

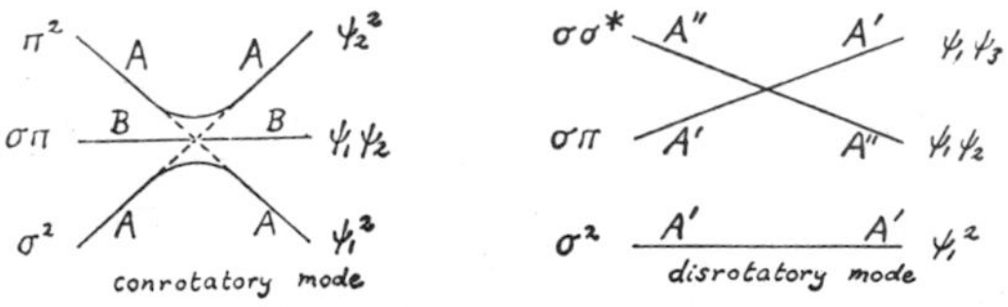

Figure 3. Interconversion of cyclopropyl cation and allyl cation.

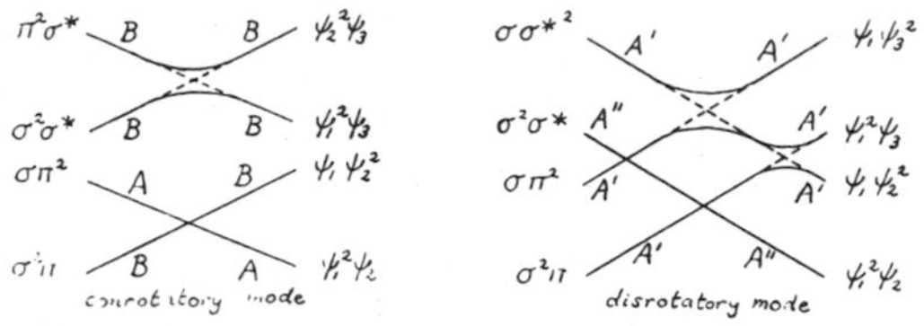

Figure 4. Interconversion of cyclopropyl radical and allyl radical.

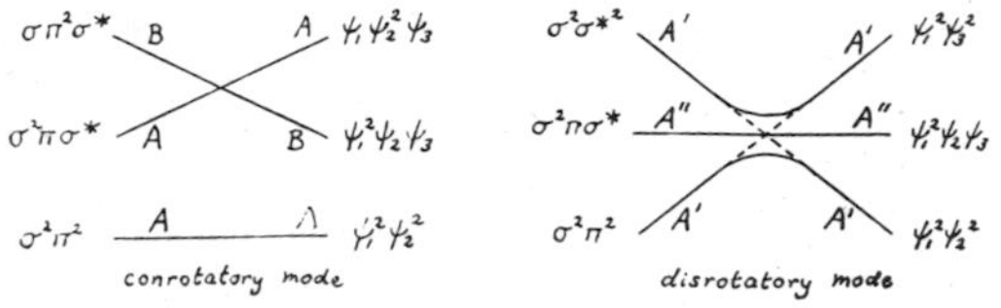

Figure 5. Interconversion of cyclopropyl anion and allyl anion.

Hoffmann. But for the radicals the situation is more complex. In both the conrotatory and disrotatory modes the ground state of each radical is correlated

Table II

		Cyclopropyl	Allyl
Conrotatory mode	A	σ	ψ_2
	B	π, σ^*	ψ_1, ψ_3
Disrotatory mode	A′	σ, π	ψ_1, ψ_3
	A′′	σ^*	ψ_2

with an excited state of the other. If the excited states occur in the order shown (as one would suppose), the thermal reaction should be slightly less hindered in the conrotatory than in the disrotatory mode, but should show a much larger activation energy than the thermal transformation of the ions. One would therefore expect the thermal isomerization of the cyclopropyl radical to be much slower and probably much less stereospecific than for either the positive or the negative ion. As to the photochemical transformations, the first excited state of either radical should also isomerize by the conrotatory mode, but in the second excited state, which lies very close above the first, disrotatory transformations are to be expected.

We have thought it worthwhile to supplement the discussion of Woodward and Hoffman in this way because orbital and state correlation diagrams enable one to follow a reaction along its entire course and draw qualitative conclusions without necessarily engaging in numerical computations.

(2) E. W. A., at present on leave from the Case Institute, Cleveland, Ohio, wishes to thank the National Institute of Neurological Diseases and Blindness for a Fellowship.

H. C. Longuet-Higgins, E. W. Abrahamson[2]
Department of Theoretical Chemistry
University Chemical Laboratory, Cambridge, England
Received March 20, 196

Reprinted from *J. Amer. Chem. Soc.*, **87**, 2046–2048 (1965)

41

Selection Rules for Concerted Cycloaddition Reactions

Sir:

Recently we characterized orbital symmetry relationships as stereochemical determinants in electrocyclic reactions, which may be regarded as intramolecular cycloadditions, involving net interconversions of one π-bond and one σ-bond.[1] We now examine the general problem of concerted *intermolecular* cycloaddition reactions from a similar point of view.

Our procedure consists in the construction of correlation diagrams for the molecular orbitals involved in reaction, classifying the levels with respect to the symmetry elements of the transition state.[2] The method is illustrated first in the case of the addition of two ethylenes to form cyclobutane. The assumed geometry of approach places the two ethylene molecules in parallel planes directly above each other, as in Figure 1. The four reactant π-levels and the four corresponding σ-levels in the product are shown in Figure 2, classified as symmetric (S) or antisymmetric (A) with respect to σ_1, a plane bisecting the ethylenes, and σ_2, a plane parallel to and midway between the planes of the approaching molecules. The form of the molecular orbitals, projected upon the plane passing through the four carbon atoms, is shown in Figure 3. The original

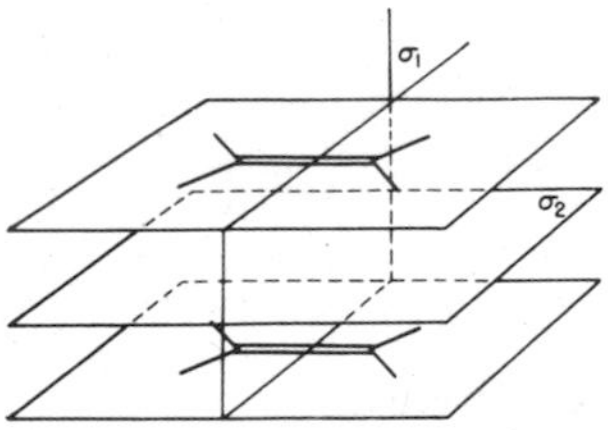

Figure 1.

(1) R. B. Woodward and R. Hoffmann, *J. Am. Chem. Soc.*, **87**, 395 (1965).

(2) This method has been independently and elegantly used in a discussion of electrocyclic reactions by Professor H. C. Longuet-Higgins (Cambridge), who very kindly communicated his results to us privately prior to publication.

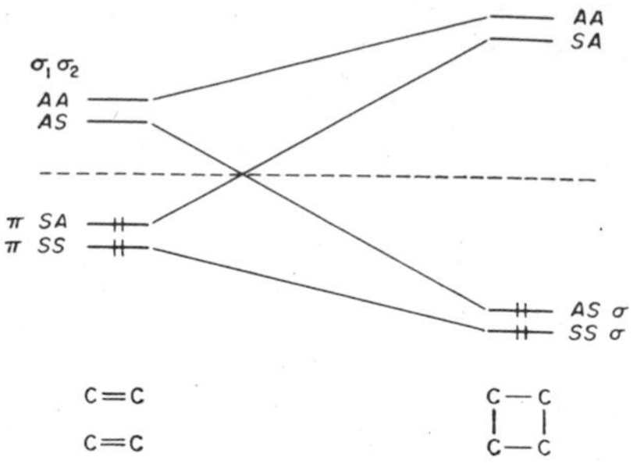

Figure 2.

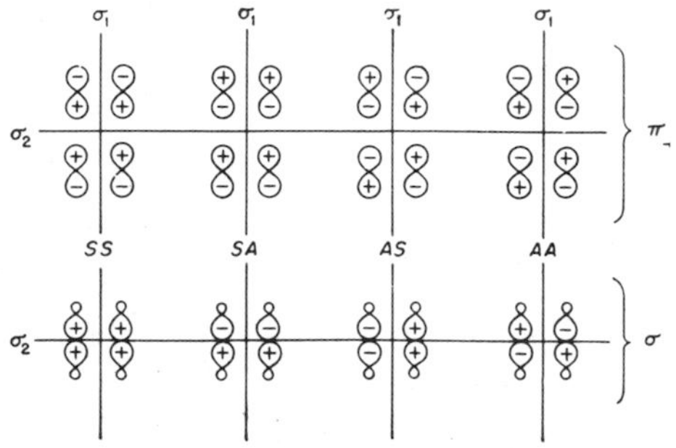

Figure 3.

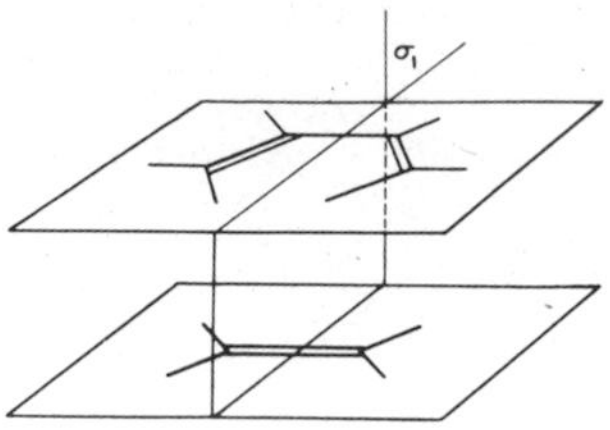

Figure 4.

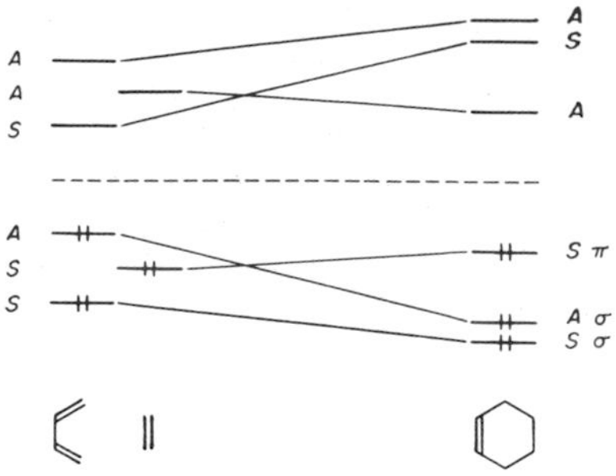

Figure 5.

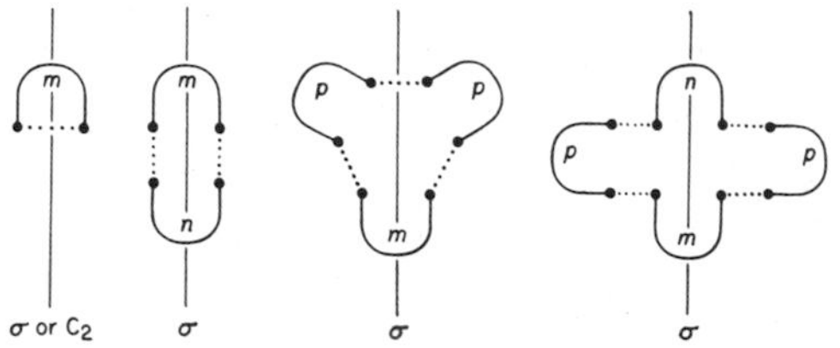

VI VII VIII IX

Figure 6.

ethylene π-bonding orbitals combine into the SS- and SA-orbitals of the interacting system, but the final σ-bonding orbitals are SS and AS. Note that πSS and πAS are bonding in the region of reaction and so decrease in energy along the reaction coordinate, while πSA and πAA are antibonding and so increase in energy. A level crossing results and is confirmed by extended Hückel calculations.

We consider next the combination of ethylene and butadiene. The transition state is formulated with diene and dienophile in parallel planes, and a perpendicular plane of symmetry σ_1, bisecting the molecules, as in Figure 4. Again a correlation diagram is drawn (Figure 5), relating the six reactant π-levels (four butadiene plus two ethylene) and the two resultant π- and four σ-orbitals. The three S-orbitals interact so that the lowest one (originally the diene most bonding π) decreases in energy, the highest S-level (originally the first unoccupied diene level) increases, while the middle S component (initially the ethylene π-level) remains at nearly constant energy; a similar pattern of interaction obtains for the A-orbitals. An extended Hückel calculation confirms the level scheme and shows some charge transfer from diene to olefin in the transition state, resulting primarily from mixing of the antibonding ethylene A level into the highest occupied diene A.

When the diagrams for the 1,2 and 1,4 additions are compared, it is clear that the bonding–antibonding level correlation in 1,2 addition makes this a highly unfavorable ground-state process, while no comparable factor mitigates against the 1,4 addition. On the other hand, if in each case one electron is excited to the next higher level, the energy along the reaction path is decreased for the 1,2 addition and increased for the 1,4 process. If now we make the reasonable postulate that, in general, correlation diagrams resembling that of Figure 5, *i.e.*, having no correlations of bonding with antibonding orbitals, are characteristic of permitted thermal reactions, while schemes similar to that of Figure 2, with bonding–antibonding correlations, are typical of photochemical processes, one can derive the following selection rules for the concerted cycloaddition reactions illustrated here (VI–IX) (m, n, p are numbers of π-electrons, q is an integer 0, 1, 2, 3 . . . .).

(1) Electrocyclic reactions (VI) (one π-bond → one σ-bond) will be thermally disrotatory and photochemically conrotatory for $m = 4q + 2$, thermally conrotatory and photochemically disrotatory for $m = 4q$. (2) The $2\pi \rightarrow 2\sigma$ process (VII) will be allowed thermally for $m + n = 4q + 2$, photochemically for $m + n = 4q$. (3) The $3\pi \rightarrow 3\sigma$ reaction, with a plane of symmetry bisecting m (VIII), will be allowed thermally for $m = 4q + 2$ (any p), photochemically for $m = 4q$ (any p). (4) The $4\pi \rightarrow 4\sigma$ process (IX) will be allowed thermally for $m + n = 4q + 2$ (any p), photochemically for $m + n = 4q$ (any p).

The proofs consist of enumeration of, say, bonding S-levels in reactants and products. If the number of bonding S-levels does not change, there will be no bonding → antibonding correlations, and the process is thermally allowed. If the number of bonding S-levels is one less or one greater in the product, a crossing must occur. As an example, consider the $2\pi \rightarrow 2\sigma$ case. In one component, containing, for example, m π-orbitals in the reactant, there will be $m/4$ bonding π-orbitals of S symmetry if $m/2$ is even, $(m + 2)/4$ if $m/2$ is odd. In the same component of the product, now containing $(m - 2)$ π-orbitals, there will be $m/4$ bonding S-levels if $m/2$ is even, $(m - 2)/4$ if $m/2$ is odd. The product σ-levels will yield one bonding level of S symmetry; thus, the total number of bonding S-levels from the reactant π's must exceed by one that from the product π's for a thermally allowed reaction. There are three cases: (1) $m = 4q_1$, $n = 4q_2$; then there will be $(q_1 + q_2)$ π-bonding S-levels before, $(q_1 + q_2)$ after. (2) $m = 4q_1 + 2$, $n = 4q_2$; this implies $(q_1 + q_2 + 1)$ before, $(q_1 + q_2)$ after. (3) $m = 4q_1 + 2$, $n = 4q_2 + 2$; this implies $(q_1 + q_2 + 2)$ before, $(q_1 + q_2)$ after. Case 2 satisfies the condition for the thermal process, and it is therefore required that $m + n = 4q_1 + 4q_2 + 2 = 4q + 2$; cases 1 and 3 will lead to bonding → antibonding correlations.

Certain special points concerning our selection rules deserve mention: (i) they apply to all concerted cycloaddition reactions, even though there may be considerable asymmetry in the rate at which the various different newly forming σ-bonds are established.[3] (ii) They need not apply to multistep cycloaddition reactions which proceed through discrete diradical or dipolar *intermediates*, containing a single newly formed σ-bond.

We may now tabulate the smaller ($\Sigma\pi \leqslant 10$) allowed concerted cycloaddition reactions (Table I). In point of fact, the agreement with recorded experience is outstanding. Examples are well known which very probably follow the allowed thermal 4 + 2 (Diels–Alder reaction) and 2 + 2 + 2[4] processes, and the allowed photochemical 2 + 2[5] and 4 + 4[6] cases, though the relevant concerted reactions may well be substantially electronically unsymmetrical. But it must be emphasized that the nature of our relations as *selection rules* makes the allowed cases *permissive*, but not *obligatory;* consequently, the rules do not *per se* exclude multistep alternative mechanisms involving discrete intermediates. By the same token, it must be concluded that those cycloadditions which do occur in cases prohibited by our selection rules for concerted reactions must proceed through multistep mechanisms (*e.g.*, formation of cyclobutane derivatives by dimerization of allenes and ketenes, and additions of perhalo- and percyanoethylenes to olefins, as well as dimerizations of *p*-xylylenes and the rarely observed photochemical Diels–Alder reaction), a conclusion in good accord with emerging experience.[7]

Table I

Type	Thermal	Photochemical
$2\pi \rightarrow 2\sigma$ (VII)	m n	m n
	4 2	2 2
	6 4	4 4
	8 2	6 2
$3\pi \rightarrow 3\sigma$ (VIII)	m p p	m p p
	2 2 2	4 2 2
	2 4 4	
	6 2 2	
$4\pi \rightarrow 4\sigma$ (IX)	m n p p	m n p p
	4 2 2 2	2 2 2 2

In assessing the predictive power of our selection rules in discerning possible new reactions, the following special points should be kept in mind: (i) (*restrictive*) the activation energies for thermal reactions should increase with the total number of π-electrons, as the total of bond elongations and contractions increases in the transition state; (ii) (*restrictive*) unless special geometrical constraints are present, entropy factors can place severe barriers in the way of realization of more complicated cases; (iii) (*extensive*) the relationships apply to ionic components as well as to neutral molecules, *e.g.*, allyl cation + cyclopentadiene → bicyclo[1.2.3]octenyl cation is permitted, while allyl cation + ethylene → cyclopentyl cation is prohibited.

The theoretical method used here has very wide applicability. We have already extended it to include the Cope and related rearrangements, and in studies of valence tautomers of $(CH)_n$; these results will form the subject of future communications.

(3) As in the mechanism earlier proposed for the Diels–Alder reaction by one of us [R. B. Woodward and T. J. Katz, *Tetrahedron*, **5**, 70 (1959).

(4) Norbornadiene + olefins, A. T. Blomquist and Y. C. Meinwald, *J. Am. Chem. Soc.*, **81**, 667 (1959); 1,3,5,7-tetramethylenecyclooctane + tetracyanoethylene, J. K. Williams and R. E. Benson, *ibid.*, **84**, 1257 (1962).

(5) Numerous cases cited in the review by R. Huisgen, R. Grashey, and J. Sauer in "The Chemistry of the Alkenes," S. Patai, Ed., John Wiley and Sons, Inc., New York, N. Y., 1964, p. 739.

(6) α-Pyridones: L. A. Paquette and G. Slomp, *J. Am. Chem. Soc.*, **85**, 765 (1963); α-pyrones: P de Mayo and R. W. Yip, *Proc. Chem. Soc.*, 84 (1964); anthracenes: D. E. Applequist and R. Searle, *J. Am. Chem. Soc.*, **86**, 1389 (1964); naphthalenes: J. S. Bradshaw and G. S Hammond, *ibid.*, **85**, 3953 (1963).

(7) *Cf.* P. D. Bartlett, L. K. Montgomery, and B. Seidel, *ibid.*, **86**, 616 (1964); L. K Montgomery, K. Schueller, and P. D. Bartlett, *ibid.*, **86**, 622 (1964); P. D. Bartlett and L. K. Montgomery, *ibid.*, **86**, 628 (1964), and references there cited.

(8) Junior Fellow, Society of Fellows, Harvard University.

Roald Hoffmann,[8] R. B. Woodward
Department of Chemistry, Harvard University
Cambridge, Massachusetts 02138
Received March 22, 1965

Reprinted from *J. Amer. Chem. Soc.*, **87**, 2511–2513 (1965)

42

Selection Rules for Sigmatropic Reactions

Sir:

We define as a *sigmatropic change of order* [*i,j*] the migration of a σ-bond, flanked by one or more π-electron systems, to a new position whose termini are $i - 1$ and $j -$ atoms removed from the original bonded loci, in an uncatalyzed intramolecular process. Thus, the well-known Claisen and Cope rearrangements are sigmatropic changes of order [3,3]. It is our purpose here to point out that orbital symmetry relationships must play a determinative role in the course of sigmatropic transformations similar to that operative in electrocyclic reactions[1] and concerted cycloadditions.[2]

Consider first the [1,*j*] sigmatropic migration of hydrogen within an all-*cis* polyolefin framework (I → II). There are two conceivable ways in which

$\overset{1}{R_2C}{=}\overset{2}{C}H{-}(CH{=}CH)_k{-}\overset{j}{C}HR'_2$

I

$\overset{1}{R_2C}H{-}(CH{=}CH)_k{-}CH{=}\overset{j}{C}R'_2$

II

hydrogen might be transferred from C-*j* to C-1. In each case the transition state for the change may be envisaged as made up by the combination of a hydrogen atom with a radical containing $2k + 3$ π-electrons. In the first process, here designated *suprafacial*, the hydrogen atom is associated at all times with the same face of the π-system, and the transition state possesses a plane of symmetry, σ. In the second, *antarafacial* process, the migrating atom is passed from the top face of one carbon terminus to the bottom of the other, through a transition state characterized by a twofold axis of symmetry, C_2. We note now that the highest occupied orbital of the framework system possesses the symmetry shown in III. Consequently, in order that positive overlap between this framework

III

orbital and the migrating hydrogen orbital be main-

(1) R. B. Woodward and R. Hoffmann, *J. Am. Chem. Soc.*, **87**, 395 (1965).
(2) R. Hoffmann and R. B. Woodward, *ibid.*, **87**, 2046 (1965).

tained, the isomerization I → II must occur thermally by the suprafacial route when k is odd, and by the antarafacial path when k is 0 or even. Analogous considerations lead to the conclusion that these relationships are precisely reversed for sigmatropic changes taking place within first-excited-state species. Our calculations by the extended Hückel method[3] fully support the generalizations deduced from simple orbital symmetry arguments.

The symmetry-allowed [1,j] sigmatropic transformations for $j \leqslant 7$ are summarized in Table I.[4] In applying these generalizations, several special factors should be borne in mind: (i) Antarafacial processes are obviously impossible for transformations which occur within small or medium-sized rings. (ii) In all cases, the carbon framework must not be so distorted during reaction as to cause serious impairment of coupling within the π-electron system; for example, this factor makes the antarafacial process difficult or impossible for $j = 3$, but is not a serious impediment to the similar process for $j = 7$. (iii) Sigmatropic reactions which violate the selection rules may take place through multistep processes, involving diradical intermediates, but such transformations may be expected to require relatively vigorous conditions.

Table I

[1,j]	Thermal	Excited state
[1,3]	Antarafacial (C_2)	Suprafacial (σ)
[1,5]	Suprafacial (σ)	Antarafacial (C_2)
[1,7]	Antarafacial (C_2)	Suprafacial (σ)

For sigmatropic reactions of order [i,j], in which both i and $j > 1$, proceeding through transition states possessing a plane of symmetry,[5] it is easy to show, following the lines laid down above, that thermal changes are symmetry-allowed when $i + j = 4n + 2$, while excited-state transformations are permitted when $i + j = 4n$.

The selection rules deduced here for sigmatropic reactions are in striking agreement with experience accumulated so far. To our knowledge there are no established examples of thermal, uncatalyzed [1,3] hydrogen shifts. On the other hand, specific [1,5] thermally induced hydrogen migrations, uncomplicated by competing [1,3] or [1,7] processes, have been observed in both cyclopentadienes[6] and cycloheptatrienes,[7] and in acyclic cases.[8] An important [1,7] thermally induced hydrogen shift in an open-chain system has been known for some time in the precalciferol–calciferol equilibrium.[9] By contrast [1,3] hydrogen shifts, as well as [1,3] carbon shifts, have been reported in photochemically activated cyclic systems,[10] and irradiation of cycloheptatrienes brings about highly specific [1,7] migrations.[11] Light-induced [1,5] hydrogen shifts have been observed in open-chain but *not* in cyclic systems.[12] Thermal sigmatropic reactions of order [3,3] have already been alluded to as the best known of all such processes. The [3,5] process has very probably been observed in a photoactivated system,[13] but if it occurs thermally at all it does so only with relatively great difficulty.[14]

In some of the cases cited above, the selection rules are permissive of alternatives which have not been observed. Thus, in the thermal rearrangement of precalciferol, a suprafacial [1,5] hydrogen migration could compete with the observed antarafacial [1,7] shift, while [1,3] migrations could accompany the photochemically induced [1,7] isomerizations in cycloheptatriene. The fact that only the transformations with the higher values of j are observed may be explicable in terms of a preference for processes with the maximum degree of linear conjugation in the transition state.

Of special interest is the fact that a cyclopropane ring may replace a π-bond in the framework system for a sigmatropic change. The symmetry-allowed, suprafacial, [1,5] hydrogen shift in the transformation of *cis*-1-methyl-2-vinylcyclopropane to *cis*-hexa-1,4-diene, which occurs very readily at 160° [$E^* \sim 31$ kcal.],[15] exemplifies this process.

It is also worthy of note that orbital symmetry arguments are applicable to sigmatropic changes within ionic species. Thus, the suprafacial [1,2] shift within a carbonium ion is symmetry-allowed and is very well known. The as yet undetected [1,4] migration within a but-2-en-1-yl cation must proceed through an antarafacial transition state, which may be difficult of access because of serious uncoupling within the framework π-system. By contrast, it may be predicted that the [1,6] shift within a hexa-2,4-dien-1-yl cation should take place through a readily accessible suprafacial transition state.

(3) R. Hoffmann, *J. Chem. Phys.*, **39**, 1397 (1963), and subsequent papers.

(4) In the above discussion of [1,j] sigmatropic changes, we have assumed throughout that a σ-orbital of the migrating group interacts with a π-system in the transition state. If the migrating group possesses an available relatively low-lying π-orbital and is not so substituted as to create an impossible steric situation in the transition state, alternative processes with relationships reversed from those in Table I can be envisaged (*cf.* i).

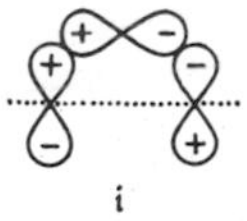

i

(5) Additional symmetry elements may be present and can be definitive in determining the preferred *conformation* for such sigmatropic changes. This point will be discussed in a subsequent communication.

(6) W. R. Roth, *Tetrahedron Letters*, 1009 (1964); S. McLean and R. Haynes, *ibid.*, 2385 (1964).

(7) A. P. ter Boorg, H. Kloosterziel, and N. van Meurs, *Rec. trav. chim.*, **82**, 717, 741, 1189 (1963); E. Weth and A. S. Dreiding, *Proc. Chem. Soc.*, 59 (1964).

(8) J. Wolinski, B. Chollar, and M. D. Baird, *J. Am. Chem. Soc.*, **84**, 2775 (1962).

(9) *Cf.* J. L. M. A. Schlatmann, J. Pot, and E. Havinga, *Rec. trav. chim.*, **83**, 1173 (1964).

(10) W. G. Dauben and W. T. Wipke, *Pure Appl. Chem.*, **9**, 539 (1964); J. J. Hurst and G. H. Whitham, *J. Chem. Soc.*, 2864 (1960).

(11) A. P. ter Boorg and H. Kloosterziel, *Rec. trav. chim.*, **84**, 241 (1965).

(12) R. Srinivasan, *J. Am. Chem. Soc.*, **84**, 3982 (1962); K. J. Crowley, *Proc. Chem. Soc.*, 17 (1964).

(13) K. Schmid and H. Schmid, *Helv. Chim. Acta*, **36**, 687 (1953).

(14) P. Fahrni and H. Schmid argue persuasively, but perhaps not entirely conclusively, that they have observed such a case [*ibid.*, **42**, 1102 (1959)].

(15) R. J. Ellis and H. M. Frey, *Proc. Chem. Soc.*, 221 (1964); *cf.* also W. Grimme, *Chem. Ber.*, **98**, 756 (1965).

(16) Junior Fellow, Society of Fellows, Harvard University.

R. B. Woodward, Roald Hoffmann[16]
Department of Chemistry, Harvard University
Cambridge, Massachusetts 02138
Received April 30, 1965

References

1. G. N. Lewis, *J. Amer. Chem. Soc.,* **38,** 762 (1916).
2. R. E. Rundle, *Rec. Chem. Progr.,* **23,** 195 (1962).
3. R. J. Gillespie, *J. Chem. Educ.,* **47,** 18 (1970) and references therein.
4. E. P. Wigner, *Group Theory,* Academic Press, New York, 1960, preface to the English translation.
5. E. Elial, *Stereochemistry of Carbon Compounds,* McGraw-Hill, New York, 1962.
6. G. M. Richardson, ed., *The Foundations of Stereochemistry,* American Book, New York, 1901.
7. J. H. van't Hoff, *Bull. Soc. Chim. France,* **23,** 295 (1875).
8. J. A. LeBel, *Bull. Soc. Chim. France,* **22,** 337 (1874).
9. E. Ruch and A. Schönhofer, *Theoret. Chim. Acta (Berlin),* **19,** 225 (1970).
10. F. A. Cotton, *Chemical Applications of Group Theory,* Wiley-Interscience, New York, 1963; 2nd ed., 1971.
11. E. P. Wigner, *Gruppentheorie,* Vieweg, Braunschweig, 1931, translation by J. J. Griffin, Academic Press, New York, 1960.
12. A. Speiser, *Die Theorie der Gruppen von endlichen Ordnung,* Zte. Aufl. Berlin, Springer, 1927.
13. R. W. G. Wyckoff, *The Analytical Expression of the Results of the Theory of Space Groups,* 2nd ed., The Carnegie Institution of Washington, Washington, 1930.
14. *International Tables for X-Ray Crystallography,* Lynch Press, Breminthaven, England, 1965.
15. J. P. Fackler, Jr., *Symmetry in Coordination Chemistry,* Academic Press, New York, 1971.
16. R. G. Pearson, *Pure Appl. Chem.,* **27,** 145 (1971); R. G. Pearson, *Accounts Chem. Res.,* **4,** 152 (1971).
17. F. D. Mango and J. H. Schachtschneider, *J. Amer. Chem. Soc.,* **89,** 2484 (1967).
18. J. Halpern, *Accounts Chem. Res.,* **3,** 386 (1970).
19. L. H. Hall, *Group Theory and Symmetry in Chemistry,* McGraw-Hill, New York, 1969; M. Hammermesh, *Group Theory and Its Applications to Physical Problems,* Addison-Wesley, Reading, Mass., 1962; M. Tinkham, *Group Theory and Quantum Mechanics,* McGraw-Hill, New York, 1964.
20. J. E. Rosenthal and G. M. Murphy, *Rev. Mod. Phys.,* **8,** 317 (1936).
21. H. A. Kramers, *Comm. Leiden,* **60;** *Proc. Amsterdam Acad.,* **33,** 959 (1930).
22. J. H. Van Vleck, *The Theory of Electric and Magnetic Susceptibilities,* Oxford University Press, New York. 1932

23. H. A. Jahn, *Proc. Royal Soc.,* **164,** 117 (1937).
24. J. H. Van Vleck, *J. Chem. Phys.,* **7,** 72 (1939).
25. W. L. Clinton and B. Rice, *J. Chem. Phys.,* **30,** 542 (1959).
26. W. Heitler and F. London, *Z. Physik,* **44,** 455 (1927); J. C. Slater, *Phys. Rev.,* **37,** 481 (1931); *ibid,* **38,** 1109 (1931); L. Pauling, *J. Amer. Chem. Soc.,* **53,** 1367 (1931).
27. L. Pauling, *J. Chem. Phys.,* **1,** 280 (1933).
28. R. S. Mulliken, *Phys. Rev.,* **41,** 49 (1932).
29. R. S. Mulliken, *Phys. Rev.,* **41,** 751 (1932).
30. F. Seitz, *Physics Today,* Oct. 1972, pp. 40–43.
31. E. L. Muetterties, *Rec. Chem. Progr.,* **31,** 51 (1970); *J. Amer. Chem. Soc.,* **91,** 1636 (1969); *Accounts Chem. Res.,* **3,** 266 (1970).

Author Citation Index

Subject Index